Sustainability with Changing Climate and Extremes

Sustainability with Changing Climate and Extremes

Editors

Xiaodong Yan
Jia Yang
Shaofei Jin

MDPI • Basel • Beijing • Wuhan • Barcelona • Belgrade • Manchester • Tokyo • Cluj • Tianjin

Editors
Xiaodong Yan
Beijing Normal University
China

Jia Yang
Oklahoma State University
USA

Shaofei Jin
Minjiang University
China

Editorial Office
MDPI
St. Alban-Anlage 66
4052 Basel, Switzerland

This is a reprint of articles from the Special Issue published online in the open access journal *Sustainability* (ISSN 2071-1050) (available at: https://www.mdpi.com/journal/sustainability/special_issues/extreme_climate).

For citation purposes, cite each article independently as indicated on the article page online and as indicated below:

LastName, A.A.; LastName, B.B.; LastName, C.C. Article Title. *Journal Name* **Year**, *Volume Number*, Page Range.

ISBN 978-3-0365-5849-3 (Hbk)
ISBN 978-3-0365-5850-9 (PDF)

Contents

sustainability

MDPI

Editorial

Sustainability with Changing Climate and Extremes

Shaofei Jin [1], Jia Yang [2] and Xiaodong Yan [3,*]

1 Department of Geography, College of Geography and Oceanography, Minjiang University, Fuzhou 350108, China
2 Department of Natural Resource Ecology and Management, Oklahoma State University, Stillwater, OK 74078, USA
3 State Key Laboratory of Earth Surface Processes and Resource Ecology, Faculty of Geographical Science, Beijing Normal University, Beijing 100875, China
* Correspondence: yxd@bnu.edu.cn

Climate change and extreme events are receiving increasingly more attention in the global sustainable development sphere. Identifying the impacts of climate change and extreme events is not only important in terms of natural processes, such as heat waves and earthquakes, but also in terms of societal processes and the societal consequences of natural disasters. Over the past two years (2021 and 2022), extreme events have occurred across the globe. The February 2021 North American cold wave led to widespread power outages for millions of people in Texas, USA. Another remarkable extreme event, the recent Coronavirus disease in 2019 (COVID-19), is shaping the entire environmental and societal sustainability situation worldwide. In 2022, persistent heatwaves have been affecting Europe, causing evacuations and heat-related deaths with a maximum temperature of 47.0 °C reported in Pinhão, Portugal, on 14 July. The flood in Pakistan, the deadliest one in worldwide since 2017, has killed more than 1300 people since 14 June 2022. With the intensity and magnitude of climate change and extreme events being unknown, neither the changes themselves nor the corresponding impacts are clear under the current circumstances. Therefore, we organized this Special Issue under the theme of sustainability with respect to the changing climate and extremes.

We will briefly discuss the contributions of the 34 published papers in this Special Issue in the following four sections.

Citation: Jin, S.; Yang, J.; Yan, X. Sustainability with Changing Climate and Extremes. *Sustainability* **2022**, *14*, 11830. https://doi.org/10.3390/su141911830

Received: 13 September 2022
Accepted: 15 September 2022
Published: 20 September 2022

1. Natural Disasters in Agriculture

As the most widely cultivated fruit genus, Wang Shuangshuang et al. (Appendix A, Contributor 1) evaluated changes in the quality and yields of citrus under the shared socioeconomic pathways scenario. In their study, first, they proposed statistical models of the relationship between daily meteorological observation variables and the yields and quality of citrus. In short, with respect to predicting the quality of citrus, the monthly mean diurnal temperature range in July was identified as the optimal variable; for predicting yields, a group of variables in October and September was identified as comprising the best predictors. Then, they analyzed the future changes in the quality and yields of citrus in the period from 2021–2060 using the ensemble mean of nineteen Coupled Model Intercomparison Project 6 (CMIP6) models. Finally, they found that the quality of citrus will increase in 3 provinces, while the yield will increase in all 11 provinces. The results can support the design of citrus plantations in the future; further, the statistical models can be coupled with the ecological process model for predicting the yield and quality of citrus fruits.

Potatoes play a significant role in global food security and human diets. As the leading potato producer, China accounts for 22% of the world's potato yield [1]. In China, the potato plant has been the fourth most important crop after rice, wheat, and maize, and is facing the negative impacts of the changing climate. Yang Li-Tao et al. (Contributor 2) analyzed the climatic production potential of potatoes from 1961 in Inner Mongolia, where they occupied the largest plantation area of potatoes in China. They obtained the annual average climate production potential for potatoes in Inner Mongolia, which was 19,318 kg/hm^{-2}. Over the past 61 years, the climatic production potential exhibited an insignificant linear decreasing trend under the changed climate. The main finding of their study lies in their identification of precipitation as the main impact factor for the productive potential for potatoes in Inner Mongolia.

Maize (*Zea mays* L.) is one of the most important crops in the global and national economies and is the third highest yielding form of crop production in China. In China, the Huanghuaihai plain is the largest summer maize-producing area with a wheat–maize cropping system. The ongoing process of global warming increases the risk of high-temperature injury to maize. In addition, previous studies have shown that the inter- and mixed-cropping of maize varieties with different genotypes is one way to effectively alleviate high-temperature injuries during the flowering period. In this study, Li Shuyan et al. (Contributor 3) investigated the response of five main genotypes of maize to high temperatures and different crop models during the flowering period. The main findings revealed that inter- and mixed cropping effectively reduced the impact of high temperatures during flowering, demonstrating that even the genotypic varieties can make significant differences in the yields.

In addition, the diurnal temperature range (DTR) is also an important meteorological variable affecting maize yields. Xie Wenqiang et al. (Contributor 4) evaluated the ability of 26 CMIP6 models to simulate DTR from 1961–2014 and projected the DTR under different shared socioeconomic pathway (SSP) scenarios from 2015 to 2050. The main findings of their study included: (1) CMIP6 models can generally reproduce spatial distribution, especially in the maize cultivation areas; (2) The DTR remains stable from 2015–2050 under SSP245, while a significant decreasing trend is found under SSP585. This study implied the efficiency of the application of DTR in the accurate prediction of maize yield prediction.

The dry-hot wind is a type of severe agricultural wind disaster with high a temperature and low humidity. Dry-hot winds can reduce wheat production by up to 30%. In China, Shandong Province has the greatest agricultural growth value among all of the provinces and is also the most affected by dry-hot winds. Wang Nan et al. (Contributor 5) assessed the spatial distribution of dry-hot wind disasters in Shandong Province. The main finding in their study demonstrated that the more developed areas in the east of Shandong province show high disaster prevention and mitigation capabilities, whereas these same aspects are weak in the west, where the economy is still behind eastern Shandong province. Their study implies the roles of economical input in preventing natural disasters.

Further, Jiang Meiyi et al. (Contributor 6) evaluated the effect of drought on the production of peanuts in Shandong province, which accounts for 16% of the total peanut cultivation area and 20% of the total peanut production in China. Drought has been considered the most severe natural disaster with respect to agricultural production. Therefore, identifying the high drought risk regions will be a benefit to the yields and quality of peanuts in Shandong. As the main finding of their study, they found that the drought risk was concentrated on the Jiaodong Peninsula, covering 20.7% of the province. Their study can support the acquisition of data for developing and carrying out peanut drought prevention, relief policies, and related decisions.

Further, in another study (Contributor 7), Pan Mingxi et al. investigated the effects of snow cover on the spring soil moisture content, which will impact the occurrence of spring drought in high-latitude areas. To investigate the role of winter snow cover with respect to soil moisture, the authors chose the main agricultural areas of Northeast China—the Songnen Plain and the Sanjiang Plain—to address these issues. Their main finding was that they found that snow cover has a strong effect on soil moisture conservation in more arid areas. This work can support the scientific basis for the early warning of spring drought, the development of more efficient irrigation schemes, and crop yield prediction based on the snow cover in Northeast China.

To predict droughts and floods in agriculture in the future, in their study (Contributor 8), Li Jiannan et al. constructed a vulnerability model based on "sensitivity–exposure–adaptability" and "vulnerability-risk, source-risk receptor" drought and flood risk models and established multi-index prediction systems. They predict the vulnerability and risk of droughts and floods in China's agroecosystem under the SSP1-2.6, SSP2-4.5, and SSP5-8.5 scenarios from 2020 to 2050 using a combination of AHP and the entropy-weighting method. They found that the levels drought and flood vulnerability intensify, and the drought or flood vulnerability area expands to southern China from SSP1-2.6 to SSP5-8.5.

2. Urban/Rural Ecosystem, Tourism, and Ecosystem Service

In an urban ecosystem, rainfall plays a crucial role in human mobility and urban management and planning. After four decades of urbanization, China is still facing the rapid developmental period of urbanization in the future. One urgent need involves understanding the impact of rainfall on residents' mobility in the city. Taxi GPS data support a large amount of spatiotemporal information about human activities and mobility in urban areas. Guo Peng et al. (Contributor 9) combined geospatial complex networks with multiscale geospatial analysis to extend the empirical research on human travel patterns using Taxi GPS data to analyze the impact of rainfall on human mobility. They proved that taxi GPS data are highly informative and exploitable in the field of human mobility analysis. The main findings based on their results were the following: (1) rainfall has a reducing effect on trip flow, trip distance, and trip duration on both weekdays and weekends; (2) rainfall has a significant effect on the network indicators; and (3) more mobile communities were detected on weekends than on weekdays, while the number of communities on weekdays and weekends did not change significantly because of rainfall.

Further, climate-related risks threaten urban safety, infrastructure stability, and socioeconomic sustainability. China is facing various climate hazards due to its diverse geomorphic conditions. In this study (Contributor 10), Sun Shao et al. propose a comprehensive analysis of the spatial pattern of major climate hazards in China from 1991 to 2020. The climate hazard types include rainstorms, droughts, heatwaves, cold waves, typhoons, and snowstorms. They found that cities are hotspots affected by intensified climate hazards in a warmer climate and they proposed an urgent need to incorporate a climate adaptation strategy into future city construction to improve social resilience and mitigate climate impacts in the rapid urbanization process in China. For the climate risks of the three major urban agglomerations of Eastern China (Beijing–Tianjin–Hebei, the Yangtze River Delta, and the Pearl River Delta), in this study (Contributor 11), Chou Jieming et al. constructed one vulnerability degree using the Gray model (GM (1,1)), and calculated the drought, heat wave, and flood hazards under different emission scenarios based on the CMIP6 model. They found that the Beijing–Tianjin–Hebei urban agglomeration has a good level of urban resilience, the Yangtze River Delta's urban agglomeration has slightly higher overall risk, and the Pearl River Delta's urban agglomeration has the highest relative risk overall.

Rural regions are sensitive to climate disasters. China has achieved its goal of eliminating absolute poverty in China. However, meteorological disasters can promote rural populations' return to poverty. In this study (Contributor 12), Li Aiwei et al. analyzed the spatiotemporal characteristics of rainstorms and droughts and their socioeconomic impacts on China's contiguous poverty-stricken areas from 1984 to 2019. The main findings are as follows: (1) rainstorms showed a significant increasing trend of 0.075 days/decade, while there is no significant trend for drought days; (2) the average annual loss rate due to disasters in the poverty-stricken areas reached 1.6%, which is 0.6% higher than the national level. Their results suggest that to obtain the realization of the United Nations Sustainable Development Goals, it is necessary to improve the capacity for meteorological disaster prevention and reduction in China's contiguous poverty-stricken areas.

The tourism markets are deeply impacted by the air pollution in China due to the necessity of pursuing fresh air in the tourism transformation process. Finding one air index to reflect the air freshness of destinations can support referable data for the tourists. Yang

Xiaoyan et al. (Contributor 13) propose a new comfort index for climate tourism: the fresh air–natural microclimate comfort index (FAI-NMCI). This index connects the fresh air index with the natural microclimate comfort index of scenic spots from transdisciplinary and multidisciplinary perspectives. Under contemporary China's high-quality tourism development background, this could provide accurate information not only for the demand and supply sides of the tourism market but could also become an important and comprehensive index for related governmental management.

Further, to explore the influence of changes in climate comfort on Arctic tourism, in this study (Contributor 14), Huang Yutao et al. analyzed the spatial-temporal changes in Arctic summer climate comfort zones from 1979 to 2019 using the ERA5-HEAT (Human thErmAl comforT) dataset from the European Center for Medium-range Weather Forecasts (ECMWF). Their results suggest that global warming increases the Arctic summer climate comfort level and may provide favorable conditions for the further development of regional tourism resources.

In the ecosystem service section, the annual terrestrial gross primary productivity (GPP) was taken as the representative ecological indicator of the ecosystem. In this study (Contributor 15), Zhang Chi et al. analyzed the GPP changes using three earth system models (ESMs) from CMIP6 under 1.5 and 2 °C global warming targets in the Shared Socioeconomic Pathway 4.5 W m^{-2} (SSP245) scenario. In their main findings, they found that Under 1.5 °C of global warming, GPP in four climate zones (temperate continental, temperate monsoonal, subtropical–tropical monsoonal, and high-cold Tibetan Plateau) increased significantly with a minimum growth of 12.3% and the increase was greater under 2 °C of global warming that at 1.5 °C. Their results implied that global warming poses no ecological risk in China from the perspective of ecosystem productivity.

The China–US trade conflict can affect the achievement of climate change goals. In this study (contributor 16), Chou Jieming et al. assessed the impact of the trade conflict on China's climate policy by combining the model from the Global Trade Analysis Project (GTAP) and the input–output analysis method. Their results showed that changes in trade will challenge China's balancing of climate and trade exigencies, implying that China–US cooperation based on energy and technology will help China cope with climate change after the trade conflict. In this Special Issue, another study by (contributor 17) Feng Qiang et al. contributed the response mechanisms of the ecosystem service's trade-offs with respect to land use changes along precipitation gradients in the Loess Plateau of China.

In one study (contributor 18), Deng Xiaofang et al. investigated human resource allocation with respect to adaptation to climate change in a state-owned forest farm in China using the questionnaire method. They found that the human resource professional and industrial structure changes in the context of climate change are the main limiting factors for the key state-owned forest farms of China. In their paper, they suggested that increasing the investment in education on climate change and the income of employees is one way to promote the adaptation to climate change in China.

Extreme events threaten human health. In one paper (contributor 19), Li Junrong et al. used big data to explore the dynamic changes regarding population exposure during a heat wave incident in Zhuhai based on real-time mobile phone data and meteorological data. The main findings are as follows: (1) the overall population exposure shows a trend of first decreasing and then increasing; (2) a high degree of population exposure is concentrated in areas such as primary and secondary schools, colleges and universities, office buildings, and residential areas; and (3) the population exposure changes in the last two days of the heat wave cycle are mainly affected by the combined influence of population factors and climate factors based on the studied heatwave events from 6 to 12 September 2021. This study has a certain practical guiding significance with respect to advancing high-temperature warning and high-temperature disaster risk prevention methods.

In addition, another study (Contributor 20) assessed China's future heatwave population exposure under 1.5 °C and 2.0 °C warming scenarios with respect to climate change adaptation using models from CMIP6. The main findings of their study are as follows:

(1) an additional 20.15% increase in the number of annual heatwave days would occur with an additional temperature increase of 0.5 °C to 2.0 °C, an over 1.5 °C increase in temperature by the mid-century; (2) from three influencing elements, namely, climate, population, and interaction (e.g., as urbanization affects population redistribution), climate explained more than 70% of the variance in different warming scenarios. Their study can support helpful insights for developing mitigation strategies for climate change.

Tropical marine fisheries provide high-quality protein for human diets and play significant roles in food security. Tuna fisheries are one of the four most highly valued fisheries worldwide and have contributed 5.2 million tons of the total capture amount. The tuna fishery in the entire western and central Pacific Ocean (WCPO) supports major industrial tuna fisheries and contributes almost 30% of the total global tuna supply. However, the changing climate is threatening tropical tuna fisheries via the increasing sea surface temperature. In this study (Contributor 21), Zhou Weifeng et al. explored the impact of climate change on tuna fishery resources by investigating the temporal and spatial characteristics of the thermocline in the main yellowfin tuna purse seine fishing grounds in the western and central Pacific Ocean during La Niña and El Niño years from 2008 to 2017. The main findings of their study include: (1) in La Niña years, the catch per unit effort (CPUE) moved westward where the high-value zone of the upper boundary contracted westward to 145° E, while in the El Niño years, this boundary moved eastward to 165° E; (2) changes in the thermocline caused by abnormal climate events significantly affected the CPUE. This study can provide additional thermocline distribution information and serve as a reference for tuna production in this area.

3. Extreme Climate Indices

Extreme climate events are more frequent under ongoing climate change. Central China, a key area for the transition between China's east–west and north–south borders and a hub for land and water transportation, is facing a complex weather system. In this study (Contributor 22), Li Yan et al. analyzed the temporal and spatial dynamics of climate events from 1988 to 2017 using nine indices: six extreme temperature indices and three extreme precipitation indices. The six extreme temperature indices are icy days (ID), frost days (FD), the duration of warm periods (WSDI), the duration of cold periods (CSDI), the lowest Tmin (TNn), and the highest Tmax (TXx). The three extreme precipitation indices are very wet days (R95), consecutive dry days (CDD), and consecutive wet days (CWD). The main findings are as follows: (1) the Jiangxi region was at greater risk of extreme climate events in central China, and (2) the drought events in central China around 2025 will be more significant. Following their results, disaster prevention and mitigation projects can be suggested to be prepared in advance for the policy-makers.

In another study (Contributor 23), Yan Weixiong et al. evaluated the distribution of extreme temperature seasonality trends in mainland China by using the following indices: the number of hot days (HD) and frost days (FD), as well as the frequency of warm days (TX90p), cold days (TX10p), warm nights (TN90p), and cold nights (TN10p). They highlighted that extreme temperatures have increased over mainland China from 1979–2020 with obvious seasonal variations, and the increase in the minimum temperature is more distinct than in the maximum temperature.

Heatwave events (HWEs) have strong negative impacts on human health, ecosystems, and sustainable social development. In one study (Contributor 24), Gao Zhibo and Yan Xiaodong analyzed the characteristics of HWEs over the Yangtze River Basin (YRB) in eastern China during the historical period and projected the changes in HWEs over the YRB in the future using a gridded observation dataset and a high-resolution regional climate model. Their main findings include: (1) Short-lived ($\geq$3 days and <6 days) HWEs are projected to increase rapidly under SSP585, but to rise slowly overall under SSP245, and (2) the increase in HWEs over the YRB region is likely to be associated with the enhancement of the western-Pacific subtropical high (WPSH) and South-Asian high (SAH). This study can support solid references for disaster prevention and mitigation in the future for these developed regions of China.

In another study (contributor 25), Li Kaiwen et al. defined compound drought and heatwave events (CDHEs) using the monthly scale standard precipitation index and evaluated the spatial and temporal variations in CDHEs in China from 1961 to 2018. The main findings of their study are as follows: (1) the mean frequency of CDHEs takes on a non-significant decreasing trend, and the mean magnitude of CDHEs takes on a non-significant increasing trend in China; (2) the significant trends in the annual frequency and annual magnitude of CDHEs are attributed to maximum temperature and precipitation changes.

The surface albedo of pure fresh snow is generally between 60% and 95%. When black carbon (BC) aerosols deposit into the surface snow through dry and wet deposition, the albedo of the snow surface will significantly reduce and thereby increase the absorption of solar radiation on the snow surface. Therefore, understanding the BC concentration on the surface of snow will play an important role with respect to studying climate change. In this study (contributor 26), Zheng Yanjiao et al. simulated the black carbon concentration on the surface snow of northeast China using an asymptotic radiative transfer model. The main findings of their study include: (1) human activities played an important role in snow black carbon pollution, and (2) the snow surface albedo will decrease by 16.448% due to the BC pollution of snow in northeast China. These results suggest that the increase in radiative forcing caused by black carbon via snow reflectivity cannot be ignored.

As one part of the earth system and one of the most sensitive regions of climate warming, cold regions are areas with low temperatures and with the presence of ice and snow for at least part of the year. In their study (Contributor 27), Wang Yumeng et al. studied the spatial distribution and variations of cold regions in China from 1961 to 2019. They found that the area of China's cold regions decreased by 49.32×104 km^2 in the period from 1991–2019 compared with that in the period of 1961–1990. In addition, in another paper (Contributor 28), Li Hao et al. studied the distribution and assessment of snow-disaster risk zoning in Heilongjiang Province. As the main findings of their study, they found that economically developed regions had strong disaster prevention and mitigation capabilities. This implied that the economic input for preventing snow disasters is required in Heilongjiang Province. For another cold region in China, in this Special Issue, Sun Shao et al. (Contributor 29) and Jia Yiru et al. (Contributor 30) evaluated and mapped the meteorological hazards and the locomotion of slope geohazards in the Qinghai-Tibet Plateau (QTP), respectively, in response to climate change.

Extreme cooling (EC) events are also a major challenge to socioeconomic sustainability and human health. In this study (Contributor 31), Song Shuaifeng and Yan Xiaodong analyzed the temporal and spatial distributions of EC events in China using the relative threshold and the relationship between EC events and the Arctic Oscillation (AO) index during the period of 1961–2017. The main findings of their study are as follows: (1) the frequency of EC events in China decreased by 0.730 d from 1961 to 2017, and (2) EC events are significantly negatively correlated with the AO index, which can explain approximately 21% of the EC event variance. Their study can help to improve the prediction and simulation of EC events in China.

To overcome the overestimation for the light precipitation and underestimation of heavy precipitation due to low model resolution, Luo Neng and Yan Guo (Contributor 32) studied the impact of model resolution on simulating the precipitation extremes over China from 1995–2014 with six extreme indices based on five models from CMIP6. All these models include low- and high-resolution versions. Six extreme indices were employed: simple daily intensity index (SDII), wet days (WD), total precipitation (PRCPTOT), extreme precipitation amount (R95p), heavy precipitation days (R20mm), and consecutive dry days (CDD). The main findings of their study are as follows: (1) models with a high resolution demonstrated better performance in reproducing the pattern of climatological precipitation extremes over China, (2) decreased biases of precipitation exist in all high-resolution models over D1, and (3) Improvement could be attributed to fewer weak precipitation events (0–10 mm/day) in high-resolution models in comparison with their counterparts with low resolutions. Their solid work proved that models with improved resolution show an obvious advantage with respect to the simulation of precipitation extremes, thereby increasing confidence in the simulation of precipitation extremes.

4. Newly Created Dataset for Climate Change

The forest ecosystem is one of the most concentrated systems under the climate change background. Identifying the forest type plays a crucial role in supporting information for forest managers, conservationists, and forest ecologists. In this study (Contributor 33), Xu Chen et al., generated a forest type distribution using an unsupervised cluster analysis method by combining climate variables with normalized difference vegetation index data. Their work will improve the depth of research in biodiversity preservation, forest management, and ecological and forestry research.

Modern meteorological methods with high-resolution and high-quality precipitation data are urgently required in the monitoring of mountain flood geological disasters as well as hydrological monitoring and prediction. In this study (Contributor 34), Wang Zheng et al. tested the newly created 0.01° multi-source fusion precipitation product developed by the National Meteorological Information Center. Their study proved that the 0.01° multi-source fusion precipitation product has better spatial continuity, a more detailed description of precipitation's spatial distribution and a more accurate reflection of extreme precipitation values, and can provide precipitation data support for refined meteorological services, major activity support, and disaster prevention and reduction.

Author Contributions: Conceptualization, S.J. and X.Y.; writing—original draft preparation, S.J., J.Y. and X.Y.; writing—review and editing, S.J., J.Y. and X.Y.; supervision, X.Y.; funding acquisition, S.J. All authors have read and agreed to the published version of the manuscript.

Funding: This research was funded by the Central Government Guides Local Projects (Grant NO. 2020L3024), and funded by Fujian Province Forestry Science and Technology Research Project (Grant NO. 2022FKJ02).

Institutional Review Board Statement: Not applicable.

Informed Consent Statement: Not applicable.

Data Availability Statement: Not applicable.

Acknowledgments: We would like to thank all authors for their excellent papers in this Special Issue: *Sustainability with Changing Climate and Extremes*. We also thank the reviewers for their valuable recommendations leading to the improvement of all manuscripts before publication. We are also grateful to all members of the *Sustainability* Editorial Office for providing us with this opportunity and for continuous support in managing and organizing this Special Issue.

Conflicts of Interest: The authors declare no conflict of interest.

Appendix A. List of Contributions

1. Wang, Shuangshuang, Wenqiang Xie, and Xiaodong Yan. 2022. "Effects of Future Climate Change on Citrus Quality and Yield in China" *Sustainability* 14, no. 15: 9366. https://doi.org/10.3390/su14159366.
2. Yang, Li-Tao, Jun-Fang Zhao, Xiang-Ping Jiang, Sheng Wang, Lin-Hui Li, and Hong-Fei Xie. 2022. "Effects of Climate Change on the Climatic Production Potential of Potatoes in Inner Mongolia, China" *Sustainability* 14, no. 13: 7836. https://doi.org/10.3390/su14137836.
3. Li, Shuyan, Junfang Zhao, Junling Li, Ruixin Shao, Hongping Li, Wensong Fang, Liting Hu, and Tianxue Liu. 2022. "Inter- and Mixed Cropping of Different Varieties Improves High-Temperature Tolerance during Flowering of Summer Maize" *Sustainability* 14, no. 12: 6993. https://doi.org/10.3390/su14126993.

4. Xie, Wenqiang, Shuangshuang Wang, and Xiaodong Yan. 2022. "Evaluation and Projection of Diurnal Temperature Range in Maize Cultivation Areas in China Based on CMIP6 Models" *Sustainability* 14, no. 3: 1660. https://doi.org/10.3390/su14031660.
5. Wang, Nan, Xiaoping Xue, Lijuan Zhang, Yue Chu, Meiyi Jiang, Yumeng Wang, Yiping Yang, Xihui Guo, Yufeng Zhao, and Enbo Zhao. 2022. "Spatial Zoning of Dry-Hot Wind Disasters in Shandong Province" *Sustainability* 14, no. 7: 3904. https://doi.org/10.3390/su14073904.
6. Jiang, Meiyi, Xiaoping Xue, Lijuan Zhang, Yuying Chen, Cheng Zhao, Haiyan Song, and Nan Wang. 2022. "Peanut Drought Risk Zoning in Shandong Province, China" *Sustainability* 14, no. 6: 3322. https://doi.org/10.3390/su14063322.
7. Pan, Mingxi, Fang Zhao, Jingyan Ma, Lijuan Zhang, Jinping Qu, Liling Xu, and Yao Li. 2022. "Effect of Snow Cover on Spring Soil Moisture Content in Key Agricultural Areas of Northeast China" *Sustainability* 14, no. 3: 1527. https://doi.org/10.3390/su14031527.
8. Li, Jiangnan, Jieming Chou, Weixing Zhao, Yuan Xu, Yidan Hao, and Yuanmeng Li. 2022. "Future Drought and Flood Vulnerability and Risk Prediction of China's Agroecosystem under Climate Change" *Sustainability* 14, no. 16: 10069. https://doi.org/10.3390/su141610069.
9. Guo, Peng, Yanling Sun, Qiyi Chen, Junrong Li, and Zifei Liu. 2022. "The Impact of Rainfall on Urban Human Mobility from Taxi GPS Data" *Sustainability* 14, no. 15: 9355. https://doi.org/10.3390/su14159355.
10. Sun, Shao, Zunya Wang, Chuanye Hu, and Ge Gao. 2021. "Understanding Climate Hazard Patterns and Urban Adaptation Measures in China" *Sustainability* 13, no. 24: 13886. https://doi.org/10.3390/su132413886.
11. Chou, Jieming, Mingyang Sun, Wenjie Dong, Weixing Zhao, Jiangnan Li, Yuanmeng Li, and Jianyin Zhou. 2021. "Assessment and Prediction of Climate Risks in Three Major Urban Agglomerations of Eastern China" *Sustainability* 13, no. 23: 13037. https://doi.org/10.3390/su132313037.
12. Li, Aiwei, Shuyuan Gao, Miaoni Gao, Xueqing Wang, Hongling Zhang, Tong Jiang, and Jing Yang. 2022. "Spatiotemporal Evolution and Socioeconomic Impacts of Rainstorms and Droughts in Contiguous Poverty-Stricken Areas of China" *Sustainability* 14, no. 16: 9927. https://doi.org/10.3390/su14169927.
13. Yang, Xiaoyan, Changshun Li, Muhammad Bilal, and Shaofei Jin. 2022. "Fresh Air–Natural Microclimate Comfort Index: A New Tourism Climate Index Applied in Chinese Scenic Spots" *Sustainability* 14, no. 3: 1911. https://doi.org/10.3390/su14031911.
14. Huang, Yutao, Xuezhen Zhang, Dan Zhang, Lijuan Zhang, Wenshuai Zhang, Chong Ren, Tao Pan, Zheng Chu, and Yuying Chen. 2021. "Spatial-Temporal Characteristics of Arctic Summer Climate Comfort Level in the Context of Regional Tourism Resources from 1979 to 2019" *Sustainability* 13, no. 23: 13056. https://doi.org/10.3390/su132313056.
15. Zhang, Chi, Shaohong Wu, Yu Deng, and Jieming Chou. 2021. "How the Updated Earth System Models Project Terrestrial Gross Primary Productivity in China under 1.5 and 2 °C Global Warming" *Sustainability* 13, no. 21: 11744. https://doi.org/10.3390/su132111744.
16. Chou, Jieming, Fan Yang, Zhongxiu Wang, and Wenjie Dong. 2021. "The Impact on Carbon Emissions of China with the Trade Situation versus the U.S." *Sustainability* 13, no. 18: 10324. https://doi.org/10.3390/su131810324.
17. Feng, Qiang, Siyan Dong, and Baoling Duan. 2021. "The Effects of Land-Use Change/Conversion on Trade-Offs of Ecosystem Services in Three Precipitation Zones" *Sustainability* 13, no. 23: 13306. https://doi.org/10.3390/su132313306.
18. Deng, Xiaofang, Junkui Li, Lijuan Su, Shan Zhao, and Shaofei Jin. 2022. "Human Resource Allocation in the State-Owned Forest Farm of China for the Changing Climate" *Sustainability* 14, no. 15: 9667. https://doi.org/10.3390/su14159667.
19. Li, Junrong, Peng Guo, Yanling Sun, Zifei Liu, Xiakun Zhang, and Xinrui Pei. 2022. "Population Exposure Changes to One Heat Wave and the Influencing Factors Using Mobile Phone Data—A Case Study of Zhuhai City, China" *Sustainability* 14, no. 2: 997. https://doi.org/10.3390/su14020997.

20. Wang, L.; Rohli, R.V.; Lin, Q.; Jin, S.; Yan, X. Impact of Extreme Heatwaves on Population Exposure in China Due to Additional Warming. *Sustainability* 2022, 14, 11458. https://doi.org/10.3390/su141811458.
21. Zhou, Weifeng, Huijuan Hu, Wei Fan, and Shaofei Jin. 2022. "Impact of Abnormal Climatic Events on the CPUE of Yellowfin Tuna Fishing in the Central and Western Pacific" *Sustainability* 14, no. 3: 1217. https://doi.org/10.3390/su14031217.
22. Li, Yan, Junfang Zhao, Rui Miao, Yan Huang, Xiaoqing Fan, Xiaoqing Liu, Xueqi Wang, Ye Wang, and Yuyang Shen. 2022. "Analysis of the Temporal and Spatial Distribution of Extreme Climate Indices in Central China" *Sustainability* 14, no. 4: 2329. https://doi.org/10.3390/su14042329.
23. Yan, Weixiong, Junfang Zhao, Jianping Li, and Yunxia Wang. 2021. "Assessment of Seasonal Variability of Extreme Temperature in Mainland China under Climate Change" *Sustainability* 13, no. 22: 12462. https://doi.org/10.3390/su132212462.
24. Gao, Zhibo, and Xiaodong Yan. 2022. "High-Resolution Regional Climate Modeling and Projection of Heatwave Events over the Yangtze River Basin" *Sustainability* 14, no. 3: 1141. https://doi.org/10.3390/su14031141.
25. Li, Kaiwen, Ming Wang, and Kai Liu. 2021. "The Study on Compound Drought and Heatwave Events in China Using Complex Networks" *Sustainability* 13, no. 22: 12774. https://doi.org/10.3390/su132212774.
26. Zheng, Yanjiao, Lijuan Zhang, Wenliang Li, Fan Zhang, and Xinyue Zhong. 2022. "Spatial-Temporal Variation of Snow Black Carbon Concentration in Snow Cover in Northeast China from 2001 to 2016 Based on Remote Sensing" *Sustainability* 14, no. 2: 959. https://doi.org/10.3390/su14020959.
27. Wang, Yumeng, Jingyan Ma, Lijuan Zhang, Yutao Huang, Xihui Guo, Yiping Yang, Enbo Zhao, Yufeng Zhao, Yue Chu, Meiyi Jiang, and Nan Wang. 2022. "Spatial Distribution of, and Variations in, Cold Regions in China from 1961 to 2019" *Sustainability* 14, no. 1: 465. https://doi.org/10.3390/su14010465.
28. Li, Hao, Wenshuang Xi, Lijuan Zhang, and Shuying Zang. 2021. "Snow-Disaster Risk Zoning and Assessment in Heilongjiang Province" *Sustainability* 13, no. 24: 14010. https://doi.org/10.3390/su132414010.
29. Sun, Shao, Qiang Zhang, Yuanxin Xu, and Ruyue Yuan. 2021. "Integrated Assessments of Meteorological Hazards across the Qinghai-Tibet Plateau of China" *Sustainability* 13, no. 18: 10402. https://doi.org/10.3390/su131810402.
30. Jia, Yiru, Jifu Liu, Lanlan Guo, Zhifei Deng, Jiaoyang Li, and Hao Zheng. 2021. "Locomotion of Slope Geohazards Responding to Climate Change in the Qinghai-Tibetan Plateau and Its Adjacent Regions" *Sustainability* 13, no. 19: 10488. https://doi.org/10.3390/su131910488.
31. Song, Shuaifeng, and Xiaodong Yan. 2021. "Changes in the Frequency of Extreme Cooling Events in Winter over China and Their Relationship with Arctic Oscillation" *Sustainability* 13, no. 20: 11491. https://doi.org/10.3390/su132011491.
32. Luo, Neng, and Yan Guo. 2022. "Impact of Model Resolution on the Simulation of Precipitation Extremes over China" *Sustainability* 14, no. 1: 25. https://doi.org/10.3390/su14010025.
33. Xu, Chen, Xianliang Zhang, Rocío Hernandez-Clemente, Wei Lu, and Rubén D. Manzanedo. 2022. "Global Forest Types Based on Climatic and Vegetation Data" *Sustainability* 14, no. 2: 634. https://doi.org/10.3390/su14020634.
34. Wang, Zheng, Yang Pan, Junxia Gu, Yu Zhang, and Jianrong Wang. 2022. "Quality Evaluation of the 0.01° Multi-Source Fusion Precipitation Product and Its Application in Extreme Precipitation Event" *Sustainability* 14, no. 2: 616. https://doi.org/10.3390/su14020616.

Reference

1. Jansky, S.H.; Jin, L.P.; Xie, K.Y.; Xie, C.H.; Spooner, D.M. Potato production and breeding in China. *Potato Res.* **2009**, *52*, 57–65. [CrossRef]

sustainability

MDPI

Article

Effects of Future Climate Change on Citrus Quality and Yield in China

Shuangshuang Wang, Wenqiang Xie and Xiaodong Yan *

State Key Laboratory of Earth Surface Processes and Resource Ecology, Faculty of Geographical Science, Beijing Normal University, Beijing 100875, China; wangss@mail.bnu.edu.cn (S.W.); wenqiangxie@mail.bnu.edu.cn (W.X.)
* Correspondence: yxd@bnu.edu.cn; Tel.: +86-13511027765

Abstract: As the world's most widely cultivated fruit, citrus in China is increasingly suffering from ongoing climate change, which affects the sustainability of agricultural systems and social economy. In this study, we linked climate factors to citrus quality and yield and established projection models to elucidate the impact of future climate change. Then, we used the ensemble mean of 19 Coupled Model Intercomparison Project 6 (CMIP6) models to project the 2021–2040 and 2041–2060 climate changes relative to the historical baseline 1995–2014 period under different shared socioeconomic pathways scenarios (SSP2-4.5, SSP5-8.5). The results show that the monthly mean diurnal temperature range in July had the greatest influence on quality, and monthly mean temperature in October, monthly mean relative humidity in October, monthly mean minimum temperature in November and monthly mean maximum temperature in September had the greatest influence on yield at the growth and ripening stages. Moreover, the quality and yield of citrus present different characteristics in terms of change in cultivation areas in the future. The quality of Sichuan, Zhejiang and Fujian Provinces in China will become significantly better, however, Hubei, Guangdong and Guangxi Provinces it will become worse. Surprisingly, yield will increase in all plantations due to future suitable climate conditions for citrus growth and ripening.

Keywords: citrus; climate change; quality; yield; future projection

Citation: Wang, S.; Xie, W.; Yan, X. Effects of Future Climate Change on Citrus Quality and Yield in China. *Sustainability* **2022**, *14*, 9366. https://doi.org/10.3390/su14159366

Academic Editor: Teodor Rusu

Received: 11 June 2022
Accepted: 26 July 2022
Published: 30 July 2022

1. Introduction

Climate change brings great challenges to natural resources and affects the sustainable development of human society [1]. It is now widely recognized as the greatest global threat of the 21st century [2]. Among many aspects of impacts, agriculture is the most sensitive sector [3,4]. In this changing environment, the production of many crops is affected, which is related to world food security and global stability [5–10]. As a result, researchers in agriculture have made it a priority to understand the relationship between crops and climate variables [11–13] and to predict crop yield and quality under climate change scenarios [14]. By correctly recognizing the contribution of climate change to crops and adopting effective adaptation measures to agriculture, human beings can make better use of improving production and resistance to adverse effects, maximizing the increase in output, reducing losses, and realizing potential benefits [15–17].

Citrus is the world's largest cultivated fruit crop, with an annual output of approximately 158 million tons, accounting for approximately 18% of the total fruit output [18,19]. As one of the most important crop types, citrus is an important source of income for farmers and is favoured by consumers for nutritional value [20,21], playing a significant role in people's livelihood, not to mention citrus juice. China has the largest population and is the largest citrus producer in the world, with an annual output of 44 million tons, accounting for almost 28% of global citrus production [18]. Thus, the production of citrus in China plays a vital role in the world citrus pattern and needs to be duly considered. However, China is increasingly suffering from ongoing climate change, and no part of the Earth is immune to this vulnerability, which could have a major effect on citrus [22,23]. Admittedly, the direction and degree of climate change in terms of influencing citrus varies locally due to regional differences in natural responses and anthropogenic factors.

The growth, development, flowering and fruiting of citrus are sensitive to climate conditions, especially in their yield and chemical quality [24–27]. Previous studies revealed that climate change in China affects the yield of citrus. Under the background of global warming, the citrus yield may be affected by climate risk in subtropical regions of China [28]. Soil erosion in citrus orchards caused by increasingly frequent and intense extreme precipitation is the main cause of productivity decline [29]. Additionally, the climate suitability of citrus affects the growth and final yield of citrus due to the change of temperature and precipitation suitability [30]. However, studies on the effect of citrus quality are rare. With advances in agricultural technology [31,32], climate change may have a greater impact on quality than yield. The contents of Vitamin C, naringin and hesperidin decreased significantly, while chlorogenic acid and caffeic acid increased during frozen temperatures [33]. The highest content of peel was observed from October to March and the Vitamin C content decreased during the ripening process [34]. Additionally, essential oils vary in content in different months [35]. Therefore, the development of this study is necessary to reveal the impact of future climate change on citrus quality, which is a factor that is considered to be as important as yield.

To project future citrus production, Tubiello [36] used two different global circulation model (GCM) scenarios to simulate climate change effects on US citrus production and the result showed that simulated fruit production benefited greatly from the projected climate change, as yields will increase by 20–50%. In this study, we combined empirical regression models based on climate factors and citrus quality or yield with datasets in different CMIP6 (Coupled Model Intercomparison Project 6) models in response to different shared socioeconomic pathways (SSPs) of future climate changes (SSP2-4.5, SSP5-8.5) on citrus. The aim of this study was to investigate the changes in citrus quality and yield in China in the near future (2021–2040) and medium future (2041–2060) relative to the historical baseline period (1995–2014) [37,38]. These results may provide useful information for perennial horticultural crops to meet the challenge of climate change and can be generalized to other parts of the world.

2. Materials and Methods

2.1. Study Area

The study areas include 79 meteorological stations in 11 major cultivation provinces in China, which are Sichuan, Chongqing, Hubei, Yunnan, Guizhou, Guangxi, Hunan, Jiangxi, Guangdong, Fujian and Zhejiang as shown in Figure 1 [39]. Typically, citrus production in these 11 provinces account for 98% of the total citrus production (45.85 million tons), and the cultivation areas account for 98% of the total citrus cultivation areas (2.62 million hectares) across China based on the National Bureau of Statistics of China (NBSC) [40].

2.2. Data Sources

2.2.1. Statistical Data

Citrus quality refers to the exterior quality and internal quality. The most important quality, in addition to fruit size, shape, colour and other economic values based on appearance, is chemical composition, mainly related to the sweet, sour or bitter taste of fruit, as well as the of active ingredient contents. The chemical composition of citrus fruit includes the following two categories in addition to water: the first is insoluble in water, such as cellulose, hemicellulose, and propectin, and the other includes water soluble substances, called total soluble solid (TSS). TSS is one of the main factors determining fruit quality [41–43], which reflects the flavor and can directly determine the commercial value of fruit. In this study, we searched relevant published articles containing citrus TSS data from the China National Knowledge Infrastructure (CNKI) and finally selected 125 available TSS data. TSS data combined with information from the China Meteorological

Station (CMS) are listed in the Table S1. Therefore, the climate of citrus cultivation areas was characterized by the local CMS from the China Meteorological Administration. From the NBSC data, the citrus production and planting area of each province were obtained to calculate the yield.

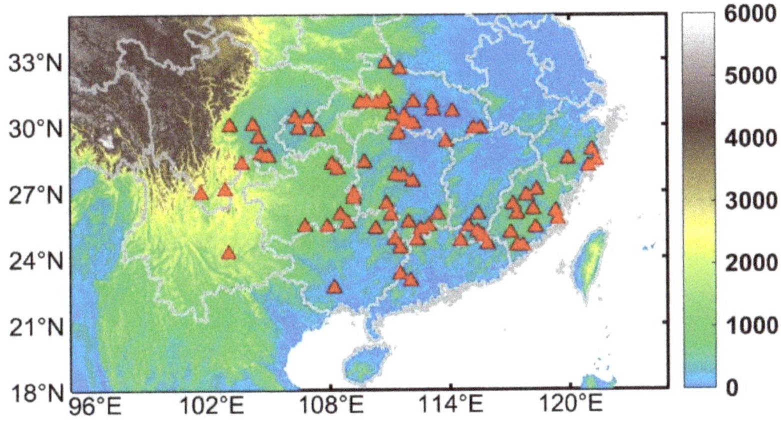

Figure 1. 79 meteorological stations in citrus cultivation areas.

2.2.2. Meteorological Observation Data

CMS (V3.0) contains the Chinese benchmark and general weather stations, including the main information of 2474 sites and the basic meteorological observation data since January 1961. We extracted meteorological data from 11 provinces in major citrus producing areas, including mean temperature, maximum temperature, minimum temperature, relative humidity and other variables on a daily scale. Finally, 79 sites were collected from papers and the data were downloaded from China Meteorological Administration as observational data and used to construct citrus quality and yield regression models.

2.2.3. CMIP6 Data

The 19 CMIP6 global climate models were used to simulate climate change in the near future (2021–2040) and medium future (2041–2060) relative to the historical base period (1995–2014) under SSP2-4.5 and SSP5-8.5 scenarios. The relevant information of 19 CMIP6 global climate models is shown in Table 1. Among them, EC-Earth3 and EC-Earth3-Veg models have the highest spatial resolution ($0.7^\circ \times 0.7^\circ$), and the CanESM5 model has the lowest spatial resolution ($2.8^\circ \times 2.8^\circ$). The relative humidity, maximum temperature, minimum temperature and mean temperature variables of the CMIP6 model were extracted (there was no simulation of relative humidity variables in the future period in the BCC-CSM2-MR model; therefore, the relative humidity in the future period was replaced by the results of 18 model sets).

2.3. Methods

Citrus is an important fruit crop with high economic value and in this study, it was assumed that when citrus lacks water, artificial irrigation is provided; therefore, the impact of precipitation was not considered. The distribution of climate factors such as the mean temperature, minimum temperature, maximum temperature and relative humidity in different growing areas is significantly different, which often results in different effects of climate change on citrus fruit quality and yield [25,44]. Diurnal temperature range (DTR) is the difference between daily maximum temperature and daily minimum temperature, which can reflect the change characteristics of the interaction and provides comprehensive information between the two [45,46]. Changes in DTR can convey climate change information, which will have an impact on human health, the circulation of the ecosystem, the growth of animals and plants, and the use of renewable energy [47]. The DTR described in this paper is the maximum temperature minus the minimum temperature in 24 h. By using a correlation analysis, regression models were established for the relationship between citrus fruit quality and yield and the key climate factors during fruit growth; furthermore, the crop models of citrus fruit quality and yield were obtained. Lobell [48] considered that all process models contain some degree of experience or statistical rules, and all statistical models also contain some hypothesis of crop processes and mechanisms [49–51]. The change in climate elements has a nonnegligible impact on the growth and development of crops. Therefore, when crop quality and yield are only determined by climate factors, the response characteristics of crops to climate elements should be understood. Statistical models can be used to predict crop responses to climate change [52,53]. By incorporating CMIP6 climate data into crop models, changes in citrus fruit quality and yield were obtained for different future periods. See the supplementary materials for the flow chart.

Table 1. 19 CMIP6 models used in the study.

CMIP6 Models	Institution	Spatial Resolution (Lat * lon)	Variables
ACCESS-CM2	CSIRO-ARCCSS, Australia	144 * 192	Tas, Tasmax, Tasmin, Hurs
ACCESS-ESM1-5	CSIRO, Australia	145 * 192	Tas, Tasmax, Tasmin, Hurs
BCC-CSM2-MR	BCC, China	160 * 320	Tas, Tasmax, Tasmin
CanESM5	CCCma, Canada	64 * 128	Tas, Tasmax, Tasmin, Hurs
CAS-ESM2-0	CAS, China	128 * 256	Tas, Tasmax, Tasmin, Hurs
CMCC-ESM2	CMCC, Italy	192 * 288	Tas, Tasmax, Tasmin, Hurs
EC-Earth3	EC-Earth-Consortium, European Union	256 * 512	Tas, Tasmax, Tasmin, Hurs
EC-Earth3-Veg	EC-Earth-Consortium, European Union	256 * 512	Tas, Tasmax, Tasmin, Hurs
EC-Earth3-Veg-LR	EC-Earth-Consortium, European Union	160 * 320	Tas, Tasmax, Tasmin, Hurs
FIO-ESM-2-0	FIO-QLNM, China	192 * 288	Tas, Tasmax, Tasmin, Hurs
FGOALS-g3	CAS, China	80 * 180	Tas, Tasmax, Tasmin, Hurs
GFDL-ESM4	NOAA-GFDL, United States	180 * 288	Tas, Tasmax, Tasmin, Hurs
INM-CM4-8	INM, Russia	120 * 180	Tas, Tasmax, Tasmin, Hurs
INM-CM5-0	INM, Russia	120 * 180	Tas, Tasmax, Tasmin, Hurs
IPSL-CM6A-LR	IPSL, France	143 * 144	Tas, Tasmax, Tasmin, Hurs
MIROC6	MIROC, Japan	128 * 256	Tas, Tasmax, Tasmin, Hurs
MPI-ESM1-2-HR	MPI-M, Germany	192 * 384	Tas, Tasmax, Tasmin, Hurs
MPI-ESM1-2-LR	MPI-M, Germany	96 * 192	Tas, Tasmax, Tasmin, Hurs
MRI-ESM2-0	MRI, Japan	160 * 320	Tas, Tasmax, Tasmin, Hurs

2.3.1. Meta-Analysis

A meta-analysis is a method that can be used to conduct a quantitative and comprehensive analysis of research results [54]. In this study, the quality data of fruit was finally determined by summarizing the research results in the relevant published literature and conducting repeated screening and averaging of the sample data. The operation process includes the following four elements: (1) keywords: TSS of citrus, fruit origin and picking year; (2) unity: recording the data of different varieties of fruit, delimiting the research area and the research benchmark period so that the data of different producing areas and different varieties of fruit have uniformity; (3) match external information: according to the origin of fruit, the relevant information from China meteorological Station was matched, and the corresponding climate data were extracted; (4) obtain results: data of TSS of citrus in different years, different producing areas and different climate conditions. The scientific database used in this study was obtained from China National Knowledge Network (CNKI), with a focus on papers published on climate change, and the exclusion of the effects of extreme climate events such as drought, flood and frost and human activities such as technological progress on fruit quality data.

2.3.2. Correlation Analysis

Correlation analysis is a method used to evaluate the relationship between two variables and the correlation coefficient indicates the strength of the relationship between variables [55]. In this study, Pearson's correlation coefficient calculation method was used to calculate the correlation degree between the two variables, i.e., citrus quality and yield with climate factors, for which the formula utilized is as follows:

$$R = \frac{\sum_{i=1}^{n}(x_i - \overline{x})(y_i - \overline{y})}{\sqrt{\sum_{i=1}^{n}(x_i - \overline{x})^2}\sqrt{\sum_{i=1}^{n}(y_i - \overline{y})^2}}$$

where R represents the correlation coefficient, n is the number of samples, x_i and y_i are the values of the ith sample in the sequence of two variables, and $\overline{x}$ and $\overline{y}$ are the average values of the sequence of two variables.

2.3.3. Least Square Estimation

In this study, a unary linear regression equation was used to fit the relationship between citrus quality and climate factors, and the trend of quality change was defined as the slope of least square estimation, for which the formula is [55] as follows:

$$\hat{y}_i = kx_i + b, i = 1, 2, \ldots, n$$

$$k = \frac{\sum_{i=1}^{n} x_i y_i - \frac{1}{n}\left(\sum_{i=1}^{n} x_i\right)\left(\sum_{i=1}^{n} y_i\right)}{\sum_{i=1}^{n} x_i^2 - \frac{1}{n}\left(\sum_{i=1}^{n} x_i\right)}$$

$$b = \overline{y} - k\overline{x}$$

$$\overline{y} = \frac{1}{n}\sum_{i=1}^{n} y_i$$

$$\overline{x} = \frac{1}{n}\sum_{i=1}^{n} x_i$$

where y_i represents citrus quality with sample size n, x_i corresponds to climate factor, k represents the regression coefficient term, and b represents the regression constant term.

2.3.4. Bilinear Interpolation and Multimodel Ensemble

To avoid the uncertainties that may occur in a single model simulation, we selected 19 models from the CMIP6 GCMs. Using the bilinear interpolation method, we interpolated the models with different resolutions on a unified $1° \times 1°$ grid and obtained the average result of the multimodel ensemble [56], for which the formula is as follows:

$$Ensemble = \frac{\sum_{i=1}^{19} M_i}{19}$$

where *Ensemble* represents the average value of the multimodel result, and M_i represents the ith model value. Then, by using the method of bilinear interpolation, the average model data of multiple model sets were uniformly interpolated to the China meteorological stations outlined in the Supplementary Materials Table S1 to obtain the local climate information of the meteorological stations simulated by the model [57].

3. Results

3.1. Changes of Quality in the Climate Change Factors

3.1.1. Quality in Relation to Climate Variables

The critical period for the growth and ripening of citrus fruits is from July to December. During this period, the change in DTR is the main climatic factor affecting the quality content of fruits [58,59]. Admittedly, a higher day temperature is preferred for the accumulation of sugar and degradation of organic acids in the fruit-ripening process, and a lower night temperature favours the same under suitable fruit growth conditions [60,61]. DTR can comprehensively reflect the information of maximum and minimum temperature, which has a considerable influence on fruit quality. To quantitatively compare the time period with the maximum correlation with TSS, the maximum temperature and minimum temperature observed at meteorological stations in different months were used to calculate the mean values of DTR and conduct a correlation analysis with TSS, as shown in Figure 2. The results show that DTR is positively correlated with TSS in citrus fruit growth in all periods and passes the significance test at 0.05. The correlation between the mean value of the DTR in July and TSS is highest (R = 0.44), and the correlation coefficient of the mean values of the DTR in the periods from July to August, July to September, July to October, July to November and July to December remains above 0.3, which is higher than that in other periods.

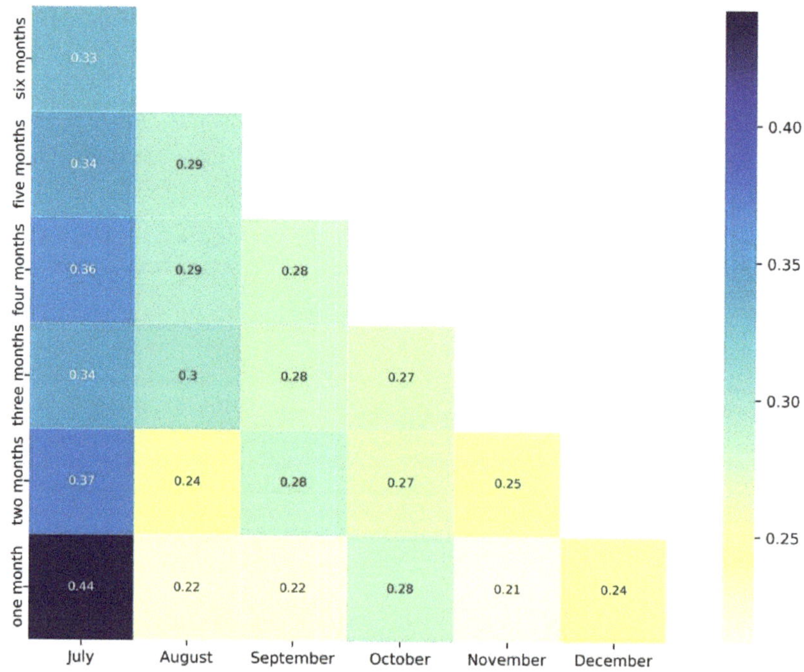

Figure 2. Correlation coefficient diagram between the TSS and DTR for each period from July to December. The abscissa represents the beginning month, and the ordinate represents different time spans of the monthly mean DTR. All data passed the significant correlation coefficient at the 0.05 significance level.

Based on the correlation analysis in Figure 2, the maximum significant positive correlation between the mean for the DTR value in July and the TSS content was obtained to select the climate factors that most affected the change in TSS content in the key period. A linear regression equation was obtained, and the quality prediction model was established as follows:

$$\mathbf{Q = 0.35 \times \mathit{X}_1 + 9.021\ (\mathit{R}_1 = 0.442, \mathit{P}_1 < 0.001)}$$

where **Q** indicates the TSS content (%); X_1 indicates the mean DTR value (°C) in July; R_1 is the correlation coefficient; and P_1 is the significance test index.

The citrus quality prediction model indicates that if the DTR becomes larger, the TSS content also increases, and the citrus quality is good; otherwise, it becomes worse under the background of future climate change.

3.1.2. Changes in the Future DTR

The CMIP6 multimodel ensemble data were used to estimate future DTR changes in July in 11 provinces in China where citrus cultivation areas are located, as shown in Figure 3. Compared with the 1995–2014 historical baseline period, the DTR in July of citrus producing areas in China presents an overall spatial distribution of increase-decrease-increase from west to east in the 2021–2040 future climate state under the two scenarios, SSP2-4.5 and SSP5-8.5. Sichuan Province in the west and Zhejiang and Fujian Provinces in the east show signs of increasing DTRs; however, the DTRs of Hubei, Hunan, Guangxi and other central regions show decreasing trends. Similarly, in the 2041–2060 mid-future period, the DTR shows almost the same spatial distribution, and the range of change is larger than that in the 2021–2040 period. The DTR in the northern part of Sichuan Province increases above 0.4 °C under the SSP5-8.5 scenario. The central regions, such as Guizhou, Hunan and Jiangxi Provinces, also show signs of increasing DTR trends, but in addition, the other regions show signs of decreasing DTR changes within 0.2 °C under the SSP2-4.5 scenario.

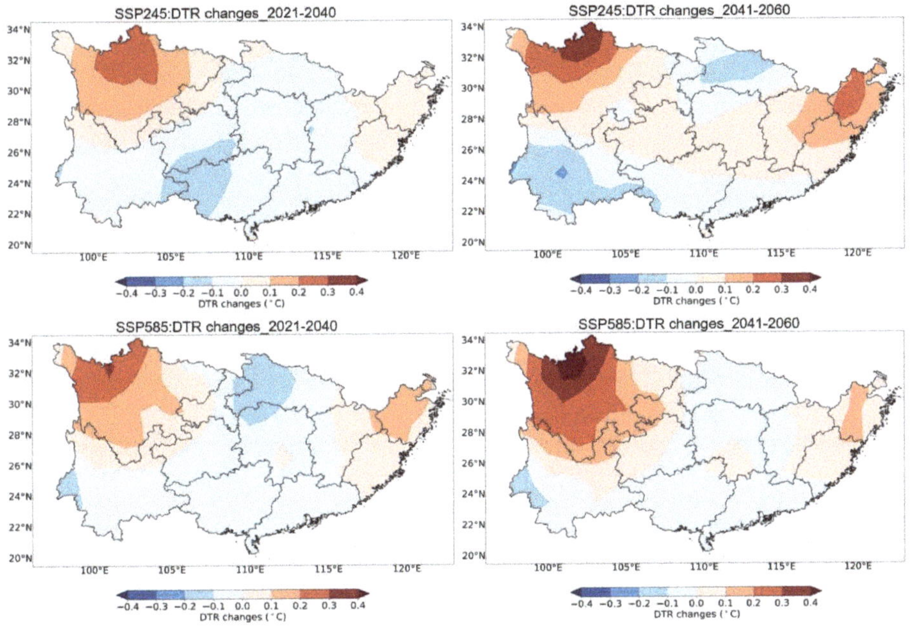

Figure 3. Under the two scenarios, SSP2-4.5 and SSP5-8.5, the DTR in July changes in the near future (2021–2040) and mid future (2041–2060) compared with the historical period (1995–2014).

3.1.3. Changes in the future TSS

The change in DTR in July in the whole region of citrus producing areas obtained in Figure 3 can be entered into the quality prediction model to calculate the change in TSS content. Moreover, the spatial grid DTR data calculated by the CMIP6 multimodel were interpolated to the local meteorological stations, and the spatial distribution map of TSS content in actual citrus-producing areas is depicted in Figure 4. Consistent with the spatial distribution in Figure 3, the change in TSS content in citrus also presents a spatial distribution characteristic of increase-decrease-increase from west to east in the 2021–2040 future climate state under the SSP2-4.5 and SSP5-8.5 scenarios. This phenomenon indicates that the quality of citrus has a tendency to become better-worse-better in these corresponding regions. In the 2041–2060 mid-future climate state, the TSS content of citrus shows a ribbon-like distribution of increasing TSS content in the western, central and eastern regions and decreasing TSS content in the southern and northern regions under the SSP2-4.5 and SSP5-8.5 scenarios. It is suggested that the quality of citrus will change in the same trend. Sichuan, Zhejiang and Fujian Provinces in terms of citrus quality present good trends for the entirety of the future situation. Citrus quality in Guizhou, Hunan, Jiangxi, northern Guangdong and northern Guangxi Provinces will change in the near future to worse and in the mid-future to good. Citrus produced in Hubei Province is affected by climate change and shows quality deterioration under different scenarios and in different future climate states.

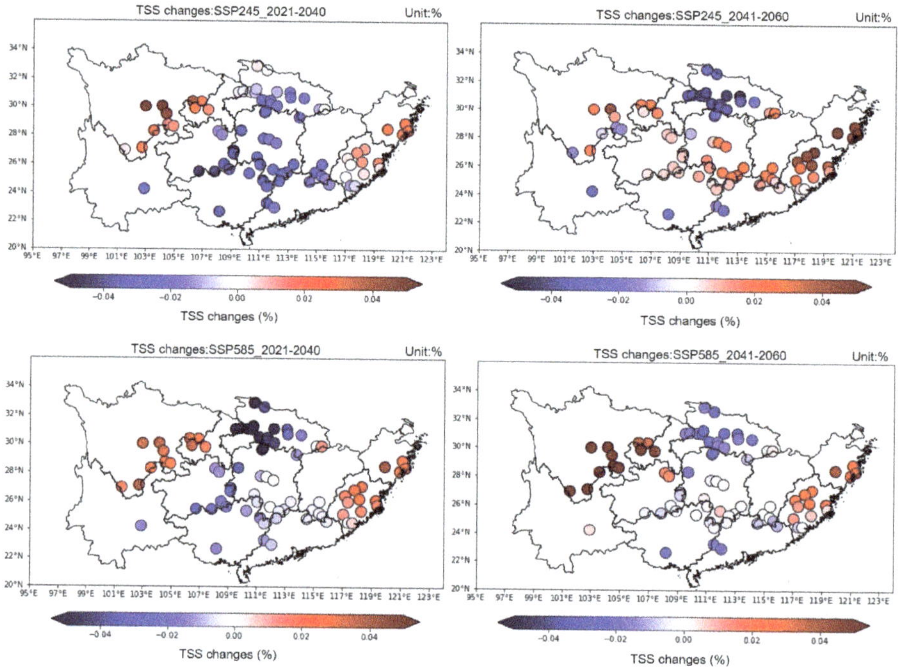

Figure 4. Under the two scenarios, SSP2-4.5 and SSP5-8.5, the TSS content of citrus at the meteorological site scale of Chinese citrus-producing areas changes in the near future (2021–2040) and in the mid future (2041–2060) compared with the historical period (1995–2014). The dots indicate citrus TSS content variations. Blue indicates decreasing change, and red indicates increasing change.

3.2. Changes of Yield in the Climate Change Factors

3.2.1. Yield in Relation to Climate Variables

Ahmad [62] confirmed that citrus fruit growth requires certain temperature conditions, and the yield is particularly sensitive to temperature. Citrus fruit grows and ripens from July to December each year, and the climate factors during this period have great effects on the yield. Consequently, the correlation between climate factors and yield in the above months was examined, and the correlation coefficient diagram was obtained, as shown in Figure 5. As seen from the figure, mean temperature in October, relative humidity in October, minimum temperature in November and maximum temperature in September have the highest correlation with yield; as a result, the particular climatic conditions were selected as independent variables to establish a multiple linear regression model and obtain the prediction of the yield model as follows:

$$\mathbf{Y = 0.007 \times X_2 + 0.108 \times X_3 - 0.025 \times X_4 - 0.03X_5 + 2.482\ (R_2 = 0.437,\ P_2 < 0.001)}$$

where **Y** indicates the yield (10 t/ha); X_2 indicates the mean temperature (°C) in October; X_3 indicates the mean minimum temperature (°C) in November; X_4 indicates the mean maximum temperature (°C) in September; X_5 indicates the mean relative humidity (%) in October; R_2 is the correlation coefficient; and P_2 is the significance test index. The citrus yield prediction model indicates that mean temperature and minimum temperature have a positive contribution to yield, while maximum temperature and relative humidity have a negative contribution under the synergistic effect of various climate factors.

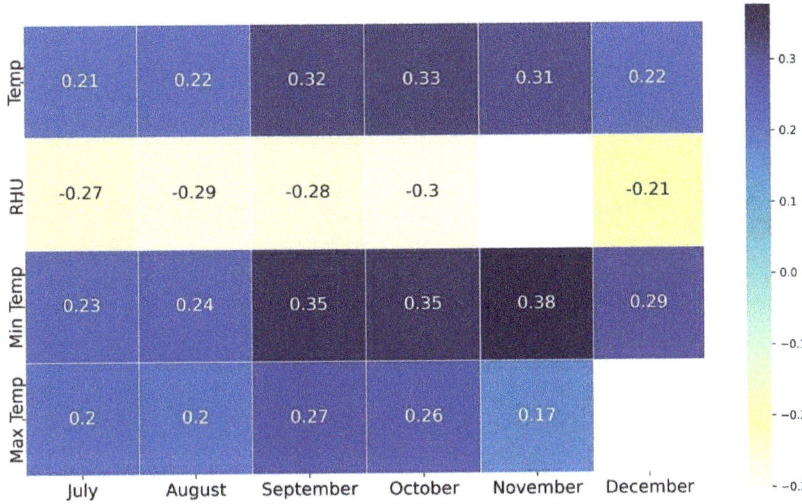

Figure 5. Correlation coefficient diagram between the monthly mean maximum temperature, minimum temperature, relative humidity and mean temperature for each month from July to December and yield. The abscissa represents each month, and the ordinate represents the monthly mean climate variables mentioned above. The correlation coefficients in the figure all passed the significance test at 0.05, and those that did not pass are displayed as blanks.

3.2.2. Changes in the Future Climate Factors

The CMIP6 multimodel ensemble data were used to predict the future changes in four key climate factors affecting citrus yield under different scenarios and different periods in the future as shown in Figure 6 (Figure 6a–d). The mean temperature in October shows an overall increase, as shown in Figure 6a. There is little difference between the simulation results of the two different scenarios, and the mean temperature range is between 0.5 °C and 1.5 °C in 2021–2040. Undoubtedly, the mean temperature range of the SSP5-8.5 scenario is significantly higher than that of SSP2-4.5 in 2041–2060. The range of mean temperature shows a phenomenon of increasing with latitude moving northwards in the citrus-growing areas. Figure 6b show that the mean minimum temperature in November presents signs of increasing in all regions, and the greatest warming trend is observed for northwestern Sichuan. In addition, the increasing range of most regions is approximately 0.5 °C higher under the SSP5-8.5 scenario than under the SSP2-4.5 scenario in 2021–2040, and the spatial distribution of the warming gradient is consistent with that in 2041–2060. In Figure 6c, the warming effect of the mean maximum temperature in September is approximately the same as that of the mean temperature in October in Figure 6a, except that the greatest warming effect shifts from the northwestern to the central and eastern parts of Sichuan. The mean relative humidity in October increases in the western region and decreases in the central and eastern regions, as shown in Figure 6d. The spatial pattern is approximately similar under the two scenarios during different future periods, while the decreasing trend is almost 0.5% higher in 2041–2060.

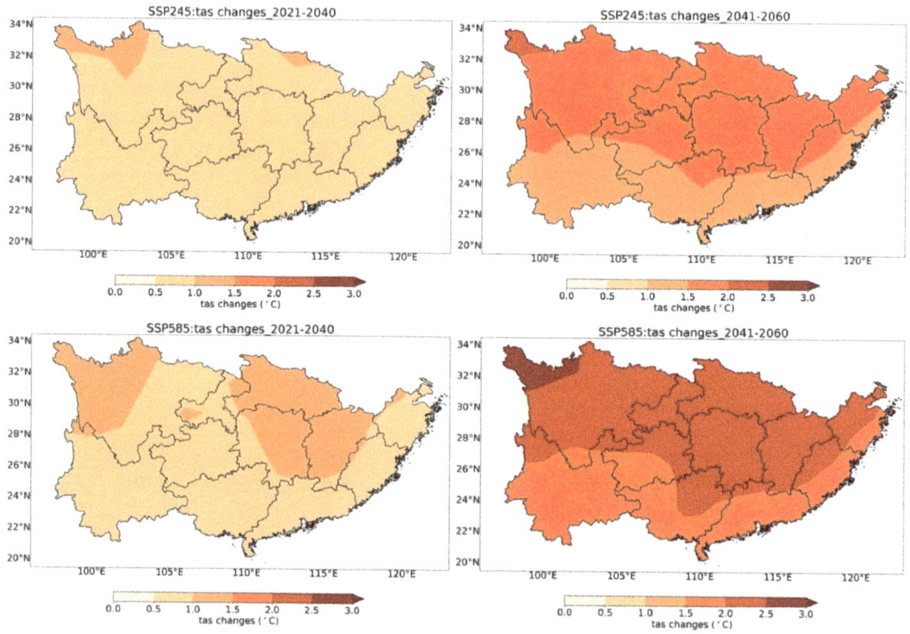

(**a**)

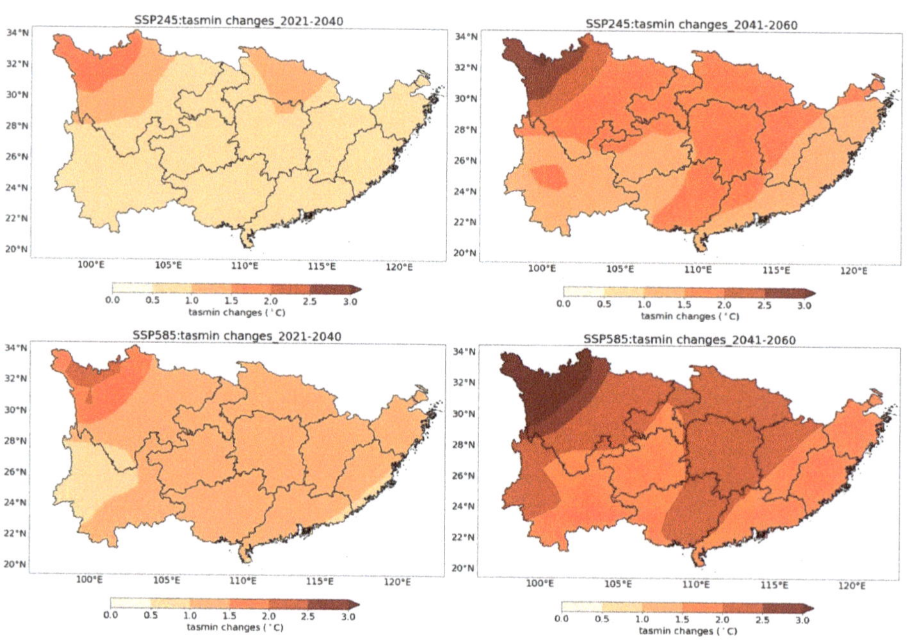

(**b**)

Figure 6. *Cont.*

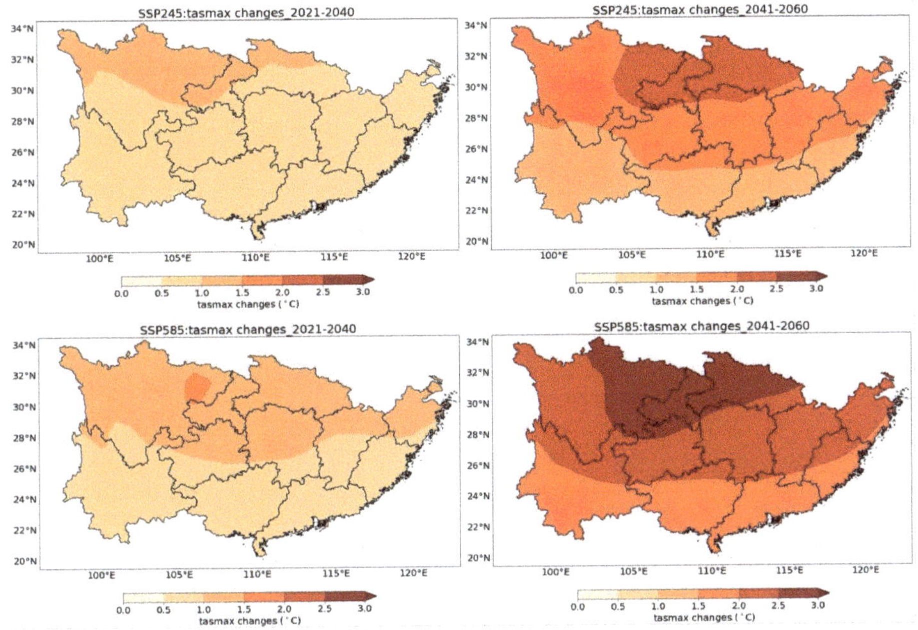

(**c**)

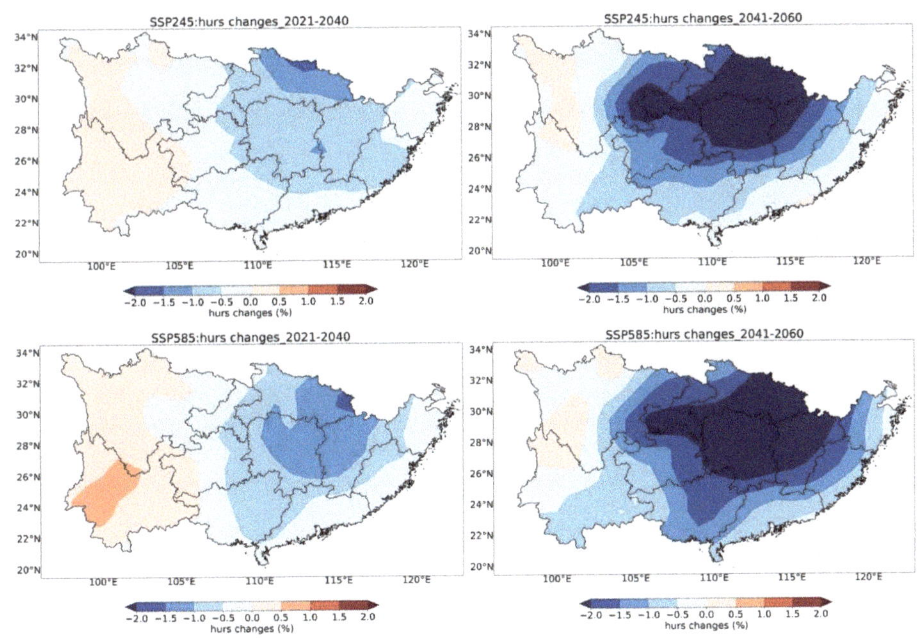

(**d**)

Figure 6. Under the two scenarios, SSP2-4.5 and SSP5-8.5, the various climate factors change in the near future (2021–2040) and mid future (2041–2060) compared with the historical period (1995–2014). (**a**) Mean temperature (°C) in October. (**b**) Mean minimum temperature (°C) in November. (**c**) Mean maximum temperature (°C) in September. (**d**) Mean relative humidity (%) in October.

3.2.3. Changes in the Future Yield

According to Figure 6, we can see key climate factor changes to some extent simulated by CMIP6 models in the future period under two different scenarios, which were entered into the citrus yield prediction model. Compared with the historical period, the change in citrus yield at the meteorological site scale in the future is shown in Figure 7. Citrus yield shows an increasing trend under different conditions, and the yield-increasing effect is more obvious in 2041–2060 than in 2021–2040, almost doubling the change. The citrus yields of Hubei, Hunan, northern Guangxi and northern Guangdong Provinces increase with an increase of 1–2 t/ha under the SSP2-4.5 scenario; however, almost all citrus-producing areas show an increase of 1–2 t/ha with little difference in yield increase among regions under the SSP5-8.5 scenario in 2021–2040. In regard to the 2041–2060 period, the yields in southeastern Sichuan, Hubei and central Hunan Provinces increase by more than 2 t/ha, and those of Zhejiang and Fujian Provinces almost double, while those of southern Hunan, northern Guangxi and northern Guangdong Provinces have almost no difference compared with the 2021–2040 period under the SSP2-4.5 scenario. A yield increase of above 2 t/ha is observed in most citrus-producing areas except those in Fujian, Yunnan and southern Sichuan Provinces and is double that from 2021–2040 under the SSP5-8.5 scenario.

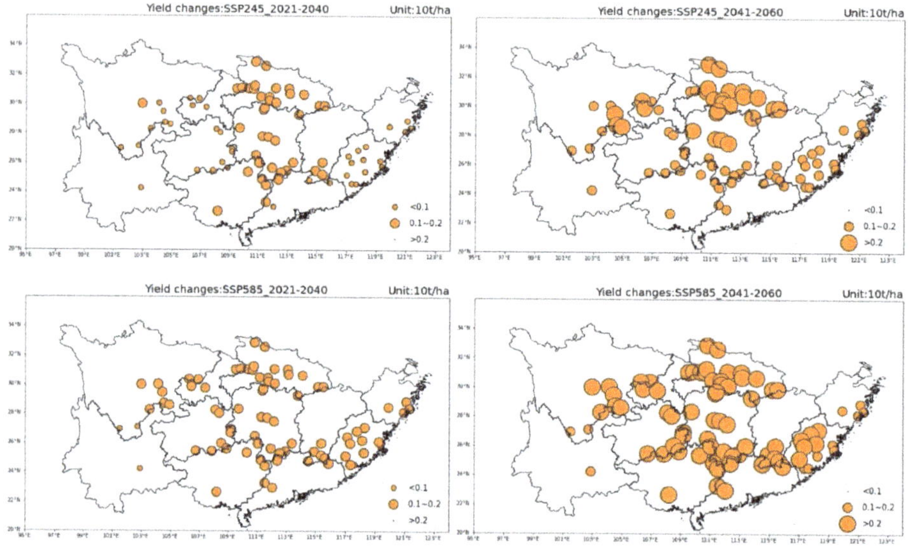

Figure 7. Under the two scenarios, SSP2-4.5 and SSP5-8.5, the yield of citrus at the meteorological site scale of Chinese citrus-producing areas changes in the near future (2021–2040) and in the mid future (2041–2060) compared with the historical period (1995–2014). The orange dots indicate yield changes whose sizes show the degree of change.

3.3. *Changes of Quality-Yield in the Climate Change Factors*

Quality-Yield (Q-Y) is defined as the total amount of TSS content contained in the yield, which reflects comprehensive information between the quality and yield changes in the future in t/ha. The change in Q-Y is shown in Figure 8 combined with the forecast for TSS content in Figure 4 and the phenomenon of all conditions increasing yield in Figure 6, which reflects the general change in citrus effective composition TSS content in yield. Although the quality of citrus in some producing areas is worsening, affected by climate change, Q-Y still shows an increasing trend with the increase in the citrus yield, which indicates that the amount of TSS content will improve in the future under the two different scenarios. The spatial distribution of the Q-Y increase is approximately 0.1–0.2 t/ha despite the obvious quality reduction in the central region and almost the same in all planting regions except in the Zhejiang and Fujian Provinces under the SSP2-4.5 and SSP5-8.5 scenarios in 2021–2040. Because of the effect of yield increase, Q-Y almost doubles compared with 2021–2040 under the SSP2-4.5 scenario in 2041–2060. Although the quality obviously declines in Hubei Province, the Q-Y increases by approximately 0.2 t/ha. The increasing effect of Q-Y is weak in Zhejiang and Fujian Provinces and some other places at approximately 0.1–0.2 t/ha. The decline in quality in the central and southern regions does not affect the Q-Y increase by more than 0.2 t/ha under the SSP5-8.5 scenario in 2041–2060.

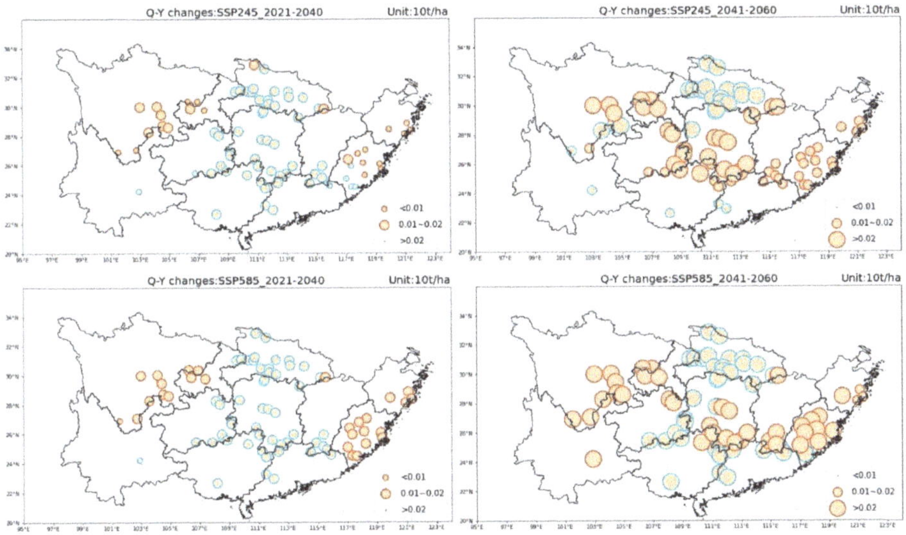

Figure 8. Under the two scenarios, SSP2-4.5 and SSP5-8.5, the quality-yield of citrus at the meteorological site scale of Chinese citrus-producing areas will change in the near future (2021–2040) and in the mid future (2041–2060) compared with the historical period (1995–2014). The orange dots indicate quality-yield changes, whose sizes show the degree of change. The red circles on the outer layer of the dots indicate the increase in TSS content, and the blue circles indicate the decrease in TSS content shown in Figure 4.

4. Discussion

4.1. Impact Mechanisms of Empirical Models for Predicting Citrus Fruit Quality and Yield

Many studies revealed the impact of climate change on the yield of crops, indicating that changes in climate factors such as temperature and precipitation will increase or decrease the yield [9,63,64]. However, fruits have not been given the same attention, and it is necessary to study the impact of climate change on fruits and what will happen to fruits in the future, especially in relation to their yield and quality. The climate factors used in this study are also relevant to temperature and precipitation factors, such as maximum temperature, minimum temperature, DTR, mean temperature and relative humidity, on a daily scale, as shown in Figures 2 and 5. Maximum temperature is beneficial for the accumulation of active substances in the fruit-ripening process, and minimum temperature favours the same under suitable fruit growth conditions [59,61]. DTR can comprehensively reflect the information of maximum and minimum temperature, which has a considerable influence on fruit quality. Mean temperature and relative humidity have been proven to be very important in the growth of citrus and have certain effects on phenology and yield [65–67]. The above climate factors in the key growth periods calculated from meteorological station data in citrus producing areas have a strong correlation with quality and yield. It has been proven that local climate change has a direct impact on citrus production, in contrast to large-scale warming conditions. This method was also applied to study climate-crop yield relationships [68]. In this study, statistical models were used to establish the relationship between climate factors and citrus quality and yield, which have certain reference values because some researchers used statistical models to predict crop yield [69,70]. Mechanism models are so complex that few suitable models can be used to predict the quality and yield changes of citrus in the future, which means that it was difficult to select a model in our study. However, the comprehensive use of mechanism models and statistical models will be of great significance for the prediction of the quality and yield of citrus and even for fruit once the mechanism model is developed and perfected [71,72].

4.2. Sensitive Areas Affected by Climate Factors in the Future

The key period of citrus fruit growth and maturation is from July to December, as shown in Figures 2 and 5, and the climate change in citrus-producing areas in this period has a direct influence on the change in citrus fruit quality and yield. This conclusion shows that the climate factors in the study area have the same trend of change under the two scenarios, but the intensity of change is greater under the SSP5-8.5 scenario, which is also in line with the simulation setting of future emission scenarios. In addition, the prediction of different future time periods presents different spatial distributions. In the 2021–2040 period, the DTR in July is projected to increase in Sichuan, Zhejiang and Fujian Provinces and decrease in most studied areas; the mean temperature in October, maximum temperature in September and minimum temperature in November seem to increase in all areas; and the relative humidity in October is projected to increase in Sichuan and Yunnan Provinces and decrease in other places. In the 2041–2060 period, the DTR in July is projected to decrease in only some parts of Hubei, Guangdong, Guangxi and Yunnan Provinces and increase in other areas; the mean temperature in October, maximum temperature in September and minimum temperature in November will increase in all areas; and the relative humidity in October is projected to decrease largely in the study areas.

4.3. Some Measures May Improve Citrus Quality and Yield

According to Figure 2, the DTR will decrease in most cultivation regions except for Sichuan, Zhejiang and Fujian Provinces, which indicates that the quality of citrus is projected to worsen. Some artificial adaptation measures may be taken to prevent negative situations. The DTR can be obtained by subtracting the daily maximum temperature from the daily minimum temperature, and both will increase significantly under the background of global warming; therefore, the reason for the decrease in the DTR is that the warming effect has a more significant enhancement effect on the daily minimum temperature. Suggested coping strategies include a lower night temperature and providing enough day warming conditions. On the other hand, changing cultivation regions is a contributing factor. Admittedly, there is a decreasing degree of DTR of between 0.1 °C and 0.2 °C, which has seldom effected TSS content as shown in Figure 3. Quality decline does not represent a serious concern based on the results under climate warming conditions. Sichuan Province may have the best natural DTR conditions for the accumulation of active substances in the future. Under the joint action of various climate factors, the changing climate is beneficial for citrus fruit growth and ripening, and the yield of citrus is projected to increase in all producing regions, not to mention the improvement of agricultural technology. Based on our hypothesis, offering irrigation is critical.

4.4. Limitations of this Study

(1) Without the support of specific citrus quality and yield data at the grid scale, the research areas of this study were limited to all provinces, and the locations of meteorological stations were used to represent the local climate conditions, which included certain errors. (2) Due to the inability of the author and the research group to undertake relevant experiments, the research data on TSS contents of citrus in this study were obtained from

other published papers, and the consistency of the data were not guaranteed. (3) In this study, the ensemble mean of 19 CMIP6 models was adopted to reduce the uncertainty of single-model simulations of climate change impacts. However, the GCM climate models also have some system errors in terms of the observations, which lead to the uncertainty in future changes of projected climate variables. (4) This study did not consider the effect of artificial technological progress on the results.

5. Conclusions

The key climate factors from July to December of citrus fruit growth and maturation have a good relationship with citrus quality and yield. The monthly mean DTR in July has the greatest influence on quality, and monthly mean temperature in October, monthly mean relative humidity in October, monthly mean minimum temperature in November and monthly mean maximum temperature in September have the greatest influence on yield. Moreover, the monthly mean DTR in July is projected to increase in Sichuan, Zhejiang and Fujian Provinces and decrease in other regions; the monthly mean temperature in October, monthly mean minimum temperature in November and monthly mean maximum temperature in September are projected to increase in all studied areas; and the monthly mean relative humidity in October is projected to increase in small regions of Sichuan and Yunnan Provinces and decrease in other places. Thus, the quality and yield of citrus presented different characteristics of change in cultivation areas when using the established prediction model for the 2021–2040 and 2041–2060 future periods relative to the 1995–2014 baseline period. The quality of western cultivation areas in Sichuan Province and eastern cultivation areas in Zhejiang and Fujian Provinces in China will become significantly better; however, that of Hubei, Guangdong and Guangxi Provinces will worsen. Surprisingly, yield will increase in all plantations due to future suitable climate conditions for citrus growth and ripening.

Supplementary Materials: The following supporting information can be downloaded at: https://www.mdpi.com/article/10.3390/su14159366/s1, Figure S1: Flow chart; Table S1: TSS data used in this study [73–85].

Author Contributions: X.Y. and S.W. conceived of the idea and method, W.X. and S.W. participated in the prepared materials, and S.W. wrote the paper with input from all authors. All authors have read and agreed to the published version of the manuscript.

Funding: This research was funded by the National Key R&D Program of China, grant number 2019YFA0606904.

Institutional Review Board Statement: Not applicable.

Informed Consent Statement: Not applicable.

Data Availability Statement: All relevant data can be found within the paper and its supporting materials. CMIP6 models data are available at link https://esgf-node.llnl.gov/search/cmip6 (accessed on 10 June 2022).

Acknowledgments: We would like to thank the authors who did the experiments to get the citrus data in the supplementary material, which enabled our work to proceed smoothly. We acknowledge the China Meteorological Administration for meteorological data and information.

Conflicts of Interest: The authors declare no conflict of interest.

References

1. Stocker, T.F.; Qin, D.; Plattner, G.-K.; Tignor, M.; Allen, S.K.; Boschung, J.; Nauels, A.; Xia, Y.; Bex, V.; Midgley, P.M. (Eds.) *IPCC Climate Change 2013: The Physical Science Basis. Contribution of Working Group I to the Fifth Assessment Report of the Intergovernmental Panel on Climate Change*; Cambridge University Press: Cambridge, UK, 2013.
2. Piao, S.; Ciais, P.; Huang, Y.; Shen, Z.; Peng, S.; Li, J.; Zhou, L.; Liu, H.; Ma, Y.; Ding, Y.; et al. The impacts of climate change on water resources and agriculture in China. *Nature* **2010**, *467*, 43–51. [CrossRef] [PubMed]

3. Adams, R.M.; Rosenzweig, C.; Peart, R.M.; Ritchie, J.T.; McCarl, B.A.; Glyer, J.D.; Curry, R.B.; Jones, J.W.; Boote, K.; Allen, L.H. Global climate change and US agriculture. *Nature* **1990**, *345*, 219–224. [CrossRef]
4. Salinger, M. Climate variability and change: Past, present and future: An overview. *Clim. Change* **2005**, *70*, 9–29. [CrossRef]
5. Rosenzweig, C.; Phillips, J.; Goldberg, R.; Carroll, J.; Hodges, T. Potential impacts of climate change on citrus and potato production in the US. *Agric. Syst.* **1996**, *52*, 455–479. [CrossRef]
6. Furuya, J.; Koyama, O. Impacts of Climatic Change on World Agricultural Product Markets: Estimation of Macro Yield Functions. *Jpn. Agric. Res. Q. JARQ* **2005**, *39*, 121–134. [CrossRef]
7. Parry, M.; Rosenzweig, C.; Livermore, M. Climate change, global food supply and risk of hunger. *Philos. Trans. R. Soc. B Biol. Sci.* **2005**, *360*, 2125–2138. [CrossRef] [PubMed]
8. Lobell, D.B.; Schlenker, W.; Costa-Roberts, J. Climate Trends and Global Crop Production since 1980. *Science* **2011**, *333*, 616–620. [CrossRef] [PubMed]
9. Lesk, C.; Rowhani, P.; Ramankutty, N. Influence of extreme weather disasters on global crop production. *Nature* **2016**, *529*, 84–87. [CrossRef] [PubMed]
10. Xie, Q.; Li, J. Effect of Indian Ocean-Pacific SST Pattern in Autumn on Winter Wheat Climatic Yield in the North China Plain in the Following Year and a Possible Mechanism. *Sci. Rep.* **2019**, *9*, 19016. [CrossRef] [PubMed]
11. Ceglar, A.; Toreti, A.; Lecerf, R.; Van der Velde, M.; Dentener, F. Impact of meteorological drivers on regional inter-annual crop yield variability in France. *Agric. For. Meteorol.* **2016**, *216*, 58–67. [CrossRef]
12. Peña-Gallardo, M.; Vicente-Serrano, S.M.; Quiring, S.; Svoboda, M.; Hannaford, J.; Tomas-Burguera, M.; Martín-Hernández, N.; Domínguez-Castro, F.; El Kenawy, A. Response of crop yield to different time-scales of drought in the United States: Spatio-temporal patterns and climatic and environmental drivers. *Agric. For. Meteorol.* **2019**, *264*, 40–55. [CrossRef]
13. Feng, S.; Hao, Z.; Zhang, X.; Hao, F. Changes in climate-crop yield relationships affect risks of crop yield reduction. *Agric. For. Meteorol.* **2021**, *304–305*, 108401. [CrossRef]
14. Tao, F.; Hayashi, Y.; Zhang, Z.; Sakamoto, T.; Yokozawa, M. Global warming, rice production, and water use in China: Developing a probabilistic assessment. *Agric. For. Meteorol.* **2008**, *148*, 94–110. [CrossRef]
15. Campbell, B.M.; Vermeulen, S.J.; Aggarwal, P.K.; Corner-Dolloff, C.; Girvetz, E.; Loboguerrero, A.M.; Ramirez-Villegas, J.; Rosenstock, T.; Sebastian, L.; Thornton, P.K.; et al. Reducing risks to food security from climate change. *Glob. Food Secur.* **2016**, *11*, 34–43. [CrossRef]
16. Lobell, D.B.; Asner, G.P. Climate and management contributions to recent trends in US agricultural yields. *Science* **2003**, *299*, 1032. [CrossRef]
17. Lu, J.; Carbone, G.J.; Gao, P. Detrending crop yield data for spatial visualization of drought impacts in the United States, 1895–2014. *Agric. For. Meteorol.* **2017**, *237–238*, 196–208. [CrossRef]
18. Food and Agriculture Organization of the United Nations. FAO Statistical Databases 2019. Available online: https://www.fao.org/faostat/ (accessed on 10 July 2022).
19. World Production of Citrus Fruits in 2020, by Region. Available online: https://www.statista.com/statistics/264002/production-of-citrus-fruits-worldwide-by-region/ (accessed on 10 July 2022).
20. Qi, L.; Qi, C.J. Status quo and development trend of world's citrus Industry. *Agric. Outlook* **2016**, *12*, 46–52. (In Chinese)
21. Balfagón, D.; Arbona, V.; Gómez-Cadenas, A. El futuro de los cítricos: Impacto del cambio climático en la citricultura. *Mètode Rev. Difus. Investig.* **2021**, *3*, 60–67. [CrossRef]
22. Fitchett, J.M.; Grab, S.W.; Thompson, D.I.; Roshan, G. Spatio-temporal variation in phenological response of citrus to climate change in Iran: 1960–2010. *Agric. For. Meteorol.* **2014**, *198–199*, 285–293. [CrossRef]
23. Lahlali, R.; Jaouad, M.; Moinina, A.; Mokrini, F.; Belabess, Z. Farmers' knowledge, perceptions, and farm-level management practices of citrus pests and diseases in Morocco. *J. Plant Dis. Prot.* **2021**, *128*, 1213–1226. [CrossRef]
24. Mittler, R.; Finka, A.; Goloubinoff, P. How do plants feel the heat? *Trends Biochem. Sci.* **2012**, *37*, 118–125. [CrossRef] [PubMed]
25. Abobatta, W.F. Drought adaptive mechanisms of plants—A review. *Adv. Agric. Environ. Sci. Open Access* **2019**, *2*, 42–45. [CrossRef]
26. Nawaz, R.; Abbasi, N.A.; Ahmad Hafiz, I.; Khalid, A. Impact of climate variables on fruit internal quality of Kinnow mandarin (*Citrus nobilis* Lour x *Citrus deliciosa* Tenora) in ripening phase grown under varying environmental conditions. *Sci. Hortic.* **2020**, *265*, 109235. [CrossRef]
27. Knowledge Centre for Food Fraud and Quality. Available online: https://knowledge4policy.ec.europa.eu/food-fraud-quality/topic/food-quality_en/ (accessed on 10 July 2022).
28. Duan, H.L.; Qian, H.S.; Li, M.; Du, Y.D. Changes of citrus climate risk in subtropics of China. *J. Geogr. Sci.* **2010**, *20*, 818–832. (In Chinese) [CrossRef]
29. Duan, J.; Liu, Y.; Yang, J.; Tang, C.; Shi, Z. Role of groundcover management in controlling soil erosion under extreme rainfall in citrus orchards of southern China. *J. Hydrol.* **2020**, *582*, 124290. [CrossRef]
30. Du, Y.D.; Duan, H.L.; Tang, L.S. Adaptability of citrus in subtropics of China under future climatic scenarios. *Chin. J. Ecol.* **2010**, *29*, 833–839. (In Chinese)
31. Li, H. Citrus tree abiotic and biotic stress and implication of simulation and modeling tools in tree management. *Tree For. Sci. Biotechnol.* **2009**, *3*, 66–78.
32. Abobatta, W.F. Influence of climate change on citrus growth and productivity (effect of temperature). *Adv. Agric. Technol. Plant Sci.* **2019**, *2*, 180036.

33. Giuffrè, A.M.; Zappia, C.; Capocasale, M. Physicochemical stability of blood orange juice during frozen storage. *Int. J. Food Prop.* **2017**, *20*, 1930–1943.
34. Giuffrè, A. Bergamot (*Citrus bergamia*, Risso): The Effects of Cultivar and Harvest Date on Functional Properties of Juice and Cloudy Juice. *Antioxidants* **2019**, *8*, 221. [CrossRef]
35. Di Rauso Simeone, G.; Di Matteo, A.; Rao, M.A.; Di Vaio, C. Variations of peel essential oils during fruit ripening in four lemon (*Citrus limon* (L.) Burm. F.) cultivars. *J. Sci. Food Agric.* **2019**, *100*, 193–200. [CrossRef] [PubMed]
36. Tubiello, F.N.; Rosenzweig, F.N.; Goldberg, C.R.; Jagtap, A.S.; Jones, J.W. Effects of climate change on US crop production: Simulation results using two different GCM scenarios. Part I: Wheat, potato, maize, and citrus. *Clim. Res.* **2002**, *20*, 259–270. [CrossRef]
37. Masson-Delmotte, V.; Zhai, P.; Pirani, A.; Connors, S.L.; Péan, C.; Berger, S.; Caud, N.; Chen, Y.; Goldfarb, L.; Gomis, M.I.; et al. (Eds.) *IPCC Climate Change 2021: The Physical Science Basis. Contribution of Working Group I to the Sixth Assessment Report of the Intergovernmental Panel on Climate Change*; Cambridge University Press: Cambridge, UK, 2021.
38. Zhou, T.J.; Chen, Z.M.; Chen, X.L.; Zuo, M.; Jiang, J.; Hu, S. Interpreting IPCC AR6: Future global climate based on projection under scenarios and on near-term information. *Clim. Change Res.* **2021**, *17*, 652–663. (In Chinese)
39. Wang, L.K.; Qi, C.J. Research on the comparative advantage and its influencing factors of in Chinese citrus main producing region—Empirical analysis based on inter-provincial panel data. *Chin. J. Agric. Resour. Reg. Plan.* **2018**, *39*, 121–128. (In Chinese)
40. National Bureau of Statistics of China (NBSC). *China Statistical Yearbook 2019*; China Statistics Press: Beijing, China, 2019. (In Chinese)
41. Rai, P.; Rai, C.; Majumdar, G.C.; Dasgupta, S.; De, S. Storage study of ultrafiltered mosambi (*Citrus sinensis* (L.) Osbeck) juice. *J. Food Process. Preserv.* **2008**, *32*, 923–934. [CrossRef]
42. Manickavasagan, A.; Ganeshmoorthy, K.; Claereboudt, M.; AlYahyai, R.; Khriji, L. Non-destructive measurement of total soluble solid (TSS) content of dates using near infrared (NIR) imaging. *Emir. J. Food Agric.* **2014**, *26*, 970. [CrossRef]
43. Wu, S.; Li, M.; Zhang, C.; Tan, Q.; Yang, X.; Sun, X.; Pan, Z.; Deng, X.; Hu, C. Effects of phosphorus on fruit soluble sugar and citric acid accumulations in citrus. *Plant Physiol. Biochem.* **2021**, *160*, 73–81. [CrossRef]
44. Nawaz, R.; Abbasi, N.A.; Hafiz, I.A.; Khalid, A. Impact of climate variables on growth and development of Kinnow fruit (*Citrus nobilis* Lour x *Citrus deliciosa* Tenora) grown at different ecological zones under climate change scenario. *Sci. Hortic.* **2020**, *260*, 108868. [CrossRef]
45. Sun, X.; Ren, G.; You, Q.; Ren, Y.; Xu, W.; Xue, X.; Zhan, Y.; Zhang, S.; Zhang, P. Global diurnal temperature range (DTR) changes since 1901. *Clim. Dyn.* **2019**, *52*, 3343–3356. [CrossRef]
46. Wang, S.; Xie, W.; Yan, X. Evalution on CMIP6 Model Simulation of the Diurnal Temperature Range over China. *Clim. Environ. Res.* **2022**, *27*, 79–93. (In Chinese)
47. Xie, W.; Wang, S.; Yan, X. Evaluation and Projection of Diurnal Temperature Range in Maize Cultivation Areas in China Based on CMIP6 Models. *Sustainability* **2022**, *14*, 1660. [CrossRef]
48. Lobell, D.B.; Burke, M. *Climate Change and Food Security: Adapting Agriculture to a Warmer World*; Springer Science & Business Media: Dordrecht, The Netherlands, 2010; Volume 37.
49. Lobell, D.B.; Burke, M.B. On the use of statistical models to predict crop yield responses to climate change. *Agric. For. Meteorol.* **2010**, *150*, 1443–1452. [CrossRef]
50. Challinor, A.J.; Simelton, E.S.; Fraser, E.D.G.; Hemming, D.; Collins, M. Increased crop failure due to climate change: Assessing adaptation options using models and socio-economic data for wheat in China. *Environ. Res. Lett.* **2010**, *5*, 34012. [CrossRef]
51. Tao, F.; Zhang, Z.; Liu, J.; Yokozawa, M. Modelling the impacts of weather and climate variability on crop productivity over a large area: A new super-ensemble-based probabilistic projection. *Agric. For. Meteorol.* **2009**, *149*, 1266–1278. [CrossRef]
52. Zhang, T.; Zhu, J.; Wassmann, R. Responses of rice yields to recent climate change in China: An empirical assessment based on long-term observations at different spatial scales (1981–2005). *Agric. For. Meteorol.* **2010**, *150*, 1128–1137. [CrossRef]
53. Li, X.; Takahashi, T.; Suzuki, N.; Kaiser, H.M. The impact of climate change on maize yields in the United States and China. *Agric. Syst.* **2011**, *104*, 348–353. [CrossRef]
54. Liu, Y.; Wang, F.; Zhang, Z.T.; Huang, C.F.; Chen, X.; Li, N. Comprehensive assessment of "climate change−Crop yield−Economic impact" in seven sub-regions of China. *Clim. Change Res.* **2021**, *17*, 455–465.
55. Wei, F.Y. *Modern Climate Statistical Diagnosis and Prediction Techniques*; Meteorological Press: Beijing, China, 2007. (In Chinese)
56. Luo, N.; Guo, Y.; Gao, Z.; Chen, K.; Chou, J. Assessment of CMIP6 and CMIP5 model performance for extreme temperature in China. *Atmos. Ocean. Sci. Lett.* **2020**, *13*, 589–597. [CrossRef]
57. Song, S.F.; Yan, X.D. Evaluation of CMIP6 Models Performance for Winter Cold Wave Frequency in China. *Clim. Environ. Res.* **2022**, *27*, 33–49. (In Chinese)
58. Soqanloo, S.S. Effect of Different Regional Climates on Persimmon Quality. *J. Civ. Eng. Environ. Sci.* **2015**, *1*, 008–012. [CrossRef]
59. Falcão, L.D.; Burin, V.M.; Sidinei Chaves, E.; Vieira, H.J.; Brighenti, E.; Rosier, J.; Bordignon-Luiz, M.T. Vineyard altitude and mesoclimate influences on the phenology and maturation of Cabernet-Sauvignon grapes from Santa Catarina State. *OENO One* **2010**, *44*, 135–150. [CrossRef]
60. Habibi, F.; Ramezanian, A.; Guillén, F.; Castillo, S.; Serrano, M.; Valero, D. Changes in Bioactive Compounds, Antioxidant Activity, and Nutritional Quality of Blood Orange Cultivars at Different Storage Temperatures. *Antioxidants* **2020**, *9*, 1016. [CrossRef]

61. Cohen, S.D.; Tarara, J.M.; Kennedy, J.A. Diurnal Temperature Range Compression Hastens Berry Development and Modifies Flavonoid Partitioning in Grapes. *Am. J. Enol. Vitic.* **2012**, *63*, 112–120. [CrossRef]
62. Ahmad, S.; Firdous, I.; Jatoi, G.H.; Rais, M.U.N.; Mohsin, A.Q. Economic impact of climate change on the production of citrus fruit in Punjab province of the Pakistan. *Sci. Int.* **2017**, *29*, 413.
63. Li, Y.; Guan, K.; Schnitkey, G.D.; DeLucia, E.; Peng, B. Excessive rainfall leads to maize yield loss of a comparable magnitude to extreme drought in the United States. *Glob. Change Biol.* **2019**, *25*, 2325–2337. [CrossRef] [PubMed]
64. Xu, R.; Li, Y.; Guan, K.; Zhao, L.; Peng, B.; Miao, C.; Fu, B. Divergent responses of maize yield to precipitation in the United States. *Environ. Res. Lett.* **2021**, *17*, 14016. [CrossRef]
65. Hirpo, F.H. Review on the Effects of Climate Change Variability on Horticultural Productivity. *Int. J. Environ. Sci. Nat. Resour.* **2019**, *17*, 555969. [CrossRef]
66. El Yaacoubi, A.; El Jaouhari, N.; Bourioug, M.; El Youssfi, L.; Cherroud, S.; Bouabid, R.; Chaoui, M.; Abouabdillah, A. Potential vulnerability of Moroccan apple orchard to climate change-induced phenological perturbations: Effects on yields and fruit quality. *Int. J. Biometeorol.* **2020**, *64*, 377–387. [CrossRef]
67. Cervantes, L.; Ariza, M.T.; Miranda, L.; Lozano, D.; Medina, J.J.; Soria, C.; Martínez-Ferri, E. Stability of Fruit Quality Traits of Different Strawberry Varieties under Variable Environmental Conditions. *Agronomy* **2020**, *10*, 1242. [CrossRef]
68. Tao, F.; Yokozawa, M.; Liu, J.; Zhang, Z. Climate-crop yield relationships at provincial scales in China and the impacts of recent climate trends. *Clim. Res.* **2008**, *38*, 83–94. [CrossRef]
69. Zhao, C.; Piao, S.; Huang, Y.; Wang, X.; Ciais, P.; Huang, M.; Zeng, Z.; Peng, S. Field warming experiments shed light on the wheat yield response to temperature in China. *Nat. Commun.* **2016**, *7*, 13530. [CrossRef]
70. Shi, W.; Tao, F.; Zhang, Z. A review on statistical models for identifying climate contributions to crop yields. *J. Geogr. Sci.* **2013**, *23*, 567–576. [CrossRef]
71. Feng, L.; Wang, H.; Ma, X.; Peng, H.; Shan, J. Modeling the current land suitability and future dynamics of global soybean cultivation under climate change scenarios. *Field Crops Res.* **2021**, *263*, 108069. [CrossRef]
72. Zhang, Z.; Chen, Y.; Wang, C.; Wang, P.; Tao, F. Future extreme temperature and its impact on rice yield in China. *Int. J. Climatol.* **2017**, *37*, 4814–4827. [CrossRef]
73. Liu, Y.C.; Liu, S.S.; Deng, J.H. Effects of climate Factors on fruit quality of Mandarin oranges. *Chin. J. Agrometeorol.* **1998**, *19*, 29–31. (In Chinese)
74. Hu, C.S.; Zhao, S.D.; Xing, W.Y.; Li, J.A. A Study on Factors Influencing the Quality of Mandarin Oranges "Good Place" Green Food and Evaluation of the Quality of Fruits. *Acta Agric. Univ. Jiangxiensis* **2000**, *22*, 70–74. (In Chinese)
75. Wen, T.; Xiong, Q.E.; Zeng, W.G.; Liu, Y.P. The Gray System Analysis of Climatic Factors to the Organic Acid and Sugar Content of Navel Orange (*Citrus sinesis* Osbeck) Fruit. *J. Sichuan Agric. Univ.* **2001**, *19*, 3. (In Chinese)
76. Bao, J.F.; Xia, R.X.; Deng, X.X.; Peng, S.A.; Liu, Y.Z.; Ma, X.T.; Zhang, H.Y. Preliminary Study on Fruit Quality of Newhall Orange from Various Spot with Different Latitude. *Subtrop. Plant Sci.* **2006**, *35*, 13–16. (In Chinese)
77. Bao, J.F. *Study and Division of Fruit Quality of Newhall Navel Orange and Main Citrus Varieties in Hubei Province*; HZAU (Huazhong Agricultural University): Wuhan, China, 2005. (In Chinese)
78. Zeng, X.G. *Study on Sugar and Acid Content and Composition of Citrus from Different Species and Producing Areas*; HZAU (Huazhong Agricultural University): Wuhan, China, 2005. (In Chinese)
79. Bao, J.F.; Xia, R.X.; Deng, X.X.; Peng, S.A.; Liu, Y.Z.; Ma, X.T.; Zhang, H.Y. Research on the Fruit Quality of Newhall Navel Orange of Hubei Province. *J. Wuhan Bot. Res.* **2005**, *23*, 583–587. (In Chinese)
80. Lu, X.P.; Li, J.; Xiong, J.; Cao, X.J.; Xie, S.X. Soil nutrition status at orchards of Newhall navel orange and the fruit quality at different latitudes of Hunan province. *J. Hunan Agric. Univ. (Nat. Sci.)* **2014**, *40*, 615–620. (In Chinese)
81. Liao, Q.H. *Study on Fruit Quality Characteristics and Origin Identification of Navel Orange in Newhall*; Southwest University: Chongqing, China, 2014. (In Chinese)
82. Li, J.Q.; Li, W.Y.; Yang, S.A.; Wu, C.M.; Long, D.C.; Luo, F.X. Analysis on fruit quality characteristics of main citrus varieties in Guizhou. *Jiangsu Agric. Sci.* **2013**, *41*, 170–171. (In Chinese)
83. Liang, X.; Zhao, X.L.; Tang, Z.P.; Wang, M.Z.; Wu, R.C.; Qin, F.L. Study on the Horticulture Features and Climatic Factors in Miyamoto SpecialPrecocious Satsuma. *J. Anhui Agri. Sci.* **2013**, *41*, 5684–5686. (In Chinese)
84. Jiang, J.F. Present situation and development countermeasure of citrus industry in Yunfu City. *Anhui Agri. Sci. Bull.* **2014**, *20*, 55–56. (In Chinese)
85. Chen, Y.T.; Luo, Y.; Li, J.Q. Fruit quality analysis of main citrus cultivars in Guizhou. *Seed* **2017**, *36*, 110–112. (In Chinese)

sustainability

MDPI

Article

Effects of Climate Change on the Climatic Production Potential of Potatoes in Inner Mongolia, China

Li-Tao Yang [1,2,3,4], Jun-Fang Zhao [2,*], Xiang-Ping Jiang [5], Sheng Wang [6], Lin-Hui Li [7] and Hong-Fei Xie [2]

1 Research Institute of Meteorological Science of Inner Mongolia Autonomous, Hohhot 010051, China; ylt9618@126.com
2 State Key Laboratory of Severe Weather, Chinese Academy of Meteorological Sciences, Beijing 100081, China; 17752014737@163.com
3 Key Laboratory of Weather Modification of Inner Mongolia Autonomous Region, Hohhot 010051, China
4 Climate Center of Inner Mongolia Autonomous Region, Hohhot 010051, China
5 Hohhot Vocational College, Hohhot 010051, China; jxp3068@126.com
6 Climate Center of Anhui Province, Hefei 230000, China; ws7810@163.com
7 Science, Technology and Forecast Division of Meteorological Bureau of Inner Mongolia Autonomous Region, Hohhot 010051, China; lilinhui-a@163.com
* Correspondence: zhaojf@cma.gov.cn; Tel.: +86-10-58995896 or +86-130-0198-1203

Abstract: Understanding the impacts of regional climate change on crop production will benefit strategic decisions for future agricultural adaptation in China. In this study, the climatic production potential of potato over the past 61 years in Inner Mongolia was simulated based on long-term observed data and the step-by-step correction method. The results show that the annual average potential for potato climatic production in Inner Mongolia is 19,318 kg·hm^{-2}, fluctuating between the highest value (25,623 kg·hm^{-2}) and the lowest value (15,354 kg·hm^{-2}). Over the past 61 years, the climatic production potential exhibited an insignificant decreasing trend, with large interannual fluctuation, especially since 2000. The high-value areas of the climatic production potential were mainly located in the central and southern regions. The climatic production potential of potato in most areas showed a decreasing trend. The influence of radiation changes on the potato climatic production potential was not obvious in most areas. The effects of temperature changes on the climatic production potential of potato were mostly negative, and were most obvious in the central and western regions and in the southeastern region. The change in precipitation in most parts of western Inner Mongolia, Hohhot, Chifeng and eastern Xingan League had a positive effect on the climatic production potential of potato. However, the change in precipitation in southern Ulanchabu, eastern Chifeng, Hulunbuir and western and eastern regions had a negative effect on the climatic production potential of potato. The main limiting factor for the climatic production potential of potato in Inner Mongolia is precipitation. Our findings have important implications for local potato production to cope with ongoing climate change in China.

Keywords: climate change; potato climatic productivity potential; Inner Mongolia; effect

Citation: Yang, L.-T.; Zhao, J.-F.; Jiang, X.-P.; Wang, S.; Li, L.-H.; Xie, H.-F. Effects of Climate Change on the Climatic Production Potential of Potatoes in Inner Mongolia, China. *Sustainability* **2022**, *14*, 7836. https://doi.org/10.3390/su14137836

Academic Editors: Xiaodong Yan, Jia Yang and Shaofei Jin

Received: 10 April 2022
Accepted: 24 June 2022
Published: 27 June 2022

1. Introduction

Under global warming, extreme weather occurs frequently, and meteorological disasters bring more and more risks to agricultural production. Food is an important commodity related to the national economy and people's livelihood. It is the basis of economic development, social stability and national independence. Ensuring national food security is always the top priority in governing a country. The Intergovernmental Panel on Climate Change (IPCC) officially released on 28 February the contribution of Working Group II to the Sixth Assessment Report (Ar6) on climate change in 2022, highlighting impacts, adaptation and vulnerability. The report further points out the severe food security situation faced at home and abroad against the background of global warming. Food security has once again received extensive attention all over the world. In order to cope with meteorological disasters, it is a challenge to understand the impact mechanisms of climate change on staple crops, and to adapt to and cope with the negative impact of climate change. Potato is one of the main climatic characteristic crops in Inner Mongolia, China. Therefore, it is urgent to acquire a more in-depth understanding of how climate change affects potato production for agricultural disaster prevention and mitigation and the healthy development of the potato industry in Inner Mongolia, China.

The climatic production potential of crops refers to the crop yields when light, heat and water resources are optimally matched under ideal conditions [1], and it is one of the important criteria for evaluating agroclimatic resources [2]. The responses of the climatic production potential of crops to climate change differ significantly among regions and crops [3–5]. At present, the commonly used models for calculating the crop production potential include the Miami model [6,7], the Thornthwaite Memory model [8,9], the AEZ model by FAO [10,11] and the step-by-step revision model [12–14]. Among these models, the step-by-step correction model is a statistical model for calculating the climatic production potential of different crops during the growing season. It can effectively reflect the matching status of climatic resources such as light, heat, and water. Its physical meaning and the causal relationships are clear. The step-by-step correction model is one of the most extensively used research methods for analyzing the food production potential [15,16].

The impacts of climate change on the climatic production potential of crops over the past few decades in China have attracted serious concern [14,17–21]. Based on the potential attenuation method, Wang et al. [14] studied the spatial variation in climatic production potential of maize, rice and soybean in the Songnen Plain, and analyzed the utilization efficiency of natural resources. Lu et al. [17] evaluated the evolution of climatic production potential in Anhui Province in the past 50 years. Wang studied and analyzed the climatic production potential of one-season rice in Anhui Province [18]. Wang et al. [19] studied the climatic production potential of winter wheat in northern China and the influence of separated water. Duan et al. [20] estimated the climatic production potential of potato in Jixi County, Ningxia, and analyzed its stability. Lai simulated the climatic production potential of five crops in Ningxia, including soybean, corn, rice, wheat and potato [21]. However, few studies have been conducted to quantitatively assess the long-term impacts of regional climate change on the climatic production potential of potato in Inner Mongolia based on the step-by-step correction model. A better understanding of how potato responds to regional climate change is essential for mitigating the negative effects of potato production to local climate change.

The objectives of the present study were to (1) determine the main parameters in a climatic production potential model for potato based on long-term historical data in Inner Mongolia and (2) explore the advantages and disadvantages of impacts of changes in main meteorological factors (light, temperature and water) on potato production in Inner Mongolia. These findings are significant for substantially improving our understanding of the impacts of regional climate change on agriculture in China.

2. Materials and Methods

2.1. Study Area

The Inner Mongolia Autonomous Region is one of 13 key grain-producing provinces (regions) in China. As of 2017, the region has 9.271 million hectares of cultivated land, accounting for 7.8% of the region's area, and the per capita cultivated land area is 3.7 times that of the whole country. The effective irrigation area is 3.175 million hectares. The soil in this area shows obvious meridional differentiation. Black soil, dark brown soil, chernozem, chestnut soil, brown calcium soil, brown desert soil and gray desert soil are distributed from east to west [22]. In some areas, cinnamon soil, calcareous soil, meadow soil and aeolian sandy soil are distributed. The content of soil organic matter (9.75~0.56%) [22] and clay particles (12.41~3.65%) [23] decreases gradually from east to west, while the soil pH value (7.39~8.90) increases gradually from east to west [23]. The soil is mostly neutral and alkaline [23,24]. Inner Mongolia covers a vast territory, with large east–west and north–south spans, complex landforms (e.g., plateaus, mountains, hills and plains) and special geographical locations, which form complex and diverse climate conditions dominated by a temperate continental monsoon climate, with varied temperature, rainfall and heat waves. In the same season, the cold climate is more suitable for the growth of potatoes, which prefer cold and cool conditions.

2.2. Data

The data are from the climate center of the Inner Mongolia Autonomous Region. The data used consist of the daily surface meteorological observation data, including sunshine hours, average temperature, precipitation, relative humidity, average wind speed and evapotranspiration. These meteorological data were obtained from 119 meteorological stations in the Inner Mongolia Autonomous Region from 1961 to 2021 (Figure 1). The long-term field observation datasets of potato were collected from 11 agricultural meteorological data stations of the Inner Mongolia Autonomous Region from 1981 to 2020 (Figure 1). The administrative boundaries of Inner Mongolia and the cities (alliances) that are involved in this article are based on the standard map with the approval number of Mongolia S (2019) 33, from the Map Institute of the Inner Mongolia Autonomous Region.

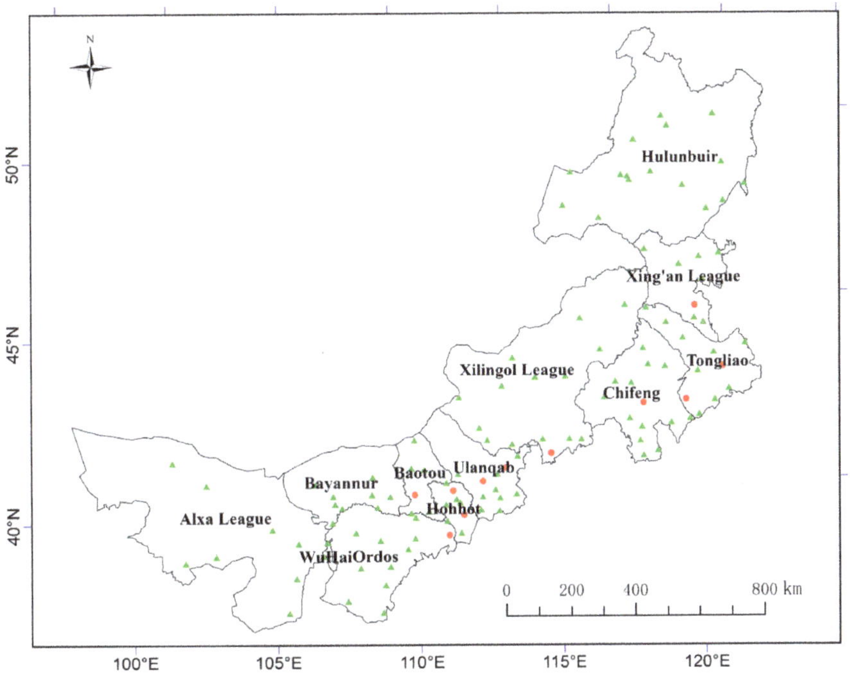

Figure 1. Distribution of 119 meteorological stations (green triangles) and 11 agrometeorological observation stations (red dots) in the Inner Mongolia Autonomous Region of China.

2.3. Research Methods

2.3.1. Calculation of the Climatic Production Potential

The step-by-step correction method was used, starting from crop photosynthesis, according to the process of crop energy conversion and yield formation, and gradually estimating the agricultural climatic production potential. It is a model that was developed from the step-by-step revision of the climatic factor function based on the photosynthetic production potential. The calculation formula is as follows [25]:

$$Y_{\mathrm{CPP}} = Y_{\mathrm{PPP}} \times f(T_{ij}) \times f(R_{ij}) = Y_{\mathrm{TPP}} \times f(R_{ij}) \quad (1)$$

where Y_{CPP}, Y_{PPP} and Y_{TPP} are the climatic production potential, photosynthetic production potential and light–temperature production potential (kg·hm^{-2}), respectively; $f(T_{ij})$ is the effective temperature coefficient of the ith growth period in the jth year; and $f(R_{ij})$ is the effective water coefficient of the ith growth period in the jth year.

(1) Photosynthetic production potential of potato

The photosynthetic production potential of potato refers to the potato yields that are uniquely determined by the local solar radiation under the assumption that the most suitable temperature, water, soil fertility, crop population and agricultural technical measures are provided. It is the theoretical upper limit of potato yield. The photosynthetic production potential model of potato adopts the state of energy utilization and loss during yield formation, takes the total solar radiation value of potato in each growth period as the basic data, and corrects it. The calculation formula is as follows [18,19]:

$$Y_{\mathrm{PPP}} = Cf(Q) \quad (2)$$

$$f(Q) = \Omega\varepsilon\varphi(1-\alpha)(1-\beta)(1-\rho)(1-\gamma)(1-\omega)(1-\eta)^{-1}(1-\xi)sq^{-1}f(L)\sum Q_i \quad (3)$$

where C is the unit conversion factor, which is 10,000; $\sum Q_i$ is the total radiation during the growing season (MJ·m^{-2}); i represents the different growth periods; Ω is the ratio of the photosynthetic ability of crops to fix CO_2, which is 0.6; ε is the ratio of photosynthetic radiation to total radiation, which is taken as 0.47; φ is the quantum efficiency of photosynthesis, which is taken as 0.224; α is the crop population reflectivity, which is taken as 0.17; β is the leakage rate of solar radiation by the crop population, which is taken as 0.1; ρ is the ineffective absorption rate of non-photosynthetic organs, which is 0.1; γ is the light saturation limitation rate, which is 0.01; ω is the respiration loss rate, which is 0.3; η is the moisture content of mature grains, which is 0.14; ξ is the proportion of crop inorganic ash content, which is taken as 0.08; s is the crop economic coefficient, which is taken as 0.75; q is the heat required to form one unit of dry matter (MJ·kg^{-1}), which is taken as 18; and $f(L)$ is the corrected value of the dynamic change in leaf area, which is taken as 0.556.

(2) Light–temperature production potential of potato

Based on the photosynthetic production potential of potato, by considering the effect of temperature on plant photosynthesis and correcting the temperature coefficient, the light–temperature production potential of potato, determined by the two factors of light and temperature, can be obtained. The calculation formula is as follows [18]:

$$Y_{\mathrm{TPP}} = Y_{\mathrm{PPP}} f(T_{ij}) \quad (4)$$

$$f(T_{ij}) = \begin{cases} \dfrac{(T_{ij} - T_1) \times (T_2 - T_{ij})^b}{(T_0 - T_1) \times (T_2 - T_0)^b} & T_1 < T < T_2 \\ 0 & T \le T_1 \text{ or } T \ge T_2 \end{cases} \quad (5)$$

$$b = \frac{T_2 - T_0}{T_0 - T_1} \quad (6)$$

where $f(T_{ij})$ is the temperature correction function of the ith growth period in the jth year; T_{ij} is the average temperature of the ith growth period in the jth year (°C); and T_1, T_2 and T_0 are the lower and the upper limits of temperature and the optimum temperature of crop growth in each growth period (°C), respectively (Table 1) [25–27].

Table 1. Three cardinal temperature points in each potato growth period (°C).

Growth Period	T_1	T_2	T_0
Sowing–emergence	5.0	29.0	14.0
Emergence–branch	9.0	32.0	18.3
Branch–inflorescence	10.0	30.0	19.5
Inflorescence–bloom	10.0	30.0	19.3
Flowering–harvestable	10.0	29.0	18.0
Full reproductive period	5.0	32.0	17.8

(3) climatic production potential of potato

The climatic production potential is the upper limit of production that can be achieved under the combined influence of radiation, temperature and precipitation. After the light–temperature production potential is corrected by the moisture correction function, the climatic production potential can be obtained, and its expression is as follows [18]:

$$Y_{\mathrm{CPP}} = Y_{\mathrm{TPP}} f(R_{ij}) \tag{7}$$

$$f(R_{ij}) = \begin{cases} R_{ij}/R_{i0} & R_{ij} < R_{i0} \\ 1 & R_{i0} \geq R_{ij} \end{cases} \tag{8}$$

where $f(R_{ij})$ is the water correction function of the ith growth period of the jth year, R_{ij} is the precipitation during the ith growth period of the jth year (mm) and R_{i0} is the physiological water demand during the ith growth period (mm). The water requirement of potatoes in the entire growth period is approximately 372.5 mm; the physiological water requirements in the sowing–seedling, seedling–branching, branching–inflorescence, inflorescence–flowering and flowering–returnable stages are 50 mm, 40 mm, 30 mm, 52.5 mm and 200 mm, respectively.

2.3.2. Impacts of Climate Change on the Climatic Production Potential of Potato

The climatic production potential of potato reflects the comprehensive impact of various meteorological factors on potato production. The photosynthetic production potential directly corresponds to the impact of light, the light–temperature production potential corresponds to the comprehensive impact of light and heat, and the climatic production potential corresponds to the comprehensive impact of the three factors of light, heat and water. The effects of light, heat, water and the comprehensive climatic conditions on the potato production potential are expressed as Y_r, Y_t, Y_p and Y_c (kg·hm^{-2}·a^{-1}), respectively. The calculation method is as follows [25,28]:

$$Y_r = (a_r/Y_1) \times Y_3 \tag{9}$$

$$Y_t = (a_t/Y_2 - a_r/Y_1) \times Y_3 \tag{10}$$

$$Y_t = (a_t/Y_2 - a_r/Y_1) \times Y_3 \tag{11}$$

$$Y_c = a_c \tag{12}$$

where a_r, a_t and a_c are the propensity rates of photosynthesis, light–temperature and the climatic production potential with time (year) (kg·hm^{-2}·a^{-1}), respectively; Y_1, Y_2 and Y_3 are the multiyear averages of photosynthesis, light–temperature and the climatic production potential (kg·hm^{-2}), respectively.

3. Results

3.1. Annual and Interdecadal Variations in the Production Potential of Potato

It can be seen from the change in the abnormal percentage of potato climatic production potential in Inner Mongolia (Figure 2) that before 2000, the abnormal percentage of potato climatic production potential in Inner Mongolia was mainly positive, and after

2000, it was mainly negative. Over the past 61 years, the climatic production potential of potato during the growth period in most of the central and western parts of Inner Mongolia and the southeast parts has shown a decreasing trend, with a decreasing range of 10~164 kg·hm^{-2}·a^{-1}; the decrease is in the areas that are above 40 kg·hm^{-2}·a^{-1} and are mainly located in most of Bayannaoer city, the southwest and northern parts of Ordos city, the southern part of Baotou city, the central part of Hohhot city, the southwest and eastern parts of Ulanqab city, west of Xilin Gol League, east of Chifeng city, most of Tongliao city, south of Xing'an League and west and southeast of Hulunbuir city. The central part of Hulunbuir city, north of Xing'an League, northeast of Xilingol League, west of Chifeng city, north parts of Baotou city, the south part of Ordos city, the northwest part of Bayannaoer city and southeast of Alxa League experienced increasing trends, with increases of 10~40 kg·hm^{-2}·a^{-1}. The increase areas greater than 20 kg·hm^{-2}·a^{-1} were mainly located in the central and northern parts of Hulunbuir city, the north part of Xing'an League, the western part of Chifeng city and the northeastern part of Xilingol League. The change trends of the climatic production potential of potato during the growth period were not obvious in other areas (Figure 3). On the whole, the climatic production potential of potato during the growth period of Inner Mongolia mainly exhibits a decreasing trend, and the increases in temperature, radiation and precipitation during the growth period are not conducive to improving the climatic production potential.

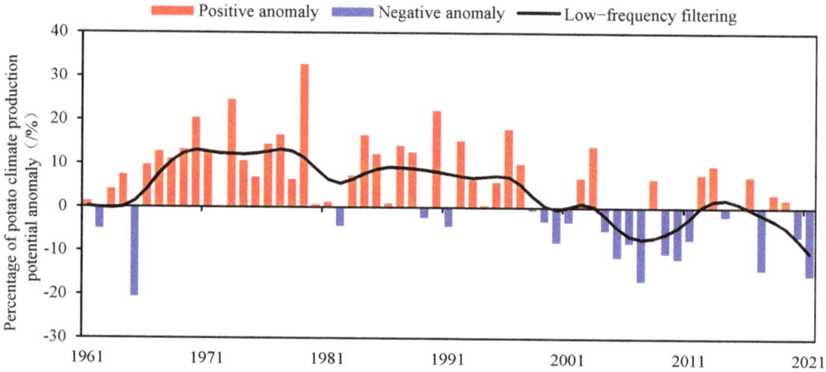

Figure 2. The anomaly (low–frequency filtering) changes in the potato potential productivity in Inner Mongolia, China, from 1961 to 2021.

Figure 4 shows the interannual and interdecadal variations in the potato production potential at all levels in Inner Mongolia. Figure 4a shows that over the past 61 years, the interannual fluctuations in the production potential of potatoes at all levels in Inner Mongolia have been relatively large, and the overall trend has decreased significantly ($p < 0.005$). The potential and climatic production potentials decreased at rates of 574 kg·hm^{-2}·(10 a)$^{-1}$, 1048 kg·hm^{-2}·(10 a)$^{-1}$ and 465 kg·hm^{-2}·(10 a)$^{-1}$. The photosynthetic production potential and light–temperature production potential were relatively close, with averages of 42,417 kg·hm^{-2} and 36,244 kg·hm^{-2} in the whole area, respectively, and the average climatic production potential was 19,318 kg·hm^{-2}, which is significantly lower than the production potential of the first two levels. The highest value of the photosynthetic production potential occurred in 1965 (45,952 kg·hm^{-2}). The highest value of the light–temperature production potential occurred in 1977 (41,001 kg·hm^{-2}). The highest value of the climatic production potential occurred in 1979 (25,623 kg·hm^{-2}). The lowest values of photosynthetic production potential, light–temperature production potential and climatic production potential occurred in 2021 (34,045 kg·hm^{-2}), 2021 (29,553 kg·hm^{-2}) and 1965 (15,354 kg·hm^{-2}), respectively.

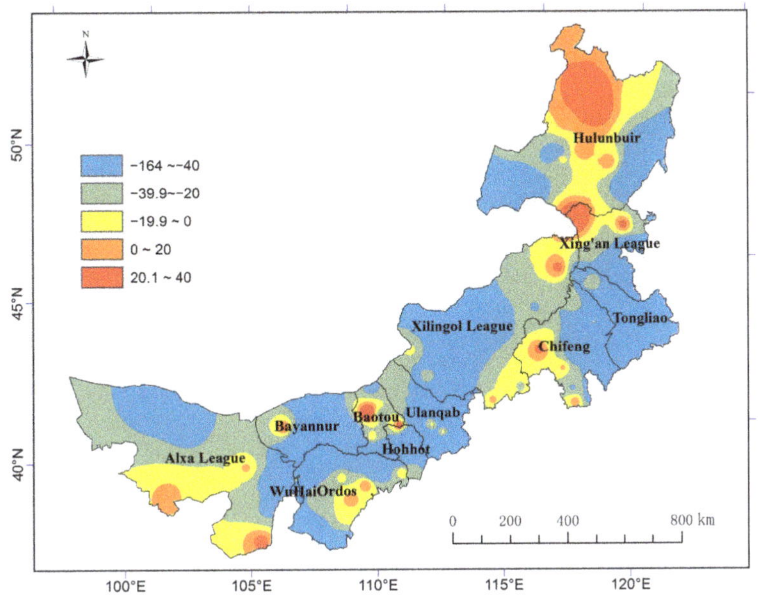

Figure 3. Change trends of climatic production potential of potato in Inner Mongolia, China ($kg{\cdot}hm^{-2}{\cdot}a^{-1}$).

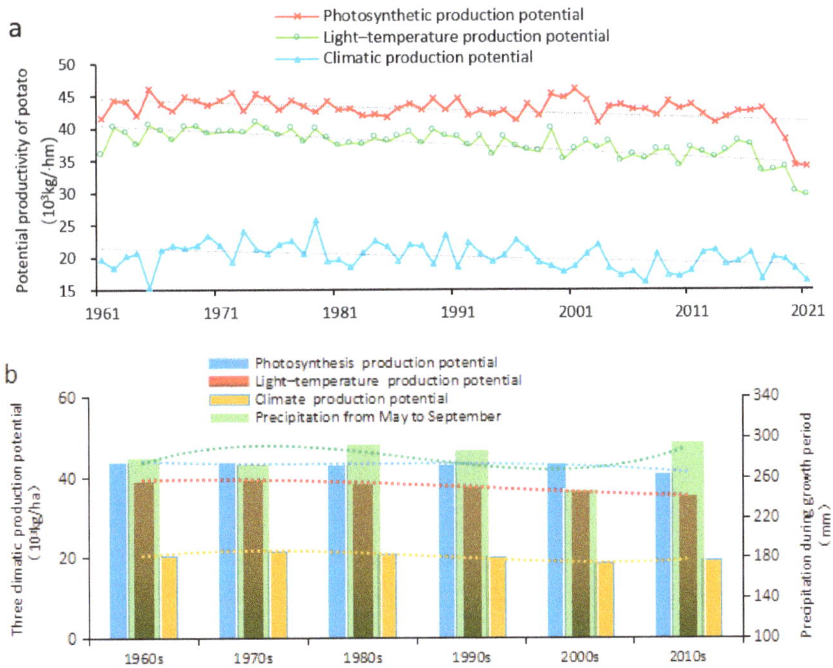

Figure 4. The annual (**a**) and decadal (**b**) changes in the potential productivity of potato of each grade in Inner Mongolia, China, from 1961 to 2021 (The red column in Figure (**b**) is brown because it overlaps with the green column.).

Judging from the interdecadal changes shown in Figure 4b, the changing trends of the production potentials at all levels are not consistent. Among them, the characteristics of the interdecadal variations in photosynthesis and the light–temperature production potential are consistent, being the largest in the 1970s and smallest in the 2010s, while the climatic production potential were the largest in the 1970s and smallest in the 2000s. The climatic production potential exhibited greater fluctuations than the previous two levels, and its variation trend had a significant positive correlation with the interdecadal variations in precipitation ($p < 0.01$), which indicated that the main limiting factor for the potato climatic production potential in Inner Mongolia was precipitation.

3.2. Characteristics of the Spatial Distribution of the Potato Production Potential

The climatic production potential of potato in Inner Mongolia decreased from the central and southern parts to the periphery, and the climatic production potential in the central and western regions had an obvious geographical distribution. The central and eastern part of Ulanqab city, the southern part of Xilin Gol League, the western part of Chifeng city and the central part of Hulunbuir city were more consistent with the dominant potato-producing areas. Among them, the highest value of the potato climatic production potential was in the Zhenglan Banner of Xilin Gol League, with the value of 27,641 kg·hm^{-2}. The second-highest value (17,601~22,600 kg·hm^{-2}) was located the eastern part of Ordos city and most of Baotou city, most of Hohhot, northern Ulanqab, central and northern Xilin Gol League and most of its eastern areas. The lowest value was located in Ejina Banner of Alxa League, being only 2594 kg·hm^{-2} (Figure 5). High temperatures and low rainfall amounts are the probable reasons for the lowest potentials of potato climatic production.

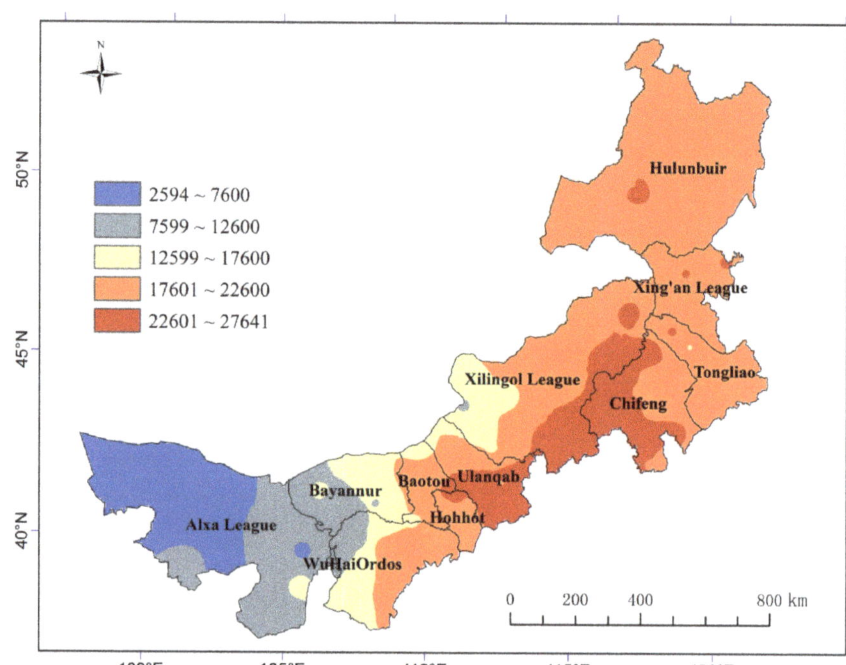

Figure 5. The spatial distribution of the climatic production potential of potatoes in Inner Mongolia, China (kg·hm^{-2}).

The annual meteorological yields and climatic production potentials of potato over the past 57 years in Wuchuan County of Hohhot city were relatively consistent, with a very significant positive correlation (0.5131 at 0.1% significance). These potato climatic production potentials are based on the results of the step-by-step revision correction method, and can better reflect the changing trends of the potato meteorological yields (Figure 6). The average annual potato yield in Wuchuan County was about 2163 kg·hm^{-2}, with 5.4% of the maximum climatic production potential and 14.2% of the minimum climatic production potential. The average application level of the potato climatic production potential in Wuchuan County was only 7.6%.

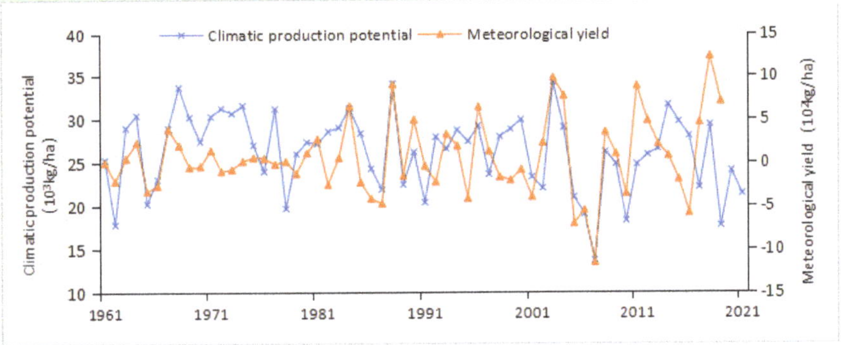

Figure 6. The relationship between the climatic production potentials and meteorological yields of potato in Wuchuan, Inner Mongolia, China, from 1961 to 2020.

3.3. *Influence of Changes in the Main Meteorological Elements on the Potato Climate Production Potential*

3.3.1. The Effect of Radiation Changes on the Potato Climatic Production Potential

Over the past 58 years, the total solar radiation in Inner Mongolia has fluctuated and decreased at a rate of 39.3 MJ·m^{-2} (10 a)$^{-1}$ ($p > 0.05$) [29]; this decreasing trend was obvious in some areas ($p < 0.05$). Over the past 61 years, the number of sunshine hours during the potato growth period in Inner Mongolia fluctuated at a rate of 18.1 h·(10 a)$^{-1}$ ($p < 0.001$) (Figure 7). Potato is a light-loving crop, and reductions in radiation and light are detrimental to its growth. Figure 8 shows that the influences of radiation changes on the climatic production potential of potato during the growth period in most of Inner Mongolia had negative effects, indicating that radiation reductions were not conducive to improving the climatic production potential of potato during the growth period. The areas experiencing negative effects are mainly distributed in the central and northeastern parts of Ordos city, southern part of Baotou city, most of Hohhot city, central and eastern parts of Ulanqab city, central and southern parts of Xilingol League, southeastern part of Chifeng city, most of Tongliao city, the southeastern part of Xing'an League, and the southeastern part of Hulunbuir city. Ulanqab city, the Chayouhou Banner in Lanchabu city and the Arong Banner in Hulunbuir city experienced the greatest impacts. The areas experiencing positive effects were relatively rare and were mainly distributed in the southwest and east parts of Xilingol League, the south part of Chifeng city and the central and eastern, central and northern Xing'an League and northwestern Hulunbuir city. However, the above-mentioned radiation changes in most of the regions had little effects on the climatic production potential of potato during the growth period, and the values were mostly between -20 MJ·m^{-2}·a^{-1} and 20 MJ·m^{-2}·a^{-1}.

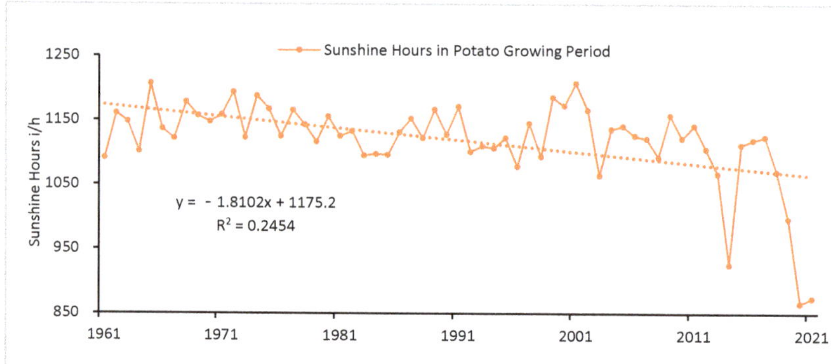

Figure 7. Changes in sunshine hours during the potato growing period in Inner Mongolia, China (h).

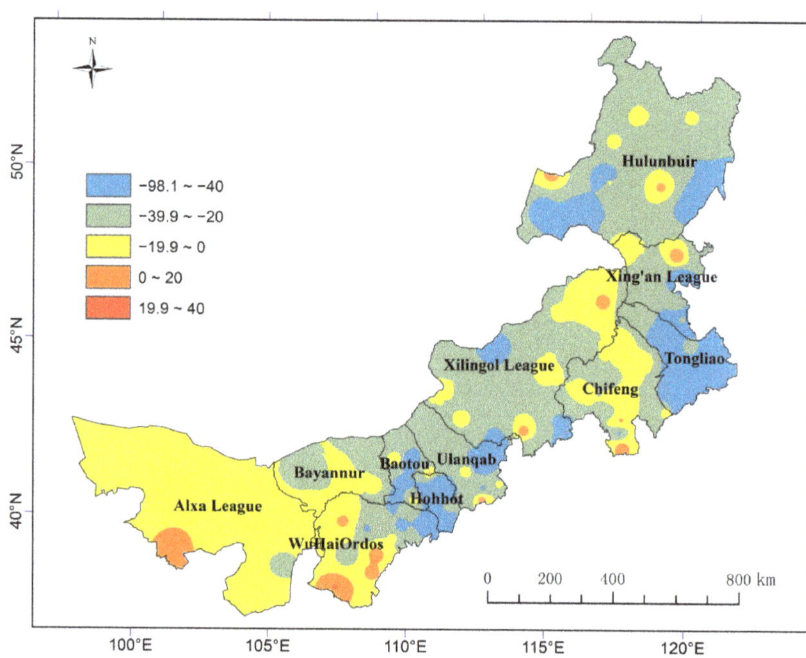

Figure 8. Impacts of radiation changes on the climatic production potential of potato in Inner Mongolia, China (kg·hm^{-2}·a^{-1}).

3.3.2. Effects of Temperature Changes on the Potato Climatic Production Potential

Over the past 61 years, the average temperatures during the potato growing season in Inner Mongolia have exhibited a significant upward trend as a whole ($p < 0.001$), with a climatic trend rate of 0.3 °C·(10 a)$^{-1}$. The temperatures increased significantly in most areas (94% of meteorological stations ($p < 0.01$) (Figure 9). Potato is a crop that prefers cool and cool conditions, and increased temperatures are not conducive to the growth and development of potato. Figure 10 shows that the impacts of the temperature changes on the potato climatic production potential were mostly negative, indicating that the temperature increases during the potato growth period in Inner Mongolia were not conducive to improving the climatic production potential. Negative effects were seen in the central and western parts of Inner Mongolia and in the southeast, while the negative effects were most obvious in the Northwest Territories, which were −102~−20 kg·hm^{-2}·a^{-1}. The greatest negative effects were found in Bayannaoer city, the north and southwest parts of Ordos city, the southwest part of Baotou city, and the central and western parts of Hohhot city. The positive effects were concentrated only in the central and northern parts of Hulunbuir city and the northwestern part of Xing'an League, with an impact range of 20~40 kg·hm^{-2}·a^{-1}; the influence range in the central and northern parts of Hulunbuir city was greater than 30 kg·hm^{-2}·a^{-1}.

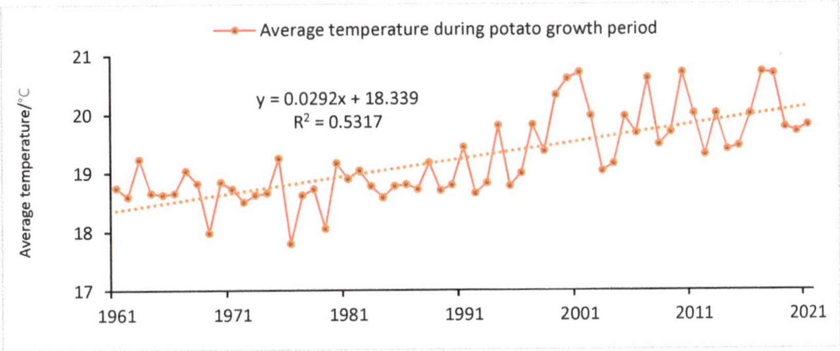

Figure 9. Variations in average temperature during the potato growth period in Inner Mongolia, China (°C).

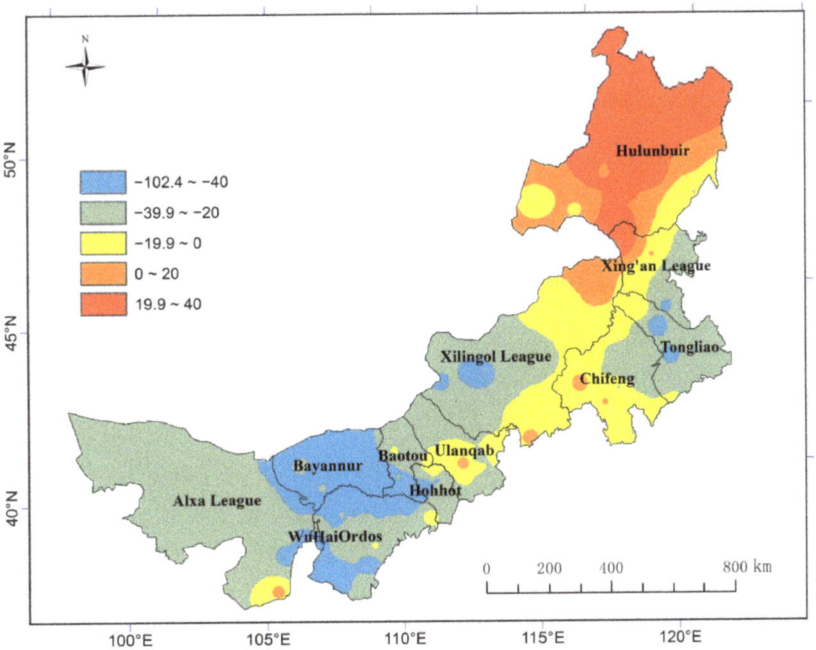

Figure 10. Impacts of temperature changes on the climatic production potential of potato in Inner Mongolia, China (kg·hm^{-2}·a^{-1}).

3.3.3. The Impact of Precipitation Changes on the Climatic Production Potential

Over the past 61 years, the precipitation levels during the potato growing season in Inner Mongolia has decreased at a rate of 0.7 mm·(10a)$^{-1}$ ($p > 0.1$), and the change trend was not significant in most areas (Figure 11). Figure 12 showed that most of western Inner

Mongolia, Hohhot, western Chifeng and eastern Xing'an League experienced positive effects on the climatic production potential of potato during the growth period of the precipitation changes, with an impact range of 20~40 kg·hm^{-2}·a^{-1}, in which the positive effects in the central and eastern parts of Ordos city and most of Baotou city were large and were greater than 30 kg·hm^{-2}·a^{-1}. The central and eastern regions experienced mostly negative effects, with an impact range of -102~ -20 kg·hm^{-2}·a^{-1}; the southeast part of Chifeng city, the southwest part of Tongliao city and the northwest part of Hulunbuir city experienced greater negative impacts, which were less than -50 kg·hm^{-2}·a^{-1}.It is worth mentioning that the change in precipitation in most areas of Wulanchabu city, which is the main potato-producing area in Inner Mongolia, had a negative impact on potato climatic production potential.

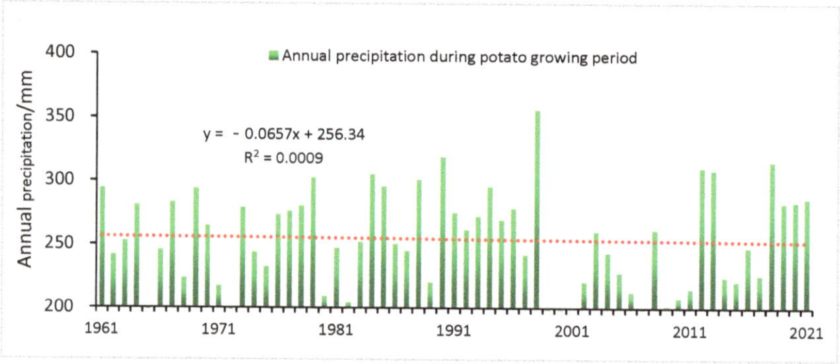

Figure 11. Variations in annual precipitation during the potato growth period in Inner Mongolia, China (mm).

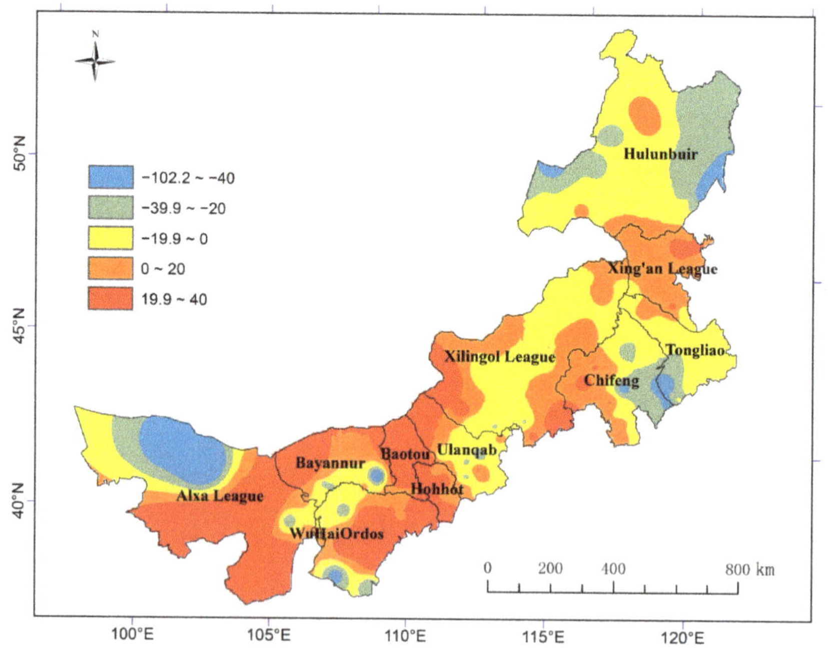

Figure 12. Impact of precipitation changes on the climatic production potential of potato in Inner Mongolia, China (kg·hm^{-2}·a^{-1}).

4. Discussion

Light, temperature and water are important drivers of climatic production potential [30,31]. Analyzing the impact of climate factors on potato climatic production potential will help people take better measures to adapt to climate change, so as to ensure the high-quality development of the potato industry. In general, the impact of climate change on potato climatic production potential in Inner Mongolia has both advantages and disadvantages, but the disadvantages outweigh the advantages. In particular, in Wulanchabu city, the main potato-producing area in the Inner Mongolia Autonomous Region, the impact of climate change on potato climatic production potential is relatively adverse. Compared with the average value of potato climatic production potential in Shanxi Province (13,428 kg·hm^{-2}), the average value of potato climatic production potential in Inner Mongolia (19,318 kg·hm^{-2}) in this study was higher. Although the average precipitation during the growing period in Shanxi Province is slightly higher than that in Inner Mongolia, the high average temperature leads to a lower climatic production potential of potato in Shanxi Province than that in Inner Mongolia. The impact of light, temperature and water changes on potato climatic production potential in Shanxi Province is also more negative; particularly, the impact of radiation and temperature changes in most areas is negative, similar to the impact of climate change on the potato climatic production potential in Inner Mongolia. The impact of precipitation change on the potato climatic production potential in Shanxi Province is positive in the north and negative in the south [28,32]. However, we found that the distribution of impacts of precipitation change on potato production potential in Inner Mongolia was more complex, spanning the main southern agricultural areas from southwest to northeast as follows: positive effect—negative effect—positive effect—negative effect—positive effect—negative effect. The average application level of the potato climatic production potential in Wuchuan County was only 7.6%, lower than the application level of the potato climatic production potential in Jixi County of Ningxia [20]. This shows that the potato yield in Wuchuan County is bound to have room for improvement. We need to pursue advantages, avoid disadvantages and make effective use of climate factors.

In order to adapt to the current climate change trend and its adverse impacts, it is necessary to vigorously promote water-saving irrigation technology and biofilm technology in agricultural areas to improve water use efficiency. In particular, most of the central and western parts of Inner Mongolia are mainly rain-fed agriculture. In Ordos, Baotou, Hohhot and other central and western regions, it is necessary to adjust the sowing date of potatoes so that the yield can occur in the critical period to avoid high temperatures and obtain high yield. The potato production in the eastern agricultural area needs to improve field management in order to improve the utilization rate of light energy and obtain high yield.

In this study, the step-by-step correction method was adopted, and the physical meaning of the model was clear. However, there are still shortcomings that need to be resolved. Because of the incompleteness and discontinuity of the data, we used the data of the average growth period of potato from agrometeorological observation stations for many years. In general, studying only the main variety of local potato, i.e., the mid-late maturing variety, without considering the early maturing variety of potato will lead to differences in climatic resources during the growing period of potato, resulting in certain differences in the calculation results of potato climatic production potential. In Inner Mongolia, water condition was the main limiting factor of potato climatic production potential. This is consistent with the previous study on various crops which found that the same precipitation was far greater than the crop water demand, or even impacted the potato climatic production potential in case of flood disaster, so there were inevitably some omissions in the calculation results [33]. The parameter determination in the step-by-step correction method was empirical, and only reflected the overall variety characteristics of current crops. With the continuous development of breeding and cultivation techniques, the parameters of better varieties might be different in the future, which is likely to lead to changes in the calculation results of potato climatic production potential [8,33]. As the precipitation in local agricultural areas was relatively low [20,34,35], the adverse impact from calculating the water correction coefficient was not considered in this study. In addition, the climatic production potential of potato was the maximum yield under ideal conditions. The actual yield of potato is affected by many factors, such as climate, soil, social economy and so on. Therefore, in the future, multiple methods can be used for comprehensive analysis to conduct in-depth comparative research on climatic production potential and actual yield, so as to explore the differences and responses between the two [35–37].

5. Conclusions

We investigated the effects of climate change on the climatic production potential of potatoes since 1961 in Inner Mongolia using the step-by-step correction method. From 1961 to 2021, the average climatic production potential of potato in Inner Mongolia decreased from the central south to the surrounding areas. Under the direct influence of local climatic factors, the average climatic production potential of potato showed a downward trend year by year, especially after 2000. The changes in three climatic factors, light, temperature and water, had different effects on the climatic production potential of potato. The impact of radiation change on the climatic production potential of potato was mainly negative, especially in the main potato-producing area of Wulanchabu. Temperature change had both advantages and disadvantages. In the cold climate region of Northeast China, the temperature change had a favorable impact on the climatic production potential of potato. Other areas were mainly adversely affected, and most areas of Ulanqab city were adversely affected by temperature changes. The change in rainfall had a favorable impact on the climatic production potential of potato west of Hohhot and in the southern and central area of Xilin Gol League. However, the change in rainfall had a negative impact on the climatic production potential of potato in most other areas.

Author Contributions: J.-F.Z. and L.-T.Y. contributed to the study conception and design. Material preparation, data collection, and analysis were performed by L.-T.Y., X.-P.J., S.W., L.-H.L. and H.-F.X. The first draft of the manuscript was written by L.-T.Y., and all authors commented on subsequent versions of the manuscript. All authors have read and agreed to the published version of the manuscript.

Funding: This research was funded by the Natural Science Foundation of Inner Mongolia, (2021MS04018) and the Climate Change Project of China Meteorological Administration (CCSF201931).

Data Availability Statement: Meteorological observation data (including surface meteorological observation data and agricultural meteorological observation data) are from the Bureau of Inner Mongolia Autonomous Region. The relevant map data comes from the map Research Institute of Inner Mongolia Autonomous Region.

Acknowledgments: We thank the reviewers for their help in improving this manuscript.

Conflicts of Interest: The authors declare no conflict of interest.

References

1. Li, F.; Zhou, M.J.; Shao, J.Q.; Chen, Z.H.; Wei, X.L.; Yang, J.C. Maize, wheat and rice production potential changes in China under the background of climate change. *Agric. Syst.* **2020**, *182*, 102853.
2. Yuan, B.; Guo, J.P.; Ye, M.Z.; Zhao, J.F. Variety distribution pattern and climatic potential productivity of spring maize in Northeast China under climate change. *Chin. Sci. Bull.* **2012**, *57*, 3497–3508. [CrossRef]
3. Jia, Q.; Li, M.F.; Dou, X.C. Climate Change Affects Crop Production Potential in Semi-Arid Regions: A Case Study in Dingxi, Northwest China, in Recent 30 Years. *Sustainability* **2022**, *14*, 3578. [CrossRef]
4. Froehlich, H.E.; Gentry, R.R.; Halpern, B.S. Global change in marine aquaculture production potential under climate change. *Nat. Ecol. Evol.* **2018**, *2*, 1745–1750. [CrossRef]
5. Yang, X.G.; Chen, F.; Lin, X.M.; Liu, Z.J.; Zhang, H.L.; Zhao, J.; Li, K.N.; Ye, Q.; Li, Y.; Lv, S.; et al. Potential benefits of climate change for crop productivity in China. *Agric. For. Meteorol.* **2015**, *208*, 76–84. [CrossRef]
6. David, P.M.Z.; Navin, R.; Carol, C.B.; Jonathan, A.F. From Miami to Madison: Investigating the relationship betweenclimate and terrestrial net primary production. *Glob. Biogeochem. Cycles* **2007**, *21*, GB3004.

7. Liu, H.; Liu, B.; Hua, L.Z.; Shi, Z.J.; Ning, G.F.; Zhang, X. Analysis of temporal and spatial pattern of vegetation NPP in Fujian Province and its driving force. *Sci. Surv. Mapp.* **2018**, *43*, 35–44. (In Chinese)
8. Zhang, X.L.; Xiao, W.H.; Wang, Y.C.; Wang, Y.; Wang, H.J.; Wang, Y.X.; Zhu, L.S.; Yang, R.X. Spatial-temporal changes in NPP and its relationship with climate factors based on sensitivity analysis in the Shiyang River Basin. *J. Earth Syst. Sci.* **2020**, *129*, 169–190. [CrossRef]
9. Chen, T.; Bao, A.M.; Jiapaer, G.L.; Guo, H.; Zheng, G.X.; Jiang, L.L.; Chang, C.; Tuerhanjiang, L. Disentangling the relative impacts of climate change and human activities on arid and semiarid grasslands in Central Asia during 1982–2015. *Sci. Total Environ.* **2018**, *653*, 1311–1325. [CrossRef]
10. Wang, X.F.; Yang, Y.Z.; You, F. Climate change in Heilongjiang Province and its impact on maize production potential. *Agric. Res. Arid Areas* **2012**, *35*, 25–29. (In Chinese)
11. Zhan, T.; Zhong, H.L.; Sun, L.X.; Fischer, G.; Velthuizen, H.V.; Liang, Z. Improving performance of Agro-Ecological Zone (AEZ) modeling by cross-scale model coupling: An application to japonica rice production in Northeast China. *Ecol. Model.* **2014**, *290*, 155–164.
12. Zhang, Y.; Yang, B.; Liu, X.H.; Wang, C.Z. Estimation of rice grain yield from dual-polarization Radarsat-2 SAR data by integrating a rice canopy scattering model and a genetic algorithm. *Int. J. Appl. Earth Obs. Geoinf.* **2017**, *57*, 75–85. [CrossRef]
13. Yu, D.; Hu, S.G.; Tong, L.Y.; Xia, C.; Ran, P.L. Dynamics and Determinants of the Grain Yield Gap in Major Grain-Producing Areas: A Case Study in Hunan Province, China. *Food* **2022**, *11*, 1122. [CrossRef]
14. Wang, Z.M.; Zhang, B.; Zhang, S.Q.; Song, K.S.; Duan, H.T. Research on agroclimatic production potential and utilization rate of natural resources in Songnen Plain. *China Agric. Meteorol.* **2005**, *26*, 1–6. (In Chinese)
15. Li, S.A.; Ju, H.; Chi, B.L. Research progress on crop production potential. *China Agric. Meteorol.* **2005**, *26*, 106–111. (In Chinese)
16. Li, F.; Zhang, S.W.; Zhang, Y.J.; Yang, H.J.; Yang, J.C. Changes of grain production potential in farming–pastoral ecotone: A case study in West Jilin, China. *J. Agric. Sci.* **2018**, *156*, 151–161. [CrossRef]
17. Lu, Y.Y.; Wang, S.; Tian, H.; Deng, H.Q.; He, D.Y. Evolution of climatic production potential and evaluation of climate carrying capacity of food security in Anhui Province in recent 50 years. *Resour. Environ. Yangtze River Basin* **2017**, *26*, 428–435. (In Chinese)
18. Wang, S.; Song, A.W.; Xie, W.S.; Tang, W.A.; Dai, J.; Ding, X.J.; Wu, R. Impact assessment of future climate change on the climatic production potential of rice in the south of Huaihe River in Anhui. *Arid Meteorol.* **2020**, *38*, 179–187. (In Chinese)
19. Wang, S.Y.; Huo, Z.G.; Li, S.K.; Xue, C.Y.; Mao, F. Water deficit and climatic production potential of Winter Wheat in northern China—A Study on dynamic changes in recent 40 years. *J. Nat. Disasters* **2003**, *12*, 121–130. (In Chinese)
20. Duan, X.F.; Qian, H.; Li, J.P.; Ma, G.F.; Yang, Y.; Cao, N.; Zhang, X.Y. Stability analysis of climatic production potential of Jixi potato under 4 sowing dates. *Agric. Res. Arid Reg.* **2011**, *29*, 266–271. (In Chinese)
21. Lai, R.S. *Analysis of the Changing Trend of Crop Production Potential and Planting Structure in Ningxia under the Background of Climate Change*; Ningxia University: Yinchuan, China, 2016. (In Chinese)
22. He, Y.N.; Yang, F. Study on background value of environmental soil organic matter in Inner Mongolia Autonomous Region. *Inn. Mong. Environ. Prot.* **1996**, *1*, 40–41. (In Chinese)
23. Xu, S.P.; Tao, S.; Cao, J. Spatial structure characteristics of soil pH value, clay and organic matter content in Inner Mongolia. *Soil Bull.* **2001**, *4*, 145–148. (In Chinese) [CrossRef]
24. Zhang, Y.M.; Liu, Y. Change trend and prediction of cultivated land soil nutrients in Inner Mongolia. *Inn. Mong. Agric. Sci. Technol.* **2004**, *6*, 45–46. (In Chinese)
25. Guo, J.P.; Zhao, J.F.; Wu, D.R.; Mu, J.; Xu, Y.H. Attribution of maize yield increase in China to climate change and technological advancement between 1980 and 2010. *J. Meteorol. Res.* **2014**, *28*, 1168–1181. [CrossRef]
26. Yang, L.T.; Jiang, X.P.; You, L.; Lin, M. Analysis and Evaluation of Potato Climate Suitability in Wuchuan County. *Inn. Mongolia. Mod. Agric.* **2017**, *11*, 96–97. (In Chinese)
27. Jiang, X.P.; Yang, L.T. Study on the Variation of Climate Suitability of Potato in Different Growth Periods–Taking Wuchuan County, Hohhot City as an Example. *Int. J. Nat. Resour. Ecol. Manag.* **2022**, *7*, 86–92. [CrossRef]
28. Ma, Y.L.; Guo, J.P.; Zhao, J.F. Distribution characteristics of crop climate production potential and its response to climate change in the agro-pastoral interlaced zone in northern Shanxi Province. *J. Ecol.* **2019**, *38*, 818–827. (In Chinese)
29. Ma, Y.F.; Li, X.C.; Song, J.H.; Gao, C.X.; Zhao, D.; Shen, Y.B. Climatological calculation of total solar radiation in Inner Mongolia Autonomous Region and its temporal and spatial distribution characteristics. *J. Meteorol. Environ.* **2013**, *29*, 102–109. (In Chinese)
30. Xie, W.Y.; Guo, X.Q.; Lu, D.R.; Yin, Y.C.; Wang, J.B. Analysis of temporal and spatial changes of climatic production potential and hydrothermal factors in the Loess Plateau in Longzhong. *Arid Meteorol.* **2015**, *33*, 861–866. (In Chinese)
31. He, Y.F.; Zhao, M.X.; Wang, J.X.; Zhang, H.S. Response of grassland productivity to climatic factors in the agro-pastoral interlaced zone of Inner Mongolia: Taking Duolun County as an example. *Arid Meteorol.* **2008**, *26*, 84–89. (In Chinese)
32. Liu, Q.; Yan, C.R.; He, W.Q. Research on the production potential of dry farming agroclimate in Shouyang County, Shanxi. *China Agric. Meteorol.* **2007**, *28*, 271–274. (In Chinese)
33. Pang, Y.M.; Chen, C.; Xu, F.X.; Guo, X.Y. Impact of climate change on potential productivities of main grain crops in the Sichuan Basin. *Chin. J. Eco-Agric.* **2020**, *28*, 1661–1672. (In Chinese)
34. Luo, H.P.; Zou, N.; Hu, X.Y.; Wang, S. Climatic potential productivity and resources utilization efficiency of major grain crops in the main grain production areas of China, 1980–2019. *Resour. Sci.* **2021**, *43*, 1234–1247. [CrossRef]

35. Lu, Y.Y.; Sun, W.; Tang, W.A.; He, D.; De, H. Climatic potential productivity and stress risk of winter wheat under the background of climate change in Anhui Province. *Chin. J. Eco-Agric.* **2020**, *28*, 17–30.
36. Wang, Z.M.; Liang, Y.L. The application of EPIC model to calculate crop productive potentialities in Loessic Yuan region. *J. Nat. Resour.* **2000**, *17*, 481–487. (In Chinese)
37. Han, S.; Liu, B.C.; Shi, C.X.; Liu, Y.; Qiu, M.J.; Sun, S. Evaluation of CLDAS and GLDAS Datasets for Near-Surface Air Temperature over Major Land Areas of China. *Sustainability* **2020**, *12*, 4311. [CrossRef]

sustainability

MDPI

Article

Inter- and Mixed Cropping of Different Varieties Improves High-Temperature Tolerance during Flowering of Summer Maize

Shuyan Li [1], Junfang Zhao [2,*], Junling Li [1,3], Ruixin Shao [4], Hongping Li [4], Wensong Fang [1,3], Liting Hu [1,3] and Tianxue Liu [4,*]

[1] China Meteorological Administration, Henan Key Laboratory of Agrometeorological Support and Applied Technique, Zhengzhou 450003, China; lsy_163com@163.com (S.L.); ljl8843@sina.com (J.L.); fws68fyf@163.com (W.F.); hlt0114@163.com (L.H.)
[2] State Key Laboratory of Severe Weather, Chinese Academy of Meteorological Sciences, Beijing 100081, China
[3] Henan Institute of Meteorological Sciences, Zhengzhou 450003, China
[4] College of Agronomy, Henan Agricultural University, Zhengzhou 450002, China; shaoruixin@henau.edu.cn (R.S.); lihongping@henau.edu.cn (H.L.)
* Correspondence: zhaojf@cma.gov.cn (J.Z.); liutianxue@henau.edu.cn (T.L.); Tel.: +86-10-58995896 (J.Z.)

Abstract: Global warming increases the risk of high-temperature injury to maize. Inter- and mixed-cropping of maize varieties with different genotypes is one way to effectively alleviate the high-temperature injury during the flowering period. However, the mitigation effect of different varieties and intercropping modes on high-temperature injury is still unclear. Based on previous years of field production, Denghai 605, which is more sensitive to high temperatures during the flowering period, was determined as the main test variety, and Zhengdan 958, Dedan 5, Weike 702, and Xianyu 335, which have great genotypic differences, were used as auxiliary varieties. The main test varieties and auxiliary varieties were intercropped and mixed cropped, respectively. Plant height, ear height, leaf area index, population light transmittance, ear characteristics, and yield were measured, and the land equivalent ratio (LER) was calculated. The plant height of Denghai 605 intercropped with Zhengdan 958 and Dedan 5 and mixed with Weike 702 and Xianyu 335 decreased significantly. The population light transmittance of the bottom or middle layer in Denghai 605 increased significantly when intercropped with other varieties. The grain number per ear increased significantly under inter- and mixed cropping with Zhengdan 958 and Weike 702. Except under intercropping with Dedan 5, the yield of Denghai 605 increased significantly, by 8.8–28.0%, under inter- and mixed cropping. Under intercropping with Zhengdan 958 and inter- and mixed cropping with Weike 702 and Xianyu 335, respectively, the group land equivalent ratio was greater than 1.1, indicating that under the combination of these varieties, inter- and mixed cropping effectively reduced the impact of high temperatures during flowering.

Keywords: summer maize; inter- and mixed cropping; high temperature; flowering period; yield

Citation: Li, S.; Zhao, J.; Li, J.; Shao, R.; Li, H.; Fang, W.; Hu, L.; Liu, T. Inter- and Mixed Cropping of Different Varieties Improves High-Temperature Tolerance during Flowering of Summer Maize. *Sustainability* **2022**, *14*, 6993. https://doi.org/10.3390/su14126993

Academic Editors: Xiaodong Yan, Jia Yang and Shaofei Jin

Received: 29 April 2022
Accepted: 1 June 2022
Published: 8 June 2022

1. Introduction

Maize (*Zea mays* L.) is one of the most important crops in the global and national economy [1,2]. The Huanghuaihai plain is the largest summer-maize-producing area in China with a wheat–maize cropping system. The growing season of summer maize is mainly in the period from June to September, which has the highest temperature in the year and two-thirds of the annual precipitation. In the context of climate change, continuous high-temperature weather occurs frequently [3–5], which increases the risk of heat damage to maize [6,7]. The flowering period of summer maize is when it is most sensitive to high temperatures. High temperatures affect male and female ear development, pollen vigor, and grain development [8–10]. Maize pollen is more susceptible to high-temperature stress than female ears [11]. Pollen vigor declines quickly when the temperature exceeds 32–35 °C [12,13]. High-temperature stress can cause abnormal pollen development or abortion, resulting in a decrease in the number of pollen grains. In addition, it can cause abnormal division of pollen mother cells, deformation, and shrinkage leading to deformed pollen grains [14–16], which in turn affects the percentage seed set, yield, and quality of maize [9,10]. Therefore, it is of great significance to study effective methods to protect maize from the harmful high temperatures during the flowering period.

Single crop varieties are more vulnerable to diseases due to their narrow genetic basis, which is particularly prominent in maize. In addition to the influence of the external environment, there are obvious genotypic differences in the tolerance of maize to high temperatures and heat injury during the flowering period [17,18]. According to the principle of ecological complementation and biodiversity, the temperature–stress resistance of a maize population can be effectively improved by intercropping or mixed cropping of different maize varieties [19,20].

Inter- and mixed cropping among different ecotypes and genotypes of maize or intercropping with other crops could improve the utilization rate of natural resources and increase the yield stability of the crop composite population [21]. The inter- and mixed cropping of maize varieties optimizes the population spatial structure and maintains a higher chlorophyll content and photosynthetic rate [22], which is more conducive to gas exchange and promotes photosynthesis [23,24], increased nutrient absorption and dry matter accumulation [25,26], enhanced crop stress resistance [27,28], and improved yield and quality of maize [29,30]. In the intercropping of multiple varieties of maize, the anthesis and pollination time of the population system are relatively prolonged. When the maize encounters adverse weather events such as continuous rain and high temperatures during the flowering period, the fertilization rate of male flowers in the compound population composed of multiple maize varieties is higher than that in the single variety planting mode, which can effectively improve the bald tip and grain shortage caused by poor pollination, so as to achieve the purpose of increasing and stabilizing yield [31–33].

An important strategy to reduce high-temperature stress is to exploit the difference in heat resistance among maize varieties using an inter- and mixed cropping system. However, few studies have considered crucial aspects of this approach, such as maize variety matching, intercropping methods, and mixed cropping ratios. High-temperature and heat damage occurred continuously during the flowering period of maize in Henan from 2013 to 2016. It was found that the yield of Denghai 605 and Xianyu 335 decreased seriously under the high temperatures, while Zhengdan 958, Weike 702, and Dedan 5 had strong heat resistance and stable yield performance. According to the previous results, the high-temperature resistance field experiment of multi-variety inter- and mixed cropping was performed in 2017. In this study, Denghai 605, which has a large planting area and is sensitive to high temperatures, was selected as the main test variety to analyze the effects of inter- and mixed cropping of Denghai 605 with Zhengdan 958, Weike 702, Dedan 5, and Xianyu 335 on the prevention and control of high-temperature and heat damage during the flowering period.

2. Materials and Methods

2.1. Study Location

Our study was located in Shizhuang (114.03° E, 34.15° N), Chencao Township, Xuchang City, Henan Province, which belongs to the north warm temperate monsoon climate zone with abundant heat resources and abundant rainfall. The total area of the experimental field was 0.7 ha, which was mechanized sowing. The heat tolerance evaluation experiment of maize was carried out from 2013 to 2016, and the high-temperature resistance experiment of multi-variety inter- and mixed cropping was performed in 2017. The soil of the experimental field was a fluvo-aquic soil. The soil chemical properties at 0–20 cm depth were as follows: 23.54 g kg^{-1} soil organic matter, 43.77 mg kg^{-1} available nitrogen, 202.04 mg kg^{-1} Olsen potassium, and 11.07 mg kg^{-1} Olsen phosphorus. The meteorological variables recorded during the summer-maize-growing season at the study site are shown in Figure 1.

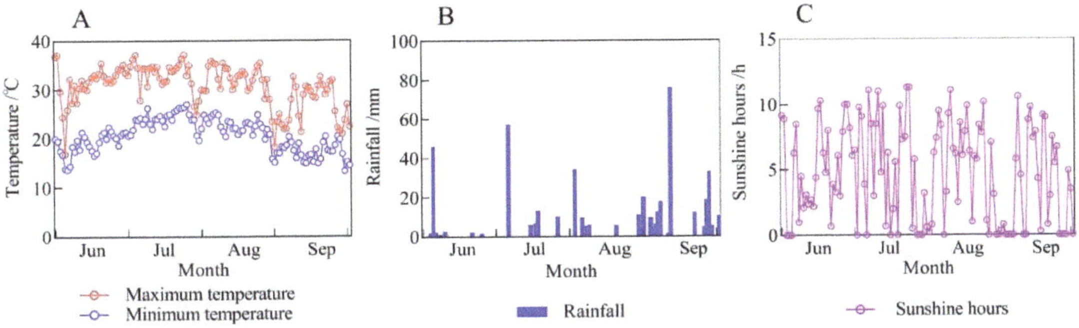

Figure 1. Meteorological variables during the summer-maize-growing season at the study site. Temperature (**A**), rainfall (**B**), and sunshine hours (**C**) at Xuchang, Henan, China from June to September in 2017.

2.2. Experimental Design

'Denghai 605' (Shandong Denghai Seed Industry Co., Ltd.; Laizhou, Shandong, China) was selected as the main test variety, and 'Zhengdan 958' (Henan Academy of Agricultural Sciences Institute of Grain Crops; Zhengzhou, Henan, China), 'Dedan 5' (Beijing Denong Seed Industry Co., Ltd.; Beijing, China), 'Weike 702' (Zhengzhou Weike Crop Breeding Technology Co., Ltd.; Zhengzhou, Henan, China), and 'Xianyu 335' (Tieling Pioneer Seed Research Co., Ltd.; Tieling, Liaoning, China) were chosen as the auxiliary varieties.

Denghai 605 was inter- and mixed cropped with the four auxiliary varieties. A monoculture of each variety was established as the control group. The experimental design and processing codes are summarized in Table 1.

Table 1. Intercropping and mixed cropping treatments of the maize varieties.

Farming Methods	Variety Combinations	Variety Codes	
		Denghai 605	**Auxiliary Varieties**
Intercropping	Denghai 605	| Zhengdan 958 [605	| 958]	(605)	| 958	605	| (958)
	Denghai 605	| Dedan 5 [605	| 005]	(605)	| 005	605	| (005)
	Denghai 605	| Weike 702 [605	| 702]	(605)	| 702	605	| (702)
	Denghai 605	| Xianyu 335 [605	| 335]	(605)	| 335	605	| (335)
Mixed cropping	Denghai 605 × Zhengdan 958 [605 × 958]	(605) × 958	605 × (958)
	Denghai 605 × Dedan 5 [605 × 005]	(605) × 005	605 × (005)
	Denghai 605 × Weike 702 [605 × 702]	(605) × 702	605 × (702)
	Denghai 605 × Xianyu 335 [605 × 335]	(605) × 335	605 × (335)
Monoculture	Denghai 605	CK605	
	Zhengdan 958	CK958	
	Dedan 5	CK005	
	Weike 702	CK702	
	Xianyu 335	CK335	

"| |" represents the intercropping of two varieties, and "×" represents the mixed cropping of two varieties. The notation also indicates the variety combination used in the experiment. For example, intercropping of Denghai 605 and Zhengdan 958 is indicated as "605||958", "Denghai 605" in the combination is indicated by the form "(605)||958", and "Zhengdan 958" is indicated by "605 | | (958)".

It has been proven that in a relay intercropping system, narrow–wide row planting improves the light environment and seed yields of intercrop species [34]. Thus, the narrow–wide row planting pattern with a wide row of 70 cm and a narrow row of 50 cm was used in this experiment. The mechanical precision seeding method was applied to fertilize simultaneously. The seeder sowed two rows concurrently. The five varieties were sown on 11 June. The intercropping sowing design was illustrated in Figure 2. For mixed cropping, varieties were sown according to a 1:1 ratio of the number of seeds. Seedlings were thinned at the three-leaf stage, weeds were chemically controlled, and a commercial compound fertilizer (N-P_2O_5-K_2O: 29-5-6, 750 kg ha^{-2}; Luxi, 0303050000001, Liaocheng, Shandong, China) was applied between rows at the jointing stage. Irrigation was applied in accordance with the soil moisture content to ensure that the entire growth period was free from drought stress. Other management measures were identical to those of local farmer practices.

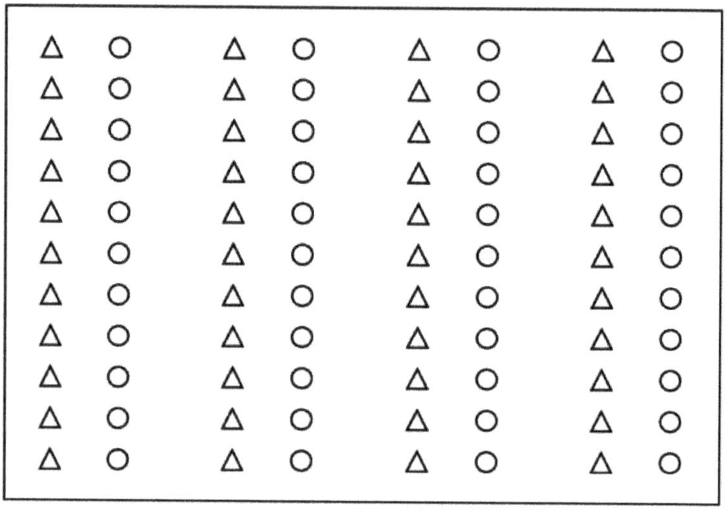

Figure 2. Schematic diagram of the intercropping sowing design. Note: △ and ○ indicate different varieties.

2.3. *Measurement of Parameters and Methods*

2.3.1. Main Reproductive Assessment Period

The timing of the tasseling, silking, flowering, and pollination stages of maize plants growing in the monoculture area was recorded, which for each stage was expressed as the number of days after sowing.

2.3.2. Population Density Determination

Eleven consecutive plants in a row were selected to measure the spacing of ten plants, and eleven consecutive rows (five wide and five narrow rows each) were selected to measure the spacing of ten rows. The measurements were used to calculate the average plant spacing (from three repetitions). Based on these data, the population density of the intercropping mode was 61,215 plant ha^{-2} and that of the mixed cropping mode was 64,815 plant ha^{-2}.

2.3.3. Main Agricultural Characters Determination

During the tasseling period of maize, five plants of relatively uniform growth were selected for each variety in each treatment area, and the plant height and ear height were measured. The leaf area index was calculated with a length–width coefficient method. The leaf length and maximum leaf width were measured in situ with a ruler. During the tasseling period of the maize, three representative sites were randomly selected in the middle of a narrow row for each treatment. The light transmittance of the middle and bottom layers in the canopy of the maize population was measured with an LAI-2000 plant canopy analyzer (LI-COR, Inc; Lincoln, NE, USA).

2.3.4. Ear Traits and Yield Determination

In the monoculture of the test variety, ten consecutive ears were selected from plants growing in a middle row of uniform growth. For the mixed cropping mode, in a relatively uniform middle row, ten ears were selected from consecutive plants of each of the two combination varieties (note that ears of the same variety were not collected from a row once ten ears had been sampled). For the intercropping treatment, ten consecutive ears were selected from one row (a non-side row) of each variety. Three replicates were collected; thus, a total of 30 ears of each variety in each treatment was sampled. Each variety was sampled 3 times for the determination of ear traits and yield, and 10 consecutive ears were measured for each replicate, for a total of 30 ears. Among them, 3 ears were selected for each repetition (9 ears were selected in total), and ear length, ear thickness, bald length, and grain number per ear were measured. The 100-grain weight and yield were determined by threshing and mixing once we were finished measuring every ten ears.

The ear length and bald length were measured with a ruler, the ear thickness was measured with a vernier caliper at the thickest portion in the middle of the ear, the 100-grain weight was measured with an electronic balance (sensitivity 0.01 g), and grain moisture content was measured using a grain moisture meter.

The population yield was calculated according to the number of grains per ear, 100-grain weight, and population density of each variety in the inter- and mixed cropping combination. The formulas used were as follows:

$$Y_i = (Y_{S1} + Y_{S2})/2 \tag{1}$$

$$Y_{S1} = W_{S1} \times N_{S1} \times D_S \tag{2}$$

$$Y_{S2} = W_{S2} \times N_{S2} \times D_S \tag{3}$$

where Y_i is the population yield of the varieties under inter- and mixed cropping mode per unit area, Y_{S1} and Y_{S2} are the yield of two varieties in the inter-and mixed cropping mode, respectively, W_{S1} and W_{S2} are the grain weight of the two varieties equivalent to 14% moisture, N_{S1} and N_{S2} are the number of grains per ear of the two varieties, respectively, and D_S is the population density.

2.3.5. Calculation of the Land Equivalent Ratio

The land equivalent ratio is the ratio of the income of two or more mixed crops (varieties) to the income of each crop in the same farmland. The land equivalent ratio (LER) was calculated using the following formula:

$$\mathrm{LER} = \frac{Y_i}{Y_{ii}} \tag{4}$$

$$Y_{ii} = (Y_{c1} + Y_{c2})/2 \tag{5}$$

where Y_{ii} is the average yield of the two varieties and Y_{c1} and Y_{c2} are the yields of the two varieties grown in the monoculture method, respectively.

2.4. Determination of Maize Flowering Period and High-Temperature Stress Threshold

The date of the onset of flowering for each variety was recorded from plants growing in the monoculture. Flowering of the entire ear of summer-sown maize usually lasts 7–10 d; therefore, the onset of tasseling is the date for the start of flowering, and the date of the end of flowering is 10 d after tasseling. The maximum temperature $\geq$ 35 °C was used as the critical threshold of high-temperature stress during the flowering period [17,35].

2.5. Statistical Analysis

Data processing and graphing were performed using Microsoft Excel 2016 and GraphPad Prism 8. All data were expressed as the mean ± standard deviation (SD), and n refers to the number of samples in each group. Statistical analyses were performed using SPSS 17.0 (SPSS, IBM Corp., Armonk, NY, USA). Analysis of variance (ANOVA) was performed to determine the significance of differences between treatments. Means for different treatments were compared using the Bonferroni test at the significance level $\alpha = 0.05$.

3. Results

3.1. Differences in the Onset of Flowering and the Occurrence of High Temperatures

As shown in Figure 3, Dedan 5 was the first variety to start flowering (54 d after sowing) and Xianyu 335 was the last variety to flower (57 d after sowing). Two main periods of high temperatures were recorded in 2017. The first period was from 3 to 6 August, in which the highest temperature exceeded 35 °C for 4 consecutive days. The second period was from 9 to 11 August, in which the highest temperature exceeded 35 °C for 3 consecutive days. The first high-temperature period coincided with the beginning of tasseling of the varieties. The second high-temperature period occurred 2–6 d after tasseling and had a greater impact on flowering and pollination.

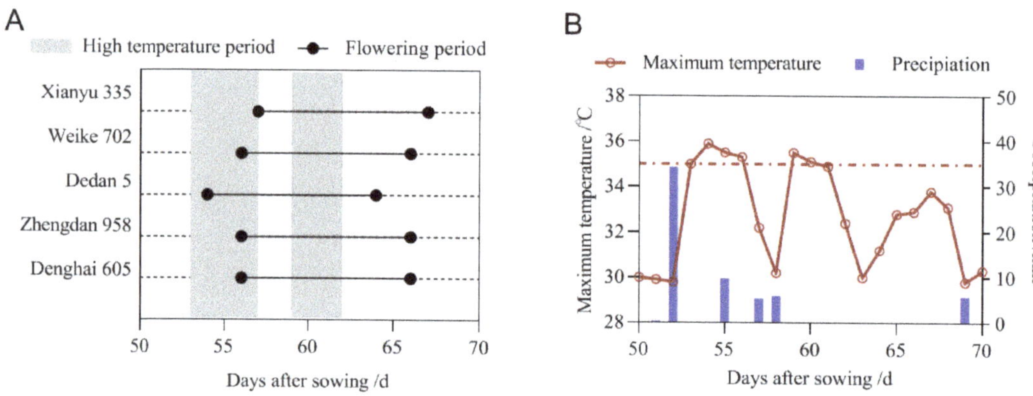

Figure 3. Flowering period (**A**) and meteorological conditions (**B**) of each variety.

3.2. Plant Height and Ear Height

As shown in Figure 4, aside from the increase in plant height of Denghai 605 intercropped with Xianyu 335, the plant height of Denghai 605 showed different degrees of decline in other inter- and mixed cropping modes. Among these treatments, intercropping with Zhengdan 958 and Dedan 5 and mixed cropping with Weike 702 and Xianyu 335 resulted in a significant decrease in the plant height of Denghai 605. The plant heights of Zhengdan 958, Dedan 5, and Weike 702 decreased under inter- and mixed cropping with Denghai 605, and the plant heights of the mixed cropping modes all decreased significantly. Given that Dedan 5, Weike 702, and Zhengdan 958 have genetically similar parents, the trends for changes in plant height under inter- and mixed cropping with Denghai 605 were generally similar. However, the plant height of Xianyu 335 increased significantly under inter- and mixed cropping with Denghai 605.

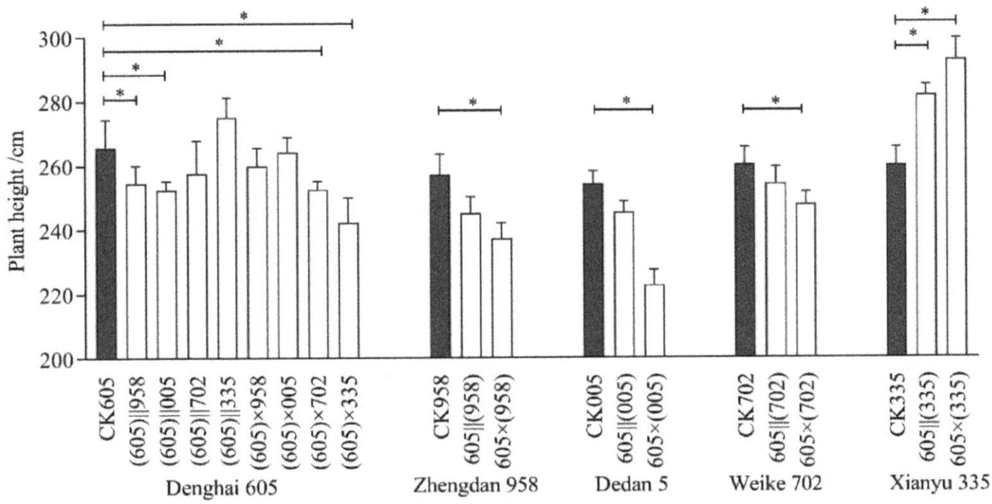

Figure 4. Plant height of summer maize varieties under different inter- and mixed cropping modes. Values represent the mean ± SD, n = 5 per group. * $p < 0.05$.

As shown in Figure 5, the changes in ear height of most varieties under the inter- and mixed cropping modes were similar to the observed changes in plant height. The main difference was that the ear height of Zhengdan 958 did not decrease significantly under the mixed cropping mode. The ear height of Weike 702 under intercropping was significantly lower than that of the monoculture. The ear height of Xianyu 335 increased significantly under the inter- and mixed cropping modes, and the ear height under intercropping exceeded that observed in the mixed cropping mode.

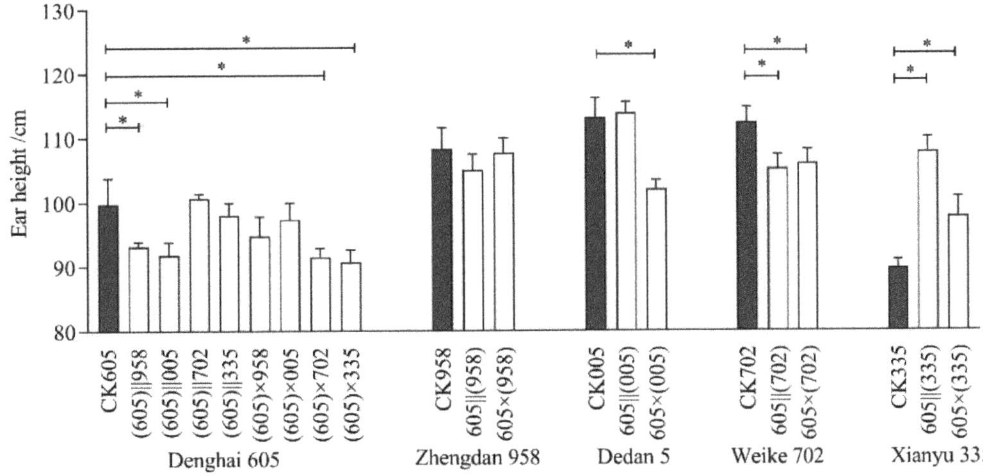

Figure 5. Ear height of summer maize varieties under different inter- and mixed cropping modes. Values represent the mean ± SD, n = 5 per group. * $p < 0.05$.

3.3. Leaf Area Index

As shown in Figure 6, except for a nonsignificant difference in leaf area between Denghai 605 and Weike 702, the leaf area index of Denghai 605 decreased significantly under the other inter- and mixed cropping modes. Under the intercropping mode, the leaf area index of the four auxiliary varieties showed no significant change compared with that of the monoculture. Under the mixed cropping mode, the leaf area index of Zhengdan 958 and Dedan 5 mixed with Denghai 605 decreased significantly compared with that of the monoculture, whereas the leaf area index of Weike 702 and Xianyu 335 mixed with Denghai 605 increased significantly compared with that of the monoculture.

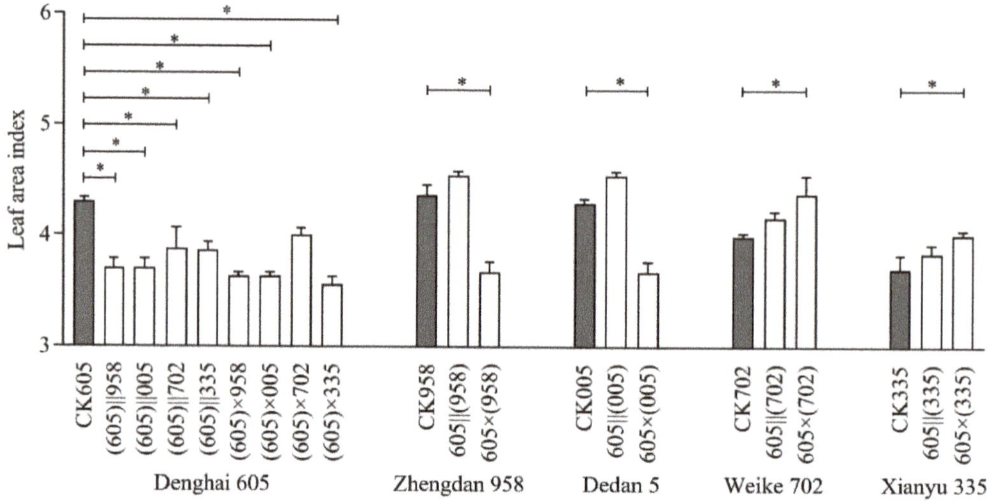

Figure 6. Leaf area index of summer maize varieties under different inter- and mixed cropping modes. Values represent the mean ± SD, n = 3 per group. * $p < 0.05$.

3.4. Population Light Transmittance

As shown in Figure 7, under the intercropping mode, the mid-canopy-level population light transmittance of Denghai 605 and Zhengdan 958 increased significantly compared with that of each monoculture. The mid-canopy-level population light transmittance between Denghai 605 and the other three varieties was not significantly affected by intercropping. Under the mixed cropping mode, the population light transmittance of the mid-canopy of the three varieties was significantly lower than that of the Denghai 605 monoculture.

Under intercropping of the other four varieties, the population light transmittance of the lower canopy increased significantly. The population light transmittance of the lower canopy under all mixed cropping modes showed no significant change compared with that of the Denghai 605 monoculture.

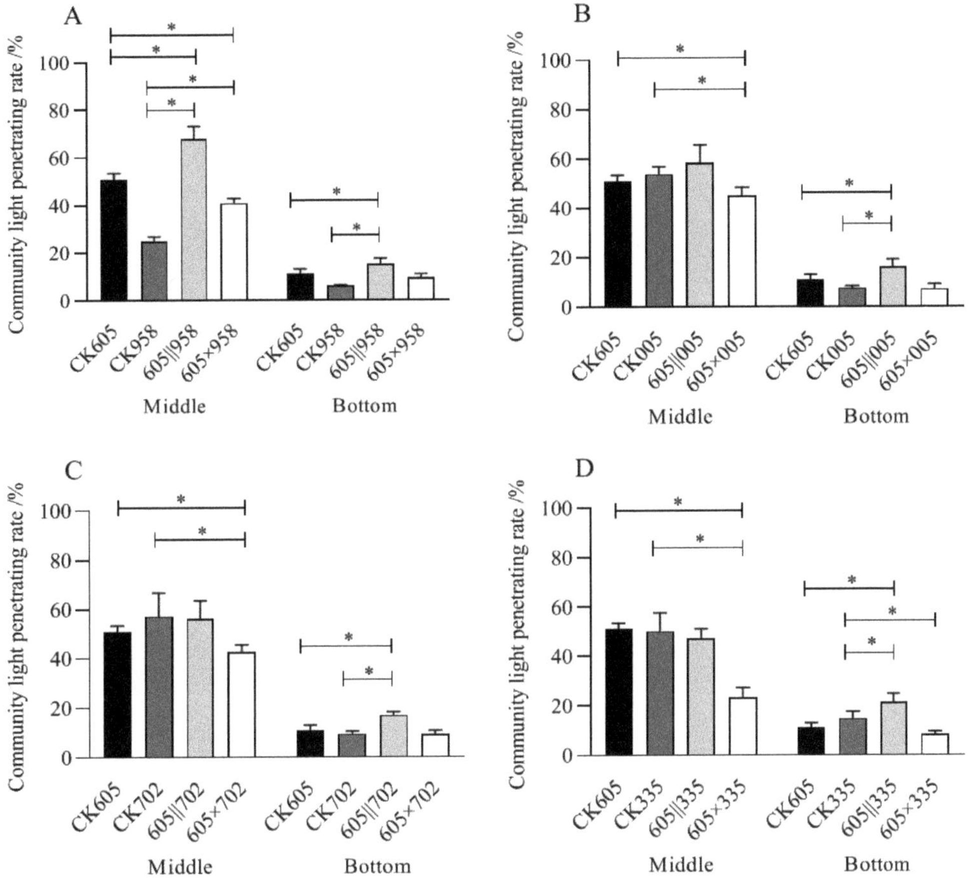

Figure 7. Population light transmittance under different inter- and mixed cropping modes. Population light transmittance in the middle and lower canopy levels under inter- and mixed cropping of (**A**) Denghai 605 and Zhengdan 958, (**B**) Denghai 605 and Dedan 5, (**C**) Denghai 605 and Weike 702, and (**D**) Denghai 605 and Xianyu 335. Values represent the mean ± SD, n = 3 per group. * $p < 0.05$.

3.5. Effects of Inter- and Mixed Cropping Modes on Ear Traits

The effects of the different inter- and mixed cropping modes on ear morphology were mainly manifested as changes in ear length (Table 2). The ear length of Denghai 605 increased significantly under inter- and mixed cropping with Zhengdan 958. The ear lengths of Weike 702 and Xianyu 335 under inter- and mixed cropping with Denghai 605 increased significantly compared with that of the monoculture. The number of grains per ear of Denghai 605 increased significantly under inter- and mixed cropping modes with Zhengdan 958 and Weike 702. Moreover, under inter- and mixed cropping, the number of grains per ear of the four auxiliary varieties increased to varying degrees compared with that of the corresponding monoculture. Specifically, intercropping of Dedan 5 and inter- and mixed cropping of Weike 702 and Xianyu 335 significantly increased the number of grains per ear. In addition, 100-grain weight Denghai 605 increased significantly under intercropping with the four varieties and under mixed cropping with Dedan 5 and Weike702. The 100-grain weight of the auxiliary varieties showed no significant differences except for that of Weike 702 under intercropping with Denghai 605.

Table 2. Ear traits of summer maize varieties under different inter- and mixed cropping modes.

Variety	Ear Length /cm	Ear Thickness /cm	Bald Length /cm	Grain Number per Ear	100-Grain Weight/g
CK 605	18.4 ± 0.8	5.1 ± 0.6	1.7 ± 0.6	503 ± 4	32.8 ± 0.3
(605)| |958	21.5 ± 1.1 *	5.1 ± 0.5	0.9 ± 0.3	615 ± 33 *	34.5 ± 0.3 *
(605)| |005	17.8 ± 0.5	5.2 ± 1.1	1.8 ± 0.2	488 ± 32	34.1 ± 0.5 *
(605)| |702	19.7 ± 2.6	5.1 ± 0.9	1.5 ± 0.2	551 ± 59 *	34.9 ± 0.7 *
(605)| |335	18.6 ± 1.1	5.0 ± 1.5	1.2 ± 0.4	523 ± 5	34.3 ± 0.7 *
(605) × 958	20.8 ± 0.6 *	4.8 ± 0.9*	1.4 ± 0.4	546 ± 16*	33.6 ± 0.4
(605) × 005	18.5 ± 0.7	5.1 ± 0.4	1.7 ± 0.4	516 ± 20	34.8 ± 0.7 *
(605) × 702	19.3 ± 0.6	5.1 ± 0.5	1.3 ± 0.4	560 ± 13 *	34.2 ± 0.8 *
(605) × 335	18.8 ± 0.9	4.9 ± 0.9	2.2 ± 0.4	488 ± 24	33.6 ± 0.1
CK958	16.9 ± 0.5	5.4 ± 1.6	0.3 ± 0.1	523 ± 49	33.5 ± 0.3
605| |(958)	17.4 ± 0.5	5.3 ± 0.3	0.6 ± 0.2	531 ± 12	34.2 ± 1.2
605 × (958)	17.1 ± 0.5	5.2 ± 0.9	0.9 ± 0.3 *	528 ± 59	33.6 ± 0.5
CK005	15.7 ± 1.0	5.0 ± 0.6	0.0 ± 0.0	504 ± 20	30.9 ± 0.3
605| |(005)	15.3 ± 0.3	5.1 ± 0.6	0.1 ± 0.1	557 ± 7 *	30.0 ± 0.3
605 × (005)	15.4 ± 0.3	5.1 ± 0.9	0.0 ± 0.0	533 ± 33	30.5 ± 0.1
CK702	17.6 ± 0.8	5.4 ± 0.1	1.2 ± 0.3	524 ± 30	35.1 ± 1.0
605| |(702)	19.7 ± 0.4 *	5.5 ± 0.3*	0.6 ± 0.3 *	577 ± 34 *	37.3 ± 0.6 *
605 × (702)	19.4 ± 0.8 *	5.3 ± 0.3	0.8 ± 0.2	565 ± 20 *	35.3 ± 0.5
CK335	16.2 ± 0.9	5.2 ± 0.5	2.0 ± 1.1	426 ± 11	34.5 ± 0.6
605| |(335)	18.9 ± 0.8 *	5.1 ± 1.2	1.7 ± 0.5	526 ± 46 *	35.4 ± 0.5
605 × (335)	18.5 ± 0.8 *	5.2 ± 0.2	2.3 ± 0.3	529 ± 42 *	35.4 ± 0.3

* Significant at the 0.05 probability level; Values are the mean ± SD (n = 9 for lines 2-5 and n = 3 for line 6).

3.6. Effect of Inter- and Mixed Cropping Modes on Yield

The yield of Denghai 605 increased significantly under inter- and mixed cropping with the auxiliary varieties except for intercropping with Dedan 5, with the yield increase ranging from 8.8% to 28.0% (Table 3). When inter- and mixed cropped with Zhengdan 958, the yield of Denghai 605 had the largest increase of 28.0% (intercropping) and 16.7% (mixed cropping). This increase was then followed by an increase of 14.9% (intercropping) and 16.2% (mixed cropping) when inter- and mixed cropped with Weike 702.

Table 3. Yield of summer maize varieties under different inter- and mixed cropping modes.

Variety	Yield /kg ha^{-2}	Yield Variety Compared with Monoculture/%
CK605	9075 ± 63	—
(605)| |958	11,615 ± 641 *	28.0
(605)| |005	9328 ± 609	2.8
(605)| |702	10,426 ± 405 *	14.9
(605)| |335	9878 ± 96 *	8.8
(605) × 958	10,587 ± 664 *	16.7
(605) × 005	9967 ± 383 *	9.8
(605) × 702	10,549 ± 381 *	16.2
(605) × 335	9988 ± 493 *	10.1
CK958	10,812 ± 318	—
605| |(958)	11,028 ± 760	2.0
605 × (958)	9846 ± 387 *	−8.9
CK005	9622 ± 379	—
605| |(005)	9842 ± 806	2.3
605 × (005)	9983 ± 624	−7.7

Table 3. *Cont.*

Variety	Yield /kg ha^{-2}	Yield Variety Compared with Monoculture/%
CK702	10,722 ± 520	—
605 I I (702)	12,861 ± 747 *	19.9
605 × (702)	12,068 ± 361 *	12.6
CK335	9144 ± 597	—
605 I I (335)	10,800 ± 606 *	18.1
605 × (335)	10,738 ± 595 *	17.4

* Significant at the 0.05 probability level; Values are the mean ± SD (n = 3).

The yield of the auxiliary varieties increased under inter- and mixed cropping with Denghai 605. The yield of Weike 702 and Denghai 605 increased by 19.9% (intercropping) and 12.6% (intercropping), respectively, and that of Xianyu 335 and Denghai 605 increased by 18.1% (intercropping) and 17.4% (intercropping), respectively. However, the yield of Zhengdan 958 decreased significantly after mixed cropping with Denghai 605.

3.7. Population Yield and Land Equivalent Ratio

The land equivalent ratio was consistently more than 1.0 under inter- and mixed cropping of Denghai 605 and four auxiliary varieties, indicating that the different inter- and mixed cropping methods were beneficial to increase the population yield (Table 4). Intercropping with Zhengdan 958 and inter- and mixed cropping with Weike 702 resulted in a population yield exceeding 11,000 kg ha^{-2}. Under intercropping with Zhengdan 958 and mixed cropping with Weike 702 and Xianyu 335, the population yield increased significantly, and the land equivalent ratio was greater than 1.1. Inter- and mixed cropping with Dedan 5 resulted in the lowest increase in population yield.

Table 4. Population yield and land equivalent ratio of summer maize varieties under different inter- and mixed cropping modes.

Inter-and Mixed Cropping Patterns	Population Yield /kg ha^{-2}	Average Yield of Monoculture/kg ha^{-2}	Land Equivalent Ratio	Population Yield Variety Compared with Denghai 605 Monoculture/%	Population Yield Variety Compared with Other Varieties' Monoculture/%
605 I I 958	11,572	9944	1.16	24.8	4.7
605 I I 005	9585	9349	1.03	5.6	−0.4
605 I I 702	11,644	9899	1.18	28.3	8.6
605 I I 335	10,339	9110	1.13	13.9	13.1
605 × 958	10,217	9944	1.03	12.6	−5.5
605 × 005	9975	9349	1.07	9.9	3.7
605 × 702	11,309	9899	1.14	24.6	5.5
605 × 335	10,363	9110	1.14	14.2	13.3

Comparing the population yield with the monoculture yield of Denghai 605, the population yield increased by more than 20% under intercropping of Denghai 605 with Zhengdan 958 and inter- and mixed cropping with Weike 702. The yield of Denghai 605 increased by 10%, compared with that of the monoculture, under mixed cropping with Zhengdan 958 and inter- and mixed cropping with Xianyu 335.

Comparing the population yield with the monoculture yield of the auxiliary varieties, the population yield was higher than that of the Xianyu 335 monoculture, with the yield increased by 13.1% (605 I I 335) and 13.3% (605 × 335), respectively.

4. Discussion

The maize growing season is in summer, with high temperatures and humidity, which brings the increasing risk of high-temperature stress [36]. High-temperature stress can cause abnormal pollen development or abortion, resulting in a decrease in the number of pollen grains; in addition, it can cause abnormal division of pollen mother cells, deformation, and shrinkage leading to deformed pollen grains [14,15]. High temperatures during flowering causes shrinkage of pollen grains and depression of the germination pore, which significantly reduces pollen vitality; the higher the temperature, the greater the reduction of pollen vitality [37]. High-temperature stress can also thicken the anther wall and hinder dehiscence, resulting in the release of fewer pollen grains and lower vitality [38]. Pollen metabolic activity is associated with the starch content of the pollen grain; a significant decrease in the starch content causes a corresponding decrease in pollen metabolic activity [17,39]. Distinct differences in pollen viability are observed among maize hybrids, which are derived from genetic differences among the parents [18,40]. Different varieties differ in their gene sources, and thus heat tolerance can differ significantly. Varieties that produce a greater number of branches, full glume, and higher number of pollen grains usually show stronger resistance to high-temperature stress [21,41]. Sowing two or more varieties with different genotypes increases the probability of cross pollination as well as the yield [42]. Therefore, changing single cropping to inter- and mixed cropping among multiple varieties without increasing cost is one of the new strategies to improve the high-temperature tolerance of maize during the flowering period by making full use of the heat tolerance differences of different genotypes.

Inter- and mixed cropping, which creates a multi-level and multi-functional composite group through different combinations of crops and varieties [43], can improve the canopy structure of the population, improve efficiency in the use of light energy and land area, and overcome the harmful impacts of diseases, insects, and grasses on monocultures, so as to increase the yield per unit area [44,45]. Intercropping forms a wavy canopy, while mixed cropping forms a concave–convex canopy of crops in order to change plane light into three-dimensional light in the upper part of the population crops. It has been shown that the photosynthetic potential of intercropping during the big bell mouth period was 76% and 78% higher, and the field light transmittance was 54.0% higher than that of the single cropping, respectively [46]. The photosynthetic intensity of the intercropping population increased by 37.2% and 28.8% compared with that of the monoculture, and the light energy utilization rate in the whole growth period increased by 58.6% compared with that of the monoculture during the jointing stage and filling stage, respectively. In addition, the yield of Jundan 20 and Dedan 5 increased by 5.6% and 7.9% compared with their monoculture after intercropping treatment, respectively [47,48]. In our study, the results showed that the light transmittance of the bottom population increased after intercropping Denghai 605 with Dedan 5, Weike 702, and Xianyu 335, and the light transmittance of the middle layer increased after intercropping with Zheng Dan 958, which is consistent with the previous findings.

The advantages of inter- and mixed cropping among different maize varieties mainly reflect the contemporary heterosis [49], including resistance complementarity and fertility complementarity. Complementary resistance means that inter- and mixed cropping of two varieties that differ in disease resistance effectively improves the disease resistance of the population after planting [27,45]. Sterility complementary refers to the similarity of the male–female interval between two varieties. The pollen of each variety is used to extend the duration of pollination and fertilization, which can enhance the utilization of heterosis in maize cross-pollination and increase yields by enhancing the number of grains per ear [50]. It had been demonstrated that intercropping was beneficial to the increase of grain number per ear and grain weight compared with monoculture, and the number of grains per ear in the intercropping of Yedan 12 and Yedan 13 increased by 9.28% and 15.66%, respectively, compared with that of the monoculture [51]. In addition, the average number of grains per ear of free pollination in different combinations increased by 40.7 grains, and 100-grain weight increased by 1.1 g, which manifested the existence of heterosis among different varieties [52]. Making full use of flowering and pollen complementary advantages of varieties and improving fertilization and seed setting rate while avoiding high temperature stress is one of the focuses of our study. The flowering period of Zhengdan 958, Weike 702, and Denghai 605 were basically the same. After inter-and mixed cropping, the flowering and pollination of Denghai 605 were compensated, and the grain number per ear was significantly increased compared with that of single cropping. It indicated that the selection of varieties with a consistent flowering period was helpful to give full play to the advantages of inter-and mixed cropping planting mode. The number of grains per panicle of Xianyu 335, a variety with a small amount of pollen and intolerant to high temperatures during the flowering period, also increased significantly after inter-and mixed cropping with Denghai 605 compared with that of single cropping. This shows that inter-and mixed cropping can make better use of heterosis of maize varieties.

In previous studies, extensive research has been performed on the physiological and ecological effects of maize intercropping. The yield-increasing benefits of maize in inter- and mixed cropping have been discussed in detail. In terms of morphological structure, the maize inter- and mixed cropping system facilitated the formation of a three-dimensional canopy structure and different intercropping row ratios and planting densities brought differences in light distribution, which improved the ventilation and light transmittance and increased CO_2 concentration in the growth space [53–55]. In terms of physiology, inter- and mixed cropping mode increased the chlorophyll content, leaf area index, and photosynthetic rate of maize [19,22]. Meanwhile, it enhanced the antioxidant enzyme activity and Rubisco's carboxylation efficiency and improved soil quality and nutrient absorption [23,25,55]. In terms of stress resistance, intercropping of different genotypes of maize profoundly strengthened the resistance to the disease [27], and the appropriate cultivar collocation effectively reduced the lodging resistance of the population. This study focused on the mitigation effect of multi-variety- inter- and mixed cropping modes on population resistance to high temperatures during the flowering period, and demonstrated that the tolerance to high temperatures during flowering was significantly enhanced by choosing reasonable variety-matching and row-spacing ratios, which were hardly studied in the past.

Previous studies have fully confirmed that the yield-increasing effect of inter- and mixed cropping was the result of multiple compound effects. However, the mechanism for increasing the yield is extremely complicated, as it is affected by diverse factors, such as genotype, phenotype, population structure, population physiology, field microclimate, and soil microecology [44]. Therefore, no simple combination of varieties will necessarily increase production, and in-depth prior analysis is required. For the selection of maize varieties, the consistency of variety traits, plant type, plant height, resistance, and other characteristics should be considered so as to increase the yield of dominant varieties, stabilize the yield of other varieties, and finally, increase the population yield. This study demonstrated that the land equivalent ratio of Denghai 605 inter- and mixed cropped with four other auxiliary varieties was more than 1.0, among which the LER intercropping with Zhengdan 958 and inter- and mixed cropping with Weike 702 and Xianyu 335, respectively, was greater than 1.1, further confirming the yield-increasing effect. Based on the previous studies, the technical key point is that the heat resistance of the combined varieties should be complementary and the growth period should be consistent. The flowering period especially should be consistent if possible. In addition, varieties with basically the same plant height and plant type should be selected if possible. If there are large differences in plant height or plant type, the 2:2 or 2:4 row ratio intercropping mode can be adopted, which is for making use of the spatial advantages of high-stalk varieties without reducing the yield of short-stalk varieties and having strong complementarity among varieties in heat resistance [56–58]. The production technology of complementary resistance enhancement of maize varieties has been listed as the main technology in 2021 by the Ministry of Agriculture and Rural Affairs of the People's Republic of China and issued and implemented as the agricultural industry standard in China [59].

5. Conclusions

In this study, Denghai 605, which produces low amounts of pollen and is intolerant of high temperatures during flowering, was used as the main test variety, and Zhengdan 958, Dedan 5, Weike 702, and Xianyu 335 were used as auxiliary varieties to determine the effect of inter- and mixed cropping of maize on high-temperature tolerance during flowering and on yield. The population light transmittance of the bottom or middle layer of crops increased in the intercropping of Denghai 605 with other varieties, which was conducive to the formation of an efficient canopy structure. The number of grains per ear increased after intercropping Denghai 605 with Zhengdan 958 and Weike 702, indicating that inter- and mixed cropping enhanced pollen vigor and improved tolerance to high temperatures during the flowering period. Moreover, the yield increased significantly by 8.8–28.0% after inter- and mixed cropping Denghai 605 with other varieties (except intercropping with Dedan 5). Among them, the LER was greater than 1.1 when Denghai 605 was intercropped with Zhengdan 958 and inter- and mixed cropped with Weike 702 and Xianyu 335, which indicated that under the combination of these varieties, inter- and mixed cropping effectively reduced the impact of high temperatures during flowering and improved the maize population yield.

Author Contributions: Conceptualization, J.Z. and T.L.; methodology, J.Z. and S.L.; validation, J.L. and S.L.; formal analysis, R.S.; investigation, H.L., L.H. and W.F.; data curation, H.L. and R.S.; writing—original draft preparation, S.L.; writing—review and editing, J.Z. and T.L.; project administration, T.L. All authors have read and agreed to the published version of the manuscript.

Funding: This research was funded by the National Key Research and Development Program of China(2018YFD0300704).

Institutional Review Board Statement: Not applicable.

Informed Consent Statement: Not applicable.

Data Availability Statement: Not applicable.

Conflicts of Interest: The authors declare no conflict of interest.

References

1. Seleiman, M.F.; Al-Suhaibani, N.; El-Hendawy, S.; Abdella, K.; Alotaibi, M.; Alderfasi, A. Impacts of Long- and Short-Term of Irrigation with Treated Wastewater and Synthetic Fertilizers on the Growth, Biomass, Heavy Metal Content, and Energy Traits of Three Potential Bioenergy Crops in Arid Regions. *Energies* **2021**, *14*, 3037. [CrossRef]
2. Seleiman, M.F.; Selim, S.; Jaakkola, S.; Mäkelä, P.S. Chemical composition and in vitro digestibility of whole-crop maize fertilized with synthetic fertilizer or digestate and harvested at two maturity stages in Boreal growing conditions. *Agric. Food Sci.* **2017**, *26*, 47–55. [CrossRef]
3. Ren, Y.J.; Wan, S.Q.; Xiao, Y.; Liu, M.; Sun, S.L. Simulative evaluation and projection of air temperatures over central China in the SRES scenarios. *Acta Meteorol. Sina* **2012**, *70*, 1098–1106. (In Chinese) [CrossRef]
4. Yang, J.; Chen, Z.; Yan, C.; Ju, H.; Mei, X.; Liu, Q.; Xu, J. Spatio-temporal characteristics and jump features of air temperature in huang-huai-hai plain during recent 50 years. *J. Agrometeorol.* **2013**, *1*, 1–7. (In Chinese) [CrossRef]
5. Guan, Y.; He, Q.J.; Liu, J.H.; Li, R.C.; Hu, Q.; Hang, B.X.; Pan, X.B. Variation characteristics of extreme temperature and its earliest and latest day sequence in Huang-Huai-Hai region during the period 1961 to 2015. *Res. Soil Water Conserv.* **2021**, *1*, 147–152. (In Chinese) [CrossRef]
6. Chen, X.; Bao, Y.J.; Li, Q.; Ding, J.K.; Yang, M.Z.; Wang, D.Y.; Cao, Z.H.; He, L.; Song, Y.H. Review on characteristics of high temperature and its damage, and prevention measures of summer maize in Huang-Huai-Hai area. *Anhui Nong Ye Da Xue Xue Bao* **2020**, *2*, 304–308. (In Chinese) [CrossRef]
7. Lv, Z.; Li, F.; Lu, G. Adjusting sowing date and cultivar shift improve maize adaption to climate change in China. *Mitig. Adapt. Strateg. Glob. Chang.* **2020**, *25*, 87–106. [CrossRef]
8. Singletary, G.; Bnisadr, R.; Keling, P. Heat-stress during grain filling in maize—Effects on carbohydrate storage and metabolism. *Aust. J. Plant Physiol.* **1994**, *6*, 829–841. [CrossRef]
9. Wilhelm, E.P.; Mullen, R.E. Heat Stress during Grain Filling in Maize. *Crop Sci.* **1999**, *6*, 1733–1741. [CrossRef]
10. Zhao, L.F.; Li, C.H.; Liu, T.X.; Wang, X.P.; Pan, X. Genotypic responses and physiological mechanisms of maize (*Zea mays* L.) to high temperature stress during flowering. *Acta Agron. Sin.* **2012**, *5*, 857–864. (In Chinese) [CrossRef]
11. Rattalino, E.J.; Budakli, C.E.; Sammarro, D.; Otegui, M.E. Heat stress effects around flowering on kernel set of temperate and tropical maize hybrids. *Field Crop Res.* **2011**, *2*, 62–73. [CrossRef]

12. Jiang, Z.B.; Tao, H.B.; Tuo, W.U.; Wang, P.; Song, Q.F. Effects of high temperature on maize pollen viability. *J. China Agric. Univ.* **2016**, *3*, 25–29. (In Chinese) [CrossRef]
13. Singh, A.; Antre, S.H.; Ravikumar, R.L.; Kuchanur, P.H.; Lohithaswa, H.C. Genetic evidence of pollen selection mediated phenotypic changes in maize conferring transgenerational heat–stress tolerance. *Crop Sci.* **2020**, *4*, 1907–1924. [CrossRef]
14. Djanaguiraman, M.; Perumal, R.; Jagadish, S.V.K.; Ciampitti, I.A.; Welti, R.; Prasad, P.V.V. Sensitivity of sorghum pollen and pistil to high–temperature stress. *Plant Cell Environ.* **2018**, *5*, 1065–1082. [CrossRef]
15. Yadav, S.K.; Tiwari, Y.K.; Pavan, K.D.; Shanker, A.K.; Jyothi, L.N.; Vanaja, M.; Maheswari, M. Genotypic variation in physiological traits under high temperature stress in maize. *Agric. Res.* **2016**, *2*, 119–126. [CrossRef]
16. Mukhtar, T.; Rehman, S.U.; Smith, D.; Sultan, T.; Seleiman, M.F.; Alsadon, A.A.; Shafaqat Ali, A.; Chaudhary, H.J.; Solieman, T.H.I.; Ibrahim, A.A.; et al. Mitigation of heat stress in *Solanum lycopersicum L.* by acc-deaminase and exopolysaccharide producing *Bacillus cereus*: Effects on biochemical profiling. *Sustainability* **2020**, *12*, 2159. [CrossRef]
17. Wang, Y.; Tao, H.; Tian, B.; Sheng, D.; Xu, C.; Zhou, H.; Huang, S.; Wang, P. Flowering dynamics, pollen, and pistil contribution to grain yield in response to high temperature during maize flowering. *Environ. Exp. Bot.* **2019**, *158*, 80–88. [CrossRef]
18. Djanaguiraman, M.; Vara, P.V.; Murugan, M.; Perumal, R.; Reddy, U.K. Physiological differences among sorghum (*Sorghum bicolor L.* Moench) genotypes under high temperature stress. *Environ. Exp. Bot.* **2014**, *100*, 43–54. [CrossRef]
19. Liu, T.X.; Wang, Z.H.; Dong, P.F.; Li, C.H. Research Progress of Physiological and Ecological Effects in Maize Intercropping System. *J. Maize Sci.* **2007**, *5*, 114–116. (In Chinese) [CrossRef]
20. Zhao, Y.L.; Kang, J.; Liu, T.X.; Li, C.H. Optimum stripe arrangement for inter–cropping and mixed–cropping of different maize (*Zea mays* L.) genotypes. *Acta Ecol. Sin.* **2013**, *12*, 3855–3864. (In Chinese) [CrossRef]
21. Sinsawat, V.; Leipner, J.; Stamp, P.; Fracheboud, Y. Effect of heat stress on the photosynthetic apparatus in maize (*Zea mays* L.) grown at control or high temperature. *Environ. Exp. Bot.* **2005**, *52*, 123–129. [CrossRef]
22. Hu, D.D.; Li, R.F.; Zhang, J.W.; Zhao, B.; Liu, P.; Dong, S.T. Mixed Cropping of Different Hybrids of Maize Optimizes Canopy Structure and Promotes Higher Grain Yield. *Agron. J.* **2019**, *111*, 2692–2702. [CrossRef]
23. Brtnicky, M.; Elbl, J.; Kintl, A.; Dokulilova, T.; Kucerova, J. Effect of maize and legume mixed cropping on soil quality in relation to planting density. In Proceedings of the 19th International Multidisciplinary Scientific GeoConference Surveying Geology and Mining Ecology Management, Albena, Bulgaria, 30 June–6 July 2019; SGEM: Sofia, Bulgaria, 2019; Volume 19, pp. 221–227. [CrossRef]
24. Cheng, B.; Wang, L.; Liu, R.J.; Wang, W.B.; Yu, R.W.; Zhou, T.; Ahmad, I.; Raza, A.; Jiang, S.J.; Xu, M.; et al. Shade-Tolerant Soybean Reduces Yield Loss by Regulating Its Canopy Structure and Stem Characteristics in the Maize-Soybean Strip Intercropping System. *Front. Plant Sci.* **2022**, *13*, 48893. [CrossRef]
25. Zhao, X.H.; Dong, Q.Q.; Han, Y.; Zhang, K.Z.; Shi, X.L.; Yang, X.; Yuan, Y.; Zhou, D.Y.; Wang, K.; Wang, X.G.; et al. Maize/peanut intercropping improves nutrient uptake of side-row maize and system microbial community diversity. *BMC Microbiol.* **2022**, *22*, 14. [CrossRef] [PubMed]
26. Astiko, W.; Ernawati, N.M.L.; Silawibawa, I.P. Effect of Intercropping on Mycorrhizal Populations, Growth, and Yield on Several Varieties of Maize (*Zea mays* L.) and Soybeans [*Glycine max* (L.) Merr.] in Dryland North Lombok, Indonesia. *IOP Conf. Ser. Earth Environ. Sci.* **2021**, *913*, 012008. [CrossRef]
27. Li, C.H.; Su, X.H.; Sun, D.L. Ecophysiological Characterization of Different Maize (*Zea mays* L.) Genotypes under Mono- or Inter-cropping Conditions. *Acta Ecol. Sin.* **2002**, *12*, 2096–2102. (In Chinese) [CrossRef]
28. Weng, W.H.; Chen, L.G.; Sun, J.Y.; Fu, Y.J. Mechanism and Application of Mixed Cropping in Crop Protection: A Review. *J. Agric.* **2017**, *2*, 15–19. [CrossRef]
29. Zaeem, M.; Nadeem, M.; Pham, T.H.; Ashiq, W.; Ali, W.; Gillani, S.S.M.; Moise, E.; Elavarthi, S.; Kavanagh, V.; Cheema, M.; et al. Corn-soybean intercropping improved the nutritional quality of forage cultivated on Podzols in boreal climate. *Plants* **2021**, *10*, 1015. [CrossRef]
30. Raza, M.A.; Cui, L.; Khan, I.; Din, A.; Yang, W. Compact maize canopy improves radiation use efficiency and grain yield of maize/soybean relay intercropping system. *Environ. Sci. Pollut. Res. Int.* **2021**, *28*, 41135–41148. [CrossRef]
31. Gajghate, R.; Chourasiya, D.; Harikrishna; Sharma, R.K. Plant Morphological, Physiological Traits Associated with Adaptation Against Heat Stress in Wheat and Maize. In *Plant Stress Biology*; Springer: Singapore, 2021; pp. 51–81. [CrossRef]
32. Rattalino, J.I.; Otegui, M.E. Heat stress in temperate and tropical maize hybrids: Differences in crop growth, biomass partitioning and reserves use. *Field Crop. Res.* **2012**, *130*, 87–98. [CrossRef]
33. Thayamini, H.S.; Brintha, I. Review on maize based intercropping. *J. Agron.* **2010**, *3*, 135–145. [CrossRef]
34. Zhao, R.B. Effects of row spacing and intercropping on canopy photosynthetic characteristics and yield of Xianyu series maize varieties. *Hebei Agric. Univ.* **2021**, *5*, S513. (In Chinese) [CrossRef]
35. Chen, Z.H.; Wang, A.L.; Wang, J.J.; Xue, J.B.; Dong, X.C.; Wei, G.Y. Influence of High Temperature on Growth and Development of Maize. *Crops* **2008**, *4*, 90–92. (In Chinese) [CrossRef]
36. Xu, Y.H.; Liu, T.X.; Fang, W.S.; Li, S.Y. Risk analysis of high temperature disaster during summer maize flowering period in henan province. *Chin. J. Agrometeorol.* **2021**, *42*, 879–888. (In Chinese) [CrossRef]
37. Naveed, S.; Aslam, M.; Maqbool, M.A.; Bano, S.; Ahmad, R.M. Physiology of high temperature stress tolerance at reproductive stages in maize. *J. Anim. Plant Sci.* **2014**, *4*, 1141–1145.

38. Kumar, S.; Thakur, P.; Kaushal, N.; Malik, J.A.; Gaur, P.; Nayyar, H. Effect of varying high temperatures during reproductive growth on reproductive function, oxidative stress and seed yield in chickpea genotypes differing in heat sensitivity. *Arch. Agron. Soil Sci.* **2013**, *6*, 823–843. [CrossRef]
39. Fonseca, A.E.; Westgate, M.E. Relationship between desiccation and viability of maize pollen. *Field Crop Res.* **2004**, *94*, 114–125. [CrossRef]
40. Song, F.W.; Peng, W.U.; Xing, J.M.; Zhou, X.Y.; Cui, X.R.; Yu, X.P.; Wang, J. Influences of high temperature stress on viability of pollen grain inbred lines of male parent. *J. Maize Sci.* **2014**, *3*, 153–158. [CrossRef]
41. Sheng, D.C.; Wang, Y.Y.; Huang, S.B.; Tao, H.B.; Wang, P. Effects of high temperature on morphology and function, yield components and grain nutrients of maize plants. *J. Maize Sci.* **2020**, *5*, 86–92. (In Chinese) [CrossRef]
42. Rasheed, A.; Seleiman, M.F.; Nawaz, M.; Mahmood, A.; Rizwan Anwar, M.; Ahsin Ayub, M.; Aamer, M.; El-Esawi, M.A.; El-Harty, E.H.; Batool, M.; et al. Agronomic and genetic approaches for enhancing tolerance to heat stress in rice: A review. *Not. Bot. Horti Agrobot. Cluj Napoca* **2021**, *49*, 1–27. [CrossRef]
43. Yang, X.Q.; Wang, Y.; Qi, X.N.; Sun, L.Y.; Song, F.B.; Liu, S.Q.; Li, X.N.; Raza, X.C.; Tian, C. Photosynthetic physio-ecological characteristics of maize intercropping system. *Soil Crop* **2019**, *8*, 70–77. (In Chinese) [CrossRef]
44. Zhu, M.; Shi, Z.S.; Li, F.H.; Wang, Z.B. Summary of Different Maize Variety Inter-planting and Mixed Cultivation. *J. Maize Sci.* **2007**, *1*, 100–103. (In Chinese) [CrossRef]
45. Liu, T.X.; Li, C.H.; Fu, J.; Yan, C.H. Population quality of different maize (*Zea mays* L.) genotypes intercropped. *Acta Ecol. Sin.* **2009**, *11*, 6302–6309. (In Chinese) [CrossRef]
46. Zhao, B.R.; Liang, Y.C.; Zhang, S.Q.; Liu, Y.T.; Yang, D.R. Study on a new model of three-dimensional intercropping of high and dwarf maize. *J. Maize Sci.* **1999**, *3*, 51–53. (In Chinese)
47. Yu, G.X. Preliminary study on Intercropping of Maize with high and short stalks. *Tillage Cultiv.* **1999**, *2*, 10–27. (In Chinese)
48. Ping, X.S.; Xing, R.R.; Liu, T.X. Complementary effect of anti-adversity and yield in different maize (*Zea mays* L.) genotypes intercropping system. *J. Gansu Agric. Univ.* **2020**, *55*, 62–67. (In Chinese) [CrossRef]
49. Jia, Y.F.; Zhang, X.S.; Zhao, M. Genetic analysis on chemical compositions of kernel of f1 embryo stage in common corn single hybrids and high oil corns. *J. Maize Sci.* **2004**, *12*, 26–29. (In Chinese) [CrossRef]
50. Ren, H.; Liu, P.; Dong, S.T.; Zhang, J.J.; Zhao, B. Research advancements of effect of high temperature stress on growth and development of maize. *J. Maize Sci.* **2019**, *27*, 109–115. (In Chinese) [CrossRef]
51. Su, X.H.; Li, C.H.; Sun, D.L.; Zhang, H.Z. First report of different genotype maize's intercropping. *J. Maize Sci.* **2000**, *8*, 57–60. (In Chinese) [CrossRef]
52. Meng, Q.P.; Wu, F.L.; Liu, Y.; Miao, S.R.; Li, C.X.; Guo, J.; Guo, F.; Yin, Y.H. Preliminary study on the utilization of contemporary heterosis in three-dimensional cultivation of Maize. *J. Maize Sci.* **1997**, *3*, 43–45. (In Chinese)
53. Jiao, N.Y.; Ning, T.Y.; Yang, M.K.; Fu, G.Z.; Yin, F.; Xu, G.W.; Li, Z.J. Effects of maize | | peanut intercropping on photosynthetic characters and yield forming of intercropped maize. *Acta Ecol. Sin.* **2013**, *33*, 4324–4330. (In Chinese) [CrossRef]
54. Cui, J.M.; Song, C.J.; Lu, D.W.; Yang, H.Y.; Guo, S.Y.; Pei, Z.Q.; Liu, Z.P.; Lu, L.Y.; Sun, H.C.; Niu, Y.F.; et al. Planting techniques of long and short-stalked multistorey intercropping of different type maize hybrid. *Rain Fed. Crops* **2005**, *25*, 253–257. (In Chinese) [CrossRef]
55. Tao, J.J.; Wang, H.B.; Zhu, Z.Y.; Tan, J.F.; Wang, Y.L. Effect of Different Genotype Summer Maize Intercropping on Yield and Nitrogen Absorption and Utilization. *Acta Agric. Boreali Sin.* **2016**, *31*, 185–191. (In Chinese) [CrossRef]
56. Bai, X.H.; Wang, T.Q.; Liu, Z.L.; Liu, S.M.; Zhang, S.H. Breeding process and high-efficiency cultivation technology of new summer maize variety weiyu 6 with high temperature resistance and high-yield. *Mod. Agric. Sci. Technol.* **2019**, *13*, 38–41. (In Chinese) [CrossRef]
57. Liu, T.X.; Wang, Y.K.; Jiang, X. High temperature and heat damage in maize blooming period and its alleviating cultivation techniques. *Henan Agric.* **2017**, *3*, 45–46. (In Chinese)
58. Liu, T.X.; Duan, P.F.; Wang, Y.L.; Zhao, C.L.; Hu, X.L. Maize Planting Method for Ecologically Mitigating Flowering-Stage High-Temperature Heat Damage. CN104996127A, 20 October 2017. (In Chinese).
59. *NY/T 3841—2021*; Technical Specification for Maize Complementary Resistance Enhancement Production. Department of Crop Management, Ministry of Agriculture and Rural Affairs of the People's Republic of China, Henan Agricultural University: Zhengzhou, China, 2021. Available online: https://hbba.sacinfo.org.cn/attachment/online?pk=293e6dac6d065349efac0ad6c5cc39045b7b258990738862da863ef34317ad30 (accessed on 28 April 2022).

sustainability

Article

Evaluation and Projection of Diurnal Temperature Range in Maize Cultivation Areas in China Based on CMIP6 Models

Wenqiang Xie, Shuangshuang Wang and Xiaodong Yan *

State Key Laboratory of Earth Surface Processes and Resource Ecology, Faculty of Geographical Science, Beijing Normal University, Beijing 100875, China; wenqiangxie@mail.bnu.edu.cn (W.X.); wangss@mail.bnu.edu.cn (S.W.)
* Correspondence: yxd@bnu.edu.cn

Abstract: The diurnal temperature range (DTR) is an important meteorological component affecting maize yield. The accuracy of climate models simulating DTR directly affects the projection of maize production. We evaluate the ability of 26 Coupled Model Intercomparison Project phase 6 (CMIP6) models to simulate DTR during 1961–2014 in maize cultivation areas with the observation (CN05.1), and project DTR under different shared socioeconomic pathway (SSP) scenarios. The root mean square error (*RMSE*), standard deviation (*SD*), Kling-Gupta efficiency (*KGE*) and comprehensive rating index (*CRI*) are used in the evaluation of the optimal model. The results show that CMIP6 models can generally reproduce the spatial distribution. The reproducibility of the annual average DTR in the maize cultivation areas is better than that in China but lower for the maize-growing season. The optimal model (EC-Earth3-Veg-LR) is used in the projection. Under the two SSPs, the DTR decreases compared with the historical period, especially in Northwest and North China. The DTR under SSP245 remains unchanged (annual) or increases slightly (growing season) during 2015–2050, while a significant decreasing trend is observed under SSP585. This highlights the importance of evaluating DTR in maize cultivation areas, which is helpful to further improve the accuracy of maize yield prediction.

Keywords: CMIP6; maize; diurnal temperature range; projection; China

Citation: Xie, W.; Wang, S.; Yan, X. Evaluation and Projection of Diurnal Temperature Range in Maize Cultivation Areas in China Based on CMIP6 Models. *Sustainability* **2022**, *14*, 1660. https://doi.org/10.3390/su14031660

Academic Editor: Mohammad Valipour

Received: 30 December 2021
Accepted: 26 January 2022
Published: 31 January 2022

1. Introduction

Many studies have addressed the future changes in temperature and precipitation under a climate change background [1–3]. These studies have provided a solid basis for assessing the risk of climate change to human health, agriculture, natural resources, water resources, etc. In agriculture, some equivalent climate variables independently play important roles. Maize is one of the main food sources of humans and is widely cultivated worldwide. The changes in maize yield are related to the food security of human society. The yield of maize is known to be influenced by the environment and other factors [4,5]. Air temperature is one of the main factors affecting maize growth and production, and maize showed a greater yield decrease than wheat and rice at the same warming level [6,7]. A large number of studies have analyzed the effects of accumulated temperature and temperature on maize production. The diurnal temperature difference (DTR) also plays an important role in maize yield. DTR is defined as the difference between the maximum and minimum 2-m temperature during a 24-h period, and it has been proven to impact multipole crop production aspects, including yield, quality, and market commercial price [8]. It has been reported that changes in daily maximum air temperature and minimum air temperature may have different influences on crops [9–11], and lower daily minimum air temperature helps to improve crop quality. Lobell and Field [4] explored the relationship between the DTR and crop yield and found that the influence of daytime warming on crops was greater than that of nighttime warming. Jiang [12] indicated that a decrease in DTR in spring and summer was beneficial for crop yield increases, while an increase in DTR in fall was beneficial. Therefore, to better predict the future maize yield, the accurate simulation of DTR via global climate models has become increasingly important.

Unfortunately, there have been few studies addressing DTR future changes, which can obviously limit our understanding of the impact of climate change on crop production. There are large uncertainties in the trends and possible interpretations of DTR [13,14]. Lobell and Field [4] evaluated 12 global climate models that participated in the Coupled Model Intercomparison Project (CMIP) phase 3 and found future changes in DTR to be inconsistent among the models. Lindvall and Svensson [15] evaluated the simulation ability of 20 Coupled Model Intercomparison Project phase 5 (CMIP5) models in simulating the terrestrial DTR of recent and future projections using HadGHCND and CRU and found that DTR varies considerably between CMIP5 models and that DTR is often underestimated. This uncertainty causes substantial confusion when applying climate model results to estimate the magnitude of crop exposure to climate change. Wang et al. [16] indicated that Coupled Model Intercomparison Project phase 6 (CMIP6) models have not improved their ability to simulate temporal DTR changes consistently during 1901–2005 relative to CMIP5. In conclusion, although CMIP6 models already have good simulation capability for air temperature and precipitation, the simulation capability for important agrometeorological elements such as DTR still needs to be improved.

Most evaluations of climate variables simulated by global climate models (GCM) are based on the annual scale and the whole region. It may be questioned whether those evaluations are suitable for agricultural applications. The spatial scales of previous studies have focused on global or nationwide scales [17], and few studies have been conducted at the scale of crop-cultivation areas. Scholars have conducted numerous studies to quantitatively assess climate model simulations of key meteorological elements and to estimate future climate [18–20]. Knox et al. [21] evaluated the impact of climate change on yield projections for maize and other crops in Africa and South Asia using several global climate models, noting that the increase in the number of climate models could reduce the uncertainty in projections. However, there is a lack of discussion on the impact of spatial scale differences in climate model performance on prediction results. Lindvall and Svensson [15] evaluated the ability of CMIP5 models to simulate the DTR over land in historical simulations and future projections and indicated that although models had good simulation ability for global DTR changes, there were uncertainties in simulations for smaller spatial scales. GCMs with good performance at global or continental scales may exhibit differences in simulation performance within the actual maize cultivation areas. Therefore, when estimating maize yield and quality, using the most suitable climate model in maize cultivation areas will improve the credibility of the prediction results.

In addition, previous studies have mainly focused on the ability to simulate the interannual variability of meteorological components [22], and few studies have been conducted on the growing season of crops. Wang et al. [16] pointed out that most individual CMIP6 models overestimated the DTR changes from December to February, especially in the high latitudes of the Northern Hemisphere. The model showed significant differences in land and did not completely capture the observed temporal and spatial evolution of the DTR. Fan et al. [23] found that the simulation stability of the annual average temperature model is higher than that of the seasonal average temperature model. There is a large difference between the model and observations of temperature from January to May, and the simulation of temperature from June to September is more stable. There are great differences between the simulation of interannual and seasonal variations in climate models. The good performance of climate models at annual or seasonal scales does not mean that climate models perform equally well in actual crop-growing seasons. Differences in model performance at different time scales may affect crop yield and quality prediction [24,25]. Therefore, a climate model with better performance in the maize-growing season should be used when predicting maize yield and quality.

The retrospective analysis of systematic biases in current climate models as well as their correction is one of the scientific issues that CMIP6 focuses on [26]. The questions that agricultural scientists are more interested in include: How well do climate models simulate the meteorological elements in crop cultivation areas? How will climate change in the future? This suggests that the ability of climate models to simulate key meteorological components that are critical in agriculture needs to be carefully evaluated to truly provide a more credible understanding and perception of the agricultural impacts of climate change. Therefore, we quantitatively evaluated the ability of the CMIP6 model to simulate the spatial and temporal characteristics of the DTR in the main maize cultivation areas from several perspectives and used the optimal model to predict and analyze the changes in the DTR in the main maize cultivation areas in China under different future scenarios. This study will contribute to further improving the ability of global climate models to simulate DTR in the maize cultivation areas of China and will serve research on the impact of climate change on maize yield and quality.

2. Materials and Methods

2.1. CMIP6 Model Output

The CMIP6 is the latest experiment to simulate global climate through climate models. It collects the best models in the world and conducts the most colorful experiments, covering the world at a high-spatial and temporal resolution. CMIP6 considers the effects of external forcing, including natural factors and human activities, over time in the simulation of historical periods. Global near-surface maximum air temperature (Tasmax) and minimum air temperature (Tasmin) data simulated by 26 CMIP6 models from 1961 to 2014 were retrieved from the CMIP6 website [27]. The DTR was calculated as the difference between the maximum and minimum near-surface temperature during a 24 h period simulated by the CMIP6 models in this study. The focus of this paper is the DTR during 1961–2050. The names of these models analyzed are listed in Table 1, together with the institution and resolution. We only considered the first ensemble simulation (CMIP6: 'r1i1p1f1') if a model had multiple ensemble simulations.

Table 1. Information of CMIP6 models.

No.	Model Name	Institution (Country)	Resolution (Lat × Lon)	Realization
1	ACCESS-CM2	CSIRO-ARCCSS (Australia)	1.875° × 1.25°	r1i1p1f1
2	ACCESS-ESM1-5	CSIRO (Australia)	1.875° × 1.24°	r1i1p1f1
3	AWI-CM-1-1-MR	AWI (Germany)	0.9375° × 0.9375°	r1i1p1f1
4	AWI-ESM-1-1-LR	AWI (Germany)	1.875° × 1.875°	r1i1p1f1
5	BCC-CSM2-MR	BCC (China)	1.125° × 1.125°	r1i1p1f1
6	BCC-ESM1	BCC (China)	2.8125 × 2.8125	r1i1p1f1
7	CanESM5	CCCma (Canada)	2.8125° × 2.8125°	r1i1p1f1
8	EC-Earth3	EC (European Community)	0.703° × 0.703°	r1i1p1f1
9	EC-Earth3-Veg	EC (European Community)	0.703° × 0.703°	r1i1p1f1
10	EC-Earth3-Veg-LR	EC (European Community)	1.125° × 1.125°	r1i1p1f1
11	FGOALS-f3-L	CAS (China)	1.25° × 1.25°	r1i1p1f1
12	FGOALS-g3	CAS (China)	2.0° × 2.0°	r1i1p1f1
13	GFDL-CM4	NOAA-GFDL (America)	1.25° × 1.25°	r1i1p1f1
14	GFDL-ESM4	NOAA-GFDL (America)	1.25° × 1.0°	r1i1p1f1
15	GISS-E2-1-G	NASA-GISS (America)	2.5° × 2.0°	r1i1p1f1
16	INM-CM4-8	INM (Russia)	2.0° × 1.5°	r1i1p1f1
17	INM-CM5-0	INM (Russia)	2.0° × 1.6°	r1i1p1f1
18	IPSL-CM6A-LR	IPSL (France)	2.5° × 1.25°	r1i1p1f1
19	KIOST-ESM	KIOST (Korea)	1.875° × 1.875°	r1i1p1f1
20	MIROC6	MIROC (Japan)	1.40625° × 1.40625°	r1i1p1f1
21	MPI-ESM-1-2-HAM	MPI-M (Germany)	1.975° × 1.975°	r1i1p1f1
22	MPI-ESM1-2-HR	MPI-M (Germany)	0.9375° × 0.9376°	r1i1p1f1
23	MPI-ESM1-2-LR	MPI-M (Germany)	1.875° × 1.875°	r1i1p1f1
24	MRI-ESM2-0	MRI (Japan)	1.125° × 1.126°	r1i1p1f1
25	NESM3	NUIST (China)	1.875° × 1.875°	r1i1p1f1
26	NorESM2-MM	NCC (Norway)	1.25° × 0.9375°	r1i1p1f1

For future emission scenarios, the shared socioeconomic pathway (SSP) provided in CMIP6 is the combination of the representative concentration pathway (RCP) and other pathways. SSP scenarios provided in CMIP6 include the updated versions from CMIP5 (SSP126, SSP245, SSP460, and SSP585) and the new combinations (SSP119, SSP370, and SSP434). In addition, SSP245 and SSP585 represent radiative forcing stabilized at 4.5 and 8.5 W m^{-2} by the end of the 21st century. The SSP245 scenario is subjective for most countries pursuing sustainable development, while the SSP585 scenario represents the worst scenario (fossil-based energy-intensive economy), reflecting the impact of unconventional development [1,26]. Therefore, we selected SSP245 and SSP585 in CMIP6 for future projections.

To facilitate the analysis, a bilinear interpolation method was used to uniformly interpolate the model data to the $0.25° \times 0.25°$ grid, corresponding to the grid position and resolution of the observed dataset. Due to the different periods of the model data and observational data, only China's land area was considered in this study.

2.2. Observation Data

To evaluate the simulated results of the CMIP6 models, the daily maximum and minimum temperature data of the China high-resolution dataset CN05.1 released by the Open Laboratory for Climate Research of China Meteorological Administration [28] were used as observational data in this study. The available starting and ending times of these data were 1961–2018, with a high-spatial resolution of $0.25° \times 0.25°$. This dataset has a long timescale and high spatial resolution. The generation process of this dataset only used the actual observational data of observation stations for statistical interpolation, covering the entire land area of China (Taiwan Province is missing statistical data). Interpolation of this dataset is based on an "anomaly approach" using over 2400 stations [29], which is similar to the method used to create the Climatic Research Unit dataset [30]. Compared with the reanalysis data, the CN05.1 data have greater reliability. This dataset has been widely used for regional and global climate model validation [29,31].

2.3. Methods

According to the harvested area and yield of 175 crops [32], grid points with production greater than the 5th percentile of maize production in China were extracted from maize yield data as the maize cultivation areas. This dataset was created by combining national, state, and county level census statistics with a recently updated global data set of croplands on a 5 min by 5 min latitude/longitude grid. The resulting land use datasets depict circa the year 2000 the area (harvested) and yield of 175 distinct crops of the world [32]. All assessments were conducted on these grid points. Continental China was divided into seven regions according to Wu et al. [33] and the distribution of the maize cultivation areas. The regional division of China is shown in Figure 1.

2.3.1. Climatology and Interannual Variability

The historical simulation experiments in CMIP6 were conducted from 1850 to 2014, while different future scenarios were divided from 2015 onward. The historical period from 1961 to 2014 was used to evaluate the ability of models to simulate the interannual variability of the DTR. The 1995–2014 period can better represent the current climate state and help us understand the simulation capability of the CMIP6 model for the current climate state [34]. Therefore, the 20-year period from 1995 to 2014 was selected to evaluate the simulation ability of CMIP6 models regarding the spatial distribution of the DTR in maize cultivation areas in China.

According to the ten-day dataset of crop growth and soil moisture in China and the research results of Cao [35], March to June of the current year was selected as the maize-growing season. Fifty-four years of data from a historical period (1961 to 2014) were selected for analysis in this study. The annual mean DTR and the maize-growing season mean DTR simulated using CMIP6 models in maize cultivation areas were calculated and compared.

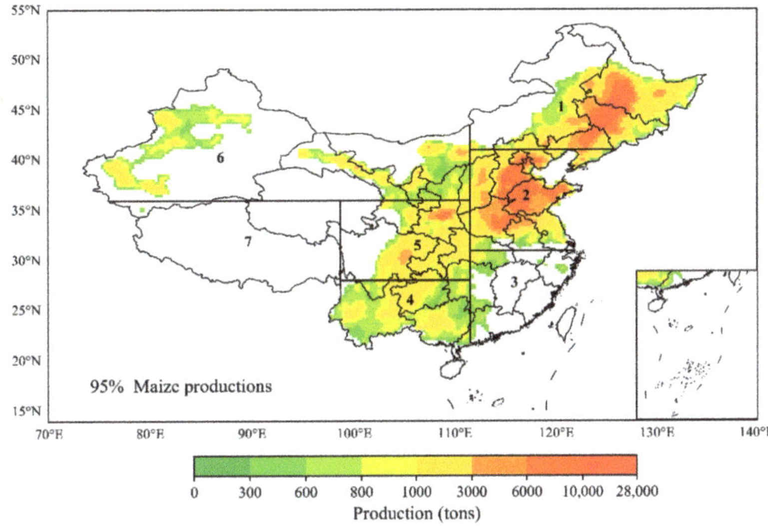

Figure 1. The regional divisions in China (subregion 1: Northeast China (NEC), subregion 2: North China (NC), subregion 3: Southeast China (SEC), subregion 4: Southwest China (SWC), subregion 5: Chuanyu (CY), subregion 6: Northwest China (NWC), and subregion 7: Tibetan Plateau (TP)).

The future period from 2015 to 2050 was selected for future projection. CMIP6 models performed more stable in this period than for the late 21st century [34]. The period from 2031 to 2050 was selected to project the future climatological DTR, as it represents the middle of the 21st century. The spatial distribution of the climatological DTR in 2031–2050 under two SSP scenarios was calculated, and the changes relative to the historical climatology (1995–2014) simulations were analyzed according to different subregions.

The spatial distribution of the linear trends of the annual average and growing season mean DTR in the main maize cultivation areas from 2015 to 2050 was calculated. The yearly linear trend analysis was conducted on each grid of the study area. Meanwhile, the linear trends of annual and growing season average DTR were calculated to study the interannual variation trend of DTR.

2.3.2. Performance METRICS

In the evaluation of the CMIP6 model simulation capability for DTR and future projection in maize cultivation areas in China, the following indices were used.

To evaluate the simulation ability of CMIP6 models for the spatial distribution of the DTR in China's maize cultivation areas, root mean square error ($RMSE$) was used. It was widely employed to illustrate the bias between the simulations and observations [36]. The $RMSE$ is:

$$RMSE = \sqrt{\frac{1}{N}\sum_{i=1}^{N}(M_i - O_i)^2} \quad (1)$$

where M_i and O_i are the simulated value and observation of the ith grid, respectively. N is the grid number. The $RMSE$ is larger than 0. When the $RMSE$ is 0, it indicates that the pattern matches the observation perfectly; a smaller $RMSE$ indicates that the pattern has better performance.

The standard deviations (SD) of the yearly growing season and annual anomalies of the maize cultivation area DTR were calculated as indicators to quantitatively express the ability of the models to simulate interannual variability. The DTR anomalies were linearly detrended before calculating the SD [15,22,37]. The SD is:

$$SD = \sqrt{\frac{1}{N}\sum_{i=1}^{n}(M_i - \overline{M})^2} \tag{2}$$

where $\overline{M}$ and $\overline{O}$ denote the mean of simulations and observations. The SD value is equal to 0 when M_i is identical to $\overline{M}$, and the closer the SD value is to SD_O (standard deviation of observations), the greater the skill in simulating the interannual variability.

To further evaluate the ability of models to simulate DTR in China's maize cultivation areas, the Kling-Gupta efficiency (KGE) was used [36,38,39]. KGE is:

$$KGE = 1 - \sqrt{(r-1)^2 + \left(\frac{\overline{M}}{\overline{O}} - 1\right)^2 + \left(\frac{SD_M/\overline{M}}{SD_O/\overline{O}} - 1\right)^2} \tag{3}$$

where SD_M and SD_O denote the standard deviation of simulations and observations. The KGE value varies between 1 and -∞, where 1 represents a complete match. There is no specific meaning attached to the KGE value when it equals zero [40].

To evaluate the ability of models to simulate both climatic state DTR and interannual variability, this study comprehensively ranked the simulation ability of each CMIP6 model based on $RMSE$ and SD. The comprehensive rating index (CRI) enables efficient ranking of model simulation values [41]. The CRI is:

$$CRI = 1 - \frac{1}{n \times m}\sum_{i=1}^{n} rank_i \tag{4}$$

where n is the number of evaluation indicators, m is the number of models, and $rank_i$ is the ranking of the i th indicator of the model. The closer the CRI value is to 1, the better the model simulation.

Previous studies have revealed that the multi-model ensemble (MME) usually shows higher reliability in reproducing the present Chinese climate relative to an individual model [42]. Therefore, the multi-model arithmetic mean ensemble with the same weights was used in this study. The MME is:

$$MME = \frac{1}{N}\sum_{j=1}^{N} M_j \tag{5}$$

where N is the number of models, M_j is the simulations of the j th model.

Time series analysis was carried out to determine the interannual variability of the DTR. The linear trend analysis was used to investigate trends in DTR variability. The linear relationship between climate variable x_i and time series t_i was established [43].

$$x_k = a + bt_k,\ \ k = 1,\ 2, \ldots\ m \tag{6}$$

where x_k is the yearly average DTR and m is the number of years of data used. t_k is the time series. a is the linear regression constant, b is the linear tendency coefficient, and $b \times 10$ is defined as the climatic tendency rate (°C/10a), which can be calculated using the least square method. b is:

$$b = \frac{\sum_{k=1}^{m} x_k t_k + \frac{1}{m}\left(\sum_{k=1}^{m} x_k\right)\left(\sum_{k=1}^{m} t_k\right)}{\sum_{k=1}^{m} {t_k}^2 - \frac{1}{m}\left(\sum_{k=1}^{m} t_k\right)^2} \tag{7}$$

a is:

$$a = \overline{x} + b\overline{t} \tag{8}$$

where $\overline{x}$ and $\overline{t}$ are the mean of the yearly average DTR and the time series. $b > 0$ represents an increasing trend of DTR over time, and vice versa. Significance levels of the b are estimated according to the two-tailed Student t-test.

3. Model Evaluation

3.1. Climatology

The CMIP6 models can reproduce the spatial characteristics of climatological DTR in maize cultivation areas (Figure 2). The DTR increased gradually from low to high latitudes and ranged from 6 °C to 16 °C. The DTR gradually increased from coastal to inland regions, with higher DTR in NWC and TP than in other regions.

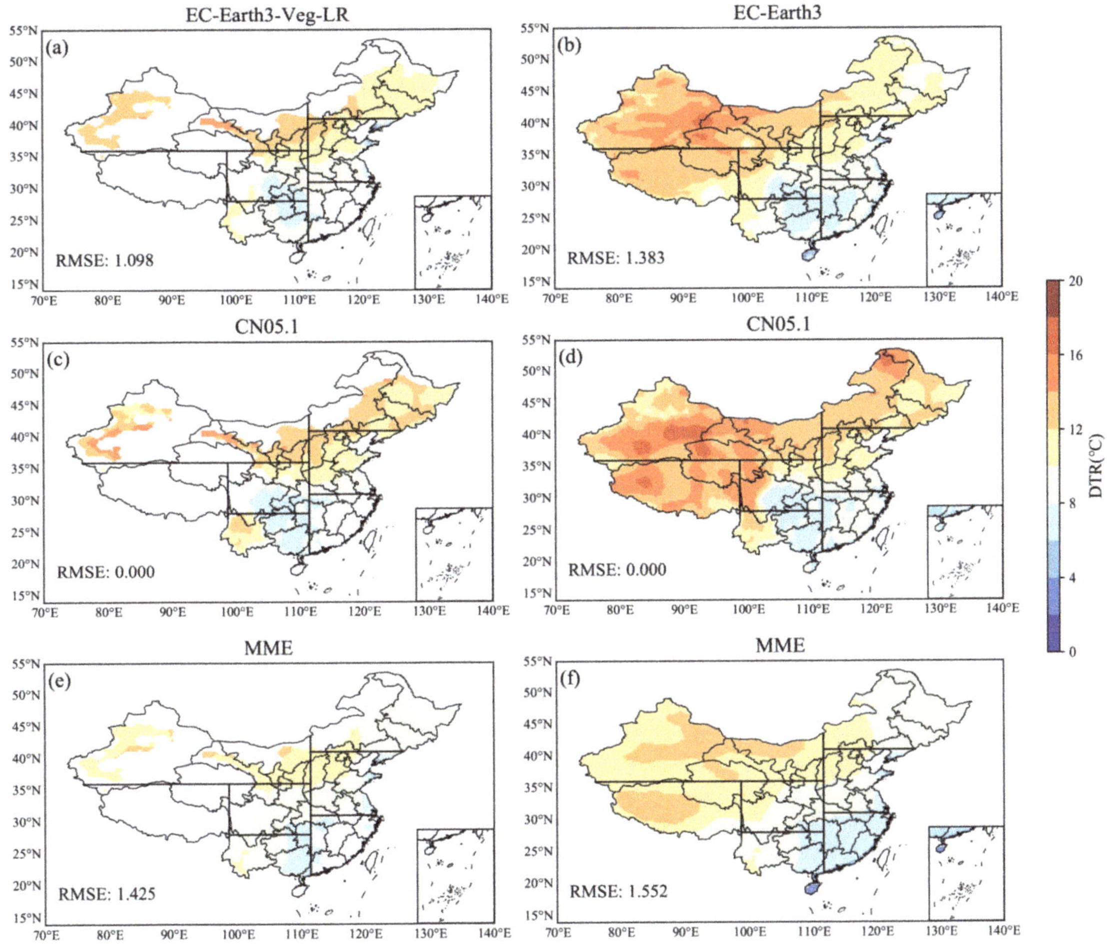

Figure 2. Observation and simulations of climatological DTR spatial distribution during 1995–2014 in China (left column displays simulated results of DTR in maize cultivation areas; the right column shows the nationwide results); (**a**,**b**) are the models with the highest *RMSE*. Observation (**c**,**d**) multi-model ensemble (*MME*) data are shown in (**e**,**f**).

When compared with observations, the multi-model ensemble data were approximately 3 °C lower than the observations nationwide and 6 °C lower in NWC. In addition, the DTR in CY was 2 °C higher than the observation.

EC-Earth3-Veg-LR had the best simulation ability among the 26 CMIP6 models for simulating the climatological DTR in maize cultivation areas ($RMSE$ = 1.098). The reproducibility of the DTR averaged over China was lower than that of the maize cultivation area averages, but it was still acceptable. The DTR simulated using the MME was not as good as EC-Earth3 in either China or maize cultivation areas. The same conclusions could be drawn across the country. The simulation of EC-Earth3 was relatively greater in CY compared with the observation, while in other regions, they were approximately 1 °C lower.

There was geographic variability in the distribution of SDs among models (Figure 3). The average SD was 2.33 in maize cultivation areas and 2.73 in China, which indicates that the consistency within maize cultivation areas was higher than that in China. The SDs of simulations in NEC and TP were approximately 2 °C higher than those in other regions, and there were significant differences between models. These results indicated that the CMIP6 models have good simulation capability in eastern China. There were great differences between the simulated results of different models in NEC and TP. CMIP6 models were still deficient in their ability to simulate the climatological DTR on the TP, which is consistent with CMIP5. Improving the model to make the simulation more reliable has become a new challenge for model developers.

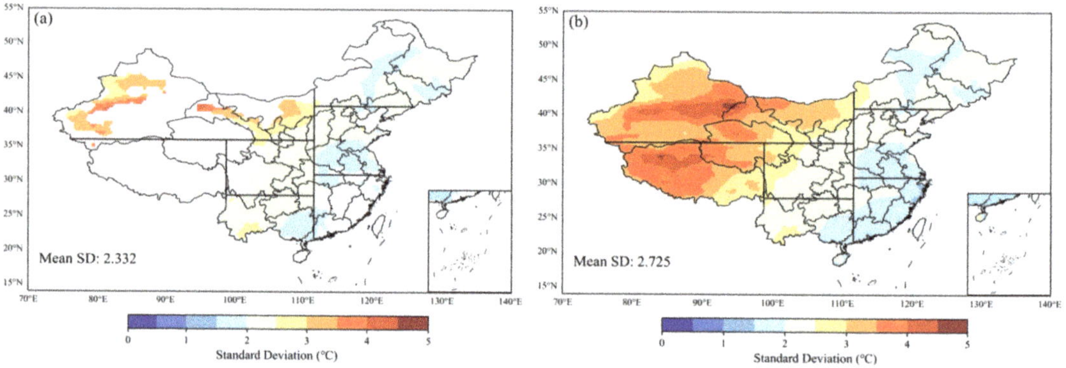

Figure 3. Standard deviations (SD) between CMIP6 models in maize cultivation areas (**a**) and China (**b**).

We evaluated each region and calculated the $RMSE$ for each model separately, and the evaluation results are presented in Figure 4 to show the performance of the models more visually.

In general, the CMIP6 models could reproduce the spatial distribution of climatological DTR (Figure 4), and most models had $RMSE$s less than 2. CMIP6 models performed better in simulating DTR in maize cultivation areas than in China ($RMSE_M$ was smaller than $RMSE_C$). EC-Earth3-Veg-LR had better simulation effects for maize cultivation areas ($RMSE_M$ = 1.098) than the other models, and the MME showed the same characteristics. Models with good performance (smaller $RMSE$) had high KGEs; the KGE (0.82) of EC-Earth3-Veg-LR ranked in the top 3 among models. However, the difference between the MME ($RMSE$ = 1.43) and observations was slightly greater than that of EC-Earth3-Veg-LR. There were large differences in model performance across regions. Moreover, the results show that the models had better simulations in SWC and NC. However, EC-Earth3-Veg was not the best in other regions. EC-Earth3 had the smallest $RMSE$ and the highest ranking in NWC, which is consistent with the results in China evaluations. These results suggested that even the model with the best performance among the 26 models may not have the best simulation capability in all regions.

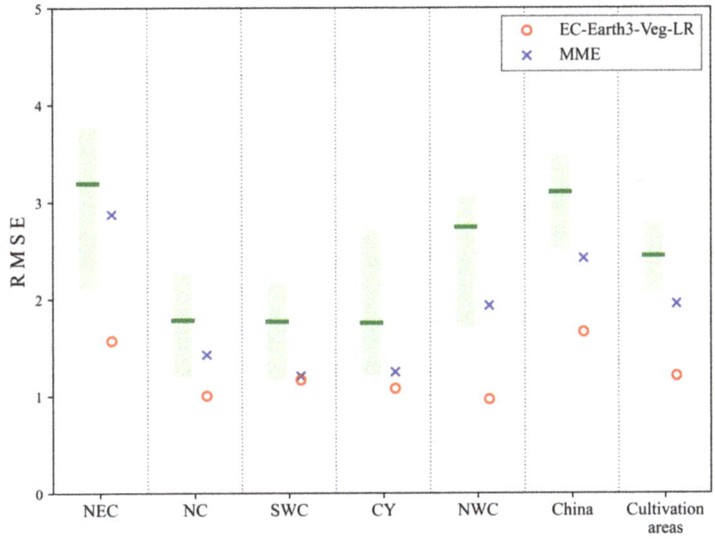

Figure 4. The *RMSE*s of climatological DTR for the historical simulation in each region during 1961–2014. Green indicates the range from the 25th to 75th percentile of the *RMSE* s for the simulations.

3.2. Interannual Variability

According to the observations, the growing season DTR in maize cultivation areas showed a decreasing trend at a rate of −0.151 °C/10 a (Figure 5). The same trend was observed for the annual DTR, with a decreasing rate of −0.178 °C/10 a. The annual DTR showed a greater rate of decline relative to the maize-growing season. CMIP6 models could better simulate these trends: among the 26 CMIP6 models, EC-Earth3 had the best simulation for the annual DTR trend (−0.119 °C/10 a) and performed best in the maize-growing season (−0.161 °C/10 a). The *MME* could simulate these trends with slower rates.

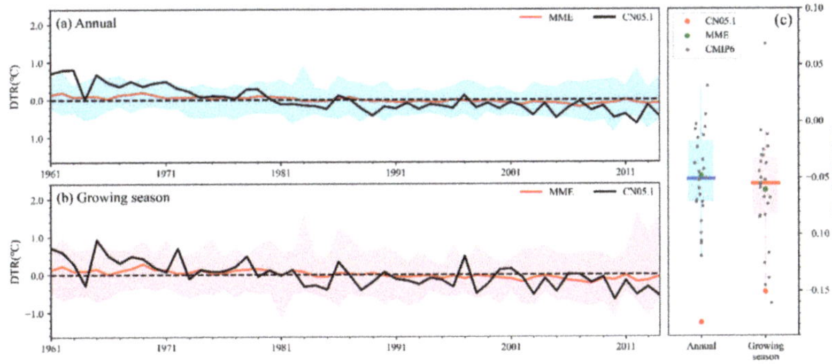

Figure 5. Annual (**a**) and growing season (**b**) DTR anomalies simulated via CMIP6 models in maize cultivation areas show patterns of fluctuations that reflect annually varying correlations of DTR in China. Blue and red shadings indicate the range of simulations. DTR trends (**c**) of annual and growing season simulated using CMIP6 models and observations. Blue and red indicate the range from the 25th to 75th percentile of the trends simulated by the models. The red and blue lines are the medians of the model-simulated trends, and the gray dots are the results of the CMIP6 models. The red and green dots represent the results of CN05.1 and *MME*, respectively.

In general, CMIP6 models were able to simulate the interannual variation in the annual DTR and growing season mean DTR in maize cultivation areas. Most models had *SD*s less than 0.7. The maize-growing season DTR simulated using CMIP6 models performed differently from that of the annual DTR, and some models simulated the annual DTR better than the simulation of the growing season DTR. Similarly, it has been reported that CMIP5 simulates the interannual variability in annual surface air temperature better than monthly and seasonal temperatures. It is worth noting that models with good simulations of annual DTR were slightly inferior in simulating the growing season mean DTR, while models with good simulations of the growing season mean DTR showed unsatisfactory simulations for interannual variation of annual DTR (Figure 6). A model with a good simulation of annual DTR does not imply a good simulation of the maize-growing season; therefore, targeted evaluation for the maize-growing season becomes essential. AWI-CM-1-1-MR had the best ability to simulate DTR for the maize-growing season, and it was more suitable for studies targeting the maize-growing season compared with the interannual variation. Models with smaller SDs also had higher KGEs; the KGE (0.36) of AWI-CM-1-1-MR ranked in the top 5 among models. The performance of the CMIP6 model varied greatly across regions (Figure 6). The MME was more suitable for the simulation of annual DTR.

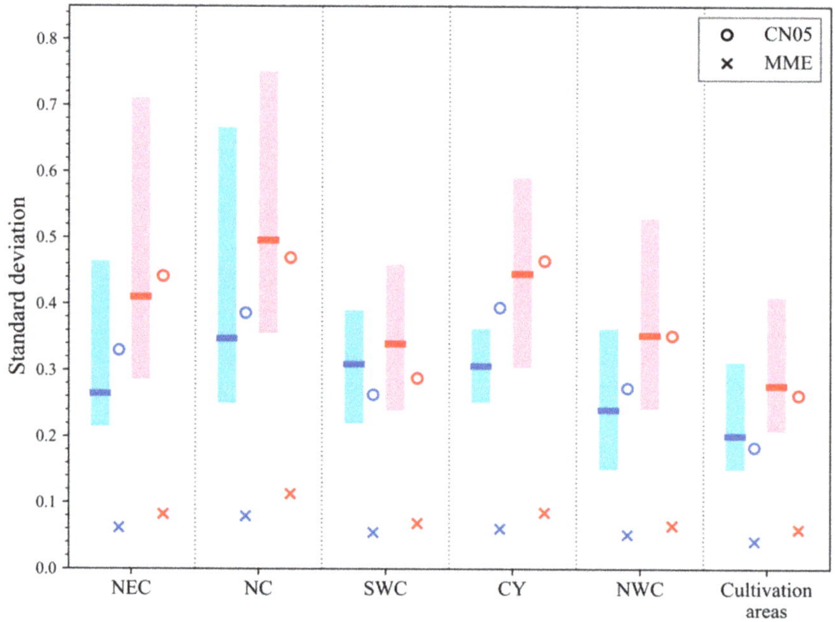

Figure 6. The standard deviations of linearly detrended annual and yearly growing season mean anomalies of DTR for the historical simulation during 1961–2014. Blue indicates the range from the 10th to 90th percentile of the annual averages, and red indicates the maize-growing season means. Each row represents a region of China.

3.3. Comprehensive Evaluation

To comprehensively evaluate the ability of each model to simulate the spatial and temporal characteristics of the DTR in maize cultivation areas, the CRI was used to evaluate each model in this study. According to Table 2, EC-Earth3-Veg-LR had the best simulation of DTR in maize cultivation areas (CRI = 0.92), which is more suitable for the simulation of DTR in maize cultivation areas of China.

Table 2. Scores of the top three models with the best performance.

	Index	EC-Earth3-Veg-LR	EC-Earth3	GFDL-CM4
Climatology	*RMSE*	1.098	1.116	0.865
	KGE	0.818	0.821	0.455
Interannual variability	*SD*	0.248	0.278	0.260
	KGE	0.358	0.404	0.262
Overall	*CRI*	0.865	0.808	0.808

4. Future Projections

Acceptable performances of climate models are the basis for developing credible data of future climate through CMIP6 scenario simulations. In this section, we explore the future changes in the diurnal temperature range of maize cultivation areas of China simulated using EC-Earth3-Veg-LR. The period is focused on two SSPs (SSP245 and SSP585): the period from 2031 to 2050 was selected to project the future climatological DTR, as it represents the middle of the 21st century. The linear trends of the annual average and growing season mean DTR in the main maize cultivation areas from 2015 to 2050 were calculated.

4.1. Climatology

The DTR and future changes in maize cultivation areas of China under different future scenarios are shown in Figure 7. The future DTR spatial distributions shared the same characteristics as historical observations. The DTR increased gradually from low to high latitudes, ranged from 6 °C to 16 °C, and gradually increased from coastal to inland regions.

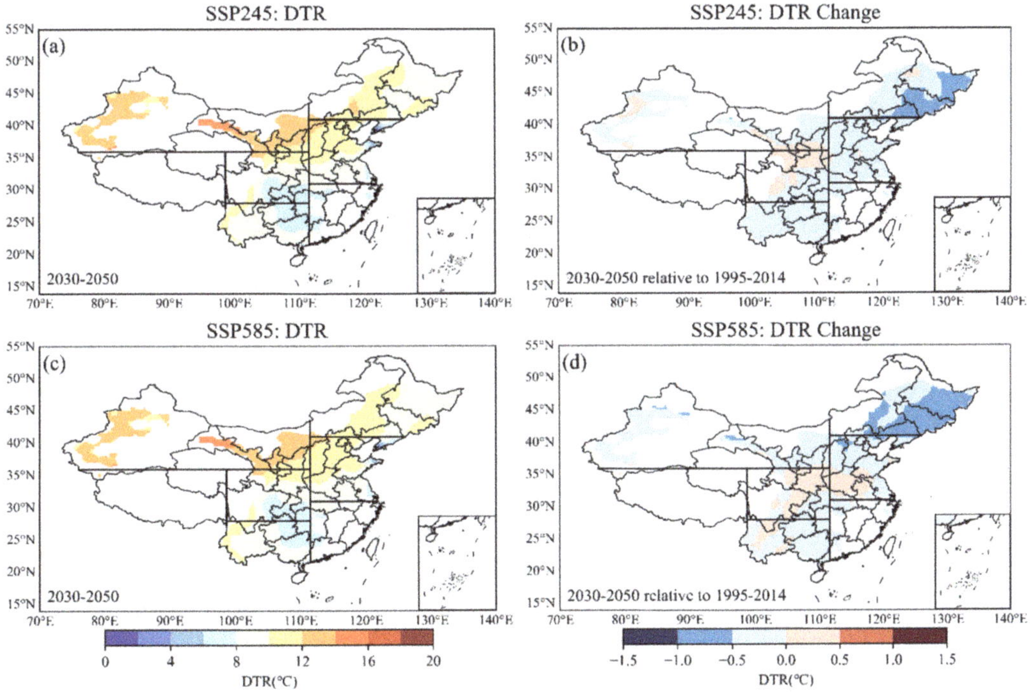

Figure 7. The climatological DTR spatial distribution (**a**,**c**) during 2030–2050 and changes (**b**,**d**) relative to 1995–2014 in maize cultivation areas of China under different scenarios simulated via EC-Earth3-Veg-LR.

Compared with historical (1995–2014) simulations, reductions were widely detected: under the SSP245 scenario, there was a decline in 86.90% of grid points within China and 81.2% under the SSP585 scenario. The climatological DTR of maize cultivation areas decreased by 0.151 °C (SSP245) and 0.207 °C (SSP585). The reduction proportion of the DTR under the SSP585 scenario was slightly smaller than that under the SSP245 scenario (Table 3). Under the SSP245 scenario, the reduction was mainly distributed in NWC, NC, and SWC, and the magnitude of change was concentrated at approximately 0.20 °C. The climatological DTR during 2030~2050 in NEC was approximately 0.38 °C higher than that of historical simulations. Almost all subregions showed a reduction under the SSP585 scenario, with a magnitude of 0.19 °C.

Table 3. Changes in the climatological DTR (°C) during 2030–2050 relative to 1995–2014 in regions of China under different scenarios.

Scenario	NEC	NC	SWC	CY	NWC	Cultivation Areas
SSP245	0.382	−0.149	−0.223	−0.043	−0.094	−0.151
SSP585	−0.556	−0.097	−0.092	0.005	−0.212	−0.207

4.2. Interannual Variability

Compared with the historical period (1961–2014), the annual DTR remained essentially unchanged under the SSP245 scenario in maize cultivation areas (Figure 8), while the growing season DTR had an increasing trend (0.014 °C/10 a). Under the SSP585 scenario, the growing season DTR in maize cultivation areas exhibited a decreasing trend at a rate of −0.052 °C/10 a, and the annual DTR showed a greater decreasing trend (−0.069 °C/10 a). The decreasing trends of DTR under the SSP585 scenario were greater than those under the SSP245 scenario.

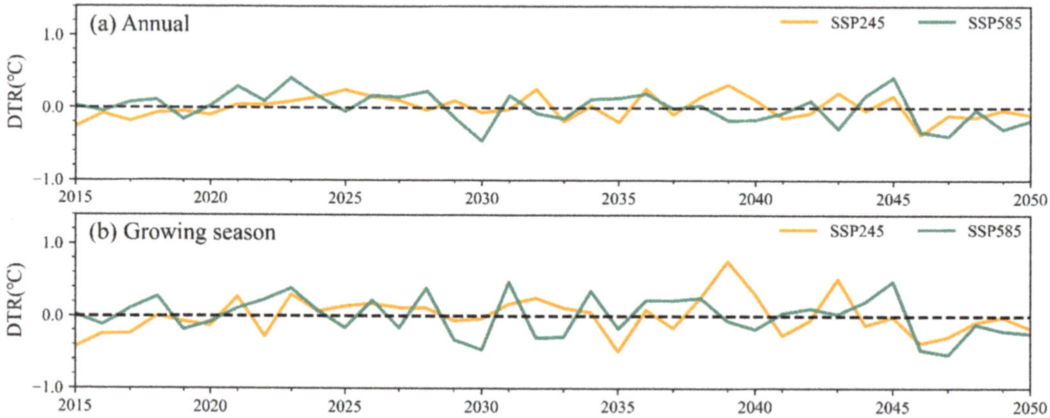

Figure 8. Annual (**a**) and growing season (**b**) DTR anomalies under different scenarios simulated via EC-Earth3-Veg-LR in maize cultivation areas show patterns of fluctuations that reflect annually varying correlations of DTR in China.

The annual DTR always showed a greater rate of decline relative to the growing season DTR in maize cultivation areas in both scenarios (Table 4). The annual DTR in the maize cultivation areas under the SSP245 scenario showed a decreasing trend at 45.90% of the grid points, while in NEC, SWC, and NWC, it was dominated by a decreasing trend. There was a significant increase in NC and CY. For the growing season DTR, more grid points showed decreasing trends in both NEC and CY. The proportion of grid points with decreasing trends in CY increased significantly, while the proportion of significant grid points increased and was more concentrated (Figure 9). In NWC and NC, the proportion of decreasing grid points decreased and that of significant grid points increased, mainly in the eastern area of Northwest China. The annual (76.28%) and growing season (70.09%) DTR under the SSP585 scenario showed decreasing trends at most grid points in maize cultivation areas, with decreasing rates exceeding 0.2 °C/10 a in NEC, SWC, and NWC. It showed increasing trends in most NC grid points and had a high percentage of significance (23.99%). The proportion of grid points with decreasing trends in DTR was greater under the SSP585 scenario than under the SSP245 scenario. Although the growing season DTR followed an increasing trend in NC and CY under the SSP245 scenario in general, the increasing trend was not significant in grid points with high maize cultivation in China (Figure 1), even with a large decreasing trend (NEC). Since Chinese maize production is mainly concentrated in NC and NEC, and the growing season DTR in these two subregions showed opposite changes, this may cause some impact on the quality and yield of maize in China.

Table 4. The proportion of grid points with linear decreasing trends of annual and growing season DTR during 2015–2050 in different subregions of China. ("Sig. proportion" indicates the proportion of grid points that were significant at $p < 0.1$).

	Region	NEC	NC	SWC	CY	NWC	Cultivation Areas
SSP245	Annual	63.78%	16.48%	60.38%	20.43%	57.13%	45.90%
	Sig. proportion	1.15%	6.99%	0.32%	15.10%	9.81%	6.07%
	Growing season	80.02%	7.28%	61.97%	46.50%	24.28%	45.36%
	Sig. proportion	3.44%	16.70%	0.00%	19.55%	27.30%	12.92%
SSP585	Annual	100.00%	40.62%	53.01%	87.36%	93.62%	76.28%
	Sig. proportion	0.00%	0.00%	0.00%	0.00%	0.00%	0.00%
	Growing season	100.00%	25.02%	88.27%	25.27%	91.77%	70.09%
	Sig. proportion	0.00%	23.99%	0.00%	1.28%	0.00%	4.96%

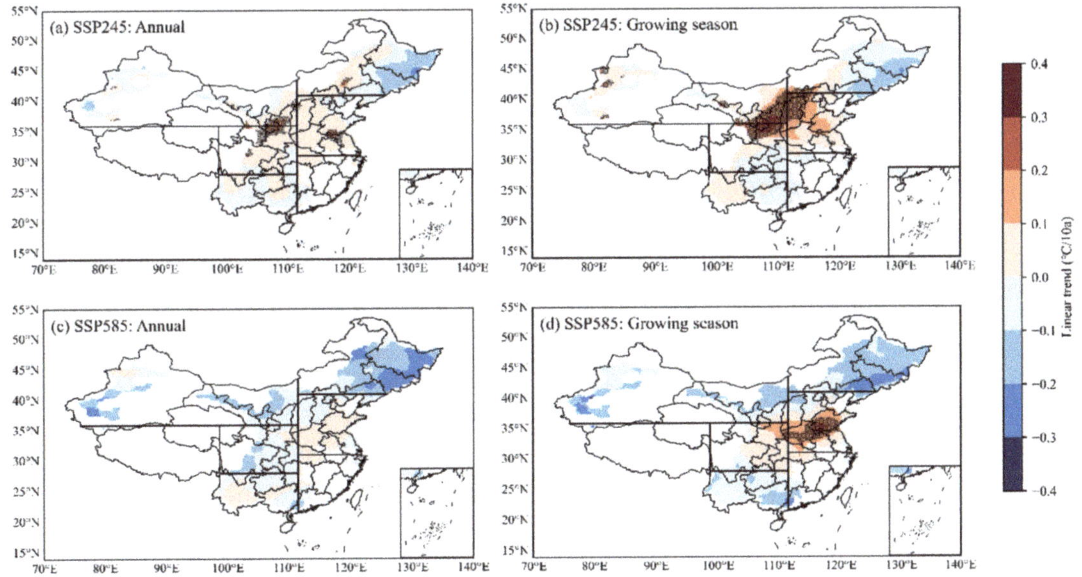

Figure 9. Spatial distribution of linear trends (°C/10 a) of annual (**a**,**c**) and growing season (**b**,**d**) DTR during 2015–2050 in the main maize cultivation areas under different scenarios simulated via EC-Earth3-Veg-LR. The black dot indicates the grid point with the trend that was significant at $p < 0.1$.

5. Discussion

To explore the applicability of the CMIP6 model to maize research, an assessment was first conducted to evaluate the ability of CMIP6 models to simulate DTR in maize cultivation areas in China. The results show that CMIP6 models can reproduce the spatial and temporal characteristics of DTR in historical periods. CMIP6 models still have obvious deviations in simulating DTR in western China, which shares the same characteristics with other meteorological components [37,44,45]. The complex terrain in NWC may lead to simulation uncertainty [46]. The simulation effect of CMIP6 models on the historical climate DTR of the main maize cultivation areas is better than that of China. The decreasing DTR trend has been universally observed since the 1950s [47,48]. CMIP6 models can also capture the slow decreasing trend in annual DTR and growing season average DTR in maize cultivation areas during the historical period. CMIP6 models simulate the annual DTR better than that for the growing season DTR, and the same characteristics can be found in the simulations of meteorological components such as temperature and precipitation [49,50]. This difference makes it critical and essential to perform targeted model evaluations of the maize-growing season. As a result of the comprehensive evaluation, EC-Earth3-Veg-LR has the best simulation ability for the spatial and temporal distribution characteristics of DTR in maize cultivation areas.

We projected the spatial and temporal characteristics and changes in the DTR in the maize cultivation areas under different future scenarios with the selected optimal model. Compared with the historical simulation, the climatic DTR of the main maize-producing areas in the middle 21st century will be reduced by 0.151°C (SSP245) and 0.207 °C (SSP585). SSP585 showed a greater proportion of DTR reduction than SSP245. The DTR in the main maize cultivation areas under SSP245 is expected to remain unchanged (annual) or increase slightly (growing season). Under SSP585, DTR is expected to decrease both annually and during the growing season. The projected DTR changes shared the same characteristics with previous studies [15,48,51]. The downward trend of DTR in future scenarios is consistent with the historical period [52,53]. At the same time, the DTR trends in maize cultivation areas are also similar to the global DTR trends. It is worth noting that there are significant increasing trends of growing season DTR in NC and NEC, and the impact on the quality and yield of crops in China needs to be further explored.

A considerable number of studies have been conducted to evaluate the ability of climate models to simulate key meteorological components [18–23]. However, the spatial scales of previous evaluations were primarily global or national. There have been few studies on maize cultivation areas and maize-growing seasons. When estimating the effects of future climate on maize yield and quality, using inappropriate climate model data will greatly reduce the credibility of the results. We quantitatively evaluated the ability of CMIP6 models to simulate the temporal and spatial characteristics of the DTR in the main maize production areas from multiple perspectives and used the optimal model to predict and analyze the changes in the DTR in the main maize production areas in China under different scenarios in the future. This will help to further improve the ability of global climate models to simulate the DTR in China's maize cultivation areas and to serve research on the effects of climate change on maize yield and quality.

This study only evaluates the DTR of the maize cultivation areas in China and projected the DTR changes during 2015–2050 under different SSP scenarios. The scope of our study is limited in China, and the simulation capability of the climate model in other regions needs further evaluation and projection. The yield of crops is known to be influenced by the environment and other factors [1–5]. Thus, the ability of the climate model to simulate other climate components in the maize cultivation areas in China needs to be further evaluated. Meanwhile, the evaluation and prediction of other meteorological elements are also worthy of further study and are of great significance and necessity.

6. Conclusions

The ability of 26 CMIP6 models to simulate the DTR during 1961–2014 is examined using model output together with the high-resolution CN05.1 dataset. Based on the overall rankings of the historical simulation capacities of 26 CMIP6 models in maize cultivation areas, the optimal model is screened to serve future estimates. Then, with the application of the optimal model, a future projection of DTR in the main maize cultivation areas during 2015–2050 under two SSPs is presented in this study. The main conclusions are summarized as follows:

1. CMIP6 models can generally reproduce the spatial distribution and interannual variation in the DTR in the main maize cultivation areas. The reproducibility of the DTR averaged over the main maize cultivation areas is better than that of China ($RMSE_M$ is smaller than $RMSE_C$). The DTR varies substantially between the models, and the intermodel spread is particularly large in NWC.
2. Based on the comprehensive evaluation, EC-Earth3-Veg-LR is more suitable for the simulation of DTR in the main maize cultivation areas in China. It is essential to pertinently evaluate global climate models. The reproducibility of the maize-growing season DTR averaged over the main maize cultivation areas is lower than that of the annual DTR, but it is still acceptable.
3. Compared with historical simulations, reductions are widely detected: the climatological DTR of the main maize cultivation areas decreases by 0.151 °C (SSP245) and 0.207 °C (SSP585). Under the SSP245 scenario, the reduction is mainly distributed in NWC, NC, and CY. All subregions show a reduction under SSP585. The reduction proportion of DTR under SSP245 is slightly smaller than that under SSP585.
4. The DTR in the main maize cultivation areas under SSP245 is expected to remain unchanged (annual) or to increase slightly (growing season). Under SSP585, DTR is expected to decrease both annually and during the growing season. The annual and growing season DTRs are dominated by decreasing trends in NEC, NWC, and SWC under the two scenarios, while in CY and NC, the growing season DTR shows a significant increase.

Author Contributions: Conceptualization, X.Y. and W.X.; methodology, W.X.; software, W.X.; validation, W.X., S.W. and X.Y.; formal analysis, W.X.; investigation, W.X.; resources, W.X. and S.W.; data curation, W.X. and S.W.; writing—original draft preparation, W.X.; writing—review and editing, W.X. and X.Y.; visualization, W.X.; supervision, X.Y.; project administration, X.Y.; funding acquisition, X.Y. All authors have read and agreed to the published version of the manuscript.

Funding: This research was funded by the National Key R&D Program of China, grant number 2019YFA0606904.

Institutional Review Board Statement: Not applicable.

Informed Consent Statement: Not applicable.

Data Availability Statement: Not applicable.

Acknowledgments: We would like to thank the China Meteorological Administration for temperature data and information. We also acknowledge the Coupled Model Intercomparison Project (CMIP) launched by the World Climate Research Program (WCRP) for CMIP6 data.

Conflicts of Interest: The authors declare no conflict of interest. The funders had no role in the design of the study; in the collection, analyses, or interpretation of data; in the writing of the manuscript, or in the decision to publish the results.

References

1. Arunrat, N.; Sereenonchai, S.; Chaowiwat, W.; Wang, C. Climate change impact on major crop yield and water footprint under CMIP6 climate projections in repeated drought and flood areas in Thailand. *Sci. Total Environ.* **2022**, *807*, 150741. [CrossRef] [PubMed]
2. Arunrat, N.; Sereenonchai, S.; Wang, C. Carbon footprint and predicting the impact of climate change on carbon sequestration ecosystem services of organic rice farming and conventional rice farming: A case study in Phichit province, Thailand. *J. Environ. Manag.* **2021**, *289*, 112458. [CrossRef]
3. Arunrat, N.; Pumijumnong, N.; Sereenonchai, S.; Chareonwong, U.; Wang, C. Assessment of climate change impact on rice yield and water footprint of large-scale and individual farming in Thailand. *Sci. Total Environ.* **2020**, *726*, 137864. [CrossRef] [PubMed]
4. Lobell, D.B.; Field, C.B. Global scale climate–crop yield relationships and the impacts of recent warming. *Environ. Res. Lett.* **2007**, *2*, 014002. [CrossRef]
5. Moriondo, M.; Giannakopoulos, C.; Bindi, M. Climate change impact assessment: The role of climate extremes in crop yield simulation. *Clim. Change* **2011**, *104*, 679–701. [CrossRef]
6. Xiong, W.; Holman, I.; Lin, E.; Conway, D.; Li, Y.; Wu, W. Untangling relative contributions of recent climate and CO_2 trends to national cereal production in China. *Environ. Res. Lett.* **2012**, *7*, 044014. [CrossRef]
7. Wang, L.; Xiong, W.; Wen, X.; Feng, L. Effect of climatic factors such as temperature, precipitation on maize production in China. *Trans. Chin. Soc. Agric. Eng.* **2014**, *30*, 138–146. [CrossRef]
8. Wang, Y.; Yang, H.; Liu, R.; Cai, H. Try to discuss the relationships between the seed qualities of wheat and meteorological conditions. *Chin. J. Agrometeorol.* **1990**, *11*, 1–7.
9. Dhakhwa, G.B.; Campbell, C.L. Potential effects of differential day-night warming in global climate change on crop production. *Clim. Change* **1998**, *40*, 647–667. [CrossRef]
10. Tubiello, F.N.; Rosenzweig, C.; Goldberg, R.A.; Jagtap, S.; Jones, J.W. Effects of climate change on US crop production: Simulation results using two different GCM scenarios. Part I: Wheat, potato, maize, and citrus. *Clim. Res.* **2002**, *20*, 259–270. [CrossRef]
11. Smika, D.E.; Greb, B.W. Protein Content of Winter Wheat Grain as Related to Soil and Climatic Factors in the Semiarid Central Great Plains. *Agron. J.* **1973**, *65*, 433–436. [CrossRef]
12. Jiang, L.X.; Lv, J.J.; Wang, L.L.; Yang, X.Q.; Li, S. Variation of Diurnal Temperature Range and its Relationship with Crop Yield in Heilongjiang Province. *Chin. J. Agrometeorol.* **2013**, *2*, 179–185.
13. Fall, S.; Watts, A.; Nielsen-Gammon, J.; Jones, E.; Niyogi, D.; Christy, J.R.; Pielke, R.A., Sr. Analysis of the impacts of station exposure on the US Historical Climatology Network temperatures and temperature trends. *J. Geophys. Res. Atmos.* **2011**, *116*, D14. [CrossRef]
14. Zhou, Y.; Ren, G. Change in extreme temperature event frequency over mainland China, 1961−2008. *Clim. Res.* **2011**, *50*, 125–139. [CrossRef]
15. Lindvall, J.; Svensson, G. The diurnal temperature range in the CMIP5 models. *Clim. Dyn.* **2015**, *44*, 405–421. [CrossRef]
16. Wang, J.; Chen, Y.; Tett, S.F.; Yan, Z.; Zhai, P.; Feng, J.; Xia, J. Anthropogenically-driven increases in the risks of summertime compound hot extremes. *Nat. Commun.* **2020**, *11*, 528. [CrossRef]
17. Wang, B.; Feng, P.; De, L.; Garry, J.O.; Macadam, I.; Waters, C.; Asseng, S.; Cowie, A.; Jiang, T.; Xiao, D.; et al. Sources of uncertainty for wheat yield projections under future climate are site-specific. *Nat. Food* **2020**, *1*, 720–728. [CrossRef]
18. Slafer, G.A.; Rawson, H.M. Rates and cardinal temperatures for processes of development in wheat: Effects of temperature and thermal amplitude. *Funct. Plant Biol.* **1995**, *22*, 913–926. [CrossRef]
19. Larmure, A.; Salon, C.; Munier-Jolain, N.G. How does temperature affect C and N allocation to the seeds during the seed-filling period in pea? Effect on seed nitrogen concentration. *Funct. Plant Biol.* **2005**, *32*, 1009–1017. [CrossRef]
20. Tang, B.; Hu, W.; Duan, A. Future Projection of Extreme Precipitation Indices over the Indochina Peninsula and South China in CMIP6 Models. *J. Clim.* **2021**, *34*, 8793–8811. [CrossRef]
21. Knox, J.; Hess, T.; Daccache, A.; Wheeler, T. Climate change impacts on crop productivity in Africa and South Asia. *Environ. Res. Lett.* **2012**, *7*, 034032. [CrossRef]
22. Lewis, S.C.; Karoly, D.J. Evaluation of historical diurnal temperature range trends in CMIP5 models. *J. Clim.* **2013**, *26*, 9077–9089. [CrossRef]
23. Fan, X.; Duan, Q.; Shen, C.; Wu, Y.; Xing, C. Global surface air temperatures in CMIP6: Historical performance and future changes. *Environ. Res. Lett.* **2020**, *15*, 104056. [CrossRef]
24. Tao, F.; Yokozawa, M.; Liu, J.; Zhang, Z. Climate–crop yield relationships at provincial scales in China and the impacts of recent climate trends. *Clim. Res.* **2008**, *38*, 83–94. [CrossRef]
25. Zhang, T.; Zhu, J.; Wassmann, R. Responses of rice yields to recent climate change in China: An empirical assessment based on long-term observations at different spatial scales (1981–2005). *Agric. For. Meteorol.* **2010**, *150*, 1128–1137. [CrossRef]
26. Zhou, T.; Zou, L.; Chen, X. Commentary on the Coupled Model Intercomparison Project Phase 6 (CMIP6). *Clim. Change Res.* **2019**, *15*, 445–456. [CrossRef]
27. CMIP6 Search Interface. Available online: https://esgf-node.llnl.gov/search/cmip6 (accessed on 29 December 2021).
28. Wu, J.; Gao, X.J. A gridded daily observation dataset over China region and comparison with the other datasets. *Diqiu Wuli Xuebao* **2013**, *56*, 1102–1111. [CrossRef]

29. Xu, Y.; Gao, X.J.; Yan, S.Y.; Xu, C.H.; Shi, Y.; Giorgi, F. A daily temperature dataset over China and its application in validating a RCM simulation. *Adv. Atmos. Sci.* **2009**, *26*, 763–772. [CrossRef]
30. New, M.; Lister, D.; Hulme, M.; Makin, I. A high-resolution data set ofsurface climate over global land areas. *Clim. Res.* **2002**, *21*, 1–25. [CrossRef]
31. Li, Q.X.; Huang, J.; Jiang, Z.; Zhou, L.; Chu, P.; Hu, K. Detection of urbanizationsignals in extreme winter minimum temperature changes over northern China. *Clim. Change* **2014**, *122*, 595–608. [CrossRef]
32. Monfreda, C.; Ramankutty, N.; Foley, J.A. Farming the planet: 2. Geographic distribution of crop areas, yields, physiological types, and net primary production in the year 2000. *Glob. Biogeochem. Cycles* **2008**, *22*, GB1022. [CrossRef]
33. Wu, Y.; Miao, C.; Duan, Q.; Shen, C.; Fan, X. Evaluation and projection of daily maximum and minimum temperatures over China using the high-resolution NEX-GDDP dataset. *Clim. Dyn.* **2020**, *55*, 2615–2629. [CrossRef]
34. IPCC. Summary for policymakers. In *Climate Change 2021: The Physical Science Basis. Contribution of Working Group I to the Sixth Assessment Report of the Intergovernmental Panel on Climate Change*; Cambridge University Press: Cambridge, UK, 2021.
35. Cao, Y.; Li, W.; Zhao, B. Water requirements of spring maize in Northwest Liaoning Province under climate change. *Resour. Sci.* **2018**, *40*, 150–160. [CrossRef]
36. Kobayashi, K.; Salam, M.U. Comparing simulated and measured values using mean squared deviation and its components. *Agron. J.* **2000**, *92*, 345–352. [CrossRef]
37. Zhou, T.; Yu, R. Twentieth-century surface air temperature over China and the globe simulated by coupled climate models. *J. Clim.* **2006**, *19*, 5843–5858. [CrossRef]
38. Gupta, H.V.; Kling, H.; Yilmaz, K.K.; Martinez, G.F. Decomposition of the mean squared error and NSE performance criteria: Implications for improving hydrological modelling. *J. Hydrol.* **2009**, *377*, 80–91. [CrossRef]
39. Kling, H.; Fuchs, M.; Paulin, M. Runoff conditions in the upper Danube basin under an ensemble of climate change scenarios. *J. Hydrol.* **2012**, *424*, 264–277. [CrossRef]
40. Hamed, M.M.; Nashwan, M.S.; Shahid, S.; bin Ismail, T.; Wang, X.J.; Dewan, A.; Asaduzzaman, M. Inconsistency in historical simulations and future projections of temperature and rainfall: A comparison of CMIP5 and CMIP6 models over Southeast Asia. *Atmos. Res.* **2022**, *265*, 105927. [CrossRef]
41. Jiang, Z.; Li, W.; Xu, J.; Li, L. Extreme precipitation indices over China in CMIP5 models. Part I: Model evaluation. *J. Clim.* **2015**, *28*, 8603–8619. [CrossRef]
42. Jiang, D.; Wang, H.; Lang, X. Evaluation of East Asian climatology as simulated by seven coupled models. *Adv. Atmos. Sci.* **2005**, *22*, 479–495. [CrossRef]
43. Wei, F.Y. *Modern Climate Statistical Diagnosis and Prediction Technology*; Meteorological Press: Beijing, China, 1999; p. 269.
44. Guo, Y.; Dong, W.; Ren, F.; Zha, Z.; Huang, J. Surface air temperature simulations over China with CMIP5 and CMIP3. *Adv. Clim. Change Res.* **2013**, *4*, 145–152. [CrossRef]
45. Xu, Y.; Xu, C. Preliminary assessment of simulations of climate changes over China by CMIP5 multi-models. *Atmos. Ocean. Sci. Lett.* **2012**, *5*, 489–494. [CrossRef]
46. Hu, Q.; Jiang, D.; Fan, G. Evaluation of CMIP5 models over the Qinghai-Tibetan Plateau. Chin. *J. Atmos. Sci.* **2014**, *38*, 924–938. [CrossRef]
47. Easterling, D.R.; Horton, B.; Jones, P.D.; Peterson, T.C.; Karl, T.R.; Parker, D.E.; Salinger, M.J.; Razuvayev, V.; Plummer, N.; Folland, C.K. Maximum and minimum temperature trends for the globe. *Science* **1997**, *277*, 364–367. [CrossRef]
48. Wang, K.; Clow, G.D. The Diurnal Temperature Range in CMIP6 Models: Climatology, Variability, and Evolution. *J. Clim.* **2020**, *33*, 8261–8279. [CrossRef]
49. Xia, J.; Yan, Z.; Jia, G.; Zeng, H.; Jones, P.D.; Zhou, W.; Zhang, A. Projections of the advance in the start of the growing season during the 21st century based on CMIP5 simulations. *Adv. Atmos. Sci.* **2015**, *32*, 831–838. [CrossRef]
50. Kamworapan, S.; Surussavadee, C. Evaluation of CMIP5 global climate models for simulating climatological temperature and precipitation for Southeast Asia. *Adv. Meteorol.* **2019**, *2019*, 1067365. [CrossRef]
51. Hamal, K.; Sharma, S.; Talchabhadel, R.; Ali, M.; Dhital, Y.P.; Xu, T.; Dawadi, B. Trends in the Diurnal Temperature Range over the Southern Slope of Central Himalaya: Retrospective and Prospective Evaluation. *Atmosphere* **2021**, *12*, 1683. [CrossRef]
52. You, Q.; Wang, D.; Jiang, Z.; Kang, S. Diurnal temperature range in CMIP5 models and observations on the Tibetan Plateau. *Q. J. R. Meteorol. Soc.* **2017**, *143*, 1978–1989. [CrossRef]
53. Sun, X.; Ren, G.; You, Q.; Ren, Y.; Xu, W.; Xue, X.; Zhang, P. Global diurnal temperature range (DTR) changes since 1901. *Clim. Dyn.* **2019**, *52*, 3343–3356. [CrossRef]

sustainability

MDPI

Article

Spatial Zoning of Dry-Hot Wind Disasters in Shandong Province

Nan Wang [1], Xiaoping Xue [2,3,*], Lijuan Zhang [1,*], Yue Chu [1], Meiyi Jiang [1], Yumeng Wang [1], Yiping Yang [1], Xihui Guo [1], Yufeng Zhao [1] and Enbo Zhao [1]

1 Heilongjiang Province Key Laboratory of Geographical Environment Monitoring and Spatial Information Service in Cold Regions, Harbin Normal University, Harbin 150025, China; wangnan199710@163.com (N.W.); cy254654@163.com (Y.C.); jiangmi199608@163.com (M.J.); wangyumeng19980603@163.com (Y.W.); ep_yangyiping@163.com (Y.Y.); 18235889220@163.com (X.G.); janekabesilsen@163.com (Y.Z.); zeb030410@163.com (E.Z.)

2 Key Laboratory for Meteorological Disaster Prevention and Mitigation of Shandong, Jinan 250000, China

3 Shandong Provincial Climate Center, Jinan 250000, China

* Correspondence: xxpdhy@163.com (X.X.); zlj@hrbnu.edu.cn (L.Z.)

Abstract: As a major agricultural province of China, Shandong province has long ranked first in agricultural growth value among all of the provinces; at the same time, it is also the province that is most affected by dry-hot wind. Therefore, it is of great significance to study the spatial zoning of the risks of dry-hot wind in this province. Based on meteorological, slope, and altitude data, and the principle of disaster risk assessment, this study uses a weighted comprehensive evaluation method, analytic hierarchy process, and ARC-GIS spatial analysis to study the spatial zoning of the risks of dry-hot wind in Shandong province. The results show that the high-risk regions of dry-hot wind are concentrated in the north-central portion of the province, the medium-risk regions are in the peripheral areas, and the low-risk regions are located mainly in the west, southwest, and east. Exposure of disaster-bearing bodies is high in the south and low in the north, while vulnerability to disaster-bearing bodies is high in the west and low in the east. The more developed areas in the east show high disaster prevention and mitigation capability, whereas this is weak in the west. In summary, dry-hot wind risk in Shandong province varies significantly by area. The medium- and high-risk areas are mainly in the west and central portions of the province.

Keywords: dry-hot wind disaster; risk zoning; Shandong province; natural disaster risk assessment principle

Citation: Wang, N.; Xue, X.; Zhang, L.; Chu, Y.; Jiang, M.; Wang, Y.; Yang, Y.; Guo, X.; Zhao, Y.; Zhao, E. Spatial Zoning of Dry-Hot Wind Disasters in Shandong Province. *Sustainability* **2022**, *14*, 3904. https://doi.org/10.3390/su14073904

Academic Editors: Salvador García-Ayllón Veintimilla and Hariklia D. Skilodimou

Received: 20 December 2021
Accepted: 22 March 2022
Published: 25 March 2022

1. Introduction

Dry-hot wind is a type of severe agricultural wind disaster with high temperature and low humidity. The late spring and early summer are the seasons when the direct sunlight in the northern hemisphere is the greatest, and the weather is sunny and drier before the arrival of the northern rainy season. Under the control of the dry air mass, the sky is fine, dry, and windy, and there are few opportunities for cloud formation to cause rain, so it is easy to form dry-hot winds. It causes low humidity in the air and water in the soil to evaporate, severely impacting crop yields and economic development [1,2]. The dry-hot wind disaster has become an important factor restricting the growth and development of crops by intensifying plant transpiration, resulting in insufficient water supply to the roots, causing an imbalance of water and nutrients in the plant. Typically, the leaves change color and normal physiological activities of the plant are damaged or inhibited, resulting in a significant shortening of the filling period of the crop, the high temperature brought by the dry hot wind after flowering will shorten the growing period of the seeds, and damage to protein and starch structure [3–8]. When the risk of dry-hot wind is low, wheat yield can decline by 5–10%, and in severe cases, by 20–30% or more [9].

In the 1950s, the Soviet Union conducted preliminary research on dry-hot wind disasters, which was at the leading level compared with other countries, mainly on the formation indicators and causes of dry-hot wind, spatial distribution, and that of disaster prevention and mitigation measures [10]. In recent years, more scholars have paid attention to the occurrence regularity of the number and intensity of dry-hot winds. Tavakol et al. [11] analyzed the spatial patterns and temporal changes of hot, dry, and windy events (HDWs) in the central United States for two time periods: 1949 to 2018 (70 years) and 1969 to 2018 (50 years). рокопец et al. [12] evaluated the dynamic variation characteristics of the total number and intensity of dry-hot winds over the lower Volga River. In contrast, Chinese scholars have conducted relatively more studies on the occurrence of hot-dry wind. Hou et al. [13] considered that the occurrence characteristics of dry-hot wind in the Hexi region of Gansu province and surrounding areas during June and July from 1960 to 2017 showed a tendency to decrease slowly and then increase rapidly. Cheng et al. [14] pointed out the occurrence frequency of different graded of the hazard in Henan province is on the rise. You et al. [15] analyzed the temporal-spatial distribution of dry-hot wind in the Hebei province winter wheat region during the past 35 years. The Huang-Huai-Hai region is the region with the most frequent occurrence of hot-dry winds in China, so it has attracted the attention of scholars. Li et al. [16] considered the annual average number of regional DHW events in the Huang-Huai-Hai Region showed a decreasing trend from 1961 to 2010 and increased in 2011–2018. Shi et al. [17] also pointed out the annual average of light and serve dry-hot wind in the Huang-Huai-Hai plain declined from 1963 to 2012. Zhao et al. [18] studied the spatial-temporal changes of dry-hot wind of winter wheat in the Huang-Huai-Hai plain under climate change. Wheat is the crop most affected by hot and dry wind, so scholars have focused on the analysis of the impact of hot and dry wind on wheat. Chen et al. [19] analyzed the influence of dry-hot wind on the wheat in Henan province and proposed that the number of dry-hot wind days has a significantly negative correlation with wheat meteorological yield. Shi et al. [20] proposed that the total number of dry-hot wind days had a negative correlation with thousand grain weight of winter wheat in Hebei province. Yang et al. [21] constructed a dry-hot wind risk assessment index system, including a dry-hot wind intensity risk index and a comprehensive disaster resistance index by using the meteorological data, yield and structure data, and development period data of the winter wheat observation station, and established a dry-hot wind risk assessment model. The risk of the dry-hot wind in the main winter wheat producing areas in North China was assessed, and the results showed that southeastern Hebei and northwestern Shandong were high-risk areas, while southern Henan, eastern Shandong, and eastern Hebei were low-risk areas. Based on the theory of agrometeorological disaster risk analysis, Chen et al. [22] analyzed the influence degree and risk probability of dry-hot wind on wheat yield in wheat production in Henan province by constructing a disaster function and using EOF and probability analysis methods. The results showed that dry-hot wind was the main disaster that influenced the high and stable yield of wheat in most of Henan province. In order to further study the resistance to dry-hot wind of wheat, Juraev et al. [23] planted varieties and lines in November in the late sowing period. The daily temperature, wind speed, and relative humidity were selected to compare and study the changes of plant height, ear length, and grain number per ear of wheat varieties in Casdalia and Surcandalia during their developmental stages. The results showed that dry-hot wind had a significant effect on the traits of wheat varieties and lines. Wang et al. [24] proposed a framework (DID) to quantify the impact of dry-hot wind on winter wheat in northern China and the framework can effectively detect winter wheat growing areas affected by dry-hot wind hazards. The estimated damage showed a notable relationship ($R^2 = 0.903$, $p < 0.001$) with the dry-hot wind intensity calculated from meteorological data. Deng et al. [25] comprehensively summarized the causes, protecting technology and answering tactics of dry-hot wind disasters.

With the deepening of research, scholars' research on hot-dry wind has shifted from occurrence to a disaster defense system, and disaster risk zoning is the basis for establishing a disaster prevention and mitigation system. Natural disaster zoning is the division of regions based on the temporal and spatial distribution of the occurrence and development of natural disasters; it can provide a scientific basis for regional disaster prevention and mitigation. In fact, strengthening the research on the comprehensive zoning of natural disasters is listed in China's 21st Century Agenda [26]. For example, Cheng et al. [27] established the index of yield loss risk of dry-hot wind and integrated to zone the comprehensive hazard risk in Henan province. Wu et al. [28] developed a new regionalization method, wherein type one is high temperature and low humidity and type two is immature death after rain, for the dry-hot windy days regionalization in the NCP. According to the risk assessment theory of natural disasters, some scholars have performed fine zoning and assessment of dry-hot wind risks in the winter wheat region of Henan province [29] and spring wheat region of Inner Mongolia [30] from four aspects: risk of disaster-causing factors, vulnerability of disaster-pregnant environment, exposure of disaster-bearing body, and ability of disaster prevention and mitigation. As a major agricultural province in China, Shandong province has a winter wheat planting area up to 4.003×10^6 hm^2 and an annual yield of 2.472×10^7 T [31]. Shandong province is located in the Huang-Huai-Hai region of China, which is the region most affected by hot-dry wind disasters. However, there are few studies on the risk of hot-dry wind in Shandong province. Therefore, it is urgent to carry out the research on the risk regionalization of hot-dry wind in Shandong province. Based on disaster risk theory, we analyze dry-hot wind disaster from four perspectives: risk, exposure, vulnerability, and disaster prevention and mitigation capability. A dry-hot wind disaster risk index model is established, and spatial zoning of dry-hot wind disasters in Shandong is examined using ARC-GIS spatial analysis, weighted comprehensive evaluation method, and analytic hierarchy process. The aims of this study are to comprehensively assess and zone the risk of dry-hot wind, so as to ensure the safety of wheat production. In order to reduce disaster risk and provide reference for agricultural production layout and scientific decision-making, dry-hot wind risk zoning is of great significance to regional agricultural management and production, people's lives, and food security.

2. Materials and Methodology

2.1. Study Area

Shandong province is located on the east coast of China and the lower reaches of the Yellow River (114°48′ E–122°42′ E and 34°23′ N–38°17′ N), as shown in Figure 1. The total land area is 157,900 km^2. The climate type is warm temperate monsoon. Precipitation is concentrated, and rain and heat occur in the same season. Spring and autumn are short, while winter and summer are long. The annual average temperature range is 11–14 °C and the annual average precipitation range is 550–950 mm. The rainfall season is unevenly distributed, with 60–70% of annual precipitation in summer. Landform types include plains, terraces, hills, and mountains. There is a dense river network in the region, including the Yellow River, Huaihe River, Haihe River, and smaller rivers in the central and southern mountainous area.

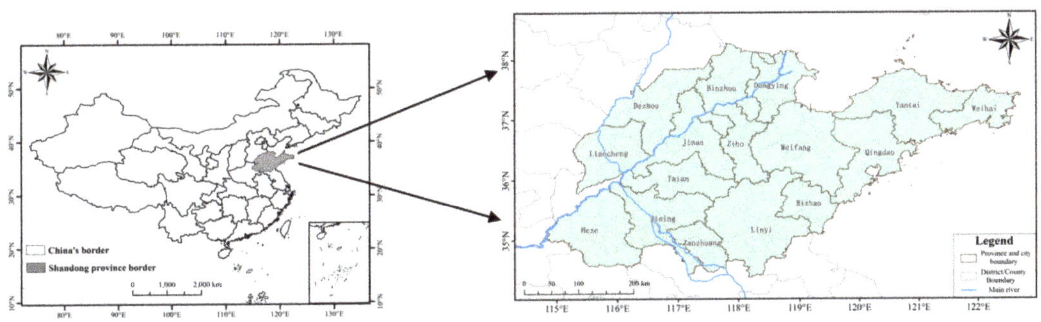

Figure 1. Administrative division of Shandong province, China.

2.2. Data Sets

The meteorological data used here include wind speed, temperature, and relative humidity from 1991 to 2020. The aspect, altitude, slope, river network density, and land-use type are obtained from the Meteorological Information Center of Shandong Meteorological Bureau. Total GDP, total population, administrative area, wheat planting area, crop planting ratio, population density, education level, per capita GDP, and crop planting area are from the 2018–2020 Statistical Yearbook of Shandong province.

2.3. Methods

Based on the natural disaster risk theory, this paper constructs a scientific framework as follows (Figure 2):

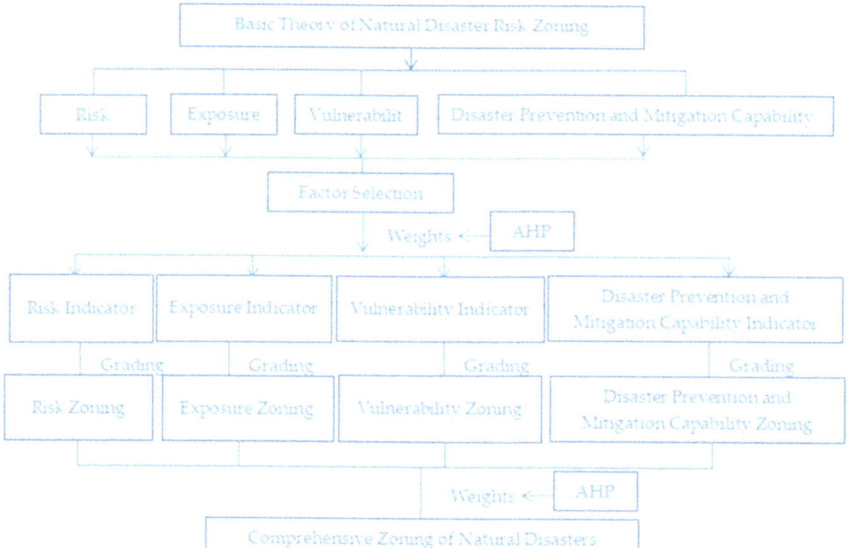

Figure 2. Scientific framework or spatial zoning of hot-dry wind disasters in Shandong province.

2.3.1. Basic Theory of Disaster Risk Assessment

Based on the theory of natural hazard risk formation [32], meteorological hazard risk is formed by the combination of four components: hazard (causative factor), exposure (carrier), vulnerability (carrier), and prevention and mitigation capacity. Each factor is in turn composed of a series of subfactors. The expressions are:

$$\text{Disaster risk index} = \text{f (hazard, exposure, vulnerability, disaster prevention and mitigation capacity)} \quad (1)$$

Hazardous factors: Hazardous factors include meteorological factors and environmental sensitivity. All meteorological factors that may lead to disasters can be called meteorological factor hazards; the sensitivity of the pregnant environment refers to the degree of strengthening or weakening of meteorological factors in the natural surface environment.

Exposure of disaster-bearing body: Disaster-bearing body is the object of disaster-causing factors and is the entity that bears the disaster. Exposure of the hazard-bearing body is the result of the interaction between the hazard-causing factor and the hazard-bearing body, and the exposure of the hazard-bearing individual to the hazard-causing factor.

Vulnerability of the disaster-bearing body: A disaster can be formed only when it acts on the corresponding object, i.e., human beings and their socioeconomic activities. Specifically, it refers to the degree of hazard or loss caused by the potential risk factors for all objects that may be threatened by the disaster-causing factors that exist in a given hazard area, and its combination reflects the degree of loss from meteorological disasters.

Prevention and mitigation capacity: This refers to various management measures and countermeasures used to prevent and mitigate meteorological hazards, including management capacity, mitigation input, and resource preparation. The more proper management measures and strong management capacity, the less potential losses may be suffered and the less risk of meteorological disasters.

Based on the above theory, the hierarchical structure model of dry-heat wind risk assessment was constructed. The risk index values of hazard factor, exposure factor, vulnerability factor, and disaster prevention and mitigation capacity of dry-heat wind disaster are calculated as *Ya*, *Yb*, *Yc*, and *Yd*, respectively, by the weighted comprehensive evaluation method. Through the natural disaster risk index formula, combined with the dry-heat wind disaster assessment index system of Shandong province, its disaster risk index model is:

$$F = W_a \times Y_a + W_b \times Y_b + W_c \times Y_c + W_d \times Y_d \quad (2)$$

In which, F is the dry heat wind hazard risk index, which indicates the degree of dry heat wind hazard; the larger the value of F, the higher the risk and the opposite the lower. y is the risk index value of hazard factor, exposure factor, vulnerability factor, and prevention and mitigation capacity of dry heat wind hazard. w is the weight of each index.

Based on the composition of meteorological data, topography, and socioeconomic elements, Figure 3 below shows the hierarchical structure model of dry-heat wind disaster risk assessment in Shandong province.

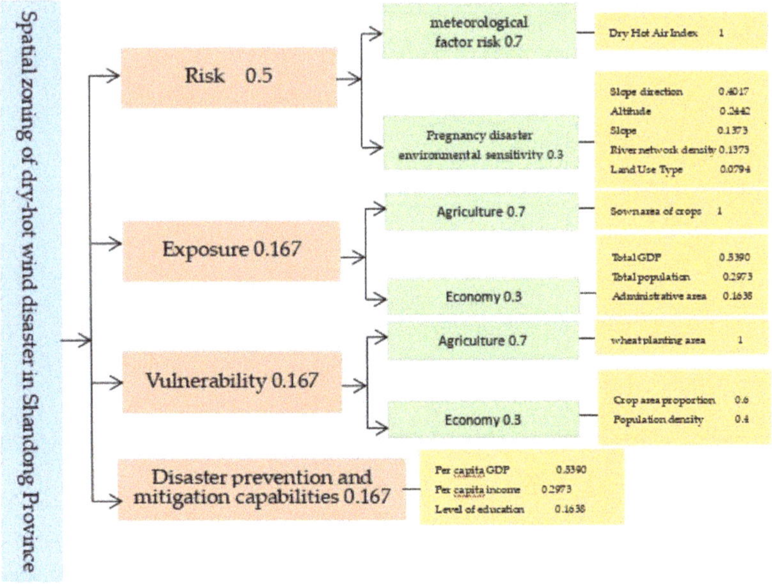

Figure 3. Hierarchical model of dry hot air risk assessment.

2.3.2. Weighted Comprehensive Evaluation Method

The weighted comprehensive evaluation method is a method that solves the "bottom-up" indexes in the risk hierarchy analysis and evaluation model, which considers the degree of influence of each factor on the overall object and integrates the strengths and weaknesses of each specific index and uses a numerical index to focus on the strengths and weaknesses of the entire evaluation object. This method is especially suitable for comprehensive analysis and evaluation of technology, strategy, or programs and is one of the most commonly used calculation methods. Its expression is:

$$Y_i = \sum_{j=1}^{m} \lambda_j X_j \quad i = 1,\ 2,\ \ldots,\ m;\ j = 1,\ 2,\ \ldots,\ m \tag{3}$$

where Y_i denotes the disaster risk index, i denotes hazard, exposure, vulnerability, and disaster prevention and mitigation capacity, respectively; X_j is the factor affecting hazard, exposure, vulnerability, and disaster prevention and mitigation capacity, and λ_i is the weight value ($0 \leq \lambda_j \leq 1$).

For the comprehensive risk index of natural disasters, the expressions are:

$$Y = \sum_{i=1}^{n} W_i Y_i \quad i = 1 \tag{4}$$

where Y denotes the comprehensive disaster risk index; Y_i is the hazard index, exposure index, vulnerability index, and disaster prevention and mitigation capacity index, and W_i is the weight value.

The stronger the disaster prevention and mitigation capacity is, the smaller the comprehensive risk index is, so the "negative sign" is used.

Where λ_j and W_i are determined using hierarchical analysis, as described in research Section 2.3.3, each factor in the formula needs to be standardized because of different dimensions; see research Section 2.3.4 for details.

2.3.3. Analytic Hierarchy Process

Analytic Hierarchy Process (AHP) is a simple method for making decisions on some more complex and vague problems, especially for those problems that are difficult to fully quantitatively analyze [33]. This paper uses the operation principle of the analytic hierarchy process and uses the 1–9 scale method given by Saaty to construct the judgment matrix for the pairwise relationship of the influence factors. The pairwise comparison of all influence factors determines the weight of each influence factor, which avoids the result error caused by the subjectivity of the expert. The qualitative comparison scale values between the two influencing factors are shown in Table 1 below:

Table 1. Scale of AHP analysis method.

Scale b_{ij}	Definition
1	The i factor is as important as the j factor.
3	The i factor is slightly more important than the j factor.
5	The i factor is more important than the j factor.
7	The i factor is much more important than the j factor.
9	The i factor is absolutely more important than the j factor.
2, 4, 6, 8	Between the noted levels.

Solve the maximum eigenvector value of the judgment matrix and its corresponding eigenvector by the sum-product method and check the consistency of the matrix (the following formula): After passing, solve it by the sum-product method.

$$CI = \frac{\lambda_{\max} - n}{n - 1} = \frac{-\sum_{i=1}^{n} \lambda_i}{n-1} \tag{5}$$

$$CR = \frac{CI}{RI} < 0.1 \tag{6}$$

In the formula, CI is the consistency index of the judgment matrix, $\lambda_{\max}$ is the largest characteristic root of the matrix, n is the order of the discrimination matrix, CR is the random consistency index of the judgment matrix, and *RI* is the average random consistency index of the discrimination matrix. The values of *RI* are shown in Table 2:

Table 2. Numerical values of random consistency index *RI*.

M	**1**	**2**	**3**	**4**	**5**	**6**	**7**	**8**	**9**	**10**	**11**
RI	0.00	0.00	0.58	0.90	1.12	1.24	1.32	1.41	1.45	1.49	1.51

This paper adopts the Analytic Hierarchy Process (AHP), taking the sensitivity of the disaster-pregnant environment as an example, and constructs the judgment matrix of each index and the calculation results are shown in Table 3. Since $CR < 0.1$, the matrix passed the consistency test.

Table 3. Judgment matrix and weights of various perceptual factors.

	I	**II**	**III**	**IV**	**V**	**Weights (W)**	**Matrix Product (AW)**	**AW/W**	λ_{max}	***CI***	***CR***
I	1	2	3	3	4	0.402	2.03	5.05	5.033	0.008	0.007
II	1/2	1	2	2	3	0.244	1.23	5.04	$\lambda = \sum(AW/W)/n$		
III	1/3	1/2	1	1	2	0.137	0.689	5.03	$CI = (\lambda - n)/n - 1$		
IV	1/3	1/2	1	1	2	0.137	0.689	5.03	$RI = 1.12$		
V	1/4	1/3	1/2	1/2	1	0.079	0.399	5.05	$CR = CI/RI$		
								25.18			

Note: In the table, I. Slope direction, II. Elevation, III. Slope, IV. River network density, V. Land-use type.

Similarly, the weights of each factor of hazard, exposure, vulnerability, disaster prevention and mitigation capacity, and combined disaster risk were obtained as shown in Figure 3.

2.3.4. Standardization

In the process of zoning, the different magnitudes of the selected factors lead to a large difference in order of magnitude; for example, the total population is 3,923,000 people, while the total GDP is about 302.22 billion yuan, so when calculating the hazard index of disaster-causing factors, normalization is required so that the values of each factor are between 0 and 1. Furthermore, when assessing the hazard of disaster-causing factors, exposure of disaster-bearing bodies, vulnerability of disaster-bearing bodies, and disaster prevention and mitigation capacity, the larger the number of influencing factors, the larger the hazard of disaster-causing factors, exposure of disaster-bearing bodies, vulnerability of disaster-bearing bodies, and disaster prevention and mitigation capacity, while some factors are the opposite. Therefore, in the assessment process, the criterion of a great value or the criterion of a very small value should be standardized first, and the formula is as follows. For example, the greater the dry and hot wind index, the greater the hazard of disaster-causing factors, so choose the great value standardization for the dry and hot wind index, and choose Equation (7); for example, the greater the slope, the less the sensitivity of disaster-preventing environment, so standardize the slope for the small value, and choose Equation (8).

Maximum standardization:

$$X'_{max} = \frac{|X_{ij} - X_{min}|}{X_{max} - X_{min}} \tag{7}$$

Minimum standardization:

$$X'_{min} = \frac{|X_{max} - X_{ij}|}{X_{max} - X_{min}} \tag{8}$$

where X_{ij} is the index number of the *j*-th factor of the *x* factor; X'_{max} and X'_{min} are the dimensionality of X_{ij}; X_{max} and X_{min} are the minimum and maximum values in the index sequence.

2.3.5. Arc-GIS Spatial Analysis

The spatial analysis methods involved in this paper include Kriging interpolation, spatial reclassification, spatial raster calculation, and slope extraction. The raster resolution is 100 m × 100 m.

Kriging interpolation is a method of unbiased optimal estimation of regionalized variables in a limited region based on the variogram theory and structural analysis. Not only can it reflect the spatial structure characteristics of variables, but it can also reflect the random distribution characteristics of variables [34]. There are many factors that affect the spatial change of natural geographical elements. The comprehensive action of these factors forms the zonal regularity on the Earth's surface, and the natural geographical elements are also disturbed by various random factors. Geographers try to explore the regional regularity of their natural geographical elements and strive to minimize the interference of random factors. Kriging interpolation can minimize the interference of random factors with the help of the optimal method. Therefore, this method can be used to analyze the changes of elements in the interpolation space, such as the change of temperature, the regional distribution law of water quality, vegetation, soil, and other elements with zonal distribution law [35]. Based on the above research conclusions, this paper uses the Kriging interpolation method to interpolate the zonal geographical elements. According to these research results, this paper applies the Kriging interpolation method to zoning elements and zoning results.

In this paper, the natural breakpoint method is used for classification in risk, exposure, vulnerability, disaster prevention and mitigation capacity and comprehensive risk. In fact, there are many classification methods, such as equal division, standard deviation classification, and so on. Fu et al. [36] and others have concluded that the natural breakpoint method can adequately extract the useful information contained in the index, so as to establish a more reasonable and accurate index evaluation space. At the same time, the risk, exposure, vulnerability, disaster prevention, and reduction ability and comprehensive risk are divided into low, medium, and high levels, mainly because if there are too many levels, the spatial expression effect is not clear enough.

3. Results

3.1. Spatial Distribution of Dry-Hot Wind Risk

The risk of dry-hot wind includes the risk of meteorological factors and the disaster environment sensitivity, and the risk of meteorological factors is the main factor constituting the risk of dry-hot wind. Disaster environment sensitivity can aggravate or reduce the risk of meteorological factors. At the same time, this paper also consulted relevant experts, such as Shandong Meteorological Bureau and the Department of Agriculture, and gave the risk of meteorological factors and the disaster environment sensitivity the weights of 0.7 and 0.3, respectively. The weight in the composition of dry-hot wind meteorological factor risk index is explained in the fourth question (Figure 3).

3.1.1. Zoning of Meteorological Factor Risk

According to the ground meteorological observation specification of the People's Republic of China (Table 4), soil relative humidity at 20 cm, daily maximum temperature (°C) air relative humidity at 14:00 (%) and wind speed at 14:00 (m/s) were selected as the grade indicators of dry-hot wind. Since the meteorological station does not observe 20 cm soil relative humidity, there is no 20 cm soil relative humidity in the meteorological observation records. Considering that precipitation is the main factor affecting 20 cm soil relative humidity, the maximum process precipitation in early and middle May was used to replace 20 cm soil relative humidity in this study. The classification standard of dry-hot wind used in this study is shown in Table 5.

Table 4. High temperature and low humidity type dry-hot wind grade indicators.

Area	20 cm Soil Relative Humidity	Mild			Medium			Severe		
		Daily Maximum Temperature (°C)	Air Relative Humidity at 14:00 (%)	Wind Speed at 14:00 (m/s)	Daily Maximum Temperature (°C)	Air Relative Humidity at 14:00 (%)	Wind Speed at 14:00 (m/s)	Daily Maximum Temperature (°C)	Air Relative Humidity at 14:00 (%)	Wind Speed at 14:00 (m/s)

Table 5. Dry-hot wind generation index.

Maximum Process Precipitation in Early and Mid-May	The Time Period is Mid to Late May								
	Mild			Medium			Severe		
	Daily Maximum Temperature (°C)	Air Relative Humidity at 14:00 (%)	Wind Speed at 14:00 (m/s)	Daily Maximum Temperature (°C)	Air Relative Humidity at 14:00 (%)	Wind Speed at 14:00 (m/s)	Daily Maximum Temperature (°C)	Air Relative Humidity at 14:00 (%)	Wind Speed at 14:00 (m/s)
<25 mm	31	≤30	≥2	≥32	≤25	≥3	≥35	≤25	≥3
≥25 mm	≥33	≤30	≥3	≥35	≤25	≥3	≥36	≤25	≥3

The harmful degrees of the days of mild, medium, and severe dry-hot wind are different. The more instances of severe dry-hot wind that happen, the stronger the influence of dry-hot wind is in the area. Therefore, when constituting the dry-hot wind index, different weights must be given to the days of mild, medium, and severe dry-hot wind. According to the basic principle of AHP, when calculating the weight, first, the ratio matrix is constructed according to the scale grade table. Since the severe dry-hot wind is very important compared with the mild dry-hot wind, the ratio is assigned as one. The severe dry-hot wind is slightly more important than the medium dry-hot wind, so the ratio is assigned as two. The medium dry-hot wind is slightly more important than the mild dry-hot wind, so the ratio is assigned as three, so the ratio matrix is formed (as shown in the Table 6 below). The consistency test index *CR* of the matrix was calculated as *CR* = 0.08. Because *CR* < 0.1, the matrix passed the consistency test. Using the sum product method, the weights of the times of mild, medium, and severe dry-hot wind are 0.2, 0.3, and 0.5, respectively.

Table 6. Judgment matrix and weights of various perceptual factors.

	I	II	III	Weights (W)	Matrix Product (AW)	AW/W	λ_{max}	*CI*	*CR*
I	1	2	3	0.5	1.62	3.01	3.01	0.004	0.008
II	1/2	1	2	0.3	0.89	3.00	$\lambda = \sum(AW/W)/n$ $CI = (\lambda - n)/n - 1$		
III	1/3	1/2	1	0.2	0.49	2.99 9	$RI = 0.58$ $CR = CI/RI$		

Note: I. Severe dry-hot wind days, II. Moderate dry-hot wind days, III. Mild dry-hot wind days.

According to Table 5, the number of hot-dry days at all levels from 1991 to 2020 was calculated. In the risk zoning of dry-hot wind days, the harm degrees of mild, medium, and severe dry-hot wind days are different. The more severe dry-hot wind days, the stronger the impact of dry-hot wind. Therefore, the mild, medium, and severe dry-hot wind days constitute a comprehensive risk index, and different weights are assigned to the mild, medium, and severe dry-hot days, which are 0.2, 0.3, and 0.5, respectively.

$$R = 0.2D_1 + 0.3D_m + 0.5D_s \tag{9}$$

where R is the comprehensive index of dry-hot wind (d), D_1 is average number of days of mild dry-hot wind disaster during the 30-year study period (d), D_m is the average number of days of medium dry-hot wind disaster during the 30-year study period (d), and D_s is the average number of days of severe dry-hot wind disaster during the 30-year study period (d).

To sum up, mild, medium, and severe dry-hot wind refers to the results obtained by combining different meteorological indicators. The comprehensive index is a linear addition of the days of mild, medium, and severe dry-hot wind, which is formed for the risk zoning index.

The risk of dry-hot wind is composed of two parts: the risk of meteorological factors and the pregnancy disaster environmental sensitivity. Dry-hot wind is a type of meteorological disaster with high temperature, low humidity, and a certain wind force. The topographic factors in the pregnancy disaster environmental sensitivity affect the hazard degree of dry-hot wind. Since meteorological factors are more important compared with topographic factors, the risk of meteorological factors is given a higher weight. According to the scale of the AHP analysis method, when the *i* factor is more important than the *j* factor, the weight of the *i* factor is given to 0.7, and the weight of the *j* factor is given to 0.3. Therefore, the weight of the risk of meteorological factors is assigned 0.7, and the weight of the pregnancy disaster environmental sensitivity is assigned 0.3.

The spatial distribution of the average number of days of dry-hot wind in Shandong province over the past 30 years is shown in Figure 4. The spatial distribution of the days of mild dry-hot wind is similar to that of moderate dry-hot wind. High values are distributed mainly in the central area, and low values are found in the east and west. The highest values are 1.1 d and 0.4 d, and the lowest value is 0. Areas with a high number of days with severe dry-hot wind disasters are concentrated in the north, and the highest figure is 0.4 d. In the remaining areas, the number of days with severe dry-hot wind disaster is significantly reduced, with the lowest value at 0. The spatial distribution of the index varies significantly. The high-value areas are concentrated in Weifang, Zibo, Jinan, Binzhou, and Dongying. The highest number of days is 1.9, and the lowest number of days is 0.

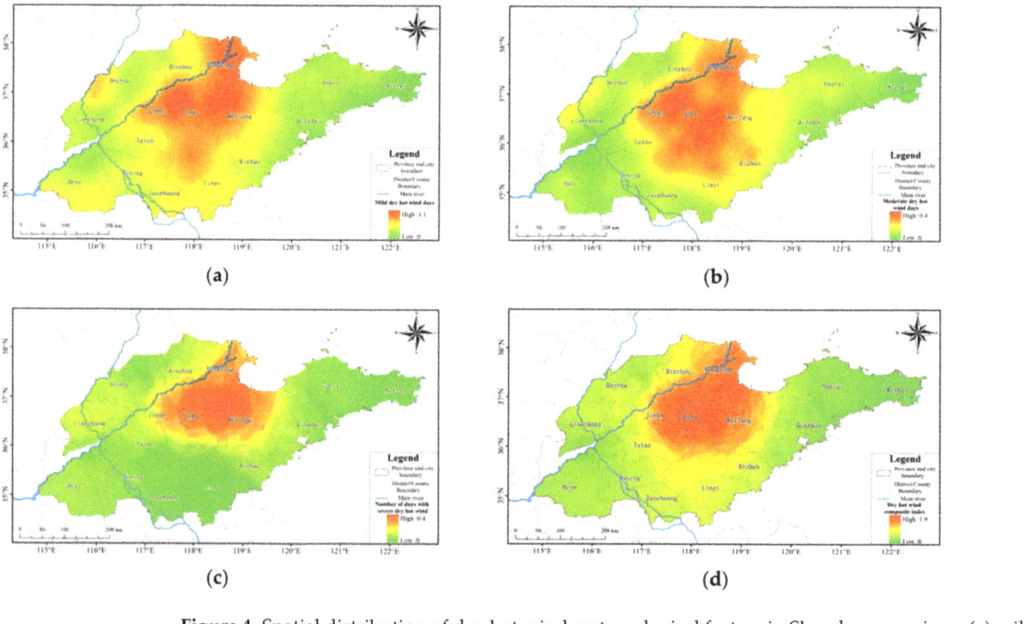

Figure 4. Spatial distribution of dry-hot wind meteorological factors in Shandong province: (**a**) mild dry-hot wind days; (**b**) moderate dry-hot wind days; (**c**) number of days with severe dry-hot wind; (**d**) dry-hot wind composite index.

3.1.2. Zoning of Disaster Environment Sensitivity

Aspect, altitude, slope, river network density, and land-use type are selected as the zoning indices for disaster environment sensitivity (Figure 5). The south slope has longer sunshine duration and higher temperature, so the closer to the south slope, the more dangerous the hot-dry wind will be. The southwest slope and southeast slope also receive more solar radiation relatively, so the sensitivity is also higher. The east slope warms faster than the west slope, so the sensitivity is higher than that of the west slope; thus, that of the northeast slope is also slightly higher than that of the northwest slope. The temperature of the north slope is the lowest, so the sensitivity is the lowest. The sensitivity of slope-free area is slightly lower than that of the south slope and higher than that of the east slope. Therefore, the ordering and scoring of the slope aspect are shown in Table 7. Temperature decreases with increased altitude; therefore, the higher the altitude, the lower the temperature. The greater the slope is, the less solar radiation per unit an area receives, and therefore the lower the sensitivity would be [37]. The slope range in Shandong province is 0–49.5°, so every 10° is assigned a grade, as shown in Table 8. In addition, the higher the river network density, the higher the air humidity, and the less the impact from dry-hot wind. Compared with unused land, woodland and grassland have better water conserving capacity and higher air humidity, which helps to reduce the influence of dry-hot wind disasters. The scores for different land use types are shown in Table 9.

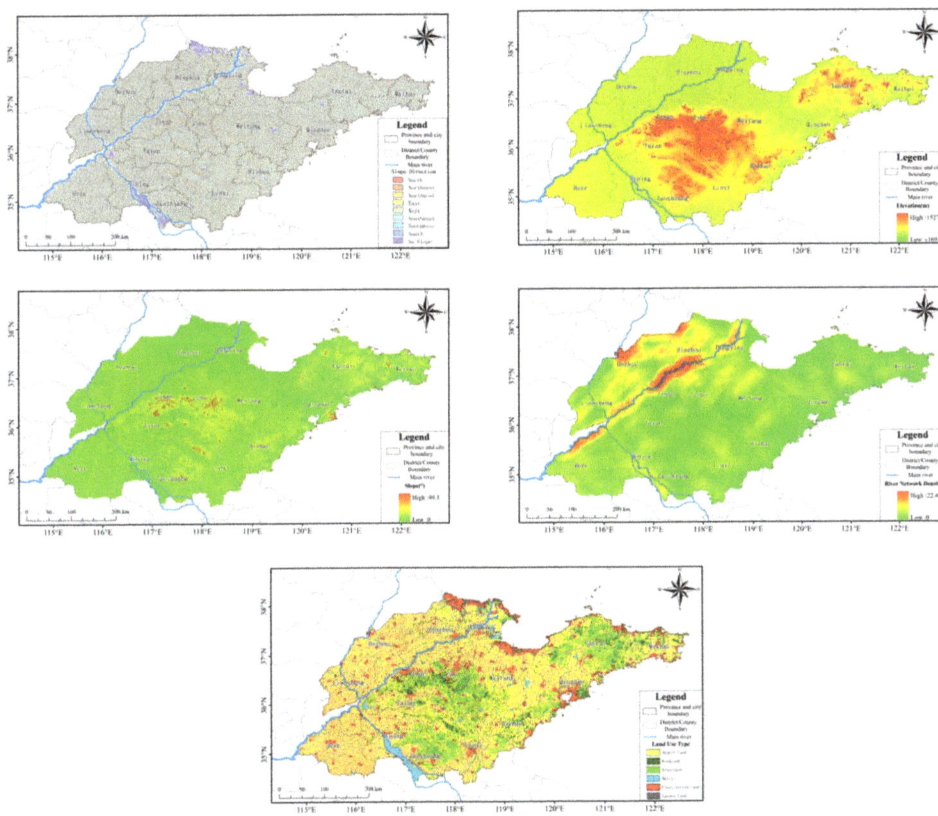

Figure 5. Spatial distribution map of slope aspect, altitude, slope, river network density, and land-use type in Shandong province.

Table 7. Grading and score of slope direction.

Aspect	South	South West	South East	No Slope	East	West	Northeast	North West	North
Score	8	7	6	5	4	3	2	1	0

Table 8. Grading and score of the slope.

Slope	**50–40°**	**40–30°**	**30–20°**	**20–10°**	**10–0°**
Score	5	4	3	2	1

Table 9. Land-use type scores.

Land Use Type	**Arable land**	**Woodland**	**Grassland**	**Waters**	**Construction Land**	**Unused Land**
Score	3	2	4	1	5	6

Adding the value of each factor according to its weight, the spatial distribution of the environmental sensitivity risk in Shandong province is obtained, as shown in Figure 6. There is little spatial difference in the environmental sensitivity from dry-hot wind disasters in Shandong province, yet the spatial distribution is uneven and shows a high degree of fragmentation.

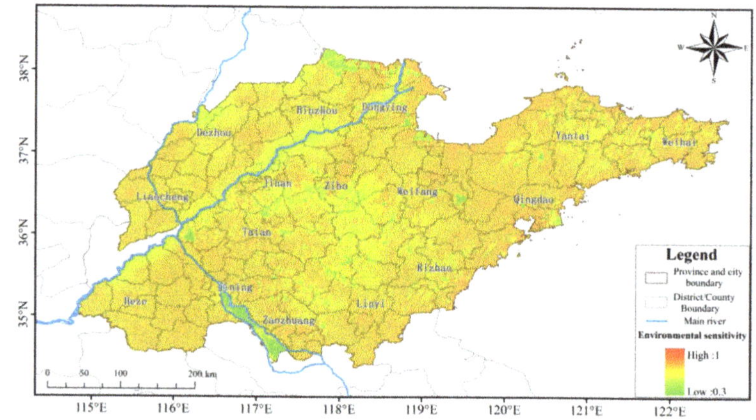

Figure 6. Spatial distribution of environmental sensitivity to dry-hot wind pregnancy in Shandong province.

3.1.3. Zoning of Dry-Hot Wind Risks

Meteorological factor risk and disaster environmental sensitivity are added according to their weight, and then classified to obtain the spatial distribution map of dry-hot wind disasters in Shandong province, as shown in Figure 7. Note that dry-hot wind disasters in Shandong province have clear spatial characteristics. The dry-hot winds at all levels in the north central area of Shandong province are all high-value areas, and the slope in the north central area is dominated by the southern slope with low altitude, small terrain slope, low river network density, and a large proportion of construction land. The sensitivity of the north central is higher. To sum up, the risk of dry-hot wind is higher in the north central area of Shandong province, with a distribution area of 32,587.6 km^2, accounting for 20.6% of the total land area. These areas include Dongying, Weifang, Zibo, east of Jinan, and south of Binzhou. The medium-risk regions are distributed in the peripheral area of the high-risk regions, including northwest of Binzhou, west of Jinan, Tai'an, Linyi, Rizhao, and east of Weifang. The medium-risk area comprises 6649.7 km^2, accounting for 42.1% of the total land area. The low-risk regions are in the west, southwest, and east, including Weihai, Yantai, and Qingdao on the Jiaodong Peninsula, and Heze, Jining, and Zaozhuang in the southwest. The low-risk area comprises 28,820.6 km^2, accounting for 37.3% of the total land area.

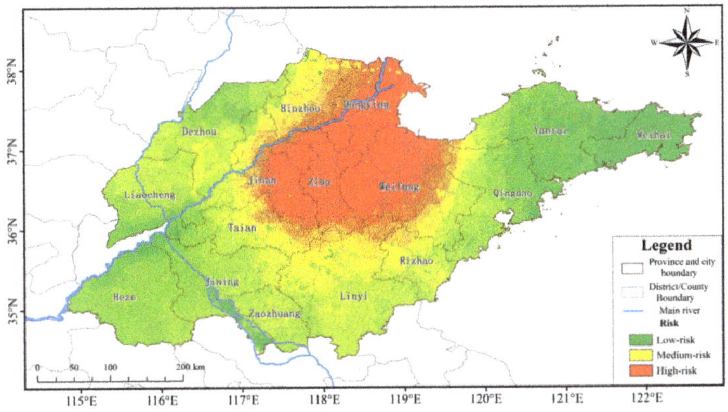

Figure 7. Spatial distribution of dry-hot wind hazards in Shandong province.

3.2. Spatial Distribution of Exposure of Disaster-Bearing Bodies

Exposure includes agricultural exposure and economic exposure. Crops are directly impacted by dry-hot wind disasters, and planting area directly reflects the degree of exposure. Thus, the sown area of crops is selected as the index of agricultural exposure. When a dry-hot wind disaster occurs, the higher the total GDP, the greater the total population, and the larger the administrative area, the stronger exposure will be from the disaster. Therefore, total GDP, total population, and administrative area are selected as economic exposure indices and added according to their weights in Figure 3 to obtain a spatial distribution map of economic exposure. Then, agricultural exposure and economic exposure are added with a weight of 0.7 and 0.3, respectively, and then classified to obtain the spatial zoning results of dry-hot wind exposure in Shandong province (Figure 8). Note that exposure to dry-hot wind disasters shows a clear spatial distribution pattern: generally high in the south and low in the north. High-exposure areas include Heze, Jining, Linyi, and Weifang, and medium-exposure areas are mainly in Dezhou, Liaocheng, Tai'an, Jinan, Yantai, and Qingdao. Low-exposure areas include Binzhou, Dongying, Zibo, Zaozhuang, Rizhao, and Weihai. The areas of high, medium, and low exposure are 56,581.8 km^2, 62,815.1 km^2, and 38,503.2 km^2, accounting for 35.8%, 39.8%, and 24.4% of the total land area, respectively.

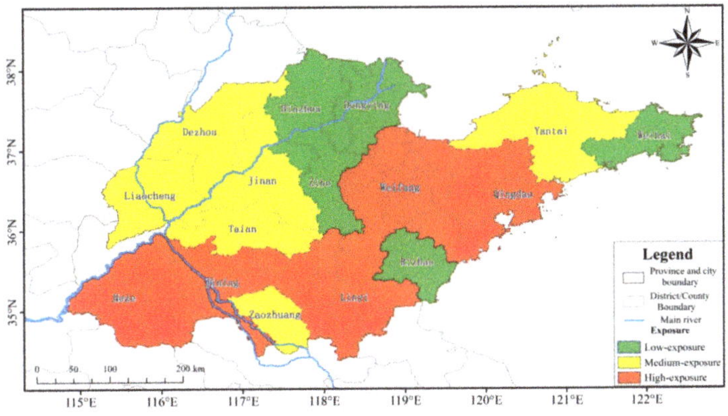

Figure 8. Spatial distribution of exposure of dry-hot wind-bearing bodies in Shandong province.

3.3. Spatial Distribution of Vulnerability of Disaster-Bearing Bodies

Vulnerability includes agricultural vulnerability and economic vulnerability. Dry-hot wind disasters hinder the grain filling of wheat and forces it to ripen, affecting its maturation, and the thousand-grain weight is significantly reduced, resulting in a serious reduction in wheat production [38]. Therefore, the larger the wheat planting area, the greater the impact of dry-hot wind disasters. The present study uses the wheat planting area as the index of agricultural vulnerability. As for economic vulnerability, crop area proportion and population density are selected as indices of economic vulnerability. Crop area proportion refers to the ratio of crop planting area to administrative area. The larger the crop area proportion, the higher the vulnerability. The higher the population density, the higher the vulnerability. Crop area proportion and population density are spatially superimposed, as shown in Figure 3, to obtain the distribution map of economic vulnerability (not shown). Agricultural vulnerability and economic vulnerability are then added with a weight of 0.7 and 0.3, respectively, and classified to obtain the spatial zoning results of the vulnerability of disaster-bearing bodies (Figure 9). Note that vulnerability is high in the west and low in the east. The high-vulnerability areas include Heze, Liaocheng, and Dezhou in the west. The low vulnerability areas are Dongying, Zibo, Rizhao, Yantai, and Weihai. The remaining cities show medium vulnerability. The areas of high, medium, and low vulnerability are 31,503.3 km^2, 88,001.4 km^2, and 38,395.3 km^2, accounting for 20.0%, 55.7%, and 24.3% of the total land area, respectively.

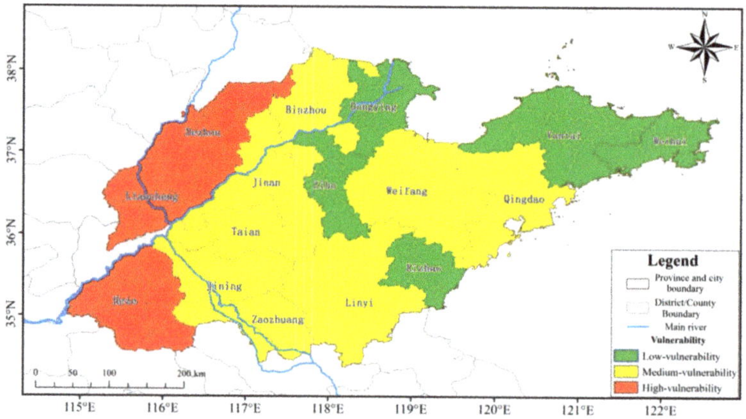

Figure 9. Spatial distribution of vulnerability of dry-hot wind-bearing bodies in Shandong province.

3.4. *Spatial Distribution of Disaster Prevention and Mitigation Capability*

Disaster prevention and mitigation capability refer to management measures and countermeasures used to prevent and reduce meteorological disasters. The higher the economic level of a place, the higher the disaster prevention and mitigation capability [39]. The higher the per capita GDP, per capita income, and level of education, the stronger the ability to defend against and respond to dry-hot wind disasters. Per capita GDP, per capita income, and education level are selected as indices of disaster prevention and mitigation capability. In accordance with the weights shown in Figure 3, the indices are added and then classified to obtain the spatial zoning results of the disaster prevention and mitigation capability in Shandong province (Figure 10). Note that the disaster prevention and mitigation capability of dry-hot wind disasters is high in the east and low in the west. Areas with high disaster prevention and mitigation capability are located mainly in Yantai, Weihai, Jinan, Qingdao, Jinan, and Dongying. Regions with low disaster prevention and mitigation capability include Dezhou, Liaocheng, Heze, Zaozhuang, and Linyi. The remaining areas are medium-risk regions. Areas with high, medium, and low disaster prevention and mitigation capability comprise 48,613.6 km^2, 56,086.5 km^2, and 53,199.8 km^2, accounting for 30.8%, 35.5%, and 33.7% of the total area, respectively.

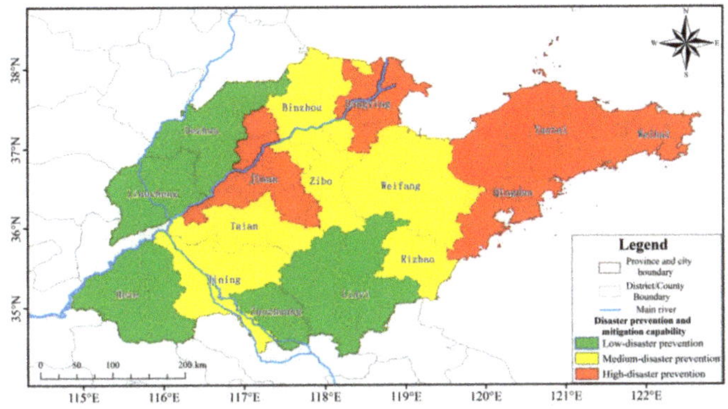

Figure 10. Spatial distribution of disaster prevention-mitigation capacity of dry and hot winds in Shandong province.

3.5. Spatial Distribution of Comprehensive Risk of Dry-Hot Wind Disasters

The four factors of the zoning results of risk, exposure of disaster-bearing bodies, vulnerability of disaster-bearing bodies, and disaster prevention and mitigation capability are spatially superimposed according to the weights shown in Figure 3. This obtains the zoning results of the comprehensive risk of dry-hot wind disaster in Shandong province, as shown in Figure 11. Note that the comprehensive risk of dry-hot wind disaster differs substantially in different areas. The medium- and high-risk regions are located mainly in the west and central areas, with low-risk regions in the east. The overall spatial distribution shows a strong degree of fragmentation. Table 10 shows the areas of the high-, medium-, and low-risk regions in each city. In Jining, Weifang, Heze, and Linyi, the area of high risk is largest. In Dongying, Qingdao, Rizhao, Weihai, Yantai, and Zaozhuang, there are no high-risk regions. Jining, Jinan, Tai'an, Binzhou, Linyi, and Dongying are medium-risk regions. The area of medium risk in Heze and Weihai is zero. In Yantai, Qingdao, and Weihai, the area of low risk is largest. In summary, there are no low-risk regions in Jining, Dezhou, Heze, Liaocheng, Linyi, and Weifang, only medium- and high-risk areas. The area of high-, medium-, and low-risk regions for dry-hot wind disasters is 64,076.7 km^2, 58,474.3 km^2, and 35,349.0 km^2, accounting for 40.6%, 37.0%, and 22.4% of the total land area, respectively.

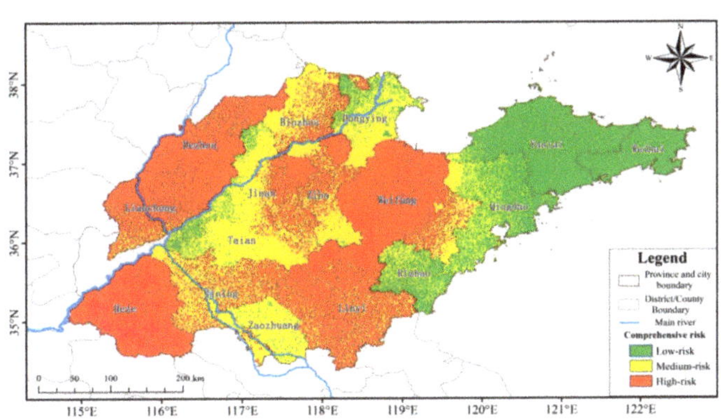

Figure 11. Spatial distribution of integrated riskiness of dry-hot wind in Shandong province.

Table 10. Area of high-, medium-, and low-risk areas in Shandong province by city.

	High-Risk Area		Medium-Risk Area		Low-Risk Area	
	Area (km^2)	Ratio (%)	Area (km^2)	Ratio (%)	Area (km^2)	Ratio (%)
Binzhou City	4028.6	2.6	5582.1	3.6	1	0
Dezhou City	9594.5	6.1	1019.1	0.7	0	0
Dongying City	0	0	5104.5	3.3	2064.5	1.3
Heze City	12,083.8	7.7	0	0	0	0
Jinan City	1955	1.2	7069.6	4.5	1443.9	0.9
Jining City	297,514	1.9	8184	5.2	0	0
Liaocheng City	6488.6	4.1	2263.2	1.4	0	0
Linyi City	11,891.2	7.6	5329.6	3.4	0	0
Qingdao City	0	0	3915.1	2.5	7276.5	4.6
Rizhao City	0	0	1668	1.1	3679.8	2.3
Taian City	375.2	0.2	6609.2	4.2	875.1	0.6
Weihai City	0	0	0	0	5687.5	3.6
Weifang City	13,089.7	8.4	3027.8	1.9	0	0
Yantai City	0	0	323.5	0.2	13,652.5	8.7
Zaozhuang City	0	0	3958.7	2.5	581	0.4
Zibo City	1595	1	4420	2.8	87.2	0.1

4. Discussion

(1) This study evaluates and classifies the risk of dry-hot wind disasters in Shandong province from the perspectives of risk, exposure, vulnerability, disaster prevention and mitigation capability, and comprehensive risk. Similar studies are rare. Thus, the results of this study provide a framework for related research in this area. The comprehensive risk of dry-hot wind disasters obtained here is compared with existing studies. Li et al. [40] use daily maximum temperature, relative humidity at 14:00, and wind speed at 14:00 from 1961 to 2017, as well as winter wheat growth period data to analyze the spatiotemporal characteristics of the disaster-causing factors of dry-hot wind disasters in the Huanghai and Huaihai areas. Their results show that dry-hot wind disasters occur frequently in northern and western Shandong. In comparison, the present study shows that the comprehensive high-risk area of dry-hot wind disasters is located mainly in Binzhou, Zibo, and Weifang, which is north of Shandong province, Dezhou and Liaocheng in the west, and Heze, Jining, and Linyi in the south. Note that these results are consistent with the previous study.

(2) In the present study, based on observation data from meteorological stations, the dry-hot wind index R of each station is calculated. Then, the Kriging interpolation method in ARC-GIS is used to obtain the spatial distribution map. In addition to the Kriging interpolation method, there are also other spatial interpolation methods, such as the inverse distance weight interpolation method, spline function method, and trend surface analysis. The results of different interpolation methods are shown in Figure 12. The spatial distribution of the dry-hot wind index obtained by different interpolation methods differs. The results of the Kriging interpolation, inverse distance weight interpolation, and spline function methods are similar, whereas the results of trend surface analysis are quite different. According to the results from the Kriging interpolation, inverse distance weight interpolation, and spline function methods, the areas with a high dry-hot wind index are located in the north-central area of Shandong province. Existing studies show that the occurrence of hot-dry wind disaster is frequent in the central part of Shandong province, and less frequent in the west [18]. Huimin County (Binzhou City) and Yangjiaogou Town (Weifang City) are two high-incidence areas of dry-hot wind disasters. Dezhou, Liaocheng, Jining, and Heze are low-risk areas. Yanzhou (Jining City) has a significantly higher occurrence frequency than Heze. In addition, the frequency of dry-hot wind disasters in Tai'an City increases from southwest to northeast [28]. Lin et al. [41] analyzed the weather data and hazard symptom information during the later stage of wheat growth, weather data for dry-hot wind days in past years, and field test data. These authors found that dry-hot wind

disasters occurred in Dezhou, Heze, Weifang, and Jining. The results obtained by Kriging interpolation are consistent with the results of previous studies.

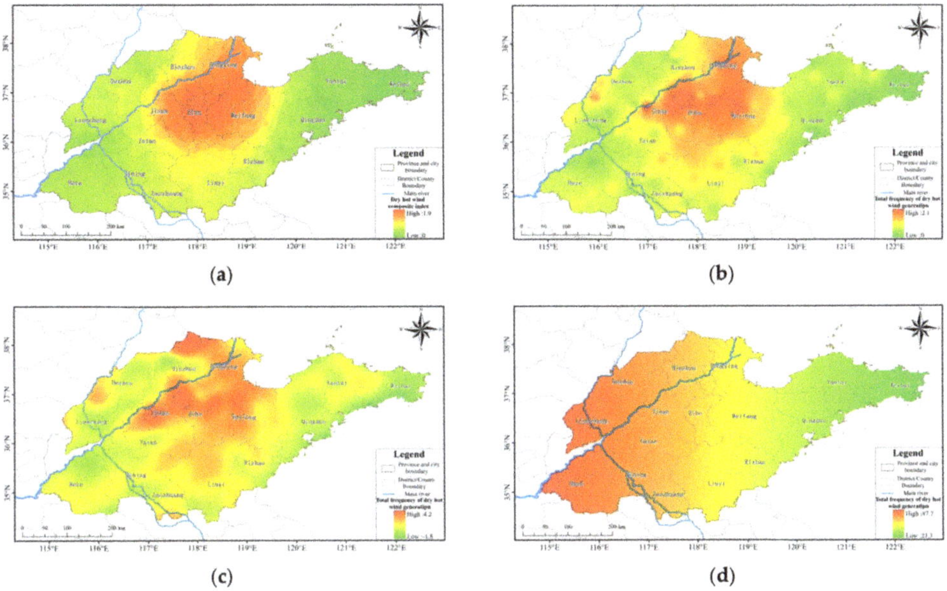

Figure 12. Spatial distribution of dry-hot wind index: (**a**) Kriging interpolation method; (**b**) inverse distance weight interpolation method; (**c**) spline function method; (**d**) trend surface analysis method.

(3) To perform the zoning of exposure, vulnerability, and disaster prevention and mitigation capability, only social and economic indicators are used in this study. However, the indicators in different counties are not standardized. For example, some indicators may not be present for certain counties. For the purpose of standardization, adjustments or replacements are made based on specific conditions, with varying effects on the zoning results. Moreover, some indicators are limited to the city level, and there is no county-level data, which affects the spatial resolution of the zoning. In addition, some indicators are not included in the statistical yearbooks; these are replaced with similar indicators. For instance, education level is supposed to be the percentage of graduation at each level, yet in the statistical yearbook, there is no relevant information. Thus, the number of school students is used to calculate education level. Apparently, such treatment affects the zoning results.

(4) The zoning results for dry-hot wind disasters include not only comprehensive risk, but also zoning for risk, exposure, vulnerability, and disaster prevention and mitigation capabilities. Therefore, in practical application, the results can be analyzed from various perspectives. For example, from the perspective of comprehensive risk zoning of dry-hot wind disasters (Figure 11), the comprehensive risk is high in Dezhou, Liaocheng, and Heze. However, when the zoning results of risk, exposure, vulnerability, and disaster prevention and mitigation capability are considered separately, we find that for areas with high comprehensive risk, such as Weifang, Jinan, Zibo, and Binzhou, disaster prevention awareness should be enhanced, and scientific disaster prevention and relief plans should be formulated. For areas with high exposure, such as Heze, Weifang, Linyi, and Jining, investment in disaster relief facilities should be increased. For Dezhou, Liaocheng, and Heze, which have high vulnerability and weak disaster prevention and mitigation capability, the focus should be on strengthening economic development and increasing investment in education, so as to reduce losses caused by dry-hot disaster. In conclusion, the risk assessment and zoning of dry-hot wind disasters in Shandong province can not only improve our understanding

of such disasters, but they can also provide a framework for government to formulate disaster prevention and relief policies.

(5) After calculating the risk index, exposure index, vulnerability index, disaster prevention and mitigation capabilities index, and comprehensive risk index, the risk space should be divided according to the size of the index. There are many methods to divide the index, such as the equal interval method, defined interval method, natural breakpoint classification method, standard deviation method, and so on. In the classification, we take the risk as an example and select four methods for comparison (Figure 13). It can be seen that the spatial differences of the zoning results obtained by the equal interval method (a), defined interval method (b) and standard deviation method (d) are not obvious, which means a smaller high-risk area in Figure 13a, a too-large risk area in Figure 13b, and the uncleared risk boundary in Figure 13d. The natural breakpoint classification method, compared with other methods, is the method with zoning results where spatial distribution of each grade is clear and is more consistent with the reality. Therefore, the natural breakpoint method is selected. The results of various division methods are as follows, and the classification scope of each index is added to the corresponding research content of the article.

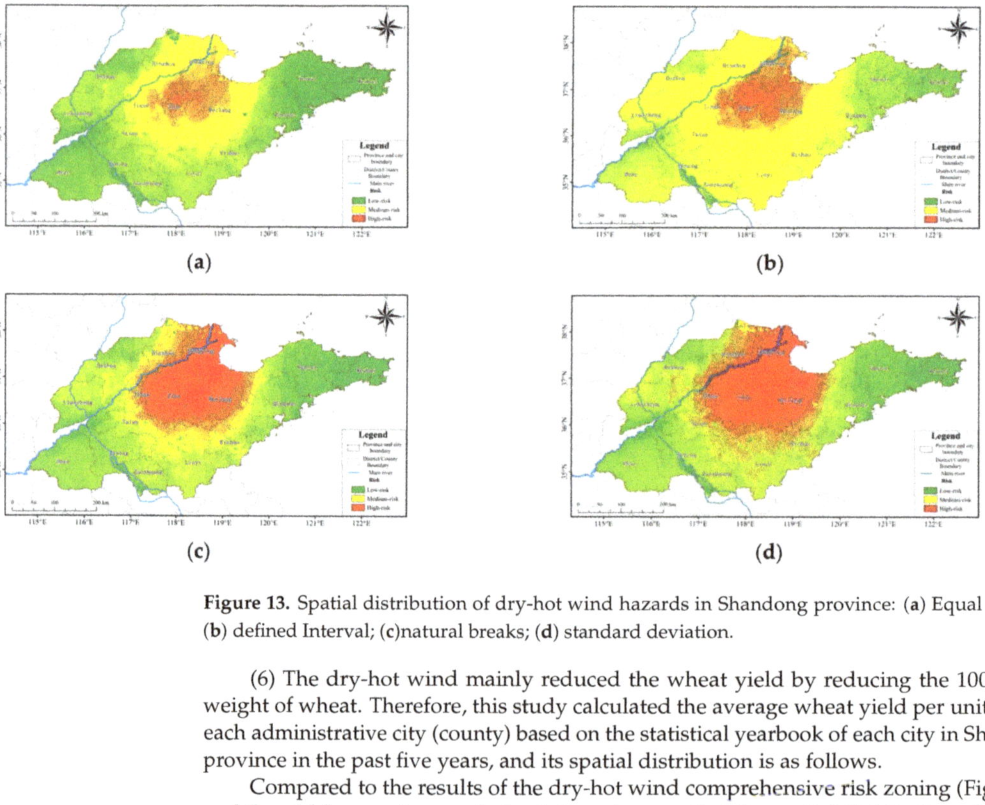

Figure 13. Spatial distribution of dry-hot wind hazards in Shandong province: (**a**) Equal Interval; (**b**) defined Interval; (**c**)natural breaks; (**d**) standard deviation.

(6) The dry-hot wind mainly reduced the wheat yield by reducing the 1000-grain weight of wheat. Therefore, this study calculated the average wheat yield per unit area of each administrative city (county) based on the statistical yearbook of each city in Shandong province in the past five years, and its spatial distribution is as follows.

Compared to the results of the dry-hot wind comprehensive risk zoning (Figure 11) and the yield per unit area of wheat, it can be seen that the spatial distribution of the two figures is basically similar (Figure 14). In areas with a high comprehensive risk of dry-hot wind, such as Dezhou, Liaocheng, Heze in the west, the yields per unit area of wheat are also the lowest; Dongying, Jinan, Tai'an, Rizhao, Qingdao, Weihai and other counties (cities) with low comprehensive risk of dry-hot wind, and the yields per unit area of wheat are also higher. However, there are also individual areas that do not match. For example, in the southwest of Dongying, the comprehensive risk of dry-hot wind is the highest, and the average yield per unit area of wheat is also high, which may be caused by several factors, such as artificial irrigation.

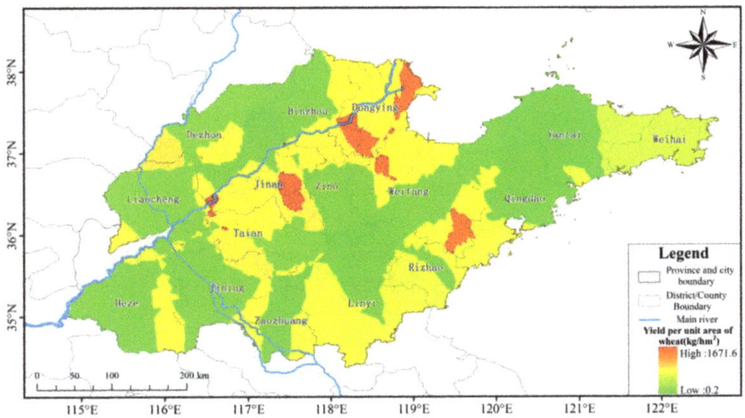

Figure 14. Spatial distribution of wheat yield per unit area in Shandong province.

5. Conclusions

The dry-hot wind, at all levels in Shandong province, mainly occurs in the central area of Shandong province, and the number of dry-hot wind in a year is at most 1.9 days. Severe dry-hot wind mainly occurs in the northern part of the central region, specifically in Zibo City, Weifang City, and Dongying City, where it occurs, at most, for 0.4 d. Considering the topographic factors, the high-value area of dry-hot wind risk index is located in the north-central area, with an area of 32,587.6 km^2, accounting for 20.6% of the province's area, and other areas are gradually decreasing. The high-value area of the exposure index is located in the southwest of Shandong province, covering an area of 56,581.5 km^2, accounting for 35.8% of the province's area. The high-risk areas in Zibo and Dongying are all low-exposure, and only Weifang is high-exposure. The high-value area of the dry-hot wind disaster-affected body vulnerability index in Shandong province is located in the west of Shandong province, with an area of 31,503.3 km^2, accounting for 20.0% of the province's area, while the high-value area of dry-hot wind risk has relatively low vulnerability. The high-value areas of the dry-hot wind disaster prevention and mitigation capabilities index in Shandong province are located in the eastern and central parts of Shandong province, with an area of 48,613.6 km^2, accounting for 30.8% of the province's area, and the lowest are in the southern and western regions. The disaster prevention and mitigation capabilities of the high-risk dry-hot wind area belongs to the medium area, among which Dongying City has the strongest disaster prevention and mitigation capabilities for dry-hot wind. To sum up, the high-value areas of dry-hot wind comprehensive risk in Shandong province are located in the western and central parts of Shandong province, with an area of 64,076.7 km^2, accounting for 40.6% of the province's area. Due to its strong disaster prevention and mitigation capabilities, Dongying City has become a low-value area in terms of comprehensive risk. Both Zibo City and Weifang City belong to areas with high comprehensive risk of dry-hot wind.

The study results showed that the comprehensive risk zoning results of dry-hot wind proposed in this paper were basically consistent with the spatial distribution of wheat yield per unit area in Shandong province, indicating that the results of dry-hot wind zoning in this study were of high accuracy. The results of this paper have important guiding significance for the formulation of disaster prevention and reduction planning of dry-hot wind in Shandong province. It is suggested to strengthen the construction of dry-hot wind early warning systems in the central and western regions, including improving the accuracy of dry-hot wind prediction, cultivating wheat varieties resistant to high temperature and low humidity in the central and western regions, especially strengthening the water conservancy construction in the central and western regions, and improving the disaster prevention and reduction capacity of dry-hot wind resistance, so as to reduce the losses caused by strong exposure and vulnerability. In particular, Weifang City, Dezhou City, and Heze City should formulate different disaster prevention and reduction plans according to different mechanisms leading to high comprehensive risk of dry-hot wind. For example, Weifang is in a high-risk area.

Author Contributions: Writing—original draft preparation, N.W.; resources, X.X.; writing—review and editing, L.Z.; data curation, Y.C.; investigation, M.J.; software, Y.W. and Y.Z.; validation, X.G. and Y.Y.; visualization, E.Z. All authors have read and agreed to the published version of the manuscript.

Funding: This research was funded by the National Natural Science Foundation of China (Grant No. 41771067) and the Key projects of Natural Science Foundation of Heilongjiang province of China (No. ZD2020D002).

Institutional Review Board Statement: Not applicable.

Informed Consent Statement: Not applicable.

Data Availability Statement: Not applicable.

Conflicts of Interest: The authors declare no conflict of interest.

References

1. Ma, Y.L.; Luan, Q.; Li, W.W.; Liu, W.P. Distribution Characteristics of Dry-hot Wind and Its Effect on Yield of Winter Wheat in Shanxi province. *J. Shanxi Agric. Sci.* **2017**, *45*, 1134–1138. [CrossRef]
2. Li, J.H.; Zhai, B.L. Harm of hot dry wind to wheat and its preventive measures. *Henan Agric.* **2021**, *7*, 39. [CrossRef]
3. Zhang, Z.H.; Cheng, L.; Li, S.L.; Peng, J.Y.; Fang, W.S. Dry-hot wind effects on physiology of winter wheat. *Chin. J. Ecol.* **2015**, *34*, 712–717. [CrossRef]
4. Xu, Y.N.; Wu, Y.; Liu, B.H.; Song, J.Q.; Lv, G.H.; Ji, B.Y.; Zhang, W.Y.; Yang, Z.L.; Bai, W.B. Effect of Dry-Hot Wind on Grain Weight of Winter Wheat at Different Spikelet and Grain Positions. *Chin. J. Agrometeorol.* **2021**, *42*, 583–595. [CrossRef]
5. Jin, F.G. Impact of dry and hot winds on wheat and countermeasures. *Mod. Agric. Sci. Technol.* **2021**, *10*, 19.
6. Reeves, T.G.; Cassaday, K. History and past achievements of plant breeding. *Aust. J. Agric. Res.* **2002**, *53*, 851–863. [CrossRef]
7. Dhonde, S.H.; Kute, N.S.; Kanawade, D.G.; Sarode, N.D. Variability and characters association in wheat (*Triticum aestivum*). *Agric. Sci. Dig.* **2000**, *20*, 99–101.
8. Sofield, I.; Evans, L.; Cook, M.; Wardlaw, I. Factors Influencing the Rate and Duration of Grain Filling in Wheat. *Aust. J. Plant Physiol.* **1977**, *4*, 785–797. [CrossRef]
9. Huo, Z.G.; Shang, Y.; Wu, D.R.; Wu, L.; Fan, Y.X.; Wang, P.J.; Yang, J.Y.; Wang, C.Z. Review on Disaster of Hot Dry Wind for Wheat in China. *J. Appl. Meteorol. Sci.* **2019**, *30*, 129–141. [CrossRef]
10. Ivanov, B.K. Conditions and characteristics of dry hot air in West Siberia. In *M. Literature Data of Dry Hot Air in the Soviet Union*; Shanxi Provincial Meteorological Bureau: Jinzhong, China, 1982; pp. 34–58.
11. Tavakol, A.; Rahmani, V.; Harrington, J. Temporal and spatial variations in the frequency of compound hot, dry, and windy events in the central United State. *Sci. Rep.* **2020**, *10*, 15691. [CrossRef]
12. Viktorovich, P.R.; Aleksandrovna, V.A.; Vladimirovich, S.K. Recurrence of hot dry winds on the territory of Lower Volga region. *J. Sci. Rev.* **2014**, *5*, 41–46.
13. Hou, Q.; Zhang, B.; He, H.; Li, S. Effects of Climate Change on Characteristics of Dry-hot Wind in Hexi Region of Gansu province. *Plateau Meteorol.* **2020**, *39*, 162–171. [CrossRef]
14. Cheng, L.; Zhang, Z.H.; Chang, J. Analysis on the Change of Dry-hot Wind Hazard for Winter Wheat in Last 47 Years in Henan province. *Chin. J. Agrometeorol.* **2011**, *32*, 455–456. [CrossRef]
15. You, F.C.; Hao, L.S.; Shi, Y.S.; Duan, S.L.; Sun, F.C. Causation analysis of Dry-Hot Wind Formation in Hebei province Winter Wheat Region. *Meteorol. Mon.* **2007**, *3*, 95–100. [CrossRef]
16. Li, S.; Zhang, L.; Huang, B.X.; He, L.; Zhang, J.F.; Guo, A.H. A comprehensive index for assessing regional dry-hot wind events in Huang-Huai-Hai Region, China. *Phys. Chem. Earth Parts A/B/C* **2020**, *116*, 102860. [CrossRef]
17. Shi, B.L.; Zhu, X.Y.; Li, H.Z.; Hu, Y.C.; Zhang, Y. Analysis of how dry-hot wind hazard has changed for winter wheat in the Huang-huai-hai plain. *Nat. Hazards Earth Syst. Sci. Discuss.* **2016**, 1–20, *preprint*. [CrossRef]
18. Zhao, J.F.; Zhao, Y.X.; Guo, J.P.; Fang, S.B. Spatial-Temporal Changes of Dry-Hot Wind for Winter Wheat in Huanghuaihai Plain During the Past 50 Years. *Sci. Agric. Sin.* **2012**, *45*, 2815–2825. [CrossRef]
19. Chen, J.Z.; Wang, Y.Q.; Zhang, J. Climatic Features of Dry-hot Wind and Its Influence on the Wheat Yield in Henan province. *J. Anhui Agric. Sci.* **2012**, *40*, 7152–7154. [CrossRef]

20. Shi, Y.S.; You, F.C.; Wei, R.J.; Hao, L.S.; You, H.L. Impact Analysis of Dry-Hot Wind on Weight of Thousand Grain in Hebei province. *Meteorol. Sci. Technol.* **2007**, *5*, 699–702. [CrossRef]
21. Yang, F.Y.; Zhu, Y.J.; Liu, W.C. Occurrence rules and risk zoning of dry-hot wind in winter wheat producing areas of north China. *Nat. Disasters* **2013**, *22*, 112–121.
22. Chen, H.L.; Deng, W.; Zhang, X.F.; Zou, C.H. Analysis and zoning of agrometeorological disasters risk for wheat growing in Henan province. *J. Nat. Disasters* **2006**, *15*, 135.
23. Juraev, D.T.; Amanov, O.A.; Dilmurodov, S.D.; Boysunov, N.B.; Kayumov, N.S.; IshankuLlova, G.N.; Togaeva, K.R. The Influence of Hot-Dry Wind on Farm Valuable Traits of Wheat Genotypes in Southern Regions of Uzbekistan. *Plant. Cell Biotechnol. Mol. Biol.* **2021**, *22*, 34–49.
24. Wang, S.; Rao, Y.H.; Chen, J.; Liu, L.C.; Wang, W.Q. Adopting "Difference-in-Differences" Method to Monitor Crop Response to Agrometeorological Hazards with Satellite Data: A Case Study of Dry-Hot Wind. *Remote Sens.* **2021**, *13*, 482. [CrossRef]
25. Deng, Z.Y.; Zhang, Q.; Xu, J.F.; Chen, M.; Qin, S.J.; Zhang, S.Y. Comparative Studies on the Harm Characteristic of Hot-dry wind and High Temperature Heat Waves. *Adv. Earth Sci.* **2009**, *8*, 865–873. [CrossRef]
26. Wang, P. Research on Method of Natural Disaster Regionalization Based on Geogeraphical Information System. *J. Beijing Norm. Univ.* **2020**, *36*, 410–416. [CrossRef]
27. Cheng, L.; Zhang, Z.H.; Fang, W.S. Comprehensive risk zoning of dry-hot wind on winter wheat based on yield loss. *Agric. Res. Arid. Areas* **2019**, *37*, 238–244. [CrossRef]
28. Wu, D.R.; Liu, J.D.; Liu, L.; Jiang, C.Y. Regionalization Study on Dry-hot Windy Days in North China Plain Using CAST Model. *Sci. Technol. Rev.* **2012**, *30*, 19–23. [CrossRef]
29. Li, X.Y.; Zhang, J.P.; Chen, M. Risk assessment and zoning of winter wheat dry-hot wind in Henan province based on GIS technology. *J. Nat. Disasters* **2017**, *26*, 63–70. [CrossRef]
30. Yang, Z.J.; Jin, L.X.; Wu, R.S.; Wang, H.Z.; Wu, R.F. Refined Risk Division of Dry—Hot Wind Disaster for Spring Wheat in Inner Mongolia Based on GIS. *J. Arid. Meteorol.* **2019**, *37*, 866–872. [CrossRef]
31. Zhang, X.Y.; Wang, L.L.; Lu, D.C.; Liu, J.J.; Liu, F. Analysis and Outlook of Wheat Market Quotation in Shandong province in 2018–2019. *Shandong Agric. Sci.* **2020**, *52*, 151–155. [CrossRef]
32. Zhang, J.Q.; Liang, J.D.; Zhou, D.W. Risk assessment of ecological disasters in Jilin province based on GIS. *Chinses J. Appl. Ecol.* **2007**, *8*, 1765–1770. [CrossRef]
33. Beynon, M. An analysis of distrbutions of priority values from altemative comparison scales within AHP. *Eur. J. Oper. Res.* **2002**, *140*, 104–117. [CrossRef]
34. Tang, G.A.; Yang, X. *ArcGIS GIS Spatial Analysis Experiment Tutorial*; Science Press: Beijing, China, 2012.
35. Wang, G.D.; Guo, C.L. Application of the Kricing Technique in Geography. *Acta Geogr. Sin.* **1987**, *4*, 366–375. [CrossRef]
36. Fu, Z.; Zhou, Y.X.; Liu, D.W.; Liu, W.S. Research on Spatial Fuzzy Comprehensive Assessment of Eco- environmental Quality. *J. Arid. Land Resour. Environ.* **2005**, *5*, 97–105. [CrossRef]
37. Li, C.Q. A brief analysis of the relationship between the natural geographical environment and the slope aspect in the mountains. *City Geogr.* **2016**, *4*, 225. [CrossRef]
38. Hu, L.; Wang, Q.C.; Liu, B.K.; Su, W.J.; Zhou, W.F. Distribution Characteristics of Dry Hot Wind and Its Response to Climate Change in Qinghai Plateau. *Meteorol. Mon.* **2014**, *40*, 450–457. [CrossRef]
39. Liu, X.R.; Kang, J.; Wang, Y.; Li, Y.H. Assessment and regionalization on drought disaster risk of different season in Chongqing area based on GIS. *J. Nat. Disasters* **2019**, *28*, 92–100. [CrossRef]
40. Li, S.; Han, L.J.; Zhang, L.; Guo, A.H. Spatial-temporal characteristics of dry-hot wind factors in Huang-Huai-Hai region. *J. Nat. Disasters* **2020**, *29*, 183–192. [CrossRef]
41. Lin, M.Y. Climatic Regionalization of Wheat Dry-Hot Wind in Shandong province. *J. Shandong Meteorol.* **1982**, *1*, 18–22. [CrossRef]

sustainability

MDPI

Article

Peanut Drought Risk Zoning in Shandong Province, China

Meiyi Jiang [1], Xiaoping Xue [2,*], Lijuan Zhang [1,*], Yuying Chen [1], Cheng Zhao [1], Haiyan Song [1] and Nan Wang [1]

1 Heilongjiang Province Key Laboratory of Geographical Environment Monitoring and Spatial Information Service in Cold Regions, Harbin Normal University, Harbin 150025, China; jiangmi199608@163.com (M.J.); chenyuying000@gmail.com (Y.C.); zhao957175130@163.com (C.Z.); songhaiyanya@163.com (H.S.); 18845565244@163.com (N.W.)

2 Shandong Provincial Meteorological Bureau, Jinan 250000, China

* Correspondence: xxpdhy@163.com (X.X.); zlj19650205@163.com (L.Z.)

Abstract: Peanut growth in Shandong Province, a major peanut-producing area in China, is greatly affected by drought. The present study uses hierarchical analysis, weighted comprehensive evaluation, and ArcGIS spatial analysis to conduct spatial zoning of peanut drought risk in Shandong Province based on daily precipitation data for the province acquired from 1991 to 2020, the per capita GDP, and the peanut planting area of Shandong Province, so as to quantify the disaster risk of peanut drought and formulate disaster prevention and resilience planning accordingly. The results show the high-drought-risk zone was mainly distributed in the northwestern part of Shandong Province and on the Jiaodong Peninsula, covering 32.4% of the province. Drought risk was concentrated on the Jiaodong Peninsula, covering 20.7% of the province. The high-vulnerability zone was mainly distributed in the cities of Yantai, Weihai, Linyi, and Rizhao, accounting for 26.8% of the total area. The low-disaster-prevention and low-mitigation-capacity zone was mainly distributed in the western part of Shandong Province, covering 38.7% of the province. Medium- and high-risk areas for drought affecting peanuts were widely distributed, while the overall comprehensive risk index was high, covering 76.2% of the province. Spatial analysis to conduct risk zoning and assessment of peanut drought in Shandong Province, so as to provide a basis for peanut drought disaster prevention and safe peanut production in Shandong Province.

Keywords: peanut drought; risk zoning; Shandong Province; natural disaster risk assessment principles

Citation: Jiang, M.; Xue, X.; Zhang, L.; Chen, Y.; Zhao, C.; Song, H.; Wang, N. Peanut Drought Risk Zoning in Shandong Province, China. *Sustainability* **2022**, *14*, 3322. https://doi.org/10.3390/su14063322

Academic Editors: Xiaodong Yan, Jia Yang, Shaofei Jin and Tommaso Caloiero

Received: 20 December 2021
Accepted: 7 March 2022
Published: 11 March 2022

1. Introduction

According to the World Meteorological Organization, meteorological disasters cause about 85% of the total losses caused by various types of natural disasters, and drought in turn accounts for about 50% of meteorological disasters losses [1]. The peanut is an important food, source of oil, and cash crop in China [2,3]. Drought can seriously affect the plants during flowering and the quality of peanut kernels during the podding period [4–6], leading to a decline in peanut quality and yield [7,8]. Therefore, drought is an important factor limiting peanut growth and yield. Shandong Province ranks among the top peanut-growing areas in China, with the peanut cultivation area accounting for about 15–16% of the total domestic peanut growing area; total production in Shandong accounts for about 18–20% of the total domestic production [9,10]. The critical period of water demand for peanut growth is concentrated from June to August, which coincides with the occurrence of summer drought in Shandong Province; therefore, drought disasters are one of the major types of disasters affecting the growth and yield of peanuts in Shandong Province [11].

Natural disaster risk refers to the possibility of loss from the impact of a certain disaster in a certain area after considering the natural and social attributes together [12]. Therefore, the purpose of natural disaster risk research is to provide a scientific basis for regional disaster prevention and mitigation, and strengthening the research of comprehensive natural disaster zoning has been listed as one of the actions of disaster prevention and mitigation in China's Agenda 21 [13]. Significant work has been carried out in various countries for drought risk zoning research. Araya, A. et al. [14] developed a suitable drought assessment technique by analyzing long-term climate data from four sites in northern Ethiopia; Moumita Palchaudhuri et al. [15] used a combination of AHP and GIS to conduct a drought zoning study in Puruliya, West Bengal, India; Nazarifar Mohammadhadi et al. [16] assessed and zoned drought risk in the Karkheh basin for different years and return periods; Zhongyi Sun et al. [17] proposed a methodology for integrated risk analysis, assessment, combination, and regionalization of droughts and floods in Anhui Province; and Luo D et al. [18] assessed the drought hazard by constructing a gray predictive incidence model (GPIM). Additionally, for peanut drought, risk zoning has also attracted the attention of scholars, but up to now the relevant research literature has been relatively scarce. Wei S. Cheng et al. [19] conducted a risk evaluation of peanut drought in the Yellow and Huaihai Sea region and concluded that high-risk areas for peanut drought disaster were scattered and mainly concentrated in the northwestern part of the Yellow River Basin. Additionally, more scholars have studied the impacts of drought on peanut growth and yield. For example, Celikkol Akcay U. et al. [20] concluded that the growth retardation of peanuts under drought stress conditions was mainly due to drought-induced oxidative damage and antioxidant responses; Jiang, C.J. et al. [21] proposed that drought inhibited different varieties of peanut and the drought resistance of different peanut varieties varied; and Zhang, K. et al. [22] selected 16 peanut varieties for drought resistance testing and concluded that geological drought can start and end quickly, while meteorological drought takes longer to develop and recover. These research results provide important reference values for conducting peanut drought risk assessment and zoning studies.

This paper selected Shandong Province, China, as the study area, and conducted a spatial zoning study on peanut drought risk in Shandong Province based on natural disaster risk theory, considering four aspects, hazard, exposure of disaster-affected bodies, vulnerability of disaster-affected bodies, and disaster prevention and mitigation capacity, by establishing a peanut drought risk index model, combined with Arc-GIS spatial analysis, a weighted comprehensive evaluation method, and hierarchical analysis method. Compared with the existing studies, in addition to the study of peanut drought risk, the exposure of disaster-affected bodies, vulnerability of disaster-affected bodies, and disaster prevention and mitigation capacity were also evaluated and zoned, providing a reference for carrying out peanut drought risk assessment and zoning studies. Research results provide a basis for the prevention of peanut drought and the safe production of peanuts in Shandong Province and provide a quantitative basis for the scientific formulation of disaster prevention and mitigation policies and planning by relevant departments.

2. Materials and Methodology

2.1. Study Area

Shandong Province is located on the east coast of China and the lower reaches of the Yellow River (114°48′ E–122°42′ E and 34°23′ N–38°17′ N), as shown in Figure 1. Total land area is 157,900 km^2. The climate type is warm temperate monsoon. Precipitation is concentrated, and rain and heat occur in the same season. Spring and autumn are short, while winter and summer are long. The annual average temperature range is 11 °C–14 °C, and the annual average precipitation range is 550–950 mm. The rainfall season is unevenly distributed, with 60–70% of annual precipitation in summer. Landform types include plains, terraces, hills, and mountains. There is a dense river network in the region, including the Yellow River, Huaihe River, Haihe River, and smaller rivers in the central and southern mountainous area.

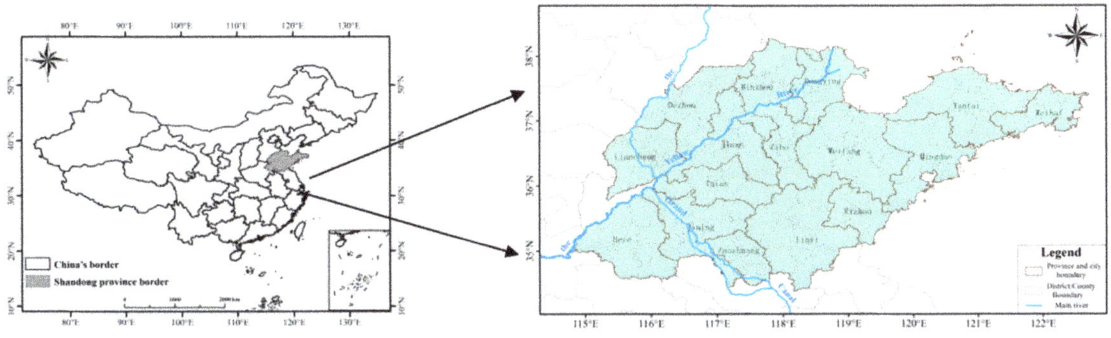

Figure 1. Administrative division of Shandong Province, China.

2.2. Data Sets

This paper covers a total of 122 meteorological stations in Shandong Province from 1991 to 2019, and the precipitation data for each month were obtained based on the daily precipitation data of each station. The daily precipitation data were obtained from Shandong Meteorological Center. According to the ground meteorological observation specification of China Meteorological Administration, the daily precipitation data were reviewed by the stations, their municipal meteorological bureaus, and the data review department of the provincial meteorological bureau before being entered into the database of Shandong Meteorological Center, and the anomalous values were eliminated to ensure the uniformity and accuracy of the data. Total GDP, total population, peanut planting area, percentage of peanut cultivation area, population density, per capita GDP, and water facilities are from the 2018–2020 Statistical Yearbook of Shandong Province.

2.3. Methods

This paper constructs a theoretical model for peanut drought risk assessment based on the basic theory of disaster risk assessment. The trend analysis method is used to analyze the temporal characteristics of the disaster-causing factors; the hierarchical analysis method is used to determine the weights of the factors constituting the risk impact; and the comprehensive weighted evaluation method is used to constitute the risk index model.

2.3.1. Basic Theory of Disaster Risk Assessment

Based on the theory of natural hazard risk formation [12], meteorological hazard risk is formed by the combination of four components: hazard (causative factor), exposure (carrier), vulnerability (carrier), and prevention and mitigation capacity. Each factor is in turn composed of a series of subfactors. The expressions are:

Disaster risk index = f (hazard, exposure, vulnerability, disaster prevention and mitigation capacity) (1)

Hazardous factors: Hazardous factors include meteorological factors and environmental sensitivity. All meteorological factors that may lead to disasters can be called meteorological factor hazards; the sensitivity of the pregnant environment refers to the degree of strengthening or weakening of meteorological factors in the natural surface environment.

Exposure of disaster-bearing body: Disaster-bearing body is the object of disaster-causing factors and is the entity that bears the disaster. Exposure of the hazard-bearing body is the result of the interaction between the hazard-causing factor and the hazard-bearing body, and the exposure of the hazard-bearing individual to the hazard-causing factor.

Vulnerability of the disaster-bearing body: A disaster can be formed only when it acts on the corresponding object, i.e., human beings and their socioeconomic activities. Specifically, it refers to the degree of hazard or loss caused by the potential risk factors for all objects that may be threatened by the disaster-causing factors that exist in a given hazard area, and its combination reflects the degree of loss from meteorological disasters.

Prevention and mitigation capacity: It refers to various management measures and countermeasures used to prevent and mitigate meteorological hazards, including management capacity, mitigation input, and resource preparation. The more proper the management measures and the stronger the management capacity, the less potential losses that may be suffered and the less risk of meteorological disasters.

Based on the above theory, a hierarchical analysis model for peanut drought disaster risk assessment in Shandong Province was built (Figure 2). Figure 2 shows the model of peanut drought risk zoning. The risk was divided into meteorological factor risk and pregnancy disaster environmental sensitivity. The risk of meteorological factor selected was the precipitation anomaly percentage. The environmental sensitivity of pregnant disaster referred to the environmental factors that can enhance or weaken the risk of peanut drought. In this paper, factors such as elevation and slope were selected. Exposure was selected as the peanut planting area; vulnerability referred to the percentage of the peanut cultivation area; and the factors for disaster prevention and mitigation capabilities selected were the aspects of per capita GDP, level of education, and so on. However, when selecting indicators, they will be selected or replaced according to the factors in the Statistical Yearbook. Please refer to Sections 3.1–3.4 for the selection basis of specific indicators.

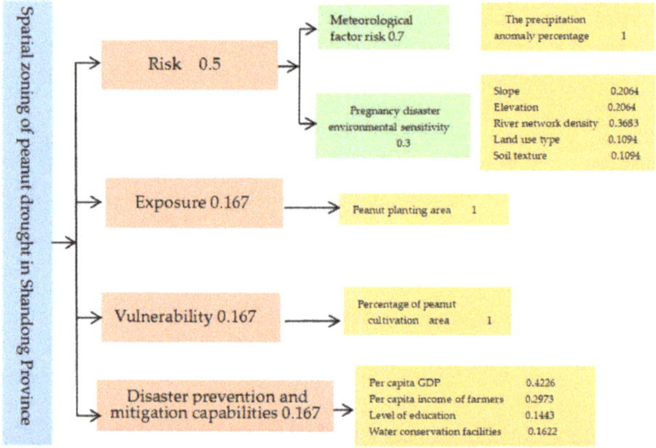

Figure 2. Hierarchical model for peanut drought risk assessment.

2.3.2. Weighted Comprehensive Evaluation Method

The weighted comprehensive evaluation method is a method that solves the "bottom-up" indexes in the risk hierarchy analysis and evaluation model, which takes into account the degree of influence of each factor on the overall object and integrates the strengths and weaknesses of each specific index and uses a numerical index to focus on the strengths and weaknesses of the entire evaluation object. This method is especially suitable for comprehensive analysis and evaluation of technologies, strategies, or programs and is one of the most commonly used calculation methods. Its expression is:

$$Y_i = \sum_{i=1}^{m} W_{ij} Y_{ij} \; i = 1, 2, 3, 4; \; j = 1, 2, \ldots, m \quad (2)$$

In the formula, Y_i represents the disaster risk index, and i, respectively represents the risk, susceptibility, vulnerability, and disaster prevention and mitigation capabilities; Y_{ij} is the factor that affects the risk, susceptibility, vulnerability, and disaster prevention and mitigation capabilities, and W_{ij} is the weight value of risk, susceptibility, vulnerability, and disaster prevention and mitigation capabilities ($0 \leq W_{ij} \leq 1$), while j represents the number of factors affecting i.

For the comprehensive risk index of natural disasters, the expressions are:

$$Y = \sum_{i=1}^{n} W_i Y_i \; i = 1,2,3,4 \tag{3}$$

In the formula, Y represents the comprehensive disaster risk index; Y_i is the risk index, susceptibility index, vulnerability index, and disaster prevention and mitigation capability index, and W_i is the weight value. The stronger the disaster prevention and mitigation capacity, the smaller the comprehensive risk index, so the "negative sign" is used.

Among them, W_{ij} and W_i are determined by the analytic hierarchy process; see research method Section 2.3.3 for details. Each factor in the formula needs to be standardized because of different dimensions; see research method Section 2.3.4 for details.

2.3.3. Analytic Hierarchy Process

Analytic hierarchy process (AHP) is a simple method for making decisions on more complex and vague problems, especially for those problems that are difficult to fully quantitatively analyze [23]. This paper used the operation principle of the analytic hierarchy process and used the 1–9 scale method given by Saaty to construct the judgment matrix for the pairwise relationship of the influence factors. The pairwise comparison of all influence factors determines the weight of each influence factor, which prevents the result error caused by the subjectivity of the expert. The qualitative comparison scale values between the two influencing factors are shown in Table 1 below.

Table 1. Scale of AHP analysis method.

Scale b_{ij}	Definition
1	The i factor is as important as the j factor.
3	The i factor is slightly more important than the j factor.
5	The i factor is more important than the j factor.
7	The i factor is much more important than the j factor.
9	The i factor is absolutely more important than the j factor.
2, 4, 6, 8	Between the noted levels.

The maximum eigenvector value of the judgment matrix and its corresponding eigenvector need to be solved by the sum-product method, and the consistency of the matrix (the following formula) should be solved; then, this should be solved by the sum-product method.

$$CI = \frac{\lambda_{max} - n}{n-1} = \frac{-\sum_{i=1}^{n} \lambda_i}{n-1} \tag{4}$$

$$CR = \frac{CI}{RI} < 0.1 \tag{5}$$

In the formula, CI is the consistency index of the judgment matrix; λ_{max} is the largest characteristic root of the matrix; n is the order of the discrimination matrix; CR is the random consistency index of the judgment matrix; and RI is the average random consistency index of the discrimination matrix. The values of RI are shown in Table 2.

Table 2. Numerical values of random consistency index *RI*.

M	1	2	3	4	5	6	7	8	9	10	11
RI	0.00	0.00	0.58	0.90	1.12	1.24	1.32	1.41	1.45	1.49	1.51

We summarize the calculation process in Table 3, as follows.

Table 3. Judgment matrix and weights of various perceptual factors.

	I	II	III	IV	V	Weight (W)	Matrix Product (AW)	AW/W	λ_{max}	CI	CR
I	1	2	2	3	3	0.368	1.851	5.03	5.013	0.009	0.003
II	1/2	1	1	2	2	0.206	1.035	5.02			
III	1/2	1	1	2	2	0.206	1.035	5.02	$\lambda = \sum(AW/W)/n$		
IV	1/3	1/2	1/2	1	1	0.109	0.548	5.03	$CI = (\lambda - n)/n - 1$		
V	1/3	1/2	1/2	1	1	0.109	0.548	5.03	$RI = 1.12$		
								25.13	$CR = CI/RI$		

Note: In the table: I. river network density, II. slope, III. elevation, IV. land use type, and V. soil texture.

This paper adopted the analytic hierarchy process (AHP), taking the sensitivity of the pregnancy disaster environmental sensitivity as an example, and constructed the judgment matrix of each index; the calculation results are shown in Table 3. Since $CR < 0.1$, the matrix passed the consistency test.

Similarly, the weights of each factor of risk, exposure, vulnerability, disaster prevention and mitigation capacity, and comprehensive risk were obtained as shown in Figure 2.

2.3.4. Standardization

Because the dimensions of the selected factors are different, the values of each factor vary greatly. It is necessary to normalize each factor so that the values of each factor are between 0 and 1. When evaluating the risk of disaster-causing factors, the exposure of the disaster-affected body, the vulnerability of the disaster-affected body, and the disaster prevention and mitigation capacity, the correlations between the selected factors and each evaluation index are different; some are positively correlated, and some are negatively correlated. Therefore, when normalizing the factors with a positive correlation, if the cause subvalue is large, the normalized value is also large, so the maximum value should be selected for standardization. On the contrary, if the factors with a negative correlation are standardized, the minimum standardization is selected. For example, the larger the peanut planting area, the greater the exposure, so the great value standardization is chosen for the peanut planting area, and Equation (6) is selected; for example, the greater the river network density, the smaller the sensitivity of the pregnant environment, so the very small value standardization is performed for the river network density, and Equation (7) is selected.

Maximum standardization:

$$X'_{max} = \frac{|X_{ij} - X_{\min}|}{X_{max} - X_{\min}} \tag{6}$$

Minimum standardization:

$$X'_{\min} = \frac{|X_{max} - X_{ij}|}{X_{max} - X_{\min}} \tag{7}$$

where X_{ij} is the index number of the j-th factor of the x factor; X'_{max} and X'_{min} are the dimensionality of X_{ij}; and X_{max} and X_{min} are the minimum and maximum values in the index sequence.

2.3.5. Arc-GIS Spatial Analysis

In this paper, the meteorological elements and geographic environment elements were interpolated using the Kriging method to obtain spatial distribution maps with a spatial resolution of 100 m × 100 m. The socioeconomic factors were resampled by administrative units to obtain spatial distribution maps with a spatial resolution of 100 m × 100 m. Then, according to the weights of each element, the raster calculation method was used to superimpose each element spatially to obtain the spatial distribution map of risk index, exposure index, vulnerability index, and disaster prevention and mitigation capability index. Finally, each index of risk was spatially superimposed by weights to obtain the spatial distribution map of comprehensive risk index. The natural grading discontinuity method was used to grade each index, and the zoning map of each index of risk zoning and the comprehensive risk zoning map were obtained.

2.3.6. Drought Classification

A precipitation anomaly percentage indicator can visually reflect the degree of drought caused by precipitation anomalies; therefore, the precipitation anomaly percentage of peanuts for the entire growing period was selected as the risk indicator of a peanut drought meteorological factor. The precipitation anomaly percentage for a certain period was calculated according to Equation (8):

$$Pa = \frac{P - \overline{P}}{\overline{P}} \times 100\% \tag{8}$$

where Pa is the precipitation anomaly percentage (%); P is the precipitation for a certain time period (mm); and $\overline{P}$ is the multiyear average precipitation for the corresponding time period (mm), and the average value of 30 years was generally calculated.

In this paper, all grades of drought were calculated for each station according to (QX/T 82-2019) (Table 4) [24], and their frequencies were calculated and integrated in the formula of the danger index of meteorological factors of drought, which was calculated as:

$$\mathrm{R} = 0.0960\mathrm{D_l} + 0.1611\mathrm{D_m} + 0.2771\mathrm{D_s} + 0.4658\mathrm{D_e} \tag{9}$$

where R is the meteorological risk index of peanut drought; D_l is the average number of days in 30 years of light drought (d); D_m is the average number of days in 30 years of moderate drought (d); D_s is the average number of days in 30 years of severe drought (d); and D_e is the average number of days in 30 years of exceptional drought (d).

Table 4. The precipitation anomaly percentage drought classification table (based on meteorological drought rating criteria).

Level	Types	The Precipitation Anomaly Percentage P_a (%)		
		Monthly Scale	Quarterly Scale	Annual Scale
1	Drought-free	$-40 < P_a$	$-25 < P_a$	$-15 < P_a$
2	Light drought	$-60 < P_a \leq -40$	$-50 < P_a \leq -25$	$-30 < P_a \leq -15$
3	Moderate drought	$-80 < P_a \leq -60$	$-70 < P_a \leq -50$	$-40 < P_a \leq -30$
4	Severe drought	$-95 < P_a \leq -80$	$-80 < P_a \leq -70$	$-45 < P_a \leq -40$
5	Exceptional drought	$P_a \leq -95$	$P_a \leq -80$	$P_a \leq -45$

3. Results

3.1. Spatial Distribution of Peanut Drought Risk

Peanut drought risk includes meteorological factor risk and disaster formative environmental sensitivity, which were assigned weights of 0.7 and 0.3, respectively, based on the AHP method (Figure 2).

3.1.1. Meteorological Factor Risk Zoning

In this paper, the daily precipitation data of 122 stations in Shandong Province from 1991 to 2020 were used for meteorological data, and the precipitation anomaly percentage was calculated according to Equation (8), and the frequency of different drought levels in 122 stations in Shandong Province was calculated using the precipitation anomaly percentage according to the drought grade (Table 4); the drought grade was determined according to the meteorological drought standard of the people's Republic of China (QX/T

82-2019), and the spatial distribution map of the frequency of different drought levels in peanut in Shandong Province during the whole reproductive period was obtained, and the spatial distribution of the frequency of different levels of drought in peanut growing areas in Shandong Province during the entire growing period was obtained (Figure 3). These findings show that the frequency of light drought was higher mainly in the southwestern part of Shandong Province, with a frequency of about 1.72%, with local high values in the northwestern and some eastern areas; the frequency of light drought in the northwestern area was the lowest, at 0. The frequency of moderate drought was higher in Dezhou, Zibo, Dongying, and Weifang, with the highest value being 2.14%, and lower in other areas. Severe drought was mainly concentrated in the Jiaodong Peninsula, with a frequency of about 2.59%, and it was also higher in some areas in western Shandong and lower in other areas; exceptional drought was widely distributed in the province, with a higher frequency in both northern and eastern areas at about 2.55% in the northern and eastern parts of the province, while the frequency was the lowest in the southwest.

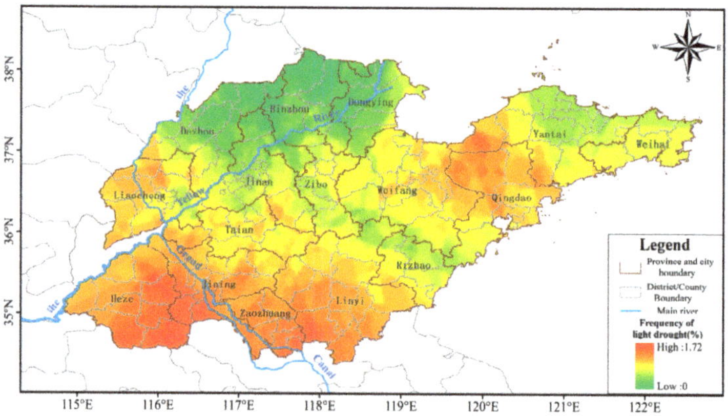

(a)

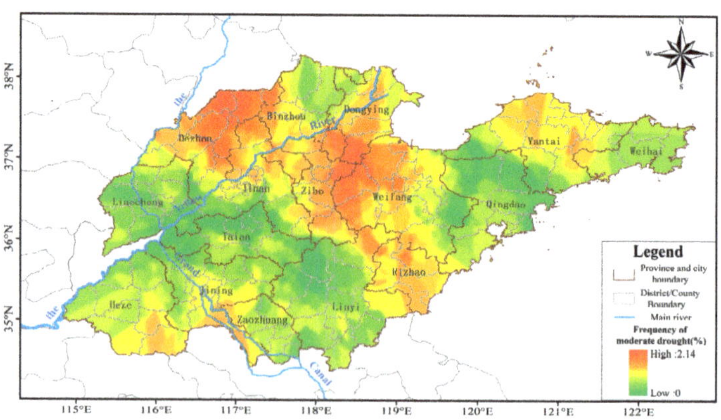

(b)

Figure 3. *Cont.*

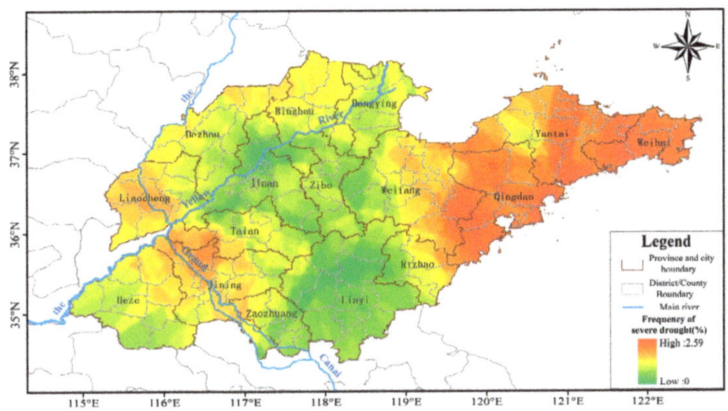

(c)

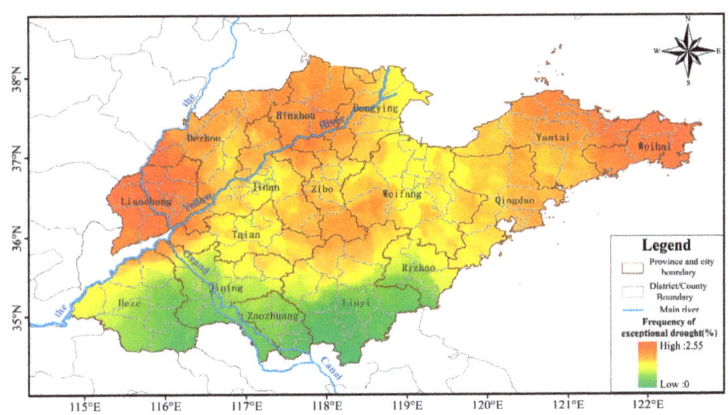

(d)

Figure 3. Spatial distribution of the frequency of different drought levels for peanuts in Shandong Province during the whole reproductive period: (**a**) frequency of light drought; (**b**) frequency of moderate drought; (**c**) frequency of severe drought; and (**d**) frequency of exceptional drought.

The frequencies of light, moderate, severe, and exceptional drought were assigned weights of 0.0960, 0.1611, 0.2771, and 0.4658, respectively, to construct the meteorological factor risk index and obtain the spatial distribution of peanut drought meteorological factor risk in Shandong Province, as shown in Figure 4. The spatial distribution of the peanut drought meteorological factor risk in Shandong Province varied significantly. Specifically, the high-value areas were mainly distributed in Liaocheng, Dezhou, Binzhou, Zibo, Weifang, Qingdao, Yantai, and Weihai; the low-value areas were mainly distributed in Jinan, Zaozhuang, Linyi, and Rizhao.

3.1.2. Zoning of Disaster Environment Sensitivity

Slope, elevation, river network density, land use type, and soil texture were selected as zoning indicators for the disaster-formative environmental-sensitivity analysis. Generally, the higher the elevation, the greater the chance of a drought occurring and the greater the sensitivity of an area to drought. Areas with greater slopes experience faster runoff and less infiltration so less moisture is stored in the slope body, making an area more prone to drought. Areas with a less dense river network had lower atmospheric humidity, making them more prone and sensitive to drought. Urbanization has resulted in arable land, grassland, and woodland being replaced by buildings and hardened ground that blocks rainwater from infiltrating into the soil, resulting in a worsening of drought conditions. Therefore, areas different land use types were given unique scores (Table 5). Areas with more clayey and heavier soil texture were less permeable to water, resulting in weaker vertical infiltration, and they were more likely to retain ground water, which is not conducive to the occurrence of drought disasters. Different soil textures were thus scored as in Table 6. In the calculation of the environmental sensitivity index, all factors except the river network density were normalized by the maximum value.

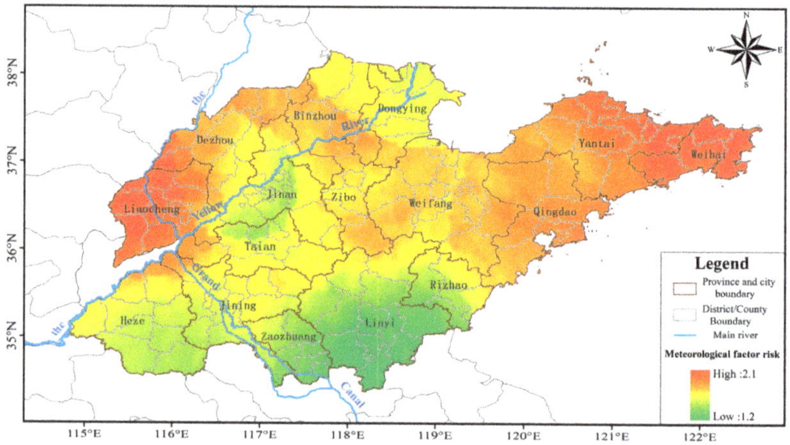

Figure 4. Spatial distribution of peanut drought meteorological factor risk in Shandong Province.

Table 5. Land use type scores.

Land Use Type	Arable land	Woodland	Grassland	Waters	Construction Land	Unused Land
Score	4	2	3	1	5	6

Table 6. Soil texture composite score.

Soil Texture	Overall Score	Soil Texture	Overall Score
Clay	1	Clay loam	2
Chalky loam	5	Loam	4
Sandy clay loam	3	Sandy loam	6
Loamy sand	6	Sand	7

The spatial distribution of elevation, slope, river network density, land use type, and soil texture in Shandong Province are shown in Figure 5. It can be seen that the elevation of Shandong Province ranges from −169 to 1527 m, with an average elevation of about 90 m. The slope ranges from 0 to 49.1°, and the slope is below 1.5° in most areas, and the spatial distribution of elevation and slope is roughly the same. Areas with higher elevation and slope are concentrated in mountainous areas, such as Tai'an, Zibo, Jinan, and parts of Linyi in central Shandong, while the peninsula areas such as Yantai are also relatively high. The density of the river network in the northwest of Shandong Province is obviously higher than in the rest of the province; In particular, along the Yellow River where the river network density is high, high-value areas are mainly distributed in Dezhou, Binzhou, Jinan, and other parts of the region. The land use types in Shandong Province are mainly arable land and construction land, and arable land is distributed in a large area in all cities; forest land and grassland are more concentrated in the south and east of Shandong Province;

the distribution of water is not concentrated; construction land is mainly distributed in the center of each city; and the area occupied by unused land is very small. Loam and sandy clay loam are widely distributed in Shandong Province, and clay soil is scattered in the province.

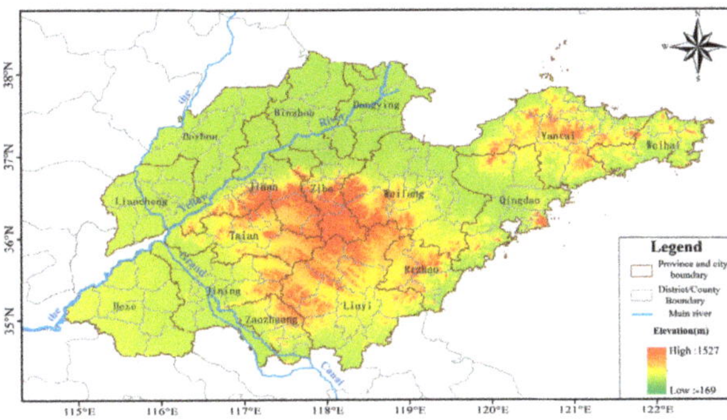

(a)

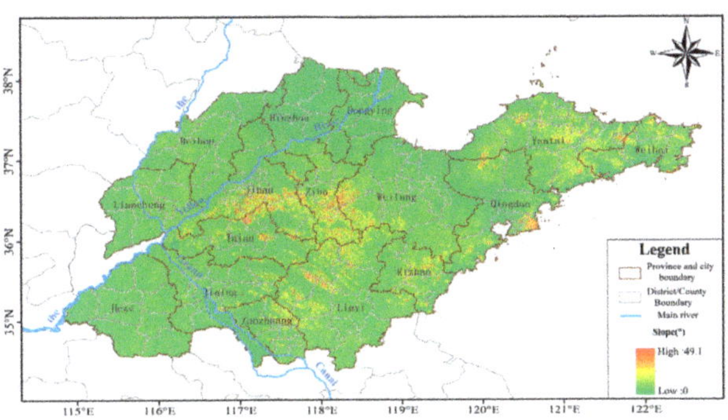

(b)

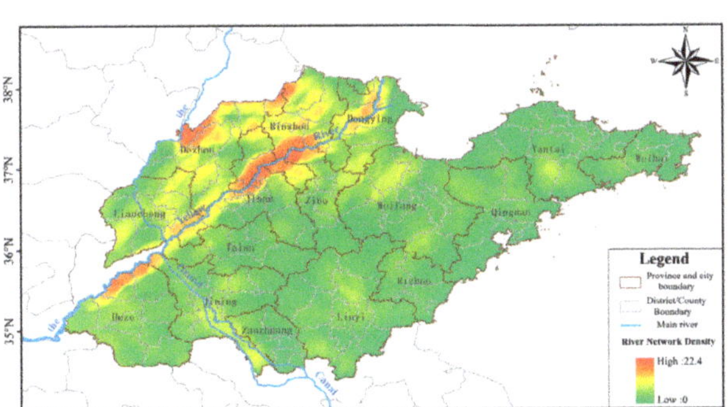

(c)

Figure 5. *Cont.*

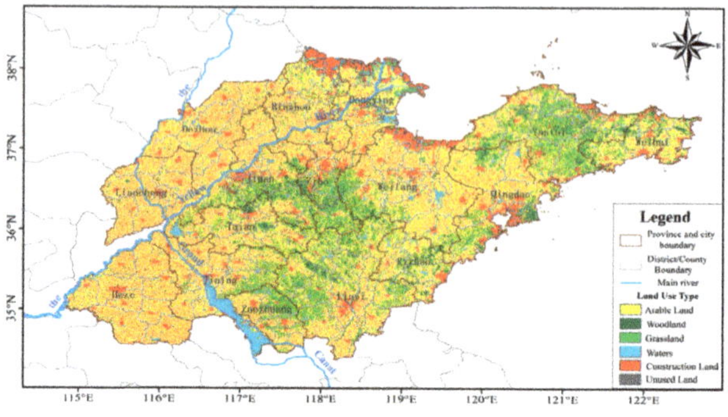

(d)

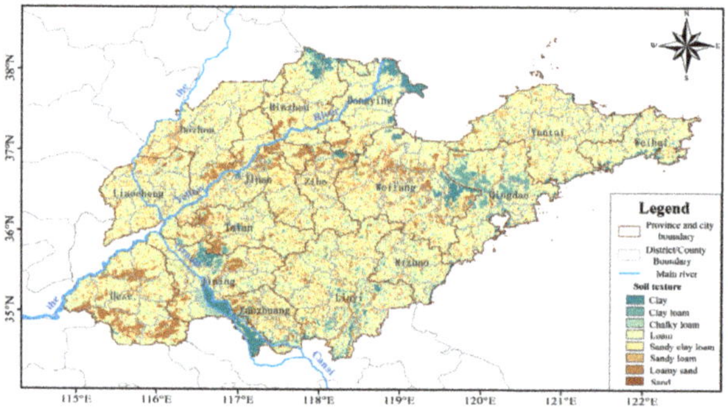

(e)

Figure 5. Spatial distribution of elevation (**a**), slope (**b**), river network density (**c**), land use type (**d**), and soil texture (**e**) in Shandong Province.

The five factors of the zoning results for elevation, slope, river network density, land use type, and soil texture were spatially superimposed according to the weights of 0.2064, 0.2064, 0.3683, 0.1094, and 0.1094. This resulted in the zoning results for the environmental sensitivity risk of peanut drought pregnancy in Shandong Province, as shown in Figure 6. Figure 6 shows the spatial distribution of this type of risk in Shandong Province had relatively little variability, with the low-value areas concentrated along the Yellow River and the rest sporadically distributed in Dezhou, Binzhou, Zaozhuang, and Jining. The high-risk areas were distributed in the province, but mainly in Jinan and the cities of Weifang, Qingdao, Yantai, Weihai, Heze, and Linyi.

3.1.3. Drought Risk Zoning for Peanuts in Shandong Province

According to the calculation from (2) and using the natural discontinuity point method to classify the areas with risks into low, medium, and high levels of risk, we obtained a spatial distribution map of peanut drought risk in Shandong Province (Figure 7).

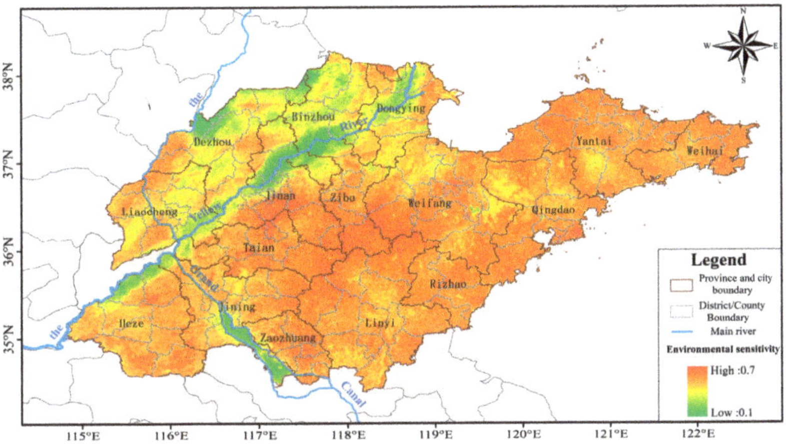

Figure 6. Spatial distribution of environmental sensitivity of peanut drought pregnancy in Shandong Province.

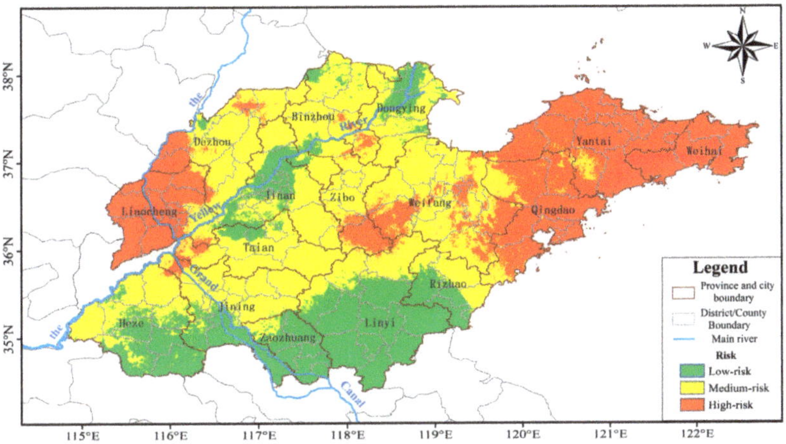

Figure 7. Spatial distribution of peanut drought risk in Shandong Province.

As Figure 7 shows, peanut drought risk in Shandong Province has an obvious spatial distribution trend, showing the spatial distribution characteristics from high in the northwest to low in the southeast. Low-, medium-, and high-risk areas accounted for 22.6%, 50%, and 27.4% of the province's area, with areas of 35,685.4 km^2, 79,013.8 km^2, and 43,227.8 km^2, respectively. The entire area of Weihai City was a high-risk area while the cities of Qingdao, Yantai, and Liaocheng all had relatively larger areas identified as high-risk areas; meanwhile, the cities of Dezhou, Binzhou, Zibo, and Weifang had sporadically distributed areas with a high risk for drought that affects peanuts. The cities of Weifang and Dongying had a relatively large area with a moderate risk. The cities of Jining, Heze, Tai'an, Jinan, Zaozhuang, Linyi, and Rizhao were mainly exposed to low and moderate risks. In particular, Zaozhuang was basically in a low-risk area.

3.2. Spatial Distribution of Exposure of Peanut-Drought-Affected Bodies

In this paper, the disaster-affected body is the peanut, so the peanut planting areas were selected as the exposure index. The larger the peanut planting area, the higher the chance of being affected by the drought, so when calculating the exposure index, the factor was standardized by the maximum value, and the natural discontinuity point method was used to classify the exposure of the disaster-affected body into low, medium, and high exposure, and the spatial distribution of the exposure of the peanut-drought-affected body in Shandong province was obtained (Figure 8).

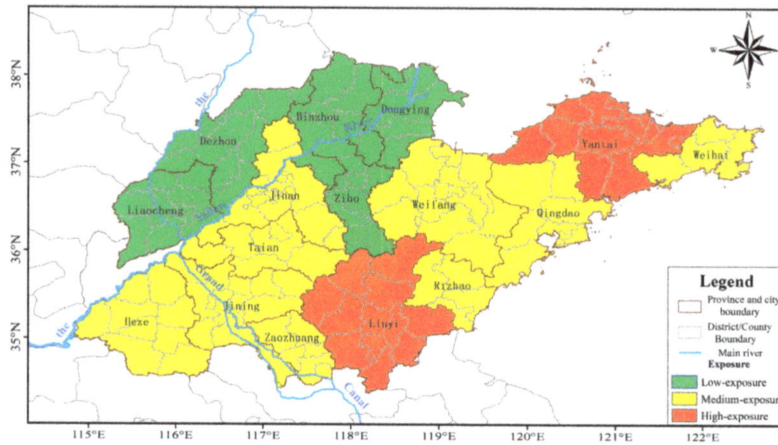

Figure 8. Spatial distribution of exposure of peanut-drought-bearing bodies in Shandong Province.

Figure 8 shows that the medium- and high-exposure areas for peanut drought in Shandong account for 76.8% of the total area of the province, with a high degree of exposure overall. Among these areas, the high-exposure areas covered 32,669.1 km^2, or 20.7%, of the province; these were mainly distributed in Yantai and Linyi. The medium-exposure areas were mainly distributed in Heze, Jining, Zaozhuang, Tai'an, Jinan, Weifang, Rizhao, Weihai, and Qingdao with an area covering 88,538.9 km^2, or 56.1%, of the province. The low-exposure area covered only 36,692 km^2, mainly including the cities of Liaocheng, Dezhou, Binzhou, Zibo, and Dongying, or 25% of the province.

3.3. Spatial Distribution of Vulnerability of Peanut-Drought-Bearing Bodies

Since this paper took administrative districts as the unit for zoning, if the planting areas of peanuts in two administrative districts were the same, but two administrative districts were different, the percentages of peanut cultivation area were different, as obviously their vulnerabilities were not the same. Therefore, the percentage of peanut cultivation area was chosen as a vulnerability indicator in this paper. The larger the percentage of the peanut cultivation area, the stronger the vulnerability. This factor was normalized to the maximum value in the calculation of the vulnerability index. The natural discontinuity point method was used to classify the vulnerability of disaster-affected bodies into low, medium, and high vulnerability, and the spatial distribution of vulnerability of peanut-drought-affected bodies in Shandong province was obtained (Figure 9).

Figure 9 shows the vulnerability of drought-affected areas where peanuts are grown in Shandong Province has obvious spatial regional differences. The low-, medium-, and high-vulnerability areas accounted for 58.3%, 14.9%, and 26.8% of the province's area, with areas of 92,059.1 km^2, 23,566.5 km^2, and 42,274.4 km^2, respectively. Among these, the cities of Linyi, Rizhao, Yantai, and Weihai were areas with a generally high vulnerability to drought. Medium-vulnerability areas were mainly distributed in the cities of Tai'an, Zaozhuang, and Qingdao. Low-vulnerability areas were widely distributed in the cities of Heze, Jining, Liaocheng, Dezhou, Jinan, Zibo, Binzhou, Dongying, and Weifang.

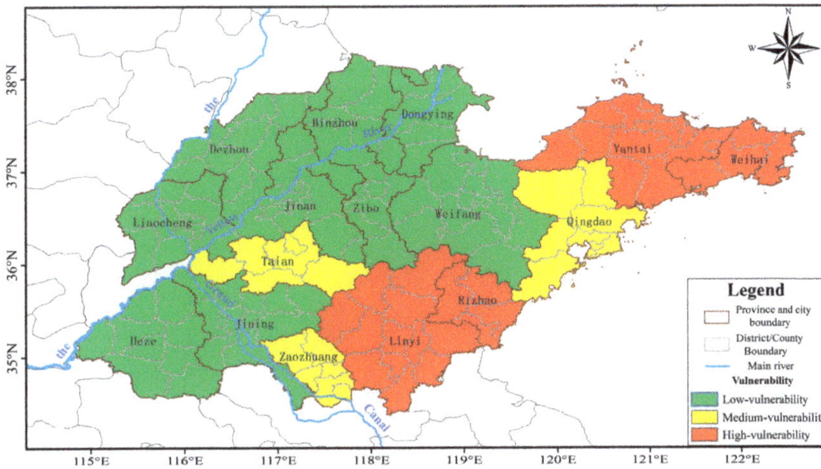

Figure 9. Spatial distribution of vulnerability of peanut-drought-bearing bodies in Shandong Province.

3.4. Spatial Distribution of Disaster Prevention and Mitigation Capability

The capacity to prevent and mitigate disasters refers to various management measures and countermeasures designed to prevent and mitigate meteorological disasters. The higher the economic level of a location, the stronger the capacity of humans to prevent and mitigate disasters. The higher the per capita GDP and income of farmers as and the higher their education level, the better ability they will have to develop policies designed to prevent and respond to disasters when peanut drought occurs. Improved construction of water conservation facilities is a direct manifestation of a stronger capacity to prevent and mitigate drought disasters; to reflect this reality, the per capita GDP and income of farmers, their level of education, and the existence of water conservation facilities were selected as indicators of the capacity of an area to prevent and mitigate disasters. Additionally, all factors are normalized to the maximum value in the calculation of the disaster prevention and mitigation capacity index. These selected indicators were given weights of 0.4226, 0.2708, 0.1443, and 0.1622, respectively, by using the AHP method. The calculations were carried out according to (2), and the natural discontinuity point method was used to classify the disaster prevention and mitigation capacity into low, medium, and high levels to obtain the spatial distribution of the capacity to prevent and mitigate drought disasters related to growing peanuts in Shandong Province (Figure 10).

Figure 10 shows the spatial distribution of the capacity to prevent and mitigate drought disasters related to peanuts in Shandong Province is highly variable. The area with a high level of this type of capacity covered 64,581.1 km^2 and was mainly distributed in Jinan, Dongying, Weifang, Qingdao, Yantai, and Weihai, or 40.9% of the province's area. The area with a medium level of this type of capacity was mainly distributed in Jining, Binzhou, Zibo, and Rizhao and covered 32,211.6 km^2 or 20.4% of the province's area. The area with a low level of this type of capacity covered 61,107.3 km^2 or 38.7% of the province's area; this area was mainly in the cities of Liaocheng, Dezhou, Tai'an, Heze, Zaozhuang, and Linyi.

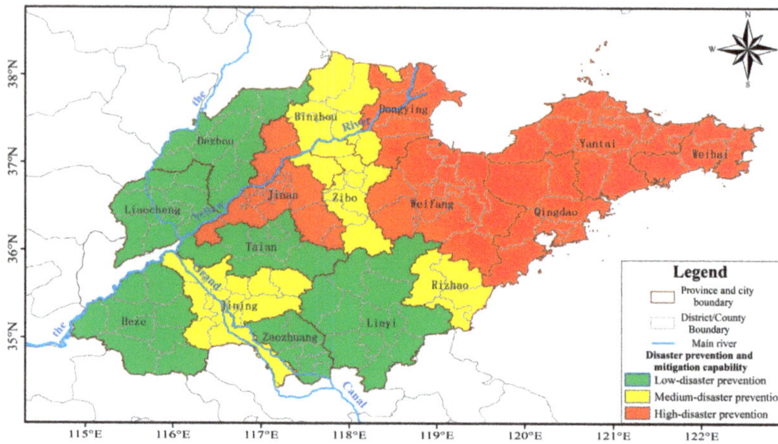

Figure 10. Spatial distribution of peanut drought prevention and mitigation capacity in Shandong Province.

3.5. Spatial Distribution of Comprehensive Peanut Drought Risk

The four factors of the zoning results, risk, exposure of disaster-bearing bodies, vulnerability of disaster-bearing bodies, and disaster prevention and mitigation capability, were spatially superimposed according to the weights of 0.5, 0.167, 0.167, and 0.167. This allowed us to obtain the zoning results of the comprehensive risk of peanut drought in Shandong Province, as shown in Figure 11. Note that the comprehensive risk of peanut drought differs substantially in different areas. The medium- and high-risk regions were located mainly in the west and central areas, with low-risk regions in the east. The overall spatial distribution shows a strong degree of fragmentation. Table 7 shows the areas of the high-, medium-, and low-risk regions in each city. In Yantai, Linyi, Liaocheng, and Weihai, high-risk regions had the largest area. In Binzhou, Dongying, and Jinan, there were no high-risk regions. In Tai'an, Weifang, and Heze, the medium-risk regions had the largest area. The number of areas with medium risk in Weihai was zero. In Jinan, and Dongying, low-risk regions had the largest area. In summary, there were no low-risk regions in Qingdao or Yantai, only medium- and high-risk areas. High-, medium-, and low-risk regions for peanut drought amounted to 36,833.4 km^2, 83,441.1 km^2, and 37,625.5 km^2, accounting for 40.6%, 37.0%, and 22.4% of the total land area, respectively.

Table 7. Low-, medium-, and high-risk areas in Shandong Province by city.

	Low-Risk Area		Medium-Risk Area		High-Risk Area	
	Area (km^2)	**Ratio (%)**	**Area (km^2)**	**Ratio (%)**	**Area (km^2)**	**Ratio (%)**
Binzhou	5098.2	53.1	4498.6	46.9	0	0
Dezhou	3104.1	29.3	6953.8	65.6	541.1	5.1
Dongying	7082.6	98.1	135.8	1.9	0.0	0.0
Heze	1178.3	9.7	10,921.3	90.1	18.3	0.2
Jinan	10,330.0	98.9	110.3	1.1	0.0	0.0
Jining	5342.9	47.6	5862.8	52.2	16.3	0.1
Liaocheng	18.3	0.2	3003.9	34.3	5737.3	65.5
Linyi	13.1	0.1	10,004.1	58.2	7173.1	41.7
Qingdao	1.0	0.0	11,068.9	98.9	121.7	1.1
Rizhao	9.1	0.2	1940.8	36.3	3392.6	63.5
Taian	147.6	1.9	7484.3	95.4	214.4	2.7
Weihai	0.0	0.0	0.0	0.0	5714.6	100.0
Weifang	887.0	5.5	15,168.6	94.4	12.2	0.1
Yantai	0.0	0.0	15.3	0.1	13,961.3	99.9
Zaozhuang	3345.6	73.8	1175.7	26.0	9.1	0.2
Zibo	1076.9	17.7	5005.8	82.3	3.0	0.0

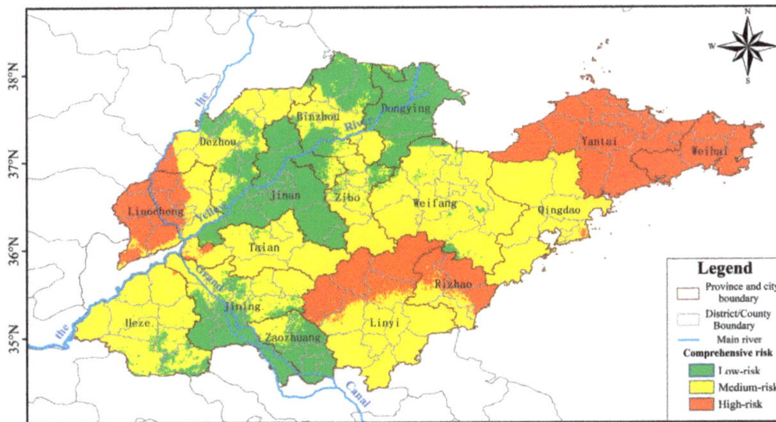

Figure 11. Spatial distribution of peanut drought comprehensive risk index zoning result in Shandong Province.

4. Discussion

Droughts are divided into meteorological drought, climatic drought, atmospheric drought, agricultural drought, hydrological drought, and socioeconomic drought [25], which correspond to different monitoring methods and indicators for different drought types. These include precipitation (P) and precipitation anomaly percentages (Pa), a standardized precipitation index (SPI), relative wetness index, integrated drought index, soil moisture drought index, Palmer drought index, soil moisture remote sensing model, vegetation water supply index, and so on [26]. Among them, atmospheric drought, agricultural drought, hydrological drought, and socioeconomic drought involve several meteorological factors, soil property factors, and socioeconomic factors, which are easier to obtain at smaller spatial scales but more difficult to obtain at larger spatial scales. Therefore, this study focused on whether precipitation during the critical growing period of peanuts met its growth demand and therefore selected meteorological drought index and considered only precipitation. In the future, with more and more basic data at high spatial resolution, agricultural drought will be further considered, and parameters such as field moisture capacity and soil weight will be added to more accurately target the water deficit in peanuts caused by soil water deficiency during the peanut reproductive period.

In this paper, the meteorological factor risk was based on the observational results of meteorological observation sites; the frequency of drought occurrence in the entire growth cycle of peanuts at each site was calculated. A spatial distribution map of drought risk was obtained using Kriging interpolation in Arc-GIS. However, spatial interpolation methods other than Kriging exist, such as inverse distance weight interpolation, the spline function method, and trend surface analysis. This paper further compared the results of the centralized interpolation methods, as shown in Figure 12, which shows the spatial distribution of the frequency of drought occurrence obtained using different interpolation methods. The comparison shows that the spatial distribution of drought occurrence frequencies obtained by different interpolation methods varied in that the results of Kriging interpolation, inverse distance weight interpolation, the spline function method, and trend surface analysis all differed greatly. These four interpolation methods show that the areas with a high frequency of drought affecting peanuts during the entire growing period were located in the northwestern and eastern areas of Shandong Province. The results of existing studies show that the areas with drought frequency from 20% to 25% in the Yellow River and Huaihai regions include certain planting areas in the western, northwestern, northern, and central Yellow River basin, as well as in the northern and northeastern planting areas in the Huaihe River basin. The planting areas in the northwestern Huaihe River basin have a lower drought frequency of less than 15% and are not prone to drought [19]; after image comparison, the results obtained with Kriging interpolation method were more consistent with the results of related studies.

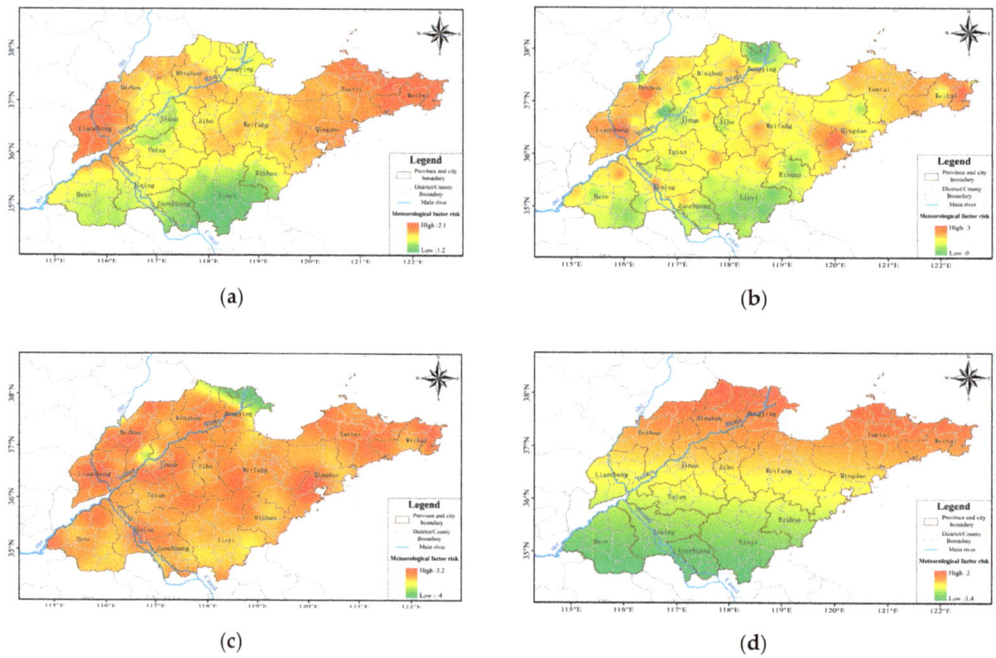

Figure 12. Spatial distribution of peanut drought meteorological factors risk in Shandong Province: (**a**) Kriging interpolation method; (**b**) inverse distance weight interpolation method; (**c**) spline function method; and (**d**) trend surface analysis method.

In the process of risk zoning, the main indicators that were selected were exposure, zoning of the vulnerability, and the capacity of an area to prevent and mitigate disasters, along with social and economic indicators; however, the influence of the statistical yearbook data based on the indicators tended to not be uniform across counties, with some indicators available in some counties and not in others. In order to achieve uniformity, there were cases in which adjustments or replacements were mad according to specific situations, which would more or less affect the zoning results. Nevertheless, the influence of the statistical yearbook information caused some indicators to be limited to the city level, so there were no indicators at the county level in some places, which would affect the spatial resolution of the zoning. In addition, the influence of statistical yearbook data caused the indicators selected to not be available, such as indicators based on the physical significance of exposure, and the vulnerability and capacity of an area to prevent and mitigate disaster because the data were not available in the statistical yearbooks, so similar indicators had to be substituted for these. For example, the education level of peanut farmers should be expressed by the percentage of farmers with a diploma at each level, but some statistical yearbooks lacked relevant information, and the number of school students had to be used to express the local level of education. This type of treatment also affected the final risk zoning results.

In this paper, when constructing the risk index, exposure index, vulnerability index, disaster prevention and mitigation index, and comprehensive risk index, weights needed to be assigned to different influencing factors. Since the above indices were used as dependent variables in this study and no specific data were available, methods such as multiple regression, principal components, and neural networks could not be used. The main advantage of the hierarchical analysis method is that it is a more appropriate method to determine the weights of the respective variables when the specific value of the dependent variable is not known and only the independent variable is available. It constructs a discriminant matrix by the relative importance of two factors and solves the weight of each factor on the premise that the matrix passes the heterogeneous consistency test. The hierarchical analysis method organically combines qualitative and quantitative methods of evaluation, neither one-sidedly pursuing high mathematical logic, nor simply focusing on subjective behavior and conscious judgment, so it is widely used when the dependent variables lack quantitative data and the weights of their influence factors need to be determined.

This paper assessed and zoned the risk of peanut-related drought in Shandong Province in terms of risk, exposure, vulnerability, prevention, mitigation capacity, and integrated risk assessment. Because little in the way of related research results exist, the analysis of this paper is comprehensive, and the zoning indices used proved to be good. The present study provides a research example for related studies, has certain reference value, and can promote future studies. The Jiaodong Peninsula is one of the major peanut-production areas in Shandong Province, so the actual peanut yields per unit in Yantai and Weihai of Shandong Province were used to verify the comprehensive peanut drought risk result obtained in this paper, as shown in Figure 13. The comprehensive peanut drought zoning results of the selected areas show that the high-risk areas were mainly distributed in Weihai and the northeastern areas of Yantai, and the spatial distribution of peanut yields per unit showed that peanut yields per unit in Weihai and the northeastern areas of Yantai were low. This shows that peanut yields per unit were lower in areas with higher integrated risks, indicating that the results obtained in this paper have a certain degree of accuracy and credibility.

Comprehensive analysis of peanut drought risk zoning results showed that the spatial distribution of peanut drought hazard, exposure, vulnerability, and disaster prevention and mitigation capacity in Shandong Province varied widely. Yantai in Shandong Province is a high-risk zone and also the area with the largest peanut cultivation area, the highest exposure zone, and highest vulnerability zone, and the comprehensive risk of Yantai determined it as a high-value zone. While Liaocheng in western Shandong Province is a high-value area for comprehensive risk of peanut drought, which is mainly caused by high risk and low disaster prevention and mitigation capacity; Weihai in eastern Shandong Province is a high-risk area with high vulnerability, which eventually forms a high-comprehensive-risk area. Therefore, for different regions, targeted measures can be proposed. For example, Liaocheng should focus on improving disaster prevention and mitigation capacity, while Yantai and Weihai should focus on improving the drought resistance of peanut varieties. The risk assessment and zoning of peanut drought in Shandong Province can not only improve people's understanding of peanut drought, but also provide a reference for relevant departments to develop and carry out peanut drought prevention and relief policies and decisions.

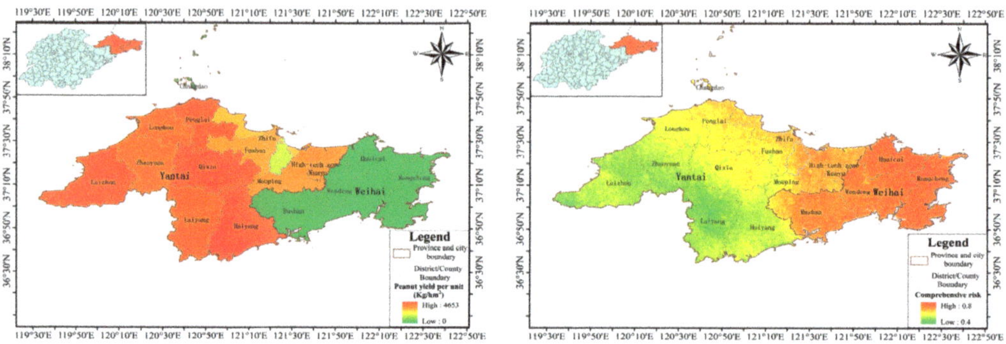

(**a**) (**b**)

Figure 13. Spatial distribution of peanut yields per unit in Yantai and Weihai (**a**); spatial distribution of comprehensive peanut drought risk in Yantai and Weihai (**b**).

5. Conclusions

The proportion of areas at medium and high risk for peanut drought in Shandong Province was 80.6%, with the cities of Weihai, Yantai, and Qingdao having a significantly higher risk than other cities. The proportion of medium- and high-exposure areas was 76.8%, which is high overall and distributed in patches in the cities of Yantai and Linyi. The medium- and high-vulnerability areas accounted for 41.7% of the province's area. Except for the cites of Weihai, Yantai, Linyi, and Rizhao, other cities had low vulnerability. Areas with a strong, medium, and relatively strong capacity to prevent and mitigate drought accounted for 61.3% of the province's area. The cities of Qingdao, Yantai, and Dongying had the highest capacity to prevent and mitigate drought disasters among the cities of the entire region. In addition, 76.2% of the area had medium and high integrated drought risk for peanuts in Shandong Province; the overall integrated risk was high.

Peanut drought risk, exposure, vulnerability, disaster prevention and mitigation capacity, and integrated risk in Shandong Province all showed spatial variability, with an inconsistent distribution of these factors across cities. The cities of Liaocheng had a significantly higher risk, poorer disaster prevention and mitigation capacity, and an overall higher integrated risk. The cities of Yantai had a higher risk, exposure, and overall higher integrated risk while Yantai and Weihai had a higher risk, vulnerability, and integrated risk. Dongying City had lower exposure, vulnerability, and integrated risk.

Author Contributions: Writing—original draft preparation, M.J.; resources, X.X.; writing—review and editing, L.Z.; methodology, Y.C. and H.S.; visualization, C.Z.; data curation, N.W. All authors have read and agreed to the published version of the manuscript.

Funding: This research was funded by the National Natural Science Foundation of China (grant No. 41771067) and the Key projects of Natural Science Foundation of Heilongjiang Province of China (grant No. ZD2020D002).

Institutional Review Board Statement: Not applicable.

Informed Consent Statement: Not applicable.

Data Availability Statement: Not applicable.

Conflicts of Interest: The authors declare no conflict of interest.

References

1. Sarkar, T.; Thankappan, R.; Kumar, A.; Mishra, G.P.; Dobaria, J.R. Stress inducible expression of AtDREB1A transcription factor in transgenic peanut (*Arachis hypogaea* L.) conferred tolerance to soil-moisture deficit stress. *Front. Plant Sci.* **2016**, *7*, 935. [CrossRef]

2. Tang, S.; Yu, S.L.; Liao, B.S.; Zhang, X.Y.; Sun, H.Y. Industry Status Existing Problems and Development Strategy of Peanut in China. *J. Peanut Sci.* **2010**, *39*, 3538. (In Chinese) [CrossRef]
3. Subrahmanyam, P.; Reddy, L.J.; Gibbons, R.W.; McDonald, D. Peanut Rust: A Major Threat to Peanut Production in the Semiarid Tropics. *Plant Dis.* **1985**, *69*, 813–819. [CrossRef]
4. Miao, H.C.; Li, Q.; Hou, X.F.; Jia, D.H.; Shi, B.X.; Ding, H.; Li, L.M.; Zhang, Z.M. Effects of Drought at Different Growth Stages on Growth and Yield of Peanut. *Xinjiang Agric. Sci.* **2021**, *58*, 441–449. (In Chinese) [CrossRef]
5. Wang, X.; Li, Y.; Zhu, H.; Chi, X.Y.; Wu, L.R.; Zhao, L.G.; Wang, J.S.; Yu, S.L. Directional Screening on Peanut Mutant with High Oil Content and Correlation between Oil Content and Leaf Water Potential. *J. Qingdao Agric. Univ.* **2019**, *36*, 30–33. (In Chinese) [CrossRef]
6. Jiang, H.F.; Ren, X.P. The Effect on SOD Activity and Protein Content in Groundnut Leaves by Drought Stress. *Acta Agron. Sin.* **2004**, *30*, 169–174. (In Chinese)
7. Girdthai, T.; Jogloy, S.; Vorasoot, N.; Akkasaeng, C.; Wongkaew, S.; Holbrook, C.C.; Patanothai, A. Heritability of, and Genotypic Correlations between, Aflatoxin Traits and Physiological Traits for Drought Tolerance under End of Season Drought in Peanut (*Arachis hypogaea* L.). *Field Crops Res.* **2010**, *118*, 169–176. [CrossRef]
8. Jongrungklang, N.; Toomsan, B.; Vorasoot, N.; Jogloy, S.; Boote, K.J.; Hoogenboom, G.; Patanothai, A. Drought Tolerance Mechanisms for Yield Responses to Pre-flowering Drought Stress of Peanut Genotypes with Different Drought Tolerant Levels. *Field Crops Res.* **2013**, *144*, 34–42. [CrossRef]
9. Wan, S.B.; Wang, C.B.; Zhu, J.H. Advantages, Problems and Countermeasures of Peanut Industry in Shandong Province. *Shandong Agric. Sci.* **2004**, *5*, 5–8. (In Chinese) [CrossRef]
10. Liao, B.S. A Review on Progress and Prospects of Peanut Industry in China. *Chin. J. Oil Crop Sci.* **2020**, *42*, 161–166. (In Chinese) [CrossRef]
11. Li, Z.D. Study on the Pricing of Peanut Drought Index Insurance in Shandong Province. Ph.D. Thesis, Guizhou University of Finance and Economics, Guiyang, China, 2016. (In Chinese).
12. Zhang, J.Q.; Liang, J.D.; Zhou, D.W. Risk Assessment of Ecological Disasters in Jilin Province Based on GIS. *Chin. J. Appl. Ecol.* **2007**, *18*, 1765–1770. (In Chinese) [CrossRef]
13. Wang, P. Research on Method of Natural Disaster Regionalization Based on Geographical Information System. *J. Beijing Norm. Univ. Nat. Sci.* **2000**, *36*, 410–416. (In Chinese)
14. Araya, A.; Stroosnijder, L. Assessing Drought Risk and Irrigation Need in Northern Ethiopia. *Agric. For. Meteorol.* **2011**, *151*, 425–436. [CrossRef]
15. Palchaudhuri, M.; Biswas, S. Application of AHP with GIS in Drought Risk Assessment for Puruliya District, India. *Nat. Hazards* **2016**, *84*, 1905–1920. [CrossRef]
16. Mohammadhadi, N.; Amir, S. Drought Risk Assessment And Zoning Using The Standardized Precipitation Index (Spi) (Case Study: Karkheh Basin). *Desert Ecosyst. Eng. J.* **2017**, *6*, 87–100.
17. Sun, Z.Y.; Zhang, J.Q.; Zhang, Q.; Hu, Y.; Yan, D.H.; Wang, C.Y. Integrated Risk Zoning of Drought and Waterlogging Disasters Based on Fuzzy Comprehensive Evaluation in Anhui Province, China. *Nat. Hazards* **2014**, *71*, 1639–1657. [CrossRef]
18. Luo, D.; Ye, L.; Zhai, Y.; Zhu, H.Y.; Qian, Q.C. Hazard Assessment of Drought Disaster Using a Grey Projection Incidence Model for the Heterogeneous Panel Data. *Grey Syst. Theory Appl.* **2018**, *8*, 509–526. [CrossRef]
19. Wei, S.C.; Li, K.W.; Zhang, J.Q.; Yang, Y.T.; Liu, C.; Wang, C.Y. Hazard Assessment of Peanut Drought and Flood Disasters in Huang-Huai-Hai Region. *J. Appl. Meteorol. Sci.* **2021**, *32*, 629–640. (In Chinese) [CrossRef]
20. Celikkol Akcay, U.; Ercan, O.; Kavas, M.; Yildiz, L.; Yilmaz, C.; Oktem, H.A.; Yucel, M. Drought-induced Oxidative Damage and Antioxidant Responses in Peanut (*Arachis hypogaea* L.) Seedlings. *Plant Growth Regul.* **2010**, *61*, 21–28. [CrossRef]
21. Jiang, C.J.; Li, X.L.; Zou, J.X.; Ren, J.Y.; Jin, C.Y.; Zhang, H.; Yu, H.Q.; Jin, H. Comparative Transcriptome Analysis of Genes Involved in the Drought Stress Response of Two Peanut (*Arachis hypogaea* L.) Varieties. *BMC Plant Biol.* **2021**, *21*, 64. [CrossRef] [PubMed]
22. Zhang, K.; Liu, Y.; Luo, L.; Zhang, X.; Li, G.; Wan, Y.; Liu, F. Root Traits of Peanut Cultivars with Different Drought Resistant under Drought Stress at Flowering and Pegging Phase. *Acta Agric. Scand. Sect. B Soil Plant Sci.* **2021**, *71*, 363–376. [CrossRef]
23. Beynon, M. An Analysis of Distrbutions of Priority Values from Altemative Comparison Scales within AHP. *Eur. J. Oper. Res.* **2002**, *140*, 104–117. [CrossRef]
24. Gu, Y.C.; Yu, L.W.; Dai, Q. Overview of Drought Classification and Common Calculation Methods. *Northeast Water Power* **2011**, *29*, 37–39. (In Chinese) [CrossRef]
25. Zhao, M.; Huang, S.; Huang, Q.; Wang, H.; Leng, G.; Xie, Y. Assessing Socio-economic Drought Evolution Characteristics and Their Possible Meteorological Driving Force. *Geomat. Nat. Hazards Risk* **2019**, *10*, 1084–1101. [CrossRef]
26. Keyantash, J.; Dracup, J.A. The Quantification of Drought: An Evaluation of Drought Indices. *Bull. Am. Meteorol. Soc.* **2002**, *83*, 1167–1180. [CrossRef]

sustainability

MDPI

Article

Effect of Snow Cover on Spring Soil Moisture Content in Key Agricultural Areas of Northeast China

Mingxi Pan [1,2], Fang Zhao [3,4], Jingyan Ma [1,*], Lijuan Zhang [1], Jinping Qu [5], Liling Xu [2] and Yao Li [6]

1 Heilongjiang Province Key Laboratory of Geographical Environment Monitoring and Spatial Information Service in Cold Regions, Harbin Normal University, Harbin 150025, China; mhjpmx@126.com (M.P.); zhlj@hrbnu.edu.cn (L.Z.)
2 Mohe Meteorological Bureau, Mohe 165300, China; mhxll@126.com
3 Harbin Meteorological Bureau, Harbin 150036, China; zhaofang213022@163.com
4 Heilongjiang Province Institute of Meteorological Sciences, Harbin 150036, China
5 Tieli Meteorological Bureau, Tieli 152500, China; ziyu-571@126.com
6 Daxing'anling District Meteorological Bureau, Daxing'anling 165000, China; lz4818@163.com
* Correspondence: 18686818800@163.com

Abstract: As an important source of soil moisture content during spring in high-latitude areas, snow cover affects the occurrence of spring drought and crop yield and quality. There has not been sufficient research on the effect of winter snow cover on spring soil moisture content. This paper focuses on the main agricultural areas of Northeast China—the Songnen Plain and the Sanjiang Plain. Using meteorological data of both spring soil moisture content and snow cover at 19 agricultural meteorological stations from 1983 to 2019, the effect of snow cover on spring soil moisture content in the Sanjiang Plain and Songnen Plain is studied by variance analysis, spatial analysis, and correlation analysis. The results show that: (1) Compared to the Sanjiang Plain, the Songnen Plain has a significantly lower content of soil moisture at the surface (0–10 cm) and deep layer (10–20 cm, 20–30 cm) during the entire spring and every month of spring ($p < 0.05$), and a greater interannual variation of soil moisture. (2) Snow cover has a significant effect on spring soil moisture in the Songnen Plain, but not as much as one in the Sanjiang Plain. For the Songnen Plain, snow-cover duration and the snow-cover onset date has a lasting influence on spring soil moisture until May, which can extend to as deep as 20–30 cm. As months go by, its influence on shallow-layer soil gradually wears off. Maximum snow depth and the snow-cover end date only influence the April surface soil. (3) Snow cover has a strong effect on soil moisture conservation in more arid areas. Delayed snow-cover onset date, earlier snow-cover end date, and significantly shortened snow-cover duration all contribute to a spring drought soil condition in the Songnen Plain.

Keywords: snow cover; spring soil moisture; impact mechanism; Songnen Plain; Sanjiang Plain

Citation: Pan, M.; Zhao, F.; Ma, J.; Zhang, L.; Qu, J.; Xu, L.; Li, Y. Effect of Snow Cover on Spring Soil Moisture Content in Key Agricultural Areas of Northeast China. *Sustainability* **2022**, *14*, 1527. https://doi.org/10.3390/su14031527

Academic Editors: Jia Yang, Xiaodong Yan and Shaofei Jin

Received: 15 December 2021
Accepted: 25 January 2022
Published: 28 January 2022

1. Introduction

Snow is the most active constituent of the cryosphere [1]. In high-latitude areas, water is released in the form of melted snow in just a few days [2] and comprises an important source of soil moisture in spring [3,4]. At the same time, snow cover also reduces the change in soil moisture content and temperature by hindering the energy exchange between soil moisture, temperature, and the environment [5], thus conserving soil moisture. However, due to global warming, the area of snow cover has dropped significantly in the northern hemisphere in the past few decades [6,7]. Snow duration in the northern hemisphere decreases at a rate of 5.3 d/10 a [8]. Particularly prominent changes in snow cover are observed in spring. Snow cover area has seen a significant reduction during spring in the northern hemisphere [9], and snow starts to melt at a significantly earlier time in Eurasia during spring [10–13]. The decline in snow reserves and the rapid, earlier disappearance of snow cover, which causes the spring warming and soil aridness, have been the center of widespread concern in the scientific community, and they have become an important influencing factor on the degree and duration of soil aridity in spring [14,15]. Researchers both in China and overseas have studied the impact of snow cover on soil moisture through the one-point method, actual observations, remote sensing, and simulation. These studies are mainly carried out from three aspects. First, some researchers have studied the effect of snow cover on the level of spring soil moisture. For example, Shinoda [16] studied the relationship between snowmelt and soil moisture in Central Eurasia with data collected at meteorological stations and found that greater annual maximum snow depth and delayed snow-cover end date correspond to higher level soil moisture, and vice versa. Ren et al. [17] studied the effect of snowmelt on soil water and heat conditions. Their results indicate that snowmelt significantly increases the water content of shallow-layer soil. Niu et al. [18] conducted field experiments to observe the change in soil moisture content as snow melting takes place, and they concluded that snowmelt infiltration acts to a certain extent to replenish the water content of the soil. Qi [19] found through simulation that without snow, the soil moisture level in Northeast China will drop at least 20% in the March to May period. Second, some researchers study the effect that the duration of the impact snowmelt has on soil moisture. For example, Douville [20] performed simulations with the Meteor-France GCM and suggested that the effect of spring snowmelt on soil moisture can last until summer. McNamara [21] simulated and analyzed the variational characteristics of soil moisture using observed data and modeling, and concluded that snowmelt, rainfall, and evaporation jointly drive the water and heat balance in soil moisture during late spring. Zhang et al. [22] reported that snow cover has an impact on soil moisture mainly as it melts. Third, some researchers have investigated how the depth of snowmelt affects soil moisture content. For example, Jan [23] proposed that 200 mm of snowmelt water had a very small effect on spring soil moisture content for soil layers below the 90-cm horizon. Zhang [10] used soil moisture data in conjunction with snow cover stage to analyze the influence of snowmelt on the humidity of seasonal frozen soil; the research shows that the maximum depth of snow cover that influences soil moisture content is 20 cm. Flerchinger et al. [24] conducted experiments to simulate the physical process of groundwater recharge by infiltrated snowmelt. It can be seen that snow cover has a certain replenishment effect on spring soil moisture. Greater snow-cover depth and longer snow-cover duration tend to have a more significant impact on soil moisture content. However, there is no consistent conclusion on that how much time the effect of snow cover can last and how deep the snow cover can affect. Different researchers have drawn different conclusions, which may be related to differences of research areas, but there is no further study. In addition, it is no certain answer about how much spring soil moisture is from snow cover. All these problems require further research, which are also the scientific problems that this study wants to solve.

As the largest crop production site in China, Heilongjiang Province often suffers from particularly serious spring droughts. This affects crop yield and quality. However, research on the factors influencing spring soil moisture in Heilongjiang Province has mostly been focused on temperature and precipitation [25–28]. As a region with a stable amount of snow cover, the effect of winter snow cover on soil moisture conservation in spring has not been sufficiently studied. In this paper, two agricultural bases of Heilongjiang Province, i.e., the Songnen Plain and Sanjiang Plain, which have clearly different soil moisture contents, are selected as the study sites. Using the soil moisture and meteorological data (1983–2019) from 19 agricultural meteorological stations, the impact of snow cover on spring soil moisture is analyzed for both sites. The objectives of present study were to (1) determine the variance analysis of two agricultural bases in Heilongjiang Province. (2) discuss the influence of snow cover on spring soil moisture content (3) reveal the contribution by snow-cover conditions on spring soil moisture. The results of this study form the scientific basis for the early warning of spring drought, the development of more efficient irrigation schemes, and crop yield prediction.

2. Materials Methods

2.1. Study Sites

The two main agricultural bases of Heilongjiang Province, the Songnen Plain and Sanjiang Plain (Figure 1), were chosen as the study sites. The Songnen Plain is in the western part of Heilongjiang Province. With a crop production area of 5.35 million hm^2, this region makes up 52% of the total crop production area of the province. It has a high-latitude continental monsoon climate, with high temperature and frequent rain in summer, severe cold and little rain in winter, and a short spring and autumn. Temperature in this region tends to have sharp fluctuations. The summer temperature is high, with high temperature and frequent rain occurring in the same season. The Sanjiang Plain is in the northeastern part of Heilongjiang Province, and it has a cultivated land area of 2.696 million hm^2. It has a temperate humid and semi-humid monsoon climate, with large annual temperature variation and abundant rainfall. Heilongjiang Province is one of the regions in China that has a stable amount of snow cover. The winter snow-cover period is primarily from November to April of the next year. During this time, the entire study area is covered with snow, with the maximum snow depth occurring in February. The snow begins to melt in March and completely melts in mid-April. Crop cultivation typically starts in May. Main food crops grown in this area include rice, wheat, corn, and soybean.

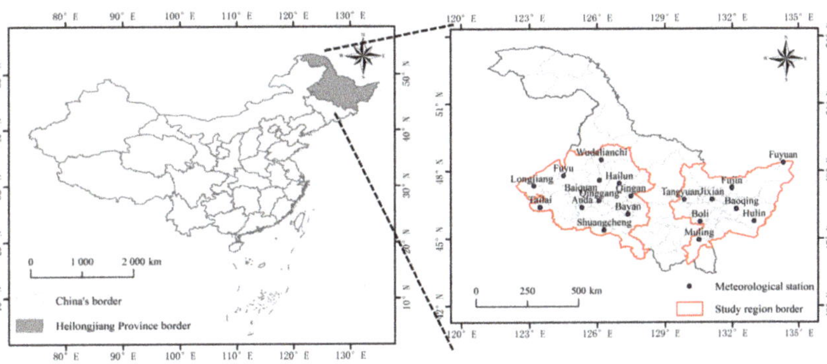

Figure 1. The study region and distribution of 19 meteorological stations in Heilongjiang Province, China.

2.2. Data Sources

Soil moisture data: spring soil moisture data for each ten days (March–May) from 19 agricultural meteorological stations (Figure 1). The time period covered spans 1983–2019. Observations were made on three soil depth layers (0–10 cm, 10–20 cm, and 20–30 cm). The data were obtained from the Meteorological Department of Heilongjiang Province. There are 31 agricultural meteorological stations in Heilongjiang Province with soil moisture data available. Provincial-wide recording of soil moisture started in 1980, but for each station, the exact commencement time is different, ranging from 1980 to 1987. To ensure continuity of data, soil moisture data from 19 agricultural meteorological stations spanning 1983–2019 were used in this study, with 11 stations on the Songnen Plain and 8 stations on the Sanjiang Plain. The data collected adequately reflect the climate characteristics of the main agricultural sites of Heilongjiang Province. Soil layers at the 0–30 cm depth range were selected for analysis, as the past study has shown that this depth interval was found to show the greatest variability in soil effective water content, and it is the cultivated layer of the soil. Soil moisture observations were all made in the observation field of agricultural meteorological stations, on grassland. No irrigation activity was carried out during the study period. The stations are located on land with different soil types, with black loam, sandy loam, clay loam, loam, yellow sandy soil, and dark brown loam being the most common ones. To eliminate the impact of soil type on soil moisture content, relative soil humidity was used to characterize soil moisture content. Relative soil humidity (RSH) is defined as the ratio between percent soil moisture (weight basis) and field capacity, and it is expressed as a percentage (%). Field capacity is the greatest amount of water that can be stably held by soil, and it is a constant value. As such, relative soil humidity can be used to characterize soil moisture and compare the degree of soil dryness and wetness between different regions. Using this value as an index, the degree of drought is defined on the following scale: severe drought: RSH $\leq$ 40%; moderate drought: 40% < RSH $\leq$ 50%; mild drought: 50% < RSH $\leq$ 60%; suitable for agriculture: 60% < RSH $\leq$ 90%; waterlogged: RSH > 90%.

Meteorological parameters: the following daily meteorological data were used: temperature, total precipitation, previous autumn precipitation, average surface temperature, average wind speed, sunshine duration, snow-cover duration, maximum snow depth, snow onset date, and snow-cover end date. Temperature, precipitation, surface temperature, wind speed, and sunshine duration were all taken in March–May of 1983–2019. The previous autumn precipitation is the amount of rainfall in September–November of the previous year. Snow-cover duration is the number of days with snow cover from August of the previous year to July of the present year. Maximum snow depth is the greatest depth of snow recorded from August of the previous year to July of the present year.

2.3. *Data Analysis*

2.3.1. Variance Analysis

The soil moisture content was analyzed by one-way analysis of variance (ANOVA). These statistical tests determine whether the differences in the soil moisture of the Songnen Plain and Sanjiang Plain were significant. Duncan's shortest significant range method was used to test the differences in the soil moisture of the Songnen Plain and Sanjiang Plain and the level of significance. First, the sum of squares for the deviations was obtained for each set of data. The statistical independence of the data among each treatment was then tested. The probability of the events was given when the statistic was greater than the F value, i.e., $p\{>F\} = p$. When $p < 0.01$, the difference was considered extremely significant; when $0.01 < p < 0.05$, the difference was considered significant; when $p > 0.05$, the difference was considered not significant.

2.3.2. Spatial Assessment Method

The spatial distribution characteristics of soil moisture content of the Songnen Plain and Sanjiang Plain were statistically calculated via the ArcGIS Grid module and Spatial Analyst module. The Kriging interpolation method is employed to analyze the spatial distribution of soil moisture content. Kriging is a regression algorithm for spatial modeling and prediction (interpolation) of random processes/random fields based on covariance functions.

2.3.3. Correlation Analysis

Correlation analysis is statistical method used to discover if there is a relationship between two variables/datasets, and the relatedness and negative/positive correlation of this relationship. The Pearson correlation method is adopted in this paper to analyze the relationship between soil moisture content and climate indicators quantitatively. In each pair, the Pearson's correlation coefficient (r) is calculated as:

$$r = \frac{\sum_{i=1}^{n}(x_i - \overline{x})(y_i - \overline{y})}{\sqrt{\sum_{i=1}^{n}(x_i - \overline{x})^2}\sqrt{\sum_{i=1}^{n}(y_i - \overline{y})^2}} \quad (1)$$

where $\overline{x}$ represents the soil moisture content, and $\overline{y}$ represents meteorological parameters.

2.3.4. Percentage Contribution

To quantitatively study the long-term percentage contribution by snow cover and other meteorological parameters on the spring soil moisture content at different depths, multiple linear regression was performed on soil moisture content at all depths and meteorological parameters of significant relevance for different spring months. The following standardized regression equation was obtained:

$$S_i = \sum a_j \times M_j \quad (2)$$

where S_i is the soil moisture content at different depths across spring months; M_j is the meteorological parameter significantly related to S_i; and a_j is the normalization constant of M_j.

Using Equation (3), the percent contribution by each meteorological parameter on spring soil moisture across different soil depths is calculated:

$$ConM_j = \frac{a_j}{\sum_{i=1}^{n} a_i} \times R^2 \times 100\% \quad (3)$$

where $ConM_j$ is the percent contribution by M_j to spring soil moisture; a is the normalization constant for each meteorological parameter; n is the number of meteorological parameters; and R^2 is the goodness of fit of the normalized regression equation.

3. Results

3.1. Comparison of Spring Soil Moisture Content for the Songnen and Sanjiang Plains for 1983–2019

3.1.1. Comparison of Spring Soil Moisture Content at the 0–30 cm Soil Depth

Figure 2 shows the spatial distribution of spring soil moisture on both the Songnen and Sanjiang Plains for 1983–2019. The Sanjiang Plain has significantly higher soil moisture content than the Songnen Plain. Spring soil moisture content varies between 69.35% and 97.44% for the Songnen Plain during this time period, and the average value is 81.39%. The coefficient of interannual variation is 0.09, and the coefficient of spatial variation is 0.072. Spring soil moisture content varies between 81.44% and 110.0% for the Songnen Plain during this time period, and the average value is 92.37%. The coefficient of interannual variation is 0.07 and the coefficient of spatial variation is 0.101. These results show that the Songnen Plain has a greater interannual variation of soil moisture than the Sanjiang Plain, whereas the intra-region variation is smaller. Variance analysis shows that there is a significant difference in spring soil moisture content between the Songnen Plain and the Sanjiang Plain, with that of the former significantly lower than the latter ($p < 0.01$).

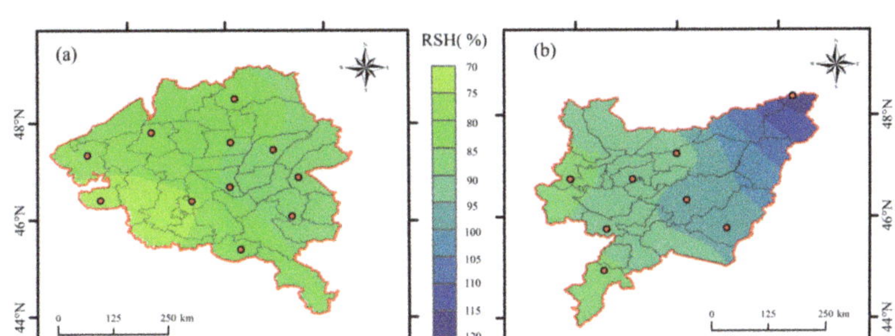

Figure 2. The spatial distribution of spring soil moisture at the 0–30 cm soil depth on both the Songnen and Sanjiang Plains for 1983–2019. ((**a**): Songnen Plain, (**b**): Sanjiang Plain).

Figure 3 shows the spatial distribution of spring soil moisture at the 0–30 cm depth in different month on the Songnen Plain and Sanjiang Plain for 1983–2019. On the Songnen Plain, the ranges of soil moisture content are 64.38–92.59%, 72.63–91.01%, and 72.30–84.73%, in March–May, respectively. On the Sanjiang Plain, the values are 89.05–129.70%, 84.96–122.54%, and 83.17–102.20% in March–May, respectively. It is obvious that the Sanjiang Plain has higher soil moisture content than the Songnen Plain. Variance analysis shows that there is a significant difference in spring soil moisture content between the Songnen Plain and the Sanjiang Plain in each month, with that of the former significantly lower than the latter ($p < 0.01$).

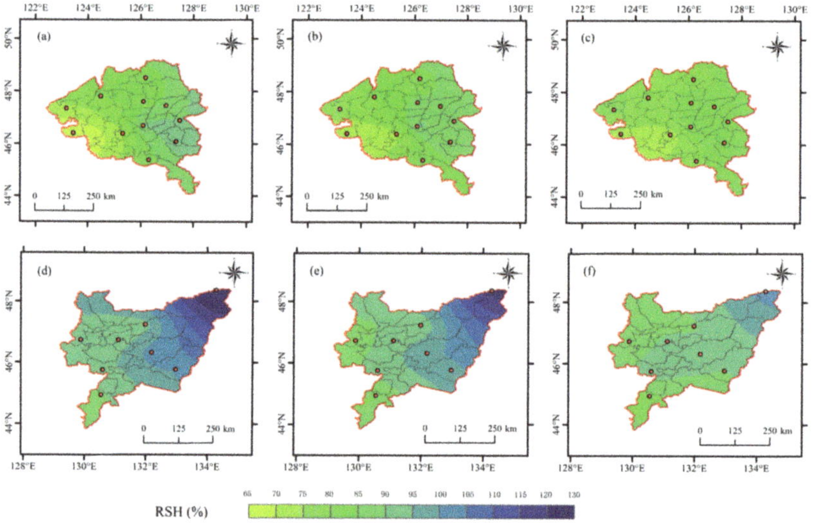

Figure 3. The spatial distribution of spring soil moisture at the 0–30 cm soil depth in different month on both the Songnen and Sanjiang Plains for 1983–2019. ((**a**): Songnen Plain in March, (**b**): Songnen Plain in April, (**c**): Songnen Plain in May, (**d**): Sanjiang Plain in March, (**e**): Sanjiang Plain in April, (**f**): Sanjiang Plain in May).

3.1.2. Comparison of Spring Soil Moisture in Across Soil Layers

Figure 4 shows the spatial distribution of spring soil moisture at different depths on the Songnen Plain and Sanjiang Plain for 1983–2019. On the Songnen Plain, the ranges of soil moisture content are 51.90–82.30%, 70.83–91.27%, and 76.34–98.69, at 0–10 cm, 10–20 cm, and 20–30 cm depths, respectively. On the Sanjiang Plain, the values are 82.94–118.37%, 88.21–130.59%, and 91.81–133.29% for 0–10 cm, 10–20 cm, and 20–30 cm depths, respectively. The soil moisture content of the Songnen Plain is lower than that of the Sanjiang Plain at each corresponding depth. Variance analysis further shows that significant differences exist in spring soil moisture at the depth ranges of 0–10 cm, 10–20 cm, and 20–30 cm between the Songnen Plain and the Sanjiang Plain. For the surface soil layer of the Songnen Plain (0–10 cm), the lowest level of soil moisture is observed in Tailai, indicating a state of mild drought, while soil conditions at other stations are suitable for farming. Conversely, 37% of the stations in the Sanjiang Plain feature waterlogged soil. These stations are primarily located in the Boli–Baoqing–Fuyuan region, while the soil moisture level at other stations is suitable for agriculture.

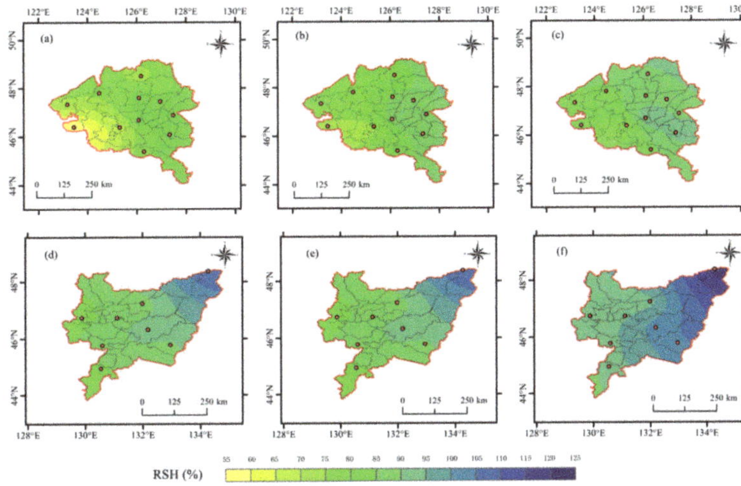

Figure 4. The spatial distribution of spring soil moisture at different depths on the Songnen Plain and Sanjiang Plain for 1983–2019. ((**a**): 0–10 cm of Songnen Plain, (**b**): 10–20 cmof Songnen Plain, (**c**): 20–30 cm of Songnen Plain, (**d**): 0–10 cm of Sanjiang Plain, (**e**): 10–20 cm of Sanjiang Plain, (**f**): 20–30 cm of Sanjiang Plain).

Figure 5 shows a comparison of spring soil moisture at different soil depths in each month for both the Songnen Plain and the Sanjiang Plain from 1983 to 2019. Variance analysis results showed that compared to the Sanjiang Plain, the Songnen Plain has significantly lower soil moisture content across various soil depths in spring months ($p < 0.05$). In some areas of the Songnen Plain, the 0–10 cm soil layer experiences drought of varying degrees in the March–May period (RSH < 60%, indicating mild drought), while on the Sanjiang Plain, the soil moisture content across different soil layers shows a mostly waterlogged state for the spring months. Compared to the Sanjiang Plain, the soil moisture content of the Songnen Plain shows greater interannual variation during spring months, with March being the largest in variation range (nearly 15%).

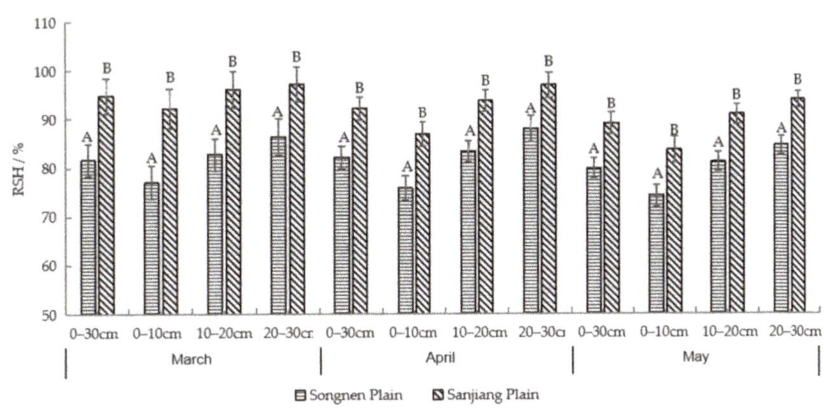

Figure 5. The spring soil moisture at different soil depths in each month for both the Songnen Plain and the Sanjiang Plain from 1983 to 2019. Note: "A", "B": Variance analysis result between the Songnen Plain and the Sanjiang Plain.

The above spring soil moisture distribution results show that the Songnen Plain has a significantly lower spring soil moisture content than the Sanjiang Plain during the entire spring season, each month of the season, across the entire soil layer, and at every depth.

3.2. Factors Influencing the Spring Soil Moisture Content of the Songnen Plain and Sanjiang Plain

Numerous studies have shown that soil moisture is closely related to meteorological factors. Taking into consideration the climate characteristics of Heilongjiang Province, 10 meteorological factors, including snow-cover conditions, were selected to perform correlation analysis with spring soil moisture at different depths and for each month of spring at different depths. The 10 factors are: daily average temperature (T), total precipitation (P), surface temperature (ST), daily average wind speed (WS), daily average sunshine duration (SD), precipitation in the previous autumn (PPA), snow-cover duration (SCD), maximum snow depth (MSD), snow-cover end date (SED), and snow-cover onset date (SOD).

3.2.1. Factors Affecting Spring Soil Moisture at Different Soil Depths

Table 1 shows the correlation analysis results for spring soil moisture at different depths. Both similarities and major differences are noticed in the meteorological factors affecting the spring soil moisture of the Songnen Plain and the Sanjiang Plain. For both locations, previous autumn precipitation has an impact on spring soil moisture at all soil depths. A strong positive correlation is observed in this case, i.e., a greater amount of previous autumn precipitation leads to higher spring soil moisture content across all the soil layers on both the Songnen Plain and the Sanjiang Plain. The only factor affecting the spring soil moisture on the Sanjiang Plain is previous autumn precipitation, whereas for the Songnen Plain, spring temperature and snow-cover conditions, including snow-cover duration, snow-cover onset date, and maximum snow depth, also play an effect. A strong negative correlation is present between spring soil moisture on the Songnen Plain and spring temperature, i.e., higher spring temperature leads to greater soil moisture evaporation in spring, and thus lower soil moisture content on the plain. There is a strong positive correlation between spring soil moisture on the Songnen Plain and snow-cover duration. Longer duration of snow cover leads to higher soil moisture content. A strong negative correlation exists between spring soil moisture on the plain and snow-cover onset date. An earlier onset of snow cover leads to higher soil moisture content. There is a significant positive correlation between surface soil moisture (0–10 cm) and the maximum snow depth on the Songnen Plain. This means that snow depth only affects surface soil moisture. A greater maximum snow depth means a higher amount of surface soil moisture in spring. The table also shows a significant positive correlation between soil moisture on the Songnen Plain at the 20–30 cm depth range and wind speed. This relationship is, however, purely numerical and has no practical significance.

Table 1. Correlation coefficient between spring soil moisture and meteorological parameters at different layers of the Songnen Plain and the Sanjiang Plain.

Area	Layer	T	P	ST	WS	SD	PPA	SCD	MSD	SDD	SOD
Songnen Plain	0–30 cm	−0.455 **	0.135	−0.093	0.168	0.179	0.656 **	0.484 **	0.28	−0.461 **	0.176
	0–10 cm	−0.420 **	0.283	0.122	−0.013	0.057	0.628 **	0.523 **	0.395 *	−0.386 *	0.153
	10–20 cm	−0.435 **	0.152	−0.05	0.119	0.149	0.660 **	0.469 **	0.284	−0.424 **	0.156
	20–30 cm	−0.454 **	−0.024	−0.309	0.352 *	0.292	0.611 **	0.414 *	0.148	−0.511 **	0.195
Sanjiang Plain	0–30 cm	−0.204	0.139	−0.189	0.122	−0.148	0.595 **	0.229	−0.047	0.013	−0.064
	0–10 cm	−0.238	0.253	−0.132	0.092	−0.205	0.566 **	0.282	0.026	−0.004	−0.100
	10–20 cm	−0.169	0.068	−0.200	0.146	−0.119	0.597 **	0.178	−0.097	0.032	−0.043
	20–30 cm	−0.182	0.074	−0.221	0.117	−0.099	0.566 **	0.200	−0.071	0.011	−0.041

Note: "*", "**": Significance at 0.05 and 0.01 levels, daily average temperature (T) (°C), total precipitation (P) (mm), surface temperature (ST) (°C), daily average wind speed (WS) (m/s), daily average sunshine duration (SD) (h), precipitation in the previous autumn (PPA) (mm), snow-cover duration (SCD) (d), maximum snow depth (MSD) (cm), snow-cover end date (SED), and snow-cover onset date (SOD).

3.2.2. Factors Affecting Soil Moisture at Different Depths for Each Month of Spring

Correlation analysis on soil moisture content at different depths for each month of spring (Table 2) shows the following: (1) previous autumn precipitation impacts the soil moisture at different depths for each month of spring; (2) temperature and precipitation affect soil moisture differently for different soil depths and spring months; and (3) a significant correlation exists between spring soil moisture on the Songnen Plain and snow-cover conditions, whereas such a relationship is barely present on the Sanjiang Plain. Specifically, previous autumn precipitation has a positive correlation with soil moisture at every depth for each month of spring (except for the surface soil moisture of May), i.e., a greater amount of precipitation in previous autumn leads to higher spring soil moisture content for both the Songnen Plain and the Sanjiang Plain. On the Songnen Plain, the soil moisture at each depth shows a significant negative correlation with temperature during April and May, and a significant positive correlation with precipitation in May. On the Sanjiang Plain, the soil moisture at each depth shows a significant positive correlation with temperature and precipitation in May. On the Songnen Plain, snow-cover duration shows a significant positive correlation with soil moisture at every depth for each month of spring; snow-cover onset date has a significant correlation with soil moisture at every depth in March. For April, this impact is demonstrated in soil layers at 10–20 cm and 20–30 cm depths; snow-cover end date shows a significant correlation with surface soil moisture; maximum snow depth shows a significant correlation with April soil moisture at the depth ranges of 0–10 cm and 10–20 cm. On the Sanjiang Plain, however, a significant correlation between surface soil moisture and snow-cover duration is observed only in March. The correlation coefficient also shows that for the Songnen Plain, as months go by, snow-cover duration and snow-cover onset date affect soil moisture to a lesser extent as soil depth increases. However, these two factors continue to influence soil moisture until May. The effect of maximum snow depth and snow cover end date is mostly observed in April.

Table 2. Correlation coefficient between spring soil moisture and meteorological parameters at different layers in each month of the Songnen Plain and the Sanjiang Plain.

Area	Month	Layer	T	P	ST	WS	SD	PPA	SCD	MSD	SDD	SOD
Songnen Plain	March	0–10 cm	−0.282	−0.195	0.062	−0.147	0.145	0.621 **	0.547 **	0.230	−0.390 *	−0.062
		10–20 cm	−0.278	−0.210	−0.095	0.000	0.138	0.657 **	0.445 **	0.152	−0.504 **	0.014
		20–30 cm	−0.295	−0.283	−0.292	0.195	0.099	0.600 **	0.372 *	0.037	−0.505 **	0.112
	April	0–10 cm	−0.380 *	0.267	0.159	0.089	−0.041	0.507 **	0.479 **	0.471 **	−0.314	0.362 *
		10–20 cm	−0.427 **	0.136	−0.083	0.179	0.022	0.558 **	0.461 **	0.363 *	−0.343 *	0.267
		20–30 cm	−0.367 *	0.010	−0.313	0.410 *	0.121	0.567 **	0.441 **	0.181	−0.540 **	0.231
	May	0–10 cm	−0.455 **	0.688 **	0.137	−0.084	−0.397 *	0.410 *	0.239	0.318	−0.247	0.155
		10–20 cm	−0.384 *	0.537 **	0.052	−0.017	−0.230	0.532 **	0.346 *	0.303	−0.235	0.216
		20–30 cm	−0.405 *	0.415 *	−0.112	0.109	−0.070	0.530 **	0.349 *	0.271	−0.364 *	0.252
Sanjiang Plain	March	0–10 cm	−0.192	0.094	−0.236	0.096	0.119	0.543 **	0.352 *	−0.050	−0.175	−0.138
		10–20 cm	−0.057	0.108	−0.165	0.076	0.002	0.590 **	0.192	−0.147	−0.020	−0.097
		20–30 cm	−0.118	0.158	−0.199	0.052	−0.069	0.576 **	0.200	−0.109	0.033	−0.046
	April	0–10 cm	−0.188	−0.070	0.056	0.040	−0.217	0.501 **	0.275	0.182	0.122	0.104
		10–20 cm	−0.092	−0.309	−0.095	0.031	0.027	0.563 **	0.160	0.004	0.052	0.053
		20–30 cm	0.044	−0.316	−0.077	0.052	0.089	0.489 **	0.173	−0.077	−0.078	−0.021
	May	0–10 cm	−0.454 **	0.616 **	0.018	0.034	−0.427 **	0.271	0.009	−0.002	0.141	−0.159
		10–20 cm	−0.513 **	0.503 **	−0.167	0.194	−0.462 **	0.363 *	0.095	−0.050	0.106	−0.020
		20–30 cm	−0.410 *	0.497 **	−0.034	−0.014	−0.433 **	0.385 *	0.152	0.079	0.106	−0.030

Note: "*", "**": Significance at 0.05 and 0.01 levels, daily average temperature (T) (°C), total precipitation (P) (mm), surface temperature (ST) (°C), daily average wind speed (WS) (m/s), daily average sunshine duration (SD) (h), precipitation in the previous autumn (PPA) (mm), snow-cover duration (SCD) (d), maximum snow depth (MSD) (cm), snow-cover end date (SED), and snow-cover onset date (SOD).

3.3. Contribution by Snow-Cover Conditions on Spring Soil Moisture of the Songnen Plain

Using meteorological parameters with significant relevance to soil moisture at each depth across all the spring months, a multiple linear regression equation is established ($p < 0.01$) (Table 3) for the Songnen Plain. Following Equation (3), the percent contribution of each meteorological factor on the spring soil moisture content of the Songnen Plain is calculated using the normalization constant (Table 4). The conclusions that follow can be drawn based on the percent contribution. (1) Snow-cover onset date and snow-cover duration affect soil moisture for the entire spring. Compared to surface soil, a longer-term effect is detected at the deeper-layer soil. For example, the effect of snow-cover onset date on the moisture level of soil at the depth range of 20–30 cm could last until May, but the percent contribution gradually drops from 20.99% to 18% and further drops to 8.12% in May. The effect of snow-cover duration on the moisture level of soil at the depth ranges of 10–20 cm and 20–30 cm also extends until May, but the percent contribution decreases gradually. For surface-layer soil (0–10 cm), the effect of snow-cover onset date is felt until March, and the percent contribution is 11.68%. The effect of snow-cover duration lasts until April, but the percent contribution declines from 17.77% to 7.84%. (2) A comparison is made between the contribution by snow-cover onset date and snow-cover duration on soil moisture. For surface-layer soil, snow-cover duration makes a greater percent contribution than snow-cover onset date. For example, for March surface soil, the percent contribution by snow-cover onset date is 11.68%, and that by snow-cover duration is 17.66%. For deeper-layer soil, the opposite trend is observed. Taking the 10–20 cm layer as an example, the percent contributions by snow-cover onset date are 20.67% and 11.25%, respectively, for March and April, while those by snow-cover duration are 9.02% and 3.74%, respectively. (3) For a single month, the impact on soil moisture by snow-cover onset date gradually increases as soil depth increases, while that by snow-cover duration gradually decreases. Taking March as an example, the percent contributions on soil moisture by snow-cover onset date at depth ranges of 0–10, 10–20, and 20–30 cm deep are 11.68%, 20.67%, and 20.99%, respectively, and the percent contributions by snow-cover duration are 17.77%, 9.02%, and 4.92%, respectively. Maximum snow depth and snow-cover end date contribute greatly to the surface soil moisture during the snow melting, but the effect lasts only for a short period. In April, they make a large contribution (up to 10.94%) to the moisture content of just the surface soil (0–10 cm).

Table 3. Multiple linear regression model equation between spring soil moisture and meteorological parameters at different layers of the Songnen Plain.

Month	Layer	Multiple Linear Regression Equation	R^2	DF	SE	SE of Regression Coefficient				
March	0–10 cm	$y = 0.469^{**}x_{PPA} + 0.324^{*}x_{SCD} - 0.213x_{SOD}$	0.552 **	34	7.87	PPA 0.039	SCD 0.064	SOD 0.159		
	10–20 cm	$y = 0.536^{**}x_{PPA} - 0.361^{**}x_{SOD} + 0.158x_{SCD}$	0.604 **	34	7.12	PPA 0.035	SOD 0.144	SCD 0.058		
	20–30 cm	$y = 0.497^{**}x_{PPA} - 0.388^{**}x_{SOD} + 0.091x_{SCD}$	0.528 **	34	8.87	PPA 0.044	SOD 0.179	SCD 0.072		
April	0–10 cm	$y = 0.370^{*}x_{PPA} + 0.285x_{SDD} + 0.240x_{MSD} + 0.172x_{SCD} - 0.012x_{T}$	0.491 **	31	6.33	PPA 0.031	SDD 0.104	MSD 0.071	SCD 0.062	T 0.863
	10–20 cm	$y = 0.408^{**}x_{PPA} - 0.285x_{T} - 0.248x_{SOD} + 0.098x_{MSD} + 0.083x_{SCD}$	0.508 **	31	5.23	PPA 0.026	T 0.63	SOD 0.109	MSD 0.059	SCD 0.054
	20–30 cm	$y = 0.377^{**}x_{PPA} - 0.369^{**}x_{SOD} - 0.243x_{T} + 0.175x_{WS} + 0.125x_{SCD}$	0.629 **	31	5.77	PPA 0.029	SOD 0.132	T 0.69	WS 1.853	SCD 0.053
May	0–10 cm	$y = 0.715^{**}x_{P} + 0.313^{**}x_{PPA} - 0.204x_{T} + 0.178x_{SD}$	0.643 **	32	4.67	P 0.047	PPA 0.022	T 0.655	SD 0.036	
	10–20 cm	$y = 0.439^{**}x_{P} + 0.427^{**}x_{PPA} - 0.134x_{T} + 0.132x_{SCD}$	0.559 **	32	4.57	P 0.034	PPA 0.023	T 0.636	SCD 0.037	
	20–30 cm	$y = 0.408^{**}x_{PPA} + 0.283^{*}x_{P} - 0.185x_{SOD} - 0.176x_{T} + 0.100x_{SCD}$	0.507 **	31	4.72	PPA 0.024	P 0.035	SOD 0.097	T 0.667	SCD 0.039

Note: "*", "**": Significance at 0.05 and 0.01 levels.

Table 4. The percent contributions on soil moisture by meteorological parameters at different layers of the Songnen Plain (%).

	March			April			May		
Layer (cm)	**0–10**	**10–20**	**20–30**	**0–10**	**10–20**	**20–30**	**0–10**	**10–20**	**20–30**
PPA	25.75	30.71	26.89	16.82	18.46	18.41	14.26	21.09	17.93
SOD	11.68	20.67	20.99	-	11.25	18.00	-	-	8.12
SCD	17.77	9.02	4.92	7.84	3.74	6.10	-	6.51	4.42
MSD	-	-	-	10.94	4.43	-	-	-	-
SDD	-	-	-	12.97	-	-	-	-	-
T	-	-	-	0.54	12.92	11.84	9.32	6.61	7.76
WS	-	-	-	-	-	8.56	-	-	-
P	-	-	-	-	-	-	32.61	21.69	12.46
SD	-	-	-	-	-	-	8.11	-	-

4. Discussion

(1) In the past, researchers have studied the effect of snow cover on spring soil moisture. The results show that snow-cover promotes the change rate of soil moisture [29] and has a noticeable impact on shallow-layer soil moisture [23]. Greater snow-cover depth and longer snow-cover duration tend to have a more significant and longer-term impact on shallow-layer soil moisture content [16]. Pan et al. [30] suggested and tested an empirical approach to estimated root-zone soil moisture in snow-dominated regions using a soil moisture diagnostic equation that incorporates snowfall and snowmelt processes. The result indicated that the soil moisture diagnostic equation is capable of accurately estimating soil moisture in snow-dominated regions after the snowfall and snowmelt processes are included in the soil moisture diagnostic equation. Qi et al. [19] have investigated snow performs similar to an important reservoir. In March–May, the soil moisture would decrease at least 20.1% when there is no snow, and the main cropland area suffers more. Shinoda et al. [16] found that the yearly maximum snow depth represents a major portion of the soil water upon snow disappearance. Potopova et al. [31] presented a detailed analysis which showed that snow-cover characteristics can significantly influence soil water saturation during the first part of the growing season, while seasonal amount of SWE can explain up to 45% of soil moisture variability during early summer (April–May–June). Liang [32] investigated the farmland of Northeast China and concluded that snow-cover depth has a strong positive correlation with April and May soil moisture. This correlation is, however, spatially differentiated, with regions showing a significant correlation concentrated mostly in the southwestern part of Heilongjiang Province. A greater correlation was also observed between snow-cover depth and April soil moisture by Liang et al. [30], which is in line with the conclusion of this paper. Taking the Songnen Plain and the Sanjiang Plain in Heilongjiang Province as examples, this study analyzes the impact of snow cover on spring soil moisture. It is noticed that snow cover plays a greater effect on the spring soil moisture content of part of the Songnen Plain located in northwestern Heilongjiang Province. This effect could last until May, but it impacts April soil moisture greatly. Yet, snow cover has less effect on the spring soil moisture content of the part of the Sanjiang Plain located in southeastern Heilongjiang Province. Our results show that previous autumn precipitation, snow-cover duration, and snow-cover onset date are the most important factors affecting the soil moisture of each layer on the Songnen Plain during the spring months. Due to the winter soil freezing, the precipitation of the previous autumn is well contained with little loss. Snow cover also acts as an insulation layer, conserving soil moisture [5]. An earlier onset of snow cover enhances the conservation effect on soil moisture. Soil memory ensures that previous autumn precipitation and snow cover could have a longer-term effect on soil moisture conservation at deeper layers, lasting until May. Snow-cover depth and snow-cover end date mainly affect the April soil moisture content at shallow layers. Our analysis shows that snow melting mainly takes place from the end of March to early April. During this time, the seeping of snowmelt has a great impact on soil moisture in the shallow layers. Greater snow-cover depth and delayed snow-cover end date lead to higher

shallow-layer soil moisture. In May, as the temperature rises and precipitation increases, the effect of snow cover on soil moisture gradually decreases.

(2) The effects of snow cover on soil moisture were also different in different study areas. Douville [20] suggested that the effect of spring snowmelt on soil moisture can last until summer. McNamara [21] found it lasts until late spring. Although Zhang et al. [22] reported that snow cover has an impact on soil moisture mainly as it melts. Our analysis points to clear differences in the effect played by snow-cover conditions for soil moisture of the Songnen Plain and the Sanjiang Plain. A further comparison is made on the climatic background of both locations. Five meteorological factors (spring temperature, precipitation, surface temperature, wind speed, and sunshine duration) are selected for comparison. The results are shown in Table 5. Variance analysis reveals significant differences in the effect of four of these meteorological factors, except for spring temperature, i.e., compared to the Sanjiang Plain, the Songnen Plain experiences significantly less spring precipitation, lower surface temperature, greater wind speed, and longer sunshine duration. In particular, the spring precipitation on the Songnen Plain is only about 70% that of the Sanjiang Plain, which means a drier climate for the former. It can thus be concluded that snow helps to conserve and replenish soil moisture in dry areas. For places of higher humidity, the contribution of snow cover to soil moisture is not as significant.

Table 5. Statistical table of spring climate conditions of the Songnen Plain and the Sanjiang Plain for 1983–2019 (average value and variance analysis result).

	Songnen Plain	Sanjiang Plain
T	5.54 °C ^{a}	5.40 °C ^{a}
P	69.48 mm ^{a}	99.16 mm ^{b}
ST	−2.88 °C ^{a}	−1.86 °C ^{b}
WS	3.63 m/s ^{a}	3.38 m/s ^{b}
SD	745.1 h ^{a}	677.7 h ^{b}

Note: "a", "b": Variance analysis result between the Songnen Plain and the Sanjiang Plain.

(3) Among all the snow-cover parameters, snow-cover duration, snow-cover onset date, maximum snow depth, and snow-cover end date are thought to have a greater contribution to soil moisture on the Songnen Plain. The variation characteristics of these four snow-cover parameters have been analyzed for the Songnen Plain, and they are shown in Figure 6. The average value of snow-cover duration is 104 d, the value of maximum snow depth is 13.6 cm, the average snow-cover onset date is November 1, and the average snow cover end date is April 2. On the Songnen Plain snow-cover duration showed declining trend, and the decline rate was −0.19 d/a while the trend was not significant. The increase rate maximum snow depth was 0.20 cm/a, which was significant ($p < 0.05$). Snow-cover onset date showed increasing trend ($p < 0.05$), and the increase rate was 0.36 d/a. Snow-cover end date showed declining trend ($p < 0.05$), and the decline rate was −0.41 d/a. Compared with those in the 1980s, maximum snow depth on Songnen Plain increased by 8.0 cm, snow-cover onset date was 14.4 days later, and snow-cover end date was 16.4 days earlier in the 2010s. Changes in any of these snow-cover parameters will negatively impact the maintenance of soil moisture. Despite the significant increase in snow-cover depth, this factor only influences the April surface soil and no noticeable influence on the deeper layers.

(4) The soil become "wet soil" due to the snow melting, which keeps snow signal for a long time and interacts with the atmosphere in the long term [33]. This soil memory can influence regional and even global climate change [34,35]. Previous studies have not studied the spatial difference of snow-cover influence, but this study result indicates that the impact of snow cover on soil moisture is different in different regions, which means that the indirect impact of snow cover on climate is different when the study area changes. In the future, more attention should be focus on the difference of snow effect in different regions. The cause of this difference also needs further research, whether it is caused by the

difference of soil properties or climate background. For agriculture, exploring the influence of snow cover on spring soil moisture can improve the accuracy of soil moisture estimation in spring and predict crop growth. Based on the estimated soil moisture, more efficient irrigation scheme can be developed, and water resources can be rationally allocated.

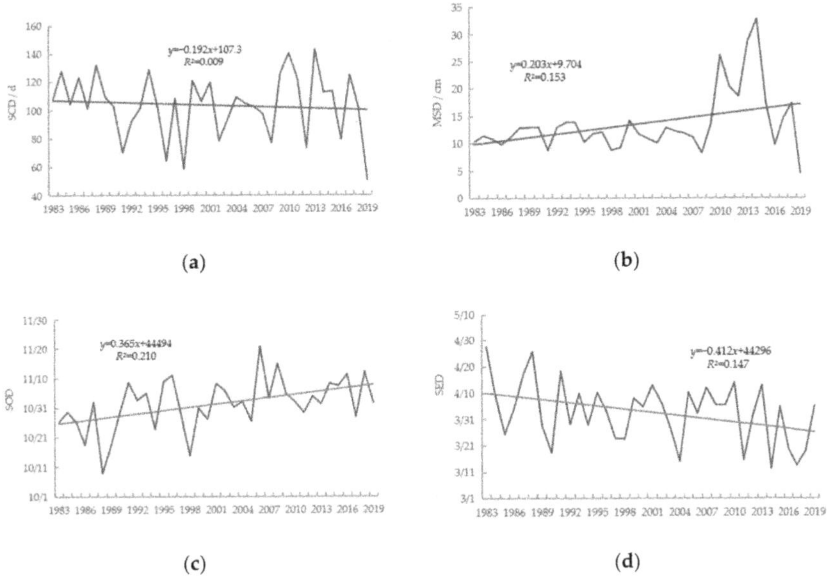

Figure 6. Interannual variation of snow-cover parameters for Songnen Plain from 1983 to 2019. ((**a**): SCD, (**b**): MSD, (**c**): SOD, (**d**): SED).

5. Conclusions

During the period of 1983–2019, the average spring soil moisture contents for the Songnen Plain and the Sanjiang Plain are 81.39% and 92.37%, respectively. Compared to the Sanjiang Plain, the Songnen Plain has significant lower spring soil moisture content and greater interannual variation of soil moisture. The Songnen Plain has a significantly lower spring soil moisture content than the Sanjiang Plain across all soil layers for the spring months.

Among all the meteorological factors, previous autumn precipitation is the main influencer of the spring soil moisture content of both the Songnen Plain and Sanjiang Plain. Snow-cover conditions have little effect on the spring soil moisture content of the Sanjiang Plain, but affects that of the Songnen Plain greatly. For the Songnen Plain, snow-cover duration and snow-cover onset date both correlate significantly with soil moisture across all the spring months. The percent contribution on soil moisture by snow-cover duration and snow-cover onset date is about 30% for March. As the months go by, the percent contribution gradually decreases. The impact on the shallow soil layer disappears in May, but a contribution of 12% is still felt at deeper layers (20–30 cm). The maximum snow depth and snow-cover end date only affect April surface soil moisture for a short while, but the percent contribution is as high as 24%.

Comparing the climate characteristics of the Songnen Plain and the Sanjiang Plain, the former is found to have a drier climate, while the different impacts snow cover has on soil moisture of the two areas could come from the differences in their climatic conditions. Snow has a stronger soil moisture conservation effect for drier areas.

Analyzing the variation characteristics of snow-cover parameters in the Songnen Plain from 1983 to 2019, it is found that the average maximum snow depth is found to increase greatly, along with greatly delayed snow-cover onset date and much earlier snow-cover end date. Snow-cover duration, however, does not change significantly. In terms of the change rate, changes in snow-cover end date and onset date happen more rapidly, and the increase in maximum snow depth happens more slowly. Overall, changes in snow-cover conditions intensify the decrease in spring soil moisture content on the Songnen Plain, which may lead to reduced grain production.

This study is only limited to two agricultural bases in Heilongjiang Province and does not conduct a detailed analysis of all the stable snow covers regions. The conclusions may have regional limitation. It has not conducted an in-depth study on the mechanism that the different impacts snow cover has on soil moisture of the two areas with different climatic conditions. It will be subject to special research in the future.

Author Contributions: M.P. analyzed the data and drafted the manuscript; F.Z., J.M. and L.Z. completed the manuscript and made major revision; J.Q. and L.X. searched references; Y.L. checked and proofread the manuscript. All authors have read and agreed to the published version of the manuscript.

Funding: This research was funded by National Natural Science Foundation of China under grant number 41771067, which is Key Projects of Natural Science Foundation of Heilongjiang Province (No. ZD2020D002). This research was also funded by National Natural Science Foundation of China under grant number 41665007, which is supported by open-ended fund of the Shenyang Institute of Atmospheric Environment of China Meteorological Administration (No. 2021SYIAEKFMS28).

Institutional Review Board Statement: Not applicable.

Informed Consent Statement: Not applicable.

Data Availability Statement: Not applicable.

Conflicts of Interest: The authors declare no conflict of interest.

References

1. Li, P. Variation of snow water resources in northwest China from 1951 to 1997. *Sci. China (Ser. D)* **1999**, *S1*, 63–69. (In Chinese)
2. Jones, H.G.; Pomeroy, J.W.; Walker, D.A.; Hoham, R.W. *Snow Ecology: An Interdisciplinary Examination of Snow-Covered Ecosystems*; University Press: Cambridge, UK, 2001; pp. 103–120.
3. Che, T.; Li, X. Spatial Distribution and Temporal Variation of Snow Water Resources in China during 1993–2002. *J. Glaciol. Geocryol.* **2005**, *27*, 64–67. (In Chinese)
4. Barnett, T.P.; Adam, J.C.; Lettenmaier, D.P. Potential impacts of a warming climate on water availability in snowdominated regions. *Nature* **2005**, *438*, 303–309. [CrossRef]
5. Fu, Q.; Hou, R.J.; Wang, Z.L.; Li, T.X. Soil moisture thermal interaction effects under snow cover during freezing and thawing period. *Trans. Chin. Soc. Agric. Eng.* **2015**, *31*, 101–107. (In Chinese) [CrossRef]
6. Brown, R.D.; Robinson, D.A. Northern Hemisphere spring snow cover variability and change over 1922–2010 including an assessment of uncertainty. *Cryosphere* **2011**, *5*, 219–229. [CrossRef]
7. McCabe, G.J.; Wolock, D.M. Long-term variability in Northern Hemisphere snow cover and associations with warmer winters. *Clim. Change* **2010**, *99*, 141–153. [CrossRef]
8. Choi, G.; Robinson, D.A.; Kang, S. Changing Northern Hemisphere snow seasons. *Climate* **2010**, *23*, 5305–5310. [CrossRef]
9. Derksen, C.; Brown, R. Spring snow cover extent reductions in the 2008–2012 period exceeding climate model projections. *Geophys. Res. Lett.* **2012**, *39*, 19504. [CrossRef]
10. Zhang, Y.; Ning, M. Spatiotemporal variability of snow cover and snow water equivalent in the last three decades over eurasia. *J. Hydrol.* **2018**, *559*, 238–251. [CrossRef]
11. Zhong, X.Y.; Zhang, T.J.; Kang, S.C.; Wang, J. Spatiotemporal variability of snow cover timing and duration over the Eurasian continent during 1966–2012. *Sci. Total Environ.* **2021**, *750*, 141670. [CrossRef]
12. Takala, M.; Pulliainen, J.; Metsamaki, S.J.; Koskinen, J.T. Detection of Snowmelt Using Spaceborne Microwave Radiometer Data in Eurasia from 1979 to 2007. *IEEE Trans. Geosci. Remote Sens.* **2009**, *47*, 2996–3007. [CrossRef]
13. Mioduszewski, J.R.; Rennermalm, A.K.; Robinson, D.A.; Wang, L. Controls on Spatial and Temporal Variability in Northern Hemisphere Terrestrial Snow Melt Timing, 1979–2012. *J. Clim.* **2015**, *28*, 2136–2153. [CrossRef]
14. Zuo, Z.Y.; Zhang, R.H. Spatial-temporal variation of soil moisture in spring in Eastern China. *Sci. China (Ser. D)* **2008**, *38*, 1428–1437. (In Chinese)
15. Vanand, L.A.F.; Laaha, G. Hydrological drought severity explained by climate and catchment characteristics. *J. Hydrol.* **2015**, *526*, 3–14.

16. Shinoda, M. Climate memory of snow mass as soil moisture over central Eurasia. *J. Geophys. Res.-Atmos.* **2001**, *106*, 33393–33403. [CrossRef]
17. Ren, X.Z.; Hu, C.S.; Zuo, H.J.; Yu, J. The effect of seasonal snow ablation on soil temperature. *J. Arid Land Resour. Environ.* **2010**, *24*, 122–125. (In Chinese)
18. Niu, C.X.; Yang, J.M.; Zhang, B.; Liu, Z.H. Influence of seasonal accumulated snow melting on the moisture and heat of shallow soil layer in northern slope of Tianshan Mountain. *J. Arid Land Resour. Environ.* **2016**, *30*, 131–136. (In Chinese) [CrossRef]
19. Qi, W.; Feng, L.; Liu, J.; Yang, H. Snow as an important natural reservoir for runoff and soil moisture in northeast China. *J. Geophys. Res. Atmos.* **2020**, *125*, e2020JD033086. [CrossRef]
20. Douville, H.; Royer, J.F. Sensitivity of the Asian summer monsoon to an anomalous Eurasian snow cover within the Météo-France GCM. *Clim. Dyn.* **1996**, *12*, 449–466. [CrossRef]
21. McNamara, J.P.; Chandler, D.; Seyfried, M.; Achet, S. Soil moisture states, lateral flow, and streamflow generation in a semi-arid, snowmelt-driven catchment. *Hydrol. Process.* **2005**, *19*, 4023–4038. [CrossRef]
22. Zhang, Y.; Haimidan, H.; Gulimire, H.; Liu, Q.Q.; Su, L.T. The influence of snowmelt on soil temperature and moisture in different freezing-thawing stages on the north slope of Tianshan mountain. *Acta. Ecol. Sin.* **2020**, *40*, 1602–1609. (In Chinese) [CrossRef]
23. Laudon, H.; Seibert, J.; Kohler, S.; Bishop, K. Hydrological flow paths during snowmelt: Congruence between hydrometric measurements and oxygen 18 in meltwater, soil water, and runoff. *Water Resour. Res.* **2004**, *40*, W03102. [CrossRef]
24. Flerchinger, G.N.; Cooley, K.R.Y.; Deng, Y. Impacts of spatially and temporally varying snowmelt on subsurface flow in a mountainous watershed: 1. Snowmelt simulation. *Hydrol. Sci. J.* **1994**, *39*, 507–520. [CrossRef]
25. Wang, M.; He, F.; Zhu, H.; Ji, Y.H. Spatial and temporal distribution of soil moisture in main agricultural areas of Heilongjiang Province under the background of climate change. *Heilongjiang Meteorol.* **2017**, *34*, 24–25. (In Chinese)
26. Zhu, H.X.; Gong, L.J.; Qu, H.H.; Lv, J.J.; Wang, L.L.; Li, B.C.; Ji, Y.H.; Yan, P.; Wang, Q.J.; Jiang, L.N. The evolution of summer soil moisture from 1981 to 2010 in Heilongjiang province. *J. Meteorol. Environ.* **2012**, *28*, 34–39. (In Chinese)
27. Wang, D.; Nan, R.; Han, J.J.; Gao, Y.G.; Li, X.F.; Sun, Y.K. Characteristics of soil humidity and its sensitivity to temperature andprecipitation in Heilongjiang province. *J. Meteorol. Environ.* **2012**, *28*, 49–53. (In Chinese)
28. Han, J.J.; Gao, Y.G.; Nan, R.; Cao, W.D. Characteristics of Soil Moisture Variation in Main Agricultural Areas of Heilongjiang Province from 1984 to 2005. *Chin. J. Agrometeorol.* **2009**, *30*, 41–44. (In Chinese)
29. Chang, J.; Wang, G.X.; Gao, Y.H.; Wang, Y.B. Impacts of snow cover change on soil water-heat processes of swamp and meadow in Permafrost Region, Qinghai–Tibetan Plateau. *Acta Ecol. Sin.* **2012**, *32*, 7289–7301. (In Chinese) [CrossRef]
30. Pan, F.F.; Nieswiadomy, M. Estimating daily root-zone soil moisture in snow-dominated regions using an empirical soil moisture diagnostic equation. *J. Hydrol.* **2016**, *542*, 938–952. [CrossRef]
31. Potopova, V.; Boroneant, C.; Mozny, M.; Soukup, J. Driving role of snow cover on soil moisture and drought development during the growing season in the Czech Republic. *Int. J. Climatol.* **2016**, *36*, 3741–3758. [CrossRef]
32. Liang, S.; Li, X.; Zheng, X.; Jiang, T.; Jiang, T.; Li, X.J.; Qiao, D.J. Effects of winter snow cover on spring soil moisture based on remote sensing data product over farmland in northeast china. *Remote Sens.* **2020**, *12*, 2716. [CrossRef]
33. Qi, D.H.; Zhou, B.T.; Xiao, C.D. Progress in studies of cryospheric changes an their impacts on climate of China. *Acta Meteorol. Sin.* **2014**, *72*, 869–879. [CrossRef]
34. Meehl, G.A. Influence of the land surface in the Asian summer monsoon—External conditions versus internal feedbacks. *J. Clim.* **1994**, *7*, 1033–1049. [CrossRef]
35. Santanello, J.A.; Peters-Lidard, C.D.; Kumar, S.V.; Alonge, C.; Tao, W.K. A modeling and observational framework for diagnosing local land-atmosphere coupling on diurnal time scales. *J. Hydrometeorol.* **2009**, *10*, 577–599. [CrossRef]

sustainability

MDPI

Article

Future Drought and Flood Vulnerability and Risk Prediction of China's Agroecosystem under Climate Change

Jiangnan Li, Jieming Chou *, Weixing Zhao, Yuan Xu, Yidan Hao and Yuanmeng Li

State Key Laboratory of Earth Surface Processes and Resource Ecology, Faculty of Geographical Science, Beijing Normal University, Beijing 100875, China

* Correspondence: choujm@bnu.edu.cn

Abstract: Droughts and floods cause serious damage to agricultural production and ecosystems, and system-based vulnerability and risk prediction are the main tools used to address droughts and floods. This paper takes the agroecosystem as the research object, uses the vulnerability model based on "sensitivity–exposure–adaptability" and "vulnerability-risk, source-risk receptor" drought and flood risk models, and establishes multi-index prediction systems covering climate change, population, agricultural technology, economy, ecology, and other factors. Using a combination of AHP and the entropy weighting method, we predict the vulnerability and risk of droughts and floods in China's agroecosystem under the SSP1-2.6, SSP2-4.5, and SSP5-8.5 scenarios from 2020 to 2050. The results show that as the scenario changes from SSP1-2.6 to SSP5-8.5 in turn, drought and flood vulnerability intensify, and the drought or flood vulnerability area expands to southern China. At the same time, future drought and flood risk patterns present the characteristics of high risk in Northeast, North, Central, and Southwest China. Therefore, major grain-producing provinces such as Heilongjiang and Henan need to do a good job of preventing and responding to agroecosystem drought and flood risks by strengthening regional structural and nonstructural measures.

Keywords: climate change; drought and flood; vulnerability; risk prediction; agroecosystem

Citation: Li, J.; Chou, J.; Zhao, W.; Xu, Y.; Hao, Y.; Li, Y. Future Drought and Flood Vulnerability and Risk Prediction of China's Agroecosystem under Climate Change. *Sustainability* **2022**, *14*, 10069. https://doi.org/10.3390/su141610069

Academic Editor: Ivo Machar

Received: 24 June 2022
Accepted: 8 August 2022
Published: 14 August 2022

1. Introduction

In the 1960s, international organizations and government agencies introduced vulnerability research into the scope of ecological research. With the continuous growth of the population, the scope of the global fragile ecological environment has increased significantly. The IPCC has officially released six scientific assessment reports on climate change [1–5], the purpose of which is to assess the scientific understanding of climate change, the impact of climate change, and possible countermeasures for the adaptation and mitigation of climate change.

A large number of scholars have successively carried out research on climate change vulnerability and risk. These studies used vegetation and ecological models and other simulation studies [6–8], indicators to assess climate change vulnerability and risk [9] or focused on adaptation measures and technological innovations for climate change risks [10–13]. Budiyono et al. [6] used vulnerability curves and flood risk assessment models; considered local factors related to hazards, exposure and vulnerability; assessing flood risk in Jakarta quantitatively, and they found that Jakarta is estimated to lose approximately US$321 million annually due to river flooding. Simane et al. [9] used the livelihood vulnerability index to study the resilience and vulnerability of five different agroecosystems in Choke Mountain communities in the Blue Nile Highlands of Ethiopia. They found that high-altitude sloping land and low-altitude steep land exhibited relatively low adaptive capacity and high vulnerability, but this study has drawbacks in regard to simplifying the internal characteristics of the community and ignoring the temporal variability of vulnerability.

Senyolo et al. [12] studied innovations in climate-smart agricultural technology at the farm level in South Africa and established a framework for classifying climate change risk, variability, and technological innovation. Drought-tolerant and early-maturing conservation agriculture, rainwater harvesting, and improved seed varieties were found to be the most suitable technologies for climate-smart agriculture in South Africa. Durowoju et al. [14] used monthly rainfall, temperature, soil moisture, vegetation condition index, normalized differential water index, and the land cover index to assess agricultural and hydrological drought vulnerability in the Kaduna River Basin in Nigeria. They found that the agricultural and hydrological drought and high vulnerability areas in this region account for about 18%. Meza et al. [15] used the drought index to calculate the comprehensive hazard index of irrigation and rain-fed systems and assessed the drought vulnerability and risk of irrigation and rain-fed agricultural systems on a global scale. The analysis shows that the drought risk of rainfed and irrigated agricultural systems presents a heterogeneous pattern on a global scale, with higher risks in southeastern Europe and Africa. Swami et al. [16] comprehensively assessed agriculture in Maharashtra, India, from 1966 to 2015 based on indicators such as monsoon and temperature changes, wasteland, scattered land holdings, human capital, physical capital, total assets, and land productivity vulnerability. The results showed that the agricultural system in the region is fragile, and regional-level variability in resource distribution, exposure, and sensitivity parameters was found, underscoring the importance of regional policy development in the region.

China is a large agricultural country, and agricultural production is vulnerable to climatic disasters, causing serious damage. Therefore, the main research object of domestic scholars is the agricultural system [17–22]. Zhou [17] analyzed the changing laws and trends of agricultural droughts, floods, and other meteorological disasters, such as climatic resources, diseases, and insect pests in China under global change. He found that with the continuous warming of the global climate and the frequent occurrence of catastrophic events, China's agricultural meteorological disasters also showed a significant trend of change. Xu et al. [20] showed that the agriculture and food system, as an important area for addressing climate change and comprehensive adaptation measures from the supply and demand side, can effectively reduce food waste and greenhouse gas emissions from agricultural sources and can increase the resilience of agricultural systems.

There are various approaches for assessing climate change vulnerability and risk, each with its own advantages and disadvantages. The biological ecological simulation method is based on theories of natural ecosystems, simulating the energy and material exchange processes between climate, soil, water, and organisms quantitatively; however, the establishment and application of comprehensive ecological models often requires interdisciplinary research by several different professional fields and work teams. The method of indicator evaluation has strong operability, but it is necessary to ensure the scientificity and rationality of the selection of evaluation indicators. At the same time, the index weighting methods of most studies are single and subjective. In addition, the main object of vulnerability and risk assessment in most studies is the ecosystem, agricultural system, economic system, or other relatively single system. Although a few studies have taken the agroecosystem as the research object, and most assessments have included static vulnerability and risk assessments, they have rarely considered different shared socioeconomic pathways (SSPs).

Therefore, this paper adopts multi-index comprehensive prediction methods based on the "sensitivity–exposure–adaptability" vulnerability framework. We established multi-level indicator evaluation systems for the drought and flood vulnerability of agroecosystems. To improve the scientificity and rationality of the evaluation indicator empowerment and evaluation, this study uses both subjective and objective weighting methods; that is, combining the AHP and entropy weight methods and assigning weights to the projection indicators of drought or flood vulnerability of agroecosystems, respectively. To explore the distribution pattern and difference in drought or flood vulnerability of agroecosystems caused by climate change under different shared socioeconomic pathways in the future, we use multiclimate model ensemble data under SSP1-2.6, SSP2-4.5 and SSP5-8.5, predicting the drought or flood sensitivity, exposure, adaptability and vulnerability of agroecosystems under three future shared socioeconomic scenarios from 2020 to 2050. On this basis, we use an agroecosystem drought or flood risk prediction model, considering vulnerability, risk source, and risk receptors, to predict the drought or flood risk of China's agroecosystem and compare the vulnerability and risk status of agroecosystems under different shared socioeconomic scenarios from 2020 to 2050.

2. Data and Methods

The research area of this paper includes 31 provincial administrative units in China, except for Hong Kong, Macao and Taiwan.

2.1. Climate Data and Population Data

The data of future climate scenarios are selected from the data output by 22 global climate models in the Sixth International Coupling Model Intercomparison Project (CMIP6) (https://esgf-node.llnl.gov/projects/cmip6)(accessed on 5 January 2022). The specific information on the selected climate model can be found in Appendix A, Table A1. The data include three SSP scenarios from 2020 to 2050: the monthly average temperature and precipitation under low-forcing scenario SSP1-2.6, medium-forcing scenario SSP2-4.5, and high-forcing scenario SSP5-8.5. We first interpolate the monthly scale data from the climate model to meteorological stations. Then, we refer to a new statistical downscaling method based on random weather generators in Liu and Zuo [23], correct the monthly scale data of the climate model based on the observation data and feed them into the random weather generator, generating climate prediction data from 699 reference weather stations in China.

The future population data includes multi-dimensional population prediction grid data ($0.5^\circ \times 0.5^\circ$) under the SSP1-2.6, SSP2-4.5, and SSP5-8.5 scenarios from 2020 to 2050 [24]. We use GIS to add population grid data of the same provincial administrative unit to the future provincial population data. Finally, the population forecast data of each provincial administrative unit from 2020 to 2050 are obtained.

2.2. Socioeconomic and Agroecological Data

The agroecological and socioeconomic data used in this study are mainly from the China Statistical Yearbook [25] by the National Bureau of Statistics of China, and the statistical data of China's forest resources inventory results [26]. The data categories include agricultural disaster-affected area, agricultural fertilizer use, gross production value of agriculture, forestry, animal husbandry and fishery, GDP per capita, total power of agricultural machinery, total reservoir capacity, flood control area, grain sown area, per capita water resources, agricultural ecological water consumption, and forest area.

2.3. Methods

2.3.1. Agroecosystem Vulnerability Prediction Model

To conduct scientific and accurate agroecosystem vulnerability predictions for each provincial administrative unit in China and build drought or flood prediction indicator systems for the vulnerability of agroecosystems, we use the "sensitivity–exposure–adaptability" vulnerability model in the IPCC [3,4,27]. This paper considers vulnerability as the degree to which a system is susceptible or unable to cope with the adverse effects of climate change, typically characterized by high sensitivity to damage, high exposure, and low adaptive capacity, as shown below. Among them, adaptability is defined as the external support of assisting a province to adapt to the hazards. It does not refer to the resilience phase of a system's adaptation to a hazard after experiencing it. We use indicators that reflect ecosystem service functions, economic and agricultural science and technology development factors in human agricultural activities, and corresponding adaptation measures to define resilience. Since the higher the vulnerability is, the smaller the adaptability is. In the data preprocessing link, this paper reversely normalizes the adaptability index data [28,29].

$$Vulnerability = f\,(Sensitivity, Exposure, Adaptability) \quad (1)$$

2.3.2. Combination of AHP and the Entropy Weight Method to Determine the Weight

AHP is a method of subjectively determining the weight of indicators. It mainly decomposes the evaluation objectives into different levels and indicators, and compares and calculates the indicators at the same level to determine the weights of different evaluation indicators [30,31].

The entropy weight method is an objective analysis method that determines the relative weight of each index in the comprehensive evaluation by the degree of dispersion between the evaluation index values. The main calculation steps are as follows:

First, the corresponding evaluation matrix is constructed.

$$A = [Y_{n1}, Y_{n2}, \ldots Y_{nm}] \tag{2}$$

Second, the data are normalized.

$$Y_{ij} = \frac{Y_{ij} - min\{Y_j\}}{max\{Y_j\} - min\{Y_j\}} \tag{3}$$

Third, the proportion P_{ij} of the i-th province under the j-th indicator is determined.

$$P_{ij} = \frac{x_{ij}}{\sum_{k=1}^{n} x_{ij}}, i = 1, 2, \ldots, n,\ j = 1, 2, \ldots m \tag{4}$$

Fourth, the entropy value e_j of the jth index is determined.

$$e_j = -1/lnn \times \sum_{i=1}^{n} P_{ij} \times ln(P_{ij}) \tag{5}$$

Fifth, the jth index difference coefficient d_j and weight w_j are determined.

$$d_j = 1 - e_j \tag{6}$$

$$w_j = \frac{d_j}{\sum_{k=1}^{m} d_j} \tag{7}$$

In this study, we use both the analytic hierarchy process and the entropy weight method to weigh the index. The AHP has been relatively and maturely applied to the determination of the weight of a multi-index system. The advantage of the entropy weight method is that it considers the objective numerical characteristics of the data. The combination of the two to determine the weight not only reflects the actual importance of each vulnerability prediction index but can also reflect the objective characteristics of each data point, making the weight of the vulnerability prediction index more scientific and reasonable [32,33].

In determining the selection method of the combination weight, we adopt the revised formula proposed by Wang et al. [34] by analyzing the problems existing in the commonly used subjective and objective combination weighting formulas. That is, the original common formula: $z_j = v_j w_j / m$ is revised to: $z_j = (v_j + w_j)/m$, and the combined weighting calculation formula is derived from this formula, as shown below:

$$z_j = (v_j + w_j)/2 \tag{8}$$

where n denotes the number of each evaluation index, m denotes the number of each province, Y_{ij} denotes the element in the ith row and jth column of matrix A, v_j is the weight determined by the analytic hierarchy process, w_j is the weight determined by the entropy weight method, and z_j is the weight determined by the combined weighting (Figures 1 and 2).

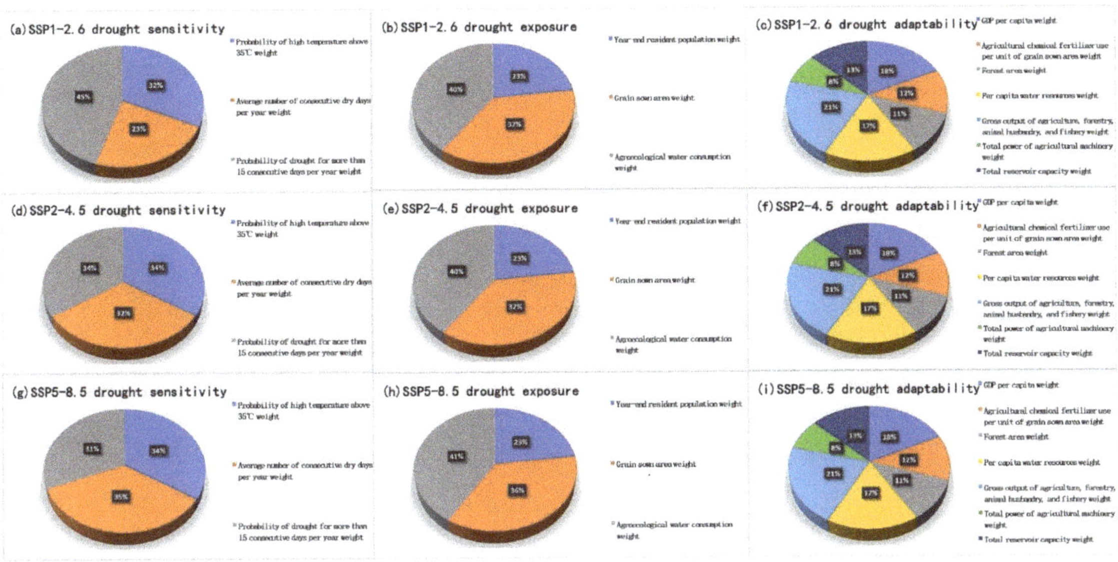

Figure 1. Estimated indicator weights of drought vulnerability in China's agroecosystem from 2020 to 2050: (**a**) SSP1-2.6 drought sensitivity; (**b**) SSP1-2.6 drought exposure; (**c**) SSP1-2.6 drought adaptability; (**d**) SSP2-4.5 drought sensitivity; (**e**) SSP2-4.5 drought exposure; (**f**) SSP2-4.5 drought adaptability; (**g**) SSP5-8.5 drought sensitivity; (**h**) SSP5-8.5 drought exposure; (**i**) SSP5-8.5 drought adaptability.

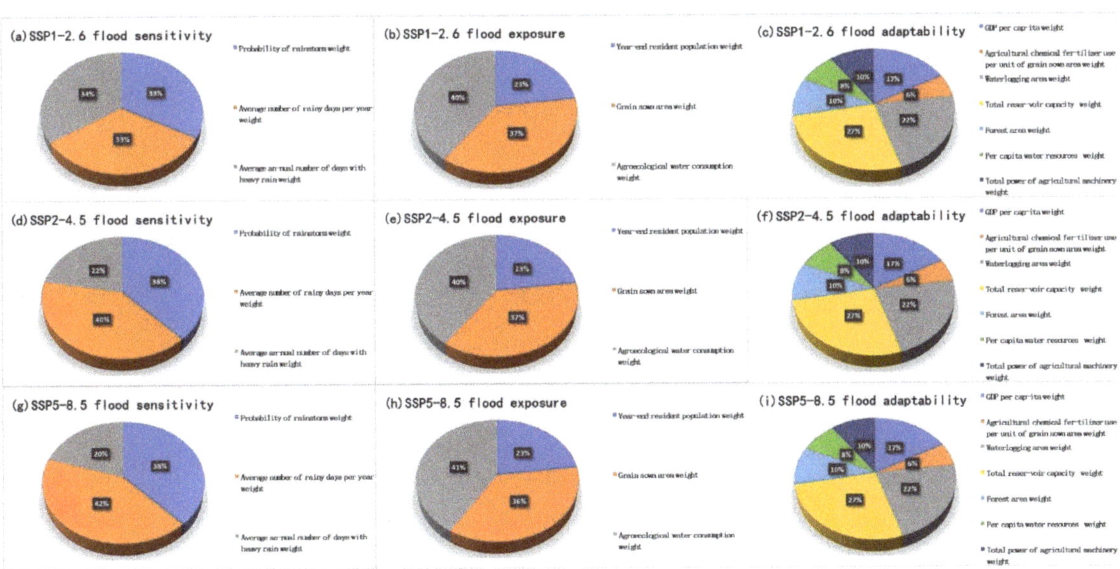

Figure 2. Estimated indicator weights of flood vulnerability in China's agroecosystem from 2020 to 2050: (**a**) SSP1-2.6 flood sensitivity; (**b**) SSP1-2.6 flood exposure; (**c**) SSP1-2.6 flood adaptability; (**d**) SSP2-4.5 flood sensitivity; (**e**) SSP2-4.5 flood exposure; (**f**) SSP2-4.5 flood adaptability; (**g**) SSP5-8.5 flood sensitivity; (**h**) SSP5-8.5 flood exposure; (**i**) SSP5-8.5 flood adaptability.

2.3.3. Agroecosystem Drought and Flood Risk Prediction Model

This study is based on the research by Xu et al. [35] and IPCC [5,27] on climate change and natural disaster risk. The drought and flood risks of agroecosystems in the context of climate change can be expressed as a functional formula of risk source, risk receptor, and vulnerability, and the multiplication of the three is the fundamental relationship. We use $R = f(H, V, E)$ to predict the drought risk or flood risk of China's agroecosystem under the SSP1-2.6, SSP2-4.5, and SSP5-8.5 scenarios from 2020 to 2050. Among them, the risk source refers to the frequent occurrence of drought or flood risk sources in China, and the intensity is expressed as the probability of occurrence over the next 30 years. The risk receptor, namely, the agroecosystem, is expressed by characterizing nine agroecosystem values, including food production, soil formation and protection, climate and gas regulation, and water conservation. Vulnerability represents drought vulnerability or flood vulnerability of agroecosystems. In addition, the product of agroecosystem value and vulnerability is the possible loss of the region and is then multiplied by the probability of drought or flooding to obtain the risk of agroecosystem drought or flooding. The equations are as follows:

$$R_y = H_y \times V_y \times E \tag{9}$$

$$R_Z = H_Z \times V_Z \times E \tag{10}$$

$$H_y = \frac{\sum n}{N} \tag{11}$$

$$H_z = \frac{\sum m}{M} \tag{12}$$

$$E = P \times A \times \alpha \tag{13}$$

where R_y is the drought risk of the agroecosystem in each province. H_y is the drought probability of the agroecosystem in each province; that is, the probability of occurrence of moderate drought, severe drought and extreme drought in a certain area from 2020 to 2050 [36]. V_y is the agroecosystem drought vulnerability in each province. E is the total value of agroecosystem services in a certain region of China. R_z is the flood risk of the agroecosystem in each province. H_z is the flood probability of the agroecosystem in each province; that is, the probability that the daily precipitation will reach a certain condition in a certain area from 2020 to 2050. V_z is the agroecosystem flood vulnerability in each province. $\sum n$ denotes the total number of months in which the drought in a certain area reaches the level of moderate drought, severe drought, and extreme drought. N denotes the total number of months in the desired year, which is 372. $\sum m$ denotes the total number of days with daily precipitation greater than or equal to the average daily precipitation in a certain area from 2020 to 2050. M denotes the total number of days in the desired year, which is 11,315. P is the total value of 9 ecological services per unit area of agroecosystem in the average state of China in the early 21st century, which is 6114.3 [37–39]. A is the area of the agroecosystem in each region, and this study uses the area of agricultural vegetation coverage; that is, the sown area of grain in each region. α denotes the ratio of the estimated annual economic price of ecological services to the economic price of ecological services in the early 21st century and uses the growth rate β of the consumer price index to calculate, which is $\alpha = (1+\beta)^{30}$. The average growth rate of China's consumer price index is assumed to be 0.03 [40], and α is 2.4.

The flow chart is shown in Figure 3.

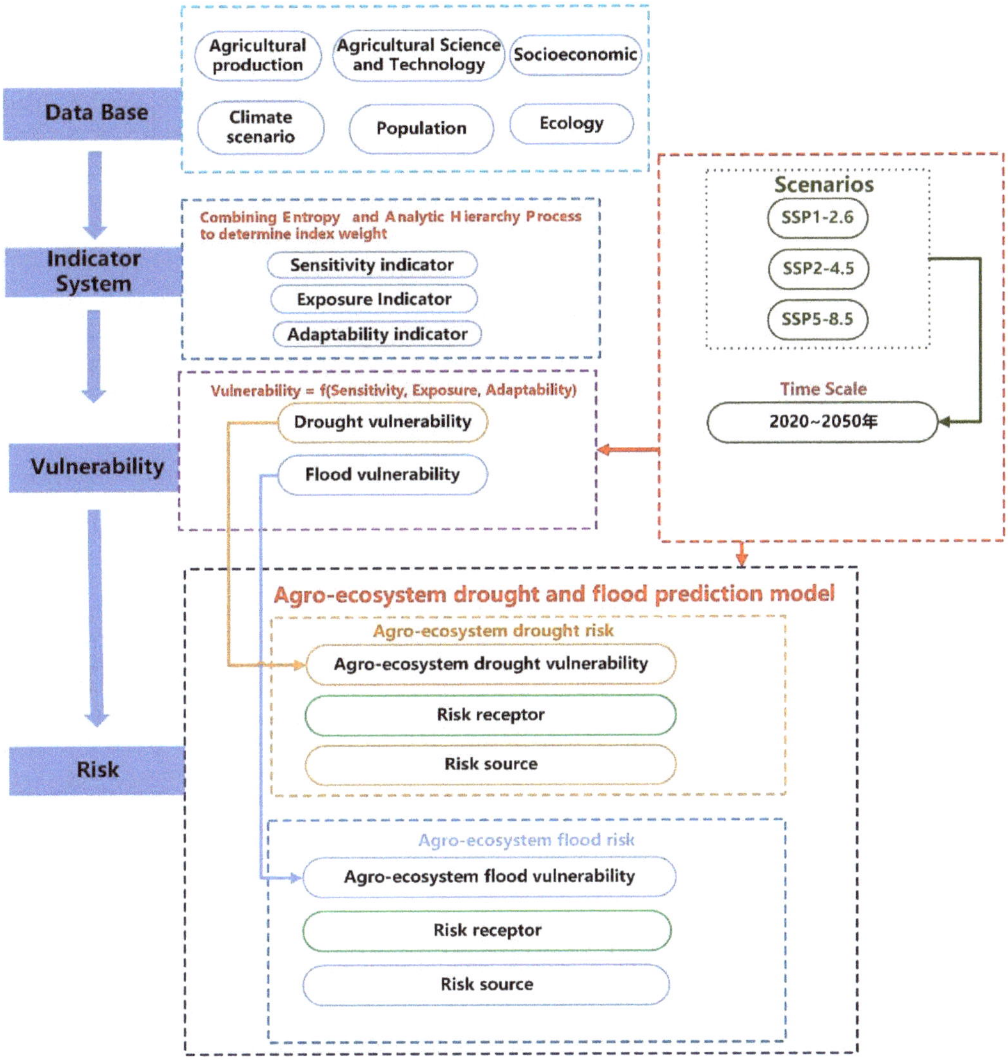

Figure 3. Flowchart of the future drought and flood vulnerability and risk prediction. (年 : year).

3. Results and Discussion

3.1. Vulnerability Estimation

In a previous study [33], we selected sensitivity, exposure, and adaptability evaluation indicators based on the "sensitivity–exposure–adaptability" vulnerability assessment framework. We built an evaluation index system separately for China's agroecosystem drought or flood sensitivity, exposure, adaptability, and vulnerability from 1991 to 2019 and conducted corresponding evaluations. The results of vulnerability assessment in the past 30 years showed that the drought–flood vulnerability of China's agroecosystem denotes a weakening trend from the central part to the surrounding areas of China, and the central provinces of Henan and Hubei are at the high drought–flood vulnerability level [33]. To ensure the accuracy of China's agroecosystem vulnerability estimation and the continuity of research in the next 30 years, we explore the distribution pattern and change in China's agroecosystem vulnerability under different shared socioeconomic scenarios. Based on previous assessment work, we predict the drought vulnerability and flood vulnerability of China's agroecosystem from 2020 to 2050.

3.1.1. Results of the Construction of the Drought Vulnerability Prediction Index System

In this study, the drought vulnerability prediction index system is constructed based on the vulnerability model (Table 1). Meteorological indicators that combine temperature and rainfall are used to construct drought sensitivity indicators for 2020~2050. The National Meteorological Administration of China is used to define the probability of occurrence of China's high temperature yellow warning at 35 °C to characterize the state of surface evaporation [41], and the surface water budget is characterized by precipitation to indirectly quantify the water budget of the agroecosystem [42]. Considering that the three major food crops in China, rice, maize, and wheat, are most vulnerable to climate change, the growth length of the critical period of water demand is usually 15–30 days [43–46]. April to September is considered to be the main growth period of most food crops in food production activities [47–50]. We use the probability of the occurrence of drought for more than 15 consecutive days and the average annual number of consecutive drought days to scientifically describe the drought status of crops and determine the drought sensitivity of agricultural ecology.

When constructing the drought exposure index in this paper, not only are the population and grain sown area considered, but three different socioeconomic path factors in the future are also included in the future population changes in different provinces. The country's basic agricultural policies and development planning factors are incorporated into the changes in the sown area of food crops, and agroecological water consumption is also included in the vulnerability analysis of agroecological drought as an assessment indicator of the degree of human participation in arid environmental exposure.

Drought adaptability is the ability to avoid or mitigate losses due to climate change risk by improving the level of science and technology and enhancing the ability to resist disasters when human beings realize the task of addressing adverse and imminent environmental changes [51]. To explore drought vulnerability in response to future extreme climate change based on the current level of drought adaptability, this study assumes that the drought resilience in 2020~2050 is the same as that in 1991–2019. Therefore, we select seven indicators, including the GDP per capita, per unit grain use of agricultural chemical fertilizers, gross production value of agriculture, forestry, animal husbandry and fishery, forest area, per capita water resources, total power of agricultural machinery, and total reservoir capacity.

3.1.2. Drought Vulnerability of China's Agroecosystem from 2020 to 2050

According to the grading threshold of drought vulnerability from 1991 to 2019 delineated by the standard deviation grading method [33,52], this paper divides the drought vulnerability in the next 30 years into three grades: low, medium, and high. The results show (Figure 4) that in the next 30 years, as the shared socioeconomic scenario increases from SSP1-2.6 to SSP5-8.5, the drought vulnerability of China's agroecosystem will gradually increase. In the SSP1-2.6 scenario, the only low-drought vulnerable province is Sichuan Province, and fourteen provinces, including Heilongjiang, Liaoning, Beijing, Guizhou, Hunan, and Guangdong, are located in drought-vulnerable areas in the agricultural ecosystem. More than half of the provinces, such as Yunnan, Fujian, and Shaanxi, which are concentrated in Northwest, Central, and East China, are located in areas with high drought vulnerability. Under the SSP2-4.5 scenario, the vulnerability of China's agroecosystem is divided into two levels: medium and high drought vulnerability. Compared with the SSP1-2.6 scenario, the range of provinces with high drought vulnerability expands to the central and southern regions of China, and the drought vulnerability of the Henan, Hunan, and Guangdong agroecosystems rises to the level of high drought vulnerability. In the SSP5-8.5 scenario, the range of provinces with high drought vulnerability further extends to southern China compared with the SSP2-4.5 scenario, and the agroecosystem drought vulnerability in Guizhou, Yunnan, and Guangxi rises to the high drought vulnerability level. In summary, as the shared socioeconomic scenarios increase from SSP1-2.6 to SSP2-4.5 and SSP5-8.5, the drought vulnerability of China's agroecosystem increases overall, and the scope of provinces with high drought vulnerability gradually expands to the central and southern regions.

Table 1. Prediction indicator system for agroecosystem drought vulnerability in China's provinces from 2020 to 2050.

Target Layer	Criterion Layer	Indicator (Unit)	Indicator Description and Calculation Method
Drought vulnerability of China's agroecosystem	Sensitivity	Probability of high temperature above 35 °C (%)	Positive indicator. Calculated by dividing the cumulative number of years with the daily maximum temperature ≥35 °C by the total number of years in the desired year
		Average number of consecutive dry days per year (d)	Positive indicator. According to the standardized precipitation index (SPI) and *Meteorological Drought Scale* [36]. The number of consecutive drought days refers to the number of consecutive days when the daily SPI reaches moderate drought, severe drought, and extreme drought. The probability of occurrence of drought for more than 15 consecutive days is calculated by dividing the cumulative number of SPI reaching moderate drought, severe drought, and extreme drought for 15 consecutive days or more in the crop growing season from April to September in the desired year by the number of years.
		Probability of drought for more than 15 consecutive days per year (%)	
	Exposure	Year-end resident population (10^4)	Positive indicator. Multidimensional population forecast data under SSP1-2.6, SSP2-4.5, and SSP5-8.5 scenarios from 2020 to 2050 [24].
		Grain sown area (10^3 hm)	Positive indicator. According to the red line policy of 1.8 billion mu of arable land in China, it is assumed that the sown area of grain will remain unchanged after 2020.
		Agroecological water consumption (10^9 m^3)	Positive indicator. Assuming that the per capita agricultural ecological water consumption from 2020 to 2050 is constant, which is the same as the situation in 2004 to 2019, then the agricultural ecological water consumption from 2020 to 2050 = $\sum$ (2004 to 2019 per capita agricultural ecological water consumption of each province × annual predicted population of each province).
	Adaptability	GDP per capita (Yuan/person)	Inverse indicator. The adaptability level is assumed to be the same as the drought adaptability status from 1991 to 2019, and the reverse normalized value of the drought adaptability data from 1991 to 2019 is used as the drought adaptability index.
		Agricultural chemical fertilizer use per unit of grain sown area (t/hm)	
		Forest area (10^4 hm)	
		Per capita water resources (m^3/person)	
		Gross output of agriculture, forestry, animal husbandry, and fishery (10^9 yuan)	
		Total power of agricultural machinery (10^4 kW)	
		Total reservoir capacity (10^9 m^3)	

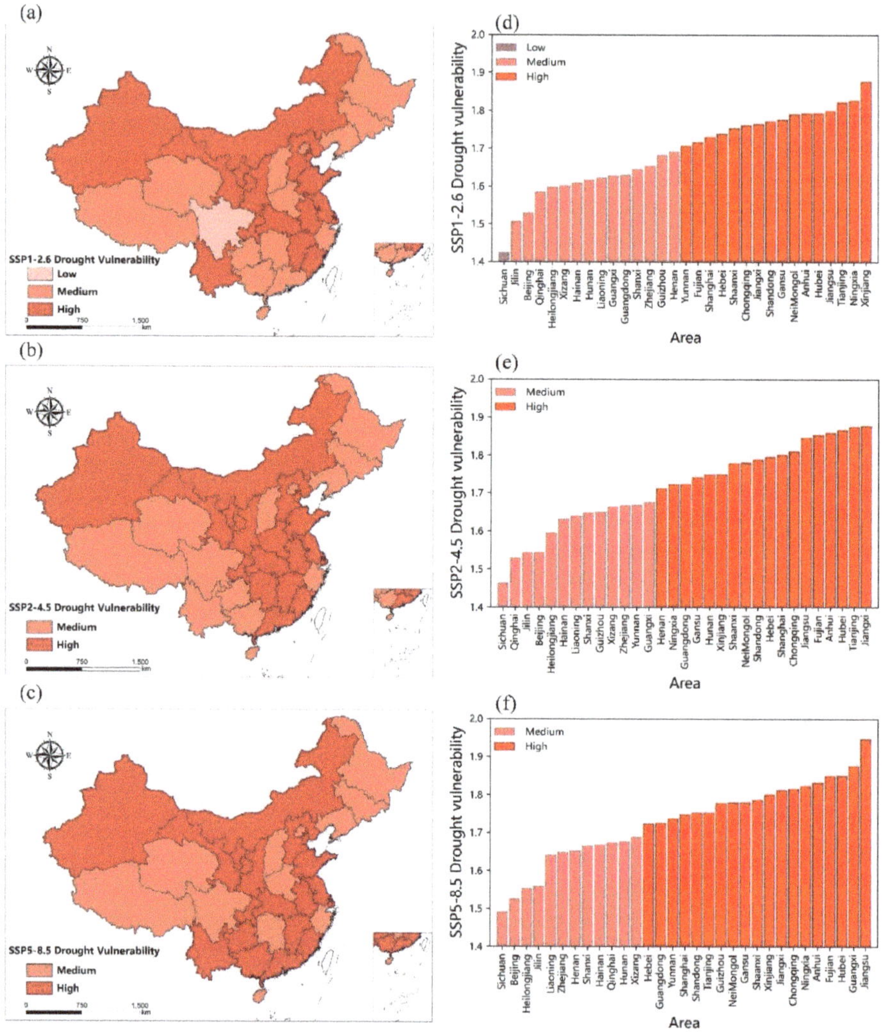

Figure 4. Drought vulnerability of China's agroecosystem under the SSP1-2.6, SSP2-4.5, and SSP5-8.5 scenarios from 2020 to 2050. Note: (**a**) SSP1-2.6 drought vulnerability classification; (**b**) SSP2-4.5 drought vulnerability classification; (**c**) SSP5-8.5 drought vulnerability classification; (**d**) SSP1-2.6 drought vulnerability index; (**e**) SSP2-4.5 drought vulnerability index; and (**f**) SSP5-8.5 drought vulnerability index.

3.1.3. Construction of the Flood Vulnerability Prediction Index System

This study builds a flood vulnerability prediction index system based on the vulnerability model (Table 2). Generally, floods are characterized by a certain order of magnitude, and long-term continuous precipitation leads to submerged or stagnant water in low-lying areas. In this study, the probability of occurrence of heavy rain, the average annual number of heavy rain days, and the average annual number of heavy rain days are used to construct the flood sensitivity index from 2020 to 2050. Flood exposure refers to the order of magnitude of population, production, and living, ecosystem life and environmental service functions, and infrastructure or economic and cultural assets that may be affected by flood disaster losses.

Table 2. Prediction index system of agroecosystem flood vulnerability in China's provinces from 2020 to 2050.

Target Layer	Criterion Layer	Indicator (Unit)	Indicator Description and Calculation Method
Flood vulnerability of China's agroecosystem	Sensitivity	Probability of rainstorm (%)	Positive indicator. Calculated by dividing the cumulative number of years with daily precipitation ≥50 mm by the total number of years in the desired year *.
		Average number of rainy days per year (d)	Positive indicator. Calculated by dividing the cumulative number of days with daily precipitation ≥ 50 mm by the number of years in the desired year *.
		Average annual number of days with heavy rain (d)	Positive indicator. Calculated by dividing the cumulative number of days with daily precipitation at (25 mm and 50 mm) by the number of years in the desired year *.
	Exposure	Year-end resident population (10^4)	Positive indicator. Multidimensional population forecast data under SSP1-2.6, SSP2-4.5, and SSP5-8.5 scenarios from 2020 to 2050 [24].
		Grain sown area (10^3 hm)	Positive indicator. According to the red line policy of 1.8 billion mu of arable land in China, it is assumed that the sown area of grain will remain unchanged after 2020.
		Agroecological water consumption (10^9 m^3)	Positive indicator. Assuming that the per capita agricultural ecological water consumption from 2020 to 2050 is constant, which is the same as the situation in 2004 to 2019, then the agricultural ecological water consumption from 2020 to 2050 = $\sum$ (2004 to 2019 per capita agricultural ecological water consumption of each province × annual predicted population of each province).
	Adaptability	GDP per capita (Yuan/person)	Inverse indicator. Adaptability is assumed to be the same as the flood adaptability from 1991 to 2019, and the reverse-standardized value of the flood adaptability data from 1991 to 2019 is used as the flood adaptability index
		Agricultural chemical fertilizer use per unit of grain sown area (t/hm)	
		Waterlogging area (10^3 hm)	
		Per capita water resources (m^3)	
		Forest area (10^4 hm)	
		Total power of agricultural machinery (10^4 kW)	
		Total reservoir capacity (10^9 m^3)	

Note: The data of per capita water resources are missing; therefore, the data time of per capita water resources is selected from 2004 to 2019 *, indicating that the calculation method is based on the *precipitation grade* [53].

Flood adaptability refers to the active coping and adaptation capabilities brought about by direct and indirect service functions of ecosystems, human-led agricultural science, technology, and economic development factors. To explore flood vulnerability in response to future extreme climate change based on the current flood adaptability level of each province, this study assumes that the flood adaptability from 2020 to 2050 is the same as that from 1991 to 2019. Therefore, this paper uses per capita GDP and unit grain sown area to utilize agricultural chemical fertilizers. This is characterized by seven indicators:

scalar volume, flood control area, forest area, per capita water resources, total power of agricultural machinery, and total reservoir capacity.

3.1.4. Flood Vulnerability of China's Agroecosystem from 2020 to 2050

The results of the study on the distribution pattern and changes in vulnerability to floods in China's agroecosystem under different SSP scenarios from 2020 to 2050 show that (Figure 5) as the shared socioeconomic scenarios increase from SSP1-2.6 to SSP5-8.5 in turn, the degree of China's agroecosystem flood vulnerability increases slightly, and the number of provinces with high flood vulnerability increases. Under the SSP1-2.6 scenario, areas with high flood vulnerability include Jiangsu, Chongqing, and Guizhou. Under the SSP2-4.5 scenario, Hebei is added to the provinces with high flood vulnerability. Under the SSP5-8.5 scenario, Guangxi and Tibet are added to the provinces with high flood vulnerability. In addition, under the three SSP scenarios, the areas with low flood vulnerability include Xinjiang, Qinghai, Gansu, and Guangdong, which may be due to the small amount of annual precipitation in the northwestern region and the insignificant fluctuation of precipitation. As a relatively developed province in South China, Guangdong has strong flood control and disaster-resistance agricultural infrastructure, as well as flood-resistance and emergency-rescue capabilities. Therefore, the agroecosystem flood vulnerability in Xinjiang, Qinghai, Gansu, and Guangdong Provinces will be lower under the three different climate scenarios in the future. At the same time, with the increase in greenhouse gas emissions, the fluctuation in precipitation in the southwestern region will intensify. In addition, the terrain of the southwestern region is complex, and extreme precipitation and flood disasters are prone to occur. Therefore, the agroecosystems in Tibet, Chongqing, Guizhou, and Guangxi in the southwestern region are more vulnerable to floods.

It is worth noting that under the three SSP scenarios, Jiangsu has a high level of flood vulnerability, which may be because Jiangsu is a coastal area, and climate warming causes high precipitation intensity and frequency. The ability to cope with heavy precipitation and floods is weak; therefore, the vulnerability of agroecosystems to floods is high. In conclusion, as the shared socioeconomic scenarios change from SSP1-2.6 to SSP5-8.5, the overall flood vulnerability of China's agroecosystem increases slightly, and the range of provinces with high flood vulnerability shows a trend of extending to the southwest.

3.2. Risk Estimation

This paper argues that the drought or flood risk of China's agroecosystem under climate change can be expressed as a functional equation of the combined action of risk source, risk receptor, and vulnerability. Among them, vulnerability is the consequence of the factors acting on risk. In the previous part of this study, we conducted drought or flood vulnerability projections of China's agroecosystem over the next 30 years. To further explore the risk distribution and changing characteristics of China's agroecosystem in the next 30 years, the accuracy of drought risk or flood risk prediction research has been enhanced. We also considered risk sources and risk receptors and assessed drought or flood risk in China's agroecosystem from 1991 to 2019, and the assessment results are shown in Appendix A Figures A1 and A2. In addition, to ensure the continuity of risk research work, we predicted the drought or flood risk in China's agroecosystem from 2020 to 2050.

3.2.1. Probability of Drought and Flooding in China's Agroecosystem from 2020 to 2050

From 2020 to 2050, the probability of drought in China's agroecosystem under the three SSP scenarios shows that (Figure 6a) as the shared socioeconomic scenario changes from SSP1-2.6 to SSP5-8.5 in turn, the probability of drought in each province continuously increases. Under the SSP1-2.6 scenario, the probability of drought occurrence in each province is 6–17%. Under the SSP2-4.5 scenario, the probability of drought occurrence in each province is 8–19%. Under the SSP5-8.5 scenario, the probability of drought occurrence in each province is 11–21%. Moreover, under the three different scenarios, the probability of the occurrence of drought in southwestern regions such as Sichuan, Chongqing, and Yunnan is slightly higher.

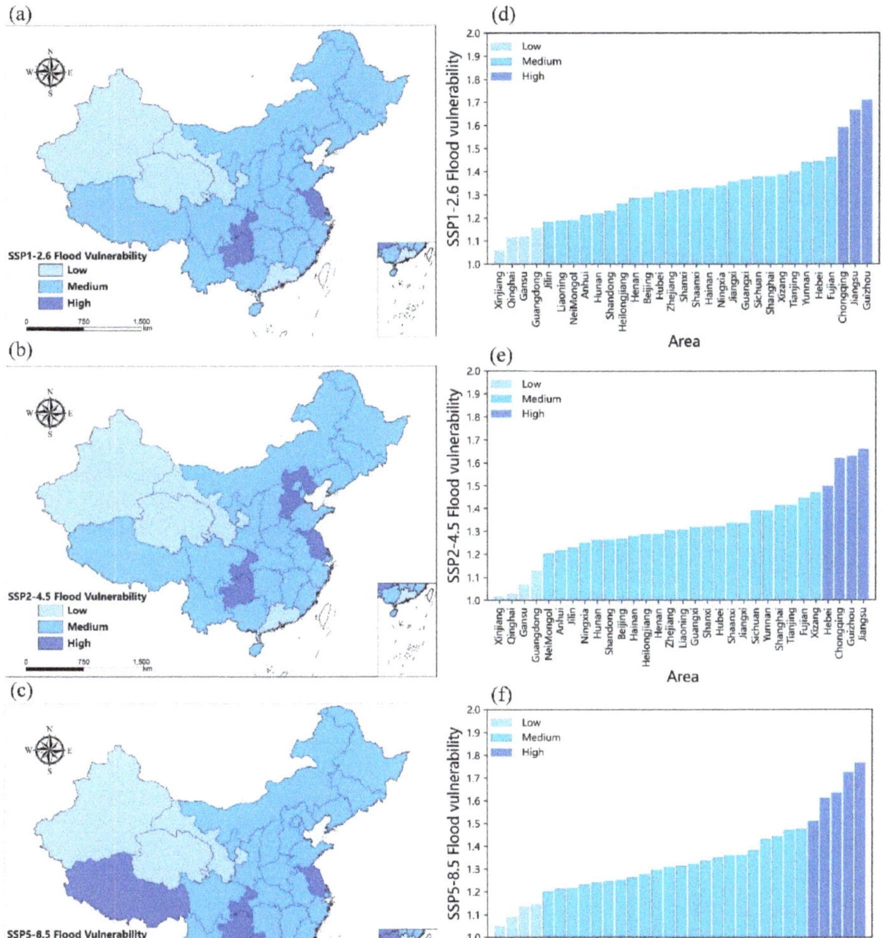

Figure 5. Flood vulnerability of China's agroecosystem under the SSP1-2.6, SSP2-4.5, and SSP5-8.5 scenarios from 2020 to 2050. Note: (**a**) SSP1-2.6 flood vulnerability classification; (**b**) SSP2-4.5 flood vulnerability classification; (**c**) SSP5-8.5 flood vulnerability classification; (**d**) SSP1-2.6 flood vulnerability index; (**e**) SSP2-4.5 flood vulnerability index; and (**f**) SSP5-8.5 flood vulnerability index.

In addition, the probability of floods in the next 30 years shows (Figure 6b) that as the scenarios change from SSP1-2.6 to SSP5-8.5 in turn, the probability of floods in each province increases slightly. In the SSP1-2.6 and SSP2-4.5 scenarios, the probability of flooding in each province is 22–34%. In the SSP5-8.5 scenario, the probability of flooding in each province is 22–35%. At the same time, under the three SSP scenarios over the next 30 years, the probability of flooding in central and southern China, such as Hunan, Fujian, Jiangxi, and Shanghai, is higher than that in northern and northwestern China.

3.2.2. The Service Value of China's Agroecosystem from 2020 to 2050

From 2020 to 2050, the evaluation results of agroecosystem service value in China's provinces show that (Figure 7) China's 31 provincial-level administrative units, except for Hong Kong, Macao, and Taiwan, have obvious differences in the value of agroecosystem services. The value of agricultural ecosystem services in the main provinces, especially Heilongjiang, Henan, and Shandong, is still relatively high. In the next 30 years, the economic value of agricultural ecosystem services will reach approximately 211.9 billion yuan and 157.6 billion yuan. In Beijing, Shanghai, and other regions that will still be dominated by finance, service manufacturing, and high-tech industries in the future, the service value of agroecosystems is relatively low, and the economic value of various services of agroecosystems is less than 2 billion yuan. In other provinces, such as Tibet, Hainan, Qinghai, and Anhui, in the next 30 years, the economic value of agroecosystem services will range from 2 billion to 110 billion yuan. In conclusion, there will be significant differences in the value of agroecosystem services in different provinces in the next 30 years, which is related to the area of agroecosystems in various provinces in China. Heilongjiang, Henan, and Shandong have always been important grain-producing areas in China in the past, mainly planting rice, wheat, and other crops. At the same time, according to *Several Opinions of the Central Committee of the Communist Party of China and the State Council on Adhering to the Prioritized Development of Agriculture and Rural Areas and Doing a Good Job in "Three Rurals"*, it clearly requires that the red line of 1.8 billion mu of arable land be strictly adhered to and the role of agricultural and rural farmers as ballast stone should be brought into play [54]. In the next 30 years or so, Heilongjiang, Henan, and Hebei may still serve as China's granaries; therefore, their agroecosystem areas will still account for a relatively high proportion, and the economic value of agroecosystem services will be high.

3.2.3. Drought and Flood Risks in China's Agroecosystem from 2020 to 2050

The drought-risk distribution patterns and changes in China's agroecosystem under different SSP scenarios from 2020 to 2050 show that (Figure 8), in the next 30 years, the drought-risk patterns of China's agroecosystem under the three SSP scenarios will denote high drought risk in Northeast, North, Central, and Southwest China. As the scenario increases from SSP1-2.6 to SSP5-8.5, in turn, the drought risk gradually increases, and the number of provinces with high drought risk also increases and shows a trend of extending to the south. In the SSP5-8.5 scenario, the overall drought risk in China is severe. Under the SSP1-2.6 scenario, half of China's provinces have a high drought-risk level, mainly distributed in the Heilongjiang River, Huaihe River, and the Yangtze River Basins, including Heilongjiang, Anhui, Sichuan, and other places. Under the SSP2-4.5 scenario, the high drought-risk provinces expand further south, adding Hunan and Jiangxi provinces. Under the SSP5-8.5 scenario, the range of provinces with high drought risk expands to the south again, adding Guizhou and Guangxi. However, under the three SSP scenarios, the western provinces of China, such as Tibet and Qinghai, are always low and medium drought-risk provinces. Regarding the spatial distribution characteristics of drought risk in China, the research of Chou et al. [55] showed that drought disasters in China have a trend of drought extending from north to south, especially in the Yangtze River Basin, where drought and extreme precipitation increase. This is consistent with the distribution of the high drought-risk areas of China's agroecosystem extending to the south as the scenario intensifies in this study.

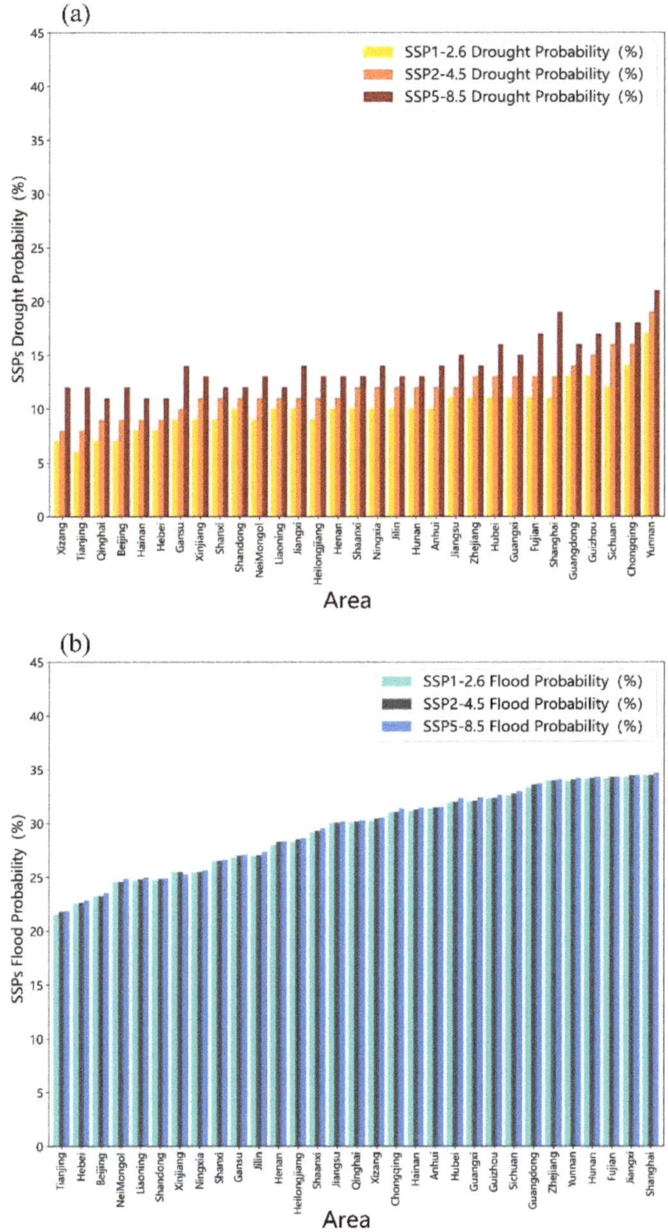

Figure 6. Probability of drought and flooding in China's agroecosystem under different SSP scenarios from 2020 to 2050. Note: (**a**) Probability of drought; and (**b**) probability of flooding.

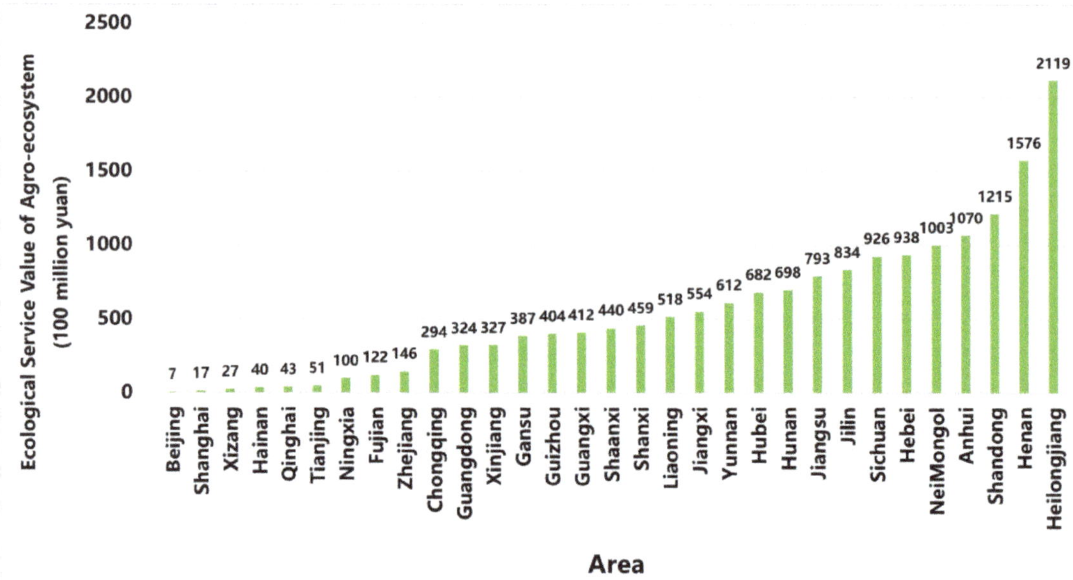

Figure 7. The value of agroecosystem services in China from 2020 to 2050.

The distribution pattern and changes in China's agroecosystem flood risk from 2020 to 2050 (Figure 9) show that the flood risk pattern of China's agroecosystem presents the characteristics of high flood risk in the northeastern, northern, central, eastern, and southwestern provinces. In the next 30 years, approximately 60% of China's provinces are at high risk of flooding, including Heilongjiang, Henan, Shandong, Sichuan, and other provinces. This may be related to the fact that Heilongjiang, Henan, Sichuan, and other provinces represent the major grain crop production provinces in China. The grain sown area occurs prior to other provinces in the country, and the service value of the agroecosystem is relatively high. These provinces are located in the Heilongjiang, Huaihe, and Yangtze River Basins. The water systems in the basins are rich, and they are prone to large floods in the whole basin. Therefore, the risk of flooding in the agroecosystem is relatively high. Under these three SSP scenarios, Tibet, Qinghai, Ningxia, Xinjiang, Gansu, Beijing, and Shanghai exhibit moderate-to-low flood risk characteristics. Alpine landforms, such as the Himalayas, Kunlun Mountains, Tianshan Mountains, and Qilian Mountains, block the transport of water vapor; therefore, there is less precipitation. In addition, hilly landforms and mountainous landforms in this area account for a large area, the climate is warm and dry, and the runoff of mountain rivers also shows a downward trend [56]. However, Beijing and Shanghai are regions represented by financial services and high-end industries. The service value of the agroecosystem is low; therefore, the risk of flooding is low.

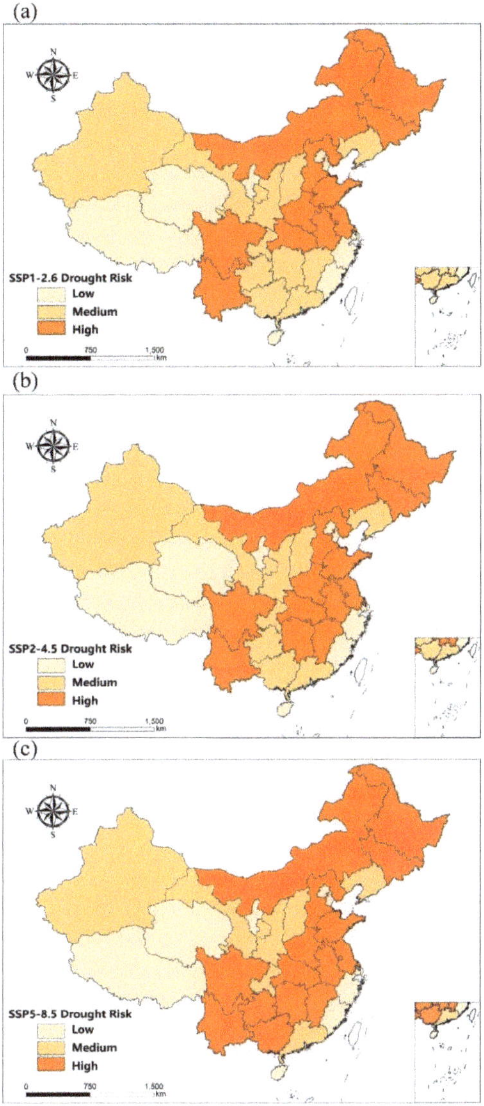

Figure 8. Drought risk of China's agroecosystem under the SSP1-2.6, SSP2-4.5, and SSP5-8.5 scenarios from 2020 to 2050. Notes: (**a**) SSP1-2.6 drought risk; (**b**) SSP2-4.5 drought risk; and (**c**) SSP5-8.5 drought risk.

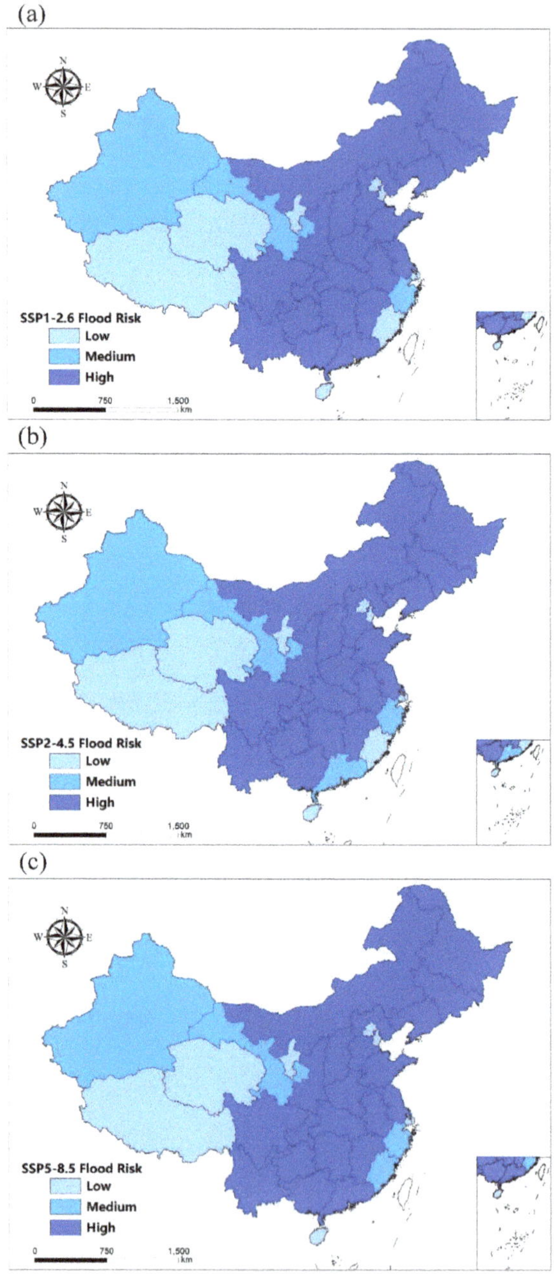

Figure 9. Flood risk of China's agroecosystem under the SSP1-2.6, SSP2-4.5, and SSP5-8.5 scenarios from 2020 to 2050. Note: (**a**) SSP1-2.6 flood risk; (**b**) SSP2-4.5 flood risk; and (**c**) SSP5-8.5 flood risk.

4. Conclusions and Discussion

This study takes China's agroecosystem as the research object and adopts a multi-index comprehensive prediction method based on the "sensitivity–exposure–adaptability" vulnerability prediction model. Taking into account factors such as climate change, society, population, agricultural science and technology, economy, ecology, etc., a multilevel indicator evaluation system was established for the drought and flood vulnerability of agroecosystems. Using a combination of subjective empowerment and objective empowerment methods, the drought and flood sensitivity, exposure, adaptability, and vulnerability of agroecosystems under three socioeconomic scenarios, SSP1-2.6, SSP2-4.5, and SSP5-8.5, in the next 30 years were estimated. On this basis, the drought and flood risk prediction model of agroecosystems was used to consider vulnerability, risk sources, and risk receptors. The drought and flood risks of the agroecosystem from 2020 to 2050 were predicted, and the vulnerability and risk status of China's agroecosystem under different shared social and economic scenarios in the future were compared and analyzed. The main findings are as follows:

(1) In the next 30 years, as the shared socioeconomic path changes from SSP1-2.6 to SSP5-8.5 in turn, the drought and flood vulnerability of China's agroecosystem will increase in general, the scope of provinces with high drought vulnerability will gradually expand to the south, and provinces with high flood vulnerability will gradually extend to the southwest.

(2) From 2020 to 2050, the regional distribution pattern of drought risk showed the characteristics of high drought risk in agroecosystems in Northeast, North, Central China, and Southwest China. The flood risk pattern showed the characteristics of high flood risk in the agroecosystem in Northeast, North, Central, East, and Southwest China. As the scenarios changed from SSP1-2.6 to SSP5-8.5 in turn, the number of provinces with high drought risk increased and showed a trend of extending to the south. Under the SSP5-8.5 scenario, the drought risk of the agroecosystem is high, and 60% of the provinces in China have a high risk level of flooding.

(3) It is worth noting that under the SSP5-8.5 scenario, Heilongjiang and Jilin in Northeast China, Henan, Hubei, Anhui, Hunan and Jiangxi in Central China, Inner Mongolia and Hebei in North China, Shandong and Jiangsu in East China, and Sichuan, Yunnan, Guizhou, and Guangxi in Southwest China, the drought and flood risk in the agroecosystem in these provinces will be higher in the next 30 years. Northeast and Central China belong to China's commodity grain bases. Crops have a long history of planting, and plant growth is easily restricted by drought and floods, resulting in a higher risk of drought and floods in the agroecosystems of Northeast and Central China. These provinces need to strengthen the use of irrigation infrastructure; promote water-saving irrigation technologies, such as sprinkler irrigation and drip irrigation; attach importance to investment; use advanced agricultural machinery and equipment; strengthen soil protection measures; and gradually develop climate-smart agriculture. At the same time, Northeast and Central China should continue to implement policies on drought, flood, and climate change adaptation; make emergency response and preparation for drought and flood risks in terms of equipment, facilities, funds, and technology; and make full use of credit, savings, markets, and other financial instruments to ensure restoration and construction in disaster-affected areas.

(4) The climate in North China is unstable and water resources are in short supply all year round. The supply of water resources mainly depends on China's South-to-North Water Diversion Project. The precipitation and reservoir capacity in this region cannot meet the regional agricultural and ecological water consumption, but floods caused by heavy rains often occur. According to the data, in July 2021, a heavy rainstorm occurred in the central and northern parts of Henan, resulting in flood disasters that affected 14.786 million people in 150 counties in Henan Province and caused a direct economic loss of 120.06 billion yuan, of which Zhengzhou City was 40.9 billion CNY, accounting for 34.1% of Henan province [57]. East China belongs to the southeastern coastal area, with a low altitude, adjacent to the Bohai Sea and the Yellow Sea, with vertical and horizontal rivers, insufficient freshwater resources, and a high risk of drought and flooding in the agroecosystem. On the one hand, these areas need to pay attention to the regular maintenance of dams, pipelines, and reservoirs

for water supply and storage systems. The southwestern provinces of China have complex topography, spanning the three steps of my country's landforms, with many mountains and ridges, and their altitudes are mostly 4000–5000 m. It is cold in winter and cool in summer, and the distribution of water and heat is uneven [58], so the risk of drought and flood in the agroecosystem is high. These areas need to advocate that farmers can diversify crops in agricultural planting and production activities, adopt intercropping, crop rotation, and other planting methods, select drought-resistant and flood-resistant crops, and adjust planting dates and planting structures. Farmers should adopt diversified livelihood strategies, actively participate in education and training on water conservation, farming methods, drought and flood awareness, and risk management, and apply them to daily agricultural production and life activities.

(5) According to the natural geographical and socio-economic background and characteristics of each province in China, we select indicators such as socio-economic development, agriculture, ecological environment, human activities, and agricultural science and technology. We combine statistics and forecast data from different repositories and establish indicator systems for predicting the vulnerability of the agroecosystem to drought and flood disasters. At the same time, this study uses the combination of AHP and the entropy weighting method to reduce the uncertainty of prediction and enhance the repeatability of this study. In addition, in future research work, a multi-indicator evaluation system can be constructed based on the local natural and human context, which can be applied not only to the agricultural system, ecosystem vulnerability, and risk assessment but also to food security, the economic system, and human health risk assessment work. Moreover, compared with the methods used in other studies such as water scarcity and similar indicator reports, the indicator system constructed in this study involves many fields such as society, nature, climate change, human activities, technological level, economic development, and so on. At the same time, based on the vulnerability and risk assessment of China's agroecosystem in the past 30 years, a comparative analysis of the vulnerability and risk distribution and characteristics of SSP1-2.6, SSP2-4.5, and SSP5-8.5 in the next 30 years is carried out in order to ensure the coherence and credibility of this research.

(6) In the next step of research, we can consider adding effective actions that farmers and agricultural production cooperatives have taken or may take and conduct adaptive analysis based on relevant driving factors and risk patterns to more accurately formulate and adjust relevant adaptation strategies and reduce ecosystem vulnerability and risk loss. Moreover, in future research, agroecosystem types or large geomorphic units can be used as the basic unit of classification for agroecosystem disaster risk estimation to take into account the influencing factors of different geomorphic units and climatic regions in China.

Supplementary Materials: The following supporting information can be downloaded at: https://www.mdpi.com/article/10.3390/su141610069/s1, Supporting Materials.

Author Contributions: J.C. and J.L. offered the idea and method, all authors participated in the prepared materials, and J.L. wrote the paper with input from all authors. Conceptualization, W.Z., Y.X., Y.H. and Y.L.; Data curation, J.L. and J.C.; Formal analysis, J.L.; Funding acquisition, J.C.; Investigation, J.L.; Methodology, J.L.; Project administration, W.Z.; Writing—original draft, J.L. All authors have read and agreed to the published version of the manuscript.

Funding: This work was supported by Project Supported by State Key Laboratory of Earth Surface Processes and Resource Ecology (Project Number: 2022-GS-01) and the Natural Science Foundation General Program of China (42075167).

Institutional Review Board Statement: Not applicable.

Informed Consent Statement: Not applicable.

Data Availability Statement: All relevant data can be found within the paper and its Supporting Materials. CMIP6 models data are available at link https://esgf-node.llnl.gov/search/cmip6.

Conflicts of Interest: The authors declare no competing interests.

Appendix A

Table A1. Brief introduction of 22 global climate models in CMIP6.

Schema Name	The Institution	Mode Resolution (Longitude × Latitude)
ACCESS-CM2	Commonwealth Scientific and Industrial Research Organization of Australia, Australian Research Council Centre of Excellence for Climate System Science	1.875° × 1.25°
ACCESS-ESM1-5	Commonwealth Scientific and Industrial Research Organization of Australia	1.875° × 1.25°
BCC-CSM2-MR	China National Climate Center	3.2° × 1.6°
FGOALS-g3	Institute of Atmospheric Physics, Chinese Academy of Sciences	1.8° × 0.8°
CanESM5	Canadian Centre for Climate Modeling and Analysis	2.8125° × 2.8125°
CanESM5-CanOE	Canadian Centre for Climate Modeling and Analysis	2.8125° × 2.8125°
CNRM-CM	French National Centre for Meteorological Research, European Centre for Computational Research and Advanced Training	1.4° × 1.4063°
CNRM-CM6-1-HR	French National Centre for Meteorological Research, European Centre for Computational Research and Advanced Training	1.4° × 1.4063°
CNRM-ESM	French National Centre for Meteorological Research, European Centre for Computational Research and Advanced Training	1.4° × 1.4063°
IPSL-CM	Pierre Simon Laplace Institute, France	1.4° × 1.4063°
EC-Earth3	European Centre for Medium-Range Weather Forecasts	0.7031° × 0.7031°
EC-Earth3-Veg	European Centre for Medium-Range Weather Forecasts	0.7031° × 0.7031°
GFDL-ESM4	NOAA Geohydrodynamics Laboratory	2.88° × 1.8°
GISS-E2-1-G	NASA Gold Institute for Space Studies	2.88° × 1.8°
INM-CM4-8	Institute of Numerical Mathematics of the Russian Academy of Sciences	2° × 1.5°
INM-CM5-0	Institute of Numerical Mathematics of the Russian Academy of Sciences	2° × 1.5°
MIROC6	Japan Marine Earth Science and Technology Agency	1.4063° × 1.3953°
MIROC-ES2L	Japan Marine Earth Science and Technology Agency	2.8125° × 2.8125°
MRI-ESM	Japan Meteorological Institute	1.125° × 1.125°
MPI-ESM1-2-HR	German Max Planck Institute for Meteorology, German Meteorological Office	0.9375° × 0.9375°
MPI-ESM1-2-LR	Max Planck Institute for Meteorology, Alfred Wegener Institute, Germany	1.875° × 1. 875°
UKESM1-O-LL	UK National Centre for Atmospheric Science, UK Met Office Hadley Centre	3.2° × 1.6°

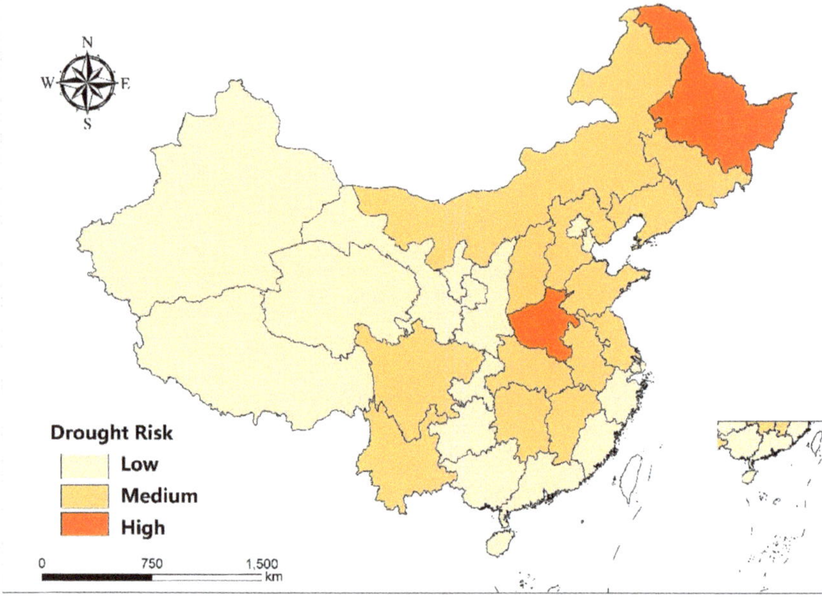

Figure A1. Drought risk in China's agroecosystem from 1991 to 2019.

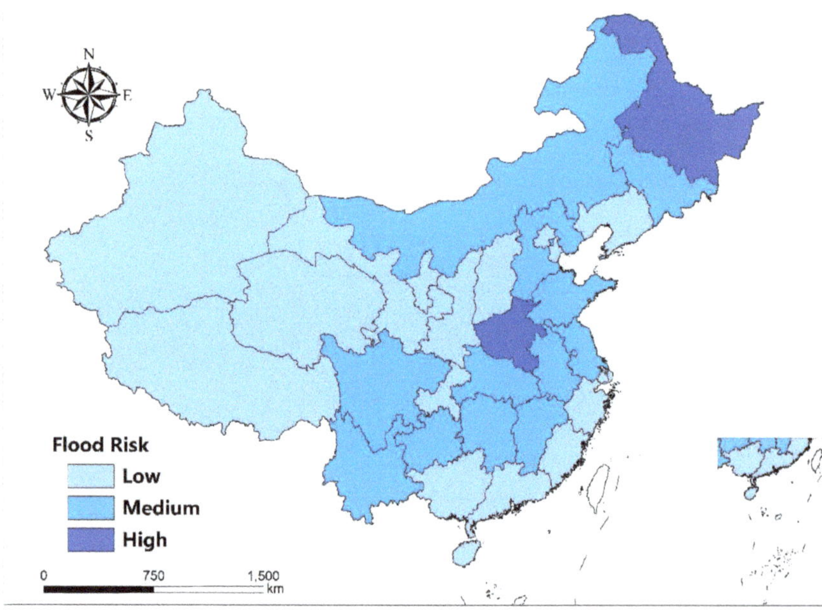

Figure A2. Flood risk of China's agroecosystem from 1991 to 2019.

References

1. Houghton, J.T.; Jenkins, G.; Ephraums, J.J. (Eds.) *The IPCC Scientific Assessment*; Cambridge University Press: Cambridge, UK, 1990.
2. Watson, R.T.; Zinyowera, M.C.; Moss, R.H. (Eds.) *Climate Change 1995. Impacts, Adaptations and Mitigation of Climate Change: Scientific-Technical Analyses*; Cambridge University Press: Cambridge, UK, 1995.
3. McCarthy, J.J.; Canziani, O.F.; Leary, N.A.; Dokken, D.J.; White, K.S. (Eds.) *Climate Change 2001: Impacts, Adaptation, and Vulnerability: Contribution of Working Group II to the Third Assessment Report of the Intergovernmental Panel on Climate Changes*; Cambridge University Press: Cambridge, UK, 2001.
4. Parry, M.L.; Canziani, O.; Palutikof, J.; Van der Linden, P.; Hanson, C. (Eds.) *Climate Change 2007—Impacts, Adaptation and Vulnerability: Working Group II Contribution to the Fourth Assessment Report of the IPCC*; Cambridge University Press: Cambridge, UK, 2007.
5. Pörtner, D.C.; Roberts, M.; Tignor, E.S.; Poloczanska, K.; Mintenbeck, A.; Alegría, M.; Craig, S.; Langsdorf, S.; Löschke, V.; Möller, A.; et al. (Eds.) *Climate Change 2022: Impacts, Adaptation, and Vulnerability.Contribution of Working Group II to the Sixth Assessment Report of the Intergovernmental Panel on Climate Change*; Cambridge University Press: Cambridge, UK, 2007.
6. Budiyono, Y.; Aerts, J.; Brinkman, J.J.; Marfai, M.A.; Ward, P. Flood risk assessment for delta mega-cities: A case study of Jakarta. *Nat. Hazards* **2015**, *75*, 389–413. [CrossRef]
7. Wang, H.M.; Chen, J.; Xu, C.Y.; Zhang, J.; Chen, H. A framework to quantify the uncertainty contribution of GCMs over multiple sources in hydrological impacts of climate change. *Earth's Future* **2020**, *8*, e2020EF001602. [CrossRef]
8. Tanure, T.; Miyajima, D.N.; Magalhaes, A.S.; Domingues, E.P.; Carvalho, T.S. Impacts of Climate Change on Agricultural Production, Land Use and Economy of the Legal Amazon Region between 2030 and 2049. *EconomiA* **2020**, *21*, 73–90. [CrossRef]
9. Simane, B.; Zaitchik, B.F.; Foltz, J.D. Agroecosystem specific climate vulnerability analysis: Application of the livelihood vulnerability index to a tropical highland region. *Mitig. Adapt. Strateg. Glob. Chang.* **2016**, *21*, 39–65. [CrossRef]
10. Vignola, R.; Harvey, C.A.; Bautista-Solis, P.; Avelino, J.; Rapidel, B.; Donatti, C.; Martinez, R. Ecosystem-based adaptation for smallholder farmers: Definitions, opportunities and constraints. *Agric. Ecosyst. Environ.* **2015**, *211*, 126–132. [CrossRef]
11. Epule, T.E.; Ford, J.D.; Lwasa, S. Projections of maize yield vulnerability to droughts and adaptation options in Uganda. *Land Use Policy* **2017**, *65*, 154–163. [CrossRef]
12. Senyolo, M.P.; Long, T.B.; Blok, V.; Omta, O. How the characteristics of innovations impact their adoption: An exploration of climate-smart agricultural innovations in South Africa. *J. Clean. Prod.* **2018**, *172*, 3825–3840. [CrossRef]
13. Ashraf, M.; Arshad, A.; Patel, P.M.; Khan, A.; Qamar, H.; Siti-Sundari, R.; Babar, J.R. Quantifying climate-induced drought risk to livelihood and mitigation actions in Balochistan. *Nat. Hazards* **2021**, *109*, 2127–2151. [CrossRef]
14. Durowoju, O.S.; Ologunorisa, T.E.; Akinbobola, A. Assessing agricultural and hydrological drought vulnerability in a savanna ecological zone of Sub-Saharan Africa. *Nat. Hazards* **2022**, *111*, 2431–2458. [CrossRef]
15. Meza, I.; Siebert, S.; Döll, P.; Kusche, J.; Herbert, C.; Rezaei, E.E.; Nouri, H.; Gerdener, H.; Popat, E.; Frischen, J.; et al. Michael Global-scale drought risk assessment for agricultural systems. *Nat. Hazards Earth Syst. Sci.* **2020**, *20*, 695–712. [CrossRef]
16. Swami, D.; Parthasarathy, D. Dynamics of exposure, sensitivity, adaptive capacity and agricultural vulnerability at district scale for Maharashtra, India. *Ecol. Indic.* **2021**, *121*, 107206. [CrossRef]
17. Zhou, G.S. Prospects for the impact of climate change on agricultural production in China. *Meteorol. Environ. Sci.* **2015**, *38*, 80–94. (In Chinese)
18. Cai, Y.L. Vulnerability and adaptation strategies of Chinese agriculture under global climate change. *Acta Geogr. Sin.* **1996**, *51*, 202–212.
19. Liu, Y.; You, M.; Zhu, J.; Wang, F.; Ran, R. Integrated risk assessment for agricultural drought and flood disasters based on entropy information diffusion theory in the middle and lower reaches of the Yangtze River, China. *Int. J. Disaster Risk Reduct.* **2019**, *38*, 101194. [CrossRef]
20. Xu, Y.L.; Zhao, Y.C.; Zhai, P.M. IPCC Special Report SRCCL New Cognition and Enlightenment on Climate Change and Food Security. *Adv. Clim. Chang. Res.* **2020**, *16*, 37–49. (In Chinese)
21. Wu, S.H.; Zhao, D.S. New Progress in Research on Climate Change Impacts, Risks and Adaptation in China. *China Popul. Resour. Environ.* **2020**, *30*, 1–9. (In Chinese)
22. Shi, W.J.; Wang, M.L.; Liu, Y.T. Crop yield and production responses to climate disasters in china. *Sci. Total Environ.* **2021**, *750*, 141147–141194. [CrossRef] [PubMed]
23. Liu, D.L.; Zuo, H. Statistical downscaling of daily climate variables for climate change impact assessment over New South Wales, Australia. *Clim. Chang.* **2012**, *115*, 629–666. [CrossRef]
24. Huang, J.; Qin, D.; Jiang, T.; Wang, Y.; Feng, Z.; Zhai, J.; Su, B. Effect of fertility policy changes on the population structure and economy of China: From the perspective of the shared socioeconomic pathways. *Earth's Future* **2019**, *7*, 250–265. [CrossRef]
25. National Bureau of Statistics of the People's Republic of China. *China Statistical Yearbook*; China Statistics Press: Beijing, China, 2021. (In Chinese)
26. National Bureau of Statistics of the People's Republic of China. Inventory Results of China's Forest Resources. Available online: http://forest.ckcest.cn (accessed on 16 June 2022).
27. Field, C.B.; Barros, V.R.; Dokken, D.J.; Mach, K.J.; Mastrandea, M.D.; Bilir, T.E.; Chatterjee Ebi, K.L.; Estrada, Y.O.; Genova, R.C. (Eds.) *Climate Change 2014: Impacts, Adaptation, and Vulnerability*; Cambridge University Press: Cambridge, UK, 2014.

28. Qin, J.; Ding, Y.J.; Zhao, Q.D.; Wang, S.P.; Chang, Y.P. Assessments on surface water resources and their vulnerability and adaptability in China. *Adv. Clim. Chang. Res.* **2020**, *11*, 381–391. [CrossRef]
29. Schardong, A.; Simonovic, S.P.; Tong, H. Use of quantitative resilience in managing urban infrastructure response to natural hazards. *Int. J. Eng. Sci.* **2019**, *9*, 13–25. [CrossRef]
30. Saaty, T.L. What is the Analytic Hierarchy Process? In *Mathematical Models for Decision Support*; Springer: Berlin/Heidelberg, Germany, 1988; pp. 109–121.
31. Alamanos, A.; Mylopoulos, N.; Loukas, A.; Gaitanaros, D. An Integrated Multicriteria Analysis Tool for Evaluating Water Re source Management Strategies. *Water* **2018**, *10*, 1795. [CrossRef]
32. Wu, J.R.; Chen, X.L.; Lu, J.Z. Assessment of Long and Short-Term Flood Risk Using the Multi-Criteria Analysis Model with the AHP-Entropy Method in Poyang Lake Basin. *Int. J. Disaster Risk Reduct.* **2022**, *75*, 102968. [CrossRef]
33. Li, J.N.; Chou, J.M.; Zhao, W.X.; Li, Q.M.; Xu, Y.; Sun, M.Y.; Yang, F. 1991–2019 drought and flood vulnerability assessment of China's agro-ecosystem. *Clim. Environ. Res.* **2022**, *27*, 19–32. (In Chinese)
34. Wang, Q. Research on the Effect of Tourism Poverty Alleviation in Yecheng County, Xinjiang Based on AHP & Entropy Weight Method. Master's Thesis, Tarim University, Aral Shehri, China, 2020. (In Chinese).
35. Xu, X.G.; Yan, L.; Xu, L.F.; Lu, Y.L.; Ma, L.Y. Ecological Risk Assessment of Natural Disasters in China. *J. Peking Univ. Chem. Technol.* **2011**, *47*, 901–908. (In Chinese)
36. *GB/T 20481–2017*; Grades of Meteorological Drought. Standardization Administration of the People's Republic of China: Beijing, China, 2017. Available online: https://kns.cnki.net/kcms/detail/detail.aspx?FileName=SCSF00050541&DbName=SCSF (accessed on 16 June 2022).
37. Costanza, R.; Arge, R.; Groot, R.; Farber, S.; Grasso, M.; Hannon, B.; Belt, M. The value of the world's ecosystem services and natural capital. *Nature* **1997**, *387*, 253–260. [CrossRef]
38. Xie, G.D.; Lu, C.X.; Leng, Y.F.; Zheng, D.; Li, S.C. Evaluation of Ecological Assets on the Qinghai-Tibet Plateau. *J. Nat. Resour.* **2003**, *18*, 189–196. (In Chinese)
39. Xu, Y.Q.; Zhou, B.T.; Yu, L.; Shi, Y.; Xu, Y. The spatiotemporal characteristics of future forest ecosystem service value in China under the background of climate change. *Chin. J. Ecol.* **2018**, *38*, 1952–1963. (In Chinese)
40. Chen, Y.B.; Liu, L.J.; Chen, X.L. Forecasting China's Inflation Rate: A Comparative Analysis Based on LSTM Model and BVAR Model. *Res. Financ. Issues* **2021**, *6*, 18–29. (In Chinese)
41. China Meteorological Administration. High Temperature Warning Signal and Defense Guide [EB/OL]. Available online: http://www.qxkp.net/qxbk/yjxhjfyzn/202103/t20210312_2948454.html (accessed on 3 August 2022).
42. Li, B.Z.; Zhou, G.S. Research progress on drought indicators. *Chin. J. Ecol.* **2014**, *34*, 1043–1052. (In Chinese)
43. Chen, D.D.; Zi, T.; Zhang, Y.F. Drought risk assessment during wheat growing period in Sichuan Province based on water gain and loss index. *J. Nat. Disasters* **2017**, *26*, 155–163. (In Chinese)
44. Yan, J. Analysis of Temporal and Spatial Variation Characteristics of Maize Drought in Liaoning Province Based on Crop Water Deficit Index. Master's Thesis, Shenyang Agricultural University, Shenyang, China, 2018. (In Chinese).
45. Cao, Y.Q.; Li, L.H.; Lu, J.; Zhang, R.N. Analysis of drought characteristics during maize growth period in Liaoning Province based on SPEI. Chinese. *J. Ecol.* **2021**, *41*, 7367–7379. (In Chinese)
46. Pan, G.F.; Wang, B.F.; Chen, B.; Fang, Z.B.; Zhao, S.S.; Tian, Y.H. Effects of sowing date on yield, growth period and utilization of temperature and light of different types of japonica rice in central and northern Hubei. *Crop. J.* **2021**, *4*, 105–111. (In Chinese)
47. Wang, H.Y.; Gong, Q. Analysis of the abnormal characteristics of precipitation in the growing season of crops in Northeast China. *Meteorol. Sci. Technol.* **2005**, *33*, 345–349. (In Chinese)
48. Zhao, C.Y.; Wang, Y.; Zhang, Y.S.; Yan, X.Y. Changes of climatic conditions during the growing season of crops in Liaoning Province in the past 50 years and their impact on agricultural production. *Disaster Sci.* **2009**, *24*, 102–106. (In Chinese)
49. Chou, J.M.; Dong, W.J.; Feng, G.L. A method to quantitatively assess the impact of climate change on economic output. *Sci. Bull.* **2011**, *10*, 21–23. (In Chinese)
50. Xiong, W.; Holman, I.P.; You, L.Z.; Yang, J.; Wu, W. Impacts of observed growing-season warming trends since 1980 on crop yields in China. *Reg. Environ. Chang.* **2014**, *14*, 7–16. [CrossRef]
51. Xu, G.C.; Kang, M.Y.; He, L.N.; Li, Y.F.; Chen, Y.R. Ecological vulnerability and its research progress. *Chin. J. Ecol.* **2009**, *29*, 2578–2588. (In Chinese)
52. Wu, S.H.; Pan, T.; Liu, Y.H.; Deng, H.Y.; Jiao, K.W.; Lu, Q.; Gao, J.B. China's comprehensive climate change risk zoning. *Acta Geogr. Sin.* **2017**, *72*, 3–17. (In Chinese)
53. *GB/T 28592-2012*; Precipitation Level. Standardization Administration of the People's Republic of China: Beijing, China, 2012. Available online: http://cmastd.cmatc.cn/bzjd/1357.jhtml (accessed on 16 June 2022).
54. Xinhua News Agency. Several Opinions of the Central Committee of the Communist Party of China and the State Council on Adhering to Prioritizing the Development of Agriculture and Rural Areas and Doing a Good Job in "Three Rurals". *China Pig Ind.* **2019**, *14*, 17–23. (In Chinese)
55. Chou, J.M.; Xian, T.; Dong, W.J.; Xu, Y. Regional temporal and spatial trends in drought and flood disasters in China and assessment of economic losses in recent years. *Sustainability* **2018**, *11*, 55. [CrossRef]
56. Shi, B.; Zhu, X.; Hu, Y.; Yang, Y. Drought characteristics of Henan province in 1961–2013 based on Standardized Precipitation Evapotranspiration Index. *J. Geogr. Sci.* **2017**, *27*, 311–325. [CrossRef]

57. The Central People's Government of the People's Republic of China. The Investigation Report on the "7.20" Heavy Rainstorm Disaster in Zhengzhou, Henan Announced [EB/OL]. Available online: http://www.gov.cn/xinwen/2022-01/21/content_5669723.htm (accessed on 3 August 2022).
58. Jing, J.L.; He, C.X.; Wang, Y.F.; Ma, B.X.; Liu, B. Analysis of temporal and spatial evolution characteristics of drought in South west China from 1902 to 2018. *Soil Water Conserv. Res.* **2022**, *29*, 220–227. (In Chinese)

sustainability

MDPI

Article

The Impact of Rainfall on Urban Human Mobility from Taxi GPS Data

Peng Guo, Yanling Sun *, Qiyi Chen, Junrong Li and Zifei Liu

College of Geography and Environmental Science, Tianjin Normal University, Tianjin 300387, China; pguo@tjnu.edu.cn (P.G.); 1830080085@stu.tjnu.edu.cn (Q.C.); lijunrong98@163.com (J.L.); 2110080002@stu.tjnu.edu.cn (Z.L.)

* Correspondence: sunyanling@tjnu.edu.cn

Abstract: Rainfall severely impacts human mobility in urban areas and creates significant challenges for traffic management and urban planning. There is an urgent need to understand the impact of rainfall on residents' travels from multiple perspectives. Taxi GPS data contains a large amount of spatiotemporal information about human activities and mobility in urban areas. For this study, we selected the central area of Zhuhai as the study area and used taxi data from August 2020 for the investigation. Firstly, we divided the taxi data into four scenarios, i.e., weekdays with and without rainfall and weekends with and without rainfall and analyzed and compared the trip characteristics for the different scenarios. Then, using the traffic analysis zone (TAZ) as the node and taxi flow between TAZs as edges, we constructed a network and compared the network indicators under the different scenarios. Finally, we used the Leiden algorithm to detect communities in different scenarios and compared the network indicators of the communities. The results showed that on days with rainfall, taxi flow and its spatial and temporal distribution pattern changed significantly, which affected transportation supply and demand. These findings may provide useful references for the formulation of urban transport policies that can adapt to different weather conditions.

Keywords: human mobility; rainfall; taxi GPS data; complex network; community; Zhuhai central areas

Citation: Guo, P.; Sun, Y.; Chen, Q.; Li, J.; Liu, Z. The Impact of Rainfall on Urban Human Mobility from Taxi GPS Data. *Sustainability* **2022**, *14*, 9355. https://doi.org/10.3390/su14159355

Academic Editors: Xiaodong Yan, Jia Yang and Shaofei Jin

Received: 27 June 2022
Accepted: 27 July 2022
Published: 30 July 2022

1. Introduction

The study of human mobility can be used to capture spatiotemporal operational patterns in urban areas and to understand the complex relationship between human activity and the urban environment. This understanding plays an important role in various aspects, such as floating population access, traffic forecasting, urban planning, and epidemic modeling [1–3].

Traditional studies of human mobility usually adapt travel diary survey data, which are expensive and labour-intensive to obtain. However, because these data often have problems, such as a small sample size, a short time span, and slow update speed, they cannot thoroughly reflect the spatiotemporal regulation of urban group activities over time. Additionally, the accuracy of the data can also be questionable as a result of the subjectivity of the survey design and the interviewees. With the development of positioning and information and communication technologies (ICTs), humans have entered the era of big data. The proliferation of various sensors and positioning technologies makes it possible to collect large-scale and high-precision big data on human mobility in a long time sequence (such as mobile phone data, bus smart card data, and taxi data). These datasets contain abundant information about individual spatiotemporal activities, which contributes to understanding human mobility patterns at a more precise spatiotemporal level [4–6]. As one of the representative types of data on traffic and travel, taxi data have become an important basis for studies of human mobility patterns [7].

As a component of the human living environment, weather conditions have significant impacts on daily trips made by inhabitants [8–11]. Taking taxi travel as an example, the rainfall not only affects the demand for taxis but also leads to changes in the taxi flow between areas. Additionally, it is likely that changes in road conditions caused by rainfall affect driving speed, travel time, and choice of route during the taxi journey from passenger pickup to the destination. [12,13]. In brief, people's travel is influenced by weather conditions, and their travel strategies usually vary significantly in different weather conditions [14,15]. As a result, understanding how weather impacts human travel patterns can contribute to improving the public transport service, and better satisfying the travel demands of passengers under different weather conditions [16].

Studies of human mobility patterns that consider weather factors generally focus on the field of transportation, usually analyzing several aspects, including traffic volume [17], the speed and density of traffic flow [18,19], and traffic jams [20]. In addition, weather changes make a big difference to vehicles, road conditions, driving behaviors (such as psychology, judgement, and reflection) and the riding environment [21].

When considering complex and changeable weather factors, current research mainly focuses on the association between different weather factors and human travel activities, such as the demand for and security of traffic travel under various weather conditions, including elevated temperatures, smog, and high winds [22,23]. These studies, for the most part, investigated the impact of weather factors on people's travel behaviors using various modes of transportation from four perspectives: number of journeys, modes of travel, travel speed, and travel time [24–28]. In addition, many studies regard rainfall as having the most significant impact on people's daily travel activities [29,30]. In particular, normal travel time is usually delayed, and elastic demand is restricted and decreased in rainstorm conditions, resulting in significant changes in the spatiotemporal distribution of traffic demand [31].

However, because they lack sufficient space-dependent visual representation, current research studies mainly focus on the overall statistical analysis and do not explore the characteristics of the spatiotemporal distribution of people's travel activities under different weather conditions. Because weather conditions have several significant impacts on human travel patterns and their spatial differences, the scarcity of study in this field needs to be addressed, with particular regard to the significant spatial impact of two weather conditions, the spatial change of people's travel behavior patterns, and their interplay [32]. As a result, analysis of the impacts of changes in weather conditions on people's travel behavior patterns requires multiscale, comprehensive analysis and visual expression.

In addition, complex geospatial networks can combine statistics on the network index with a spatial analysis based on statistical analysis and spatial visualization of network characteristics [33,34]. Mobility network statistics can thus describe and evaluate how human mobility is distributed and developed on different scales. Therefore, these complex network-based analytical methods improve the understanding of urban mobility [35–37].

In this study, taking Zhuhai City, Guangdong Province, China as the study area, we combined geospatial complex networks with multiscale geospatial analysis to extend empirical research on human travel patterns by analyzing the impact of rainfall on human mobility. By dividing the taxi data into four scenarios: weekdays without rainfall, weekdays with rainfall, weekends without rainfall, and weekends with rainfall, we aimed to: (1) explore the feasibility of using taxi data to investigate human mobility in urban areas under rainfall conditions; (2) compare the differences in basic travel characteristics and explore the changes in the spatial distribution of trips in the different scenarios; and (3) quantitatively explore the impacts of rainfall on human mobility at the whole network and community network levels using the complex networks method.

2. Study Area and Data

2.1. Case Study: Zhuhai, China

This study was conducted in Zhuhai, China, which is in southern China and borders Macao to the south, with a total area of approximately 1736 km^2. According to the seventh national population census conducted at the end of 2020, the residential population of

Zhuhai was 2.44 million. Zhuhai consists of three districts (Xiangzhou, Doumen, and Jinwan) and five economic function districts (Hengqin, Gaoxin, Baoshui, Wanshan, and Gaolan). As the location of municipal government, Xiangzhou is the most flourishing district. In Zhuhai, the central part of Xiangzhou district is called the Zhuhai Central Area, and this represents the city center of Zhuhai. The central area occupies 153.15 km^2 and has a population of 1.12 milion, which is 46.08 percent of the total residential population. The spatial map of Zhuhai and its central area is shown in Figure 1. In this study, we focused on human mobility in the central area.

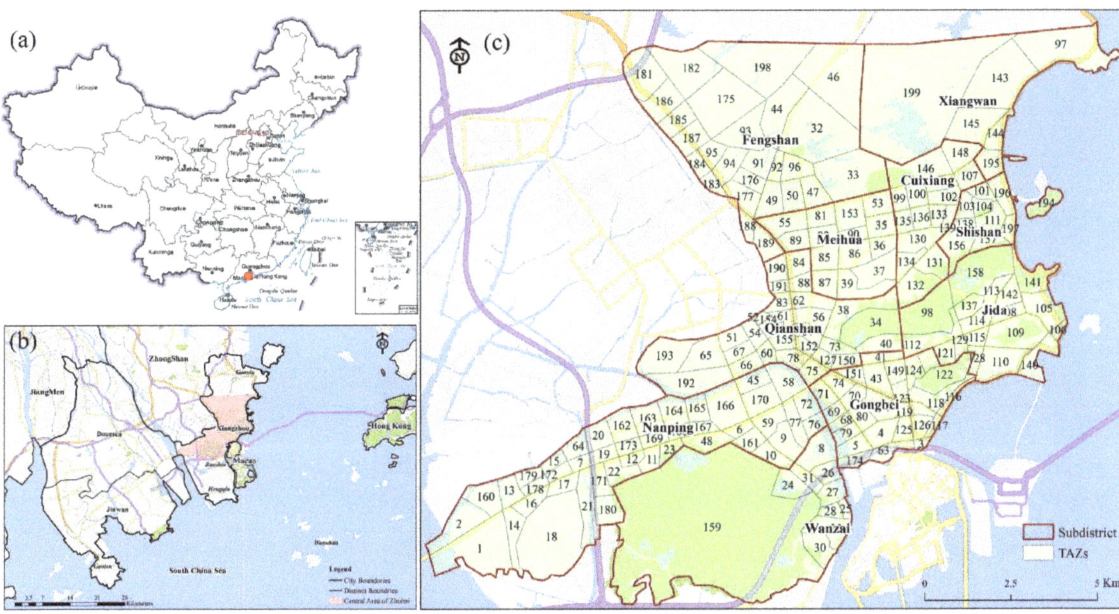

Figure 1. (**a**) Location of the study area in China; (**b**) Location of the study area in Zhuhai City; (**c**) Overview of the study area-Zhuhai Central Area.

2.2. Data Source

To address our research questions and examine the association between human mobility and rainfall, three datasets from Zhuhai, China were used in this study. The first dataset was the weather conditions dataset, which consisted of one-hour measurements of weather conditions for two weather stations from the Meteorology Bureau of Zhuhai. Since drizzle and showers have minor effects on human mobility, this study defined days with rainfall as those with a total precipitation exceeding 25 mm and a duration of more than 6 h. Using this definition, we selected ten days with rainfall in August 2020, six weekdays and four days at weekends. In order to identify differences in human mobility on days with rainfall, twelve days without rainfall were also selected, half of which were weekdays and the other half at weekends. As shown in Figure 2, the daily precipitation in Zhuhai was plotted as a bar graph, and the dates of the four scenarios are represented by different colors.

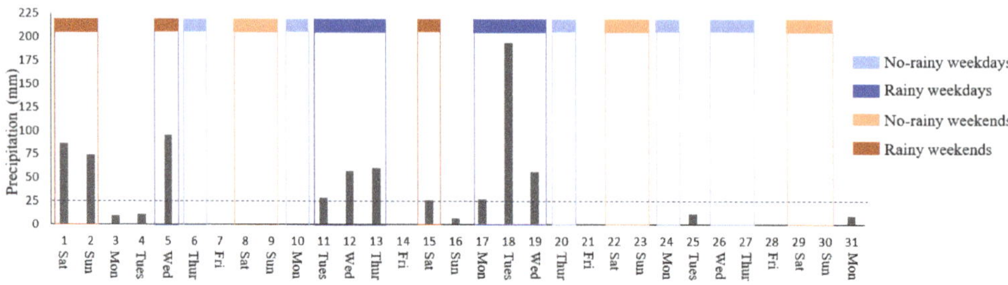

Figure 2. Zhuhai precipitation in August 2020 and the four scenarios.

The second dataset was taxi GPS trajectories for the same twenty-two days in August 2020. For study purposes, only the car IDs, pick-up and drop-off locations, and timestamps were considered (Table 1). The data covered trips taken by 2,165,106 passengers in 3284 taxis. Each trip represents a purposeful human movement from origin to destination. We used ArcGIS to calculate the TAZs (Traffic Analysis Zones) based on the pick-up and drop-off locations. In this sense, these trips can reflect spatial connections made through human movements and serve as edges to build spatial interaction networks.

Table 1. Processed trip record data.

ID	Pickup Datetime	Dropoff Datetime	Pickup Longitude	Pickup Latitude	Dropoff Longitude	Dropoff Latitude	Trip Distance (m)	Trip Duration (min)	Origin TAZ	Destination TAZ
1001	1 August 2020 19:06:02	1 August 2020 19:10:53	113.470733	22.215318	113.4896	22.224246	2700	4.8	12	165
1002	1 August 2020 17:19:50	1 August 2020 17:27:37	113.532791	22.256026	113.541893	22.240246	2900	7.4	39	41
1003	2 August 2020 18:49:03	2 August 2020 19:02:14	113.548533	22.222178	113.506586	22.226488	6300	13.1	126	170
...	...	...	...	...	...	...	...	...	...	...

In order to analyze changes in human mobility using the complex network method, taxi OD points were integrated into different TAZs. Thus, the last dataset consists of the TAZ data for the central area, consisting of 199 vector polygons (Figure 1).

3. Methodology

This section presents the methodological framework proposed in this study. Figure 3 shows our framework, which consists of two major components. In the first component, we first calculated basic trip characteristics for different scenarios, such as trip distance and trip duration, and then analyzed the spatial distribution of the trip in the four scenarios using kernel density estimation. In the second component, we constructed the entire networks for the four scenarios, and from these we detected the mobility communities in order to investigate the impact of rainfall on human mobility by comparing the whole network and community network indicators.

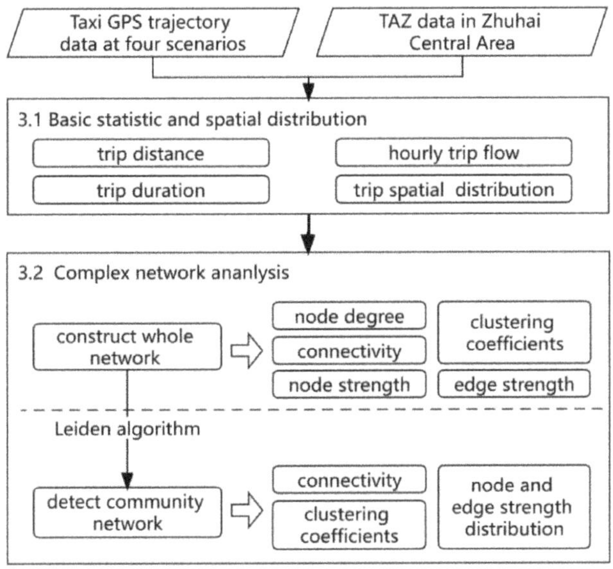

Figure 3. Workflow of the analytical framework.

3.1. *Basic Mobility Characteristics*

As each trip is simplified to a vector $<(O_x, O_y, O_m, O_t), (D_x, D_y, D_m, D_t)>$, the basic trip patterns can be analyzed from the following two perspectives: Firstly, the properties of all trips, such as distance and duration, can be computed and the associated statistical distributions of the four scenarios are thus obtained. Trip distance $Trip_{dist}$ and trip duration $Trip_{dura}$ were calculated as follows:

$$Trip_{dist} = D_m - O_m \quad (1)$$

$$Trip_{dura} = D_t - O_t \quad (2)$$

Secondly, we used kernel density estimation to investigate the spatial distribution of the trips in the different scenarios. Kernel density estimation is a nonparametric method for estimating a density function from a random sample of data. The kernel density estimation $f(x)$ was calculated as follows:

$$f(x) = \frac{1}{h^2}\Sigma_{i=1}^{n} K\left(\frac{x - X_i}{h}\right) \quad (3)$$

where h is the bandwidth, n is the number of discrete points in the bandwidth range, and $K(x)$ is the kernel function.

3.2. *Complex Network Analysis*

3.2.1. Network Construction and Community Detection

To capture a holistic picture of the urban mobility network, we extracted the travel connection relationships between each pair of origin and destination (OD) TAZs and aggregated all the taxi trips to construct a weighted directed network. This is defined as $G = (V, E)$, where $V = \{v_1, v_2, \cdots, v_n\}$ contains all distinct visited TAZs, where n is the number of TAZs in the central area. The edge set $E = \{(l_{i,j}, w_{i,j}) | i, j \in V \wedge i \neq j\}$ contains all existing directed trips, where $l_{i,j}$ represents the directed flow between pairs of TAZs. $w_{i,j}$ is the weight of edge $l_{i,j}$, which corresponds to the taxi trip flow of TAZs.

Urban mobility subnetworks were constructed using community detection. Community detection methods aim to identify partitions (structures composed of communities) which maximize the density of intragroup connections and thus find dense optimal subgraphs in large graphs. In other words, the community detection technique was used in our study to detect TAZs that had a higher quantity of interactions with each other than with the other TAZs, which means that people located in those TAZs have closer links with each other than those located elsewhere.

The Louvain method is a popular community detection algorithm that has the advantage of minimizing computation time [38]. However, it can yield arbitrarily badly connected communities. The Leiden technique is used to identify well-connected and locally optimal dynamic mobility communities in urban areas [39]. So, in this study, we used the Leiden technique to identify the dynamic urban mobility communities. This algorithm mainly consists of the following three phases: (1) local movement of nodes, (2) refinement of the partition, and (3) aggregation of the network based on the refined partition, using the nonrefined partition to create an initial partition for the aggregate network. For the network construction and analysis in this study, we used the Leidenalg package and the Python NetworkX package.

3.2.2. Statistical Indicators of Network

Degree, strength, connectivity, and clustering coefficient can be used to reflect the topology characteristics of the network. To further investigate the discrete characteristics of network indicator distribution, a standardized measure coefficient was used to represent the discrete characteristics of the network. By comparing the temporal changes of these indicator values in different weather conditions, we can obtain a better understanding of the impact of rainfall on network mobility.

Node degree is an important quantity that reveals the spatial heterogeneity of urban mobility [40]. Nodes with larger degrees represent more highly connected areas in the city. In a network, the degree of a node is the number of edges directly connected to the node, as shown in the following formula. In this study, it was the total number of passengers who were picked up or dropped off at a TAZ.

$$k_i = \sum_{j \in V} N(v_i, v_j) \tag{4}$$

The strength of edge and node are two indicators for the analysis of the network flow. Edge strength W represents the taxi flow in a specific direction between the two TAZs. Node strength S_i is employed to generalize the degree measure of weighted networks, which is defined as the sum of the taxi flow on all edges associated with node i. The calculation formula is as follows:

$$W = \sum_{i=1}^{m} r_i \tag{5}$$

$$S_i = \sum_{j \in V} W(v_i, v_j) \tag{6}$$

where r_i represents one trip in this direction, and m is the total number of such trips.

The connectivity of the network δ is quantitatively calculated as follows:

$$\delta = \frac{2 \times L}{N^2} \tag{7}$$

where N is the number of nodes and L is the number of edges. A large δ indicates that the taxi connections between TAZs were relatively denser and, thus, the network had better overall connectivity.

Clustering coefficients include the local clustering coefficient and the average clustering coefficient. The local clustering coefficient of a node describes the likelihood that the neighbors of this node are also connected. If a node has a high local clustering coefficient value, this indicates local cohesiveness and a high tendency to form groups. For node i, its local clustering coefficient is the fraction of the links that are present among the total possible links between its neighbors. There are several generalizations of clustering coefficient to weighted graphs, and the definition by [41] is a local node-level quantity. Its formula is

$$C_i^w = \frac{1}{s_i(k_i - 1)} \sum_{j,h} \frac{w_{i,j} + w_{i,h}}{2} a_{ij} a_{ih} a_{jh} \tag{8}$$

where s_i is the strength of the node i, a_{ij} are elements of the adjacency matrix, k_i is the node degree, $w_{i,j}$ are the weights. The average clustering coefficient of all nodes, $\langle C_w \rangle$, can be applied to quantify the density of the entire network. The calculation formula is as follows:

$$\langle C_w \rangle = \frac{\sum_{i \in V} C_w(i)}{N} \tag{9}$$

Closeness centrality is tightly related to the notion of distance between nodes. It is calculated as the average of the shortest path length from the node to every other node in the network. The calculation formula is as follows:

$$CC_i = \frac{N - 1}{\sum_{i \neq j} d(i,j)} \tag{10}$$

where N is the number of nodes in the network, and $d(i,j)$ is the shortest path between nodes i and j. The larger the CC_i, the higher the node's closeness centrality, and the better its connection with other nodes.

For any indicator x of the network, such as degree or strength, we use the standardized measure coefficient of variation $CV(x)$ to further investigate the discrete characteristics of network indicator distribution. The calculation formula is as follows:

$$CV(x) = \frac{[x]}{\langle x \rangle} \tag{11}$$

where $[x]$ is the standard deviation and $\langle x \rangle$ is the average value. In particular, the coefficient of variation is not affected by measurement scale and dimension.

4. Results

4.1. Basic Statistics and Spatial Distribution of Trip Data

To analyze the general distribution of trips in the different scenarios, average daily statistics of the total number of trips, trip duration, and trip distance of each taxi were calculated, as shown in Figure 4. In the figure, NRWD and RAWD represent weekdays without rainfall and weekdays with rainfall, respectively, and NRWE and RAWE represent weekends without rainfall and weekends with rainfall, respectively. These abbreviations have the same meaning in the diagram below.

Overall, the average number of trips at weekends was slightly higher than on weekdays. As expected, in terms of weekdays, the number of trips on days with rainfall was obviously lower than on days without rainfall. This is likely to be because people canceled nonessential travel. By the same token, at weekends, there were also more taxi trips on days without rainfall than on days with rainfall. From Figure 4b, we can see that trip distance did not differ significantly between weekdays and weekends and was slightly higher on days without rainfall than on days with rainfall. Figure 4c shows that the trip duration at weekends was slightly reduced compared to weekdays.

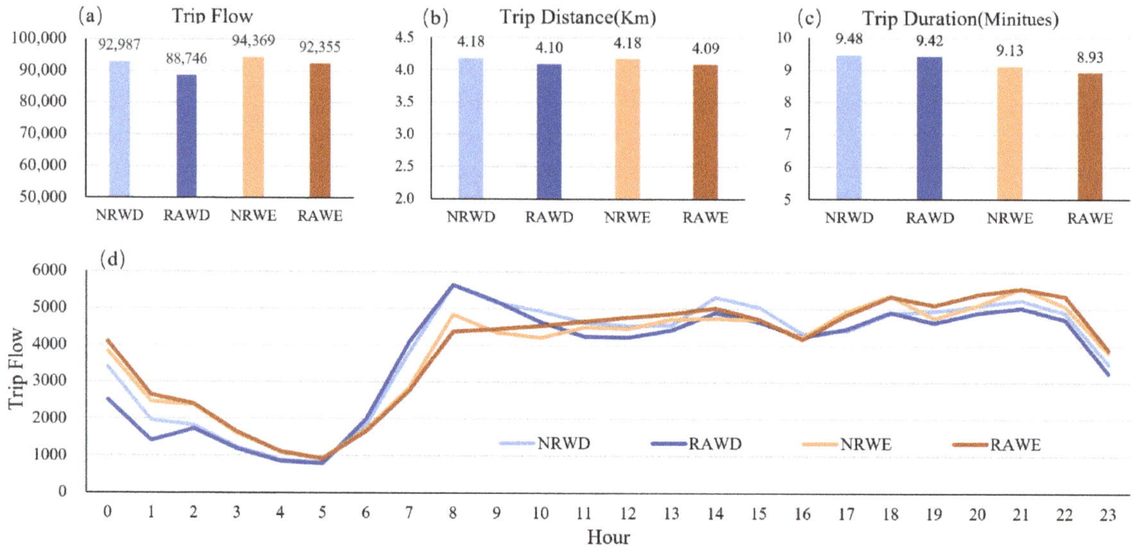

Figure 4. Average daily statistics of taxi trip in the four scenarios. (**a**) Trip flow; (**b**) Trip distance; (**c**) Trip duration; (**d**) Hourly trip flow.

The demand for taxis changes in time and space according to the travel needs of citizens. Figure 4d shows the hourly changes for the four scenarios; the x-axis indicates the time horizon of 24 h and the y-axis is the number of trips. On weekdays, three peaks can be observed: (1) a morning peak starting around 8 a.m., (2) an afternoon peak around 2 p.m., and (3) an evening peak starting around 8 p.m. During the first peak period, the trip flows were not affected by the rainfall. At the second and third peak hours, the taxi flow on days without rainfall was significantly higher than it was on days with rainfall. At weekends, the maximum trip flow occurred in the evening peak hours, in contrast to the maximum trip flow on weekdays which was during the morning peak hour.

To gain a better understanding of the patterns of taxi services, we further investigated the proportions according to different trip distances and durations, as shown in Figure 5. In general, the difference in the proportion of different distances and durations was not obvious. In Figure 5a, we can see that almost all trips were within 15 km and that 85 per cent of trips were within 7 km. Figure 5b shows that almost all trips took less than 35 min, with more than 85 percent taking less than 15 min.

When comparing trip flow by distance on weekdays and weekends, the proportion for days with rainfall was higher than for days without rainfall when the distance was less than 3 km, and the result was reversed once the distance exceeded 3 km. Compared to the trip duration in Figure 5a, the average weekend duration was shorter than it was on weekdays, and the proportion that took less than ten minutes was greater on weekends. This may be related to the fact that travel needs on the weekends are mainly leisure and close to home.

To investigate the spatial patterns of taxi passengers in the four scenarios, we interpolated the daily average pick-up location in each scenario using the kernel density estimation method. Kernel density estimation can intuitively reflect the spatial distribution of taxi passengers in the different scenarios, and can also represent the changes by comparison of the density results.

As shown in Figure 6, we identified several identical hot spots, which represent the locations with high passenger flow in the four scenarios. The highest are located near Zhuhai Railway Station and Gongbei Port. On days with rainfall, the passenger density decreased, and some hot spot areas were not obvious. Hot spots such as Mingzhu Railway

Station, Mingyang Plaza, and Huafa Plaza were not affected by rainfall on weekdays or weekends. Fuhuali Plaza was a significant hotspot on weekends without rainfall, but the hotspot disappeared on weekends with rainfall.

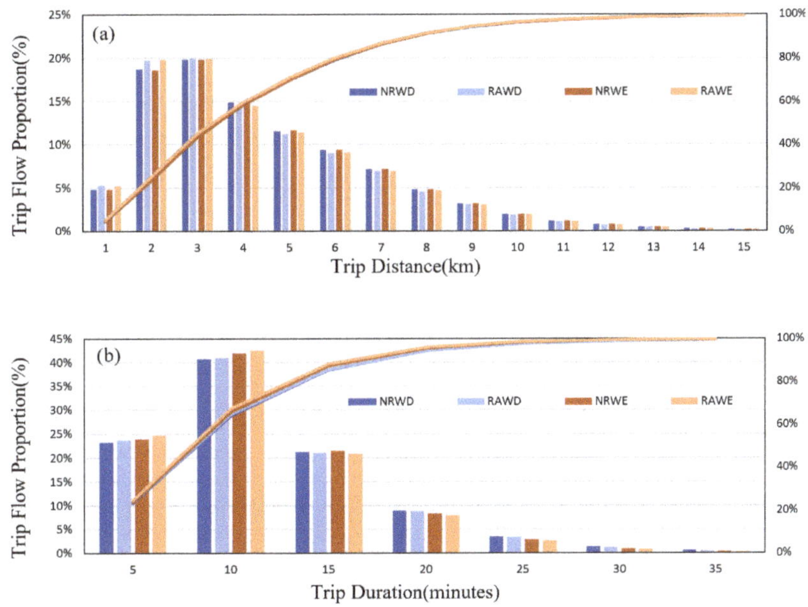

Figure 5. (**a**) The proportional distribution of trip distance in the four scenarios; (**b**) The proportional distribution of duration in the four scenarios.

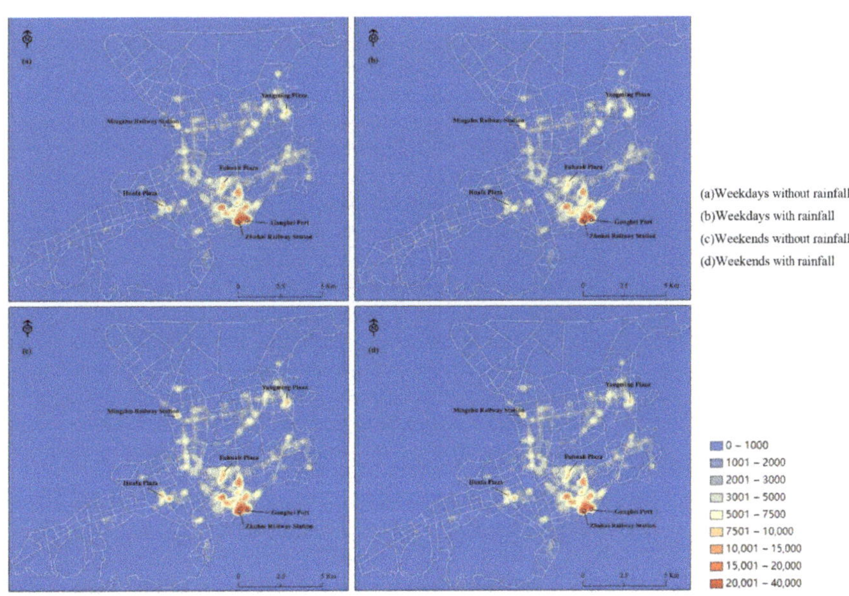

Figure 6. Spatiotemporal comparison of kernel density interpolation results.

4.2. Complex Network-Based Analytical Indicators

4.2.1. Indicator Analysis of Whole Network

The network was constructed according to the taxi trajectory data for weekdays with rainfall, weekdays without rainfall, weekends with rainfall, and weekends without rainfall, respectively. Then, the complex network indicators introduced in 2.2 were calculated, as shown in Table 2.

Table 2. The whole network indicators in four scenarios.

Indicators	Description	NRWD	RAWD	NRWE	RAWE
L	The number of edges	24194	21522	24232	20491
<K>	Node average degree	244.38	216.3	243.54	205.94
δ	Network connectivity	1.234	1.087	1.224	1.035
$<C_W>$	Average clustering	0.816	0.794	0.812	0.773
<S>	Node average strength	939.26	891.92	948.43	928.19
CV(S)	Coefficient of variation of node strength	1.39	1.43	1.41	1.43
<W>	Edge average strength	3.84	4.12	3.89	4.51
CV(W)	Coefficient of variation of edge strength	3.08	3.08	3.10	2.99

Compared to days without rainfall, the number of connections between TAZs (i.e., the number of network edges) decreased significantly during the corresponding period of days with rainfall, both on weekdays and at weekends. Additionally, the negative impact of rain at weekends was stronger than that on weekdays. On days without rainfall, the number of network edges (L) increased from weekdays to weekends, following the same trend as the average daily trip flow. However, there were fewer connections between TAZs on weekends with rainfall than on weekdays with rainfall compared to the average daily trip flow.

The change in the number of network edges affects the average degree of nodes <K>. <K> decreased as network edges decreased on days with rainfall. Overall, rainfall reduced the external contact of TAZs by 11.5% on weekdays and 15.4% on weekends. Similarly, the network connectivity (δ) of the network showed an obvious decline on days with rainfall. However, there was no significant change between weekdays and weekends. Finally, analysis of the average cluster coefficient C showed that it decreased on days with rainfall. The decrease in network edges due to rainfall reduced the connection density and the number of closed triplets between TAZs, thereby weakening the cluster connection between TAZs.

In terms of network strength, the node strength <S> and edge strength <W> of weekdays with and without rainfall, were lower than they were at weekends. As expected, days with rainfall had lower node strength than days without rainfall. Conversely, edge strength was intense on days with rainfall. That is, as the number of edges decreased on days with rainfall, the average edge strength was higher than it was on days without rainfall.

In terms of flow distribution, the coefficients of variation of node strength CV(S) on days with rainfall were higher than on days without rainfall both on weekdays and at weekends. When we compared this value for weekdays and weekends, the CV(S) did not change on weekdays with rainfall, whereas the value increased from weekdays to weekends on days with no rain. This means that the distribution of node strength on days with rainfall was more homogeneous compared to weekdays and weekends without rainfall. The same coefficient of variation of edge strength CV(W) on weekdays indicated that rainfall on weekdays had no effect on the heterogeneity of edge strength distribution, but rainfall on weekends had a significant effect on edge strength distribution.

In order to better compare the changes of various indicators on days with rainfall, we computed and visualized the indicators of each network node on weekdays and weekends. In terms of direct connection indicators, the spatiotemporal distribution of the difference in node degree K and local clustering coefficient Cw on the weekdays and weekends with rainfall is shown in Figure 7.

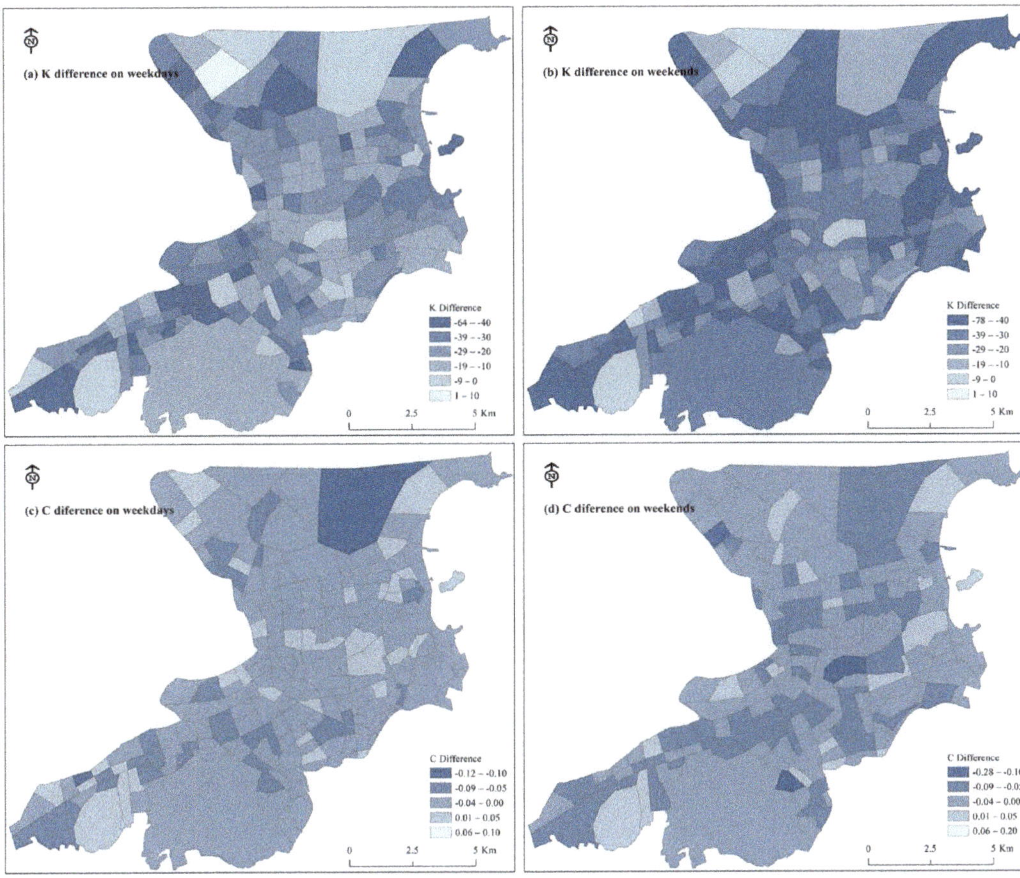

Figure 7. Visualization of the difference between K and C.

The difference in TAZ degrees between weekdays and weekends with rainfall decreased significantly at most of the TAZ nodes, spatially in the periphery of the study area. This means that the number of travel connections between TAZs decreased because of rainfall. At weekends, the node degree decreased more than on weekdays. This means that the impact of rainfall on the number of travel connections between TAZs was minor on weekdays. The large areas of dark blue in Figure 7b indicate that TAZs in these locations reduced some unnecessary travel connections on weekends with rainfall. As shown in Figure 7, the local clustering coefficient of most TAZs did not change significantly on the weekdays. However, at weekends, the number of TAZs with lower C increased significantly, and the TAZs with weakened C showed the characteristics of a large weakening range. For the spatial distribution, although some TAZs with increased C could be found both on the weekdays and weekends, they were relatively uniform and in a mixed state.

To further investigate the connection between TAZs, the differences in node strength S and closeness centrality CC on days with and without rainfall were calculated. As shown in Figure 8, we found that regardless of whether it was a weekday or the weekend, the TAZs with the highest node strength decline were mainly concentrated in commercial areas such as Huafa, Fuhua, and Yangming Plaza. Only a few TAZ nodes with slightly poor traffic conditions increased in node strength on the days with rainfall. This means that on days with rainfall, people with strong travel needs may prefer taxis. On days with rainfall, the closeness centrality of TAZs decreased overall on weekdays and weekends, which means that rainfall weakened the indirect connectivity between TAZs.

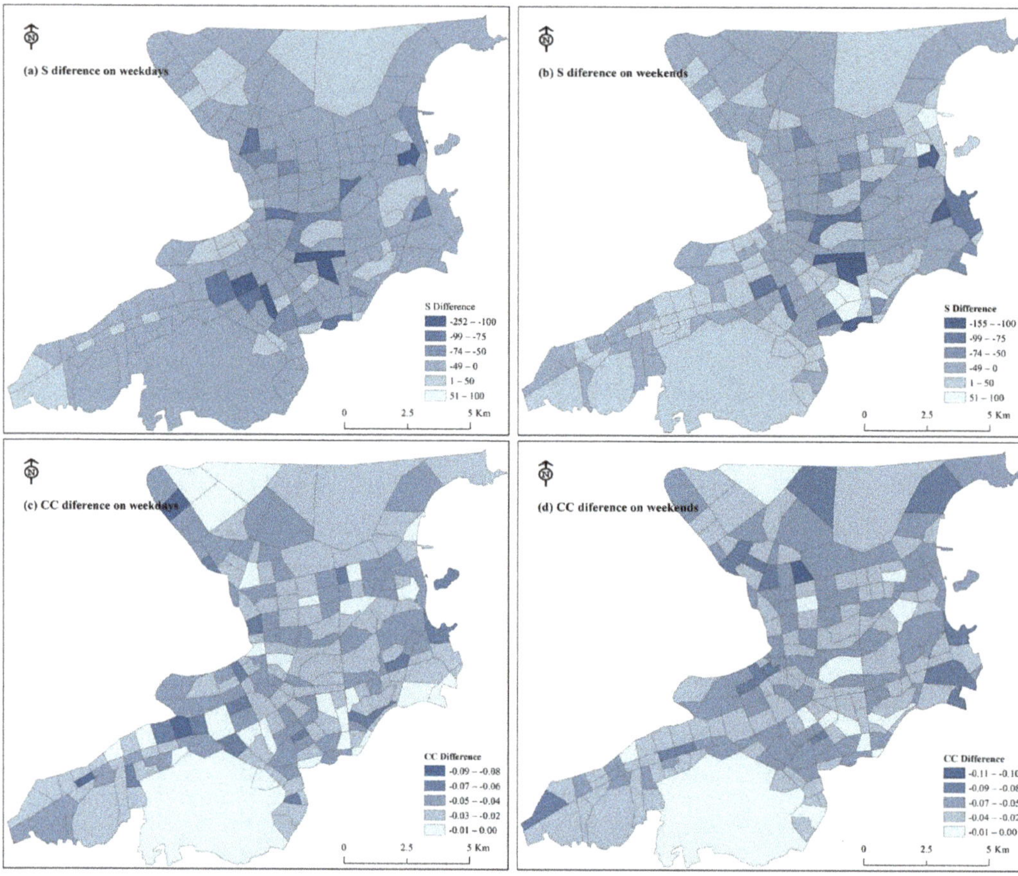

Figure 8. Visualization of the difference between S and CC.

4.2.2. Indicators Analysis of Community Network

Using the Leiden algorithms, we detected network communities from the dataset of the four scenarios. As shown in Figure 9, four communities were detected on the weekdays and five communities were detected at the weekends, and different communities are visualized in different colors.

On weekdays, the spatial pattern of the four communities was very similar. The red community C1, where Jida is located, had an enclave in its southern area on days without rainfall. The blue community C2, consisting of Shishan, Xiangwan, Cuixiang, and Meihua, extended southwards on days with rainfall. On days without rainfall, there were a few enclaves between different communities, but on days with rainfall, the enclaves disappeared, and the pattern of each community was more concentrated.

Compared with weekdays with rainfall, the structures of communities change more on weekends with rainfall. The TAZ number of the yellow community clearly decreased. By absorbing a small part of the yellow community, the red community extended to the northwest and the blue community to the east. As for the weekdays, on weekends with rainfall, each community was more compact, with no enclaves.

To carry out a more detailed exploration at the community level, we calculated the network connectivity δ, the average clustering coefficient <C>, the coefficient of variation of node strength CV(S), and the coefficient of variation of edge strength CV(W) of these four community networks. The results are shown in Table 3.

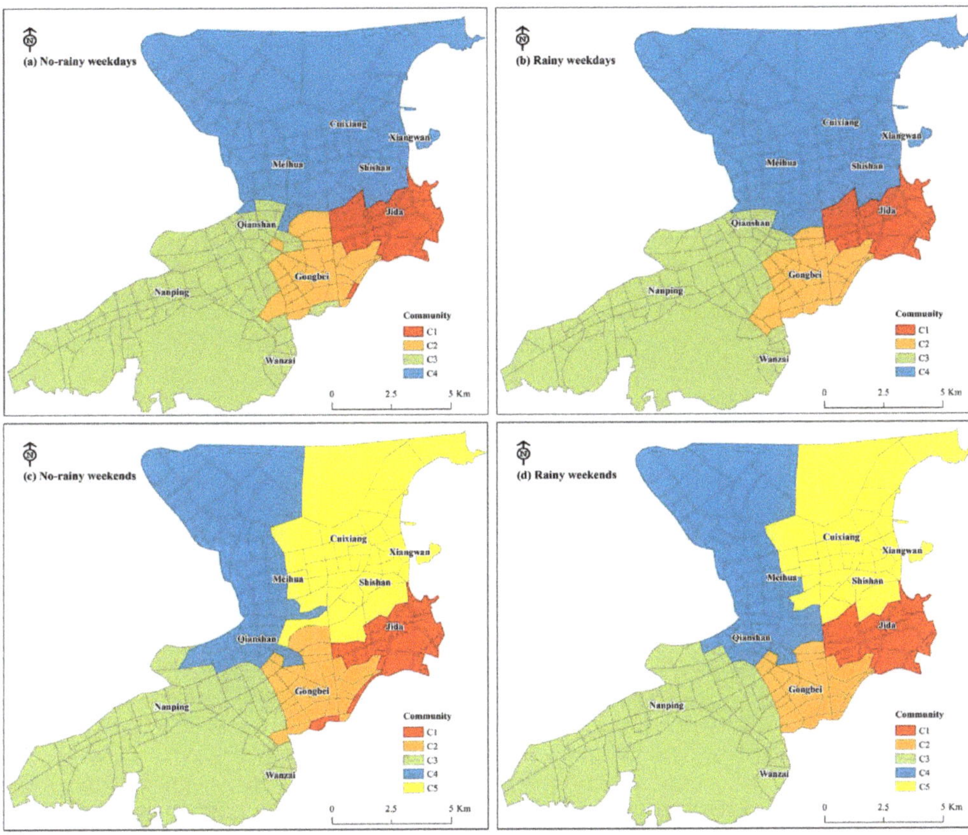

Figure 9. The detected communities of the four datasets.

Table 3. The network indicators of communities of four scenarios.

Community	Weekdays without Rainfall				Weekdays with Rainfall				Weekends without Rainfall					Weekends with Rainfall				
	C1	C2	C3	C4	C1	C2	C3	C4	C1	C2	C3	C4	C5	C1	C2	C3	C4	C5
δ	1.884	1.857	1.211	1.487	1.852	1.898	1.042	1.309	1.985	1.814	1.182	1.389	1.755	1.827	1.906	0.964	1.217	1.679
<C>	0.975	0.961	0.778	0.876	0.966	0.97	0.731	0.864	0.995	0.959	0.749	0.853	0.945	0.958	0.962	0.725	0.82	0.919
CV(S)	0.738	0.984	1.22	0.98	0.742	0.998	1.23	1.001	0.65	1.009	1.436	0.954	0.88	0.753	0.989	1.437	0.937	0.904
CV(W)	1.21	1.736	1.883	1.682	1.21	1.834	1.741	1.601	1.121	1.766	2.362	1.521	1.458	1.221	1.8	2.032	1.329	1.512

On weekdays, with the exception of community C2, the network connectivity δ and average clustering coefficient <C> of all the communities decreased on days with rainfall, as shown in Table 2. This indicates that most communities had higher clustering characteristics on weekdays with rainfall. In addition, community C3, where Nanwan is located, exhibited the lowest clustering characteristics on weekdays both with and without rainfall. In terms of network flow, the CV(S) of node strength for all communities increased on days with rainfall, whereas the CV(W) of communities C3 and C4 differed from the overall trend. As can be seen from Figure 6, the traveler density in the C3 and C4 areas was relatively small, so the coefficient of variation of edge strength was lower on the days with rainfall.

At weekends, we found that the network structure was relatively unstable, and its connectivity was more affected by rainfall. As on weekdays, only the network connectivity δ of community C2 increased on weekends with rainfall. Additionally, it can be seen from Table 2 that the average clustering coefficient <C> of community C2 increased slightly on weekends with rainfall. This means that on the days with rainfall, the community network near Zhuhai Railway Station and Gongbei Port not only maintained high connectivity but also maintained high aggregation characteristics among the TAZs. On weekends with rainfall, the CV(S) and CV(W) tend to vary by communities. The CV(S) and CV(W) of communities C1 and C5 showed the same increasing trend on weekends with rainfall, but C4 was the opposite, with both showing decreasing trends.

5. Summary and Discussion

Urban China currently faces a low-level natural disaster in terms of rainstorms and flooding, which have a significant impact on travel. Based on taxi data in different scenarios, this study analyzed the impact of rainfall on residents' travel using basic statistical analysis and complex network analysis, and conducted comparative analysis and detailed discussion from time and space dimensions.

This study benefits from the advantages of taxi data in terms of the large sample size, its accuracy, and its individual dimension. Acquiring the spatial–temporal characteristics of inhabitants' travel on various days with rainfall can help us to further understand the impact of rainfall on travel in urban areas. In addition, it contributes to a deeper understanding of the interaction between residents' daily travel and the complex geographic environment of cities and provides more detailed support for decision-making, planning, and management of urban transportation and land use systems.

However, this study also has some limitations in data and methods. For instance, it only discusses the impact of rainfall factors on residents' travel based on taxi trajectory data, and the analysis is performed on a time scale of days. In the future, other travel data (such as smart card data and mobile signaling data) could be used for further investigation, and this could be analyzed in depth by hours (such as morning and evening peak or different hours of the day). In addition, our research defined days with rainfall as those with continuous rain for 6 h and an amount of rain greater than 25 mm. Only one month's worth of data was used for the analysis. With the accumulation of data, it would be possible to thoroughly analyze the impact on human travel behavior at different levels of rainfall and extreme rainfall conditions. Moreover, rainfall has a certain spatial and temporal heterogeneity. This study did not analyze the travel impacts of changes in rainfall conditions within a day, such as how long the delayed impact of rainfall on travel lasted and what the conditions were for the recovery of mobility in different regions after rain. Neither did we consider what factors directly affected the recovery time for human mobility. In future work, we will use hourly rainfall data and analysis of travel characteristics to answer these questions. Finally, this study only focused on the impact of rainfall on residents' travel, and the impact of other weather factors (such as temperature, relative humidity, and wind speed) could be considered in the future.

6. Conclusions

In this paper, we took the central area of Zhuhai as our research area, and based on taxi data, used basic statistics and complex network analysis methods to compare and analyze human mobility in four scenarios. The research conclusions are as follows:

(1) Taxi GPS data are highly informative and exploitable in the field of human mobility analysis. Using the location and times at which passengers were picked up and dropped off in taxi trip GPS data, we can analyze activity levels across the city and the way people move around the city;

(2) Rainfall has a reducing effect on trip flow whether on weekdays or at weekends, as well as on trip distance and trip duration, but has no significant impact on the appearance and duration of peak hours. From the spatial distribution of passengers,

it is evident that rainfall has little effect on most hotspots, with the exception of a few commercial centers;

(3) From the perspective of the whole mobility network, rainfall has a significant effect on the network indicators. For instance, the edges of the network and the average degree of nodes decreased significantly on days with rainfall. Node and edge strength in some commercial areas declined significantly on the days with rainfall;

(4) There were more mobility communities were detected on weekends than on weekdays. The number of communities on weekdays and weekends did not change because of rainfall. For communities located in transportation hubs or port areas, the changes in network indicators were opposite to those of other communities.

Author Contributions: Conceptualization, P.G. and Y.S.; methodology, P.G.; formal analysis, P.G. and Y.S.; writing—original draft preparation, P.G. and Y.S.; writing—review and editing, Q.C., J.L. and Z.L. All authors have read and agreed to the published version of the manuscript.

Funding: This research was funded by the National Natural Science Foundation of China (Grant No. 41001022 and 41971306).

Institutional Review Board Statement: Not applicable.

Informed Consent Statement: Not applicable.

Data Availability Statement: Not applicable.

Acknowledgments: Special thanks to anonymous reviewers for their valuable comments. In addition, the authors gratefully acknowledge Jialing Wu, Ying Zhang, and Jie Liu for assisting with the data preparation.

Conflicts of Interest: The authors declare no conflict of interest. The funders had no role in the design of the study; in the collection, analyses, or interpretation of data, in the writing of the manuscript, or in the decision to publish the results.

References

1. Xu, M.L.; Fu, X.; Tang, J.Y.; Liu, Z.Y. Effects of weather factors on the spatial and temporal distributions of metro passenger flows: An empirical study based on smart card data. *Prog. Geogr.* **2020**, *39*, 45–55. [CrossRef]
2. Barbosa, H.; Barthelemy, M.; Ghoshal, G.; James, C.R.; Lenormand, M.; Louail, T.; Me-nezes, R.; Ramasco, J.J.; Simini, F.; Tomasini, M. Human mobility: Models and applications. *Phys. Rep.* **2018**, *734*, 1–74. [CrossRef]
3. Lu, F.; Liu, K.; Chen, J. Research on Human Mobility in Big Data Era. *J. Geo-Inf. Sci.* **2014**, *16*, 665–671.
4. Yang, X.P.; Fang, Z.X. Recent progress in studying human mobility and urban spatial structure based on mobile location big data. *Prog. Geogr.* **2018**, *37*, 880–889.
5. Ding, L.F.; Fan, H.C.; Meng, L.Q. Understanding taxi driving behaviors from movement data. *AGILE* **2015**, *2015*, 219–234.
6. Yuan, Y.H.; Raubal, M.; Liu, Y. Correlating mobile phone usage and travel behavior—A case study of Harbin, China. *Comput. Environ. Urban Syst.* **2012**, *36*, 118–130. [CrossRef]
7. Kang, C.G.; Liu, X.; Xu, X.Y.; Kun, Q. Impact of Weather Condition on Intra-Urban Travel Behavior: Evidence from Taxi Trajectory Data. *J. Geo-Inf. Sci.* **2019**, *21*, 118–127.
8. Zhan, G.; Wilson, N.H.W.; Rahbee, A. Impact of weather on transit ridership in Chicago, Illinois. *Transp. Res. Rec. J. Transp. Res. Board.* **2007**, *2034*, 3–10.
9. Koetse, M.J.; Rietveld, P. The impact of climate change and weather on transport: An overview of empirical findings. *Transp. Res. Part D Transp. Environ.* **2009**, *14*, 205–221. [CrossRef]
10. Mesbah, M.; Luy, M.; Currie, G. Investigating the lagged effect of weather parameters on travel time reliability. *WIT Trans. Ecol. Environ.* **2014**, *191*, 795–801.
11. Syeed, K.; Bunker, J.M. Adverse weather effects on bus ridership. *Road Transp. Res.* **2015**, *24*, 44–57.
12. Zhang, X.Q.; An, S.; Sheng, H.F. Discrete dynamic road network capacity under adverse weather. *J. Harbin Inst. Technol.* **2009**, *41*, 85–88.
13. Gong, D.P.; Song, G.H.; Li, M.; Gao, Y.; Yu, L. Impact of rainfalls on travel speed on urban roads. *J. Transp. Syst. Eng. Inf. Technol.* **2015**, *15*, 218–225.
14. Cools, M.; Moons, E.; Creemers, L.; Wets, G. Changes in travel behavior in response to weather conditions: Do type of weather and trip purpose matter? *Transp. Res. Rec. J. Transp. Res. Board.* **2010**, *2157*, 22–28. [CrossRef]
15. Palma, D.A.; Rochat, D. Understanding individual travel decisions: Results from a commuters survey in Geneva. *Transportation* **1999**, *26*, 263–281. [CrossRef]

16. Arana, P.; Cabezudo, S.; Peñalba, M. Influence of weather conditions on transit ridership: A statistical study using data from Smartcards. *Transp. Res. Part A Policy Pract.* **2014**, *59*, 1–12. [CrossRef]
17. Keay, K.; Simmonds, I. The association of rainfall and other weather variables with road traffic volume in Melbourne, Australia. *Accid. Anal. Prev.* **2005**, *37*, 109–124. [CrossRef] [PubMed]
18. Lin, L.; Ni, M.; He, Q.; Gao, J. Modeling the impacts of inclement weather on freeway traffic speed: Exploratory study with social media data. *Transp. Res. Rec. J. Transp. Res. Board.* **2015**, *2482*, 82–89. [CrossRef]
19. Ma, F.M.; Liu, H. The Influence of Snow Weather Affects Fast Road Traffic Characteristics and It's Countermeasures. *Technol. Econ. Areas Commun.* **2016**, *18*, 53–56.
20. Theofilatos, A.; Yannis, G. A review of the effect of traffic and weather characteristics on road safety. *Accid. Anal. Prev.* **2014**, *72*, 244–256. [CrossRef]
21. Ao, M.; Qu, R. Analysis of Influence of Meteorological Conditions on Road Traffic. *Highw. Automot. Appl.* **2011**, *2*, 58–62.
22. Stern, E.; Zehavi, Y. Road safety and hot weather: A study in applied transport geography. *Trans. Inst. Br. Geogr.* **1990**, *15*, 102–111. [CrossRef]
23. Edwards, J.B. Weather-related road accidents in England and Wales: A spatial analysis. *J. Transp. Geogr.* **1996**, *4*, 201–212. [CrossRef]
24. Li, C.; Huang, Y.; Li, G.; Yi, C. Analysis of the influence of smog on the travel behavior. *J. Xi'an Univ. Archit. Technol.* **2015**, *47*, 728–733.
25. Trépanier, M.; Agard, B.; Morency, J. Using smart card data to assess the impact of weather on public transport user behavior. In Proceedings of the Conference on Advanced Systems for Public Transit-CASPT12, Santiago, Chile, 23–27 July 2012; pp. 1–15.
26. Tao, S.; Corcoran, J.; Rowe, F.; Hickman, M. To travel or not to travel: 'Weather' is the question. Modelling the effect of local weather conditions on bus ridership. *Transp. Res. Part C Emer. Technol.* **2018**, *86*, 147–167. [CrossRef]
27. Singhal, A.; Kamga, C.; Yazici, A. Impact of weather on urban transit ridership. *Transp. Res. Part A Policy Pract.* **2014**, *69*, 379–391. [CrossRef]
28. Stover, V.M.; Mccormack, E.D. The impact of weather on bus ridership in Pierce County, Washington. *J. Public Transp.* **2012**, *15*, 95–110. [CrossRef]
29. Eisenberg, D. The mixed effects of precipitation on traffic crashes. *Accid. Anal. Prev.* **2004**, *36*, 637–647. [CrossRef]
30. Kamga, C.N.; Yazici, M.A.; Singhal, A. Hailing in the rain: Temporal and weather-related variations in taxi ridership and taxi demand-supply equilibrium. In Proceedings of the Transportation Research Board Annual Meeting, Washington, DC, USA, 8–12 January 2013; Volume 13, p. 31.
31. Su, Y.; Zhou, L.; Cui, A. Analysis of the influence of extreme weather on urban traffic. *Transp. Enterp. Manag.* **2016**, *10*, 6–9.
32. Sui, T.; Corcoran, J.; Hickman, M.; Stimson, R. The influence of weather on local geographical patterns of bus usage. *J. Transp. Geogr.* **2016**, *54*, 66–80.
33. Saberi, M.; Mahmassani, H.S.; Brockmann, D.; Hosseini, A. A complex network perspective for characterizing urban travel demand patterns: Graph theoretical analysis of large-scale origin–destination demand networks. *Transportation* **2017**, *44*, 1383–1402. [CrossRef]
34. Zhong, C.; Arisona, S.M.; Huang, X.F.; Batty, M.; Schmitt, G. Detecting the dynamics of urban structure through spatial network analysis. *Int. J. Geogr. Inf. Sci.* **2014**, *28*, 2178–2199. [CrossRef]
35. Xin, R.; Ai, T.; Ding, L.; Zhu, R.X.; Meng, L.Q. Impact of the COVID-19 pandemic on urban human mobility—A multiscale geospatial network analysis using New York bike-sharing data. *Cities* **2022**, *126*, 103677. [CrossRef] [PubMed]
36. Cao, J.Z.; Li, Q.Q.; Tu, W.; Cao, Q.L.; Cao, R.; Zhong, C. Resolving urban mobility networks from individual travel graphs using massive-scale mobile phone tracking data. *Cities* **2021**, *110*, 103077. [CrossRef]
37. Jia, T.; Cai, C.X.; Li, X.; Luo, X.; Zhang, Y.Y.; Yu, X.S. Dynamical community detection and spatiotemporal analysis in multilayer spatial interaction networks using trajectory data. *Int. J. Geogr. Inf. Sci.* **2022**. [CrossRef]
38. Blondel, V.D.; Krings, G.; Thomas, I. Regions and borders of mobile telephony in Belgium and in the Brussels metropolitan zone. *Bruss. Stud.* **2010**, *42*, 1–12. [CrossRef]
39. Traag, V.A.; Waltman, L.; Van Eck, N.J. From Louvain to Leiden: Guaranteeing well-connected communities. *Sci. Rep.* **2019**, *9*, 5233. [CrossRef]
40. Jacob, R.; Harikrishnan, K.P.; Misra, R.; Ambika, G. Measure for degree heterogeneity in complex networks and its application to recurrence network analysis. *R. Soc. Open Sci.* **2017**, *4*, 1–15. [CrossRef]
41. Barrat, A.; Barthélemy, M.; Pastor-Satorras, R.; Vespignani, A. The architecture of complex weighted networks. *Proc. Natl. Acad. Sci. USA* **2004**, *101*, 3747–3752. [CrossRef]

sustainability

MDPI

Article

Understanding Climate Hazard Patterns and Urban Adaptation Measures in China

Shao Sun, Zunya Wang *, Chuanye Hu * and Ge Gao

Laboratory for Climate Studies, National Climate Center, China Meteorological Administration, Beijing 100081, China; sunshao@cma.gov.cn (S.S.); gaoge@cma.gov.cn (G.G.)
* Correspondence: wangzy@cma.gov.cn (Z.W.); hucy@cma.gov.cn (C.H.)

Abstract: Climate-related risks pose a great threat to urban safety, infrastructure stability and socioeconomic sustainability. China is a country that crosses diverse geomorphic and climatic regions in the world and is frequently affected by various climate hazards. In this study, we propose a comprehensive analysis on the spatial pattern of major climate hazards in China from 1991 to 2020, including rainstorms, droughts, heatwaves, coldwaves, typhoons, and snowstorms, and generate an integrated sketch map on multi-hazard zones. It is detectable that South of the Yangtze River is in danger of heatwaves, rainstorms, and typhoons, while the North China Plain is more likely to suffer droughts. Coldwaves, snowstorms, and freezing mainly affect Northeast China, Northwest China, and the Qinghai–Tibet Plateau. In the view of climate governance, cities are hotspots affected by intensified climate hazards in a warmer climate. There is an urgent need to incorporate a climate adaptation strategy into future city construction, so as to improve social resilience and mitigate climate impacts in rapid urbanization process. Specific adaptation measures have been developed from the perspectives of land-use planning, prevention standard, risk assessment, and emergency response to facilitate the understanding of climate resilience and urban sustainability.

Keywords: climate hazards; geospatial analysis; urban adaptation; risk management; China

Citation: Sun, S.; Wang, Z.; Hu, C.; Gao, G. Understanding Climate Hazard Patterns and Urban Adaptation Measures in China. *Sustainability* **2021**, *13*, 13886. https://doi.org/10.3390/su132413886

Academic Editors: Xiaodong Yan, Jia Yang and Shaofei Jin

Received: 28 October 2021
Accepted: 30 November 2021
Published: 15 December 2021

1. Introduction

Climate action has emerged as one of the defining issues attracting great attention from scientists, governments, and the public. A warming climate is believed to boost the frequency of extreme events and hence aggravate climate risks in the future, endanger the sustainable development of human society [1,2]. China, located in the southeast of the Eurasian continent and the west of the Pacific Ocean, is one of the countries with the most severe climate hazards in the world [3,4]. Climate hazards in China are characterized by various kinds, high frequency, long duration, distinct seasonality, and regional differentiations. In monsoon regions, the hazards of coldwaves, strong winds, and snowstorms concentrate in winter, while the hazards of rainstorms, heatwaves, droughts, and typhoons occur frequently in summer. Moreover, the interaction of compound hazards can lead to the nonlinear amplification in hazard intensity, resulting in more serious socioeconomic impacts [5,6]. Since the 21st century, climate risks in China remain high due to the elevated exposure caused by rapid economic growth and urbanization process, and climate-related economic losses have been climbing in recent years (Figure 1).

Since the impacts of climate hazards are experienced locally, it is understandable that certain cities located in hazard zones have the needs to occupy a crucial position in adaptation agenda [7–10]. Allied to the urbanization trend in China, the pressing nature of adaption in cities becomes apparent. Cities create unique microclimates, complex topographies, and plentiful emission of heat and mass of buildings and allied with their heavy reliance on interconnected networked infrastructure, high population densities, and multifarious population constitutions, increase exposure to climate hazards, while poor governance structures or inadequate urban design exacerbate climate risks [11–13].

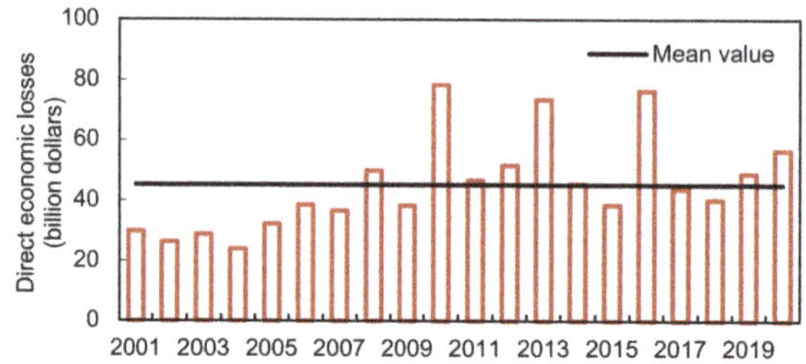

Figure 1. Direct economic losses caused by meteorological disasters in China from 2001 to 2020.

At present, actions on climate adaptation are mainly concentrated in national and supranational levels, while the potential for climate adaptation in urban level remains grossly underestimated [14]. Scholars points out that urban planning, as an important policy tool to optimize urban land-use and arrange public infrastructures, should become one of the main ways for implementing climate adaptation strategies [15–17]. However, due to some practical problems such as the inadequate understanding of climate hazards and the lack of technical standards, traditional urban planning in China has not played the leading role in climate actions. A literature review shows that many attempts have been made to focus on the impacts of climate hazards on urban areas [18–20] and incorporate climate adaptation an important part of urban planning and city expansion management [21–24].

Given that previous studies on climate hazards in China are scattered in a certain hazard type or a certain region, this study first conducts a comprehensive analysis on the spatial patterns of multiple climate hazards in China, then clarifies the difficulties and challenges that urban development faces in mitigating climate impacts. On this basis, specific suggestions on building climate-resilient cities in China are proposed from the perspectives of land-use planning, prevention standards, risk assessment, and emergency response in order to facilitate collaborations between economic, social, and climate policies and provide scientific reference for policy makers dealing with climate risks.

2. Data and Method

Daily observations of 2419 national meteorological stations in China from 1 January 1991 to 31 December 2020 are adopted in the research. The datasets are compiled and issued by the Meteorological Information Center (MIC) of the China Meteorological Administration (CMA), and it can be accessed from http://data.cma.cn (28 October 2021). The preliminary quality control has been conducted by the MIC, through checking spatial consistency, temporal consistency, and internal consistency and adjusting the suspicious records [25]. The variables used include daily precipitation, daily mean temperature, daily maximum/minimum temperature, and weather phenomena. There are some missing records in the dataset. To obtain reliable climatic statistics, two steps are processed. First, the annual mean value is taken as a missing one when the total missing records exceed 20% of a year. Second, the station with the consecutive annual mean values less than 30 years is removed from the calculation.

The daily records of weather phenomena are used to identify the snowfall day.

The CMA TC database is also adopted in the analysis [26]. The TC best-tracks are applied to derive major typhoon tracks affecting China.

The hazard data from 2001 to 2020 are collected from "Annuals of Meteorological disaster (2020)", including direct economic losses, affected population, mortality, affected croplands, crop failure areas, and collapsed buildings.

In this analysis, a hot day is defined when the daily maximum temperature reaches 35 °C and above. A rainstorm is identified with the daily precipitation reaches 50 mm and above. The scientific basis of these definitions are from *Warning Signals for Meteorological Hazard* issued by the CMA. Drought is defined based on the meteorological drought composite index (MCI) [27,28], which is adopted to monitor drought operationally by the CMA. The MCI is calculated by precipitation and mean temperature in each station. A snowstorm day is identified when the daily snowfall reaches 10 mm and above.

3. Spatial Pattern of Multiple Climate Hazards in China

Figure 2 shows the spatial pattern of the major climate hazards in China from 1991 to 2020. It is detectable that Southeastern China and Northwestern China are the two hazard zones that are frequently affected by heatwaves (Figure 2a). The number of annual hot days reaches 20 to 30 in the south of the Yangtze River, Hainan, and Chongqing. They are even greater in southern Xinjiang, Junggar Basin, and Western Inner Mongolia, being generally 30 to 50 hot days on annual average. Turpan Basin is the region with the highest temperature in China, with more than 60 hot days on annual average. In addition, a historical extreme temperature of 49.0 °C has been recorded in Turpan city, Xinjiang. The heatwaves in China usually occur from May to September. It has negative effects on human health and agriculture production, while long-lasting heatwaves can also strain energy supplies by leading to a surge in demand for water and electricity. Due to accelerated climate warming, the areas affected by heatwaves in mainland China expanded from 468 km^2 in the 1990s to 515 km^2 in the 2010s (Figure 3a), which indicates that the heatwave risk will continue to aggravate in the future.

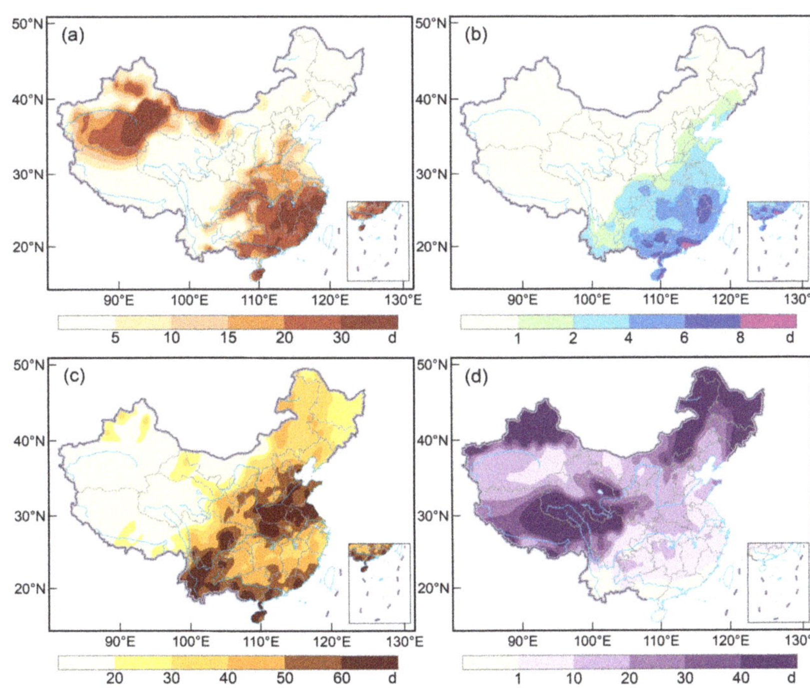

Figure 2. Climatological spatial pattern of (**a**) hot days, (**b**) rainstorm days, (**c**) drought days, and (**d**) snowfall days in China from 1991 to 2020.

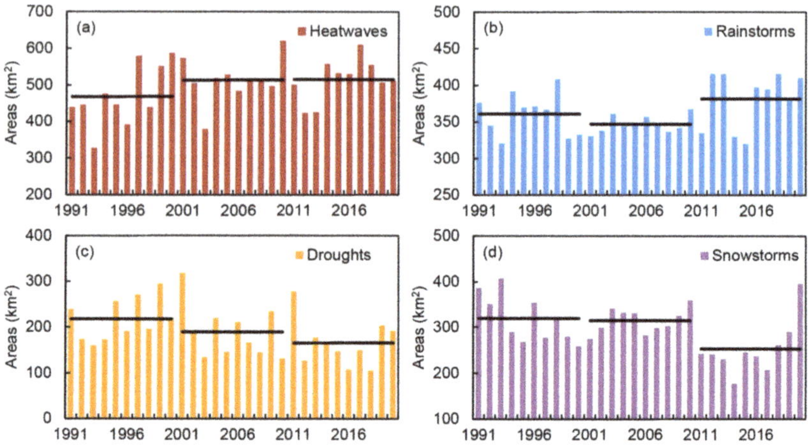

Figure 3. Areas affected by (**a**) heatwaves, (**b**) rainstorms, (**c**) droughts, and (**d**) snowstorms in China from 1991 to 2020.

China's rainfall is greatly influenced by the East Asian monsoon. With the northward movement of the East Asian monsoon, the monsoon rain belt experiences three notable stationary stages and forms the pre-summer rainy season in South China, the Meiyu in the Yangtze River, and the rainy season in North China [29–35]. Accompanying the southward retreat of the East Asian monsoon, a flood season caused by tropical cyclones affects South China again [36,37]. Rainstorms occur frequently in the rainy seasons and tend to result in floods. As shown in Figure 2b, the annual number of days with rainstorm decreases from southeast to northwest in China. There are generally four to eight rainstorm days in the south of the Yangtze River and more than eight days in coastal South China, but it is almost rare in Western China. Compared with the 2010s and 1990s, the areas affected by rainstorms in mainland China expanded from 361 km^2 to 382 km^2 (Figure 3b), indicating the intensification of rainstorm hazard in nearly 30 years.

Meteorological droughts refer to surface water shortage due to the imbalance between evapotranspiration and precipitation in a certain period, which has serious impacts on crop growth and even causes crises of water resources. Seasonality and regionality features the meteorological droughts in China. It mainly occurs in late spring, summer, and autumn in North China, in autumn and winter in South China, and in winter and spring in Southwest China. Generally, meteorological droughts mostly occur in North China, Huang-Huai Plain, Eastern Inner Mongolia, and Southwestern China, with an annual number of drought days being more than 60 days (Figure 2c). From 1991 to 2020, the areas affected by droughts in China shows a weak upward trend, shrinking from 218 km^2 in the 1990s to 165 km^2 in the 2010s (Figure 3c). However, an increase in drought occurrence are observed in North China, eastern Northwest China, and eastern Southwest China (Figure not shown).

The winter climate in China is dominated by the East Asian winter monsoon [38–41]. The strong East Asian winter monsoon leads the active cold air generating in the polar area to China and results in chilly weather, strong winds, snowstorms, and ice freezing, etc. The cold air is usually active in late autumn, winter, and early spring, which could lead to damages in houses and infrastructures and can adversely affect agriculture, transportation, livestock, and fishery production. The cold air breaks out southward along four main paths, including the west path, the middle path, the west path, and the concurrent east-west path [42]. The active cold air tends to cause snowfall. Heavy snowfall, especially snowstorms, has a great impact on agriculture, animal husbandry, communication, energy supply, and traffic. Figure 2d shows the climatic distribution of annual snowy days. Snowfall mostly occurs in the northern Xinjiang autonomous region, Northeast China, the Tibetan Plateau, and Inner Mongolia, with the annual snowfall days reaching 30 days and above. The annual snowfall days appear the most in Southern Qinghai and Eastern Inner Mongolia, being over 60 days, and these areas are also prone to snowstorm [43,44]. Compared with the 2010s and 1990s, the areas affected by snowstorms in China shrank from 319 km^2 to 253 km^2 but varies widely on an inter-annual scale (Figure 3d).

China is heavily affected by the tropical cyclones generated in the northwest Pacific Ocean and the South China Sea. Typhoons not only bring wild winds and huge waves but are also accompanied by heavy rainfall and storm surges, causing serious socioeconomic impacts. Typhoons occur from April to December, especially from July to September, and are observed with three main moving paths to affect China: the northwest path, the westward path, and the offshore turning path. Recent decades have witnessed an average of 26 typhoons generated in the northwest Pacific Ocean every year, and about 7 of them land in China. The average length of typhoon season in China is 104 days and shows a shortening trend in recent years, but the intensity and duration of landing typhoons are increasing.

The disastrous impacts of climate hazards in China during 2001–2020 are estimated from six aspects, including affected population, mortality, affected cropland, crop failure areas, collapsed buildings, and direct economic losses (Figure 4). In terms of affected population, floods and droughts account for a high proportion, 31.9% and 34.4%, respectively. The highest proportion of deaths is caused by rainstorms and floods at 52.1%, followed by severe convective weather at 36.2%. Drought is the dominant hazard to agricultural production, accounting for 48.1% of total affected croplands and 46.2% of total crop failure areas. Rainstorms and floods cause the majority of collapsed buildings among all hazards, occupying 73.8% of the total. In terms of direct economic losses, the highest proportion of 43.1% is caused by rainstorms and floods, followed by droughts and tropical cyclones, and comparatively limited losses can be seen in snowstorms and freezing.

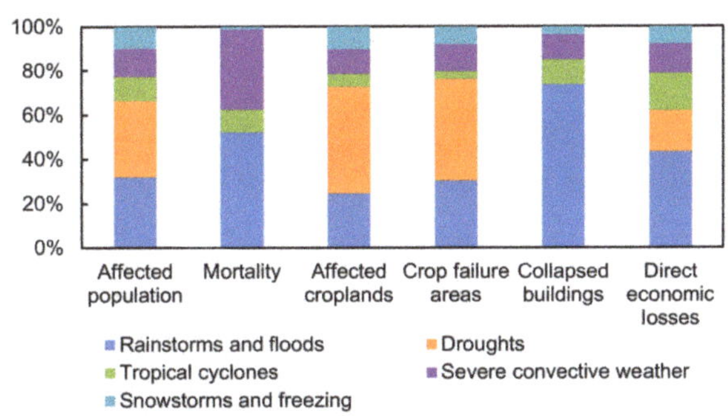

Figure 4. The proportional composition in disastrous impacts of affected population, mortality, affected croplands, crop failure areas, collapsed buildings, and direct economic losses caused by climate hazards.

To learn the integrated spatial pattern of climate hazards in China, a comprehensive hazard map is generated combining heatwaves, rainstorms, droughts, snowstorms, as well as the main moving paths of coldwaves and typhoons during 1991 to 2020. The hazard zones are identified with comparative thresholds that are at least one standard deviation above the spatial average of certain indices in China (refer to Figure 2). Specifically, the heatwave hazard zone refers to an area with more than 20 hot days per year, the rainstorm hazard zone refers to an area with more than 4 rainstorm days per year, the drought hazard zone refers to an area with more than 45 drought days per year, and the snowstorm hazard zone refers to an area with more than 40 snowfall days per year.

As shown in Figure 5, most areas of China are affected by different types of climate hazards. Northern China is most susceptible to coldwaves and snowstorms. Droughts dominate central-eastern China and coldwaves also have widespread impacts in this region. Heatwaves, rainstorms, and typhoons superimpose in Southeastern China, and coldwaves may also reach south of the Yangtze River. It is obvious that there are various types of climate hazards in China with broadly negative impacts. Southeastern China, in particular, is heavily affected by multiple hazards simultaneously, and the dense population and concentrated economic activities are expected to further amplify climate-related socioeconomic risks.

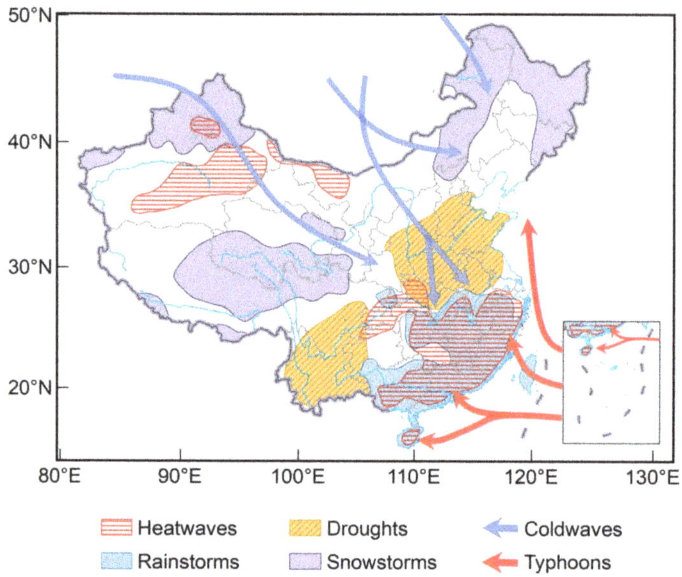

Figure 5. Mapping of multiple climate hazards in China during 1991–2020.

4. Strategies for Addressing Climate Hazards in Urban Development

4.1. Impacts of Climate Hazards on Urban Development

Climate hazards have negative impacts on urban development. Cities, where humans gather and economic activities are concentrated, have been rapidly expanding during Chinese urbanization in recent decades and hence have become hotspots affected by climate hazards [45–47]. Since climate changes are expected to drive the intensification of climate hazards, it is imperative to build climate-resilient cities to mitigate climate risks. In order to facilitate decision making on urban risk management, some issues and solutions have been discussed as follows based on the climate hazard patterns in China.

Climate hazards are proposed to be taken into consideration in urban planning from two aspects: climate change and extreme events. Climate change affects urban development through long-term changes such as rising sea levels, environmental aridification, and the intensification of urban heat/rain island effects, while extreme events have an immediate impact through heavy rainfalls, typhoons, and heat/cold waves. From the perspective of climate impacts, extreme-temperature-related impacts include increased summertime strain on materials, peak electricity loads in summer (conversely, reduced heating requirements in winter) [48–50]. Extreme-precipitation-related impacts include increased flooding (street, basement, sewer) and reduction in water quality [51,52]. Sea-level-rise-related impacts include inundation of low-lying areas, expansion of wetlands, increased structural damage, and impaired operations [53–55]. In warmer climate, heatwaves are projected to increase in frequency, intensification, and duration; inland flooding induced by precipitation extremes are likely to exacerbate, while flash droughts are also simulated to intensify in the future. Under this trend, climate risks are expected to bring greater challenges to urban development in the future.

The potential changes in climate hazards increase the complexity of urban planning for policymakers. For example, sea-level rise may cause the low-lying coastal areas, flood plains, and steep slope areas to become unaccommodated for residence; an increase in drought may result in the underground and drinkable water to be in short supply. Policymakers should fully understand the hazard exposure of residents and urban systems and consider how to reach the balance between urban developments and climate risks. Given that building of climate-resilient cities in China has a high priority, we have developed a strategic framework to achieve urban sustainability, shown in Figure 6.

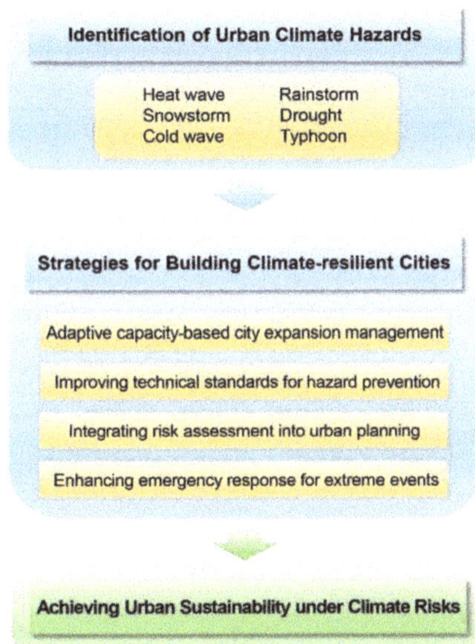

Figure 6. A Strategic framework for building climate-resilient cities in China.

4.2. Adaptive Capacity-Based City Expansion Management

City adaptive capacities are defined as the ability to absorb and recover from climate impacts [7]. Factors determining adaptive capacity include but are not limited to the following: income levels and Gross Domestic Product (GDP), natural resource availability and distribution, levels of public cognition on climate risks; the availability of technological capacity and adaptation options, the availability and quality of environmental factors (e.g., land, water, raw materials, biodiversity), infrastructure quality and provision, ability to act collectively to develop and implement adaptation responses, and public education as well as emergency skills.

As the expansion of cities aggravates the overwhelming energy consumption, transportation systems, and drainage systems, it is necessary to restrict the extent of cities according to environmental conditions. Moreover, the integrated assessment of climate risks, vulnerabilities, and adaptive capacities may provide a solution on city expansion. The types of climate hazards, levels of vulnerability and capacity, and socio-economic characteristics should be taken to promote the efficiency of adaptation planning and policy making in each city. The probability of climate extremes in the present and the foreseeable future can be estimated by climate models, and socioeconomic exposure as well as vulnerabilities are always changing with time, resulting in certain uncertainties of climate risks. Hence, the urban development path should be matched with the enhancement of adaptive capacity.

4.3. Improving Technical Standards for Hazard Prevention

Improving technical standards is good practice for preventing climate hazards at hardware level. It is necessary to develop risk assessment models for city construction based on scenario simulations. To date, urban climate assessment mainly relies on historical events or trend extrapolation but lacks the understanding of potential or unprecedented risks. In particular, specific industries and regions are rarely focused [56]. With the advances in numerical simulation, scientists obtained climate projection by using a combination of dynamic and statistical methods. Regional climate models can further improve the temporal and spatial resolution of climate simulations and enhance the performance of urban extreme weather and climate events. Given the differences in geographical location, economic level, and urban adaptability, each city in China should independently estimate the potential risks based on climate simulations and then formulate the specialized urban development plans.

The technical standards involve various industrial sectors, such as road traffic, water supply and drainage, electricity, communications, gas, flood control, and greening space. At present, climate change is limitedly considered in the technical standards referring to urban construction. Moreover, the infrastructure standards are far below the requirements of defense of climate hazards in coastal China [57,58]. The increased occurrence of compound hazards in coastal areas highlights the necessity of cross-department collaboration. For example, the prevention standards for coastal flooding caused by river flood, extreme rainstorm, and storm surge need to be jointly designed by meteorological, hydrological, and oceanic departments [5]. In the future, the formulation of technical standards will no longer be a simple technical issue but require more consideration of compound or cascading effects of multi-hazards on urban operations through multi-industry modeling and public participation [59–61].

4.4. Specifying Climate Risk Assessment as a Mandatory Part of Urban Planning

As the application of climate risk assessments in China's urban planning is not mandatory, the potential impacts of climate risk are significantly underestimated [62,63]. This neglect can directly affect the normal operation of the social activities, as the climate hazard can damage the critical infrastructure in cities (e.g., power transmission tower, signal tower, water storage and supply system, railway stations, hazardous substance installations). Appropriate climate risk assessments can minimize the negative impacts of climate hazards and protect human life, health, and property.

The expanding urban area, increasing populations and growing economic status have elevated exposure of cities. Meanwhile, warming temperatures have intensified climate hazards in urban areas. This double pressure makes it necessary to consider climate risk assessment in urban planning, and due to the close and complicated connections of infrastructure in different sectors, it is not wise to consider risk management when the construction is already completed. Oppositely, the best timing of risk assessment is the preliminary stage of urban planning. To set climate risk assessment as a mandatory issue in urban planning has been approved an effective and economic way to enhance the climate resilience in many newly developed cities around the world.

4.5. Strengthening Emergency Management for Extreme Climate Hazards

As mentioned above, many cities are vulnerable to facing future climate changes. In order to mitigate climate risks, emergency management for extreme climate events should be strengthened. Both direct and indirect interventions contribute to the improvement of emergency management capabilities. Direct intervention is to improve the existing emergency management methods. For example, flood protection strategies should be adjusted according to the precipitation intensity. Indirect intervention includes reducing the vulnerability of cities through strategic spatial planning. For example, public transport networks should be designed or improved for easier access and then to reduce the exposure to high temperature in heatwave events.

Urban emergency infrastructures are also crucial to respond to extreme climate hazards, serving as a lifeline for people exposed to extreme events. Routine maintenance and stress testing of emergency infrastructures should be strengthened to maximize rescue functions. Since emergency management cannot take effect simply through the command of planning strategies, increasing public knowledge and awareness of extreme climate events is essential for vulnerability reduction and risk mitigation. Administrative management strategy should also combine with public initiatives, and the important role of scientific researches in supporting decision making should be constantly strengthened and verified in practice.

5. Conclusions and Discussion

China is one of the countries with the highest climate risks in the world due to its monsoon phenomenon and complex climate patterns. Here, we systematically reveal the spatial pattern of multiple climate hazards in China, including heatwaves, rainstorms, snowstorms, droughts, as well as the moving paths of coldwaves and typhoons. On this basis, we summarize the impacts of climate hazards on urban development, and further propose to incorporate the strategic goals of climate adaptation into urban development of China, hoping to mitigate climate risks and enhance urban sustainability.

The Southeastern China and Northwestern China are two hazard zones with more than 30 hot days annually, and the extent of hazard zone has been expanding in recent 30 years. Cities located in the heatwave hazard zone face great risks to human health and energy supply. South of the Yangtze River, especially in coastal South China, belongs to the hazard zone of rainstorms. Cities located in rainstorm hazard zone face great risks of urban waterlogging, river flooding, and triggered geological hazards. Drought mostly occurs in North China, the Yellow River–Huaihe River basin, eastern Inner Mongolia, and Southwestern China, with more than 45 drought days annually. Cities located in the drought hazard zone face great risks of water supply and agriculture production. Coldwaves, accompanied by strong winds, snowstorms, and ice freezing, frequently invade northern Xinjiang, Northeast China, Tibetan Plateau, and Inner Mongolia, threatening public infrastructures, transportation, and energy supply in cities along the way. Cities located in southeast coastal China should be prepared for typhoon system that bring wild winds, heavy rainfall, and storm surges in summer half year.

Given that cities are hotspots affected by intensified climate hazards in a warmer world, some specific adaptation measures have been developed to facilitate the construction of climate-resilient cities. Specifically, city scale and land-use pattern should be formulated by urban carrying capacity assessment. Technical standard of public infrastructures should be improved to resist intensified climate hazards. Local risk assessment should become a mandatory part of urban planning to guide future development away from hazard zones. Emergency response capabilities should be strengthened by stress testing to reduce the disastrous impacts of extreme hazards.

Due to the current underestimations of climate impacts on urban development, there is an urgent need to improve climate adaptation. If climate-related policies and actions are still marginalized, the steady deterioration of urban sustainability could lead to irreversible economic and social issues in the foreseeable future. Several actions proposed in this study could provide a guideline for building climate-resilient cities that are better able to address climate change in the coming future.

Author Contributions: Conceptualization, S.S. and Z.W.; data curation, S.S.; methodology, S.S.; formal analysis, S.S. and Z.W.; investigation, C.H. and G.G.; resources, Z.W. and G.G.; validation, C.H. and G.G.; writing—original draft preparation, S.S.; writing—review and editing, Z.W., C.H., and G.G.; visualization, S.S.; supervision, Z.W. All authors have read and agreed to the published version of the manuscript.

Funding: This research was funded by the National Key Research and Development Program of China (Grant No. 2018YFC1509003 and 2019YFC1510202), the National Natural Science Foundation of China (Grant No. 41701103), the Major Research and Development Program of China Railway Group (P2018T006), and the UK-China Cooperation on Climate Change Risk Assessment.

Data Availability Statement: The historical observations of meteorological stations in China are compiled and issued by the Meteorological Information Center (MIC), China Meteorological Administration (CMA), and can be accessed online: http://data.cma.cn/en.

Conflicts of Interest: The authors declare no conflict of interest.

References

1. Hansen, G.; Cramer, W. Global distribution of observed climate change impacts. *Nat. Clim. Chang.* **2015**, *5*, 182–185. [CrossRef]
2. James, R.; Otto, F.; Parker, H.; Boyd, E.; Cornforth, R.; Mitchell, D.; Allen, M. Characterizing loss and damage from climate change. *Nat. Clim. Chang.* **2014**, *4*, 938–939. [CrossRef]
3. Chen, H.; Sun, J. Increased population exposure to precipitation extremes in China under global warming scenarios. *Atmos. Ocean. Sci. Lett.* **2020**, *13*, 63–70. [CrossRef]
4. Zhou, B.; Xu, Y.; Wu, J.; Dong, S.; Shi, Y. Changes in temperature and precipitation extreme indices over China: Analysis of a high-resolution grid dataset. *Int. J. Climatol.* **2016**, *36*, 1051–1066. [CrossRef]
5. AghaKouchak, A.; Chiang, F.; Huning, L.; Love, C.; Mallakpour, I.; Mazdiyasni, O.; Moftakhari, H.; Papalexiou, M.S.; Ragno, E.; Sadegh, M. Climate extremes and compound hazards in a warming world. *Annu. Rev. Earth Planet. Sci.* **2020**, *48*, 519–548. [CrossRef]
6. Carleton, T.; Hsiang, S. Social and economic impacts of climate. *Science* **2016**, *353*, aad9837. [CrossRef] [PubMed]
7. Carter, J.; Cavan, G.; Connelly, A.; Guy, S.; Handley, J.; Kazmierczak, A. Climate change and the city: Building capacity for urban adaptation. *Prog. Plan.* **2015**, *95*, 1–66. [CrossRef]
8. Larsen, K.; Gunnarsson-Östlingb, U. Climate change scenarios and citizen-participation: Mitigation and adaptation perspectives in constructing sustainable futures. *Habitat Int.* **2009**, *33*, 260–266. [CrossRef]
9. Balogun, A.; Marks, D.; Sharma, R.; Shekhar, H.; Salehi, P. Assessing the potentials of digitalization as a tool for climate change adaptation and sustainable development in urban centres. *Sustain. Cities Soc.* **2019**, *53*, 101888. [CrossRef]
10. While, A.; Whitehead, M. Cities, urbanisation and climate change. *Urban Stud.* **2013**, *50*, 1325–1331. [CrossRef]
11. IPCC. *Global Warming of 1.5 °C*; Cambridge University Press: Cambridge, UK; New York, NY, USA, 2018.
12. Trusilova, K.; Jung, M.; Churkina, G. Urbanization impacts on the climate in Europe: Numerical experiments by the PSU-NCAR Mesoscale Model (MM5). *J. Appl. Meteorol. Climatol.* **2008**, *47*, 1442–1455. [CrossRef]
13. Stone, B.; Hess, J.; Frumkin, H. Urban form and extreme heat events: Are sprawling cities more vulnerable to climate change than compact cities. *Environ. Health Perspect.* **2010**, *118*, 1425–1428. [CrossRef] [PubMed]
14. Rosenzweig, C.; Solecki, W.; Hammer, S.A.; Mehrotra, S. Cities lead the way in climate–change action. *Nature* **2010**, *467*, 909–911. [CrossRef] [PubMed]
15. Greiving, S.; Fleischhauer, M. National climate change adaptation strategies of European states from a spatial planning and development perspective. *Eur. Plan. Stud.* **2011**, *20*, 27–48. [CrossRef]
16. Cobbinah, P.B.; Asibey, M.O.; Opoku-Gyamfi, M.; Peprah, C. Urban planning and climate change in Ghana. *J. Urban Manag.* **2019**, *8*, 261–271. [CrossRef]
17. Arshad, A.; Ashraf, M.; Sundari, R.S.; Qamar, H.; Hassan, M.U. Vulnerability assessment of urban expansion and modelling green spaces to build heat waves risk resiliency in karachi. *Int. J. Disaster Risk Reduct.* **2020**, *46*, 101468. [CrossRef]
18. Caparros-Midwood, D.; Dawson, R.; Barr, S. Low carbon, low risk, low density: Resolving choices about sustainable development in cities. *Cities* **2019**, *89*, 252–267. [CrossRef]
19. Matthews, T.; Llex, A.; Byrne, J. Reconceptualizing green infrastructure for climate change adaptation: Barriers to adoption and drivers for uptake by spatial planners. *Landsc. Urban Plan.* **2015**, *138*, 155–163. [CrossRef]
20. Bulkeley, H. Cities and the governing of climate change. *Annu. Rev. Environ. Resour.* **2010**, *35*, 361–375. [CrossRef]
21. Brown, D. Making the linkages between climate change adaptation and spatial planning in Malawi. *Environ. Sci. Policy* **2011**, *14*, 940–949. [CrossRef]
22. Yang, L.; Chen, W.; Wu, J.; Sun, W.; Li, Y. Spatial planning adapting to climate change: Progress in the content and methodology. *Urban Plan. Int.* **2020**, *35*, 96–103. (In Chinese)
23. Roggema, R.; Kabat, P.; Dobbelsteen, A. Towards a spatial planning framework for climate adaptation. *Smart Sustain. Built Environ.* **2012**, *1*, 29–58. [CrossRef]
24. Wilson, E. Adapting to climate change at the local level: The spatial planning response. *Local Environ.* **2006**, *11*, 609–625. [CrossRef]
25. Cao, L.; Zhu, Y.; Tang, G.; Yuan, F.; Yan, Z. Climatic warming in China according to a homogenized data set from 2419 stations. *Int. J. Climatol.* **2016**, *36*, 4384–4392. [CrossRef]
26. Ying, M.; Zhang, W.; Yu, H.; Lu, X.; Feng, J.; Fan, Y.; Zhu, Y.; Chen, D. An overview of the China Meteorological Administration cyclone database. *J. Atmos. Ocean. Technol.* **2014**, *31*, 287–301. [CrossRef]
27. Zou, X.; Zhang, Q. Preliminary studies on variations in droughts over China during past 50 years (in Chinese). *J. Appl. Meteorol. Sci.* **2008**, *19*, 679–687.

28. Wu, Y.; Bake, B.; Zhang, L.; Rasulov, H. Spatil-temporal patterns of drought in North Xinjiang, China, 1961–2012 based on meteorological drought index. *J. Arid. Land* **2015**, *7*, 527–543. [CrossRef]
29. Tao, S.; Chen, L. A review of recent research on the East Asian summer monsoon in China. *Rev. Monsoon Meteorol.* **1987**, 60–92.
30. Ding, Y. Summer monsoon rainfall in China. *J. Meteorol. Soc. Jpn.* **1992**, *70*, 373–396. [CrossRef]
31. Ding, Y. Monsoons over China. *Adv. Atmos. Sci.* **1994**, *2*, 135–136.
32. Ding, Y. Seasonal march of the East Asian summer monsoon. *East Asian Monsoon* **2004**, 3–53.
33. Ding, Y.; Chan, J. The East Asian summer monsoon: An overview. *Meteorol. Atmos. Phys.* **2005**, *89*, 117–142.
34. Lau, K.; Yang, S. Climatology and interannual variability of the Southeast Asian summer monsoon. *Adv. Atmos. Sci.* **1997**, *14*, 141–162. [CrossRef]
35. Tanaka, M. Intraseasonal oscillation and the onset and retreat dates of the summer monsoon over east, southeast Asia and the western Pacific region using GMS high cloud amount data. *J. Meteorol. Soc. Jpn.* **1992**, *70*, 613–629. [CrossRef]
36. Qian, W.; Lee, D. Seasonal march of Asian summer monsoon. *Int. J. Climatol.* **2000**, *20*, 1371–1378. [CrossRef]
37. Chen, T.; Wang, S.; Huang, W. Variation of the East Asian summer monsoon rainfall. *J. Clim.* **2004**, *17*, 744–762. [CrossRef]
38. Zhang, Y.; Sperber, K.; Boyle, J. Climatology and interannual variation of the East Asian winter monsoon: Results from 1979–95 NCEP/NCAR reanalysis. *Mon. Weather. Rev.* **1997**, *125*, 2605–2619. [CrossRef]
39. Nakamura, H.; Izumi, T.; Sampe, T. Interannual and decadal modulations recently observed in the pacific storm track activity and East Asian winter monsoon. *J. Clim.* **2002**, *14*, 1855–1874. [CrossRef]
40. Gong, D.; Wang, S.; Zhu, J. East Asian winter monsoon and arctic oscillation. *Geophys. Res. Lett.* **2001**, *28*, 2073–2076. [CrossRef]
41. Compo, G.; Kiladis, G.; Webster, P. The horizontal and vertical structure of east Asian winter monsoon pressure surges. *Q. J. R. Meteorol. Soc.* **1999**, *125*, 29–54. [CrossRef]
42. Zhang, P.; Chen, G. A statistical analysis of the cold wave high which infulences on China. *Acta Meteorol. Sin.* **1999**, *57*, 493–501.
43. Sun, J.; Wang, H.; Yuan, W.; Chen, H. Spatial-temporal features of intense snowfall events in China and their possibel change. *J. Geophys. Res. Atmos.* **2010**, *115*, D16110. [CrossRef]
44. Wang, Z.; Zhou, B. Large-scale atmospheric circulations and water vapor transport influcing interannual variations of intense snowfalls in northern China. *Chin. J. Geophys.* **2018**, *61*, 2654–2666. (In Chinese)
45. Jiang, S.; Wang, K.; Mao, Y. Rapid local urbanization around most meteorological stations explains the observed daily asymmetric warming rates across China from 1985 to 2017. *J. Clim.* **2020**, *33*, 9045–9061. [CrossRef]
46. Shen, X.; Liu, B.; Jiang, M.; Lu, X. Marshland loss warms local land surface temperature in China. *Geophys. Res. Lett.* **2020**, *47*, e2020GL087648. [CrossRef]
47. Wu, X.; Wang, L.; Yao, R.; Luo, M.; Wang, S.; Wang, L. Quantitatively evaluating the effect of urbanization on heat waves in China. *Sci. Total. Environ.* **2020**, *731*, 138857. [CrossRef]
48. Auffhammer, M.; Baylis, P.; Hausman, C. Climate change is projected to have severe impacts on the frequency and intensity of peak electricity demand across the United States. *Proc. Natl. Acad. Sci. USA* **2017**, *114*, 1886–1891. [CrossRef] [PubMed]
49. Neumann, J.; Price, J.; Chinowsky, P.; Wright, L.; Martinich, J. Climate change risks to US infrastructure: Impacts on roads, bridges, coastal development, and urban drainage. *Clim. Chang.* **2015**, *131*, 97–109. [CrossRef]
50. Rizwan, A.; Dennis, L.; Liu, C. A review on the generation, determination and mitigation of Urban Heat Island. *J. Environ. Sci.* **2008**, *20*, 120–128. [CrossRef]
51. Hallegatte, S.; Green, C.; Nicholls, R.; Corfee-Morlot, J. Future flood losses in major coastal cities. *Nat. Clim. Chang.* **2013**, *3*, 802–806. [CrossRef]
52. Arnell, N.; Gosling, S. The impacts of climate change on river flood risk at the global scale. *Clim. Chang.* **2016**, *134*, 387–401. [CrossRef]
53. Buchanan, M.; Kopp, R.; Oppenheimer, M.; Tebaldi, C. Allowances for evolving coastal flood risk under uncertain local sea-level rise. *Clim. Chang.* **2016**, *137*, 347–362. [CrossRef]
54. Jongman, B.; Ward, P.; Aerts, J. Global exposure to river and coastal flooding: Long term trends and changes. *Glob. Environ. Chang.* **2012**, *22*, 823–835. [CrossRef]
55. Vitousek, S.; Barnard, P.; Fletcher, C.; Frazer, N.; Erikson, L.; Storlazzi, C. Doubling of coastal flooding frequency within decades due to sea-level rise. *Sci. Rep.* **2017**, *7*, 1399. [CrossRef]
56. Broto, V.; Bulkeley, H. A survey of urban climate change experiments in 100 cities. *Glob. Environ. Chang.* **2013**, *23*, 2–102.
57. Xu, Y. Development strategy of China's coastal cities for addressing climate change. *Clim. Chang. Res.* **2020**, *16*, 88–98. (In Chinese)
58. Cui, S.; Xu, L.; Huang, Y.; Huang, W. Progress and prospect of study on urban spatial patterns to cope with climate change. *Prog. Geogr.* **2015**, *34*, 1209–1218. (In Chinese)
59. Preston, B.; Westaway, R.; Yuen, E. Climate adaptation planning in practice: An evaluation of adaptation plans from three developed nations. *Mitig. Adapt. Strateg. Glob. Chang.* **2011**, *16*, 407–438. [CrossRef]
60. Xuemei, B.; Dawson, R.; Ürge-vorsatz, D. Six research priorities for cities and climate change. *Nature* **2018**, *555*, 23–25.
61. Rivera, C.; Wamsler, C. Integrating climate change adaptation, disaster risk reduction and urban planning: A review of Nicaraguan policies and regulations. *Int. J. Disaster Risk Reduct.* **2014**, *7*, 78–90. [CrossRef]

62. Wu, S.; Zhao, D. Progress on the impact, risk and adaptation of climate change in China. *China Popul. Resour. Environ.* **2020**, *30*, 1–9. (In Chinese)
63. Fünfgeld, H. Institutional challenges to climate risk management in cities. *Curr. Opin. Environ. Sustain.* **2010**, *2*, 156–160. [CrossRef]

sustainability

Article

Assessment and Prediction of Climate Risks in Three Major Urban Agglomerations of Eastern China

Jieming Chou [1], Mingyang Sun [1,2,*], Wenjie Dong [2], Weixing Zhao [1], Jiangnan Li [1], Yuanmeng Li [1] and Jianyin Zhou [3]

1 State Key Laboratory of Earth Surface Processes and Resource Ecology, Faculty of Geographical Science, Beijing Normal University, Beijing 100875, China; choujm@bnu.edu.cn (J.C.); 201921051146@mail.bnu.edu.cn (W.Z.); 201921051150@mail.bnu.edu.cn (J.L.); 202021051153@mail.bnu.edu.cn (Y.L.)
2 Atmospheric Science School, Sun Yat-Sen University, Guangzhou 510275, China; dongwj3@mail.sysu.edu.cn
3 College of Meteorology and Oceanography, National University of Defense Technology, Changsha 410073, China; 15165381462@163.com
* Correspondence: 201821051163@mail.bnu.edu.cn

Abstract: In the context of global climate change and urban expansion, extreme urban weather events occur frequently and cause significant social problems and economic losses. To study the climate risks associated with rapid urbanization in the global context of climate change, the vulnerability degree of urban agglomeration is constructed by the Grey Model (GM (1, 1)). Based on the sixth phase of the Coupled Model Intercomparison Project (CMIP6) data sets SSP1-2.6, SSP2-4.5, and SSP5-8.5, drought, heat wave, and flood hazards under different emission scenarios are calculated. The vulnerability degree of the urban agglomeration and the climate change hazard were input into the climate change risk assessment model to evaluate future climate change risk. The analysis results show regional differences, with the Beijing–Tianjin–Hebei urban agglomeration having good urban resilience, the Yangtze River Delta urban agglomeration having slightly higher overall risk, and the Pearl River Delta urban agglomeration having the highest relative risk overall. On the whole, the higher the emission intensity is, the greater the risk of climate change to each urban agglomeration under different emission scenarios.

Keywords: climate change; urban agglomeration; drought; heat wave; flood; risk assessment; GM (1, 1)

Citation: Chou, J.; Sun, M.; Dong, W.; Zhao, W.; Li, J.; Li, Y.; Zhou, J. Assessment and Prediction of Climate Risks in Three Major Urban Agglomerations of Eastern China. *Sustainability* **2021**, *13*, 13037. https://doi.org/10.3390/su132313037

Academic Editors: Xiaodong Yan, Jia Yang, Shaofei Jin and Adriana Del Borghi

Received: 17 August 2021
Accepted: 18 November 2021
Published: 25 November 2021

1. Introduction

The global climate system has significantly warmed in the past century, with the global average surface temperature increasing by approximately 0.9 °C from 1900 to 2018 [1]. China has a high incidence of extreme weather events. In the context of global warming and high incidence of extreme weather events, this paper studies the inherent relationship between the high incidence of extreme weather events and climate warming. The multi-scenario risk assessment of drought, high temperature, heat wave, and flood under climate change and urban expansion is a hot topic in urban sustainable development. In this paper, we predict the risks of drought, heat wave, and flood during the next 30 years based on the BCC-CSM2-MR data sets SSP1-2.6, SSP2-4.5, and SSP5-8.5. This paper focuses on the climate change risks to the three major urban agglomerations: Beijing–Tianjin–Hebei, Yangtze River Delta, and Pearl River Delta in China, which is significant for the study of climate change risks to large urban agglomerations in other countries.

In their research, domestic and foreign scholars have carried out many studies on the responses of urban agglomerations to climate change. According to the Intergovernmental Panel on Climate Change (IPCC)'s Fifth Assessment Report (AR5) [2], climate change risk is quantified as a function of hazard, exposure, and vulnerability. However, there are two shortcomings at present. First, previous climate change risk studies focused on the risk of a certain disaster in a particular city; however, in the context of climate change, more attention should be paid to the risk of multiple disaster types and complex disaster-bearing bodies, and the evolution characteristics of each disaster risk in time and space are also worth examining [3–5]. Second, a large number of studies have focused more on internal factors and climate risks such as hazards, exposure, and vulnerability [6,7], while ignoring the risk forcing of external factors such as urbanization and climate change [8]. According to the United Nations Strategy for Disaster Reduction (ISDR), climate change and urbanization are the two main factors that make human beings more vulnerable to disasters [9]. Though the AR5 delineates exposure as a separate component of the risk, in this study, we included exposure as an integral part of the vulnerability. Therefore, the vulnerability degree of an urban agglomeration is defined based on the comprehensive consideration of exposure and vulnerability. The concept of vulnerability involves different disciplines and professional fields such as sociology, economics, and disaster management. In this paper, the vulnerability degree of an urban agglomeration refers to the degree to which the social system affected by climate pressure and the urbanization process has been damaged because of its own sensitivity and vulnerability. Hazard, as the core of climate change risk, is used to identify the location and intensity of weather events such as drought, heat wave, and floods [10]. Based on the vulnerability of urban agglomerations and the risk of climate change, this paper establishes a basic model for climate change risk assessment.

The three major urban agglomerations along the east coast of China are important engines of regional economic and social development; however, they are also vulnerable to climate change. Edward pointed out that the impacts of sea level rise and extreme weather, such as heat waves and floods, have seriously affected the economic life of coastal cities, and it is very important to adapt to the impact of climate change on coastal cities through economic construction and emergency response [11]. Global warming, melting glaciers, and the rising sea level have a huge impact on coastal cities, making them more vulnerable to rainstorms and floods [12,13]. With the acceleration of urbanization and the impact of global warming, the frequency and intensity of extreme events such as droughts, heat waves, and floods increase, which has a huge impact on the power supply, water supply, and transportation in urban infrastructure [14,15]. In this paper, the vulnerability of urban agglomerations and the hazard of climate change are input into the climate change risk assessment model to comprehensively estimate the future climate change risks to the three major urban agglomerations in eastern China. The combination of natural ecosystems and socioeconomic systems can help to control the degree and probability of the impact of climate change on the three major urban agglomerations. This can provide a scientific basis for disaster prevention and mitigation, and emergency response in cities.

2. Research Area

With the acceleration of urbanization, population and industries have been concentrated in the eastern coastal areas of China, forming three city clusters: Beijing–Tianjin–Hebei, Yangtze River Delta, and Pearl River Delta [16]. These generally have a large population density and rapid economic development. At the same time, a nationally important transport hub exists around the port and the nuclear power plant, which means that the degree of exposure and vulnerability of these urban agglomerations to disasters and large economic losses is bigger than other cities. Therefore, eastern China's urban agglomeration, with the highest level of national economic contribution, has become a potential large-scale disaster-bearing body.

In the eastern coastal area with the largest population density distribution and the highest level of national economic development, the three urban agglomerations of Beijing, Tianjin, and Hebei, the Yangtze River Delta, and the Pearl River Delta are the most mature and competitive urban agglomerations in China [17]. Table 1 shows their location information and basic characteristics while Figure 1 shows their geographical distribution and the prefecture-level cities they contain. With the rapid growth of China's economy and the continuous evolution of its cities, urban agglomeration has gradually become the symbol of a higher stage of urban development. It can strengthen the interconnection among various cities, promote the process of economic integration, and encourage the rapid growth of the regional economy. On the other hand, because of the high concentration of resources, environment, ecosystem, and social economy in urban agglomerations, they have a stronger impact on the exposure, sensitivity, and vulnerability to extreme weather events, and the risk to them from climate change is significant. Therefore, studying the climate, economic change characteristics, and development laws of these three urban agglomerations is of great significance for promoting their social development process as the main bodies in China and the balanced and stable development of various regions.

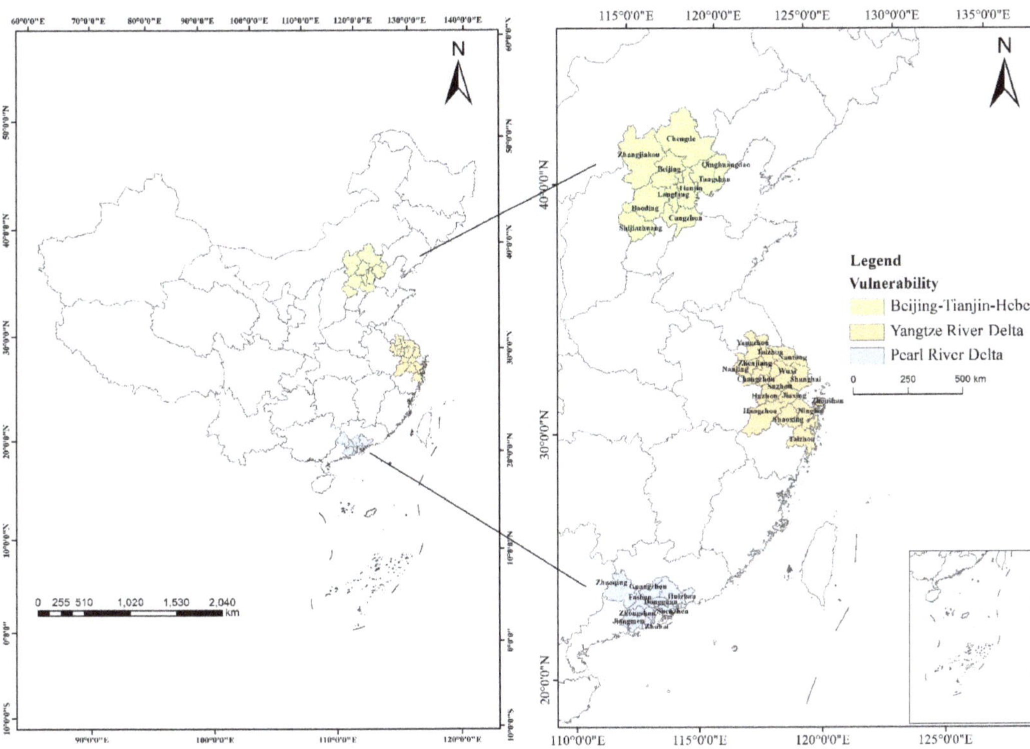

Figure 1. Distribution and names of the three major urban agglomerations in China.

Table 1. Three urban agglomerations in China and their basic characteristics.

Name	Range	Geographical Position	Basic Characteristics
Beijing–Tianjin–Hebei Urban Agglomeration	With Beijing, Tianjin as the center, including Shijiazhuang, Baoding, Zhangjiakou, Tangshan, Qinhuangdao, Langfang, Cangzhou, Chengde.	Located north of the North China Plain, the center of Northeast Asia in China, bordering on Bohai Sea.	It covers an area of 183,088 km^2, has a population of 90.09 million, and a GDP of CNY 7.70 trillion. It is the capital economic circle of China and the largest and most dynamic area.

Table 1. *Cont.*

Name	Range	Geographical Position	Basic Characteristics
Yangtze River Delta Urban Agglomeration	With Shanghai and Nanjing as the center, including Suzhou, Wuxi, Changzhou, Zhengjiang, Nantong, Yangzhou, Taizhou (in Jiangsu), Hangzhou, Ningbo, Jiaxing, Huzhou, Shaoxing, Zhoushan and Taizhou (in Zhejiang).	Located in the lower reaches of the Yangtze River, bordering on the Yellow Sea and the East China Sea, with many ports, it is an impact plain formed before the Yangtze River enters the sea.	The area is 132,065 km^2, with a population of 11.47 million and a GDP of CNY 15.02 trillion. It is an important zone between the „one belt, one road" and the „Yangtze River Economic Belt" and is the most densely distributed area and largest economic zone in China.
Pearl River Delta Urban Agglomeration	With Guangzhou, Shenzhen as the center, including Zhuhai, Foshan, Jiangmen, Dongguan, Zhongshan, Huizhou, Zhaoqing.	It is located at the Pearl River estuary in the south central part of Guangdong Province. It is adjacent to Hong Kong and Macao.	It has an area of 55,061 km^2, population of 0.63 million and a GDP of CNY 8.10 trillion. As one of the most dynamic economic groups in the Asia Pacific region, it is also the gateway for the southern region to open up to the outside world.

Note: the data are calculated according to the *China Statistical Yearbook 2018* and the statistical yearbook of relevant provinces (cities) for 2018.

3. Data and Methods

3.1. Data

Data Sources

The historical climate data were taken from the National Meteorological Information Center of China (http://data.cma.cn (accessed on 22 November 2021)). The meteorological stations of Beijing–Tianjin–Hebei, Yangtze River Delta, and Pearl River Delta were selected. To maintain the homogeneity of the data, this paper selected the daily average temperature (unit: °C), daily average precipitation (unit: mm), daily maximum temperature (unit: °C), monthly average temperature (unit: °C), and relative humidity (unit: %) from 1981 to 2019.

The social and economic data used in this paper were taken from the China National Statistical Center (http://www.bjstats.gov.cn (accessed on 22 November 2021)). The *China Statistical Yearbook* provided the GDP per capita (unit: yuan), permanent population (unit: thousand), urban population (unit: thousand), urban area (unit: thousand ha), total area (unit: thousand ha), and proportion of primary industry in the GDP (unit: %) of the Beijing–Tianjin–Hebei, Yangtze River Delta, and Pearl River Delta urban regions from 1981 to 2019.

The future climate scenario data came from the BCC-CSM2-MR climate model of the sixth phase of the Coupled Model Intercomparison Project (CMIP6). The model uses 320 (longitude) * 160 (latitude) grid points to cover the globe. It is newly developed by the National Climate Center. The ability of the BCC-CSM2-MR climate model to simulate the climate mean state and global warming trend, quasi-biennial oscillation (QBO), tropical intraseasonal oscillation (MJO), and diurnal variation in precipitation for the 20th century has significantly improved [18]. In this paper, daily temperature (unit: °C), daily precipitation (unit: mm), and daily relative humidity (%) from 2020 to 2050 were selected. Three emission scenarios, SSP1-2.6, SSP2-4.5, and SSP5-8.5, were selected for the climate prediction experiment (https://esgf-node.llnl.gov/search/cmip6/ (accessed on 22 November 2021)). Table 2 shows the main characteristics of various representative concentration paths.

Table 2. Main characteristics of various representative concentration paths.

Scenario Component	SSP1-2.6	SSP2-4.5	SSP5-8.5
Radiative forcing	2.6 Wm^{-2}	4.5 Wm^{-2}	8.5 Wm^{-2}
Greenhouse gas emission	Very low	Medium-low	High

3.2. Method

In this paper, the AHP is used to construct the vulnerability assessment system of the three major urban agglomerations. Using an integrated meteorological drought index, heat wave index, and flood index, the risk to the three urban agglomerations of meteorological disasters was evaluated. In this paper, the vulnerability of urban agglomerations and the hazard of climate change were input into the climate change risk assessment model to comprehensively estimate the future climate change risk to the three major urban agglomerations in eastern China. Figure 2 shows the flow chart of climate change risk assessment. The specific methods are as follows:

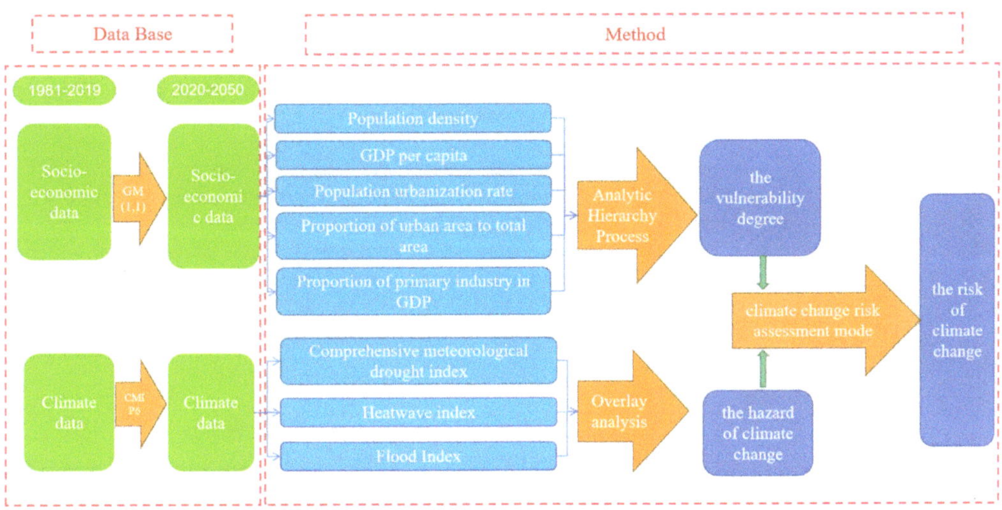

Figure 2. Flow chart of climate change risk assessment.

3.2.1. Grey System Model GM (1, 1)

The GM (1, 1) model constructed by Deng is a method to solve the lack of data and shortage of information, and create a fuzzy long-term description of the law of development of things [19]. Socioeconomic data include the GDP per capita (unit: yuan), permanent population (unit: thousand), urban population (unit: thousand), urban area (unit: thousand ha), total area (unit: thousand ha), and proportion of primary industry in the GDP (unit: %). As the data are affected by many factors, it was difficult to use a single linear trend for effective prediction. Therefore, this paper used the GM (1, 1) to predict the socioeconomic data from 2020 to 2050. The GM (1, 1) is a method to solve the problems of insufficient data, poor information, and high fitting accuracy [20]. The GM (1, 1) uses differential equations to fully explore the essence of the system. This theory takes a small amount of irregular raw data as samples and organizes them in a regular series after data accumulation [21]. Then, a reduction process is performed to obtain the forecasted data for the target year.

Let the original variable be $X^{(0)} = \left\{X^{(0)}(\mathrm{i}), \mathrm{i} = 1, 2, \ldots, \mathrm{n}\right\}, X^{(0)}(\mathrm{i}) \geq 0$. By adding $X^{(0)}$, we obtain the following results:

$$X^{(1)} = \left\{X^{(1)}(k), \quad k = 1, \quad 2, \quad \ldots, \quad \mathrm{n}\right\} \tag{1}$$

where

$$X^{(1)}(k) = \sum_{i=1}^{k} X^{(0)}(\mathrm{i}) = X^{(1)}(k-1) + X^{(0)}(k) \tag{2}$$

The following differential equations were established to create the GM (1, 1) model:

$$\frac{dX^{(1)}}{dt} + aX^{(1)} = \mathrm{u} \tag{3}$$

3.2.2. Analytic Hierarchy Process

The AHP is a decision analysis method that decomposes factors related to decisions and conducts qualitative and quantitative analyses [22]. When facing a complex problem, it conducts a deeper analysis of its internal influencing factors and uses less quantitative information to mathematically carry out the decision-making process. This method is widely used at present [23]. Its basic principle is to decompose the decision-making process into different hierarchical structures. Experts assign quantitative scores according to the relative importance of each element to construct a judgment matrix A.

The calculation results of the matrix must pass a consistency test to avoid the contradictory situation where A is more important than B, B is more important than C, and C is more important than A for each internal element [24]. CI is a consistency indicator. Smaller values of CI indicate greater consistency. RI is a random consistency index, which is related to the order of the judgment matrix. In general, the greater the order of the matrix, the greater the possibility of random consistency deviation.

The random consistency ratio further evaluates the judgment matrix.

$$\mathrm{CR} = \mathrm{CI}/\mathrm{RI} \tag{4}$$

When CR is less than 0.1, matrix A is considered to have satisfactory consistency; otherwise, the judgment matrix must be revised.

To better evaluate the vulnerability degree of urban agglomeration, this paper standardized the above data and used the AHP to determine the weight of the GDP per unit area, the population density, the population urbanization rate, the proportion of urban area in total area, and the proportion of primary industry in the GDP (see Appendix A).

3.2.3. Vulnerability of the Carrier

This paper selected the GDP per unit area, the population density, the population urbanization rate, the proportion of urban area in the total area, and the proportion of primary industry in the GDP as the indicators of the vulnerability of the carrier.

The index analysis model of the vulnerability degree evaluation index system is as follows:

$$Z = \Sigma U_i W_i \tag{5}$$

where Z is the vulnerability degree of each coastal urban area; this paper divided the GDP per unit area, the population density, the population urbanization rate, the proportion of urban area in total area, and the proportion of primary industry in the GDP into five grades: U_i is the grade of the i-th index in an urban area; W_i is the weight of index i, which was obtained with the AHP (see Table 3).

Table 3. Analytic hierarchy process.

	Population Density (People/km^2)	GDP per Capita (Thousand Yuan)	Population Urbanization rate	Proportion of Urban Area to Total Area	Proportion of Primary Industry in GDP
1	<200	<20	<0.5	<0.15	<6.4
2	200–500	20–50	0.5–0.6	0.15–0.30	6.4–12.7
3	500–1000	50–80	0.6–0.7	0.30–0.50	12.7–17.7
4	1000–1500	80–100	0.7–0.8	0.50–0.70	17.7–28.2
5	>1500	>100	>0.8	>0.7	>28.2
weight	0.31	0.29	0.12	0.13	0.15

See Appendix A for the calculation process.

3.2.4. Calculation of Meteorological Risk Index

In this paper, the hazard of climate change refers to drought, heat wave, and flood. The comprehensive meteorological drought index, heat wave index, and flood index were calculated to describe the hazard degrees of the three meteorological disasters. The following are the calculation methods of the three indexes.

A: Comprehensive Meteorological Drought Index

The comprehensive meteorological drought index (*CI*) was used to analyze the spatiotemporal characteristics of drought in the three urban agglomerations in eastern China [25]. The smaller the value of the *CI*, the greater the absolute value, indicating that the drought in this region is more serious. It reflects the drought situation of a certain region on a long time scale and has been widely used since it was first proposed [26]. The comprehensive meteorological drought index (*CI*) is calculated by the following formula:

$$CI = aZ_{30} + bZ_{90} + cM_{30} \quad (6)$$

where: a is 0.4; b is 0.4; c is 0.8. Z_{30} and Z_{90} are the standardized precipitation index (SPI) values in the last 30 and 90 days, respectively, and M_{30} is the relative humidity index in the last 30 days. The specific calculation process has been detailed in the literature [27].

B: Heat Wave Index

The heat wave index reflects the duration of extremely high temperatures because China occupies a vast area with a large regional climate difference, and its tolerance to high temperatures is variable [28]. In this paper, the local daily meteorological data from May to September from 1981 to 2019 were used to first calculate the torridity index of the samples whose maximum temperature was greater than 33 °C. After sorting the torridity index sequence into ascending order, the median was selected as the local critical value of torridity, which represents the comfort level of the human body with respect to the meteorological environment [29].

The torridity index (E_r) is calculated by the following formula:

$$\text{When } RH \leq 60\%, E_r = 1.8 \times T_a - 0.55 \times (1.8 \times T_a - 26) \times (1 - 0.6) + 32 \quad (7)$$

$$\text{When } RH > 60\%, E_r = 1.8 \times T_a - 0.55 \times (1.8 \times T_a - 26) \times (1 - RH) + 32 \quad (8)$$

where T_a represents ambient temperature/°C and RH stands for average daily relative humidity/%.

After calculating the torridity index, the heat wave index (*Hi*) is calculated as follows:

$$HI = 1.2 \times (Er - Er') + 0.35 \sum_{i=1}^{N-1} 1/nd_i (Er_i - Er') + 0.15 \sum_{i=1}^{N-1} \frac{1}{nd_i} + 1 \quad (9)$$

where i is the previous day, and the number of days from day i before the day is nd_i. The torridity index of the day is Er, the critical value of torridity is Er', and the torridity index of day i before the day is Er_i.

C: Flood Index

The flood index (FI) is based on the cumulative precipitation of a region for 3 consecutive days.

$$FI = \sum_{i=N}^{N+2} P_i \quad (10)$$

where P_i is the precipitation on day i of a city, and P_i is in units of mm.

3.2.5. Climate Change Risk Assessment Model

Disaster risk generally considers disaster-causing factors and the disaster-bearing body:

$$\text{Risk} = \text{Hazard} * \text{Vulnerability} \quad (11)$$

Disaster risk is equal to the product of disaster hazard and vulnerability of the disaster-bearing body [30].

This paper improved the disaster risk model to better describe the impact of climate change and urbanization on assessing the climate change risk.

$$R = (D + H + F) * V \quad (12)$$

where R represents the risk of climate change, D represents the drought hazard, H represents the heat wave hazard, F represents the flood hazard, and V represents the vulnerability of urban agglomerations.

4. Results and Discussion

4.1. Vulnerability of Three Urban Agglomerations from 1981 to 2019

China's three major urban agglomerations are all located in the eastern coastal areas. Most of the country's infrastructure was built in the 21st century. Science and technology are the fundamental driving forces for us to improve the level of disaster tolerance. With meteorological risk and disaster occurrence rising yearly, it is very important to evaluate the climate change tolerance of these rapidly developing urban agglomerations.

This paper considered the three major urban agglomerations in eastern China as the research area and their social conditions as the research object to analyze through the following indicators:

According to the theory of natural disaster systems, the above five indicators were selected and stratified [31]. To better evaluate vulnerability, this paper standardized the above data and used the AHP to determine the weight of the five elements in the evaluation of the vulnerability degree of the carrier (see Table 3).

The vulnerability degree was divided into five registration regions according to the values of each vulnerability degree: the lowest vulnerability, low vulnerability, medium vulnerability, high vulnerability, and the highest vulnerability, and the spatial description and expression were mapped, as shown in Figure 3.

Figure 3 shows that the vulnerability degree of each urban agglomeration was relatively low in 1985, then increased over time. The main reason is that the three major urban agglomerations were expanding with the continuous advancement of urbanization and inward population migration. As a major component of the natural ecosystem, human beings are in a fragile and vulnerable state. The rapidly increasing population density increases the vulnerability of urban agglomerations. In addition, it can be seen from the figure that the high vulnerability in 2015 was mainly concentrated in the first-tier cities Beijing, Tianjin, Nanjing, Wuxi, Suzhou, Shanghai, Foshan, Guangzhou, Shenzhen, and Dongguan. In the vulnerability assessment of urban agglomerations, a higher GDP means higher economic benefits for the city and higher risks to the city when meteorological disasters strike. On the whole, among the three major urban agglomerations in 2015, the most vulnerable urban area was the Yangtze River Delta, which is also currently the city agglomeration with the largest GDP output among the three major urban agglomerations, contributing approximately one-fifth of the national GDP. Compared with the Yangtze River Delta and the Pearl River Delta, the vulnerability degree of the Beijing–Tianjin–Hebei urban agglomeration was generally lower, which was consistent with its economic development in recent years. During this time, the economic development gap between Beijing and Tianjin, and other cities in the Beijing–Tianjin–Hebei region has become increasingly obvious. There is a significant gap between the rich and the poor in the Beijing–Tianjin–Hebei urban agglomeration.

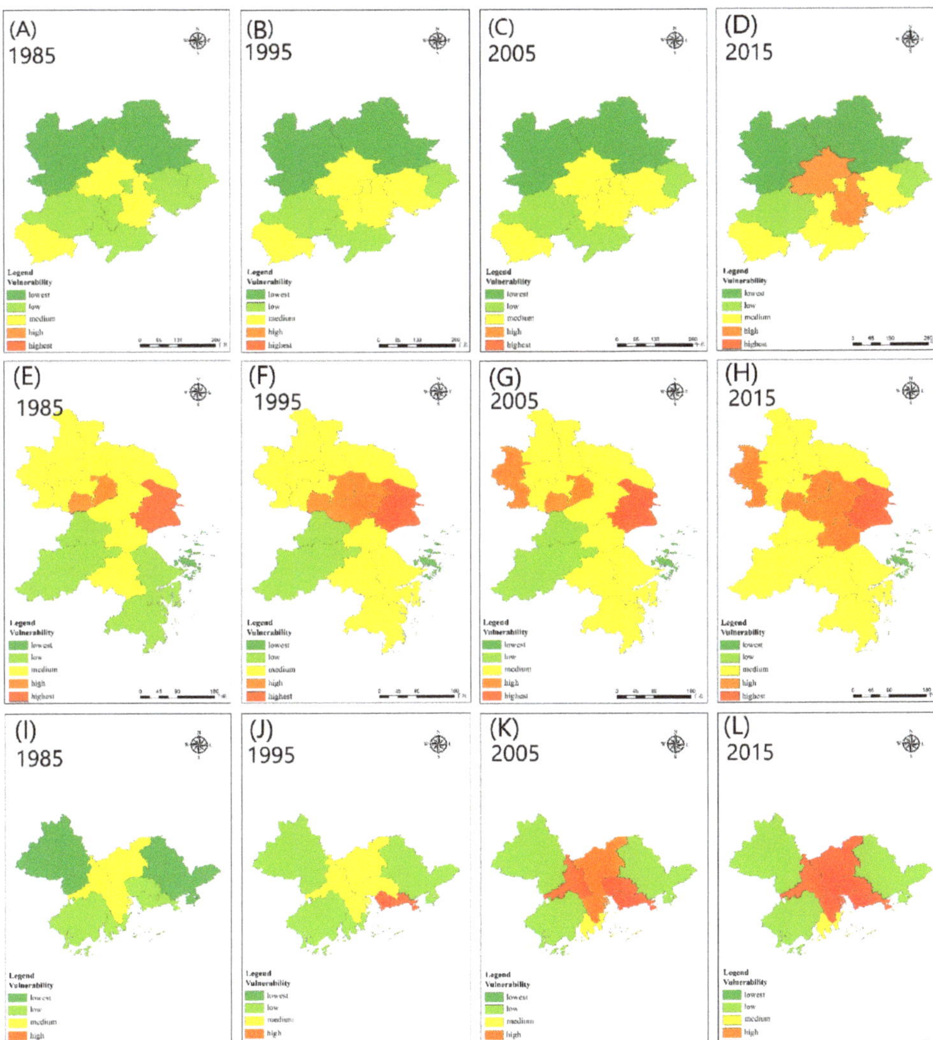

Figure 3. Evolution of the spatial pattern of the vulnerability degree of the three urban agglomerations in 1985, 1995, 2005, and 2015 (**A–D**): Beijing–Tianjin–Hebei; (**E–H**): The Yangtze River Delta; (**I–L**): Pearl River Delta.

4.2. Risk of Meteorological Disasters in Urban Agglomerations from 1981 to 2019

4.2.1. Comprehensive Meteorological Drought Index

Figure 4 shows the geographical distribution of the annual average drought index in the three eastern coastal urban agglomerations. The smaller the *CI* value is, the drier the region is. From this, we can see that the Beijing–Tianjin–Hebei region generally presented the characteristics of wetness in the west and dryness in the east [32]. The drought index in the Yangtze River Delta region gradually decreased from south to north, with the drought degree in the northern part notably stronger than in the southern part. The reason for this is that, in the plum rain season, the western Pacific Ocean brings much water vapor from south to north, and the Yangtze River Delta urban agglomeration maintains steady and sustained precipitation. The distribution of the drought index in the Pearl River Delta region was lower in the northwest and higher in the southeast. The reason for this may be that the West Pacific subtropical high brought much water vapor and precipitation to the southeast of the Pearl River Delta, therefore the drought index in the southeast direction was greater.

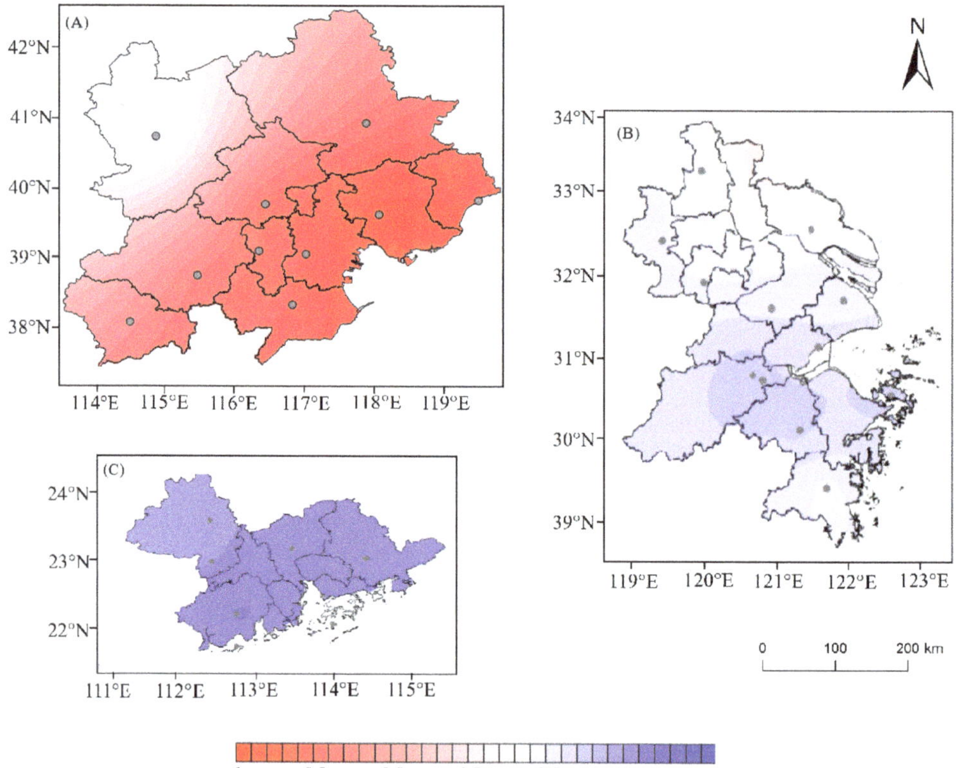

Figure 4. Geographical distribution of the annual mean comprehensive meteorological drought index in the three urban agglomerations (**A**): Beijing–Tianjin–Hebei; (**B**): Yangtze River Delta; (**C**): Pearl River Delta.

Overall, the western part of the Beijing–Tianjin–Hebei region is located on the North China Plain, and the integrated meteorological drought index *CI* was minimal. The Beijing–Tianjin–Hebei region was the driest among the three major coastal urban agglomerations in eastern China, followed by the Yangtze River Delta and the Pearl River Delta, which had the lowest drought index. Located in the southeastern coastal area of China, the Pearl River Delta urban agglomeration is closer to the ocean and has abundant water vapor. From its definition, we can see that the factors affecting the strength of the drought index were precipitation and potential evapotranspiration. Therefore, the drought index of southern coastal cities was larger than that of northern inland cities, and the drought situation of the three urban agglomerations had a spatial trend of drying from south to north.

4.2.2. Heat Wave Index

Figure 5 shows the geographical distribution of the annual average number of heat wave days in the three urban agglomerations from 1981 to 2019. According to the definition of heat waves, the factors influencing the magnitude of the heat wave index include ambient temperature and relative humidity. Summer is the season of the year's highest temperatures; during the day, temperatures rise faster inland than in coastal areas, and the relative humidity is lower. In the Beijing–Tianjin–Hebei and Yangtze River Delta urban agglomerations, the number of heat wave days was higher in the west and lower in the east, while, in the Pearl River Delta region, the number was higher in the north and lower in the south. Therefore, the geographical distribution of heat wave days in the three urban agglomerations was reasonable.

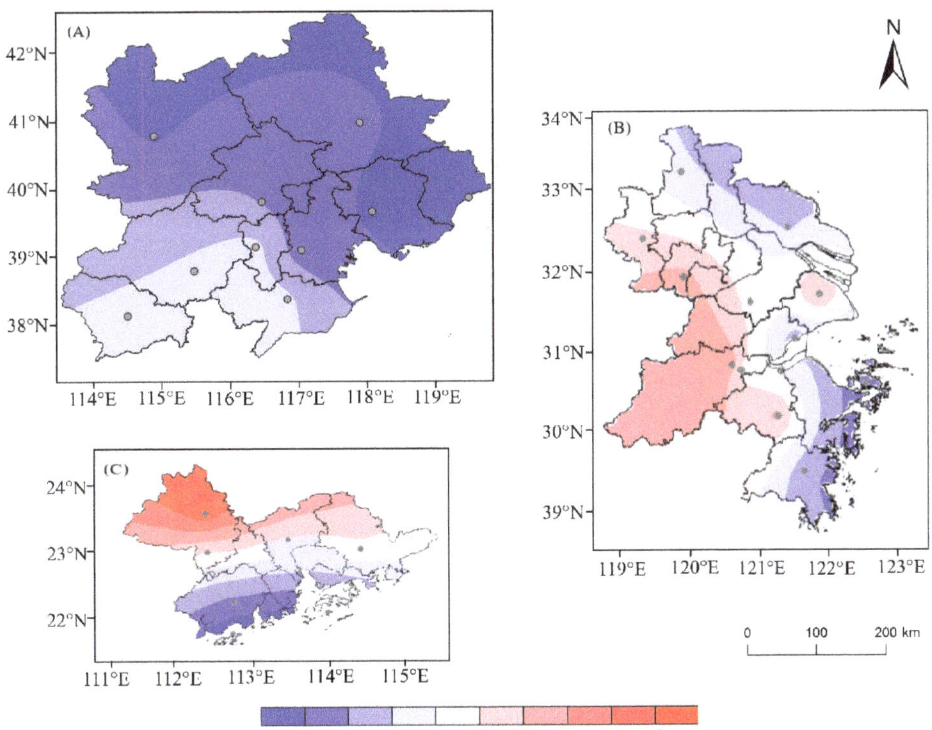

Figure 5. Geographical distribution of annual mean heat wave days in the three urban agglomerations during 1981–2019 (**A**): Beijing–Tianjin–Hebei; (**B**): Yangtze River Delta; (**C**): Pearl River Delta.

In addition, we can also see that the maximum number of heat wave days occurred in Zhaoqing in the Pearl River Delta, and the minimum number was near Qinhuangdao in the Beijing–Tianjin–Hebei area. This is because the Pearl River Delta urban agglomeration is in low latitudes and Zhaoqing is the most interior of the Pearl River Delta cities. During the year, the ambient temperature is higher, and the relative humidity is lower in summer. As a result, the Pearl River Delta had more heat wave days. As Qinhuangdao is located in northern China, the environmental temperature in summer was lower than that in southern

China. It is also located in the offshore area, which is affected by the sea breeze, and, therefore, Qinhuangdao had lower temperatures and higher ambient humidity. Therefore, Qinhuangdao had the lowest average number of heat wave days. In general, the number of heat wave days in Chinese cities mainly depends on the geographical location, with obvious spatial agglomeration and distribution [33].

4.2.3. Flood Index

The annual mean geographical distribution of the flood index in the three urban agglomerations is shown in Figure 6. In the last 39 years, the spatial pattern of precipitation in the Beijing–Tianjin–Hebei urban agglomeration was more in the east and less in the west, which was mainly because of the different terrain features and urban development in the Beijing–Tianjin–Hebei region. Zhangjiakou lies near the Taihang Mountains, a region where descending air and dry adiabatic warming causes the temperature to rise. With less atmospheric moisture and higher temperatures, the flood index near Zhangjiakou was the lowest of the regions studied [34]. The flood index value of the Yangtze River Delta urban agglomeration gradually decreased from south to north because the amount of water vapor brought from the Pacific Ocean was higher in the south, therefore there was more precipitation. The Pearl River Delta urban agglomeration is located in the subtropical monsoon climate zone, and its flood index value was the largest among the three cities, showing a trend of being higher in the south and lower in the north. Precipitation mainly occurs in the westerly belt of the subtropical high and is also affected by weather systems such as typhoons. The region is also affected by the intertropical convergence zone. The Pearl River Delta region thus received the most precipitation [35].

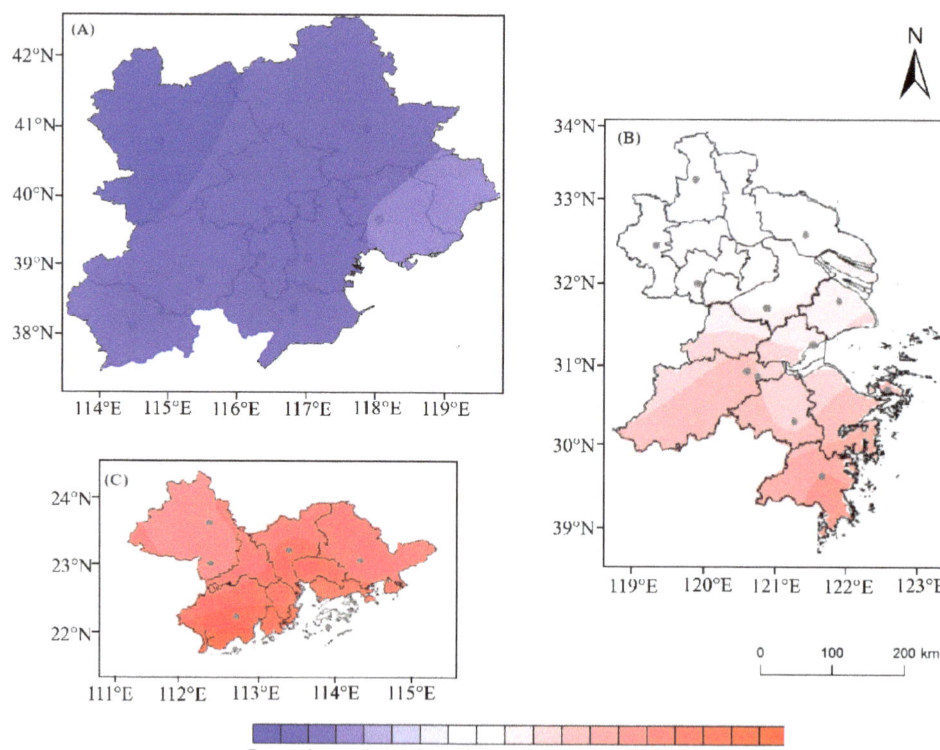

Figure 6. Geographical distribution of the annual mean flood index in the three urban agglomerations (**A**): Beijing–Tianjin–Hebei; (**B**): Yangtze River Delta; (**C**): Pearl River Delta.

4.3. Vulnerability Degree of Urban Agglomeration in 2020–2050

In the context of global warming, extreme meteorological events occur frequently, posing a serious threat to the future and human development. The impact of these meteorological disasters on human beings depends not only on the severity of the disasters themselves but also on social and economic factors such as population and social structure [36]. Therefore, predicting the vulnerability of urban agglomerations during 2020–2050 is crucial to assessing the risk of future climate change.

Based on the Grey model GM (1, 1), this paper predicted the population, GDP, and other social and economic factors from 2020 to 2050. We standardized the above data and used the AHP to determine the weight of the five elements in the evaluation of the vulnerability degree of the carrier, including population density, GDP per capita, population urbanization rate, proportion of urban area in total area, and proportion of primary industry in the GDP. The vulnerability of the urban agglomeration in 2020–2050 was calculated, and the spatial description and expression were carried out as shown in Figure 7.

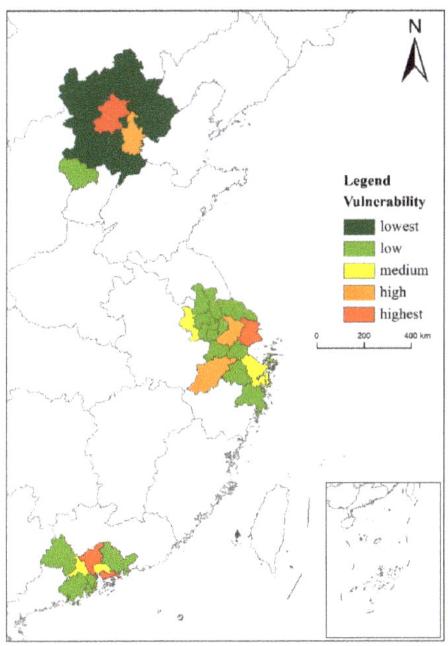

Figure 7. Vulnerability degrees of the three major urban agglomerations during 2020–2050.

Compared with the meteorological disaster risk, which mainly focuses on extreme weather events, the vulnerability degree of urban agglomerations is more concerned with the role of humans and the GDP [37]. In this paper, the vulnerability degree of urban agglomeration refers to the degree to which the social system, with regards to how it is affected by climate pressure, has been damaged because of its own sensitivity and vulnerability. As shown in Figure 6, the high vulnerability of the three urban agglomerations in the future will mainly be concentrated in the new first-tier cities: Beijing in the Beijing–Tianjin–Hebei region, Shanghai in the Yangtze River Delta region, and Shenzhen and Guangzhou in the Pearl River Delta region. With the rapid development of the economy, these cities will bear an increasing population. As the main component of the natural economy, the rapid increase in population increases the vulnerability of cities. In general, the vulnerability of the three major urban agglomerations in the future will mainly be concentrated in the Yangtze River Delta and the Pearl River Delta. In comparison, the future vulnerability of the Beijing–Tianjin–Hebei region in China is relatively low, and its resistance to extreme disaster events is stronger.

4.4. Climate Change Hazard in Urban Agglomerations under Different Emission Scenarios during 2020–2050

Based on the climate scenarios of SSP1-2.6, SSP2-4.5, and SSP5-8.5 according to CMIP6 in the near and middle periods (2020–2050), this paper obtained future meteorological elements such as daily mean temperature, daily maximum temperature, daily precipitation, and daily relative humidity under different scenarios. The comprehensive meteorological drought index (*CI*), heat wave index (*HI*), and flood index (*FI*) were calculated to quantitatively evaluate the hazard of climate change in urban agglomerations under different emission scenarios from 2020 to 2050.

Based on the comprehensive meteorological drought index, heat wave index, and flood index, this paper calculated the mild, moderate, and severe occurrence frequencies of drought, heat wave, and extreme flood events. As shown in Table 4, this paper obtained the hazard level of extreme events in different regions based on the method of overlap analysis. Overlap analysis involves the overlay of mild, moderate, and severe drought, heat wave, and flood maps to generate a new data layer with attributes.

Table 4. Extreme event hazard classification index system.

Index	Mild	Moderate	Severe
CI	$-1.2 < CI \leq -0.6$	$-1.8 < CI \leq -1.2$	$CI \leq -1.8$
HI	$2.8 \leq HI < 6.5$	$6.5 \leq HI < 10.5$	$HI \geq 10.5$
FI	30 mm < FI ≤ 50 mm	50 mm < FI ≤ 70 mm	FI ≥ 90 mm

Figure 8 shows the hazards of extreme meteorological events such as drought, heat wave, and flood under different emission scenarios. As seen in Figure 8A–C, among the three urban agglomerations, the drought hazard of Beijing–Tianjin–Hebei will be the largest, followed by the Pearl River Delta region, and, finally, the Yangtze River Delta region. The serious drought in the Pearl River Delta is mainly due to the high annual temperature in the Pearl River Delta region and the long-term absence of precipitation in a specific season, which lead to the frequent occurrence of winter and spring drought. At the same time, with the increase in greenhouse gas emission intensity, the drought hazard in Beijing and Tianjin will have an increasing trend. On the one hand, Beijing and Tianjin are located in the North China Plain at high latitudes, which makes it difficult for warm and wet air from the ocean to reach the Beijing–Tianjin–Hebei city region. On the other hand, with the increase in greenhouse gas emissions and the further aggravation of global warming, temperatures in the Beijing–Tianjin–Hebei urban agglomeration are also increasing. In addition, Beijing and Tianjin, as two super large first-tier cities in the Beijing–Tianjin–Hebei urban agglomeration, have obvious urban heat island effects. Therefore, Beijing and Tianjin will have the greatest hazard of drought in the future.

As seen from the middle row (D, E, F) of Figure 8, the hazard of heat waves will be the lowest in the Beijing–Tianjin–Hebei urban agglomeration, followed by the Yangtze River Delta, and the highest in the Pearl River Delta urban agglomeration, from 2020 to 2050. Solar radiation decreases with increasing latitude, therefore the hazard of heat waves will gradually decrease from south to north. However, the hazard of heat waves in some cities in the Yangtze River Delta will also be relatively high because the Yangtze River Delta city cluster is located in the middle and lower reaches of the Yangtze River. In summer, under the subtropical high, atmospheric subsidence prevails and the temperature rises rapidly, forming a summer drought. At the same time, the hazard of heat waves is also relatively high. With the increase in emissions, the heat wave hazard of the three major urban agglomerations on the east coast of China will continue increasing.

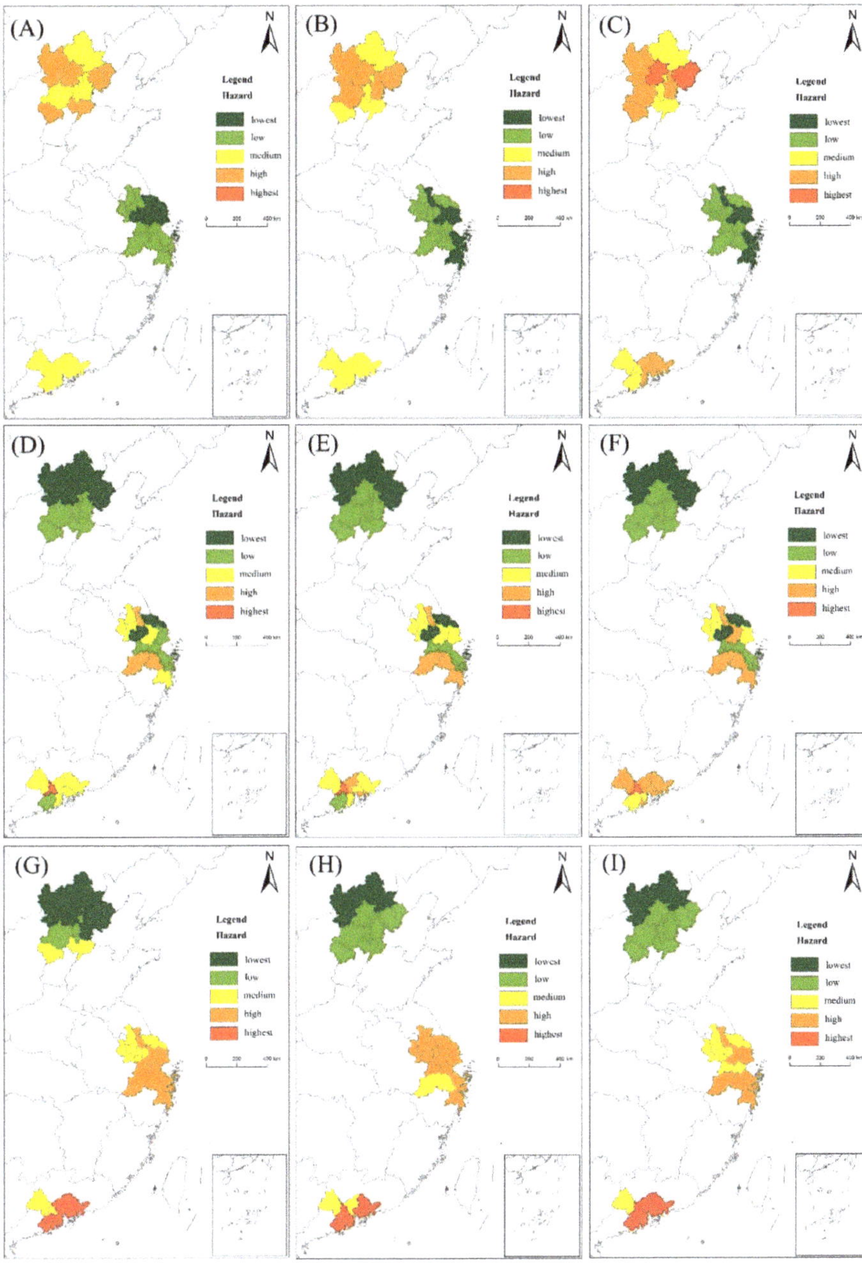

Figure 8. Hazards of drought (**A–C**), heat wave (**D–F**), and flood (**G–I**) in the three urban agglomerations under different emission scenarios: SSP1-2.6 (**A**,**D**,**G**), SSP2-4.5 (**B**,**E**,**H**), and SSP5-8.5 (**C**,**F**,**I**).

From the bottom row (G, H, I) of Figure 8, it can be seen that the flood hazard of the three major urban agglomerations in 2020–2050 will be concentrated in the middle and lower reaches of the Yangtze River. That is, the Yangtze River Delta urban agglomerations will have the greatest flood hazard, while the Pearl River Delta urban agglomerations in the low latitudes will have the middle flood hazard, and the Beijing–Tianjin–Hebei region in the middle and high latitudes will have the least flood hazard. The main reason for this is that there are many plains in the middle and lower reaches of the Yangtze River, and the terrain is relatively low and flat. Flooding mainly depends on the precipitation. The Yangtze River Delta urban agglomeration is located in the monsoon climate zone with heavy rain in summer. Huang pointed out that approximately 46% of the world's regions may suffer from moderate or high hazards of extreme precipitation changes in the future. The frequency of extreme precipitation events will increase under the RCP8.5 scenario, and most regions will show high flood hazards [38]. There was no significant difference in the spatial distribution of flood hazards under the three scenarios analyzed in this paper (SSP1-2.6, SSP2-4.5, and SSP5-8.5); however, with the increase in emission intensity, the disaster hazard of floods increased.

Based on the hazard assessment of drought, heat wave, and flood, and according to the spatial distribution pattern of meteorological disasters, this paper regionalized the hazards of climate change in the three major urban agglomerations in eastern China [39], as shown in Figure 9. It can be seen from the figure that the Pearl River Delta urban agglomeration will experience the greatest hazard of climate change, followed by the Yangtze River Delta, and the Beijing–Tianjin–Hebei urban agglomeration. The hazard of climate change will gradually decrease from south to north. With the increase in emission intensity, the climate change hazard of SSP5-8.5 in the high emission scenario will cause the largest changes.

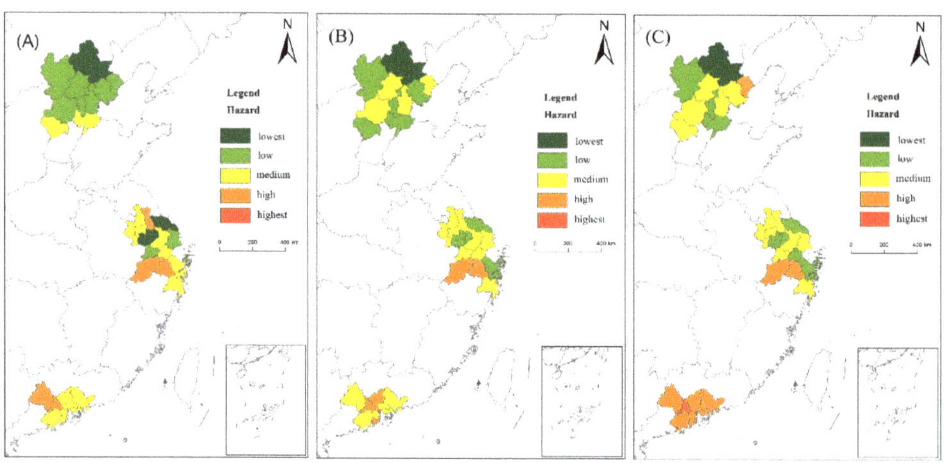

Figure 9. Climate change hazard of the three urban agglomerations under different emission scenarios (**A**): SSP1-2.6; (**B**): SSP2-4.5; (**C**): SSP5-8.5.

4.5. Climate Change Risk Zoning of Urban Agglomerations under Different Emission Scenarios during 2020–2050

In this paper, the disaster risk was divided into two parts: the disaster-causing factor and the bearing body. Drought, heat wave, and flood were regarded as disaster-causing factors, and urban agglomerations were regarded as bearing bodies. Disaster risk was equal to the product of disaster risk and the vulnerability of the disaster-bearing body. Therefore, these factores were brought into the climate change risk assessment model to obtain the climate change risk to the three urban agglomerations under different emission scenarios.

Figure 10 shows that the future climate change risk to the Beijing–Tianjin–Hebei urban agglomeration will mainly be concentrated in Beijing and Tianjin, which have higher populations and GDPs. Therefore, when large meteorological disasters occur, these two urban agglomerations will be more vulnerable to risks, while other cities in the Beijing–Tianjin–Hebei region will show good urban resilience. However, the climate change risk to the Yangtze River Delta urban agglomeration will be different, the regional coordination is not consistent and the overall risk will be slightly stronger than that to the Beijing–Tianjin–Hebei urban agglomeration. In addition, the risk of climate change will be high in the Pearl River Delta urban agglomeration, among which Guangzhou and Shenzhen are the cities with the highest risk in the entire Pearl River Delta region. Overall, the higher the emission intensity, the greater the risk of climate change in each urban agglomeration; this conclusion was also consistent with the previous research results [38,40].

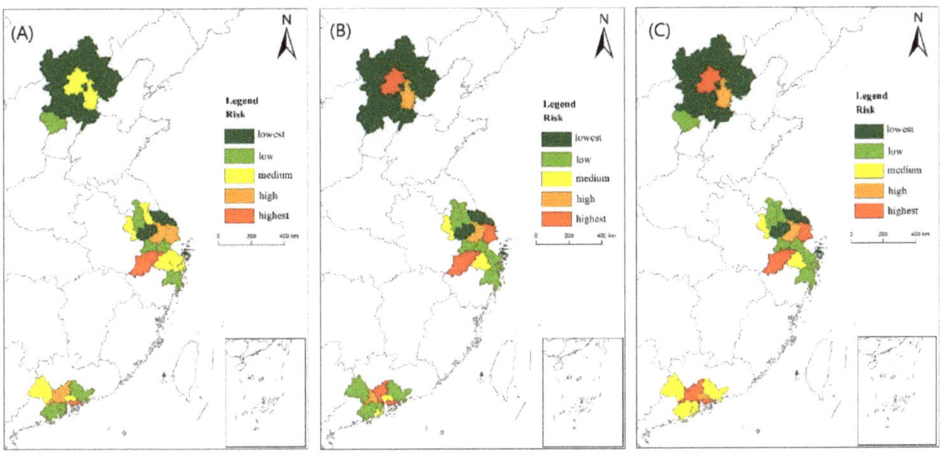

Figure 10. Climate change risks in the three urban agglomerations under different emission scenarios (**A**): SSP1-2.6; (**B**): SSP2-4.5; (**C**): SSP5-8.5.

5. Conclusions and Discussion

The risk of climate change to the three major urban agglomerations in eastern China is the result of the combined effects of climate factors and social and economic factors, which need to be cross-analyzed by multiple disciplines and fields. To solve this problem, we selected a climate change risk assessment model as a bridge between natural science analysis and humanistic socioeconomic analysis. The vulnerability degree of urban agglomerations and the hazard of climate change were input into the climate change risk assessment model. Based on the perspective of climate change economics, this paper analyzed the risk of climate change to the three major urban agglomerations in eastern China.

In terms of vulnerability, high vulnerability was mainly concentrated in Beijing, Tianjin, Nanjing, Wuxi, Suzhou, Shanghai, Foshan, Guangzhou, Shenzhen, Dongguan, and other first-tier cities as well as new first-tier cities. These cities tend to be associated with a higher GDP and population, and they are at a higher risk when meteorological disasters strike. On the whole, among the three urban agglomerations, the Yangtze River Delta region was the most vulnerable, followed by the Pearl River Delta region, and the Beijing–Tianjin–Hebei region.

With climate change, drought characteristics in the north and south have changed. In general, the Beijing–Tianjin–Hebei urban agglomeration will have the largest drought hazard in the future, followed by the Pearl River Delta urban agglomeration, and the Yangtze River Delta urban agglomeration. The hazard of heat wave in urban agglomerations in China mainly depends on geographical location, with obvious spatial agglomeration and distribution. The hazard of heat wave gradually decreases from south to north, and the hazard of heat wave in inland cities is higher than that in coastal cities in the same urban agglomerations. In China, future flood hazards will be concentrated in the middle and lower reaches of the Yangtze River; that is, the Yangtze River Delta will have the largest flood hazard, the Pearl River Delta the middle flood hazard, and the Beijing–Tianjin–Hebei region the lowest flood hazard.

Among the three coastal urban agglomerations in eastern China, the Pearl River Delta has the highest hazard of climate change, followed by the Yangtze River Delta. The Beijing–Tianjin–Hebei urban agglomeration has the lowest hazard of climate change. The hazard of climate change gradually decreases from south to north. With the increase in emission intensity, the hazard of climate change in the high emission scenario SSP5-8.5 tends to increase.

The risk of future climate change to the Beijing–Tianjin–Hebei urban agglomeration will be mainly concentrated in Beijing and Tianjin, while other cities in the Beijing–Tianjin–Hebei region will show good urban resilience. However, the risk of climate change to the Yangtze River Delta is different, and the regional coordination is not consistent. The overall risk of climate change to the Yangtze River Delta is slightly stronger than that to the Beijing–Tianjin–Hebei urban agglomeration. In addition, the risk of climate change to the Yangtze River Delta urban agglomeration is high overall, among which Guangzhou and Shenzhen are the highest risk cities in the entire Yangtze River Delta region.

Under different emission scenarios, the higher the emission intensity, the greater the future temperature rise, and the greater the risk of climate change that is faced by each urban agglomeration. Therefore, the high incidence of disaster events is closely related to climate warming. Energy saving, emission reduction, low carbon, and environmental protection are effective measures to delay the temperature rise and reduce the frequency of disasters in the future. Therefore, formulating corresponding policies and measures according to the characteristics of climate change in different urban agglomerations is essential to effectively respond to climate change, improve urban resilience, and maintain steady and rapid economic growth.

In the study of the risk of climate change, we only selected three extreme climate events, drought, heat wave, and flood, because they are the most typical and widely distributed meteorological and hydrological events. However, in fact, the risks of climate change are not limited to drought, heat wave, and flood, but also include hail, typhoons, cryogenic freezing, and other disasters. However, the timing and regional nature of these events were more obvious, therefore they were not included in this study. In this paper, five factors: GDP per unit area, the population density, the population urbanization rate, the proportion of urban area in total area, and the proportion of primary industry in the GDP, were considered in the calculation of the vulnerability index of the bearing body, and the semiquantitative method of expert scoring was also used with the AHP method, which may affect the results of risk estimation. In addition, the Grey model is only suitable for medium and short-term prediction, and the accuracy of long-term prediction is limited. In this paper, the climate data of the BCC-CSM2-MR model were selected to predict the future SSP1-2.6, SSP2-4.5, and SSP5-8.5 emission scenarios. Because of the limited space, the multi-model ensemble average requires further study.

Author Contributions: Conceptualization, J.C. and M.S.; Data curation, J.C.; Formal analysis, W.D. and W.Z.; Funding acquisition, J.C.; Investigation, Y.L.; Methodology, M.S.; Project administration, J.L.; Resources, J.Z.; Supervision, M.S.; Validation, J.C., M.S. and W.D.; Writing—original draft, M.S.; Writing—review & editing, M.S. All authors have read and agreed to the published version of the manuscript.

Funding: This research was funded by the National Key Research and Development Program of China: 2018YFC1509003; National Natural Science Foundation of China: 42075167.

Institutional Review Board Statement: Not applicable.

Informed Consent Statement: Not applicable.

Conflicts of Interest: The authors declare no conflict of interest.

Appendix A

Table A1. Experts Scores Technique.

	Population Density	GDP per Capita	Population Urbanization Rate	Proportion of Urban Area to Total Area	Proportion of Primary Industry in GDP
professor A	5	4.3	1.7	2	3
professor B	5	5	1.8	3	2.2
professor C	5	4.8	2	2.2	2.1
professor D	5	4.7	2.3	1.2	2.3
average value	5	4.7	1.95	2.1	2.4

Table A2. Analytic hierarchy Process data.

	Population Density	GDP per Capita	Population Urbanization Rate	Proportion of Urban Area to Total Area	Proportion of Primary Industry in GDP
Population density	1	1.064	2.564	2.381	2.083
GDP per capita	0.94	1	2.41	2.238	1.958
Population urbanization rate	0.39	0.415	1	0.929	0.813
Proportion of urban area to total area	0.42	0.447	1.077	1	0.875
Proportion of primary industry in GDP	0.48	0.511	1.231	1.143	1

Table A3. Analytic hierarchy Process results.

Index	Feature Vector	Weight Value	Maximum Eigenvalue	Ci
Population density	1.548	30.960%		
GDP per capita	1.455	29.102%		
Population urbanization rate	0.604	12.074%	5	0
Proportion of urban area to total area	0.65	13.003%		
Proportion of primary industry in GDP	0.743	14.861%		

Table A4. Random consistency table.

n	3	4	5	6	7	8	9	10
R	0.52	0.89	1.12	1.26	1.36	1.41	1.46	1.49

References

1. Sun, W.; Li, Q.; Huang, B.; Cheng, J.; Song, Z.; Li, H.; Dong, W.; Zhai, P.; Jones, P. The Assessment of Global Surface Temperature Change from 1850s: The C-LSAT2. 0 Ensemble and the CMST-Interim Datasets. *Adv. Atmos. Sci.* **2021**, *38*, 875–888. [CrossRef]
2. Field, C.B.; Barros, V.R.; Dokken, D.J.; Mach, K.J.; Mastrandea, M.D.; Bilir, T.E.; Chatterjee Ebi, K.L.; Estrada, Y.O.; Genova, R.C. (Eds.) *Climate Change 2014: Impacts, Adaptation, and Vulnerability*; Cambridge University Press: Cambridge, UK, 2014.
3. Liu, D.; Wang, D.; Wu, J.; Wang, Y.; Wang, L.; Zou, X.; Chen, Y.; Chen, X. A risk assessment method based on RBF artificial neural network-cloud model for urban water hazard. *J. Intell. Fuzzy Syst.* **2014**, *27*, 2409–2416. [CrossRef]
4. Yang, Y.; Peng, H. Assessment of waterlogging risk for urban drainage system. *J. Water Purif. Technol.* **2018**, *37*, 116–121. (In Chinese) [CrossRef]
5. Liu, X.; Qu, J.; Liu, L.; Li, H.; Pei, H.; Ceng, J. Research on the Assessment of Urban Climate Change Adaptation Capability in Western China. *J. Ecol Econ.* **2019**, *35*, 104–110. Available online: https://core.ac.uk/download/pdf/222802639.pdf (accessed on 18 October 2021). (In Chinese)
6. Giupponi, C.; Mojtahed, V.; Gain, A.K.; Biscaro, C.; Balbi, S. Integrated risk assessment of water-related disasters. In *Hydro-Meteorological Hazards, Risks and Disasters*; Academic Press: Cambridge, MA, USA, 2014; pp. 163–200.
7. Wu, S.; Gao, J.; Deng, H.; Liu, L.; Pan, T. Climate change risk and methodology for its quantitative assessment. *J. Prog. Geogr.* **2018**, *37*, 28–35.
8. Sahana, V.; Mondal, A.; Sreekumar, P. Drought vulnerability and risk assessment in India: Sensitivity analysis and comparison of aggregation techniques. *J. Environ. Manag.* **2021**, *299*, 113689. [CrossRef]

9. Dong, S.; Tao, S.; Yang, W.; Li, Z.; Li, Y. Impacts of climate change on urban agglomeration in central and western regions of China. *J. Arid Environ.* **2011**, *25*, 72–76.
10. Apel, H.; Aronica, G.T.; Kreibich, H.; Thieken, A.H. Flood risk analyses—How detailed do we need to be? *Nat. Hazards* **2009**, *49*, 79–98. [CrossRef]
11. Barbier, E.B. A global strategy for protecting vulnerable coastal populations. *J. Sci.* **2014**, *345*, 1250–1251. [CrossRef]
12. Hay, C.C.; Morrow, E.; Kopp, R.E.; Mitrovica, J.X. Probabilistic reanalysis of twentieth-century sea-level rise. *J. Nat.* **2015**, *517*, 481–484. [CrossRef]
13. Sun, S.; Zhai, J.; Li, Y.; Huang, D.; Wang, G. Urban waterlogging risk assessment in well-developed region of Eastern China. *J. Phys. Chem. Earth* **2020**, *115*, 102824. [CrossRef]
14. Arnell, N.W.; Lowe, J.A.; Challinor, A.J.; Osborn, T.J. Global and regional impacts of climate change at different levels of global temperature increase. *J. Clim. Chang.* **2019**, *155*, 377–391. [CrossRef]
15. Alexander, K.; Hettiarachchi, S.; Ou, Y.; Sharma, A. Can integrated green spaces and storage facilities absorb the increased risk of flooding due to climate change in developed urban environments? *J. Hydrol.* **2019**, *579*, 124201. [CrossRef]
16. Zhang, X.; Zhou, J.; Song, W. Simulating urban sprawl in china based on the artificial neural network-cellular automata-Markov model. *J. Sustain.* **2020**, *12*, 4341. [CrossRef]
17. Wang, C.; Meng, Q. Research on the sustainable synergetic development of Chinese urban economies in the context of a study of industrial agglomeration. *J. Sustain.* **2020**, *12*, 1122. [CrossRef]
18. Chou, J.; Sun, M.; Xu, Y.; Yang, F.; Li, J.; Zhao, W. Resilience of Grain Yield in China Under Climate Change Scenarios. *J. Front. Environ. Sci.* **2021**, *9*, 44. [CrossRef]
19. Deng, J. Introduction to grey system theory. *J. Grey Syst.* **1989**, *1*, 1–24.
20. Sun, M.; Chou, J.; Xu, Y.; Yang, F.; Li, J. Study on the thresholds of grain production risk from climate change in China's main grain-producing areas. *J. Phys. Chem. Earth.* **2020**, *116*, 102837. [CrossRef]
21. Yin, M. Fifteen years of grey system theory research: A historical review and bibliometric analysis. *J. Expert Syst. Appl.* **2013**, *40*, 2767–2775. [CrossRef]
22. Huang, C.; Yin, J.; Zhang, J. Calculation of Risk Assessment Index Weight by Analytic Hierarchy Process. *J. China Public Secur.* **2018**, 19–22. (In Chinese) [CrossRef]
23. Tan, R.R.; Aviso, K.B.; Huelgas, A.P.; Promentilla MA, B. Fuzzy AHP approach to selection problems in process engineering involving quantitative and qualitative aspects. *J. Process. Saf. Environ. Prot.* **2014**, *92*, 467–475. [CrossRef]
24. Chen, D.; Da, L.; Yan, W.; Deng, D.; He, L.; Wu, S.; Yu, L.; Yang, J.; Wang, J.; Zhang, L. Screening and Evaluation of Suitable Shrubs for Sand Control in Alpine Sandy Land of Northwest Sichuan Based on AHP Analysis. *J. Sichuan For. Sci. Technol.* **2021**, *42*, 65–69.
25. GB/T 20481–2017, Grades of Meteorological Drought. Available online: https://kns.cnki.net/kcms/detail/detail.aspx?FileName=SCSF00050541&DbName=SCSF (accessed on 12 November 2021).
26. He, J.; Yang, X.; Li, J.; Jin, J.; Wei, Y.; Chen, X. Spatiotemporal variation of meteorological droughts based on the daily comprehensive drought index in the Haihe River basin, China. *Nat. Hazards* **2015**, *75*, 199–217. [CrossRef]
27. Ma, J.; Zhang, L.; Li, J. Automatic Calculation of Comprehensive Meteorological Drought Index (Ic). *J. Meteorol. Res. Appl.* **2011**, *32*, 17–19. Available online: http://www.cqvip.com/qk/96341a/201104/40486752.html (accessed on 12 November 2021). (In Chinese)
28. GB/T 29457-2012, Grade of the heat wave. Available online: https://kns.cnki.net/kcms/detail/detail.aspx?FileName=SCSF00039877&DbName=SCSF (accessed on 20 November 2021).
29. Zhou, Y.; Zhu, S.; Hua, J.; Li, Y.; Xiang, J.; Ding, W. Spatio-temporal distribution of high temperature heat wave in Nanjing. *J. Geogr. Inf. Sci.* **2018**, *20*, 1613–1621. Available online: http://www.cqvip.com/qk/86408a/201811/676778597.html (accessed on 18 October 2021). (In Chinese)
30. Huang, C. Basic principles of natural disaster risk analysis. *J. Nat. Disasters* **1999**, 21–30. Available online: http://www.cqvip.com/qk/97398x/199902/3579615.html (accessed on 18 October 2021).
31. Chou, J.; Ban, J.; Dong, W.; Hu, C.; Dai, R. Characteristics analysis and assessment of economic damages caused by tropical cyclones in Guangdong Province. *Chin. J. Atmos. Sci.* **2018**, *42*, 357–366. (In Chinese) [CrossRef]
32. Miao, Z.; Xu, L.; Lu, M. Analysis of Drought Characters Based on SPEI Index in Beijing-Tianjin-Hebei Region. *J. Yellow River* **2018**, *40*, 51–57. (In Chinese) [CrossRef]
33. Huang, X.; Wang, B.; Guo, Y.; Li, Y. Characteristics of urban extreme heat and assessment of social vulnerability in China. *J. Geogr. Res.* **2020**, *39*, 1534–1547. (In Chinese) [CrossRef]
34. Lu, J.; Liu, J.; Liu, M.; Cao, Y.; Li, H.; Ning, Y. Multi-scale analysis of precipitation in the Beijing-Tianjin-Hebei urban agglomeration in the past 55 years. *Hydro Sci. Eng.* **2020**, *6*, 23–31. (In Chinese) [CrossRef]
35. Huang, G.; Chen, Y.; Yao, Z. Spatial and temporal evolution characteristics of extreme rainfall in the Pearl River Delta under high urbanization. *Adv. Water Sci.* **2021**, *32*, 161–170.
36. Roper, R.E. Book Review of "Heat Wave: A Social Autopsy of Disaster in Chicago". *Homel. Secur. Emerg. Manag.* **2011**, *1*. [CrossRef]
37. Huang, J.; Su, F. The Review and Prospect on the Hot Issues of Urban Social Vulnerability to Disasters. *Sci. Geol. Sin.* **2017**, *37*, 1211–1217.

38. Huang, H.; Cui, H.; Ge, Q. Assessment of potential risks induced by increasing extreme precipitation under climate change. *Nat. Hazards* **2021**, *108*, 2059–2079. [CrossRef]
39. Wu, S.; Pan, T.; Liu, Y.; Deng, H.; Jiao, K.; Lu, Q.; Feng, A.; Yue, X.; Yin, Y.; Zhao, D.; et al. Comprehensive climate change risk regionalization of China. *Acta Geogr. Sin.* **2017**, *72*, 3–17.
40. Dong, S.; Xu, Y.; Zhou, B.; Hou, M.; Li, R.; Li, Y.; Zhang, Y. Projected risk of extreme heat in China based on CMIP5 models. *Adv. Clim. Chang. Res.* **2014**, *10*, 365.

sustainability

MDPI

Article

Spatiotemporal Evolution and Socioeconomic Impacts of Rainstorms and Droughts in Contiguous Poverty-Stricken Areas of China

Aiwei Li [1,2], Shuyuan Gao [1,2], Miaoni Gao [1,2,*], Xueqing Wang [1], Hongling Zhang [1,2], Tong Jiang [1,2,*] and Jing Yang [3,4]

1 Institute for Disaster Risk Management, Nanjing University of Information Science and Technology, Nanjing 210044, China
2 School of Geographical Science, Nanjing University of Information Science and Technology, Nanjing 210044, China
3 State Key Laboratory of Earth Surface Processes and Resource Ecology, Beijing Normal University, Beijing 100875, China
4 Faculty of Geographical Science, Beijing Normal University, Beijing 100875, China
* Correspondence: gaomn@nuist.edu.cn (M.G.); jiangtong@cma.gov.cn (T.J.)

Abstract: To consolidate the achievements in the elimination of absolute poverty in China and prevent rural populations from returning to poverty as a result of meteorological disasters, this study analyzed the spatiotemporal characteristics of rainstorms and droughts and their socioeconomic impacts on China's contiguous poverty-stricken areas (CPSAs) from 1984 to 2019. The annual number of rainstorms and drought days in CPSAs of China reached approximately 1.9 days/year and 44.6 days/year, respectively. It gradually decreased from southeast to northwest. Rainstorms showed a significant increasing trend of 0.075 days/decade, while there is no significant trend in drought days. Due to rainstorms and droughts, the average annual number of people affected and direct economic losses in CPSAs reached 34 million people/year and 29 billion Chinese yuan/year, accounting for 22.9% and 12.6% of China's total amounts, respectively. The average annual loss rate due to disasters in this region reached 1.6%, which is 0.6% higher than the national level. Furthermore, the distinct features and socioeconomic impacts of rainstorms and droughts were identified on the southeastern and northwestern sides of the population density line (PDL) along Tengchong-Aihui in China. Droughts have often impacted the regions located along the PDL, while rainstorms and droughts have occurred more frequently in the regions to the southeast of the PDL than in the regions to the northwest of the PDL. As a result, the affected population and direct economic losses due to rainstorms and droughts in the regions to the southeast of the PDL were 8.8 and 9.2 times and 3.3 and 7.4 times higher, respectively, than those in the regions on the other side of the PDL. Although the losses were greater, the disaster resistance capabilities were significantly improved in these regions. In contrast, the regions to the northwest side of the PDL exhibited a significant increasing trend in losses with a relatively low disaster resistance capabilities. This study revealed that it is necessary to improve the capability of meteorological disaster prevention and reduction in China's CPSAs, especially in the regions to the west of the PDL, which could further contribute to the realization of United Nations Sustainable Development Goals.

Keywords: contiguous poverty-stricken areas; rainstorms and droughts; direct economic losses; disaster-affected population

Citation: Li, A.; Gao, S.; Gao, M.; Wang, X.; Zhang, H.; Jiang, T.; Yang, J. Spatiotemporal Evolution and Socioeconomic Impacts of Rainstorms and Droughts in Contiguous Poverty-Stricken Areas of China. *Sustainability* **2022**, *14*, 9927. https://doi.org/10.3390/su14169927

Academic Editors: Xiaodong Yan, Jia Yang and Shaofei Jin

Received: 24 May 2022
Accepted: 8 August 2022
Published: 11 August 2022

1. Introduction

The United Nations Sustainable Development Goals (SDGs) clearly state that by 2030, all forms of poverty are to be eradicated worldwide [1]. However, more than 700 million people (10% of the world's population) still live in extreme poverty. The Outline for Poverty Alleviation and Development in Rural China (2011–2020) delineated

14 contiguous poverty-stricken areas (CPSAs) in China based on economic levels and set the net income of farmers to 2300 yuan (constant 2010 prices) as the national standard for poverty alleviation. As of 23 November 2020, all 832 poverty-stricken counties in China have been removed from poverty. The poverty reduction target of the United Nations 2030 Agenda for Sustainable Development has been achieved 10 years ahead of schedule, contributing to more than 70% of the reduction in global poverty [2]. Most of these areas are located in plateau, mountainous and hilly areas with a harsh natural environment, low economic status and relatively lagging social development and represent high-risk areas for meteorological disasters in China [3]. Among these disasters, rainstorms and droughts are the two meteorological disasters with the highest proportion of the total socioeconomic impact, with 24% of the agricultural population endangered by rainstorms and droughts with a long duration and wide range, reaching losses of 67.35 billion yuan in 2016 alone [4,5].

At the global warming level of 1.5 °C, the droughts in CPSAs will change from mild to moderate, which will increase significantly in three prefectures of Southern Xinjiang and the Tibet Region [6]. Meanwhile, the intensity of rainstorms will increase in more than 85% of CPSAs in a 1.5 °C warming scenario [7]. Since the disaster resistance capabilities in poor areas are relatively low, natural hazards like rainstorms and droughts are a major reason why people become and stay poor [8]. Therefore, a comprehensive understanding of the spatiotemporal characteristics and socioeconomic impacts of rainstorms and droughts in CPSAs is important for preventing populations that have been lifted out of poverty from falling back into it again due to meteorological disasters, which further contributes to poverty eradication consolidation processes in China.

Previous studies have found that the frequency and intensity of rainstorms and droughts in China have increased since the 21st century, and the affected population and economic losses due to rainstorms and droughts have significantly increased [9–11]. However, these studies mainly focused on a specific region, such as exploring rainstorms and disaster-induced losses in eastern China [12]. There remains a lack of comprehensive studies of meteorological disaster conditions in CPSAs. In recent years, the demand for natural hazard data has increased, and numerous databases have emerged, e.g., NatCatSERVICE, a global-scale database launched by Munich Re (MUNICH RE), and the Emergency Events Database (EM-DAT) published by the University of Leuven, Belgium. China has also created databases such as the China Meteorological Administration Disaster Database, China Natural Disaster Database and China Meteorological Disaster Yearbook, according to different application needs, which provide a good opportunity to study the socioeconomic impacts of meteorological disasters in specific regions [9,13–16].

Hu Huanyong proposed the well-known population density line (PDL) in 1935, revealing the basic pattern of the population distribution in China. It showed that 64% of the areas northwest of the Aihui-Tengchong line in China is in habited by only 4% of the population [17,18]. The PDL is strongly correlated with the regional climate, and the PDL coincides closely with the climate dividing line between arid and humid zones in China [19]. This result is very consistent with the 400 mm isohyetal line and the 10 °C annual mean temperature line, and the temperature and precipitation characteristics of the above CPSAs also basically coincide with this dividing line [17]. The divergence of human factors on both sides of the PDL is also significant; CPSAs in China are mainly distributed on both sides of the PDL, and the economy in the regions to the east of the PDL is more developed than that of the regions to the west of the PDL [20,21].

Therefore, based on rainstorm and drought data, direct economic losses and affected population data, this study determined the spatial and temporal evolutionary characteristics of the socioeconomic impacts of rainstorms and droughts and revealed the unique characteristics which distinguish this region from the disaster conditions in China. Furthermore, this study compared and analyzed rainstorm and drought characteristics with the PDL as the boundary. The results of this study can provide a scientific basis for the CPSAs considered to overcome climate change challenges, prevent a return to poverty resulting from meteorological disasters and further contribute to the realization of SDGs.

2. Data and Methods

2.1. Study Area

CPSAs are mainly located in the mountains, hills and highlands of Central and Western China, with complex and diverse topographic conditions and fragile ecological environments (Figure 1) [22]. For the convenience of description, the regions to the east of the PDL (Regions5–11, R5–11 for short), the regions along the PDL (Regions1–4, R1–4 for short) and the regions to the west of the PDL (Regions12–14, R12–14 for short) were classified. The regions along the PDL (R1–4) include the Southern area of Daxing'anling Mountains, the Yanshan-Taihang Mountain area, the Lvliang Mountain area and Liupanshan area, which have a temperate continental climate, with an average annual precipitation of approximately 500 mm, a mean temperature of approximately 4–8 °C, a maximum temperature of 16 °C, and a minimum temperature of 0 °C, which decreases with increasing latitude. The regions to the west of the PDL (R12–14) include the three prefectures of southern Xinjiang, the Tibet region and Tibetan areas in four provinces which belong to the plateau mountain climate zone of China. The annual precipitation is very low, approximately 300 mm, and the average annual temperature reaches below 4 °C, a maximum temperature of 10 °C, and a minimum temperature of −6 °C. The regions to the east of the PDL (R5-11) have a subtropical monsoon climate, with an average annual precipitation of over 1000 mm and an average annual temperature ranging from approximately 13 to 18 °C, a maximum temperature of 23 °C and a minimum temperature of 10 °C [20].

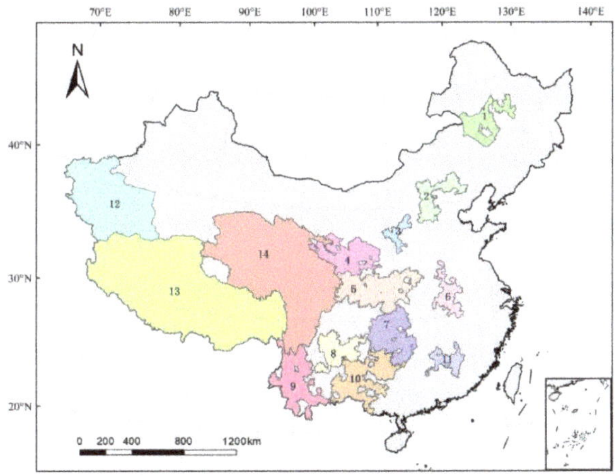

Figure 1. The contiguous poverty-stricken areas (CPSAs) (1. Southern area of Daxing'anling Mountains; 2. Yanshan-Taihang Mountain Area; 3. Lvliang Mountain Area; 4. Liupanshan Area; 5. Qinba Mountain Area; 6. Dabie Mountain Area; 7. Wuling Mountain Area; 8. Wumeng Mountain Area; 9. West Yunnan Border Area; 10. Yunnan, Guangxi and Guizhou rocky desertification area; 11. Luoxiao Mountain Area; 12. Three prefectures of Southern Xinjiang; 13. Tibet Region; 14. Tibetan Areas in Four Provinces).

During 1984–2019, the average annual total population of these CPSAs reached 210 million people, accounting for 16.7% of the total population of China. The average annual total gross domestic product (GDP) reached 1989.3 billion yuan, accounting for only 7.7% of the total GDP of China. The population of the above CPSAs grew from 180 million people to 230 million people, and the total GDP increased from 57.3 billion yuan to 7621.9 billion yuan (Figure 2a,b). The multi-year average GDP per capita in this region reached RMB 8943 per person, accounting for only 46.9% of China's per capita GDP of RMB 19,079 per person (Figure 2c). This indicates that compared to the entire country, the CPSAs considered, accounting for 40% of the total area of China, contain a relatively small population and are relatively economically under-developed.

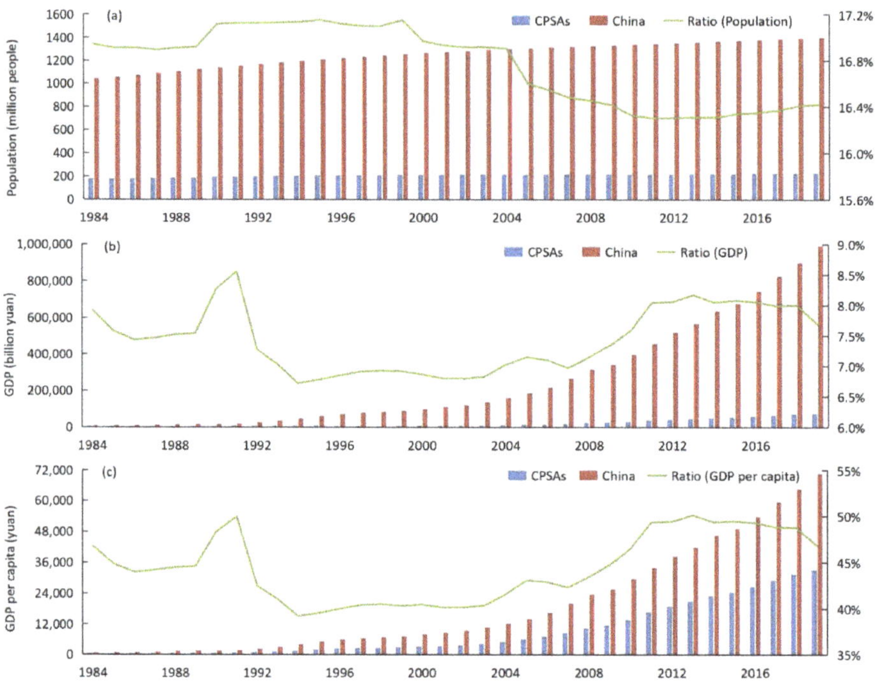

Figure 2. Time series of (**a**) total population (million people), (**b**) total GDP (billion yuan), (**c**) GDP per capita (yuan) in CPSAs, China during 1984–2019. Green lines denote their ratio compared to China's total.

2.2. Datasets

2.2.1. Observational Data

The CN05.1 dataset from the National Climate Center in China with a high spatial resolution (0.25°) for the period of 1984–2019 was used to depict climate regimes and identify rainstorms and droughts, including a daily maximum temperature, a daily minimum temperature, a daily mean temperature, precipitation and wind speed [23,24].

2.2.2. Meteorological Disaster Dataset

The Chinese meteorological disaster dataset with Chinese county units constructed by the National Climate Center based on meteorological disaster data collation standards was used in this study [25]. Two categories of disasters were selected, including rainstorms and floods and mudslides triggered by rainstorms, and droughts. This study mainly focused on the direct economic losses and affected population during the period of 1984–2019. In order to better reflect the situation of the CPSAs, we compared it with the national situation when describing the socioeconomic impacts of rainstorms and droughts. The socioeconomic impacts in CPSAs and China were averaged using the direct economic losses and affected population in the given regions, respectively.

2.2.3. Socio-Economic Data

The GDP and the population of China with a spatial resolution of 1 km were taken from the Geographic Science and Resource Environment Data Center of the Chinese Academy of Sciences [26,27]. The dataset covers the periods of 1990, 1995, 2000, 2005 and 2010. The data of the remaining years within the data range were obtained via cubic spline interpolation to ensure data continuity. Cubic spline interpolation was used to divide the data into several segments. Each segment constructed a cubic function, and each segment function smoothly connected with another [28]. Finally, the linear trend extrapolation method was used to further extrapolate the post-2010 values to 2019 [29]. The CPSAs and national values were calculated as the mean value of the grid points in the regions in which they are located.

2.3. Methods

2.3.1. Definitions of Rainstorms and Droughts

Corresponding to the two categories of disaster from the Chinese meteorological disaster database, the following definitions were used to identify the characteristics of rainstorms and droughts. Since all the socio-economic impact of rainstorm, floods and mudslides were triggered by rainstorms, we mainly focused on the characteristics of rainstorms in this study. The number of rainstorm days was defined as the number of days with a daily precipitation exceeding 50 mm, according to the China Meteorological Administration, which has been widely used in previous studies [30,31]. According to the revised national meteorological drought grade standard, the daily meteorological drought comprehensive index (MCI) was calculated by using the historical daily rainfall, the average temperature, the maximum temperature, the minimum temperature and wind speed [32]. MCI was widely used for monitoring drought and is preferable to other indices in terms of effect and monitoring capacity [33,34]. The number of drought days was defined as the number of days of medium drought conditions or above (MCI ≤ -1.5) [35,36].

2.3.2. Consumer Price Index (CPI) Standardization

The CPI reflects the movement of prices in economic operations and constitutes an important indicator of the degree of inflation [37]. To eliminate the influence of inflation and ensure comparable direct economic losses due to rainstorms and droughts over time, this study selected 2019 as the base year and converted and standardized the direct economic losses data for other years by the 2019 market value [38].

2.3.3. Assessment of the Disaster Resistance Capability

The disaster resistance capability evaluation index used in this study was the loss-to-GDP rate (the loss-to-GDP rate is the direct economic losses caused by disasters in a certain area compared to the GDP of that area in a given year, referred to as LGR) [39]. The lower the percentage of the direct economic losses in a given region is, the higher its disaster resistance capability.

2.3.4. Linear Trend Detection

In this study, the Mann-Kendall (M-K) nonparametric test method was used to assess the trend of elements, which can reduce the influence of outliers and missing measurements on trend estimation [40,41].

3. Spatiotemporal Characteristics of Rainstorms and Droughts

3.1. Temporal Characteristics

The average number of rainstorm days in CPSAs from 1984 to 2019 was 1.9 days/year, ranging from 1.5 days (1997) to 2.3 days (2016) (Figure 3a). The average number of drought days in CPSAs reached 44.6 days/year, ranging from 21.8 days (1990) to 67.3 days (2011), (Figure 3b). The rainstorm days showed a significant increasing trend of 0.075 days/decade, while there was no significant trend in drought days (Figure 3a,b).

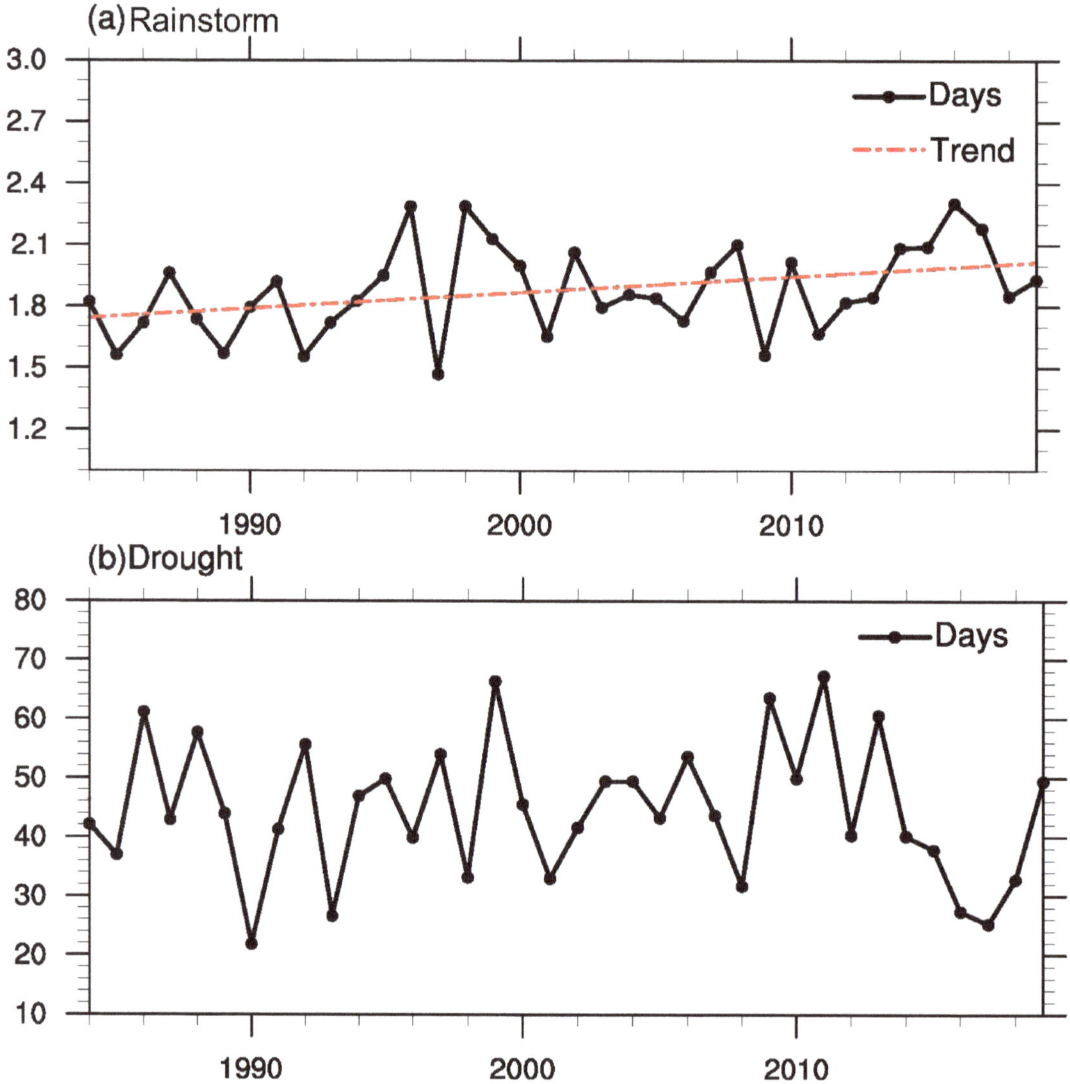

Figure 3. Time series and trend of accumulated (**a**) rainstorm days (days/year) (daily precipitation $\geq$ 50 mm) and (**b**) droughts days (MCI ≤ -1.5) (days/year) averaged over the CPSAs during 1984–2019.

3.2. Spatial Characteristics

The spatial distribution of rainstorm and drought days in the CPSAs considered revealed distinct characteristics on both sides of the PDL. From 1984 to 2019, the number of rainstorm days in CPSAs gradually decreased from southeast to northwest, with multi-year average values of 3.1 and 0.3 rainstorm days in the regions to the east of the PDL and the regions along and to the west of the PDL, respectively (Figure 4a). The number of drought days in CPSAs also gradually decreased from southeast to northwest, and the multi-year average number of drought days in the regions along and to the east of the PDL and to the west of the PDL reached 50.2 and 25.2 days, respectively (Figure 4b). It could be observed that the number of rainstorm and drought days in the regions to the east of the

PDL decreased from east to west, and only the drought conditions in the regions along the PDL were very severe, while the number of rainstorm and drought days in the regions to the west of the PDL was relatively low.

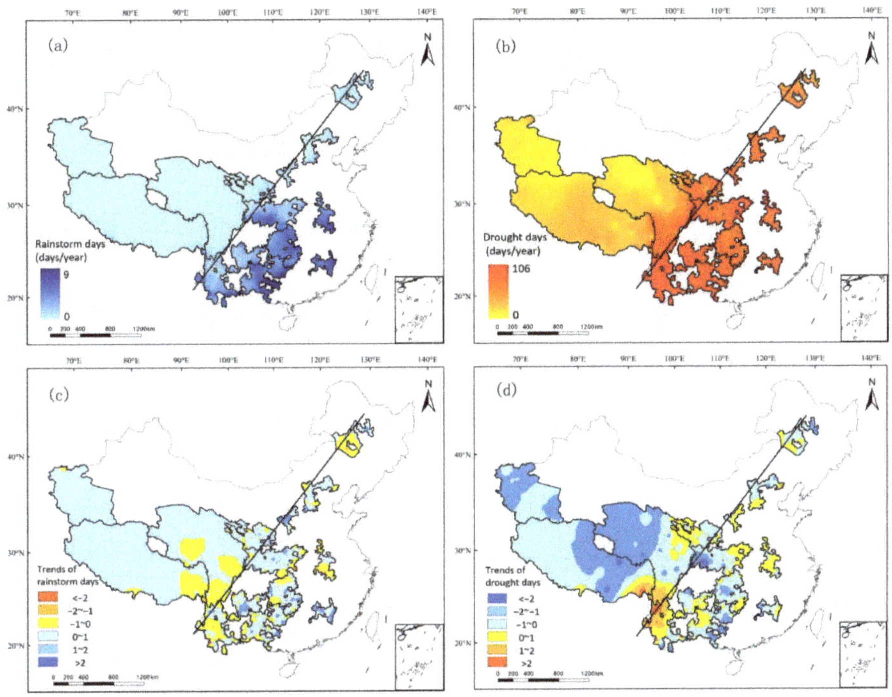

Figure 4. Spatial distribution of (**a**,**b**) annual average (days/year) and (**c**,**d**) linear trends (days/year) of (**a**,**c**) rainstorm days and (**b**,**d**) droughts days in CPSAs during 1984–2019.

During 1984–2019, the number of rainstorm days in the Southern area of Daxing'anling Mountains (R1), the Dabie Mountain area (R6), the West Yunnan Border Area (R9) and the Tibetan areas in the four provinces (R14) showed decreasing trends, while the increasing trends of rainstorm days in other regions of CPSAs were not statistically significant (Figure 4c). The number of drought days increased in certain areas of the Dabie Mountain area (R6), the Lvliang Mountain area (R3), the West Yunnan Border Area (R9), the southern Tibetan areas in four provinces (R14) and the eastern Tibet region (R13). In contrast, the drought days in other regions, especially in the three prefectures of southern Xinjiang (R12), the Tibet region (R13) and the Tibetan areas in four provinces (R14), showed a significant decreasing trend (Figure 4d).

4. Socioeconomic Impacts of Rainstorms and Droughts

In order to illustrate the affected population and direct economic losses in CPSAs, comparisons were made between the CPSAs and the whole nation.

4.1. Affected Population

In terms of the total affected population, the average population affected by rainstorms and droughts in CPSAs from 1984–2019 reached 34 million people/year, accounting for 22.9% of the national population affected by rainstorms and droughts, which is higher than the national share of the population in this region (16.7%) (Figure 5a). From 1984 to 2019, both CPSAs and China's disaster-affected population increased before 2000 and then decreased. The maximum occurred in 2010, reaching 90 million and 326 million, respectively. Moreover, to identify the affected population, the ratio of the average population affected by rainstorms and droughts to total population in CPSAs and China were 16% and 11.7%, respectively. By contrast, the ratio in CPSAs was 1.6% and 6.8% higher than that in China during 1984–1999 and 2000–2019, respectively. Notably, the affected population in CPSAs increased after 2000.

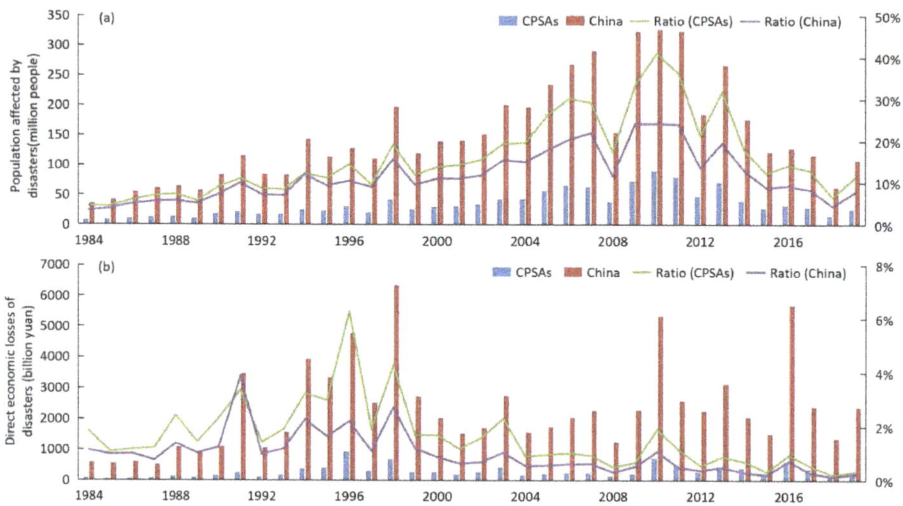

Figure 5. (**a**) Rainstorm and drought-affected population (million people) and its proportion in total population (%), and (**b**) direct economic losses caused by rainstorms and droughts (billion yuan) and its proportion in GDP (%) in CPSAs and China during 1984–2019.

As shown in Figure 6, the impact of rainstorms was more serious than droughts. Among the fourteen regions in CPSAs, the Wuling Mountain area (R7) exhibited maximum occurrence of 4.804 million people/year and 2.773 million people/year affected by rainstorms and droughts, respectively. By contrast, the Tibet region (R13) exhibited a minimum occurrence of 0.087 million people/year and 0.048 million people/year, respectively. From 1984 to 2019, the multi-year average of rainstorm and drought-affected population in CPSAs showed different characteristics between the eastern and western sides of the PDL (Figure 6a,b). Divided by PDL, the number of people affected by rainstorms and droughts in the regions to the east of the PDL reached 16.4 million people/year and 12.4 million people/year, which are 8.8 and 3.3 times higher than the population in the regions along and to the west of the PDL, respectively. In particular, the impact of rainstorms and droughts on the population in the regions to the east of the PDL was particularly severe.

4.2. Direct Economic Losses

In terms of total disaster losses, the average direct economic losses resulting from rainstorms and droughts in CPSAs reached 29 billion yuan/year from 1984–2019, accounting for 12.6% of the total national disaster losses, which is higher than the national share of the GDP of the region (7.7%) (Figure 5b). The direct economic losses due to rainstorms and droughts in CPSAs and China reached a peak of 93.1 billion yuan and 633.9 billion yuan in 1996 and 1998, respectively. The disaster-related losses in 1991, 1994, 1996, 1998, 2010, 2013 and 2016 were relatively high. To identify the extent of economic losses, the ratio of the mean direct economic losses due to rainstorms and droughts to GDP in CPSAs and China were 1.6% and 1%, respectively. Except for 1991, the ratio of disaster losses to GDP in CPSAs was higher than that of in China.

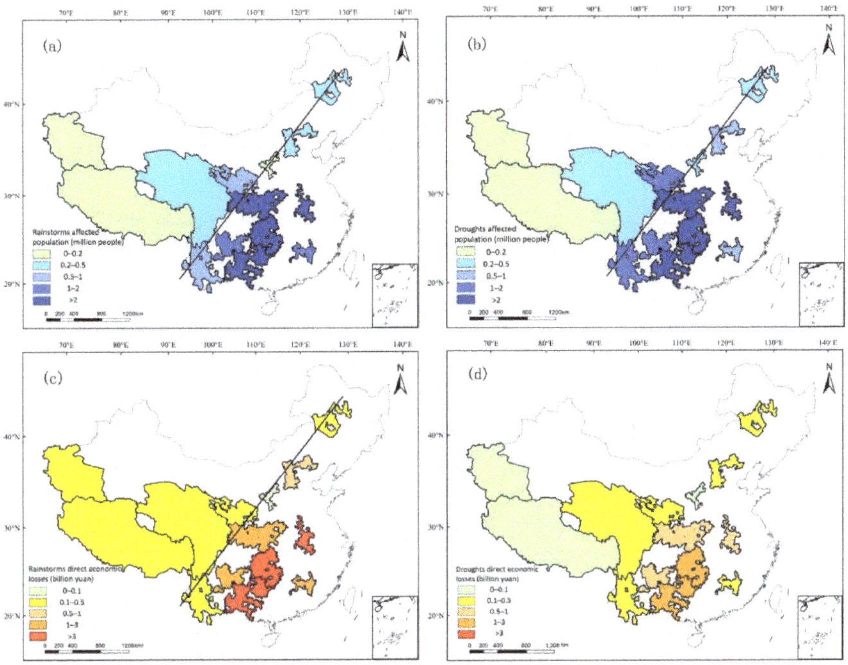

Figure 6. Spatial distribution of (**a**,**b**) disasters-affected population (million people) and (**c**,**d**) disasters-caused direct economic losses (billion yuan) due to (**a**,**c**) rainstorms and (**b**,**d**) droughts in CPSAs during 1984–2019.

From 1984 to 2019, the multi-year average direct economic losses resulting from rainstorms in CPSAs also showed significantly distinct characteristics between the eastern and western sides of the PDL. The economic losses caused by rainstorms were higher than those caused by droughts (Figure 6c,d). The region hit hardest by economic losses resulting from rainstorms and droughts was the Wuling Mountain area (R7), with direct economic losses of 7.103 billion yuan/year and 1.445 billion yuan/year resulting from rainstorms and droughts, respectively. The Lvliang Mountain area (R3) and Tibet region (R13) attained the lowest losses resulting from rainstorms and droughts, with losses of 0.098 billion yuan/year and 0.023 billion yuan/year, respectively. Divided by PDL, the direct economic losses due to rainstorms and droughts in the regions to the east of the PDL reached 20.3 billion yuan/year and 5.7 billion yuan/year, which are 9.2 and 7.4 times higher than those in the regions along and to the west of the PDL. Notably, the disaster-related losses in the regions to the east of the PDL were very severe.

As shown in Figure 7a, the trend of the total economic losses attributed to rainstorms and droughts from 1984 to 2019 increased, except in the Dabie Mountain area (R6), which showed a decreasing trend. All 13 other regions showed increasing trends, especially the regions along and to the west of the PDL, where the disaster-related losses increased faster. Among these areas, the Dabie Mountain area (R6), the Wuling Mountain area (R7) and the Luoxiao Mountain area (R11) showed statistically insignificant trends, while the increasing trends in the remaining 11 regions were significant at a confidence level of 95%.

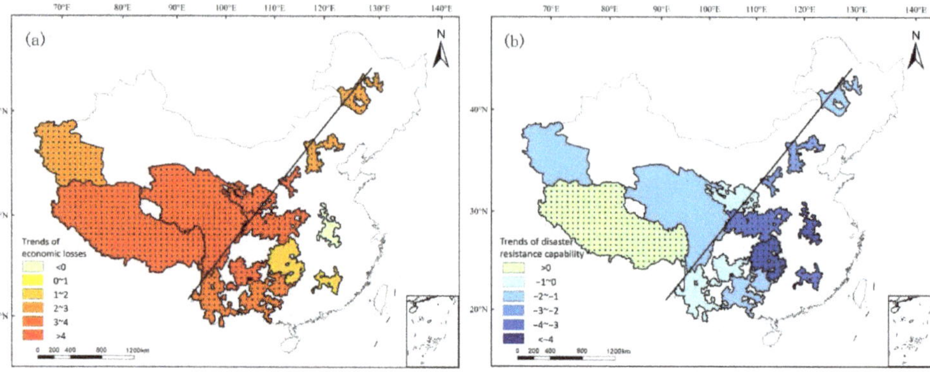

Figure 7. Spatial distribution of linear trends of (**a**) direct economic losses caused by rainstorms and droughts and (**b**) disaster resistance capability over CPSAs during 1984–2019. Black dots indicate the results are significant at a confidence level of 95%.

4.3. Disaster Resistance Capability

To identify the disaster resistance capability in each CPSA, average values of the LGR from 1984 to 2019 were calculated in this study (Figure 8). As shown by the solid line in Figure 8, the annual mean LGR in CPSAs is 1.1%. The LGR in the areas to the east of the PDL was generally higher than the average and the disaster resistance capabilities were relatively low. Among these regions, the disaster resistance capabilities in the Wuling Mountain area (R7) and the Luoxiao Mountain area (R11) were the lowest, with the LGR reaching 3% and 2.9%, respectively. The disaster resistance capabilities in the regions along and to the west of the PDL were relatively high.

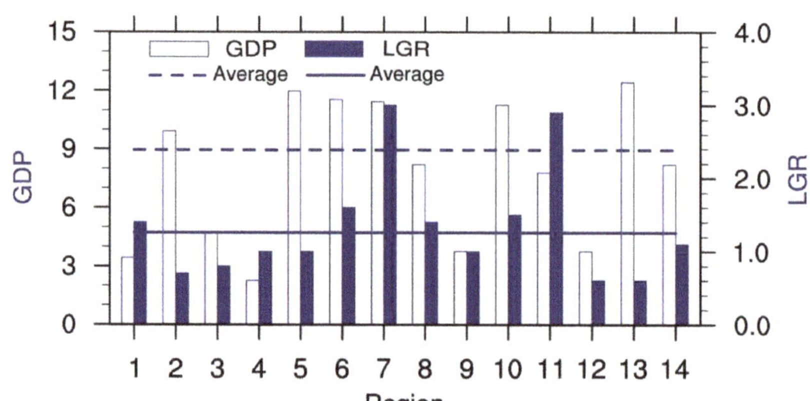

Figure 8. The average GDP per capita (yuan) and loss-to-GDP rate (LGR) (%) in CPSAs during 1984–2019. The serial numbers of CPSAs are the same as Figure 1. (1. Southern area of Daxing'anling Mountains; 2. Yanshan-Taihang Mountain Area; 3. Lvliang Mountain Area; 4. Liupanshan Area; 5. Qinba Mountain Area; 6. Dabie Mountain Area; 7. Wuling Mountain Area; 8. Wumeng Mountain Area; 9. West Yunnan Border Area; 10. Yunnan, Guangxi and Guizhou rocky desertification area; 11. Luoxiao Mountain Area; 12. Three prefectures of Southern Xinjiang; 13. Tibet Region; 14. Tibetan areas in four provinces).

The disaster resistance capabilities in CPSAs increased during 1984–2019, except for the Tibet region (R13) (Figure 7b). Among fourteen regions in CPSAs, the trends in the Yanshan-Taihang Mountain area (R2), the Lvliang Mountain area (R3), the Qinba Mountain area (R5), the Dabie Mountain area (R6), the Wuling Mountain area (R7) and the Luoxiao Mountain area (R11) are significant at a confidence level of 95%. In the regions to the east of the PDL, for example the Dabie Mountain area (R6) and the Wuling Mountain area (R7), the resistance capabilities increased faster than other regions.

5. Summary and Discussion

5.1. Summary

The population, GDP and GDP per capita in CPSAs reached 210 million people, 1989.3 billion yuan, and 8943 yuan per person, accounting for 16.7%, 7.7%, and 46.9% of China's total amounts, respectively. To consolidate the achievements in the elimination of absolute poverty in China and prevent rural populations from returning to poverty due to meteorological disasters, this study analyzed the spatiotemporal characteristics of rainstorms and droughts and their socioeconomic impacts in China's CPSAs from 1984 to 2019.

The annual average rainstorm and drought days in CPSAs were approximately 1.9 days/year and 44.6 days/year, respectively. It gradually decreased from southeast to northwest. The number of rainstorm days showed a significant increasing trend of 0.075 days/decade, while the decreasing trend of the number of drought days was not significant. Due to rainstorms and droughts, the average annual affected population and direct economic losses in CPSAs reached 34 million people/year and 29 billion yuan/year, accounting for 22.9% and 12.6% of China's total amounts, respectively. The average annual loss rate due to disasters in this region reached 1.6%, which is 0.6% higher than the national level.

Furthermore, distinct features and socioeconomic impacts of rainstorms and droughts were identified on the southeastern and northwestern sides of the PDL along Tengchong-Aihui in China. Droughts often impacted the regions located along the PDL, while rainstorms and droughts occurred more frequently in the regions to the southeast of the PDL than in the regions to the northwest of the PDL. As a result, the affected number of people and direct economic losses due to rainstorms and droughts in the regions to the southeast of the PDL reached 16.4 million people/year and 20.3 billion yuan/year, and 12.4 million people/year and 5.7 billion yuan/year, which were 8.8, 9.2, 3.3 and 7.4 times higher than those in the regions to the northwest of the PDL. Although there were more affected people and higher direct economic losses due to rainstorms and droughts in the regions to the east of the PDL, the disaster resistance capabilities were significantly improved. In contrast, the regions to the northwest side of the PDL showed a significant increasing trend of losses with relatively low disaster resistance capabilities.

5.2. Discussion

From the perspective of rainstorms and droughts, drought conditions were more severe in regions along the PDL, and both rainstorms and droughts were relatively severe in the regions to the east of the PDL. In addition, the number of rainstorm days significantly increased in the entire study area. Rainstorm-induced disasters became increasingly serious, and only the number of rainstorm days in the Southern area of Daxing'anling Mountains (R1), the Dabie Mountain area (R6), the West Yunnan Border Area (R9) and the southern part of the Tibetan areas in four provinces (R14) showed a decreasing trend. The number of drought days decreased in the entire study area, especially the drought conditions in the three prefectures of southern Xinjiang (R12), the Tibet region (R13) and the Tibetan areas in four provinces (R14), whereas the drought conditions in the West Yunnan Border Area (R9) and the southern part of the Tibetan areas in four provinces (R14) and the number of drought days in the West Yunnan Border Area (R9), the southern Tibetan areas in four provinces (R14) and the eastern Tibet region (R13) notably increased.

According to a comprehensive analysis of the direct economic losses, economic level and disaster resistance capabilities in CPSAs, the disaster-related losses significantly increased, and at a higher rate than that of the regions to the east of the PDL, while the disaster resistance capabilities did not increase significantly. The situation in the regions to the east of the PDL was the opposite, with high disaster losses but at a lower rate, and the disaster resistance capabilities were improving. The disaster resistance capabilities in the Lvliang Mountain area (R3), the Liupan Mountain region and the three prefectures of southern Xinjiang (R12) were satisfactory, although the GDP per capita was slightly lower than that in the other regions. It is worth noting that the disaster resistance capabilities in the Yanshan-Taihang Mountain area (R2), the Wuling Mountain area (R7), Yunnan, the Guangxi and Guizhou rocky desertification area (R10) and the Tibet region (R13) were relatively low. The GDP per capita in these 4 regions is higher than that of other regions, their cost in disaster prevention and reduction might be relatively low (Figure 8). These regions should enhance the disaster prevention consciousness and increase investment in disaster reduction and prevention facilities. Meanwhile, the number of affected population due to rainstorms and droughts in CPSAs showed a remarkable increase after 2000 (Figure 5a). This might be ascribed to the increase in rainstorm days and population exposed to the disasters since the population in CPSAs increased after 2000, with a growth of 39.7% relative to 1984–1999 (Figure 2a). Therefore, the government should take the occurrence of rainstorms and droughts into consideration during urban planning and land use planning in CPSAs to reduce the exposure and risk of population and economy to disasters.

For a long time, the international community has paid great attention to the relationship between poverty and natural disasters. For example, the reduction of rainfall has slowed the economic growth of sub-Saharan Africa [42]. Climate change has plunged 6.5% of the population of Latin America into poverty and vulnerability [43]. Previous studies have reported that disasters will aggravate poverty and recovery is not straightforward for poor people [44]. Under global warming, the frequency and intensity of rainstorms and droughts might increase significantly over CPSAs [6,7]. Correspondingly, the average annual direct economic losses due to rainstorms is expected to be 4 times and 17 times higher than it is currently under global warming of 1.5 °C and 4.0 °C [45]. As the relatively poor area in China, it is necessary to pay close attention to the socio-economic impact and risks of rainstorms and droughts over CPSAs in the future. Hence, the above-mentioned regions with high GDP per capita but low disaster resistance capabilities should strengthen their awareness of disaster prevention and mitigation, enhance early warnings and disaster relief inputs in response to meteorological disasters, and improve the disaster resistance capabilities to meet climate change challenges. This is an important part of consolidating China's gains in poverty eradication and further contributing to the realization of SDGs.

Author Contributions: Conceptualization, M.G., T.J. and J.Y.; Data curation, A.L. and H.Z.; Methodology, A.L. and S.G.; Supervision, M.G., T.J. and J.Y.; Validation, A.L. and X.W.; Visualization, A.L. and S.G.; Writing—original draft, A.L. and M.G.; Writing—review & editing, A.L., X.W. and M.G. All authors have read and agreed to the published version of the manuscript.

Funding: This study was funded by the National Natural Science Foundation of China (grant no.42071024, 42105158 and 42005126), the project supported by the State Key Laboratory of Earth Surface Processes and Resource Ecology, and the Startup Foundation for Introducing Talent of NUIST.

Institutional Review Board Statement: Not applicable.

Informed Consent Statement: Not applicable.

Data Availability Statement: Not applicable.

Acknowledgments: The authors would like to thank Ke Tan of Nanjing University of Information Science and Technology for the support in data.

Conflicts of Interest: The authors declare that they do not have any conflict of interest.

References

1. Kim, J.Y. *Ending Extreme Poverty by 2030*; World Bank: Washington, DC, USA, 2015.
2. IDM. Post-poverty Reduction Era and Modernization of Poverty Reduction Governance. *Leadersh. Decis. Inf.* **2020**, *25*, 2.
3. Guo, Y. Analysis of the impact of meteorological disasters in China's special types of poverty-stricken areas. *J. Environ. Dev.* **2019**, *31*, 151–152. [CrossRef]
4. Yang, D. Impact of meteorological disasters on poverty in special types of areas in China. *Abil. Wisdom* **2018**, *18*, 239.
5. China Meteorological Administration. *China Meteorological Disaster Yearbook*; China Meteorological Press: Beijing, China, 2017.
6. Wang, X.; Li, A.; Shao, X.; Gao, S.; Lin, B.; Yang-chen, X.; Jiang, T. Projected drought risk in poverty-stricken areas of China with CMIP6 models under SSPs-RCPs scenarios at 1.5 °C and 2 °C warmer levels. *J. Nanjing Univ. Inf. Sci. Technol.* **2022**, *27*, 123–133.
7. Gao, S.; Li, A.; Huang, J.; Wang, X.; Lin, B.; Yang-chen, X.; Wang, L.; Jiang, T. Projected Changes of Extreme Precipitation in Rural Revitalization Areas in China under 1.5 °C and 2.0 °C Global Warming Scenarios. *J. Clim. Environ. Res.* **2022**, *27*, 123–133.
8. Hallegatte, S.; Bangalore, M.; Bonzanigo, L.; Fay, M.; Kane, T.; Narloch, U.; Rozenberg, J.; Treguer, D.; Vogt-Schilb, A. *SHOCK WAVES: Managing the Impacts of Climate Change on Poverty*; Climate Change and Development Series; World Bank: Washington, DC, USA, 2016.
9. Lin, L. Studies on the Impact of Main Meteorological Disaster in China within Recent 30 Years. Master's Thesis, The University of Lanzhou, Lanzhou, China, 2013.
10. Ye, M.; Qian, Z.; Wu, Y. Spatiotemporal Evolution of the Droughts and Floods over China. *Wuli Xuebao/Acta Phys. Sin.* **2013**, *62*, 139203. [CrossRef]
11. Cui, M.; Chen, Y.; Ge, G. Analysis of China Meteorological Disasters Loss in Recent Years Based on Synthesis Index. In Proceedings of the IEEE 2009 First International Conference on Information Science and Engineering, Nanjing, China, 26–28 December 2009.
12. Chen, Y.; Li, J.; Chen, A. Does high risk mean high loss: Evidence from flood disaster in southern China. *Sci. Total Environ.* **2021**, *785*, 147127. [CrossRef]
13. MUNICH RE. Munich Re NatCatSERVICE. Available online: http://www.munichre.com/en/reinsurance/business/non-life/natcatservice/index.html (accessed on 26 January 2021).
14. CRED, Centre for Research on the Epidemiology of Disasters. EM-DAT: The OFDA/CRED International Disaster Database. Available online: http://www.emdat.be/database (accessed on 26 January 2021).
15. China Meteorological Science Data Sharing Service. Available online: http://www.cma.gov.cn/2011qxfw/2011qsjgx/ (accessed on 2 February 2021).
16. China's Natural Disaster Database [DB/OL]. Available online: http://www.data.ac.cn/list/tab_nature_disaster (accessed on 2 February 2021).
17. Hu, L.; Liu, Y.; Ren, Y.; Yu, L.; Qu, C. Spatial change of population density boundary in mainland China in recent 80 years. *J. Natl. Remote Sens. Bull.* **2015**, *19*, 928–934.
18. Chen, M.; Li, Y.; Gong, Y.; Lu, D.; Zhang, H. The Population Distribution and Trend of Urbanization Pattern on Two Sides of Hu Huanyong Population Line: A Tentative Response to Premier Li Keqiang. *Acta Geogr. Sin.* **2016**, *71*, 179–193. [CrossRef]
19. Huang, Y.; Yang, B. A Study of Environmental Determinism from the Perspective of "Hu Huanyong Line". *J. Yunnan Norm. Univ.* **2012**, *44*, 68–73.
20. Jing, C. Assessment of the Drought Effect in Contiguous Extreme Poverty Areas of China. Master's Thesis, Nanjing University of Information Science & Technology, Nanjing, China, 2017.
21. Teng, Y.; Zhou, F. Reread "Hu Huanyong line". *N. China's Land Resour.* **2012**, *5*. [CrossRef]
22. Wang, B.; Gao, F.; Li, H. Spatial characteristics and poverty mechanism of contiguous poverty-stricken areas in China. *J. Res. Dev.* **2016**, *6*, 59–64.
23. Xu, Y.; Gao, X.; Shen, Y.; Xu, C.; Shi, Y.; Giorgi, F. A daily temperature dataset over China and its application in validating a RCM simulation. *Adv. Atmos. Sci.* **2009**, *26*, 763–772. [CrossRef]
24. Wu, J.; Gao, X. A gridded daily observation dataset over China region and comparison with the other datasets. *J. Chin. J. Geophys. Chin. Ed.* **2013**, *56*, 1102–1111.
25. Jiang, T.; Tan, K.; Wang, Y.; Zhai, J. Spatial-temporal variation of meteorological disasters in the "Belt and Road" regions. *J. Sci. Technol. Rev.* **2020**, *38*, 59–67.
26. Huang, Y.; Jiang, D.; Fu, J. The GDP Distribution Data Set of China's KM Grid. *Acta Geogr. Sin.* **2014**, *69*, 140–143. [CrossRef]
27. Fu, J.; Jiang, D.; Huang, Y. A Data Set of Population Distribution in China's KM Grid *Acta Geogr. Sin.* **2014**, *69*, 136–139. [CrossRef]
28. McKinley, S.; Levine, M. Cubic spline interpolation. *J. Coll. Redw.* **1998**, *45*, 1049–1060.
29. Jing, C.; Su, B.; Zhai, J.; Wang, Y.; Lin, Q.; Gao, M.; Jiang, S.; Chen, Z.; Jiang, T. Gridded Value-Added of Primary, Secondary and Tertiary Industries in China under Shard Socioeconomic Pathways. *J. Sci. Data.* **2022**, *9*, 309. [CrossRef]
30. Gemmer, M.; Fischer, T.; Jiang, T.; Su, B.; Liu, L.L. Trends in Precipitation Extremes in the Zhujiang River Basin, South China. *J. Clim.* **2011**, *24*, 750–761. [CrossRef]
31. Zhao, Y.; Xu, X.; Zhao, T.; Xu, H.; Mao, F.; Sun, H.; Wang, Y. Extreme precipitation events in East China and associated moisture transport pathways. *Sci. China Earth Sci.* **2016**, *59*, 1854–1872. [CrossRef]
32. Liao, Y.; Zhang, C. Spatio-Temporal Distribution Characteristics and Disaster Change of Drought in China Based on Meteorological Drought Composite Index. *J. Meteorol. Mon.* **2017**, *43*, 8.

33. Han, L.; Zhang, Q.; Zhang, Z.; Jia, J.; Wang, Y.; Huang, T.; Cheng, Y. Drought Area, Intensity and Frequency Changes in China under Climate Warming, 1961–2014. *J. Arid. Environ.* **2021**, *193*, 104596. [CrossRef]
34. Morid, S.; Smakhtin, V.; Moghaddasi, M. Comparison of seven meteorological indices for drought monitoring in Iran. *Int. J. Climatol.* **2010**, *26*, 971–985. [CrossRef]
35. Gao, R.; Wang, S.; Gao, N.; Zuo, H. Application Comparison of CI and MCI Drought Indexes in Ningxia. *J. Arid. Meteorol.* **2021**, *39*, 185–192.
36. Yu, R.; Zhai, P. Changes in compound drought and hot extreme events in summer over populated eastern China. *Weather. Clim. Extrem.* **2020**, *30*, 100295. [CrossRef]
37. Zhang, H.; Zhang, D.; Yao, Y.; Zhang, P. Monetary Policy Transparency and Disinflation: An Empirical Evidence from China. *Econ. Res. J.* **2009**, *44*, 55–64.
38. Wen, S.; Wang, Y.; Su, B. Estimation of economic losses from tropical cyclones in China at 1.5 °C and 2.0 °C warming using the regional climate model COSMO-CLM. *Int. J. Climatol.* **2018**, *39*, 724–737. [CrossRef]
39. Research Group of Research Institute of National Conditions of Tsinghua University; Hu, A. *Main Ideas and Key Tasks of Disaster Prevention and Mitigation Construction during the "13th Five-Year" Period*; National Development and Reform Commission: Beijing, China, 2015; Volume 18.
40. Wen, S.; Zhai, J.; Thomas, F.; Su, B.; Jiang, T.; Wang, Y. Variation of normalized economic losses from influential tropical cyclones in China for 1984–2014. *J. Trop. Meteorol.* **2017**, *33*, 478–487. [CrossRef]
41. Kendall, M.; Gibbons, J. *Rank Correlation Methods*; Oxford Univerisy Press: New York, NY, USA, 1990; pp. 1–260.
42. Brown, C.; Meeks, R.; Hunu, K.; Yu, W. Hydroclimate risk to economic growth in sub-Saharan Africa. *Clim. Chang.* **2011**, *106*, 621–647. [CrossRef]
43. Kronik, J.; Verner, D. *Indigenous Peoples and Climate Change in Latin America and the Caribbean*; Directions in Development Series; World Bank: Washington, DC, USA, 2010.
44. Karim, A.; Bollino, C.A. Poverty and Natural Disasters: A Meta-Regression Analysis. *J. Rev. Econ. Inst.* **2016**, *7*. [CrossRef]
45. Jiang, T.; Su, B.; Huang, J.; Zhai, J.; Zbigniew, W. Each 0.5 °C of warming increases annual flood losses in China by more than 60 billion USD. *Bull. Am. Meteorol. Soc.* **2020**, *101*, E1464–E1474. [CrossRef]

sustainability

MDPI

Article

Fresh Air–Natural Microclimate Comfort Index: A New Tourism Climate Index Applied in Chinese Scenic Spots

Xiaoyan Yang [1], Changshun Li [1,2,*], Muhammad Bilal [3] and Shaofei Jin [1,4,*]

1 Economics and Management College, Minjiang University, Fuzhou 350108, China; Xiaoyan.Yang@mju.edu.cn
2 Fujian Meteorological Service Center, Fuzhou 350001, China
3 School of Economics and Management, Beijing University of Posts and Telecommunications, Beijing 100876, China; bilal_ashraf30@hotmail.com
4 Department of Geography, Minjiang University, Fuzhou 350108, China
* Correspondence: 2766@mju.edu.cn (C.L.); jinsf@tea.ac.cn (S.J.); Tel.: +86-188-5010-2506 (S.J.)

Abstract: Severe air pollution in China has caused significant tourism transformation for pursuing fresh air in microclimate tourism markets. Contemporary practices simply measure the air freshness of destinations and scenic spots using a single index, i.e., primarily negative oxygen ions (O_2^-). This index cannot comprehensively reveal scenic spots' air freshness degree and determine the dynamic interactions between air freshness and scenic spots' tourism development, thus inducing an illusion of air freshness for the target scenic spots. Meanwhile, the current fresh air index primarily ignores connections with the microclimate index of scenic spots and cannot provide a multidimensional index for scenic spots to take advantage of both air and microclimate resources for diverse tourism products and service production. Therefore, this study proposes a multidimensional index, the fresh air–natural microclimate comfort index (FAI-NMCI), connecting the fresh air index with the natural microclimate comfort index of scenic spots together from transdisciplinary and multidisciplinary perspectives. This study utilizes FAI-NMCI to measure four scenic spots of Fujian Province, and reveals in-depth results of scenic spots' air freshness and natural microclimate comfort degree together. The results demonstrate that the four scenic spots in Fujian province of China had different levels of air freshness degree and natural microclimate comfort degree in 2018. The natural scenic spots were mostly distributed in Healing Fresh, Very Fresh, and Super Fresh levels of FAI with the most comfortable and comfortable levels of NMCI. The cultural scenic spots were mostly distributed in Relatively Fresh and Healing Fresh levels of FAI with the most comfortable and comfortable levels of NMCI. Meanwhile, the FAI-NMCI of natural and cultural scenic spots also had significant differences within 24 Jieqi, which will promote dynamic and creative utilization of those resources in microclimate tourism development.

Keywords: fresh air index; natural microclimate comfort index; fresh air–natural microclimate comfort index; scenic spots; Fujian province

Citation: Yang, X.; Li, C.; Bilal, M.; Jin, S. Fresh Air–Natural Microclimate Comfort Index: A New Tourism Climate Index Applied in Chinese Scenic Spots. *Sustainability* **2022**, *14*, 1911. https://doi.org/10.3390/su14031911

Academic Editor: Andrei P. Kirilenko

Received: 4 December 2021
Accepted: 30 January 2022
Published: 8 February 2022

1. Introduction

Air pollution influences tourists' and residents' (potential tourists') health, which will directly or indirectly affect demand in the tourism market [1]. Exposure to air-polluted environments leads to various health problems [2], such as respiratory infections [3,4], asthma [5,6], stroke, and even lung cancers [7]. Additionally, air pollution, especially fog and haze pollution, also impacts the psychological health [8,9] and mental disorders [10] of tourists and potential tourists, which directly increases healthcare expenditures [11], health costs [12], and other socioeconomic burdens [13,14] and socioeconomic costs [15] for tourists and local residents. Meanwhile, the factors of natural microclimate and the comfort of the natural microclimate as a whole also influence tourists' decision-making for specific destinations or scenic spots, impacting tourists' behavior and the constant and dynamic tourist flow [16] from one region to another because of dynamic natural microclimate changes [17,18]. Under an air pollution background, the increasing fresh air demand of tourists in China has attracted tourists to destinations with fresh air to experience comfort in natural microclimates. Fresh air and natural microclimate together provide friendly, healthy, and comfort natural tourist attractions in the contemporary Chinese tourism market. Air freshness and microclimate comfort degree together as vital environmental information have significantly influenced tourists, destination and scenic spot management, as well as tourism market development, which has seldom been explored by contemporary research and practice.

Actually, the most popular index for measuring air freshness and air quality is the air quality index (AQI), which utilizes primary air pollutant components such as $PM_{2.5}$, PM_{10}, SO_2, NO_2, CO, or O_3 to measure air quality. AQI studies have developed according to specific research purposes or considering practical utilization; for example, many studies have explored relationships between the AQI and years of life lost [19] and the aggregated effects of internal indicators at a short-term scale [20]. However, AQI reveals the degree of air pollution situations of cities or some air-polluted areas, such as factories or streets, rather than scenic spots. It also neglects the degree of air freshness through adding indicators of O_2^- to measure the health and beneficial degree of fresh air [21]. Moreover, current studies of the fresh air index primarily utilize singular indicators, only O_2^-, to measure air freshness, without considering air pollutants' influence on the comprehensive air freshness degree [22].

For the natural microclimate of scenic spots, many studies primarily utilize one index or two indexes to analyze and reveal the natural microclimate comfort degree, which can mainly be summarized as the temperature-humidity index (THI) [23], wind-cold index (WCI) [24], index of cloth loading (ICI) [25], and human comfort index (HCI) [26]. However, it is vital to note that current indexes do not consider rain factors to measure the influence of rain volume on the comfort degree, which provides an opportunity to establish a rain comfort index and construct a new comprehensive index including all primary factors of the natural microclimate of scenic spots. Additionally, the natural microclimate comfort degree of scenic spots has seldom been integrated with the fresh air index in contemporary studies due to a lack of comprehensive air freshness data and natural microclimate factor data at the same time, along with the existence of some research barriers.

Integration embraces the power to reveal the actual distribution of real scenic spot air freshness and natural microclimate resources. Based on individual economic development trends of different places, it is essential to point out that the air freshness degree of scenic spots needs to be classified according to WHO standards [27] and actual situations together, which calls for stricter standards of fresh air, with gradual improvement in standards. For more in-depth exploration and development of scenic spots with fresh air and natural microclimate tourism resources, it is indispensable to classify different levels or add more actual levels of air freshness degree and natural microclimate degree in future research because the current classification needs to be developed based on more accurate research.

For sustainable development of destinations and scenic spots with fresh air and natural microclimate comfort resources, this paper establishes a fresh air index (FAI) based on comprehensive identification of fresh air components and the index construction methods and experiences of AQI, constructs natural and cultural FAI according to the different threshold standards, and creatively constructs a natural microclimate comfort index of scenic spots through taking advantage of the current index and adding a rain comfort index, connecting the fresh air index (FAI) and the natural microclimate comfort index (NMCI) together to assess the air freshness degree of destinations and the natural microclimate comfort degree. For in-depth air and microclimate resource distributions in the tourism market, this paper proposes an improved fresh air–natural microclimate comfort index to systematically analyze fresh air resources and natural microclimate comfort resources together in Chinese scenic spots by classifying seven levels of the fresh air index (FAI) and seven levels of the natural microclimate comfort index (NMCI). Based on the comparative advantages of embracing relatively fresher air, this paper takes four scenic spots of Fujian Province, China, as an example and analyzes the natural and cultural scenic spots' fresh air–natural microclimate comfort degree to promote the sustainable utilization and development of fresh air and natural microclimate comfort resources in the microclimate tourism market.

2. Materials and Methods

This study utilizes the comprehensive index construction method to construct one fresh air–natural microclimate comfort index (FAI-NMCI), then creatively applies this index to four scenic spots of Fujian Province to reveal the actual air freshness and microclimate comfortable degree at the same time.

2.1. Methodology of Fresh Air–Natural Microclimate Comfort Index Construction and Evaluation

2.1.1. Definition and Essence of Fresh Air

Different from polluted air, fresh air not only contains basic air components and air structures, but also contains more beneficial air components, such as negative oxygen ions (O_2^-), which are also treated as "air vitamins" for promoting human health [28,29] and include no harmful, or less harmful air components, such as $PM_{2.5}$, PM_{10}, SO_2, NO_2, CO, or O_3, that are harmless standards. Therefore, fresh air has beneficial and no harmful characteristics and can be simply defined and summarized as follows:

Fresh air (FA) = beneficial air components (BAC) + no harmful air components (NAC)

According to the aforementioned analysis, fresh air could be summarized as having health essences from protecting and promoting human health perspectives, which could be represented by a higher concentration of healthy air components and a lower concentration of polluted air components.

In this study, negative oxygen ions (O_2^-) were adopted to represent beneficial air components and were selected to reveal general nonharmful air components that severely caused air pollution in China. SO_2, NO_2, CO, or O_3 was chosen to represent local special air contaminants in different regions of China. Therefore, to illustrate fresh air in specific scenic spots of China, this study constructed a fresh air structure with selected air components, which can be summarized as follows:

Fresh Air = Beneficial Air Components (O_2^-) + No Harmful Air Components {$PM_{2.5}$+PM_{10}+(SO_2, NO_2, CO or O_3)}

2.1.2. Improved Fresh Air Index of Natural and Cultural Scenic Spots

We compared the contemporary single-factor fresh air index of Fujian, Hubei, and Zhejiang provinces of China, utilizing only negative oxygen ion (O_2^-) degree and $PM_{2.5}$, to assess scenic spots' air freshness (Table 1). The Meteorological Service Center of Fujian Province, China, and the authors of this paper have creatively proposed a comprehensive fresh air index by considering both the healthy air component factor negative oxygen ion (O_2^-) and the low degrees of $PM_{2.5}$ and PM_{10} as general polluted air components that exist in different regions of China, and NO_2, SO_2, CO, and O_3 as particular air pollutant factors in specific regions of China, determined by local geological and economic structure features.

The subindices of the beneficial air component IFAIB are expressed as the ratio of the negative oxygen ion (O_2^-) concentration to the 24 h average standard O_2^- level (1000 ions/cm^3) (Table 2) recommended by the World Health Organization Air Quality Guidelines for fresh air [30–32]. That is, $IFAI_B$ can be written as:

$$IFAI_B = \frac{C_B}{S_B} \tag{1}$$

where $IFAI_B$ represents the subindices of the beneficial air component in scenic spots, C_B represents the negative oxygen ion (O_2^-) concentration of scenic spots, and S_B represents the 24 h average standard O_2^- level (1000 ions/cm^3) of scenic spots.

Table 1. Comparison of methods and classification of air freshness in different provinces of China.

Measurement Method	Levels	$PM_{2.5}$ Concentration (μg/m³)	O_2^- Concentration (N Ions/cm³)	Level Evaluation
Measurement method of fresh air in Fujian	Level 1	≤35	$1500 \leq N$	Very Fresh
	Level 2	≤35	$1000 \leq N < 1500$	Fresh
	Level 3	≤35	$500 \leq N < 1000$	Relatively Fresh
	Level 4	≤35	$100 \leq N < 500$	General
	Level 5	≤35	$0 < N < 100$	Not Fresh
Measurement method of fresh air in Hubei	Level 1	≤35	$1200 \leq N$	High Concentration, Air Fresh
	Level 2	≤35	$500 \leq N < 1200$	High Concentration, Air Relatively Fresh
	Level 3	≤35	$300 \leq N < 500$	Middle Concentration, Air Generally Fresh
	Level 4	≤35	$100 \leq N < 300$	Relatively Low Concentration, Air Relatively Not Fresh
	Level 5	≤35	$0 < N < 100$	Low Concentration, Air Not Fresh
Measurement method of fresh air in Zhejiang	Level 1	≤35	$0 \leq N < 50$	Not Fresh
	Level 1	≤35	$50 \leq N < 200$	Not Fresh
	Level 1	≤35	$200 \leq N < 500$	Not Fresh
	Level 2	≤35	$500 \leq N < 900$	General
	Level 3	≤35	$900 \leq N < 1200$	Relatively Fresh
	Level 4	≤35	$1200 \leq N < 1800$	Fresh
	Level 5	≤35	$1800 \leq N < 2100$	Fresh
	Level 6	≤35	≥ 2100	Very Fresh

Table 2. Standard concentration degree of indicators of fresh air components.

Dimension	Indicators	Average Time	Concentration Degree		Unit
			I	II	
Beneficial Air Components	Negative Oxygen Ion (O_2^-)	24 h mean	1000	1000	ions/cm³
No Harmful Air Components	Sulfur Dioxide (SO_2)	24 h mean	50	150	μg/m³
	Nitrogen Dioxide (NO_2)	24 h mean	80	80	
	Carbon Monoxide (CO)	24 h mean	4	4	mg/m³
	Ozone (O_3)	day maximum 8 h mean	100	160	μg/m³
	Particulate Matter 10 (PM_{10})	24 h mean	50	150	
	Particulate Matter 2.5 ($PM_{2.5}$)	24 h mean	35	75	

Notes: Data source from Ambient Air Quality Standards of China (GB3095-2012). The category I region used I concentration degree, and the category II region used II concentration degree. The category I region is the place of natural conservation, famous scenic spots, and other special conservation regions that need special protection. The category II region is the residential regions, the mixed region with commercial, transportation, and residents, the cultural regions, industrial regions, and rural regions.

The subindex $IFAI_{pi}$ of nonharmful air components is expressed as the negative ratio of air pollutant concentrations, C_{pi}, to the recommended short-term Ambient Air Quality Standards of China (GB3095-2012) [33], S_{pi}. That is, $IFAI_{pi}$ can be written as:

$$IFAI_{pi} = -\frac{C_{pi}}{S_{pi}} \tag{2}$$

where $IFAI_{pi}$ represents the subindices of no harmful air components in scenic spots, C_{pi} represents air pollutant concentrations of scenic spots, and S_{pi} represents the recommended short-term air pollutant concentration standards of scenic spots.

The short-term Ambient Air Quality Standards of China [34] used in natural scenic spots are defined as concentrations with average times of 24 h means of 50 μg/m³ for PM_{10},

35 μg/m^3 for $PM_{2.5}$, 50 μg/m^3 for SO_2, 80 μg/m^3 for NO_2, and 4 mg/m^3 for CO, and running an 8 h mean of 100 μg/m^3 for O_3. The short-term Ambient Air Quality Standards of China used in cultural scenic spots are defined as concentrations with average times of 24 h means of 150 μg/m^3 for PM_{10}, 75 μg/m^3 for $PM_{2.5}$, 150 μg/m^3 for SO_2, 80 μg/m^3 for NO_2, and 4 mg/m^3 for CO, and running an 8 h mean of 160 μg/m^3 for O_3. The 24 h average standard $O_2{}^-$ level recommended by the World Health Organization Air Quality Guidelines for fresh air is 1000 ions/cm^3 in category I and II regions (Table 2).

The natural and cultural fresh air index is the sum of IFAIB and the sum of subindices of no harmful air components, $IFAI_{Pi}$, which can be written as:

$$FAI = IFAI_B + \sum IFAI_{pi} = \sum IFAI_j \tag{3}$$

where FAI refers to the fresh air index, $IFAI_B$ refers to subindices of beneficial air components, $\sum IFAI_{pi}$ indicates the sum of subindices of no harmful air components, and $\sum IFAI_j$ indicates the sum of subindices of fresh air components j.

The classifications of natural and cultural FAIs of scenic spots and their levels were explored in this paper. According to Table 1, the FAI of natural scenic spots utilizes category I region standards to further explore different threshold levels, while the FAI of cultural scenic spots utilizes category II region standards (Table 2).

According to the China Industrial Technical Regulation on Grade of Air Negative (Oxygen) Ion Concentration (QX/T 380-2017) [32] and Local Technical Regulation on Grade of Air Negative (Oxygen) Ion Concentration of Hubei Province (DB42/T 1198-2016) [30], the concentration degree of negative oxygen ion ($O_2{}^-$) increase per 500 ions/cm^3 will enhance air freshness and have a beneficial influence on human health.

According to the World Health Organization (WHO) air freshness grading standard of the negative oxygen ion index, when $1800 \leq O_2{}^- \leq 2100$, $O_2{}^-$ has natural healing power, and when $O_2{}^- > 2100$, it has therapeutic and rehabilitation effects [35]. Therefore, on the basis of a higher standard than the WHO standard for negative oxygen ion therapy, this paper proposes that when $2500 \leq O_2{}^- \leq 3000$, $PM_{2.5}$, PM_{10}, O_3, NO_2, SO_2, and CO decreased in proportion, and the air freshness degree is at the Healing Freshness Level, the resulting air freshness level of the scenic spot reaches the healing standard. Then, when $3000 \leq O_2{}^- \leq 3500$ and when $O_2{}^- > 3500$, the air could have a more active influence on human health, which could further promote the air freshness degree to very fresh and super fresh levels.

According to the World Health Organization Air Quality Guideline (WHO AQG), a $PM_{2.5}$ and PM_{10} decrease of 5 μg/m^3 will reduce mortality risk by approximately 3% [31], an O_3 decrease of 5 μg/m^3 will reduce daily mortality by approximately 0.12–0.25% [31], an SO_2 decrease of 5 μg/m^3 will reduce daily mortality by approximately 0.09–0.29% [31], a CO decrease of 0.5 mg/m^3 will also reduce mortality risk or health risk [31], and an NO_2 decrease of 5 μg/m^3 will reduce daily mortality by approximately 0.29% [36].

This paper further classifies the FAI threshold range of natural and cultural scenic spots according to the different standards of natural and cultural scenic spot air quality standards in China and the WHO standards, which are summarized in Tables 2 and 3, respectively. The air freshness degree of natural and cultural scenic spots could be classified into seven levels: Not Fresh, General, Fresh, Relatively Fresh, Healing Fresh, Very Fresh, and Super Fresh (Tables 3 and 4).

2.1.3. Improved Natural Microclimate Comfort Index

It is obvious that those indices usually assess two or three factors of natural microclimate, such as THI analyzes temperature and humidity factors, WCI analyzes wind and temperature factors, ICI analyzes wind, temperature, and solar radiation factors, and HCI analyzes temperature, humidity, and wind factors. Therefore, it is a trend in Chinese tourism research to utilize the indices mentioned above to comprehensively analyze the degree of natural microclimate comfort in tourists' destinations [37–40]. However, there is also a lack of consideration of the influence of precipitation factors on the degree of natural microclimate comfort in contemporary Chinese tourism research. According to climate comfort analysis of scenic spots, when precipitation < 1.0 mm, there is nearly no influence on tourists. When precipitation < 3.0 mm, it is within tourists' precipitation comfort range, which will provide a rainfall microclimate for tourists to experience and appreciate rainfall scenery in scenic spots, while when precipitation > 10.0 mm, tourism will decrease dramatically because large volumes of rainfall make tourists feel uncomfortable and cause them to stop tourism activities and leave scenic spots [41].

Table 3. Threshold range of natural scenic spot FAI levels.

SB	S_{pi}						**FAI (Fresh Air Index) In Natural Scenic Spot**	**Levels**
O_2 (Ion/cm^3)	**$PM_{2.5}$ (μg/m^3)**	**PM_{10} (μg/m^3)**	**O_3 (μg/m^3)**	**SO_2 (μg/m^3)**	**NO_2 (μg/m^3)**	**CO (mg/m^3)**		
≥3500	≤10	≤25	≤75	≤25	≤55	≤1.5	FAI ≥ 2	Super Fresh
3000	15	30	80	30	60	2	1.2 ≤ FAI < 2	Very Fresh
2500	20	35	85	35	65	2.5	0.4 ≤ FAI < 1.2	Healing Fresh
2000	25	40	90	40	70	3	−0.4 ≤ FAI < 0.4	Relatively Fresh
1500	30	45	95	45	75	3.5	−1.2 ≤ FAI < −0.4	Fresh
1000	35	50	100	50	80	4	−2 ≤ FAI < −1.2	General
<1000	>35	>50	>100	>50	>80	>4	FAI < −2	Not Fresh

Table 4. Threshold range of cultural scenic spot FAI levels.

SB	S_{pi}						**FAI (Fresh Air Index) In Cultural Scenic Spot**	**Levels**
O_2 (Ion/cm^3)	**$PM_{2.5}$ (μg/m^3)**	**PM_{10} (μg/m^3)**	**O_3 (μg/m^3)**	**SO_2 (μg/m^3)**	**NO_2 (μg/m^3)**	**CO (mg/m^3)**		
≥3500	≤25	≤75	≤85	≤75	≤55	≤1.5	FAI ≥ 2.14	Super Fresh
3000	35	90	100	90	60	2	1.31 ≤ FAI < 2.14	Very Fresh
2500	45	105	115	105	65	2.5	0.48 ≤ FAI < 1.31	Healing Fresh
2000	55	120	130	120	70	3	−0.35 ≤ FAI < 0.48	Relatively Fresh
1500	65	135	145	135	75	3.5	−1.17 ≤ FAI < −0.35	Fresh
1000	75	150	160	150	80	4	−2 ≤ FAI < −1.17	General
<1000	>75	>150	>160	>150	>80	>4	FAI < −2	Not Fresh

Therefore, this paper proposes a rain comfort index (RCI) to assess the comfort degree to which different volumes of rainfall influence tourists in scenic spots. The RCI refers to the influence of rainfall on the extent of human comfort in a natural microclimate environment. According to the Chinese grade of precipitation standard (GB/T 28592-2012) and considering tourism activities usually within a 12 h range, this paper chose a 12 h precipitation standard to measure rainfall levels; that is, drizzle < 0.1 mm, light rain 0.1~4.9 mm, moderate rain 5.0~14.9 mm, and heavy rain 15.0~29.9 mm. Based on the above analysis, this paper defined the comfort range of precipitation as between 0.1 (drizzle) and 10.00 mm (moderate rain) for tourists in scenic spots and then constructed the RCI as follows:

$$\mathrm{RCI} = \mathrm{R}/\mathrm{R}_0$$

where RCI refers to the rain comfort index, R refers to actual rainfall, and R_0 refers to standard rainfall, that is equal to 0.1 mm (drizzle) in this paper. Then, seven levels of RCI of different comfort degrees were classified, which are summarized in Table 5.

By considering all factors of the natural microclimate, this paper proposes the natural microclimate index (NMCI):

$$\mathrm{NMCI} = \frac{\mathrm{THI} + \mathrm{WCI} + \mathrm{ICI} + \mathrm{HCI} + \mathrm{RCI}}{5}$$

Table 5. Threshold range of RCI levels.

RCI Levels	RCI
Very Uncomfortable	≥100
More Uncomfortable	80 ≤ RCI < 100
Relatively Uncomfortable	60 ≤ RCI < 80
Comfortable	0 ≤ RCI < 30
Relatively comfortable	30 ≤ RCI < 40
Less comfortable	40 ≤ RCI < 50
Uncomfortable	50 ≤ RCI < 60

By assessing each subindex of the NMCI (THI, WCI, ICI, HCI, RCI) and defining them into their own comfort levels, the results are summarized in Table 6. For subindices of NMCI, such as THI, WCI, ICI, HCI, and RCI, which have their own levels of standards and classification patterns, it is essential to standardize each subindex of NMCI for further calculation of NMCI. This paper summed up and calculated their average comfort levels through reclassifying each subindex of NMCI into seven levels (d, c, b, A, B, C, D) and assigning values (that is, d = 2, c = 3, b = 4, A = 5, B = 6, C = 7, D = 8). Then, the average THI + WCI + ICI + HCI + RCI/5 value was calculated to determine the final levels of NMCI in Table 7.

When 5 ≤ NMCI < 6, the NMCI is at the most comfortable level; when 4 ≤ NMCI < 5, the NMCI is at the comfortable (cool comfort) level; when 6 ≤ NMCI < 7, the NMCI is at the comfortable (warm comfort) level; when 3 ≤ NMCI < 4, the NMCI is at the less comfortable (cold less comfortable) level; when 7 ≤ NMCI < 8, the NMCI is at the less comfortable (hot less comfortable) level; when 2 ≤ NMCI < 3, the NMCI is at the uncomfortable (very cold uncomfortable) level; when NMCI > 8, the NMCI is at the uncomfortable (very hot uncomfortable) level (Table 7).

2.1.4. Combined Fresh Air–Natural Microclimate Comfort Index

After the construction of the fresh air index (FAI) and the natural microclimate comfort index (NMCI), this paper comprehensively constructed a parallel fresh air–natural microclimate index for assessing the air freshness degree and natural microclimate comfort degree of specific scenic spots in China. This approach utilizes the fresh air index and natural microclimate comfort index to analyze scenic spots and then connects them together to explore fresh air and natural microclimate tourism resources in depth (Table 8).

2.2. Data Source

This study chose four fresh air index monitoring stations within four scenic spots of Fujian Province, China. That is, the Pingnan Baishuiyang-Yuanyang Brook Scenic Spot in Ningde City, the Gutian Conference Site Scenic Spot in Longyan City, the Yongding earth building (Tulou) Scenic Spot in Longyan City, and the Wuyi Mountain Scenic Spot in Nanping City (Figure 1 and Table 9).

Pingnan Baishuiyang-Yuanyang Brook Scenic Spot is a natural scenic spot. It has a total area of 66 square kilometers, integrating streams, peaks, rocks, waterfalls, caves, lakes, and other landscapes. There are more than 10 kinds of national protected animals, such as Silver Pheasant, Python, and Pangolin, which are found in the "Natural Animal and Botanical Garden" (Figure 1 and Table 9).

Gutian Conference Site Scenic Spot is a cultural scenic spot. It is located in the south of Meihua Mountain, with a planned area of 7.6 square kilometers. In October 2015, it was successfully promoted to a national 5A tourist attraction and won the honorary titles of "National Top Ten Excellent Patriotism Education Bases" and "National Red Tourism Classic Scenic Spots" (Figure 1 and Table 9).

Table 6. Threshold range of each NMCI subindex and its corresponding levels and values in the NMCI context.

Subindices 1: THI		**Subindices 2: WCI**		**Subindices 3: ICI**		**Subindices 4: HCI**		**Subindices 5: RCI**		**Levels and Values of NMCI Subindices**	
Value Range	Levels	Value Range	Levels	Value Range	Levels	Value Range	Levels	Value Range	Levels	Levels	Values
40 < THI < 45	Cold Uncomfortable	WCI ≥ 1000	Cold Wind	ICI ≥ 1.0	Cold Uncomfortable	25 < HCI < 35	Cold, Uncomfortable	RCI ≥ 100	Very Uncomfortable	d	2
45 < THI < 55	Relatively cold Relatively Uncomfortable	800 < WCI < 1000	Relatively Cold Wind	1.5 < ICI < 1.8	Thick coat sweater	35 < HCI < 45	Relatively cold Relatively Uncomfortable	80 < RCI < 100	More Uncomfortable	c	3
55 < THI < 60	Cool, Relatively Comfortable	600 < WCI < 800	Cool Wind	1.3 < ICI < 1.5	Coat, thick shin	45 < HCI < 55	Cool, Relatively Comfortable	60 < RCI < 80	Relatively Uncomfortable	h	4
60 < THI < 65	Comfortable	300 < WCI < 600	Comfortable Wind	0.7 < ICI < 1.3	Thin coal shirt	55 < HCI < 65	Comfortable	0 < RCI < 30	Comfortable	A	5
65 < THI < 70	Warm, Relatively Comfortable	200 < WCI < 300	Warm Wind	0.5 < ICI < 0.7	Shirts (long sleeves)	65 < HCI < 75	Warm, Relatively Comfortable	30 < RCI < 40	Relative Comfortable	B	6
70 < THI < 75	Slightly hot, Less Comfortable	50 < WCI < 200	Slightly Hot	0.35 < ICI < 0.5	Shirts (short sleeves)	75 < HCI < 80	Slightly hot, Less Comfortable	40 < RCI < 50	Less comfortable	C	7
75 < THI < 80	Stuffiness, Uncomfortable	−80 < WCI < 50	Hot wind	0.1 < ICI < 035	Summer clothes with short sleeves	80 < HCI < 85	Stuffiness, Uncomfortable	50 < RCI < 60	Uncomfortable	D	8

Table 7. Threshold range of NMCI Levels.

Value Range	Levels
2 ≤ NMCI < 3	Very cold uncomfortable
3 ≤ NMCI < 4	Cold less comfortable
4 ≤ NMCI < 5	Cool comfort
5 ≤ NMCI < 6	The most comfortable
6 ≤ NMCI < 7	Warm comfort
7 ≤ NMCI < 8	Hot less comfortable
NMCI > 8	Very hot uncomfortable

Table 8. Threshold range of FAI-NMCI Levels.

FAI Value Range in Cultural Scenic Spot	FAI Value Range in Natural Scenic Spot	FAI Levels	NMCI Value Range	NMCI Levels
FAI ≥ 2.14	FAI ≥ 2	Super Fresh	2 ≤ NMCI < 3	Very cold uncomfortable
1.31 ≤ FAI < 2.14	1.2 ≤ FAI < 2	Very Fresh	3 ≤ NMCI < 4	Cold less comfortable
0.48 ≤ FAI < 1.31	0.4 ≤ FAI < 1.2	Healing Fresh	4 ≤ NMCI < 5	Cool comfort
−0.35 ≤ FAI < 0.48	−0.4 ≤ FAI < 0.4	Relatively Fresh	5 ≤ NMCI < 6	The most comfortable
−1.17 ≤ FAI < −0.35	−1.2 ≤ FAI < −0.4	Fresh	6 ≤ NMCI < 7	Warm comfort
−2 ≤ FAI < −1.17	−2 ≤ FAI < −1.2	General	7 ≤ NMCI < 8	Hot less comfortable
FAI < −2	FAI < −2	Not Fresh	NMCI > 8	Very hot uncomfortable

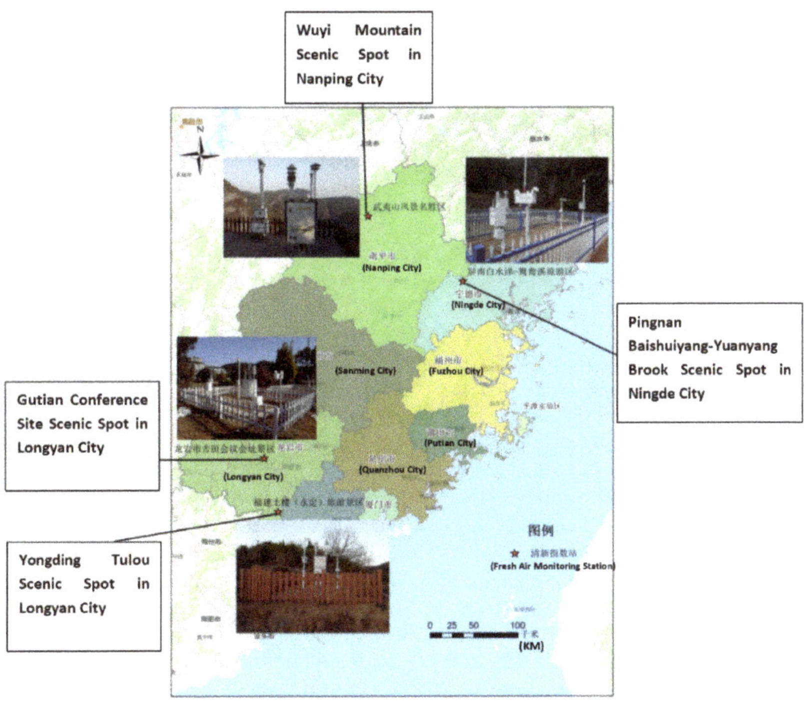

Figure 1. The four scenic spots and their monitoring stations in Fujian Province, China.

Table 9. Four scenic spots and their location information.

Scenic Spots	Site Number	Longitude	Latitude	Altitude	Located City	Located Specific Site
Wuyi Mountain Scenic Spot	F9269	117.96245	27.6677	408	Nanping City	Wuyi Mountain
Gutian Conference Site Scenic Spot	F7490	116.8376	25.2261	730	Longyan City	Gutian County
Pingnan Baishuiyang-Yuanyang Brook Scenic Spot	F3749	119.058	27.0736	679.2	Ningde City	Ping'nan County
Yongding earth building (Tulou) Scenic Spot	F7590	117.015	24.667	431	Longyan City	Yongding County

Yongding earth building (Tulou) Scenic Spot is a cultural scenic spot. It is a unique magical mountain residential building. It has a long history, unique style, large scale, and exquisite structure. In July 2008, it was successfully listed in the world heritage list (Figure 1 and Table 9).

Wuyi Mount Scenic Spot is a natural scenic spot. It is located at the junction of Jiangxi and northwest Fujian Provinces and at the southeast foot of the north section of Wuyi Mountain vein, with a total area of 999.75 square kilometers. It is a famous scenic spot and summer resort in China (Figure 1 and Table 9).

Since the $PM_{2.5}$, PM_{10}, and O_3 standards in natural and cultural scenic spots have different regulations in practice, this paper attempts to establish natural and cultural scenic spot FAIs to distinguish air freshness evaluations. In addition, natural and cultural scenic spots have different environmental backgrounds and social interaction intensities, which will significantly influence the air freshness degree of scenic spots. Meanwhile, for tourists, the fresher the air in scenic spots, the better the traveling experience, so they could utilize the FAI of natural scenic spots to measure all scenic spots in practice. For destination and scenic spot management, it is vital to distinguish the difference between natural and cultural scenic spots and to evaluate the related and compared freshness within natural scenic spots and cultural scenic spot categories. This paper chose the Wuyi Mountain scenic spot and Baishuiyang scenic spot as natural scenic spots of Fujian Province and the Yongding Tulou scenic spot and Gutian Conference Site scenic spot as cultural scenic spots of Fujian Province.

According to local air characteristics, O_3 is a special air component that influences the local air freshness of Fujian Province. Therefore, data on four indicators, O_2^-, $PM_{2.5}$, PM_{10}, and O_3, from 1 January to 31 December 2018 were collected. Temperature, relative humidity, wind speed, solar radiation, and precipitation data of the four scenic spots were collected from 1 January to 31 December 2018. Therefore, this paper utilized negative oxygen ion (O_2^-), $PM_{2.5}$, PM_{10}, and O_3 data for each day of 2018 of the four scenic spots to calculate their fresh air index (FAI). Then, temperature, relative humidity, wind speed, solar radiation, and precipitation data were utilized on each day of 2018 to calculate the natural microclimate comfort index (NMCI) of the four scenic spots.

The data for natural microclimate factors were collected from the National Meteorological Information Center of China (http://data.cma.cn/, accessed on 1 February 2022) and provided by the Fujian Meteorological Service Center, China (http://fj.cma.gov.cn/, accessed on 1 January 2022), and the data for fresh air index subindices were provided by the Fujian Meteorological Service Center, China (http://fj.cma.gov.cn/, accessed on 1 January 2022). The data were generated from the four scenic spots' microclimate and air freshness index monitoring equipment. More precisely, one set of negative oxygen ion monitoring stations (model HuaTron FR500) and one set of air index stations (model HuaTron AEI365) were built in the above four scenic spots (Figure 1 and Table 8).

3. Results

3.1. FAI-NMCI of Natural Scenic Spots

In the Wuyi Mountain scenic spot, the FAI levels were mainly distributed in the upper levels; for example, the Super Fresh level was approximately 164 days (accounting for 47%), the Very Fresh level was approximately 67 days (accounting for 19%), and the Healing Fresh level was approximately 56 days (accounting for 16%), which was approximately

287 days in 2018, accounting for almost 82% (Figure 2A). Meanwhile, the NMCI primarily included three levels. The comfortable level was approximately 246 days (accounting for 67%), which also embraced the cool comfort level (168 days, 46%) and the warm comfort level (78 days, 21%). The most comfortable level was approximately 116 days (accounting for 32%), while the less comfortable level was approximately 3 days (accounting for only 1%) (Figure 2B). Therefore, in the Wuyi Mountain scenic spot, the FAI-NMCI was mainly distributed within the upper fresh-comfort range, which was approximately 246 days of upper freshness levels and 362 days of comfort in 2018 (Figure 2A,B).

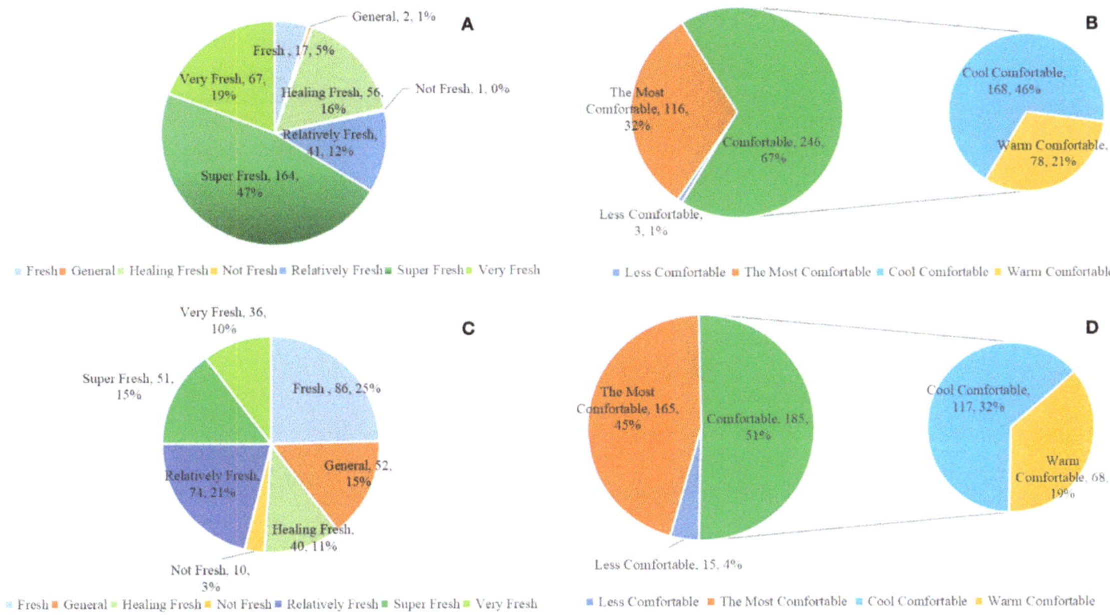

Figure 2. FAI and NMCI levels of Wuyi Mountain (**A**,**B**) and Baishuiyang scenic spot (**C**,**D**) of Fujian Province, China.

In the Baishuiyang scenic spot, the FAI levels were distributed from the Not Fresh level to the Super Fresh level. More precisely, the Fresh level was approximately 86 days (accounting for 25%), the Relatively Fresh level was approximately 74 days (accounting for 21%), the Super Fresh level was approximately 51 days (accounting for 15%), the Healing Fresh level was approximately 40 days (accounting for 11%), and the Very Fresh level was approximately 36 days (accounting for 10%), which indicated the dynamic fluctuation of FAI in the Baishuiyang scenic spot, and the FAI reached or exceeded the Fresh level for approximately 287 days, accounting for approximately 78.63% (Figure 2C). At the same time, the NMCI was mainly distributed at the comfortable level (185 days, 51%), among which the cool comfort level was 117 days (accounting for 32%) and the warm comfort level was 68 days (accounting for 19%). Meanwhile, the most comfortable level was also in a large proportion (45%) and was approximately 165 days. The less comfortable level was approximately 15 days and accounted for 4% (Figure 2D). Therefore, in Baishuiyang scenic spot, the FAI-NMCI was mainly distributed within the middle fresh-comfort range, which was approximately 160 days of middle freshness and 185 days of comfort in 2018 (Figure 2C,D).

In Wuyi Mountain scenic spot, the FAI mainly fluctuated between −2.69 and 32.7, that is, it fluctuated between Not Fresh and Super Fresh levels. The NMCI mainly fluctuated between 3.6 and 7.2, that is, it fluctuated between cold less comfortable and hot less comfortable levels. Within the whole year of 2018, from January to April, the NMCI gradually increased from 4 to 5 (in the cool comfort level), while the FAI was mostly more than 5 (in the Super Fresh level) in the second half of January, decreased dramatically to −1.87 (in General level) in early February, then slightly increased to Relatively Fresh, Healing Fresh, and Very Fresh levels, and some Super Fresh levels, which also fluctuated in this period. From May to the first half of July, the NMCI increased from 5 to 6 and 6 to 7, which generally entered into the most comfortable level and increased into the warm comfort level in this period, which fluctuated mainly within these two levels; meanwhile, the FAI mainly fluctuated between Fresh and Super Fresh levels and reached the highest FAI of approximately 3.7 in June, which was more than the threshold value of Super Fresh (FAI > 2) by approximately 16.35 times. From the second half of July to the first half of September, the NMCI remained at the warm comfort level, while the FAI decreased gradually from more than 5 to lower than 3, although it was also mainly within the Super Fresh level. Then, the FAI increased to more than 5 in the second half of August and increased to more than 20 in the first half of September. From September to October, the NMCI decreased gradually from the warm comfort level to the most comfortable level, while the FAI also decreased to lower levels, such as the Fresh level and Relatively Fresh level, although some days reached the Healing Fresh level and Very Fresh level. When entering November and December, the NMCI decreased from 5 to lower than 4, which is a decrease from the cool comfort level to the cold less comfortable level, while the FAI increased gradually to the Super Fresh level in the middle of November and then decreased dramatically in the second half of November and the first half of December. When the NMCI decreased to the cold less comfortable level in December, the FAI fluctuated, while most fluctuated between the Healing Fresh and Super Fresh levels (Figure 3A).

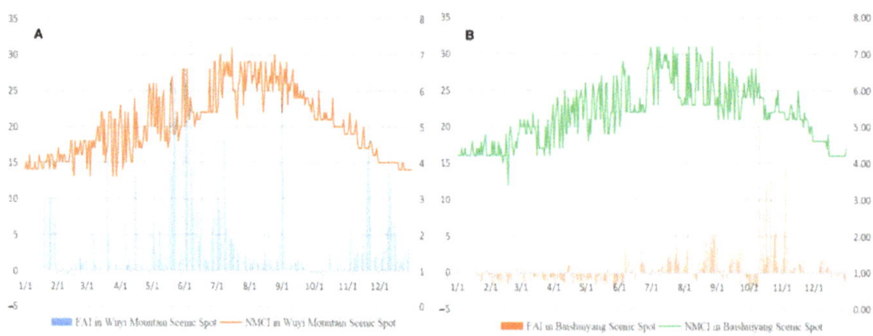

Figure 3. FAI and NMCI of Wuyi Mountain (**A**) and Baishuiyang scenic spots (**B**) of Fujian Province on each day of 2018.

In Baishuiyang scenic spot, the FAI mainly fluctuated between −2.83 and 34.27, which fluctuated between the Not Fresh level and Super Fresh levels, while the NMCI mainly fluctuated between 3.4 and 7.2, which fluctuated between cold less comfortable and hot less comfortable levels. Within the whole year of 2018, from January to early June, the NMCI gradually increased from 4 to nearly 7, namely, it increased from cool comfort to the most comfortable level and then to the warm comfort level, while the FAI was mostly below 0, namely, it fluctuated between the Not Fresh level and the Relatively Fresh level. From the middle of June to August, the NMCI fluctuated and gradually increased from 6 to more than 7, that is, from the warm comfort level to the hot less comfortable level; meanwhile, the FAI increased from 0 to between nearly 5 or more than 5 on some days, which mainly fluctuated between the General level and the Super Fresh level. From September to November, the NMCI fluctuated mainly between 6.5 and 5, which indicated that the natural microclimate fluctuated from the warm comfort level to the most comfortable level from early September to late November. The FAI slightly and gradually increased in September and then decreased drastically below 0 from late September to early October, namely, decreased from the Super Fresh level to the Not Fresh or General level, mainly during the National Holiday of China (approximately seven-day long holiday). After the seven days of the National Holiday of China, the FAI increased dramatically and reached the highest level of the year (FAI = 34.27), then fluctuated within the Super Fresh level (mainly between 2 and 14). From late November to December, the NMCI decreased below 5 but higher than 4, namely, at the cool comfort level, and the FAI also decreased to a lower level of the year, which primarily fluctuated between the General level and the Healing Fresh level (Figure 3B).

It is essential to point out that the FAI of the Wuyi Mountain and Baishuiyang scenic spots have some higher FAI values, such as more than 30 or even higher, which is primarily because of the higher O_2^- levels and the lower $PM_{2.5}$, PM_{10}, and O_3 levels in scenic spots. When scenic spots have a high O_2^- concentration, it will significantly influence the entire FAI value.

More precisely, within 24 Jieqi of Wuyi Mountain scenic spot and Baishuiyang scenic spot, the FAI-NMCI can be summarized as in Figure 4. For example, it is obvious that in Lesser Cold and Greater Cold Jieqi, both the Wuyi Mountain and Baishuiyang scenic spots were at the cool comfort level, although Wuyi Mountain was at the Super Fresh level and was fresher than Baishuiyang, which was only at the Fresh level. At the start of Spring, the FAI of both Wuyi Mountain and Baishuiyang decreased, which was not because of the decrease in negative oxygen ions (O_2^-) at the two scenic spots but because of the increase in $PM_{2.5}$ and PM_{10} due to the large number of tourists and tourism transportation entering those two scenic spots (Figure 4).

From Rain Water to Spring Equinox Jieqi, although the two scenic spots were also within the cool comfort level, the FAI both experienced a drastic increase from the General to the Super Fresh level in Wuyi Mountain and from the Not Fresh to the Relatively Fresh level in Baishuiyang. In Pure Brightness and Grain Rain Jieqi, Baishuiyang entered into the most comfortable level while Wuyi Mountain remained at the cool comfort level. Meanwhile, for more tourist travel to Baishuiyang, the FAI was at a relatively lower level, that is the General level, than Wuyi Mountain (Super Fresh and Healing Fresh level). At the Start of Summer Jieqi, Wuyi Mountain was at the Super Fresh and warm comfort levels, while Baishuiyang was only at the Fresh level and was still within the cool comfort level. In Grain Full Jieqi, Baishuiyang had increased freshness and became warm. In the Grain in Ear Jieqi, Wuyi Mountain entered the most comfortable level and was in a Super Fresh situation (Figure 4).

3.2. *FAI-NMCI of Cultural Scenic Spots*

In the Yongding Tulou scenic spot, the FAI levels were mainly distributed in Healing Fresh (159 days, 38%) and Relatively Fresh (74 days, 18%), while they were also distributed in Not Fresh (115 days, 27%). The Super Fresh (29 days, 7%), Very Fresh (31 days, 7%), and Fresh (14 days, 3%) levels were in small proportions (Figure 5A).

The NMCI was distributed in the comfort domain, that is, the most comfortable domain (149 days, 41%), and the comfort domain (216 days, 59%), which included the cool comfort domain (133 days, 36%) and the warm comfort domain (83 days, 23%) (Figure 5B). In the Gutian Conference Site scenic spot, the highest levels of FAI were at the Healing Fresh level (166 days, 48%) with a large proportion, and the Relatively Fresh level (146 days, 42%), while the Very Fresh (24 days, 7%) and Fresh (12 days, 3%) levels had a small proportion, and the Super Fresh level was not reached (Figure 5C). Meanwhile, the NMCI included three levels: the most comfortable (110 days, 30%), and the comfort (250 days, 69%), which embraced the cool comfort level (160 days, 44%) and the warm comfort level (90 days, 25%) (Figure 5D).

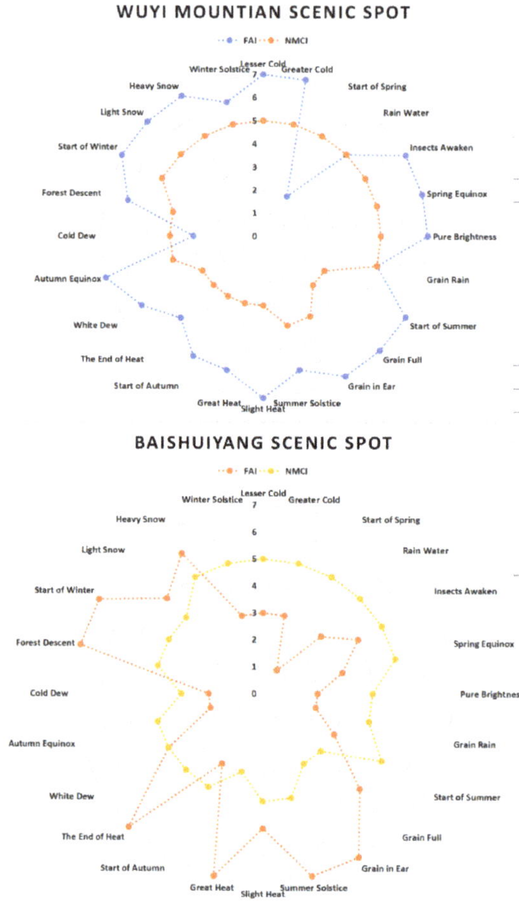

FAI Levels	Values
Super Fresh	7
Very Fresh	6
Healing Fresh	5
Relatively Fresh	4
Fresh	3
General	2
Not Fresh	1

NMCI Levels	Values
Very Cold Uncomfortable	7
Cold Less Comfortable	6
Cool Comfortable	5
The Most Comfortable	4
Warm Comfortable	3
Hot Less Comfortable	2
Very Hot Uncomfortable	1

Figure 4. 24 Jieqi FAI-NMCI of Wuyi Mountain and Baishuiyang scenic spots.

In the Yongding Tulou scenic spot, the FAI mainly fluctuated between −0.97 and 14.81, which fluctuated between Fresh and Super Fresh levels. The NMCI mainly fluctuated between 3.80 and 6.80, which fluctuated between cold less comfortable and warm comfort levels (Figure 6A). Throughout 2018, the FAI and NMCI of the Yongding Tulou scenic spot experienced dynamic fluctuations. From January to March, the NMCI increased gradually from 3 to above 4, that is from cold less comfortable to the cool comfort level, and the FAI mainly fluctuated between −0.35 and 0.48, that is mainly at the Relatively Fresh level. From late March to the middle of April, the NMCI continued to increase from 4 to above 5, that is from cool comfort to the most comfortable level. Meanwhile, the FAI increased dramatically above 2 and reached the Very Fresh level and fluctuated into the high level of Relatively Fresh. From May to September, the NMCI fluctuated from 5 to above 6, that is from the most comfortable to the warm comfort level, and the FAI mainly fluctuated within the Relatively Fresh level. When entering October, the NMCI decreased from above 6 to below 5, that is from warm comfort to the most comfortable level. The FAI increased drastically and reached the highest level (14.81) of the year, and most of FAI values were higher than 6 in the second half of October, that is it mainly fluctuated at the Very Fresh level. From November to December, the NMCI decreased from above 5 to below 4, that is from the most comfortable to the cold less comfortable level, while the FAI primarily fluctuated within the Relatively Fresh level until in the middle of December, when it increased dramatically to more than 12, namely, in the Very Fresh level, then decreased quickly and fluctuated between the Fresh and Relatively Fresh levels (Figure 6A).

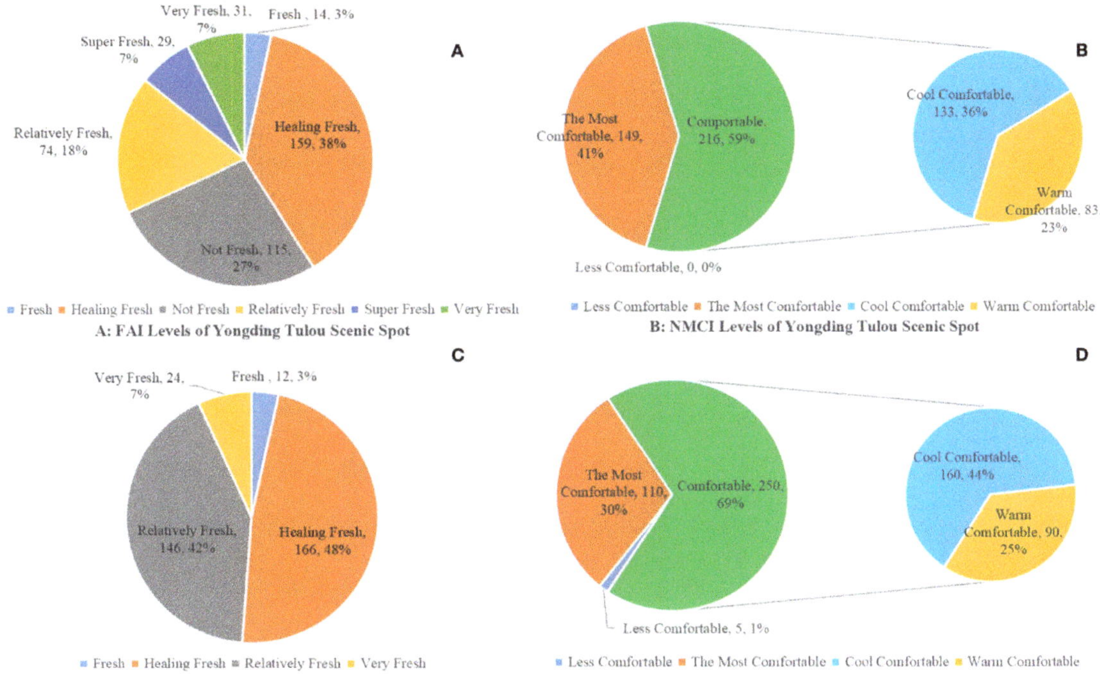

Figure 5. FAI and NMCI levels of Yongding Tulou (**A**,**B**) and Gutian Conference Site scenic spots (**C**,**D**) of Fujian Province, China.

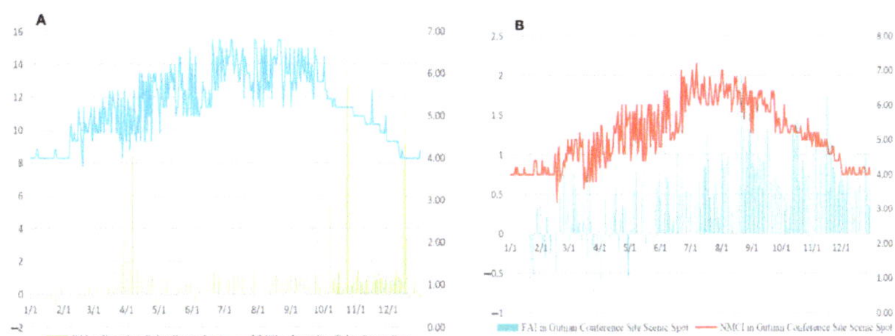

FAI and NMCI of Yongding Tulou scenic spot

FAI and NMCI of Gutian Conference Site scenic spot

Figure 6. (**A**) FAI and (**B**) NMCI of Yongding Tulou and Gutian Conference Site scenic spots of Fujian Province on each day of 2018.

In the Gutian Conference Site scenic spot, the FAI mainly fluctuated between −0.76 and 2.03, that is it fluctuated between Fresh and Super Fresh levels, while the Super Fresh level was lower than the Yongding Tulou scenic spot by approximately 7 times. The NMCI mainly fluctuated between 3.20 and 7.2, that is it fluctuated between cold less comfortable and hot less comfortable levels (Figure 5B). For the first five months of 2018, from January to May, the NMCI increased from more than 4 to more than 6, that is from cool comfort to the most comfortable level, and then to the warm comfort level. The FAI fluctuated between below −0.5 and 1, that is it fluctuated between the Fresh and Healing Fresh levels. From June to August, the NMCI increased from above 6 to above 7, that is from warm comfort to the hot less comfortable level. The FAI was mainly above 0.5 and some days even above 1, that is primarily at the Healing Fresh level. From September to November, the NMCI decreased from above 6 to 5, that is from warm comfort to the most comfortable level, while the FAI mainly fluctuated between 1 and 1.5, that is it fluctuated between Healing Fresh and Very Fresh levels. In December, the NMCI decreased and mainly varied between 4 and 4.2, that is mainly at the cool comfort level, while the FAI also decreased and mainly fluctuated above 0.5 or below 0.5, that is mainly at the Relatively Fresh and Healing Fresh levels (Figure 6B).

More precisely, within 24 Jieqi of the Yongding Tulou scenic spot and Gutian Conference Site scenic spot, the FAI-NMCI could be summarized as in Figure 7 for satisfying tourists' diverse fresh air–natural microclimate comfort preferences and demands through the provision of proper microclimate tourism products and services in each Jieqi, or different Jieqi with approximately similar air freshness and natural microclimate comfort degrees.

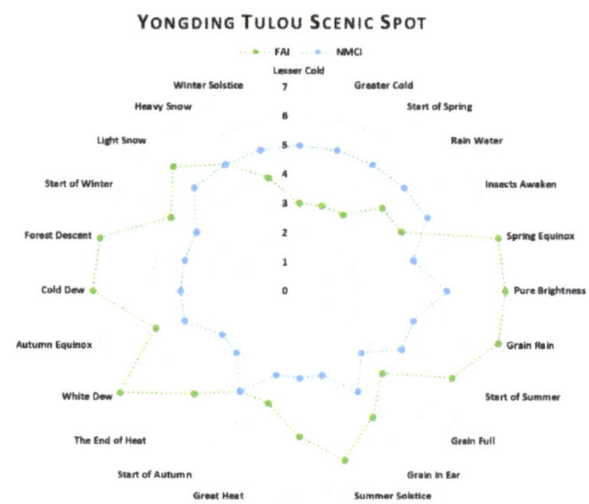

FAI Levels	Values
Super Fresh	7
Very Fresh	6
Healing Fresh	5
Relatively Fresh	4
Fresh	3
General	2
Not Fresh	1

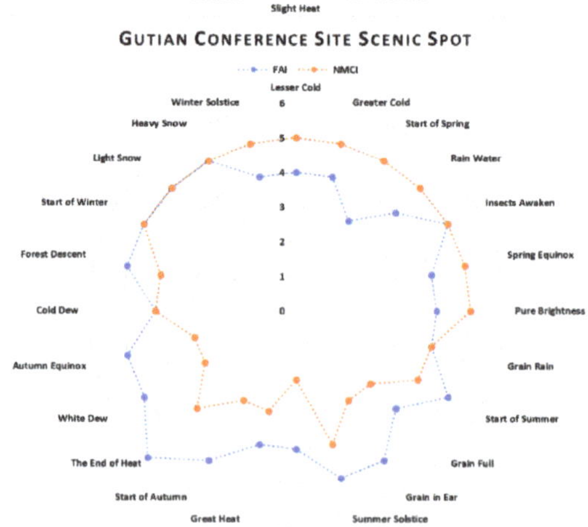

NMCI Levels	Values
Very Cold Uncomfortable	7
Cold Less Comfortable	6
Cool Comfortable	5
The Most Comfortable	4
Warm Comfortable	3
Hot Less Comfortable	2
Very Hot Uncomfortable	1

Figure 7. 24 Jieqi FAI-NMCI Index of Yongding Tulou and Gutian Conference Site scenic spots.

4. Discussion and Conclusions

Comparing FAI-NMCI in natural scenic spots of Fujian province, the Wuyi Mountain scenic spot had more days of upper-level air freshness and natural microclimate comfort degree than Baishuiyang scenic spot, which was mainly distributed within the middle fresh-comfort range. In cultural scenic spots of Fujian province, the Gutian Conference Site scenic spot had more days of middle-level air freshness and natural microclimate comfort degree than Yongding Tulou scenic spot, which had even more days of air freshness at the Not Fresh level.

Comparing fluctuating ranges of FAI of the four scenic spots in Fujian Province, the natural scenic spots had wider FAI fluctuation ranges, which was primarily determined by the higher concentration of negative oxygen ions and lower concentrations of $PM_{2.5}$, PM_{10}, and O_3, while, the cultural scenic spots had relatively shorter FAI fluctuation ranges, which was profoundly influenced by the relatively lower concentration degree of negative oxygen ions and higher concentration degree of $PM_{2.5}$, PM_{10}, and O_3.

Comparing fluctuating ranges of NMCI of four scenic spots in Fujian province, the NMCI mainly fluctuated between 3.20 and 7.2, which indicated that their natural microclimate comfort degree primarily fluctuated between cold less comfortable and hot less comfortable levels, except the Yongding Tulou scenic spot, which fluctuated between cold less comfortable and warm comfort levels.

In short, the four scenic spots in Fujian Province of China had different levels of air freshness degree and natural microclimate comfort degree in 2018. The natural scenic spots were mostly distributed in Healing Fresh, Very Fresh, and Super Fresh levels of FAI, with the most comfortable and comfortable levels of NMCI. The cultural scenic spots were mostly distributed in Relatively Fresh and Healing Fresh levels of FAI, with the most comfortable and comfortable levels of NMCI. Meanwhile, the FAI-NMCI of natural and cultural scenic spots also had significant differences within 24 Jieqi, which will promote dynamic and creative utilization of these resources in microclimate tourism development.

Therefore, the fresh air–natural microclimate of scenic spots has become a vital tourist attraction in the contemporary Chinese tourism market under the background of severe air pollution. Tourists with different physiological and psychological conditions will be more suited to different levels of air freshness and natural microclimate comfort levels, such as elderly tourists, pregnant women tourists, and children tourists will be vulnerable to polluted air environments and require more fresh and comfortable natural microclimates, such as Super Fresh and the most comfortable level of scenic spots, which has more negative oxygen ions (O_2^-) and lower $PM_{2.5}$, PM_{10}, and O_3, with a more comfortable temperature, humidity, wind speed, solar radiation, and precipitation.

Moreover, the dynamic fresh air–natural microclimate fluctuation revealed by FAI-NMCI of scenic spots will properly indicate the air freshness and natural microclimate comfort degree information for both the demand side and the supply side of the tourism market, which will profoundly promote fresh air–natural microclimate tourism resources' distribution and redistribution, accelerate fresh air–natural microclimate in-depth tourism resources combination and integration from a creative microclimate tourism development perspective, and further drive fresh-microclimate tourism development in destinations and scenic spots with comparative advantages of fresh air and a comfortable natural microclimate for triggering their in-depth transformation and improvement.

From a theoretical contribution perspective, this paper constructed the FAI-NMCI to harmoniously combine air resources and microclimate resources together, which could empirically evaluate air and microclimate tourism resource values for tourists, the tourism market, and the society as a whole. Moreover, it could figure out overlapping study areas of air tourism, climate tourism, health tourism, ecological tourism, and so forth, which have significant theoretical innovation through transdisciplinary, multidisciplinary, and interdisciplinary integration research in China's tourism market. This is a new research trend that has seldom been explored by researchers and could become an important research domain for future multidisciplinary, transdisciplinary, and interdisciplinary exploration.

From a socioeconomic practical application perspective, the creation and utilization of FAI-NMCI is significant for fresh air and comfortable microclimate tourism resources' distribution and redistribution in the tourism market. Meanwhile, it could also become a vital health tourism market indicator, which simply reveals the air freshness degree and microclimate comfort degree for tourists and some potential tourists pursuing health tourism experiences and long-term health-improving vacations. FAI-NMCI could indicate a scenic spot's air freshness and microclimate comfort levels specifically on different days of a year, which also could be calculated and summarized into 24 Jieqi, 12 months, 4 seasons, and so forth. Under contemporary China's high-quality tourism development background, this could provide accurate information not only for demand and supply sides of the tourism market but could also become an important and comprehensive index for related governmental management and NGOs.

The limitation of this paper is we only applied the FAI-NMCI to four scenic spots of China because of data collection limitations and the short time construction of the air freshness index infrastructure, with only 1–2 years of data sources. This new index has only been applied in Fujian Province rather than other provinces of China, which need to be explored in the future. This index could reveal not only the natural conditions of tourism resources but could also indicate tourism practices, tourists' and destinations' social behavior patterns, as well as socioeconomic development patterns of some destinations, which could be explored through introducing multidisciplinary methods in future research.

Author Contributions: Conceptualization, X.Y., C.L. and S.J.; methodology, X.Y. and C.L.; software, X.Y. and C.L.; validation, X.Y., M.B. and C.L.; formal analysis, X.Y. and C.L.; investigation, X.Y. and C.L.; resources, X.Y. and C.L.; data curation, X.Y. and C.L.; writing—original draft preparation, X.Y., C.L. and S.J.; writing—review and editing, X.Y., M.B., C.L. and S.J.; visualization, X.Y. and C.L.; supervision, C.L.; project administration, C.L.; funding acquisition, C.L. and S.J. All authors have read and agreed to the published version of the manuscript.

Funding: This work was financially supported by the National Science Foundation of China (42101238), the start-up funding of Minjiang University (32304307), and the Project of Central Leading Local "Fujian Mental Health Human–Computer Interaction Technology Research Center" (2020L3024).

Institutional Review Board Statement: Not applicable.

Informed Consent Statement: Not applicable.

Data Availability Statement: The data are not publicly available due to the privacy of data policy of Fujian Meteorological Service Center.

Conflicts of Interest: The authors declare no conflict of interest.

References

1. Yang, T.; Liu, W. Does air pollution affect public health and health inequality? Empirical evidence from China. *J. Clean. Prod.* **2018**, *203*, 43–52. [CrossRef]
2. Schmid, D.; Korkmaz, P.; Blesl, M.; Fahl, U.; Friedrich, R. Analyzing transformation pathways to a sustainable European energy system—Internalization of health damage costs caused by air pollution. *Energy Strat. Rev.* **2019**, *26*, 100417. [CrossRef]
3. Jones, B.A.; Fleck, J. Shrinking lakes, air pollution, and human health: Evidence from California's Salton Sea. *Sci. Total Environ.* **2020**, *712*, 136490. [CrossRef]
4. Liu, K.; Yang, B.-Y.; Guo, Y.; Bloom, M.S.; Dharmage, S.; Knibbs, L.D.; Heinrich, J.; Leskinen, A.; Lin, S.; Morawska, L.; et al. The role of influenza vaccination in mitigating the adverse impact of ambient air pollution on lung function in children: New insights from the Seven Northeastern Cities Study in China. *Environ. Res.* **2020**, *187*, 109624. [CrossRef]
5. Ding, L.; Zhu, D.; Peng, D.; Zhao, Y. Air pollution and asthma attacks in children: A case–crossover analysis in the city of Chongqing, China. *Environ. Pollut.* **2017**, *220*, 348–353. [CrossRef]
6. Li, Z.; Xu, X.; Thompson, L.A.; Gross, H.E.; Shenkman, E.A.; DeWalt, D.A.; Huang, I.-C. Longitudinal Effect of Ambient Air Pollution and Pollen Exposure on Asthma Control: The Patient-Reported Outcomes Measurement Information System (PROMIS) Pediatric Asthma Study. *Acad. Pediatr.* **2019**, *19*, 615–623. [CrossRef]
7. Raaschou-Nielsen, O.; Beelen, R.; Wang, M.; Hoek, G.; Andersen, Z.J.; Hoffmann, B.; Stafoggia, M.; Samoli, E.; Weinmayr, G.; Dimakopoulou, K.; et al. Particulate matter air pollution components and risk for lung cancer. *Environ. Int.* **2016**, *87*, 66–73. [CrossRef]

8. Wei, F.; Wu, M.; Qian, S.; Li, D.; Jin, M.; Wang, J.; Shui, L.; Lin, H.; Tang, M.; Chen, K. Association between short-term exposure to ambient air pollution and hospital visits for depression in China. *Sci. Total Environ.* **2020**, *724*, 138207. [CrossRef] [PubMed]
9. Tzivian, L.; Winkler, A.; Dlugaj, M.; Schikowski, T.; Vossoughi, M.; Fuks, K.; Weinmayr, G.; Hoffmann, B. Effect of long-term outdoor air pollution and noise on cognitive and psychological functions in adults. *Int. J. Hyg. Environ. Health* **2015**, *218*, 1–11. [CrossRef]
10. Buoli, M.; Grassi, S.; Caldiroli, A.; Carnevali, G.S.; Mucci, F.; Iodice, S.; Cantone, L.; Pergoli, L.; Bollati, V. Is there a link between air pollution and mental disorders? *Environ. Int.* **2018**, *118*, 154–168. [CrossRef]
11. Zeng, J.; He, Q. Does industrial air pollution drive health care expenditures? Spatial evidence from China. *J. Clean. Prod.* **2019**, *218*, 400–408. [CrossRef]
12. Bai, R.; Lam, J.C.; Li, V.O. A review on health cost accounting of air pollution in China. *Environ. Int.* **2018**, *120*, 279–294. [CrossRef]
13. Zhang, X.; Ou, X.; Yang, X.; Qi, T.; Nam, K.M.; Zhang, D.; Zhang, X. Socioeconomic burden of air pollution in China: Province-level analysis based on energy economic model. *Energy Econ.* **2017**, *68*, 478–489. [CrossRef]
14. Hao, Y.; Peng, H.; Temulun, T.; Liu, L.-Q.; Mao, J.; Lu, Z.-N.; Chen, H. How harmful is air pollution to economic development? New evidence from PM2.5 concentrations of Chinese cities. *J. Clean. Prod.* **2018**, *172*, 743–757. [CrossRef]
15. Xia, Y.; Guan, D.; Jiang, X.; Peng, L.; Schroeder, H.; Zhang, Q. Assessment of socioeconomic costs to China's air pollution. *Atmos. Environ.* **2016**, *139*, 147–156. [CrossRef]
16. Lopes, H.S.; Remoaldo, P.C.; Ribeiro, V.; Martín-Vide, J. Perceptions of human thermal comfort in an urban tourism destination—A case study of porto (portugal). *Build. Environ.* **2021**, *205*, 108246. [CrossRef]
17. Yang, X.; Dong, L.; Li, C. Microclimate tourism and microclimate tourism security and safety in China. *Tour. Manag.* **2019**, *74*, 110–133. [CrossRef]
18. Jeuring, J.H.G. Weather perceptions, holiday satisfaction and perceived attractiveness of domestic vacationing in The Netherlands. *Tour. Manag.* **2017**, *61*, 70–81. [CrossRef]
19. Zeng, Q.; Fan, L.; Ni, Y.; Li, G.; Gu, Q. Construction of AQHI based on the exposure relationship between air pollution and YLL in northern China. *Sci. Total Environ.* **2020**, *710*, 136264. [CrossRef]
20. Ruggieri, M.; Plaia, A. An aggregate AQI: Comparing different standardizations and introducing a variability index. *Sci. Total Environ.* **2012**, *420*, 263–272. [CrossRef]
21. Yuan, M.; Liu, X.; Guo, J.; Huang, Y.; Song, W. Analysis of Eco-Tourism Climate Resources in Xingwen, China Based on the Comfort Index and the Negative Air (Oxygen) Ion. *J. Geosci. Environ. Prot.* **2021**, *9*, 154–163. [CrossRef]
22. Tan, J.; Chen, Z.; Luo, X.; Yang, W.; Shu, S.; Xu, J. Distribution characteristics of atmospheric negative oxygen ions in scenic spots of Hubei Province and the influence of meteorological conditions. *Resour. Environ. Yangtze River Basin* **2017**, *26*, 314–323, (In Chinese with English abstract).
23. Luceño, N.L.; Angrimani, D.D.S.R.; Bicudo, L.D.C.; Szymańska, K.J.; Van Poucke, M.; Demeyere, K.; Meyer, E.; Peelman, L.; Mullaart, E.; Broekhuijse, M.L.; et al. Exposing dairy bulls to high temperature-humidity index during spermatogenesis compromises subsequent embryo development in vitro. *Theriogenology* **2019**, *141*, 16–25. [CrossRef] [PubMed]
24. Oliver, J.E. *Climate and Man's Environment: An Introduction to Applied Climatology*; John Wiley & Sons: New York, NY, USA, 1973; pp. 145–147.
25. Fischer, W.H. Human climates of northern China. *Atmos. Environ. (1967)* **1979**, *13*, 747–748. [CrossRef]
26. Yu, Y.; Wang, X.; Bründlinger, R. Improved Elman Neural Network Short-Term Residents Load Forecasting Considering Human Comfort Index. *J. Electr. Eng. Technol.* **2019**, *14*, 2315–2322. [CrossRef]
27. World Health Organization. WHO Global Air Quality Guidelines: Particulate Matter (PM2.5 and PM10), Ozone, Nitrogen Dioxide, Sulfur Dioxide and Carbon Monoxide. License: CC BY-NC-SA 3.0 IGO. 2021. Available online: https://apps.who.int/iris/handle/10665/345329 (accessed on 4 February 2021).
28. Liu, R.; Lian, Z.; Lan, L.; Qian, X.; Chen, K.; Hou, K.; Li, X. Effects of negative oxygen ions on sleep quality. *Procedia Eng.* **2017**, *205*, 2980–2986. [CrossRef]
29. Nimmerichter, A.; Holdhaus, J.; Mehnen, L.; Vidotto, C.; Loidl, M.; Barker, A.R. Effects of negative air ions on oxygen uptake kinetics, recovery and performance in exercise: A randomized, double-blinded study. *Int. J. Biometeorol.* **2013**, *58*, 1503–1512. [CrossRef]
30. *DB42/T1198-20162016*; Hubei Province Air Negative Oxygen Ion Concentration Grade. Hubei Provincial Bureau of Quality and Technical Supervision: Wuhan, China, 2016.
31. WHO. *WHO Air Quality Guidelines for Particulate Matter, Ozone, Nitrogen Dioxide and Sulfur Dioxide–Global Update 2005*; Druckpartner Moser: Rheinbach, Germany, 2005; pp. 217–415.
32. *QX/T380-2017*; Air (Negative) Oxygen Ion Concentration Grade, Meteorological Industry Standard of The People's Republic of China, Compilation of Meteorological Industry Standard of The People's Republic of China. China Meteorological Administration: Beijing, China, 2017.
33. *GB3095-2012*; Ministry of Environmental Protection and General Administration of Quality Supervision, Inspection and Quarantine, Ambient Air Quality Standard. China Environmental Science Press: Beijing, China, 2012.
34. Wu, R.; Song, X.; Bai, Y.; Chen, J.; Zhao, Q.; Liu, S.; Xu, H.; Wang, T.; Feng, B.; Zhang, Y.; et al. Are current Chinese national ambient air quality standards on 24-hour averages for particulate matter sufficient to protect public health? *J. Environ. Sci.* **2018**, *71*, 67–75. [CrossRef]

35. Yao, Y. Distribution characteristics of air anion concentration in Zhejiang Province. *Meteorol. Sci. Technol.* **2019**, *47*, 1006–1013.
36. Liu, N.M.; Liu, L.Q.; Xu, M.-M.; Liang, F.C.; Pan, X.C. Relationships between ambient nitrogen dioxide and population mortality of respiratory disease in Beijing. *J. Environ. Health* **2014**, *31*, 565–568.
37. Lam, C.K.C.; Gallant, A.; Tapper, N. Perceptions of thermal comfort in heatwave and non-heatwave conditions in Melbourne, Australia. *Urban Clim.* **2018**, *23*, 204–218. [CrossRef]
38. Martín, M.B.G. Weather, climate and tourism a geographical perspective. *Ann. Tour. Res.* **2005**, *32*, 571–591. [CrossRef]
39. Nasrollahi, N.; Hatami, Z.; Taleghani, M. Development of outdoor thermal comfort model for tourists in urban historical areas; A case study in Isfahan. *Build. Environ.* **2017**, *125*, 356–372. [CrossRef]
40. Lai, D.; Lian, Z.; Liu, W.; Guo, C.; Liu, W.; Liu, K.; Chen, Q. A comprehensive review of thermal comfort studies in urban open spaces. *Sci. Total Environ.* **2020**, *742*, 140092. [CrossRef] [PubMed]
41. Cui, X.; Liang, C.; Wu, J.; Wang, C.; Chen, L.; Liu, H. Forecasting method of tourism meteorological index in Kaiping Diaolou scenic spot. *Guangdong Meteorol.* **2018**, *40*, 57–60, (In Chinese with English Abstract).

sustainability

MDPI

Article

Spatial-Temporal Characteristics of Arctic Summer Climate Comfort Level in the Context of Regional Tourism Resources from 1979 to 2019

Yutao Huang [1], Xuezhen Zhang [2], Dan Zhang [2], Lijuan Zhang [1,*], Wenshuai Zhang [1], Chong Ren [1], Tao Pan [1], Zheng Chu [3] and Yuying Chen [1]

1 Heilongjiang Province Key Laboratory of Geographical Environment Monitoring and Spatial Information Service in Cold Regions, Harbin Normal University, Harbin 150025, China; huangyutao0128@163.com (Y.H.); vincentzhang1993@gmail.com (W.Z.); rch408785202@gmail.com (C.R.); ptaovip@gmail.com (T.P.); chenyuying000@gmail.com (Y.C.)

2 Institute of Geographic Sciences and Natural Resources Research, Chinese Academy of Sciences, Beijing 100101, China; xzzhang@igsnrr.ac.cn (X.Z.); zhangdan@igsnrr.ac.cn (D.Z.)

3 Heilongjiang Meteorological Sciences Research Institute, Harbin 150030, China; chuzheng1985@126.com

* Correspondence: zlj19650205@163.com

Citation: Huang, Y.; Zhang, X.; Zhang, D.; Zhang, L.; Zhang, W.; Ren, C.; Pan, T.; Chu, Z.; Chen, Y. Spatial-Temporal Characteristics of Arctic Summer Climate Comfort Level in the Context of Regional Tourism Resources from 1979 to 2019. *Sustainability* **2021**, *13*, 13056. https://doi.org/10.3390/su132313056

Academic Editor: Oran Young

Received: 17 October 2021
Accepted: 18 November 2021
Published: 25 November 2021

Publisher's Note: MDPI stays neutral with regard to jurisdictional claims in published maps and institutional affiliations.

Abstract: In the context of global warming, a key scientific question for the sustainable development of the Arctic tourism industry is whether the region's climate is becoming more suitable for tourism. Based on the ERA5-HEAT (Human thErmAl comforT) dataset from the European Center for Medium-range Weather Forecasts (ECMWF), this study used statistical methods such as climatic tendency rate and RAPS to analyze the spatial-temporal changes in Arctic summer climate comfort zones from 1979 to 2019 and to explore the influence of changes in climate comfort on Arctic tourism. The results showed the following: (1) With the increase in the Arctic summer temperature, the universal thermal climate index (UTCI) rose significantly from 1979 to 2019 at a rate of 0.457 °C/10a. There was an abrupt change in 2001, when the climate comfort changed from "colder" to "cool", and the climate comfort has remained cool over the past decade (2010–2019). (2) With the increase in Arctic summer temperatures, the area assessed as "comfortable" increased significantly from 1979 to 2019 at a rate of 2.114×10^5 km^2/10a. Compared with the comfortable area in the 1980s, the comfortable area increased by 6.353×10^5 km^2 over the past 10 years and expanded to high-latitude and high-altitude areas, mainly in Kola Peninsula, Putorana Plateau, and Verkhoyansk Mountains in Russia, as well as the Brooks Mountains in Alaska. (3) With the increase in Arctic summer temperatures, the number of days rated comfortable on 30% of the grid increased significantly from 1979 to 2019 (maximum increase: 31 days). The spatial range of the area with a low level of comfortable days narrowed and the spatial range of the area with a high level of such days expanded. The area with 60–70 comfortable days increased the most (4.57×10^5 km^2). The results of this study suggest that global warming exerts a significant influence on the Arctic summer climate comfort level and provides favorable conditions for further development of regional tourism resources.

Keywords: Arctic; universal thermal climate index (UTCI); spatial-temporal changes; 1979–2019

1. Introduction

According to the sixth assessment report of the Intergovernmental Panel on Climate Change (IPCC), the average global surface temperature has increased by 1.09 °C over the past century [1]. The report in the Arctic Climate Impact Assessment also pointed out that, in the context of global warming, climate change in the Arctic is mainly manifested in the melting of ice sheets in the Arctic, the continuous decrease in the volume of snow and frozen soil in the Northern Hemisphere, the frequent occurrence of extreme weather events, and the significant changes in the ecosystem in the Arctic [2]. Accordingly, climate change turned into a severe environmental problem, and soon it will have significant impacts on our lives, our world, and also on tourism [3]. Climate change has affected the length and quality of the tourist season, the choice and consumption of tourist destinations, the decision making of tourist activities, and the operating costs of the tourism industry [4]. Climate change directly causes the redistribution of tourist flows, geographically and seasonally, and climatic conditions that are attractive to tourists may shift to high latitudes and high altitudes [5]. Climate change also indirectly affects tourism by causing water shortages, floods, deterioration of water quality, changes in natural landscapes, reduction in biodiversity, forest fires, and spread of diseases [6,7]. Global warming could lead to the spatial-temporal redistribution of climate comfort [8]. The Arctic is the highest latitude region of the Northern Hemisphere, as well as a region sensitive to global climate change [9]. The heating rate of the Arctic is twice that of the global average [10]; thus, analyzing the characteristics of climate change in the Arctic is of great significance for the development of Arctic tourism resources.

In recent years, tourism in high-latitude regions has become increasingly popular. For instance, the number of tourists in Greenland increased from 26,410 in 1999 to 57,223 in 2008, which was equivalent to the total population of Greenland [11]. The number of summer tourists to Alaska also increased significantly [12]. The average annual growth rate in the number of tourists in Iceland has doubled over the past few years [13]. The increase in the climate comfort level in Norway has led to a huge increase in summer tourism; the number of tourists in the Svalbard Islands, for example, has quadrupled [14]. This significant increase in the number of tourists indicates that the climate comfort level in the Arctic has improved due to global warming, and the need to carry out quantitative studies on these changes is urgent. Roshan et al. [15] used the physiologically equivalent temperature (PET) to identify and evaluate bioclimatic conditions of 40 meteorological stations in Iran. It showed that Iran's comfortable days transferred to early spring and late autumn and an increase in the frequency of thermal comfort condition was observed in almost half of the stations. Roshana et al. [16] also used stations data to show that both SET, PET, and UTCI increased in Iran from 1995 to 2014. Eludoyin et al. [17] have examined the variations in the thermal condition of Nigeria in terms of the temperature, relative humidity, effective temperature (ET), temperature–humidity index (THI), and relative strain index (RSI) and concluded that thermal stress has increased in Nigeria from 2000 at most stations, especially in the south and north-western regions. Feng et al. [18] used the improved classic THI model to analyze the spatiotemporal evolution of annual and monthly average climate comfort in China from 2005 to 2018. The results showed that the annual average comfort level in China changed from cold to comfortable. Wu et al. [19] used 591 stations data to calculate the change of China's climate comfort index values between 1966 and 2016 and came to the same conclusion. In China, the climate comfort levels in high-altitude regions, plateaus, and mountains have improved significantly [20–24]. Zeng et al. [25] investigated the spatial and temporal variations of the universal thermal climate index (UTCI) of the China–Pakistan Economic Corridor (CPEC) from 1979 to 2018 and concluded that the UTCI generally exhibited a positive trend of 0.33 °C/10a over the past 40 years, and the seasonal variation characteristics of the UTCI showed an upward trend in all four seasons. The tourist comfort zone in North America moved from low altitude to high altitude and from low latitude to high latitude [26]. For example, southern Florida, Arizona, and particularly Mexico have decreased the warm tourist season and Canada, Georgia, and South Carolina have extended the warm tourist season. Alaska's tourist season also extended, which now starts 10 days earlier than in the 1940s [20]. Some scholars have analyzed the thermal comfort in different regions of Europe. Napoli et al. have assessed the thermal bioclimate of Europe for the summer season and concluded that an increase in heat stress up to 1 °C was observed from 1979 to 2016 [27]. Some scholars pointed out that southern Europe (Crete Island in Greece, Eastern Mediterranean, Athens, Madrid (Spain), Novi Sad (Vojvodina, Northern Serbia)) were under high temperature and pressure, and climate warming has led to a significant increase in the frequency and duration of high-temperature heat waves in southern Europe and a decline in climate comfort [28–35]. In central Europe, climate change has led to an increase in the number of strong and very strong heat stress days in Poland [36]. Tourists from Germany, Ireland, and the UK would spend more holidays in their home country, and the tourism of Ireland and Britain has moved north [37]. In northern Europe, warming has increased the length of nature and wilderness tourism and the "midnight sun" season in Finland and Sweden [5], and increased comfort in Norway has led to an increase in summer visitors [14]. Climate change could considerably enhance northwestern Europe's comfort level [38,39]. It can be seen that climate changes have led to the redistribution of tourism climate resources in time and space. The existing studies are mostly based on station data analysis, with lower spatial resolution. The area of study is mostly concentrated on the low and middle latitudes of the Northern Hemisphere, but little attention has been paid to the changes in comfort level in the Arctic, where there have been the highest rates of temperature increase.

The climate comfort index, which is widely used in public health, urban planning, tourism, and leisure, is an important index for quantifying the climatic suitability of tourism destinations [40] and as a basis for studying the effect of climate change on tourism [41]. With the measurement of the environment variables, direct indices were initially drawn by the researchers. Including wet-bulb temperature [42], Kata temperature [43], wet-bulb globe temperature (WBGT) [44], temperature–humidity index (THI) [45], effective temperature (ET) [46], and wind chill index (WCI) [47], direct indices were widely used because they are simple and easy to use. As the study of climate comfort went further, the researchers realized that besides environmental factors such as temperature and humidity, behavioral variables such as metabolic rate and the thermal insulation and moisture permeability of clothing that affected the heat comfort should also be considered [48]. Thus, some researchers raised experiential indices, such as the predicted four-hour sweat rate (P4SR), that included the behavioral variables [49]. Because of the lack of the scientific simulation of human response in experiential indices, thermodynamic theorem was used to simulate the heat exchange between the human body and environment, by which climate comfort indices such as physiological equivalent temperature (PET) [50] and the universal thermal climate index (UTCI) were raised. Comparing with the PET index, the environmental and behavioral factors that affected the climate comfort indices were more fully considered in the UTCI. The human multi-joint reaction to the external environment was in the model of climate comfort index to simulate the dynamic physiological response of the pedestrian and construct the thermal stress index the human body accepted during exposure outdoors. According to the thermal stress index, the UTCI is divided into 10 comfort levels from hot to cold [51]. The UTCI is considered the most advanced, comprehensive, and universal human comfort index [52,53], and it is applicable to various climates, seasons, temporal, and spatial scales, as well as being widely used in urban zoning and tourism planning [54,55]. The ERA5-HEAT (Human thErmAl comforT) dataset from the European Center for Medium-range Weather Forecasts (ECMWF) was used in this study. The objectives of the present study were to (1) determine the spatial pattern of Arctic summer climate comfort, (2) reveal the spatial-temporal characteristics of Arctic summer climate comfort, and (3) discuss the influence of climate change on Arctic summer climate comfort. It provides a scientific basis for sustainable development of tourism in the region.

2. Materials and Methods

2.1. Study Area

In this study, 800 km^2 of the northern land in the Arctic Circle (66°34′ N) was taken as the study area, including parts of Greenland (territory of Denmark), Canada, Alaska (the United States), Russia, Norway, Sweden, Finland, and Iceland. Located in the North Frigid Zone, the study area is covered by sea ice year-round and experiences polar weather (Figure 1). The average temperature in the coldest month reaches −40 °C–−20 °C, so the area is an important cold source for the world. The weather is characterized by a long, extreme cold winter, and a short, cool summer; the annual average precipitation is 100–250 mm, which reaches 500 mm in the Greenland sea area, and summer precipitation is usually in the form of rain. Due to its climate characteristics, the Arctic region has unique tourism resources, and current tourism in the region mainly focuses on glaciers, auroras, gorges, endemic species, skiing, underwater diving, and boating on the ice sea.

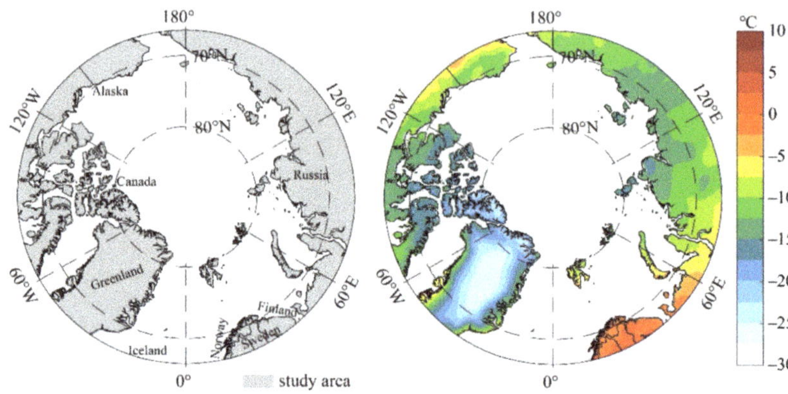

Figure 1. Location of the study area (shaded area) and the average temperature in the Arctic from 1979 to 2019.

2.2. Datasets

1. Thermal comfort data: The climate comfort data were drawn from the ERA5-HEAT dataset of ECMWF from 1979 to the present, with a spatial resolution of 0.25° × 0.25° and a time interval of one hour. The UTCI from 1979 to 2019 was analyzed in this study, and the data were downloaded from the ERA5-HEAT website Available online: https://cds.climate.copernicus.eu/cdsapp#!/dataset/ (accessed on 28 May 2020). According to the thermal physiological response of the human body, the UTCI thermal stress was divided into 10 levels (Table 1).

Table 1. UTCI equivalent temperatures categorized in terms of thermal stress and thermal perception.

UTCI (°C)	Stress Category	Comfort Level	UTCI (°C)	Stress Category	Comfort Level
>46	Extreme heat stress	Extreme hot	0~9	Slight cold stress	Cool
38~46	Very strong heat stress	Very hot	−13~0	Moderate cold stress	Colder
32~38	Strong heat stress	Hot	−27~−13	Strong cold stress	Cold
26~32	Moderate heat stress	Hotter	−40~−27	Very strong cold stress	Very cold
9~26	No thermal stress	Comfort	<−40	Extreme cold stress	Extreme cold

2. Arctic climate data: To analyze the climate background of the changes in Arctic comfort and analyze the changing factors of the UTCI, we chose temperature, wind speed, relative humidity, and sunshine duration for the analysis. These climate elements were obtained from the ERA5 dataset. Again, the time range drawn from ERA5 was from 1979 to the present, the spatial resolution was 0.25° × 0.25°, and the time interval was one hour. The temperature and wind speed data were used directly, whereas the sunshine duration and relative humidity required further processing: sunshine duration was replaced by total cloud cover (TCC) [56] and relative humidity was calculated by the average temperature and dew-point temperature. The specific equation is as follows:

$$f = \frac{E}{EW} \times 100\% \tag{1}$$

$$E = 6.11exp5417.753\{(1/\ 273.16) - [1/(\ 273.16 + Td)]\} \tag{2}$$

$$Ew = 6.11exp5417.753\{(1/\ 273.16) - [1/(\ 273.16 + T)]\} \quad (3)$$

where f is relative humidity, E is actual vapor pressure, E_w is saturated vapor pressure, Td is dew point, and T is temperature.

2.3. Methods

2.3.1. Trend Analysis Method

A univariate linear regression equation of the cold region area variable (y) and the corresponding time (x) was established:

$$y = ax + b \quad (\ i = 1, 2, \ldots, n) \quad (4)$$

where a is the linear regression coefficient indicating the rate of change in the spatial extent of the cold region. The positive or negative value of a indicates that the spatial extent of the cold region is increasing or decreasing over time. The positive value of a indicates the spatial extent of cold region is increasing while the negative value of a indicates the spatial extent of cold region is decreasing.

2.3.2. Rescaled Adjusted Partial SUMS (RAPS) Method

The rescaled adjusted partial sums (RAPS) visualization approach can be used for preliminary visual inspection of a time series, to gain a feel for the data, and/or to guide and focus subsequent statistical tests and analyses [57]. The RAPS of the X_k are defined as follows:

$$X_k = \sum_{t=1}^{k} \frac{Y_t - \overline{Y}}{S_Y};\ k = 1,\ \ldots, n \quad (5)$$

where $Y = [Yt;\ t = 1, \ldots, n]$ represent a time series of a climatic variable, $\overline{Y}$ is sample mean, ${S_Y}^2$ is variance, n is number of values in the time series, and k is counter limit of the current summation.

2.3.3. Correlation Analysis

Correlation analysis is a statistical method used to discover if there is a relationship between two variables/datasets, and the relatedness and negative/positive correlation of this relationship. The Pearson correlation method is adopted in this paper to analyze the relationship between comfortable area, comfortable days, and climate indicators, quantitatively. In each pair, the Pearson's correlation coefficient (r) is calculated as:

$$r = \frac{\sum_{i=1}^{n}(x_i - \overline{x})(y_i - \overline{y})}{\sqrt{\sum_{i=1}^{n}(x_i - \overline{x})^2}\sqrt{\sum_{i=1}^{n}(y_i - \overline{y})^2}} \quad (6)$$

where $\overline{x}$ represents the average comfortable areas or days and $\overline{y}$ is the average of a given climate variable.

The study based on the UTCI data in the ERA5-HEAT dataset provided by ECMWF used MATLAB software for spatial data calculation and overlay analysis (Figure 2). We used the Pcolor function of MATLAB to achieve graphical visualization.

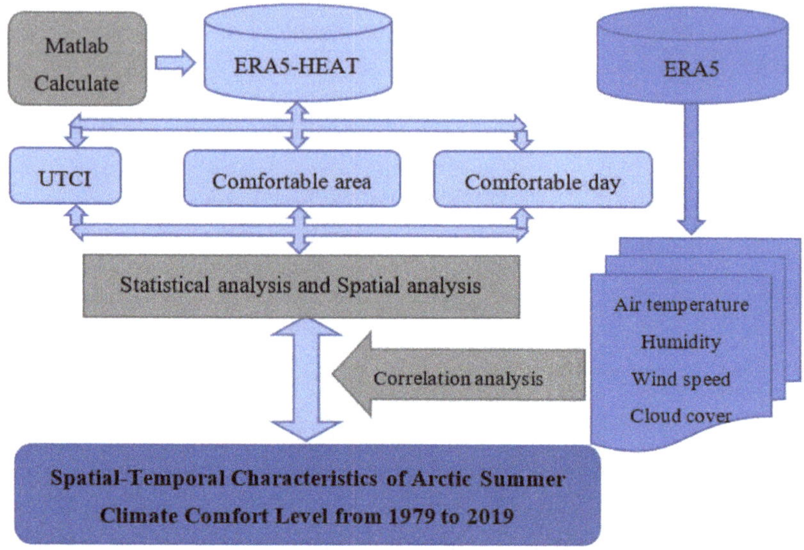

Figure 2. The framework of the study.

3. Results

3.1. Change of UTCI in the Arctic Summer (1979–2019)

The average UTCI value in the Arctic summer from 1979 to 2019 was 0.025 °C and the comfort level was "cool". The UTCI value increased significantly at a rate of 0.457 °C/10a ($p < 0.01$), for a total increase of 1.874 °C, and the thermal stress level also increased (Figure 3a). The results of RAPS analysis showed (Figure 3b) that there was an obvious change in the UTCI value in the Arctic summer in 2000, which indicates that a transformation occurred in the comfort level in the early 21st century. The UTCI anomaly map (Figure 3c) also shows that the negative anomaly changed to a positive anomaly in the early 21st century (2001) and the UTCI changed from a relatively low value stage to a relatively high value stage. Further analysis reveals that the UTCI value was −0.501 °C and the comfort level was "colder" before the abrupt change (1979–2000). After the abrupt change (2001–2019), the UTCI value was 0.634 °C and the comfort level was "cool", which indicates that there was an obvious change in the comfort level of the Arctic summer in the 21st century.

Figure 3. Change in Arctic summer UTCI from 1979 to 2019: (**a**) time series, (**b**) RAPS, (**c**) anomaly series.

3.2. Change in the Area Rated "Comfortable" in the Arctic Summer

It can be seen from Section 3.1 that the spatial distribution of comfortable areas expanded with the increase in the UTCI value for the Arctic summer from 1979 to 2019 (Figure 4a). The area of summer comfortable zones in the Arctic from 1979 to 2019 was 1.541×10^6 km^2, covering 19.152% (i.e., nearly 1/5) of the total Arctic land area. The comfortable area in the decade before and after was 1.218×10^6 and 1.864×10^6 km^2, respectively. The summer comfortable area was largest in 2019, at 2.44×10^6 km^2, and smallest in 1996, at only 0.921×10^6 km^2, with an annual variation coefficient of 23.304%. The increase rate for the summer comfortable area in the Arctic from 1979 to 2019 was 2.114×10^5 km^2/10a, which was significant ($p < 0.01$). The total increase area was 8.667×10^5 km^2, which indicates that the summer comfortable area in the Arctic has been expanding. The RAPS analysis results indicate (Figure 4b) that there was an obvious change in the size of the Arctic summer comfortable area in 2000, which was consistent with the change in the UTCI value. The comfortable area before and after the abrupt change was 1.307×10^6 and 1.811×10^6 km^2, respectively. This finding was further verified by the area anomaly (Figure 4c), which indicates that there was a significant expansion in the Arctic summer comfortable area in the early 21st century.

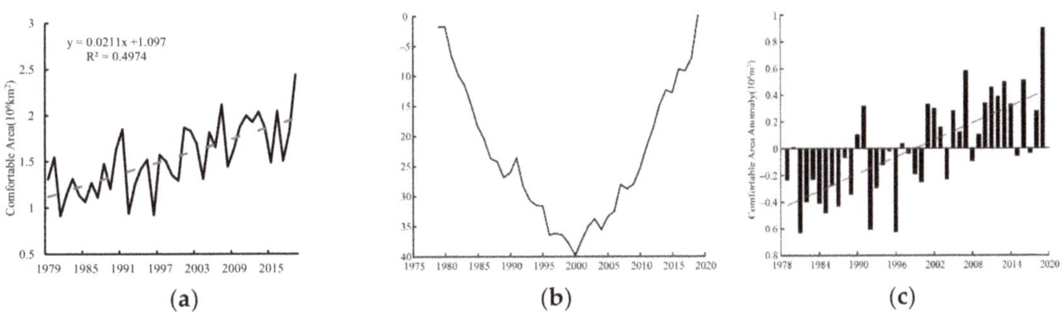

Figure 4. Change in Arctic summer comfortable area from 1979 to 2019: (**a**) time series, (**b**) RAPS, (**c**) anomaly series.

The spatial distribution of the areas with different comfort levels in the Arctic over the past ten years was further analyzed (Figure 5). The Arctic summer comfortable area from 2010 to 2019 was 1.864×10^6 km^2, accounting for 23.2% of the Arctic land area. These comfortable areas were mainly distributed in Eurasia and North America (66.5°–70 °N), mid-eastern Russia, Norway, Sweden, and Finland in northern Europe, Alaska and the Yukon territory in Canada, as well as scattered areas along the coast of Greenland and islands of Canada, north of 70 °N (Figure 5a). Compared with the ten years from 1979 to 1988, the Arctic summer comfortable area increased by 6.353×10^5 km^2, accounting for 7.9% of the Arctic land area (Figure 5b). In terms of spatial distribution, the comfortable area mainly expanded to high-latitude and high-altitude regions (66.5°–70 °N) and, specifically, in the Kola Peninsula, Putorana Plateau, Verkhoyansk Mountains in Russia, and Brooks Mountains in Alaska (Figure 5c). It is, therefore, possible for tourism to be developed in the Arctic regions rated comfortable, especially those newly designated as comfortable. Sightseeing in gorges, fishing, and watching rare species can be carried out in the peninsula regions and adventures, such as polar mountain climbing and outdoor expeditions, can be developed in the mountainous and plateau regions.

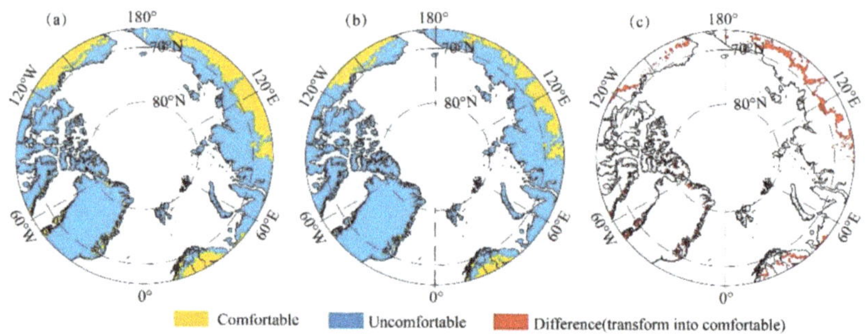

Figure 5. Spatial distribution and variation of Arctic summer comfortable area: (**a**) 2010–2019, (**b**) 1979–1988, (**c**) spatial variation of the decade before and after.

3.3. Change in the Number of Arctic Summer Comfortable Days from 1979 to 2019

There was an average of 23 Arctic summer days rated comfortable from 1979 to 2019, accounting for a quarter (25%) of the entire summer, while in the decade before and after there were 21 and 26 such days, respectively. The largest number of comfortable days was seen in 2019 (30 d), and the smallest was 18 d in 1992. The annual variation coefficient was 10.955%. The increase rate in comfortable days from 1979 to 2019 was 1.458 d/10a, which was significant ($p < 0.01$), and the total increase was 6 d, which indicates that the season for Arctic summer tourism activities has been significantly extended (Figure 6a). The RAPS analysis results showed that there was an obvious change in the number of comfortable days around 2000, which is consistent with the changes in UTCI values (Figure 6b). The number of comfortable days before and after the abrupt change was 22 d and 25 d, respectively. This finding was further verified by the anomaly of days (Figure 6c), indicating that a significant change occurred in the number of comfortable days in the early 21st century.

Figure 6. Change in the number of Arctic summer comfortable days from 1979 to 2019: (**a**) time series, (**b**) RAPS, (**c**) anomaly series.

Figure 7 shows the spatial distribution of Arctic summer comfortable days from 1979 to 2019. The number of days in all grid cells rated comfortable has clearly been increasing, the area with a statistically significant increase accounted for 30% of the total Arctic area, and the increase rate ranged −1.972–7.498 d/10a. The largest increase in the number of comfortable days was in the Qeqertarsuaq region of western Greenland (31 d) and the second largest increase was in northern and eastern Siberia of Russia (8–24 d). The summer tourism seasons were extended in the areas with a large increase in the number of comfortable days, and the tourism structure could be adjusted appropriately to invest in tourism infrastructures to meet the needs created by this extension.

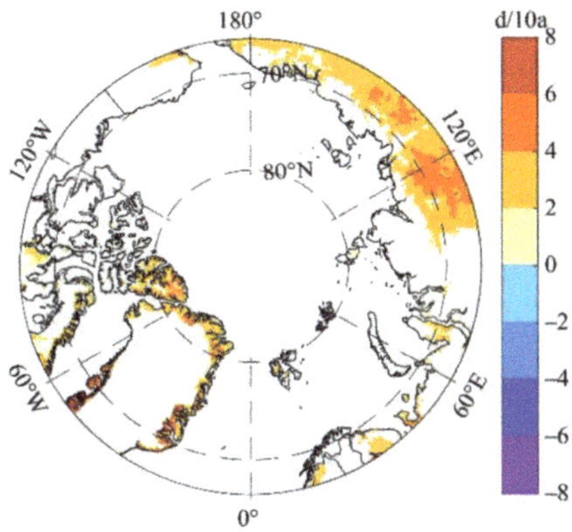

Figure 7. Rate of spatial change for the number of Arctic summer comfortable days from 1979 to 2019.

These Arctic summer comfortable days not only appear in the increase in grid cells but also in the change of space areas with the same level of comfortable days. Table 2 displays the area distribution of comfortable days at various frequency levels between the period of 1979–1988 and 2010–2019. The areas with 0–10, 10–20, 20–30, and 40–50 comfortable days decreased from 1979 to 1988 and from 2010 to 2019, with regions experiencing only 0–10 comfortable days declining most (by 2.96×10^5 km^2). Areas with 30–40, 50–60, 60–70, 70–80, and 80–90 comfortable days increased, with areas having 60–70 comfortable days increasing most (4.57×10^5 km^2).

Table 2. Areas of comfort days at different levels in 1979–1988 and 2010–2019.

(10^5 km^2)	0–10 d	11–20 d	21–30 d	31–40 d	41–50 d	51–60 d	61–70 d	71–80 d	81–92 d
2010–2019	9.74	12.79	11.90	9.57	8.07	8.93	6.66	1.67	0.04
1979–1988	12.70	15.13	12.12	9.54	9.05	7.02	2.09	0.64	0.00
Difference	−2.96	−2.34	−0.22	0.03	−0.98	1.91	4.57	1.03	0.04

In the past ten years (2010–2019), there was an average of 26 comfortable Arctic summer days, which were distributed in North America and Eurasia, excluding Greenland (Figure 8a). These basically followed a zonal distribution and gradually decreased with an increase in latitude. Mid-eastern Russia, northern Europe, Canada, and Alaska (66.5°–70 °N) had a large number of comfortable days (maximum: 90 d) and the regions north of 70 °N had a smaller number of comfortable days (average: 20 d), accounting for 21.74% of the Arctic summer. Compared with the early 1980s (Figure 8b), the number of comfortable days increased by 5 d.

In summary, there was an obvious increase in the range and number of comfortable days (between 10 d and 30 d) in the Arctic, which were mainly distributed in mid-eastern Russia (66.5°–70 °N), Greenland, and Ellesmere Island, north of 80 °N (Figure 8c). The increase in the number of comfortable days extends the summer tourism season in the Arctic, providing tourists more time to enjoy the natural scenery and creating the potential for higher economic benefits.

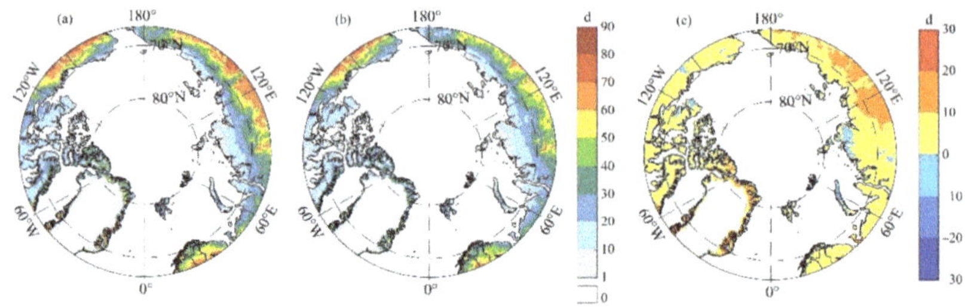

Figure 8. Spatial distribution and change in the number of Arctic summer comfortable days from 1979 to 1988 and from 2010 to 2019: (**a**) 2010–2019, (**b**) 1979–1988, (**c**) Spatial transformation.

4. Discussion

As an area sensitive to global warming, the climate comfort level in the Arctic has changed with the warming climate, which has had positive effects on the regional tourism industry. Therefore, it is of great significance to study the changes in the areas of the Arctic that can be rated comfortable during the summer, with a particular focus on the influence of climate comfort level on tourism in the context of global warming. In this study, a quantitative analysis was carried out on the spatial-temporal evolution of the Arctic summer climate comfort zones from 1979 to 2019, as well as the influence of the change in the climate comfort levels on Arctic tourism. The results of this analysis may be helpful in optimizing the Arctic tourism industry, promoting the development of tourism destinations, and providing scientific and quantitative support for the development of Arctic tourism. The following aspects were discussed in this study.

Factors that influence the climate comfort level include temperature, relative humidity, wind speed, and sunshine duration. The relationship between meteorological elements and climate comfort level was analyzed based on the ERA5-HEAT dataset. Figure 9 shows the interannual changes in Arctic summer average temperature, relative humidity, wind speed, and TCC from 1979 to 2019, which had averages of 3.809 °C, 79.461%, 4.128 m/s, and 70.249%, respectively. The temperature rose significantly from 1979 to 2019 ($p < 0.01$), with a rate of 0.418 °C/10a and a total increase of 1.714 °C. The relative humidity, wind speed, and TCC fluctuated greatly, but the change was not significant ($p > 0.05$), which indicates that the temperature increase was the main climate change factor in the Arctic.

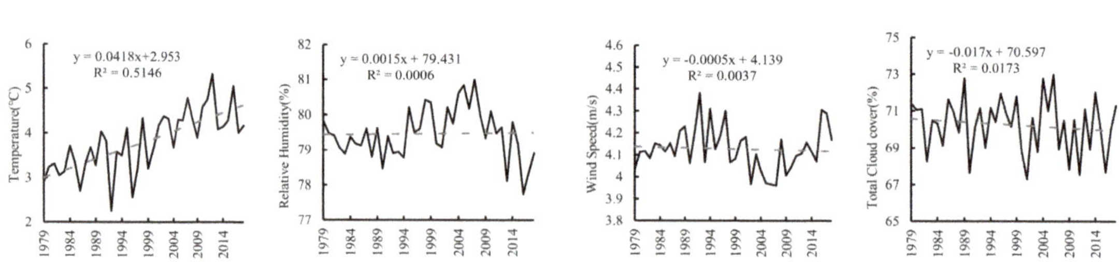

Figure 9. Changes in the average temperature, relative humidity, wind speed, and total cloud cover in the Arctic summer from 1979 to 2019.

The correlation between the comfortable area and the number of comfortable days and various meteorological factors was also analyzed (Table 3). The comfortable area and number of days were positively correlated with the average temperature ($p < 0.01$) and negatively correlated with TCC; the positive correlation coefficient was much larger than the negative correlation coefficient. Thus, it appears that the comfortable area and days are mainly affected by the temperature and TCC, with the temperature being the main factor.

Table 3. Correlation coefficients between comfortable area, comfortable days, and meteorological factors in arctic summer from 1979 to 2019.

	Temperature	Relative Humidity	Wind Speeds	TCC
Comfortable Area	0.818 **	0.07	−0.218	−0.340 *
Comfortable Days	0.944 **	0.086	−0.309	−0.514 **

Note: "*", "**": Significance at 0.05 and 0.01 levels.

The spatial correlation coefficients between the number of comfortable days and temperature, relative humidity, wind speed, and TCC in the Arctic summer are analyzed (Figure 10). It showed that there are significant spatial correlations between comfortable days and temperature, relative humidity, wind speed, and TCC, which are basically distributed between 66.5–70 °N and can reach 80 °N in areas such as Canada's Queen Elizabeth Islands and Ellesmere Island, indicating that the Arctic summer comfortable days of 66.5–70 °N is related to temperature, relative humidity, wind speed, and TCC. However, there are differences between different elements in the relevant regions. Statistics showed that the areas where summer comfortable days are significantly correlated with temperature, relative humidity, wind speed, and TCC are 5.719×10^6, 3.24×10^6, 3.974×10^6, and 5.412×10^6 km^2, respectively. The average significant correlation coefficients are 0.776, −0.447, −0.506, and −0.609, respectively, indicating that the Arctic summer comfortable days have a more significant correlation with temperature, and temperature is the most important factor affecting summer climate comfort.

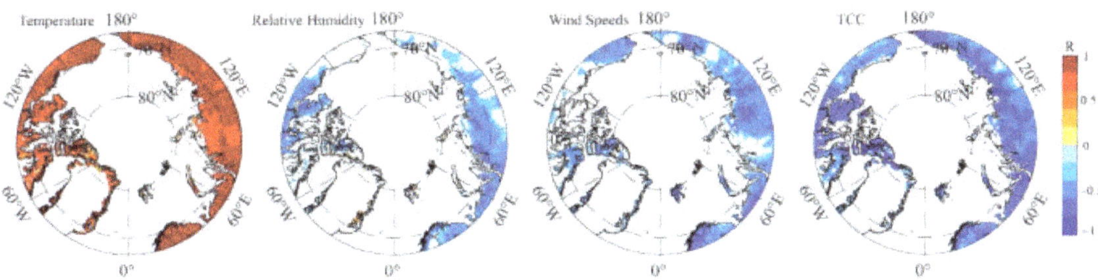

Figure 10. Spatial correlation coefficients between the number of comfortable days and temperature, relative humidity, wind speed, and TCC in Arctic summer from 1979 to 2019 (R: correlation coefficient).

The spatial pattern of and change in climate comfort levels in the peri-Arctic nations and regions has been previously studied. For instance, the spatial pattern of the summer UTCI index in Russia from 2001 to 2015 was analyzed by Vinogradova et al. [58] based on data from 500 meteorological stations in Russia. The results showed that the comfortable zones were distributed in most areas of Russia in summer, and a moderate thermal stress level was observed in daytime in southern Europe. In a study by Yu et al., the hourly weather data from meteorological stations were used to calculate the modified climate index for tourism [20]. The authors suggested that the weather conditions in parts of regions in Alaska had improved greatly from 1942 to 2005, which extended the tourism season. Based on the temperature, precipitation, TCC, and visibility data, Forland et al. found that "warm" weather in Norway, north of the Arctic Circle, increased from 1981 to 2010, the summer tourism season was extended, and the improvement in the summer climate comfort level led to an increase in the number of tourists, quadrupling in the Svalbard Islands [14]. According to statistics, from 1990 to 2005, the annual growth rate of Iceland's Arctic cruise tourists was 19.19%, from 7952 in 1990 to 266,070 in 2015, an increase of more than 30 times; in 2019, the number of overnight tourists in Greenland increased to 264,711 people, equivalent to 3.47 times the total population of the island; from 1996 to 2018, cruise tourism in Svalbard also increased greatly, and cruise tourists increased by 1.94 times [59]. These studies, to some extent, verify the findings of the present study, which showed that the Arctic summer comfortable zones have expanded and the tourism areas increased, while the number of comfortable days has significantly increased and the summer tourism season has been extended.

The results of this study showed that the Arctic summer comfortable zones have expanded and there are more comfortable days, which has created favorable conditions for tourism. In recent years, Norway, Iceland, Russia, and Canada have set up Arctic tourism routes and promoted the development of ecological tourism in the region [60]. The Arctic is the region least affected by human activities in the world, and unplanned or inadequate development of tourism resources could eventually lead to the degradation of the natural environment. Tourists are also likely to bring new species and germs into the region [61]. To achieve the sustainable development of Arctic tourism, the various countries involved should adhere to the concept of global integration and cooperate to complete legal systems for Arctic tourism development to ensure ecological balanced and sustainable growth in the region.

5. Conclusions

The average summer UTCI value in the Arctic was 0.025 °C from 1979 to 2019 and the comfort level was cool. The UTCI increased significantly at a rate of 0.457 °C/10a and there was an abrupt change in 2001. The Arctic summer comfort level changed from colder to cool after this abrupt change.

The average comfortable area in the Arctic summer from 1979 to 2019 was 1.541×10^6 km^2, accounting for a fifth of the Arctic land area. These comfortable zones were mainly distributed in mid-eastern Russia, Norway, Sweden, and Finland in northern Europe, as well as Alaska and the Yukon territory in Canada. The comfortable area increased significantly at a rate of 2.114×10^5 km^2/10a from 1979 to 2019. Compared with the 1980s, the comfortable area has increased by 6.353×10^5 km^2 in recent decades, mainly expanding to high-latitude and high-altitude regions, specifically distributed in the Kola Peninsula, Putorana Plateau, Verkhoyansk Mountains in Russia, and Brooks Mountains in Alaska.

The average number of Arctic summer comfortable days was 23 from 1979 to 2019, accounting for a quarter of the entire summer. These were distributed zonally and gradually decreased from low to high latitudes. The number of comfortable days increased significantly from 1979 to 2019, at a rate of 1.458 d/10a. Compared with the early 1980s, the number of comfortable days has increased by 5 d over the past decades. The range and number of summer comfortable days has increased significantly, and the largest number of comfortable days was seen in the Qeqertarsuaq region of western Greenland (31 d), while the area seeing 60–70 summer comfortable days expanded the most (by 4.57×10^5 km^2).

This study is only limited to the Arctic region and does not conduct a detailed analysis of the Northern Hemisphere or typical Arctic tourism cities. It has not conducted an in-depth study on how the Arctic climate comfort changes under the background of future climate change. It will be subject to special research in the future.

The improvement of Arctic summer climate comfort and the extension of suitable tourist seasons will enhance the attractiveness of Arctic tourist destinations while promoting the development and utilization of Arctic natural resources. Sightseeing in gorges, fishing, and watching rare species can be carried out in the peninsula regions and adventures, such as polar mountain climbing and outdoor expeditions, can be developed in the mountainous and plateau regions. Despite vigorously developing Arctic tourism activities, it is necessary to protect and manage Arctic tourism resources to ensure the sustainable development of Arctic tourism.

Author Contributions: Y.H. analyzed the data and drafted the manuscript; L.Z., D.Z. and X.Z. completed the manuscript and made major revisions; W.Z. completed data visualization; T.P. and C.R. downloaded data and searched references; Y.C. and Z.C. checked and proofread the manuscript. All authors have read and agreed to the published version of the manuscript.

Funding: This research was funded by National Natural Science Foundation of China (Grant No. 41771067. Supported by the National Key R&D Program of China (Grant No. 2017YFA0603301).

Institutional Review Board Statement: Not applicable.

Informed Consent Statement: Not applicable.

Data Availability Statement: The data presented in this study are available on request from the corresponding author.

Acknowledgments: Thanks to ECMWF for providing the ERA5-HEAT dataset.

Conflicts of Interest: The authors declare no conflict of interest.

References

1. IPCC. 2021: Summary for Policymakers. In *Climate Change 2021: The Physical Science Basis. Contribution of Working Group I to the Sixth Assessment Report of the Intergovernmental Panel on Climate Change*; Cambridge University Press: Cambridg, UK, 2021.
2. Chen, L.Q. Evidence of Arctic and Antarctic changes and their regulation of global climate change (further findings since the fourth IPCC assessment report released). *Adv. Polar Sci.* **2013**, *25*, 1–6. (In Chinese) [CrossRef]
3. Semenza, J.C.; Ebi, K.L. Climate change impact on migration, travel, travel destinations and the tourism industry. *J. Travel Med.* **2019**, *26*, 1–13. [CrossRef]
4. Lise, W.; Tol, R.S.J. Impact of climate on tourist demand. *Clim. Chang.* **2002**, *55*, 429–449. [CrossRef]
5. Hall, C.M.; Saarinen, J. Polar Tourism: Definitions and Dimensions. *Scand. J. Hosp. Tour.* **2010**, *10*, 448–467. [CrossRef]
6. Hansen, B.B.; Isaksen, K.; Benestad, R.E.; Kohler, J.; Pedersen, A.Ø.; Loe, L.E.; Coulson, S.J.; Larsen, J.O.; Varpe, Ø. Warmer and wetter winters: Characteristics and implications of an extreme weather event in the High Arctic. *Environ. Res. Lett.* **2014**, *9*, 114021. [CrossRef]
7. Hall, C.M.; Baird, T.; James, M.; Ram, Y. Climate change and cultural heritage: Conservation and heritage tourism in the Anthropocene. *J. Herit. Tour.* **2016**, *11*, 10–24. [CrossRef]
8. Scott, D.; Wall, G.; McBoyle, G. Chapter 7. Climate Change and Tourism and Recreation in North America: Exploring Regional Risks and Opportunities. In *Tourism, Recreation and Climate Change*; Hall, C.M., Higham, J., Eds.; Channel View Publications: Blue Ridge Summit, PA, USA; Bristol, UK, 2005; Volume 22, pp. 115–129. [CrossRef]
9. Yu, W.Y.; Wang, L.; Feng, Z.G.; Zhang, C.Y. Bibliometric analysis and development trends of global Arctic marine scientific research. *Mar. Sci.* **2018**, *42*, 64–81. (In Chinese)
10. Notz, D.; Stroeve, J. Observed Arctic sea-ice loss directly follows anthropogenic CO_2 emission. *Science* **2016**, *354*, 747–750. [CrossRef]
11. Fay, G.; Karlsdóttir, A. Social indicators for arctic tourism: Observing trends and assessing data. *Polar Geogr.* **2011**, *34*, 63–86. [CrossRef]
12. Suzanne, D.L.B.; Maher, P.; Dawson, J.; Hillmer-Pegram, K.; Huijbens, E.; Lamers, M.; Liggett, D.; Müller, D.; Pashkevich, A.; Stewart, E. Tourism and Arctic Observation Systems: Exploring the relationships. *Polar Res.* **2016**, *35*, 24980. [CrossRef]
13. Stephen, K. Societal Impacts of a Rapidly Changing Arctic. *Curr. Clim. Chang. Rep.* **2018**, *4*, 223–237. [CrossRef]
14. Forland, E.J.; Jacobsen, J.; Denstadli, J.M.; Lohmann, M.; Hanssen-Bauer, I.; Hygen, H.O.; Tommervik, H. Cool weather tourism under global warming: Comparing arctic summer tourists' weather preferences with regional climate statistics and projections. *Tour. Manag.* **2013**, *36*, 567–579. [CrossRef]
15. Roshan, G.; Yousefi, R.; Kovacs, A.; Matzarakis, A. A comprehensive analysis of physiologically equivalent temperature changes of Iranian selected stations for the last half century. *Theor. Appl. Climatol.* **2018**, *131*, 19–41. [CrossRef]
16. Roshan, G.; Yousefi, R.; Błażejczyk, K. Assessment of the climatic potential for tourism in Iran through biometeorology clustering. *Int. J. Biometeorol.* **2017**, *62*, 525–542. [CrossRef] [PubMed]
17. Eludoyin, O.; Adelekan, I.; Webster, R.; Eludoyin, A. Air temperature, relative humidity, climate regionalization and thermal comfort of Nigeria. *Int. J. Climatol.* **2014**, *34*, 2000–2018. [CrossRef]
18. Feng, L.; Liu, Y.X.; Feng, Z.Z.; Yang, S.Q. Analysing the spatiotemporal characteristics of climate comfort in China based on 2005–2018 MODIS data. *Theor. Appl. Climatol.* **2021**, *143*, 1235–1249. [CrossRef]
19. Wu, F.F.; Yang, X.H.; Shen, Z.Y.; Yi, Z.J. Long-term trends and spatiotemporal variations of climate comfort in China during 1966–2016. *Therm. Sci.* **2020**, *24*, 2445–2453. [CrossRef]
20. Yu, G.; Schwartz, Z.; Walsh, S. Effects of climate change on the seasonality of weather for tourism in Alaska. *Arctic* **2009**, *62*, 443–457. [CrossRef]
21. Yu, Z.K.; Sun, G.N.; Feng, Q.; Luo, Z.W. Spatiotemporal dynamic analysis of tourism climate comfort and climate risk in the Tibetan Plateau. *Resour. Sci.* **2014**, *36*, 2327–2336. (In Chinese)

22. Jin, A.Q.; Zhang, A.; Zhao, X.Y. Prediction of future climate comfort in eastern China under climate change scenario. *Acta Sci. Nat. Univ. Pekin.* **2019**, *55*, 887–898. (In Chinese)
23. Shi, L.; Huang, X.Q.; Ni, M.J.; Ci, R.Y.Z.; Luo, Z. Analysis on the adaptability of tourism climate in Tibet Autonomous Region. *J. Glaciol. Geocryol.* **2015**, *37*, 1412–1419. (In Chinese)
24. Sun, Y.C.; Wang, S.Y.; Li, H.; Zheng, G.F.; Wang, F.; Guan, J.D. Climate comfort analysis of summer tourism in Liu Pan Mountain Area of Ningxia. *J. Arid Meteorol.* **2018**, *36*, 1035–1042. (In Chinese)
25. Zeng, D.; Wu, J.; Mu, Y. Spatial-Temporal Pattern Changes of UTCI in the China-Pakistan Economic Corridor in Recent 40 Years. *Atmosphere* **2020**, *11*, 858. [CrossRef]
26. Scott, D.; Mcboyle, G.; Schwartzentruber, M. Climate change and the distribution of climatic resources for tourism in North America. *Clim. Res.* **2004**, *27*, 105–117. [CrossRef]
27. Napoli, C.; Pappenberger, F.; Cloke, H. Assessing heat-related health risk in Europe via the Universal Thermal Climate Index (UTCI). *Int. J. Biometeorol.* **2018**, *62*, 1155. [CrossRef] [PubMed]
28. Basarin, B.; Luki, T.; Matzarakis, A. Review of Biometeorology of Heatwaves and Warm Extremes in Europe. *Atmosphere* **2020**, *11*, 1276. [CrossRef]
29. Basarin, B.; Lukic, T.; Matzarakis, A. Quantification and assessment of heat and cold waves in Novi Sad, Northern Serbia. *Int. J. Biometeorol.* **2016**, *60*, 139–150. [CrossRef]
30. Nastos, P.T.; Matzarakis, A. The effect of air temperature and human thermal indices on mortality in Athens, Greece. *Theor. Appl. Climatol.* **2012**, *108*, 591–599. [CrossRef]
31. Bleta, A.; Nastos, P.T.; Matzarakis, A. Assessment of bioclimatic conditions on Crete Island, Greece. *Reg. Environ. Chang.* **2014**, *14*, 1967–1981. [CrossRef]
32. Katavoutas, G.; Founda, D. Response of Urban Heat Stress to Heat Waves in Athens (1960–2017). *Atmosphere* **2019**, *10*, 483. [CrossRef]
33. Katavoutas, G.; Founda, D. Intensification of thermal risk in Mediterranean climates: Evidence from the comparison of rational and simple indices. *Int. J. Biometeorol.* **2019**, *63*, 1251–1264. [CrossRef]
34. Pecelj, M.M.; Luki´c, M.Z.; Filipovi´c, D.J.; Proti´c, B.M. Summer variation of the UTCI index and Heat Waves in Serbia. *Nat. Hazards Earth Syst. Sci.* **2019**, 1–19. [CrossRef]
35. Rasilla, D.; Allende, F.; Martilli, A.; Fernández, F. Heat Waves and Human Well-Being in Madrid (Spain). *Atmosphere* **2019**, *10*, 288. [CrossRef]
36. Wereski, S.; Krzyżewska, A.; Dobek, M. Winter UTCI variability in Poland in the 21st century. *Reg. Stud. Dev.* **2020**, *24*, 128–137. [CrossRef]
37. Hamilton, J.M.; Tol, R. The impact of climate change on tourism in Germany, the UK and Ireland: A simulation study. *Reg. Environ. Chang.* **2007**, *7*, 161–172. [CrossRef]
38. Nicholls, S.; Amelung, B. Implications of climate change for rural tourism in the nordic region. *Scand. J. Hosp. Tour.* **2015**, *15*, 48–72. [CrossRef]
39. Nicholls, S.; Amelung, B. Climate change and tourism in northwestern europe: Impacts and adaptation. *Tour. Anal.* **2008**, *13*, 21–31. [CrossRef]
40. Li, S.; Sun, M.S.; Zhang, W.J.; Tan, L.; Zhu, N.N.; Wang, Y.F. Spatial pattern and evolution of climatic comfort period in mainland China from 1961 to 2010. *Geogr. Res.* **2016**, *35*, 2053–2070. (In Chinese)
41. Zhong, L.S.; Yu, H.; Zeng, Y.X. Impact of climate change on Tibet tourism based on tourism climate index. *J. Geogr. Sci.* **2019**, *29*, 2085–2100. [CrossRef]
42. Haldane, J. The Influence of High Air Temperatures No. *I. Epidemiol. Infect.* **1905**, *5*, 494–513. [CrossRef]
43. Hill, L.; Griffith, O.W.; Flack, M. The measurement of the rate of heat loss at body temperature by convection, radiation and evaporation. *Philos. Trans. R. Soc.* **1916**, *207*, 225–347. [CrossRef]
44. Yaglou, C.P.; Minard, D. Control of heat casualties at military training centers. *Am. Med. Assoc. Arch. Ind. Health* **1957**, *16*, 302–316.
45. Thorn, E.C. The discomfort index. *Weathenvise* **1959**, *12*, 57–60.
46. Houghton, F.C.; Yaglou, C.P. Determining equal comfort lines. *J. Am. Soc. Heat. Vent. Eng.* **1923**, *29*, 165–176.
47. Siple, P.; Passel, C.F. Measurements of dry atmospheric cooling in subfreezing temperature. *Proc. Am. Philos. Soc.* **1945**, *89*, 177–199. [CrossRef]
48. Fanger, P.O. Thermal comfort: Analysis and applications in environmental engineering. In *Thermal Comfort Analysis & Applications in Environmental Engineering*; Danish Technical Press: Copenhagen, Denmark, 1970; pp. 225–240.
49. McArdle, B.; Dunham, W.; Holling, H.E.; Ladell, W.S.S.; Scott, J.W.; Thomson, M.L.; Weiner, J.S.; Holing, H. The prediction of the physiological effects of warm and hot Environments. In *Renewable Northwest Project Report*; Medical Resource Council: London, UK, 1947; pp. 47–391.
50. Hoeppe, P. *Die Energiebilanz des Menschen*; Muenchen Univ. Press: München, Germany, 1984.
51. Blazejczyk, K.; Epstein, Y.; Jendritzky, G.; Staiger, H.; Tinz, B. Comparison of UTCI to selected thermal indices. *Int. J. Biometeorol.* **2012**, *56*, 515–535. [CrossRef]
52. Jendritzky, G.; De, D.R.; Havenith, G. UTCI-why another thermal index? *Int. J. Biometeorol.* **2012**, *56*, 421–428. [CrossRef] [PubMed]

53. Brode, P.; Fiala, D.; Blazejczyk, K.; Holmer, I.; Jendritzky, G.; Kampmann, B.; Tinz, B.; Havenith, G. Deriving the operational procedure for the Universal Thermal Climate Index (UTCI). *Int. J. Biometeorol.* **2012**, *56*, 481–494. [CrossRef]
54. Kong, Q.Q.; Zheng, J.Y.; Wang, X.G. Spatial pattern and temporal variation in thermal comfort in China from 1979 to 2014. *Resour. Sci.* **2016**, *38*, 1129–1139. (In Chinese)
55. Zhang, X.Y.; Jiang, C.; Sun, J.X.; Zhou, M.F. Spatiotemporal variation characteristics and influencing factors of climate comfort at different altitudes. *Chin. J. Appl. Ecol.* **2018**, *29*, 2808–2818. (In Chinese)
56. Hoyt, D.V. Percent of possible sunshine and the total cloud cover. *Mon. Weather Rev.* **1977**, *105*, 648–652. [CrossRef]
57. Garbrecht, J.; Fernandez, G.P. Visualization of trends and fluctuations in climatic records1. *J. Am. Water Resour. Assoc.* **2010**, *30*, 297–306. [CrossRef]
58. Vinogradova, V. Using the Universal Thermal Climate Index (UTCI) for the assessment of bioclimatic conditions in Russia. *Int. J. Biometeorol.* **2021**, *65*, 1473–1483. [CrossRef] [PubMed]
59. Zhang, D.; Wang, M.; Gan, M.Y.; Yang, L.S. Research progress on influencing factors of Arctic tourism. *Resour. Sci.* **2021**, *43*, 1687–1699. [CrossRef]
60. Lan, C.Z. Polar tourism: High-end tourism and sustainable development under unique resource conditions. *World Environ.* **2017**, *3*, 70–73. (In Chinese)
61. Luo, M.; Guo, Y.A. Legal dilemma and solution for the development of tourism resources in the Arctic region. *Chin. J. Marit. Law* **2020**, *31*, 58–66. (In Chinese)

sustainability

MDPI

Article

How the Updated Earth System Models Project Terrestrial Gross Primary Productivity in China under 1.5 and 2 °C Global Warming

Chi Zhang [1], Shaohong Wu [1], Yu Deng [2,*] and Jieming Chou [3]

[1] Key Laboratory of Land Surface Pattern and Simulation, Institute of Geographic Sciences and Natural Resources Research, Chinese Academy of Sciences, Beijing 100101, China; zhangchi@igsnrr.ac.cn (C.Z.); wush@igsnrr.ac.cn (S.W.)

[2] Key Laboratory of Regional Sustainable Development Modeling, Institute of Geographic Sciences and Natural Resources Research, Chinese Academy of Sciences, Beijing 100101, China

[3] State Key Laboratory of Earth Surface Processes and Resource Ecology, Beijing Normal University, Beijing 100875, China; choujm@bnu.edu.cn

* Correspondence: rain00788@163.com

Abstract: Three Earth system models (ESMs) from the Coupled Model Intercomparison Project phase 6 (CMIP6) were chosen to project ecosystem changes under 1.5 and 2 °C global warming targets in the Shared Socioeconomic Pathway 4.5 W m^{-2} (SSP245) scenario. Annual terrestrial gross primary productivity (GPP) was taken as the representative ecological indicator of the ecosystem. Under 1.5 °C global warming, GPP in four climate zones—i.e., temperate continental; temperate monsoonal; subtropical–tropical monsoonal; high-cold Tibetan Plateau—showed a marked increase, the smallest magnitude of which was around 12.3%. The increase was greater under 2 °C of global warming, which suggests that from the perspective of ecosystem productivity, global warming poses no ecological risk in China. Specifically, in comparison with historical GPP (1986–2005), under 1.5 °C global warming GPP was projected to increase by 16.1–23.8% in the temperate continental zone, 12.3–16.1% in the temperate monsoonal zone, 12.5–14.7% in the subtropical–tropical monsoonal zone, and 20.0–37.0% on the Tibetan Plateau. Under 2 °C global warming, the projected GPP increase was 23.0–34.3% in the temperate continental zone, 21.2–24.4% in the temperate monsoonal zone, 16.1–28.4% in the subtropical–tropical monsoonal zone, and 28.4–63.0% on the Tibetan Plateau. The GPP increase contributed by climate change was further quantified and attributed. The ESM prediction from the Max Planck Institute suggested that the climate contribution could range from −12.8% in the temperate continental zone up to 61.1% on the Tibetan Plateau; however, the ESMs differed markedly regarding their climate contribution to GPP change. Although precipitation has a higher sensitivity coefficient, temperature generally plays a more important role in GPP change, primarily because of the larger relative change in temperature in comparison with that of precipitation.

Keywords: GPP; climate change; CMIP6; ESM

Citation: Zhang, C.; Wu, S.; Deng, Y.; Chou, J. How the Updated Earth System Models Project Terrestrial Gross Primary Productivity in China under 1.5 and 2 °C Global Warming. *Sustainability* **2021**, *13*, 11744. https://doi.org/10.3390/su132111744

Academic Editors: Xiaodong Yan, Jia Yang, Shaofei Jin and Adriana Del Borghi

Received: 18 July 2021
Accepted: 9 October 2021
Published: 24 October 2021

1. Introduction

Gross and net primary productivity (GPP and NPP, respectively) are representative indicators that reflect ecosystem production capacity [1–3]. Many previous studies have considered future GPP/NPP change. For example, Huang et al. [4] evaluated NPP variations in the 21st century under various climate scenarios using the Lund–Potsdam–Jena dynamic global vegetation model. They found that total NPP in China is projected to increase continuously under different scenarios, with CO_2 concentration playing the dominant role. Using a machine learning model to constrain the spatial uncertainty in GPP projections, Schlund et al. [5] predicted a higher increase in GPP in northern high latitudes over the 21st century under the Representative Concentration Pathway [6] 8.5 W m^{-2} (RCP8.5) in comparison with regions closer to the equator. Under 1.5 °C of global warming, the GPP in China is expected to increase by 15.5% ± 5.4% on a stabilized pathway and by 11.9% ± 4.4% on a transient pathway [7]. Zhang et al. [3] explored the trend features of GPP/NPP in the 21st century under the Shared Socioeconomic Pathway [8] 24.5 W m^{-2} (SSP245) with the Beijing Climate Model. Their results predicted the overall trends of increase in both the near-term and long-term terrestrial GPP/NPP. However, in certain districts, the trend of GPP/NPP showed an initial increase followed by a decrease. Wang et al. [9] investigated the variation in NPP over the 21st century using the Earth system models (ESMs) of the Coupled Model Intercomparison Project phase 5. The results obtained under all four RCP scenarios suggested an increasing trend of NPP over China, especially in western areas.

In summary, GPP/NPP in China under different scenarios is expected to show a trend of increase in the 21st century. However, large uncertainties exist in the various ESMs [5,9]. Under the global warming targets of 1.5 and 2 °C above preindustrial levels set by the Paris Agreement, many regional impacts wait to be assessed. In particular, as the Coupled Model Intercomparison Project enters into the 6th phase (CMIP6), more and more ESMs have distributed their latest climate simulation under the Shared Socioeconomic Pathways (SSPs). How the latest ESMs will project the future ecosystem change in China and the corresponding climate attribution remains to be determined and revealed. In a recent study on performance in presenting historical terrestrial GPP in China, three out of seven ESMs evaluated were found to perform well in terms of climatological GPP, spatial pattern, and the ecosystem–climate relationship [10]. Consequently, these three ESMs were chosen in this study to predict ecosystem change under the warming targets. The ecological indicator of annual GPP was applied to measure the general state of the ecosystem. Changes in annual GPP predicted using the different ESMs were quantified with respect to the different climate zones in China. Furthermore, the relationship between the ecosystem and climate variables was tested and built through linear correlation and multiple regression. Relying on the model-specific parameters of the ecosystem response to the climate, the climate-related GPP changes were revealed and quantified.

2. Materials and Methods

2.1. Data

In this study, three ESMs that performed well in historical GPP reproduction [10] were chosen to project future GPP in China: (1) the Beijing Climate Center Climate System Model (BCC-CSM2-MR) [11], (2) the Euro-Mediterranean Centre on Climate Change coupled climate model (CMCC-CM2-SR5) [12], and (3) the Max Planck Institute for Meteorology Earth System Model version 1.2 (MPI-ESM1.2-HR) [13]. Specifically, BCC-CSM2-MR and MPI-ESM1-2-HR, out of seven ESMs, gave the best estimation of climatological GPP in China from 1980 to 2013. MPI-ESM1-2-HR performed best in characterizing the spatial structure. BCC-CSM2-MR and CMCC-CM2-SR5 best captured the response of the ecosystem to the climate [10]. The land surface models used for the three ESMs were BCC-AVIM2.0, CLM4.5, and JSBACH3.2. Major improvements or parameterizations have been made to these models in comparison with their predecessors; they make use of new scientific understanding to better simulate vegetation phenology [12–14]. These ESMs could provide not only the monthly GPP, but also the monthly surface air temperature and precipitation. In CMIP6, new SSPs were employed for climate modelling. The SSPs included five narratives describing alternative socio-economic developments, such as sustainable development, fossil-fueled development, etc. [8]. The middle of the road development—i.e., the SSP2 scenario—featured a continuation of the current fossil fuel-dominated energy mixed with intermediate challenges for both mitigation and adaptation, which resembled the historical pattern most [8]. SSP245, as the sole scenario of SSP2 implemented in CMIP6, was thus chosen to represent the most possible future world. Historical data (1980–2013) were applied to determine the ecosystem–climate relationship—i.e., correlation and multiple linear regression. Data from the BCC-CSM2-MR, CMCC-CM2-SR5, and MPI-ESM1.2-HR ESMs were output as 1.125° × 1.125°, 0.9375° × 1.25°, and 0.9375° × 0.9375° grids, respectively. Because the grids were not uniform, they were first transformed to a 1° × 1° grid through bilinear interpolation for comparative purposes.

A climate division map of China was applied for regional analysis. It divided the country into four climate zones—i.e., temperate continental, temperate monsoonal, subtropical–tropical monsoonal, and high-cold Tibetan Plateau, as in He et al. [15] and Zhang et al. [16].

2.2. Methods

2.2.1. Bilinear Interpolation

Bilinear interpolation can produce a smoother interpolation than that achieved using the nearest neighbor method [17]. Thus, it was applied to transform fields from various grids of the ESMs into the formal 1° × 1° grid. In this approach, g (n_1, n_2) is defined as a linear combination of the values of its four nearest neighbors. Given the four neighbors with coordinates f (n_{10}, n_{20}), f (n_{11}, n_{21}), f (n_{12}, n_{22}), and f (n_{13}, n_{23}) (i.e., the four nearest neighbors of f (n_1, n_2)), the geometrically transformed field g (n_1, n_2) is computed as:

$$g(n_1, n_2) = A_0 + A_1 n_1 + A_2 n_2 + A_3 n_1 n_2. \quad (1)$$

The bilinear weights A_0, A_1, A_2, and A_3 are found by solving:

$$\begin{bmatrix} A_0 \\ A_1 \\ A_2 \\ A_3 \end{bmatrix} = \begin{bmatrix} 1 \; n_{10} \; n_{20} \; n_{10}n_{20} \\ 1 \; n_{11} \; n_{21} \; n_{11}n_{21} \\ 1 \; n_{12} \; n_{22} \; n_{12}n_{22} \\ 1 \; n_{13} \; n_{23} \; n_{13}n_{23} \end{bmatrix}^{-1} \begin{bmatrix} f(n_{10}, n_{20}) \\ f(n_{11}, n_{21}) \\ f(n_{12}, n_{22}) \\ f(n_{13}, n_{23}) \end{bmatrix}. \quad (2)$$

2.2.2. Area Weighting

Regional and global mean variables—e.g., temperature, precipitation, and GPP—on the 1° × 1° grid are calculated through area weighting:

$$V_{reg} = \frac{\sum_i V_i \cdot \frac{\pi R}{180} \cdot \frac{\pi R}{180} \cos(\theta_i)}{\sum_i \frac{\pi R}{180} \cdot \frac{\pi R}{180} \cos(\theta_i)} = \frac{\sum_i V_i \cos(\theta_i)}{\sum_i \cos(\theta_i)}, \quad (3)$$

where θ represents the latitude of the grid, R is the Earth's radius, and V is the variable.

2.2.3. Linear Correlation and Multiple Regression

The correlation coefficient r is used to test the relationship between ecosystem productivity and climate factors. The formula can be expressed as follows:

$$r = \frac{n\sum_{i=1}^{n} E_i C_i - \sum_{i=1}^{n} E_i \sum_{i=1}^{n} C_i}{\sqrt{n\sum_{i=1}^{n} E_i^2 - (\sum_{i=1}^{n} E_i)^2} \cdot \sqrt{n\sum_{i=1}^{n} C_i^2 - (\sum_{i=1}^{n} C_i)^2}}, \quad (4)$$

where E and C represent ecosystem productivity and climate factors, respectively.

The interannual variation in GPP reflects year-to-year differences attributable mainly to climate variations [18,19]; therefore, the relationship between GPP and climate—i.e., precipitation and surface air temperature—was explored using a multiple regression approach [20]:

$$y = ax_T + bx_P + \varepsilon, \quad (5)$$

where y is the detrended anomaly of the carbon flux GPP, variable x_T is the detrended annual temperature anomaly, and x_P is the detrended annual precipitation anomaly. The fitted regression coefficients a and b define the apparent carbon flux sensitivity to interannual variations in temperature and precipitation, and ε is the residual error term. The use of the detrended time series instead of the original nonstationary time series in the above linear correlation and regression analysis provides a robust estimate of their relationship [21–23].

Some definitions set 1986–2005 as a reference period when the global surface air temperature was 0.61 °C warmer than preindustrial levels [24,25]. We adopted this definition and defined the 1.5 and 2 °C warming periods as the first time when the 20-year-moving-average global temperature was 0.89 and 1.39 °C warmer, respectively, than that from 1986–2005 in the models. The corresponding changes in ecosystem and regional climate were based upon the reference period of 1986–2005. It is also necessary to point out that 20 years is a duration that is commonly applied to represent a climate state in the scientific world [26–28].

3. Results and Discussion

3.1. GPP Distribution and Projected Changes

The climatological GPP distribution produced by each of the three ESMs from 1986 to 2005 is shown in Figure 1. The ESMs all produce a similar spatial pattern of GPP, showing high (low) values in the southeast (northwest) of China. Regionally, the GPP in the subtropical–tropical monsoonal zone is largest, followed in descending order by the temperate monsoonal zone, Tibetan Plateau, and temperate continental zone. The three ESMs produced comparable estimates in the climate zones except in the monsoonal regions, where CMCC-CM2-SR5 produced larger estimates, especially in the subtropical–tropical monsoonal zone.

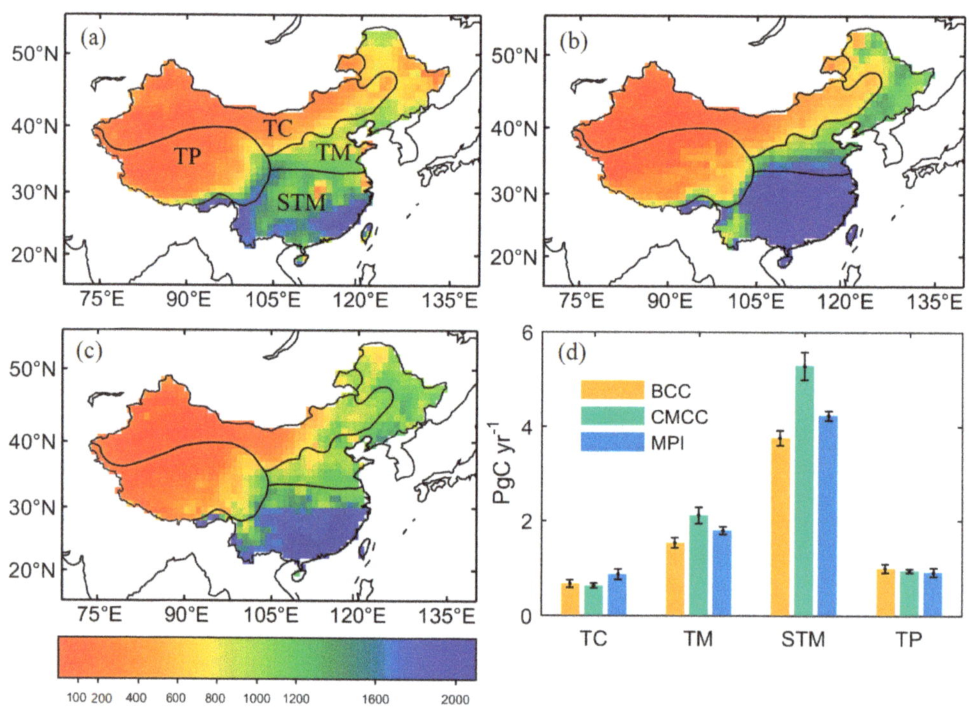

Figure 1. Climatological GPP distribution (unit: gC m^{-2} yr^{-1}) from 1986 to 2005 from (**a**) BCC-CSM2-MR, (**b**) CMCC-CM2-SR5, and (**c**) MPI-ESM1-2-HR. The four delineated regions represent the climate zones temperate continental (TC), temperate monsoonal (TM), subtropical–tropical monsoonal (STM), and high-cold Tibetan Plateau (TP). (**d**) Regional GPP from the ESMs (PgC yr^{-1}: petagram of carbon per year) (BCC: BCC-CSM2-MR; CMCC: CMCC-CM2-SR5; MPI: MPI-ESM1-2-HR). The bars and lines represent the standard deviation during the 20 years.

The GPP change under 1.5 °C of global warming is shown in Figure 2. Throughout China, the GPP of all three ESMs showed a positive anomaly except over certain individual grid points. In the subtropical–tropical monsoonal with MPI-ESM1-2-HR, the negatively changed grids tend to concentrate (Figure 2e). The areal GPP reduction may be related to reduced local precipitation. However, many factors could contribute to the GPP change in addition to precipitation and temperature, such as land-use change, soil moisture, wind speed, humidity, solar radiation, nitrogen deposition, etc. Thus, it is really hard to be conclusive. Moreover, the aggregated negative grids tend to dissipate under 2 °C global warming (Figure 3e). The GPP change patterns differ among the models (Figure 2). For example, the largest anomaly in the output of BCC-CSM2-MR appears over the southeastern Tibetan Plateau, whereas the largest anomalies in the output of CMCC-CM2-SR5 and MPI-ESM1-2-HR appear in the central and southern parts of the subtropical–tropical monsoonal zone, respectively. The absolute GPP change is largest in the subtropical–tropical monsoonal zone; however, the relative change is rather small and stable among the models (Figure 2 right). The absolute change is smallest in the temperate continental zone owing to its low base value in GPP. The relative change is large over the Tibetan Plateau, and there are strong differences in the magnitude of the GPP increments among the different models.

The GPP anomaly under 2 °C of global warming shows a spatial pattern similar to that found under 1.5 °C global warming but with a stronger intensity (Figure 3). Regional statistics indicate that the regional GPP changes will be larger under 2 °C of global warming. The projected increment of GPP in China under the different warming targets is consistent with previous findings [5,7,29]. This suggests that from the perspective of GPP, there is no ecological crisis in the projected future climate within the studied domain [3]. As with 1.5 °C warming, the subtropical–tropical monsoonal zone with the highest GPP value contributed the most to the increment in China's GPP under 2 °C of warming. However, the increase rate does not show much difference in magnitude when compared with that of other regions. It is worth noting that the rate of increase in GPP is substantial on the Tibetan Plateau—i.e., the increase is nearly 63% with regard to BCC-CSM2-MR. Thus, the Tibetan Plateau would appear to be the region most susceptible to the effects of climate warming, although the influence could be considered positive and beneficial.

The seasonal GPP anomalies under the 1.5 and 2 °C warming scenarios are shown in Figures 4 and 5, respectively. The spatial modes between the two warming scenarios are similar, noting that the magnitude in the 2 °C warming is much larger than in the 1.5 °C warming. In spring and summer, the GPP anomalies are the most prosperous, as they correspond to the growing season in China, while they drop to become the weakest in winter. In spring and winter with all ESMs, the GPP all over China generally shows a positive anomaly, with only sporadic negative points. Some negative changes occur in summer and autumn, especially with BCC-CSM2-MR and MPI-ESM1-2-HR. For BCC-CSM2-MR in summer, the negative GPP anomalies concentrate in the Huaihe River, which divides the subtropical–tropical monsoonal and temperate monsoonal regions. However, there were no negative changes in GPP at the zone scale. For MPI-ESM1-2-HR in summer and autumn, we observed some negative changes over the grassland in the temperate continental, which is similar to the results of Ma et al. [30]. They found that large areas in Northern China showed a decreasing trend in NPP under global warming, although the overall NPP increased significantly. The fact that only one ESM obtained similar results also indicates the large inter-model spread in representing the future GPP change. The negative changes in the temperate continental were weakened under the 2 °C warming scenario (Figure 5).

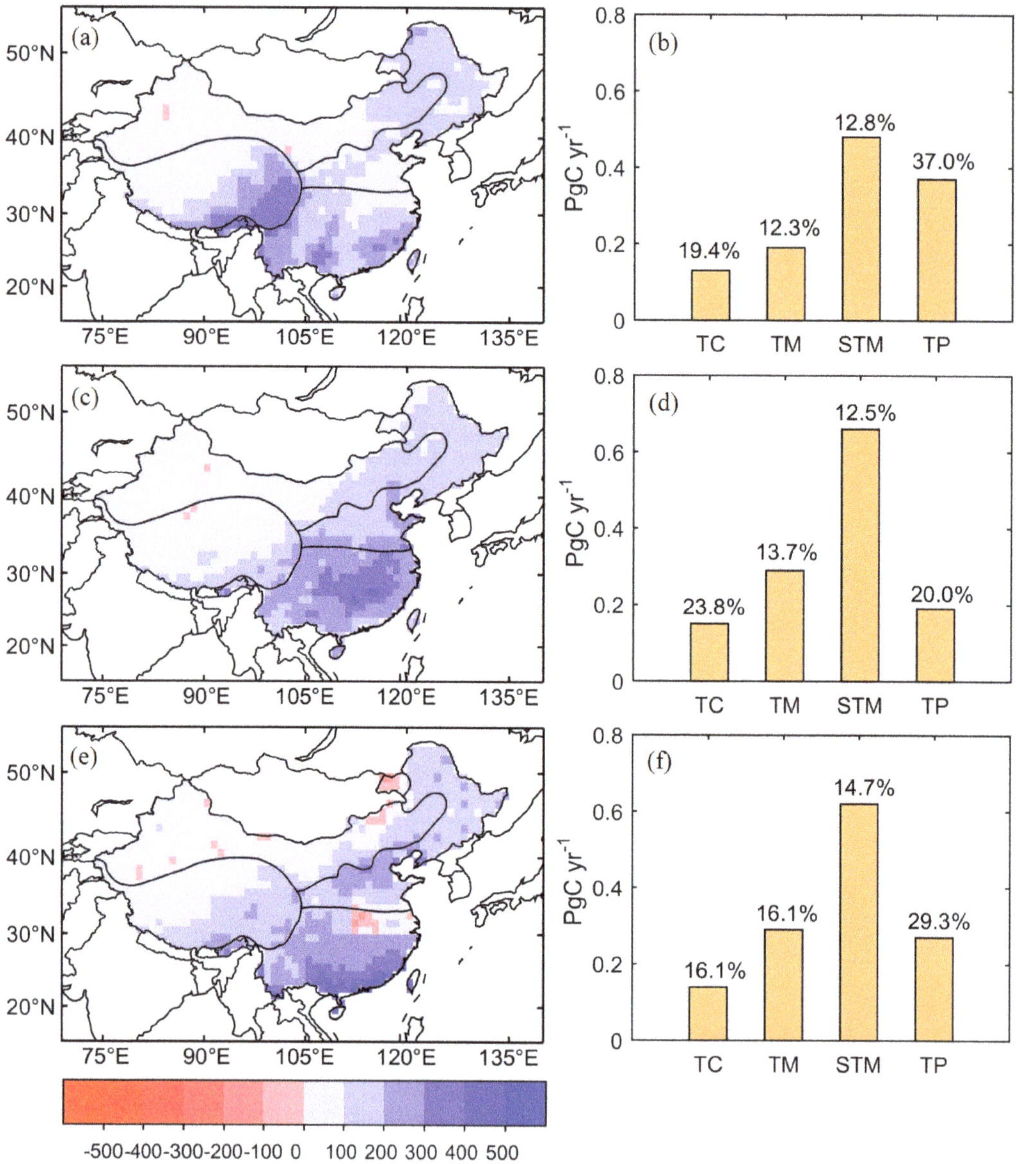

Figure 2. (Left) Spatial distribution of GPP anomaly (gC m^{-2} yr^{-1}) and (right) the respective regional GPP changes under 1.5 °C of global warming. (**a**,**b**) BCC-CSM2-MR, (**c**,**d**) CMCC-CM2-SR5, and (**e**,**f**) MPI-ESM1-2-HR. Numbers in the right-hand panels represent the relative change in comparison with regional GPP from 1986 to 2005.

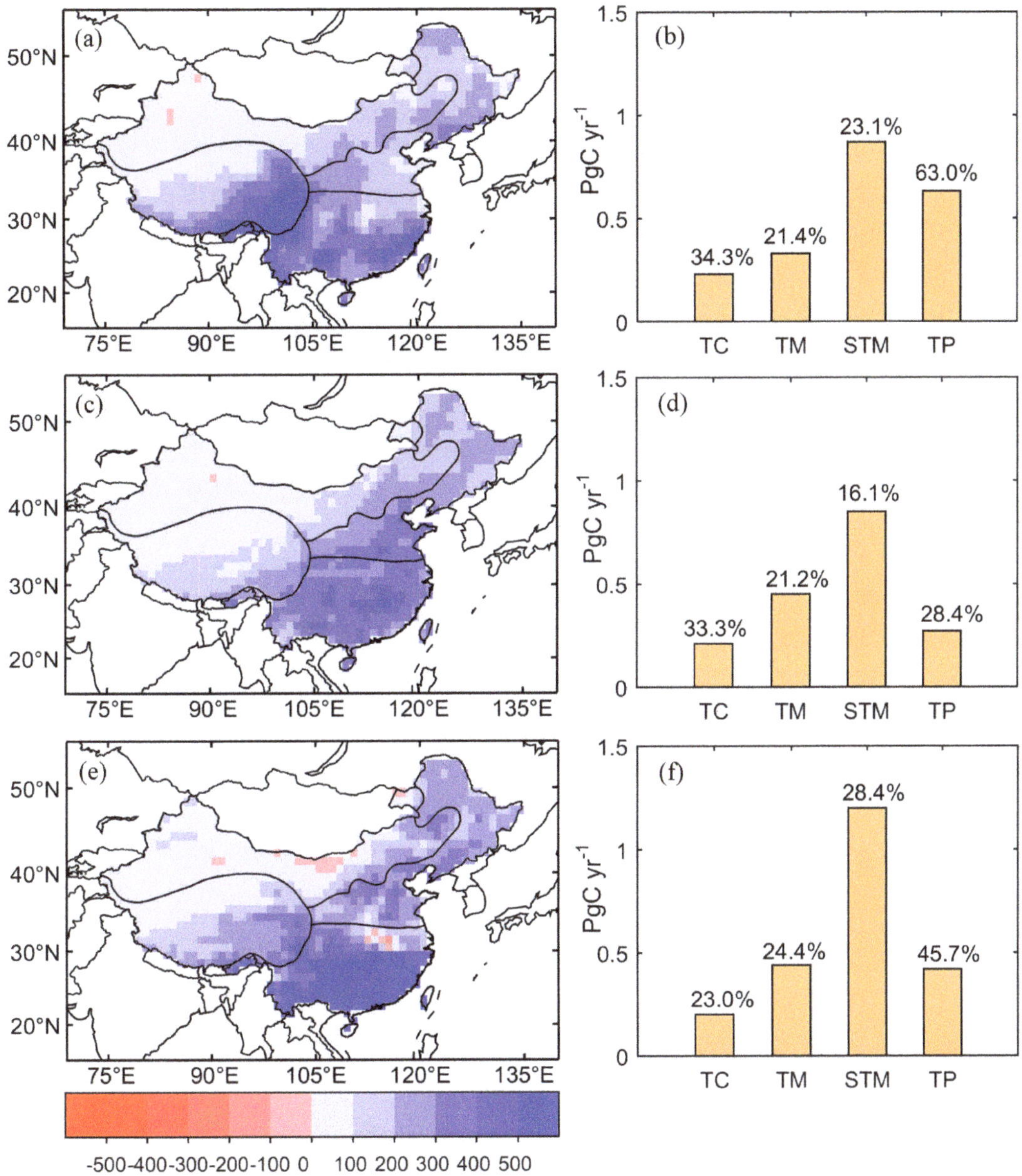

Figure 3. (Left) Spatial distribution of GPP anomaly (gC m^{-2} yr^{-1}) and (right) the respective regional GPP changes under 2 °C of global warming. (**a**,**b**) BCC-CSM2-MR, (**c**,**d**) CMCC-CM2-SR5, and (**e**,**f**) MPI-ESM1-2-HR. Numbers in the right-hand panels represent the relative change in comparison with regional GPP from 1986 to 2005.

3.2. Climate Attribution

The variation in GPP is closely related to climate [20,31], and the correlation parameters within the studied ESMs are provided in Table 1. It can be seen that GPP is correlated significantly with at least one climate variable. There are cases in which GPP correlates negatively with temperature, such as in the temperate continental zone with MPI-ESM1-2-HR and in the temperate monsoonal zone with BCC-CSM2-MR, which imply inherent differences in ecological modeling between the different ESMs [9,10]. On the Tibetan Plateau, it is unanimous within the ESMs that GPP is most closely related to temperature rather than to precipitation. Because the ESMs substantially overestimate precipitation over the Tibetan Plateau [32–34], it is possible that precipitation is not the primary climate factor constraining the regional ecosystem.

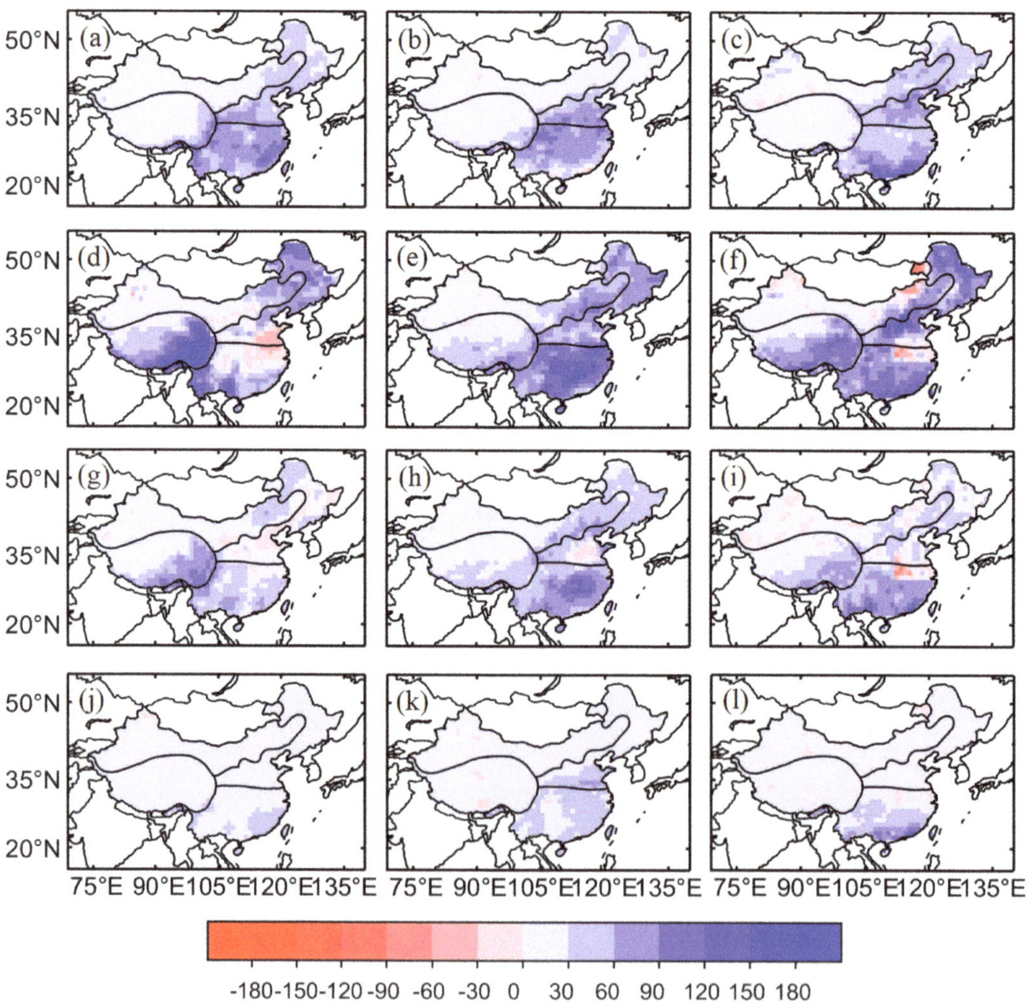

Figure 4. Seasonal GPP anomaly distribution under 1.5 °C of global warming (gC m^{-2} yr^{-1}). The four rows from up to down represent spring, summer, autumn, and winter, respectively. The three columns from left to right represent BCC-CSM2-MR, CMCC-CM2-SR5, and MPI-ESM1-2-HR, respectively.

The corresponding climate changes—i.e., precipitation and surface air temperature— under the warming targets are shown in Tables 2 and 3, respectively. The mode of temperature change in the four regions is consistent among the models. The hottest region—i.e., the subtropical–tropical monsoonal zone—increases least under the effects of warming. Both CMCC-CM2-SR5 and MPI-ESM1-2-HR produced similar estimates of temperature change, whereas the estimates from BCC-CSM2-MR were larger, especially over the monsoonal regions.

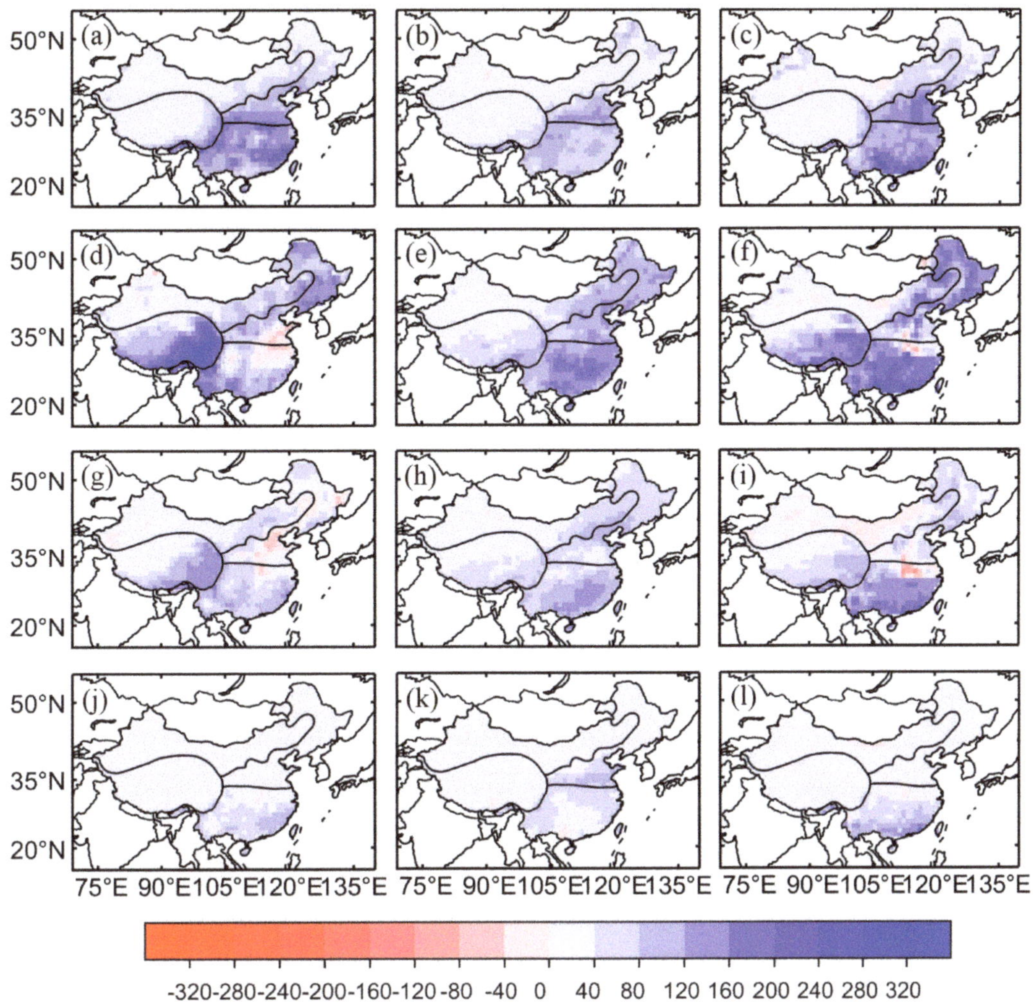

Figure 5. Seasonal GPP anomaly distribution under 2 °C of global warming (gC m^{-2} yr^{-1}). The four rows from up to down represent spring, summer, autumn, and winter, respectively. The three columns from left to right represent BCC-CSM2-MR, CMCC-CM2-SR5, and MPI-ESM1-2-HR, respectively.

There is a greater model variety regarding the change in precipitation. Under 1.5 °C of warming, there are negative changes—e.g., in the temperate monsoonal zone with CMCC-CM2-SR5 and in the subtropical–tropical monsoonal zone with both BCC-CSM2-MR and CMCC-CM2-SR5. Conversely, under 2 °C of warming, there are no negative changes, but the incremental differences for one certain region are huge. These results indicate the large uncertainty in the precipitation projections made by the ESMs. It is also worth noting that in comparison with their variabilities, the change in temperature under the warming scenarios is reasonably large, whereas the precipitation change is rather limited [10].

The apparent sensitivity of climate to the ecosystem of each of the ESMs is shown in Table 4. The response of the ecosystem to climate varies strongly among the models. For some ESM regions, climate plays a crucial role, such that the variation in climate explains more than half of the variation in GPP. However, for certain other ESM regions, the degree of explanation attributable to climate is rather small—e.g., MPI-ESM1-2-HR in the subtropical–tropical monsoonal zone and BCC-CSM2-MR on the Tibetan Plateau. For one particular region, the same climate factor might affect the ecosystem differently in the various models. Taking the subtropical–tropical monsoonal zone as an example, precipitation is the major influencing factor and affects the ecosystem positively with BCC-CSM2-MR and CMCC-CM2-SR5. However, with MPI-ESM1-2-HR, the correlation between precipitation and GPP is insignificant and negative (Tables 1 and 4). Moreover, the overall climate contribution to ecosystem variation with MPI-ESM1-2-HR is very small (Table 4). These features further reflect the inherent differences of ecological modeling within the ESMs.

Table 1. Correlation parameter *r* between GPP and climate variables during the historical period of 1980–2013. * denotes correlation that is significant at the 0.1 level; ** denotes correlation that is significant at the 0.05 level; *** denotes correlation that is significant at the 0.01 level.

GPP vs. P	TC	TM	STM	TP
BCC	0.68 ***	0.74 ***	0.62 ***	0.1
CMCC	0.72 ***	0.58 ***	0.42 **	−0.01
MPI	0.81 ***	0.36 **	−0.26	0.23
GPP vs. T				
BCC	−0.09	−0.35 **	−0.25	0.35 **
CMCC	0.17	0.42 **	0.34 *	0.74 ***
MPI	−0.42 **	0.13	0.33 *	0.62 ***

Table 2. Regional temperature change (°C) in comparison with that of the reference period of 1986–2005 under 1.5 and 2 °C (in parentheses) of global warming.

	TC	TM	STM	TP
BCC	1.37(2.21)	1.66(2.37)	1.16(1.86)	1.37(2.10)
CMCC	1.28(1.94)	1.12(1.78)	0.81(1.27)	1.19(1.71)
MPI	1.18(1.89)	1.1(1.79)	0.83(1.69)	1.14(2.02)

Table 3. Regional precipitation change (mm yr^{-1}) in comparison with that of the reference period of 1986–2005 under 1.5 and 2 °C (in parentheses) of global warming.

	TC	TM	STM	TP
BCC	12.47(28.22)	8.11(47.13)	−6.18(57.36)	10.37(32.23)
CMCC	11.31(45.68)	−20.08(16.22)	−51.89(20.50)	30.24(94.89)
MPI	2.65(8.66)	6.91(19.23)	32.83(46.13)	11.5(9.42)

Table 4. Sensitivity parameters of GPP to climate variables during the historical period of 1980–2013 through the multiple linear regression approach. The final row shows the explained variance of climate to GPP through multiple regression.

Temperate Continental	BCC	CMCC	MPI
P	0.8	0.72	0.77
T	0.26	0.14	−0.1
Var	52.20%	54.00%	66.50%
Temperate monsoonal			
P	0.78	0.54	0.66
T	0.06	0.35	0.52
Var	55.40%	45.90%	30.20%
Subtropical–tropical monsoonal			
P	0.64	0.51	−0.15
T	0.06	0.44	0.28
Var	38.10%	36.30%	13.10%
Tibetan Plateau			
P	0.16	−0.1	0.46
T	0.38	0.75	0.75
Var	15.00%	55.70%	57.50%

Based on the ecosystem–climate relationship (Table 4) and the known climate changes (Tables 2 and 3), the GPP change over the climate zones with the different ESMs is attributed quantitatively in Figure 6. As mentioned before, some contributions from the climate factors are negative, for which there are two major reasons. First, the climate change is negative—e.g., the negative precipitation change leads to a negative contribution to GPP. Second, the correlation between the climate factor and GPP is negative—e.g., temperature and GPP in the temperate continental zone in MPI-ESM1-2-HR. A positive anomaly in temperature could also lead to GPP reduction. The climate contribution to the variation in GPP changes among the studied ESMs. Even with the same model—e.g., MPI-ESM1-2-HR—it can be −12.8% in the temperate continental zone and 61.1% on the Tibetan Plateau. Generally, under the effects of global warming, the influence of temperature on the ecosystem is larger than that of precipitation. This is mainly because the relative change in temperature is much larger than that in precipitation. On the Tibetan Plateau, where temperature is the most constraining factor (Tables 1 and 4), temperature plays a more dominant role than precipitation in the increase in GPP (Figure 6). In addition, it is observed that the relative GPP increase in the Tibetan Plateau is much larger than that in other regions. On the one hand, it is related to the low baseline value of GPP in the Tibetan Plateau (Figure 1d). A light increase in the GPP of the Tibetan Plateau is salient in relative values compared to the respective change in the subtropical–tropical monsoonal region. On the other hand, this may be related to the vegetation structure on the Plateau. Demonym plants can be divided into three types based on their photosynthesis patterns (i.e., C3, C4, and crassulacean acid metabolism). C3 photosynthesis produces a three-carbon compound during the Calvin cycle, while C4 photosynthesis makes an intermediate four-carbon compound that splits into a three-carbon compound for the Calvin cycle. They favor different conditions of nature. The conditions on the frigid Tibetan Plateau are unsuitable for the growth of C4 plants [35]. Consequently, the plateau is dominant by C3 plants [36]. C3 plants are more efficient in vegetative growth than C4 plants in response to the increasing air CO_2 [37]. As a result, GPP increases more rapidly with increased air CO_2 in the Tibetan Plateau than in other regions containing both C3 and C4 plants.

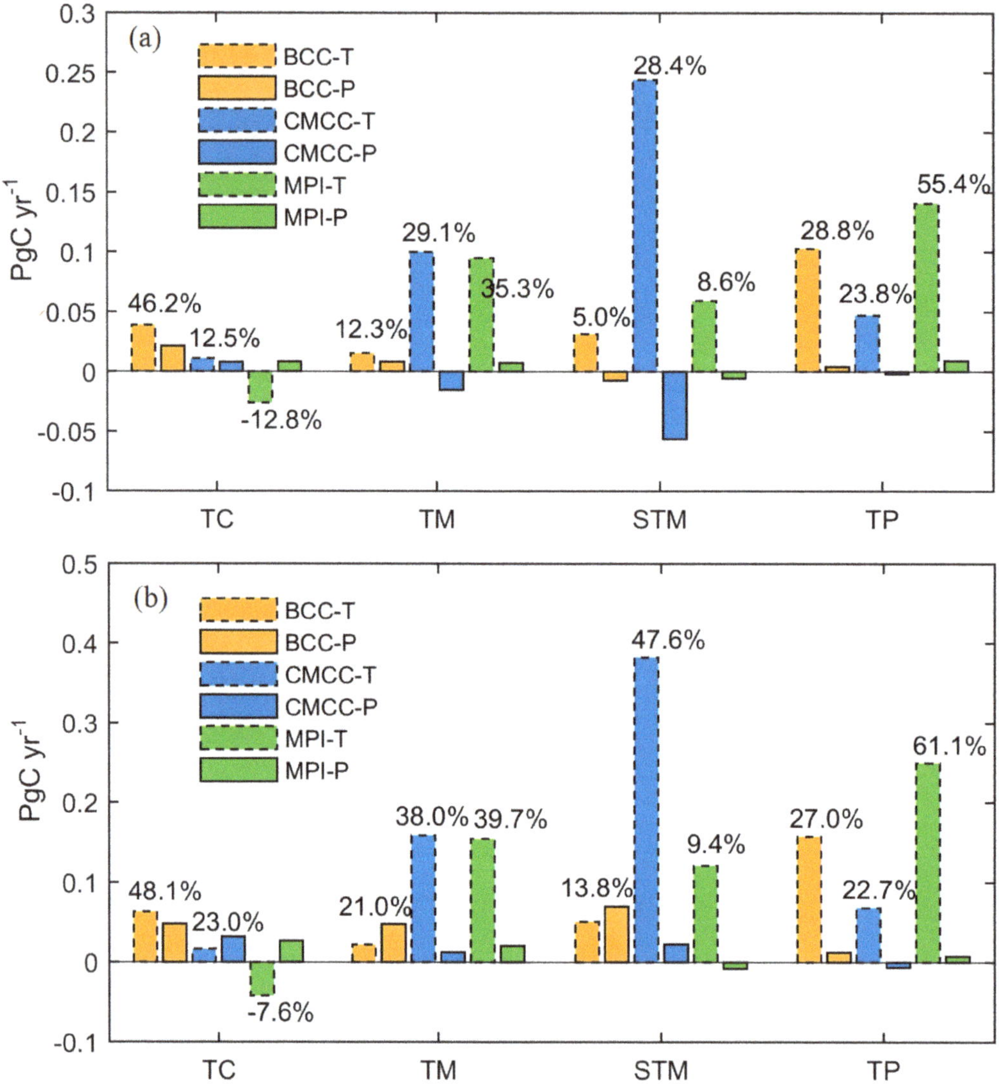

Figure 6. Climate contributions to GPP change with the different ESMs under (**a**) 1.5 and (**b**) 2 °C of global warming. Bars represent the absolute climate contributions (PgC yr^{-1}). Percentages represent the relative climate contributions to the overall GPP changes.

4. Conclusions

To evaluate the GPP change under 1.5 and 2 °C of global warming, this study selected three CMIP6 ESMs (i.e., BCC-CSM2-MR, CMCC-CM2-SR5, and MPI-ESM1-2-HR) that performed well in historical GPP modeling; the principal conclusions derived are as follows:

1. Under 1.5 and 2 °C of global warming, the projections of the ESMs indicate that global warming introduces no ecological risk in China. Although certain individual grid points showed negative GPP changes, regional GPP showed a marked increase, the smallest magnitude of which was more than 10% greater than that from 1986 to 2005.
2. Specifically under 1.5 °C warming, the GPP in the temperate continental zone is projected to increase by 16.1–23.8% in comparison with the historical value (1986–2005).

Similarly, GPP is projected to increase by 12.3–16.1% in the temperate monsoonal zone, 12.5–14.7% in the subtropical–tropical monsoonal zone, and 20.0–37.0% on the Tibetan Plateau. Under 2 °C warming, the increase in GPP is projected to be even greater—i.e., 23.0–34.3% in the temperate continental zone, 21.2–24.4% in the temperate monsoonal zone, 16.1–28.4% in the subtropical–tropical monsoonal zone, and 28.4–63.0% on the Tibetan Plateau.

3. Climate change is projected to contribute positively to GPP change, except in the temperate continental zone with MPI-ESM1-2-HR. Although precipitation has larger sensitivity parameters, temperature generally plays a more important role in GPP change because of the larger change relative to its own variability in comparison with that of precipitation.

The output of the three studied ESMs showed a marked spread, not only in GPP change but also in the accountability of climate to the ecosystem. In addition, the change in climate, especially precipitation, differed strongly within the models, which indicates the large uncertainty in the climate projections of the ESMs. All of these add to difficulties in attributing future GPP change to climate. Moreover, this study analyzed the influence of annual precipitation and temperature upon the ecosystem productivity. However, GPP variation depends not only on these, but also on wind speed, humidity, solar radiation, nitrogen deposition, etc. Future studies should be more comprehensive in building the regressed equations between GPP and the impact factors. This study failed to analyze the contribution of CO_2 to the GPP increase. This was due to the lack of gridded/regional CO_2 concentration data. Future research should take into account the CO_2 effect when analyzing the GPP change and be more specific about vegetation of C3 and C4 types. This study is more general in that it focuses on the general productivity of the climate zones. Future studies should be refined to specific vegetation covers, such as forest, grass, etc. The fact that different ESMs lack consensus in the response mechanism of the ecosystem to climate, even over one specific climate zone, highlights that there is still a long way for ecological modeling in China to go.

Author Contributions: Conceptualization, C.Z. and Y.D.; formal analysis, C.Z. and Y.D.; writing—original draft preparation, C.Z.; review and editing, Y.D., S.W. and J.C. All authors have read and agreed to the published version of the manuscript.

Funding: This work was funded by the National Key Research and Development Program of China (2018YFC1509003, 2019YFC0507802, and 2018YFC1508902), and the Strategic Priority Research Program of Chinese Academy of Sciences (XDA19040304).

Acknowledgments: This study was performed through the Institute of Geographic Sciences and Natural Resources Research (IGSNRR). We also appreciate the support from the Coupled Model Intercomparison Project (phase 6) for providing the ESM outputs.

Conflicts of Interest: The authors declare no conflict of interest.

References

1. Beer, C.; Reichstein, M.; Tomelleri, E.; Ciais, P.; Jung, M.; Carvalhais, N.; Rödenbeck, C.; Arain, M.A.; Baldocchi, D.; Bonan, G.B.; et al. Terrestrial gross carbon dioxide uptake: Global distribution and covariation with climate. *Science* **2010**, *329*, 834–838. [CrossRef] [PubMed]
2. Schimel, D.S.; House, J.I.; Hibbard, K.A.; Bousquet, P.; Ciais, P.; Peylin, P.; Braswell, B.H.; Apps, M.J.; Baker, D.; Bondeau, A.; et al. Recent patterns and mechanisms of carbon exchange by terrestrial ecosystems. *Nature* **2001**, *414*, 169–172. [CrossRef]
3. Zhang, C.; Wu, S.; Leng, G. Possible NPP changes and risky ecosystem region identification in China during the 21st century based on BCC-CSM2. *J. Geogr. Sci.* **2020**, *30*, 1219–1232. [CrossRef]
4. Huang, Q.; Ju, W.; Zhang, F.; Zhang, Q. Roles of climate change and increasing CO_2 in driving changes of net primary productivity in China simulated using a dynamic global vegetation model. *Sustainability* **2019**, *11*, 4176. [CrossRef]
5. Schlund, M.; Eyring, V.; Camps-Valls, G.; Friedlingstein, P.; Gentine, P.; Reichstein, M. Constraining uncertainty in projected gross primary production with machine learning. *J. Geophys. Res. Biogeosci.* **2020**, *125*, e2019JG005619. [CrossRef]
6. Van Vuuren, D.P.; Edmonds, J.; Kainuma, M.; Riahi, K.; Thomson, A.; Hibbard, K.; Hurtt, G.C.; Kram, T.; Krey, V.; Lamarque, J.F.; et al. The representative concentration pathways: An overview. *Clim. Chang.* **2011**, *109*, 5–31. [CrossRef]

7. Yue, X.; Liao, H.; Wang, H.; Zhang, T.; Unger, N.; Sitch, S.; Feng, Z.; Yang, J. Pathway dependence of ecosystem responses in China to 1.5 °C global warming. *Atmos. Chem. Phys.* **2020**, *20*, 2353–2366. [CrossRef]
8. Riahi, K.; Van Vuuren, D.P.; Kriegler, E.; Edmonds, J.; O'neill, B.C.; Fujimori, S.; Bauer, N.; Calvin, K.; Dellink, R.; Fricko, O.; et al. The shared socioeconomic pathways and their energy, land use, and greenhouse gas emissions implications: An overview. *Glob. Environ. Chang.* **2017**, *42*, 153–168. [CrossRef]
9. Wang, T.; Lin, X.; Liu, Y.; Dantec-Nédélec, S.; Ottle, C. Causes of uncertainty in China's net primary production over the 21st century projected by the CMIP5 Earth system models. *Int. J. Climatol.* **2016**, *36*, 2323–2334. [CrossRef]
10. Zhang, C.; Deng, Y. How the CMIP6 models project the historical terrestrial GPP in China. *Ecosystem* **2021**. in review.
11. Wu, T.; Lu, Y.; Fang, Y.; Xin, X.; Li, L.; Li, W.; Jie, W.; Zhang, J.; Liu, Y.; Zhang, L.; et al. The Beijing Climate Center Climate System Model (BCC-CSM): The main progress from CMIP5 to CMIP6. *Geosci. Model Dev.* **2019**, *12*, 1573–1600. [CrossRef]
12. Cherchi, A.; Fogli, P.G.; Lovato, T.; Peano, D.; Iovino, D.; Gualdi, S.; Masina, S.; Scoccimarro, E.; Materia, S.; Bellucci, A.; et al. Global mean climate and main patterns of variability in the CMCC-CM2 coupled model. *J. Adv. Modeling Earth Syst.* **2019**, *11*, 185–209. [CrossRef]
13. Mauritsen, T.; Bader, J.; Becker, T.; Behrens, J.; Bittner, M.; Brokopf, R.; Brovkin, V.; Claussen, M.; Crueger, T.; Esch, M.; et al. Developments in the MPI-M Earth System Model version 1.2 (MPI-ESM1.2) and Its Response to Increasing CO_2. *J. Adv. Modeling Earth Syst.* **2019**, *11*, 998–1038. [CrossRef]
14. Li, W.; Zhang, Y.; Shi, X.; Zhou, W.; Huang, A.; Mu, M.; Qiu, B.; Ji, J. Development of land surface model BCC_AVIM2.0 and its preliminary performance in LS3MIP/CMIP6. *J. Meteorol. Res.* **2019**, *33*, 851–869. [CrossRef]
15. He, H.; Wang, S.; Zhang, L.; Wang, J.; Ren, X.; Zhou, L.; Piao, S.; Yan, H.; Ju, W.; Gu, F.; et al. Altered trends in carbon uptake in China's terrestrial ecosystems under the enhanced summer monsoon and warming hiatus. *Natl. Sci. Rev.* **2019**, *6*, 505–514. [CrossRef]
16. Zhang, L.; Ren, X.; Wang, J.; He, H.; Wang, S.; Wang, M.; Piao, S.; Yan, H.; Ju, W.; Gu, F.; et al. Interannual variability of terrestrial net ecosystem productivity over China: Regional contributions and climate attribution. *Environ. Res. Lett.* **2019**, *14*, 014003. [CrossRef]
17. Bovik, A.C. *The Essential Guide to Image Processing*; Academic Press: Cambridge, MA, USA, 2009.
18. Peng, S.; Piao, S.; Ciais, P.; Myneni, R.B.; Chen, A.; Chevallier, F.; Dolman, A.J.; Janssens, I.A.; Penuelas, J.; Zhang, G.; et al. Asymmetric effects of daytime and night-time warming on Northern Hemisphere vegetation. *Nature* **2013**, *501*, 88–92. [CrossRef]
19. Yao, Y.; Wang, X.; Li, Y.; Wang, T.; Shen, M.; Du, M.; He, H.; Li, Y.; Luo, W.; Ma, M.; et al. Spatiotemporal pattern of gross primary productivity and its covariation with climate in China over the last thirty years. *Glob. Chang. Biol.* **2018**, *24*, 184–196. [CrossRef] [PubMed]
20. Piao, S.; Sitch, S.; Ciais, P.; Friedlingstein, P.; Peylin, P.; Wang, X.; Ahlström, A.; Anav, A.; Canadell, J.G.; Cong, N.; et al. Evaluation of terrestrial carbon cycle models for their response to climate variability and to CO_2 trends. *Glob. Chang. Biol.* **2013**, *19*, 2117–2132. [CrossRef]
21. Zhou, L.; Dickinson, R.E.; Tian, Y.; Vose, R.S.; Dai, Y. Impact of vegetation removal and soil aridation on diurnal temperature range in a semiarid region: Application to the Sahel. *Proc. Natl. Acad. Sci. USA* **2007**, *104*, 17937–17942. [CrossRef]
22. Zhou, C.; Wang, K.; Ma, Q. Evaluation of eight current reanalyses in simulating land surface temperature from 1979 to 2003 in China. *J. Clim.* **2017**, *30*, 7379–7398. [CrossRef]
23. Podobnik, B.; Stanley, H.E. Detrended cross-correlation analysis: A new method for analyzing two nonstationary time series. *Phys. Rev. Lett.* **2008**, *100*, 084102. [CrossRef] [PubMed]
24. IPCC. *Climate Change 2013: The Physical Science Basis*; Cambridge University Press: Cambridge, UK, 2013.
25. Zhai, P.; Yu, R.; Zhou, B.; Chen, Y.; Guo, J.; Lu, Y. Research progress in impact of 1.5 °C global warming on global and regional scales. *Clim. Chang. Res.* **2017**, *13*, 465–472. (In Chinese)
26. Rojas, M.; Lambert, F.; Ramirez-Villegas, J.; Challinor, A.J. Emergence of robust precipitation changes across crop production areas in the 21st century. *Proc. Natl. Acad. Sci. USA* **2019**, *116*, 6673–6678. [CrossRef]
27. Knutti, R.; Sedláček, J. Robustness and uncertainties in the new CMIP5 climate model projections. *Nat. Clim. Chang.* **2013**, *3*, 369–373. [CrossRef]
28. Kusunoki, S.; Arakawa, O. Are CMIP5 models better than CMIP3 models in simulating precipitation over East Asia? *J. Clim.* **2015**, *28*, 5601–5621. [CrossRef]
29. Mu, Q.; Zhao, M.; Running, S.W.; Liu, M.; Tian, H. Contribution of increasing CO_2 and climate change to the carbon cycle in China's ecosystems. *J. Geophys. Res. Biogeosci.* **2008**, *113*, G01018. [CrossRef]
30. Ma, X.; Huo, T.; Zhao, C.; Yan, W.; Zhang, X. Projection of Net Primary Productivity under Global Warming Scenarios of 1.5 °C and 2.0 °C in Northern China Sandy Areas. *Atmosphere* **2020**, *11*, 71. [CrossRef]
31. Jung, M.; Schwalm, C.; Migliavacca, M.; Walther, S.; Camps-Valls, G.; Koirala, S.; Anthoni, P.; Besnard, S.; Bodesheim, P.; Carvalhais, N.; et al. Scaling carbon fluxes from eddy covariance sites to globe: Synthesis and evaluation of the FLUXCOM approach. *Biogeosciences* **2020**, *17*, 1343–1365. [CrossRef]
32. Tong, K.; Su, F.; Yang, D.; Zhang, L.; Hao, Z. Tibetan Plateau precipitation as depicted by gauge observations, reanalyses and satellite retrievals. *Int. J. Climatol.* **2014**, *34*, 265–285. [CrossRef]
33. Zhang, C. Moisture source assessment and the varying characteristics for the Tibetan Plateau precipitation using TRMM. *Environ. Res. Lett.* **2020**, *15*, 104003. [CrossRef]

34. Pan, C.; Zhu, B.; Gao, J.; Kang, H.; Zhu, T. Quantitative identification of moisture sources over the Tibetan Plateau and the relationship between thermal forcing and moisture transport. *Clim. Dyn.* **2019**, *52*, 181–196. [CrossRef]
35. Sage, R.F.; Wedin, D.A.; Li, M. *The Biogeography of C4 Photosyn-Thesis: Patterns and Controlling Factors*; Sage, R.F., Monsoon, R.K., Eds.; C4 Plant Biology; Academic Press: San Diego, CA, USA, 1999; pp. 313–373.
36. Sitch, S.; Smith, B.; Prentice, I.C.; Arneth, A.; Bondeau, A.; Cramer, W.; Kaplan, J.O.; Levis, S.; Lucht, W.; Sykes, M.T.; et al. Evaluation of ecosystem dynamics, plant geography and terrestrial carbon cycling in the LPJ dynamic global vegetation model. *Glob. Chang. Biol.* **2003**, *9*, 161–185. [CrossRef]
37. Wand, S.J.; Midgley, G.F.; Jones, M.H.; Curtis, P.S. Responses of wild C4 and C3 grass (Poaceae) species to elevated atmospheric CO_2 concentration: A meta-analytic test of current theories and perceptions. *Glob. Chang. Biol.* **1999**, *5*, 723–741. [CrossRef]

sustainability

MDPI

Article

The Impact on Carbon Emissions of China with the Trade Situation versus the U.S.

Jieming Chou [1], Fan Yang [1,2,*], Zhongxiu Wang [3] and Wenjie Dong [4]

[1] State Key Laboratory of Earth Surface Processes and Resource Ecology, Faculty of Geographical Science, Beijing Normal University, Beijing 100875, China; choujm@bnu.edu.cn
[2] CMA Key Laboratory for Cloud Physics of China Meteorological Administration, Chinese Academy of Meteorological Sciences, Beijing 100081, China
[3] The Alliance of International Science Organizations, Beijing 100085, China; wang_zhongxiu@126.com
[4] Southern Marine Science and Engineering Guangdong Laboratory (Zhuhai), Zhuhai 519080, China; dongwj3@mail.sysu.edu.cn
* Correspondence: yangfan_bnu@mail.bnu.edu.cn; Tel.: +86-189-1052-3251

Abstract: The China–US trade conflict will inevitably have a negative impact on China's trade imports and exports, industrial development, and economic growth, and will affect the achievement of climate change goals. In the short term, the impact of the trade conflict on China's import and export trade will cause the carbon emissions contained in traded commodities to change accordingly. To assess the impact of the trade conflict on China's climate policy, this paper combines a model from the Global Trade Analysis Project (GTAP) and the input–output analysis method and calculates the carbon emissions in international trade before and after the conflict. The conclusions are as follows: (1) The trade war has led to a sharp decline in China–US trade, but for China as a whole, imports and exports have not changed much; (2) China's export emissions have changed little, its import emissions have dropped slightly, and its net emissions have increased; and (3) China's exports are still concentrated in energy-intensive industries. Changes in trade will bring challenges to China's balancing of climate and trade exigencies. China–US cooperation based on energy and technology will help China cope with climate change after the trade conflict.

Keywords: trade conflict; carbon emissions; import and export trade; cooperative emission reduction

Citation: Chou, J.; Yang, F.; Wang, Z.; Dong, W. The Impact on Carbon Emissions of China with the Trade Situation versus the U.S. *Sustainability* **2021**, *13*, 10324. https://doi.org/10.3390/su131810324

Academic Editor: Pallav Purohit

Received: 14 July 2021
Accepted: 17 August 2021
Published: 15 September 2021

1. Introduction

Currently, the world economy suffered unexpected shocks [1], affected by the epidemic COVID-19 [2]. The United States and China are the two largest economies: China relied on its institutional advantages to control the number of domestic cases [3] and the economy recovered rapidly in the US due to the popularization of vaccines. American citizens are eager for excess savings during the retaliatory consumption epidemic, and many industries are experiencing inflation [4]. Among their major suppliers, in addition to China, countries in south and southeast Asia are hardest hit by the epidemic, and it is even difficult for India to control its own situation [5,6]. The trade tensions between China and the United States tend to ease, and have been an important factor affecting international trade in recent years.

In July 2018, the United States began to impose 25% tariffs on an array of Chinese exports worth US $34 billion [7], and China and the United States began a trade war that has had an enormous impact on the economic development of the two countries as well as the world economy and global trade [8,9]. The strategic conflict between China and the US emerged at the end of 2017, when China was portrayed as a competitor in a Trump administration National Security Report [10]. The trade conflict between China and the United States reflects the strategic competition between the two countries in the new industrial revolution. In turn, future trade agreements may be conditioned on climate agreements in international negotiations. Biden's presidential campaign plan called for binding agreements on enhanced climate ambition, including shipping and aviation, and Biden may support the adoption of a carbon border adjustment [11,12].

Economic growth and rapid industrialization are considered to be the main reasons for the sharp increase in emissions [13]. Since 2006, China has been the world's largest carbon emitter [14]. At the same time, China is also the largest net exporter of carbon dioxide emissions in goods and services [15,16]. The increase in emissions embodied in China's trade has caused problems for international trade and climate policy: China and other emerging markets have a comparative advantage in manufacturing and are an essential part of international trade; however, at the same time, because China's carbon-intensive manufacturing yields much more carbon emissions than the manufacture of the same products in developed regions, trade has increased global carbon emissions [17–20].

With the rapid development of international trade, the production chain of goods and services is no longer limited to one or two countries, and more production and consumption take place in different countries. Current accounting schemes for carbon emissions are mainly based on emissions from production, with less consideration of the consumption side [21,22]. There are two principal methods for consumption-based carbon emissions accounting: life cycle assessment (LCA) and input–output analysis (IOA) [23–25]. The LCA method is typically used for relatively simple and traceable inspections of production chains such as households and enterprises. On the other hand, input–output analysis is widely used at the national and sector levels [26–28]. This method can be further divided into three model frameworks: single region input–output (SRIO), bilateral trade input–output (BTIO), and multi-regional input–output (MRIO).

The SRIO model is mostly used to study the implied energy and emissions in a country's trade, taking the country as a whole and assuming the same production technology; the BTIO model takes into account technological differences between different countries and uses separate energy consumption and emissions factors; neither of these two methods can accurately reflect the relationship between industry and trade among various sectors in each country [29]. The MRIO model distinguishes between the technical and economic structures of different countries as well as the flow of imported and exported products [30]. With the improvement of input–output tables among countries, this method is increasingly employed in research on large-scale hidden emissions in global trade. In its trade war simulation, this paper mainly focuses on changes in China's trade and the resulting changes in emissions. The single region input–output model can meet the paper's research needs with fewer data requirements than the other models, so the SRIO model is adopted.

Here, we combine existing methods to simulate the impact of the trade conflict on China's commodity trade value [31,32] and to discuss the impact on China's energy industry and the path of carbon reduction. In order to track global import and export changes caused by trade conflicts, we use the model of Global Trade Analysis Project (GTAP) [33] to simulate the trade situation of 29 sectors in 14 regions. We calculate the emissions embodied in China's trade by a single input–output (SRIO) model of emissions and trade as of the year 2018. Our calculations only include carbon emissions from China's imports and exports, and emissions from other regions are not included.

2. Materials and Methods

2.1. Materials and Data

The GTAP model data are from the GTAP v10 data package [34], which contains the input–output tables and trade volumes of countries across the world. This paper uses a recursive method to project the 2014 data in the model to 2018 [35], and the currency is US dollars. The energy statistics for China's carbon emissions accounting come from the Energy Statistics Yearbook [36–40], and the emissions factors are derived from the revised emissions factors in Liu's study [41]. Due to the slow updating of China's statistical data, energy statistics for 2018 have not been released, so energy data of 2017 are used to generate carbon emissions data. At the same time, due to the difficulty of obtaining foreign data, this paper combines the emissions data contained in the GTAP's own database and assumes that foreign countries in each region have similar technical levels and are unified into the same emissions coefficient matrix. Abbreviations for regions and departments can be found in Tables A1 and A2.

2.2. Methods

2.2.1. The GTAP Model

The model from the Global Trade Analysis Project (GTAP) is a multi-country multi-sector application general equilibrium model designed based on neoclassical economic theory (Hertel, 1997; GTAP, 2019; Walmsley et al., 2012). The GTAP, led by Thomas W. Hertel, a professor at Purdue University in the United States, was developed and has been widely used in the analysis of trade policies. In the GTAP model framework, they first establish a sub-model that can describe in detail the behavior of each country's production, consumption, government expenditure, etc., and then link the sub-models into a multi-country multi-sector general equilibrium model through international commodity trade. When we carry out policy simulations in this model framework, it is possible to simultaneously discuss the impact of the policy on factors such as production, imports and exports, commodity prices, factor supply and demand, factor compensation, gross domestic product, and social welfare levels in various countries.

The GTAP model assumes that the market is perfectly competitive, the returns to scale of production are constant, producers minimize production costs, consumers maximize utility, and all product and input factor markets clear. At the same time, each country has only one account, and all taxes, financial assets, and capital and labor income are accumulated in this account. The income in the account is divided into three parts: private consumption, deposits, and government consumption. The private expenditure equation uses the fixed difference elastic utility equation. The government's utility equation takes the form of a Cobb-Douglas equation.

GTAP establishes connections between countries (regions) through trade. Domestic products and imported products from different regions are incomplete substitutes; that is, they follow the Armington hypothesis and are characterized by a set constant elasticity of substitution. When the construction of a country's economic model is completed, the commodities and capital flows of international trade (the "global banking" sector) are added to it to form a multi-country economic model. At this time, there is a substitution relationship between imported products and domestic products, and the Armington hypothesis is adopted for product compounding; that is, imported products and domestic products are regarded as different products, and they have an incomplete substitution relationship between each other.

In the GTAP model, there are two international departments (national banks and international transportation departments). The savings of each country are aggregated into international banks and distributed among the countries according to the return on capital. The price expression of import and export commodities in the GTAP model is as follows:

$$P^{FOB} = P^{EX}\left(1 + T^{EX}\right) \tag{1}$$

$$P^{CIF} = P^{FOB}(1 + F) \tag{2}$$

$$P^{IM} = P^{CIF}\left(1 + T^{IM}\right) \tag{3}$$

where P^{FOB} represents the export port price, P^{EX} represents the domestic price of exported goods, P^{CIF} represents the import port price, P^{CIF} represents the domestic price of exported goods, P^{IM} represents the domestic price of imported goods, T^{EX} and T^{IM} represent export and import tariffs (or subsidies), and F is the freight cost.

2.2.2. Production-Based Carbon Emissions Calculation

We calculate the production-based emission according to the IPCC sectoral approach [41]. Emissions are calculated based on the sectoral consumption of different fuels, as shown in equation below.

$$CE_{ij} = AD_{ij} \times NCV_i \times CC_i \times O_{ij} \tag{4}$$

where CE_{ij} refers to the carbon dioxide emissions generated by the combustion of fossil fuel type i in sector j; AD_{ij} represents the fossil fuel consumption of the corresponding type and sector; NCV_i refers to the net calorific value, i.e., the calorific value generated by each fossil fuel combustion unit; CC_i refers to the CO_2 emissions per unit of net calorific value generated by fossil fuel i; and O_{ij} refers to the oxygenation efficiency. The fossil fuel emissions factors ($NCV_i \times CC_i$) we adopted are from a study by Liu [41], in which 602 groups of coal samples from all coal mines in China were sampled and weighted to obtain the national average emissions factor. Reference values for emission factors can be found in Table A3.

2.2.3. Input–Output Method to Calculate Trade Emissions

One method of consumption-based carbon emissions accounting is to compile an inventory based on the final consumption location of goods and services, and another including the total amount of the emissions contained in the imports used in production, and subtract the two quantities. The emissions included in exports reflect the interregional exchange of energy supply, commodities, and materials. Environmentally extended input–output analysis (EIO) can be used to calculate the emissions from regional imports and exports.

Input–output analysis is a method used to study the production balance among various sectors of the national economy. If we start from the assumption of general equilibrium, the dependence of the product volume of each sector is expressed as a system of equations. Then, based on statistical data, a matrix or checkerboard-shaped balance table is made to show the overall picture of the balance between the supply of and demand for products in various sectors of the national economy; from this is derived the total amount of products in each sector. The proportion of the product volume required by other sectors (called the technical coefficient) is used to determine the relevant parameter values in the above equations.

According to Leontief's input–output analysis method [42], the following models can be established:

$$\boldsymbol{X} = \boldsymbol{AX} + \boldsymbol{Y} \tag{5}$$

where $\boldsymbol{X}$ is the N*1 order total output column vector, N is the number of economic sectors, $\boldsymbol{Y}$ is the N*1 order final product column vector, and matrix $\boldsymbol{A}$ is the direct consumption coefficient.

After conversion, it can be transformed into:

$$\boldsymbol{X} = (\boldsymbol{I} - \boldsymbol{A})^{-1}\boldsymbol{Y} = \boldsymbol{BY} \tag{6}$$

Here, $\boldsymbol{B}$ is the Leontief inverse matrix, that is, the complete demand coefficient matrix, and $\boldsymbol{I}$ is the identity matrix.

Next, we can obtain the demand coefficient matrix $\boldsymbol{C}$ of carbon emissions in each industry,

$$\boldsymbol{C} = \boldsymbol{X^C}(\boldsymbol{1} - \boldsymbol{A})^{-1} \tag{7}$$

where $\boldsymbol{X^C}$ represents the carbon emissions on the production side of each sector.

Finally, we can obtain the carbon emissions in import and export trade,

$$\boldsymbol{C^{im}} = \boldsymbol{CY^{im}} \tag{8}$$

$$\boldsymbol{C^{out}} = \boldsymbol{CY^{out}} \tag{9}$$

where C^{im} and C^{out} represent the carbon emissions contained in imports and in exports, respectively, and Y^{im} and Y^{out} represent the import and export trade volumes, respectively.

3. Results

3.1. Goods Traded before and after the Trade Conflict

The model used in this paper is the GTAP model developed by researchers at Purdue University in the United States. It is a multi-country, multisector computable general equilibrium model and is widely used in quantitative analyses of the impact of international trade policies.

The trade conflict model setting reflects a scenario in which the United States imposes tariffs on different trade commodities to eliminate the trade deficit, and China counters with tariffs of its own. We run our simulations based on the list of 25% tariffs imposed on several key sectors. Changes in macroeconomic variables such as commodity trade variables in the process are the result of China's response to the impact of the trade war. Given the uncertainties surrounding different national policies, no scenario analysis was performed on this basis for other countries' policies (such as the EU's countermeasures to the US's increase in tariffs, countries around the world speeding up RCEP negotiations, etc.).

Table 1 shows the impact of the trade war on China's exports in various sectors. It can be seen that China's exports to the United States have fallen sharply, but its exports to other countries have increased. The total exports of most sectors have increased, mechanically leading to an increase in emissions from China's trade.

Table 1. Changes in China's exports to different countries.

Sectors	USA	Oceania	EastAsia	SEAsia	SouthAsia	Namerica	La-Amer	EU-28	MENA	SSA	Other
Transport	−0.74	0.06	0.05	0.05	0.07	0.00	0.05	0.06	0.06	0.06	0.06
ElectricalEq	−0.07	−0.05	−0.06	−0.06	−0.04	−0.13	−0.06	−0.06	−0.05	−0.05	−0.05
ElectronicEq	−0.15	−0.07	−0.07	−0.08	−0.06	−0.14	−0.08	−0.07	−0.07	−0.07	−0.07
FerrousMetal	−0.02	−0.03	−0.04	−0.04	−0.03	−0.09	−0.04	−0.03	−0.03	−0.03	−0.03
Total	−0.33	0.00	−0.04	−0.04	−0.01	−0.05	0.00	−0.02	0.00	0.00	−0.02

Table 2 shows the impact of the trade war on China's imports in various sectors. It can be seen that overall imports have been slightly reduced, and the changes are not very different across the various sectors. Imports from the United States and North America have changed significantly, mainly due to the increase in import costs caused by tariffs. Under the influence of this trend, Chinese imports from other countries have also been slightly reduced, mechanically leading to a reduction in the emissions contained in China's imported products. If we take the two together, China's trade exports have increased while its imports have decreased, and China's consumption-based carbon emissions have decreased in turn.

Table 2. Changes in China's imports from different countries.

Sectors	USA	Oceania	EastAsia	SEAsia	SouthAsia	Namerica	LatinAmer	EU-28	MENA	SSA	Other
GrainsFesFis	−0.67	0.09	0.10	0.09	0.10	0.06	0.09	0.09	0.10	0.10	0.10
ProcFood	−0.66	0.01	0.01	0.00	0.02	−0.04	0.01	0.01	0.01	0.01	0.01
Transport	−0.74	0.06	0.05	0.05	0.07	0.00	0.05	0.06	0.06	0.06	0.06
ChemicalPro	−0.76	0.02	0.02	0.01	0.03	−0.03	0.02	0.02	0.03	0.03	0.03
Total	−0.33	0.00	−0.04	−0.04	−0.01	−0.05	0.00	−0.02	0.00	0.00	−0.02

3.2. Carbon Emissions Contained in China's Trade with the US

According to the value of international trade and the emissions coefficient matrix, we calculate the emissions changes in these main sectors for the two countries and the emissions changes of other sectors.

In Figure 1a, it can be seen that the carbon emissions contained in goods imported from the United States by China in several major sectors have been reduced. Due to the difference in carbon emissions intensity, the changes in emissions contained in chemical products are obviously greater, and the decreasing imports from other sectors also have the effect of decreasing the emissions contained in those sectors. As seen in Figure 1b, the changes in carbon emissions from China's exports to the United States are different from the changes in emissions from imports. Except for those of the nonferrous metal sector, the carbon emissions of sectors with tariffs are all relatively low, while the emissions of other sectors have increased by a large margin. This is similar to the result of the trade analysis. The import shrinkage effect caused by the trade conflict has mechanically reduced China's import emissions from the United States. However, at the same time, export emissions are controlled by the trade market and have grown slightly in other sectors that do not levy tariffs, with only small changes overall. On the whole, China's net emissions to the United States have decreased.

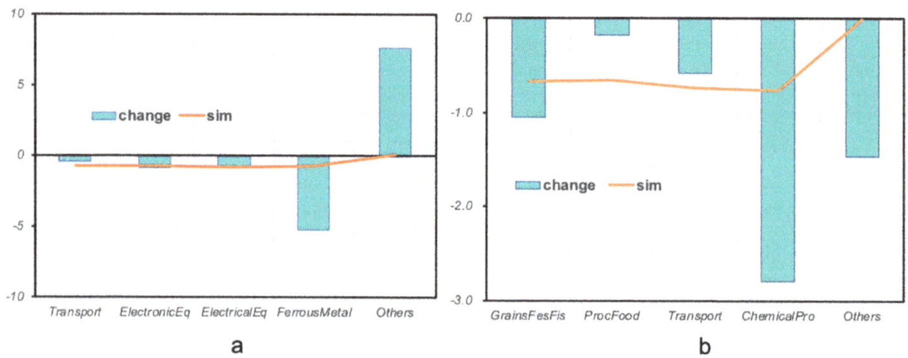

Figure 1. Changes in commodity carbon emissions from major sectors ($MtCO_2$) before and after the China–US trade conflict, where (**a**) represents emissions from China to the United States and (**b**) represents emissions from the United States to China.

3.3. *Changes in China's Trade Emissions with the Rest of the World*

Figures 2 and A1 shows the changes in China's export emissions to various countries in the world. Figure 2a shows the absolute change, and Figure 2b shows the percentage change. On the whole, China's exports to the world are mainly concentrated in the industrial and service industries at this stage, while the sectors with the largest export emissions are the electricity and water sectors, with emissions that are much higher than those of other sectors. Since the start of the trade conflict, except for in a few major sectors in which tariffs have been imposed, emissions have decreased, and those of other sectors have increased slightly.

Figures 3 and A2 shows the changes in emissions from China's imports from various countries in the world. Figure 3a shows the absolute change, and Figure 3b shows the percentage change. It can be seen that the distribution of emissions from China's imports is relatively even, with transportation services accounting for the largest share. Since the start of the trade conflict, the import emissions of all sectors have fallen, and China's import trade has been more affected.

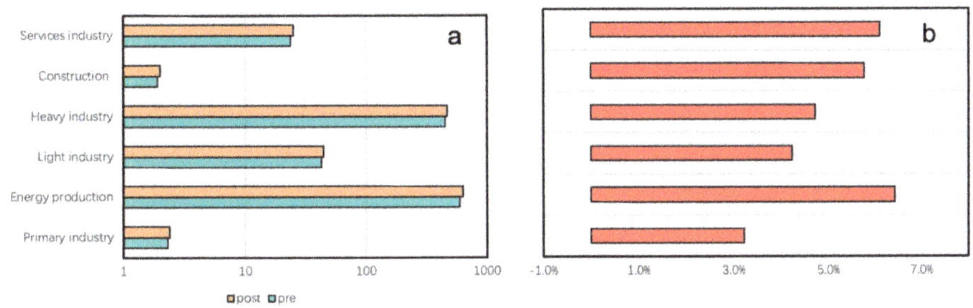

Figure 2. Changes in the world's carbon emissions from China's exports, where (**a**) represents the change in carbon emissions ($MtCO_2$) and (**b**) represents the percentage change.

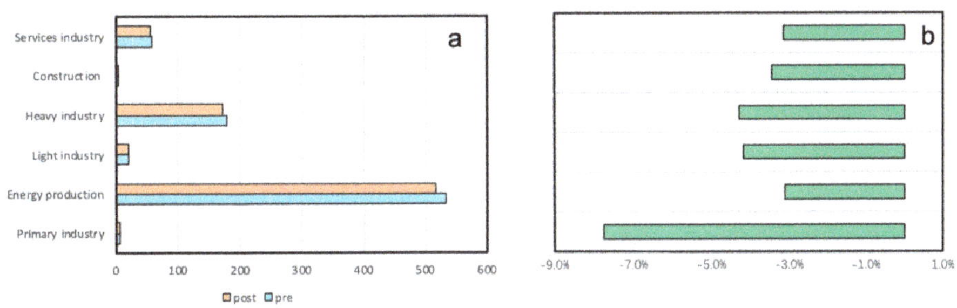

Figure 3. Changes in the world's carbon emissions from China's imports, where (**a**) represents the change in carbon emissions ($MtCO_2$) and (**b**) represents the percentage change.

4. Discussion

4.1. Spatial Distribution of Emissions Included in China's Trade

To further discuss the impact of the trade conflict on carbon emissions, this paper examines the changes in China's import and export emissions from different countries from a spatial perspective. As seen in Figure 4, whether through imports or exports, China's share of carbon emissions to the United States is smaller than the shares of other Asian countries. Due to the influence of spatial location, the countries that trade most with China are Asian countries. Whether because of transportation costs or the demand for a large number of daily necessities caused by population growth, these countries have closer trade ties with China. In contrast, China–US trade is more concentrated in certain sectors. Before the trade conflict, China's exports to the United States were electronic products, which accounted for 1/3 of all of China's exports and half of China's total exports of electronic products. Since the start of the trade conflict, the share has plummeted to approximately 1/8. On the other hand, the emissions coefficient of electronic products is so low that even before the trade conflict, the carbon emissions of electronic products accounted for only 1/50 of China's total emissions from exports to the United States.

Unlike China's exports to the United States, China's imports from the United States are the main component of the changes to China's imports. Compared with the emissions from imports from other countries and regions, which have shown only minor changes, China's emissions from imports from the United States have been reduced by nearly one-third, which has had an impact on China's overall import situation. Although China's import market is not highly dependent on the United States, the United States is an important source of imports for Chinese agricultural products and transportation equipment. China's response to the tariffs has also had a considerable impact on these two sectors, which have seen their imports reduced by nearly 70%.

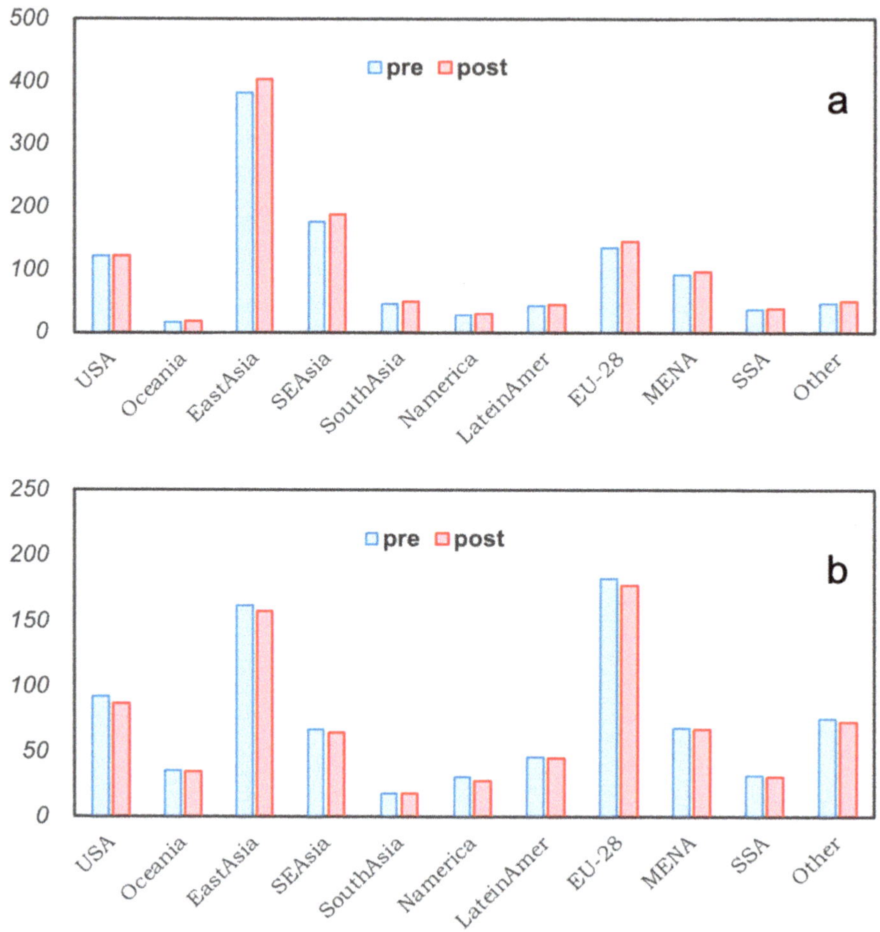

Figure 4. Spatial distribution of emissions included in China's trade ($MtCO_2$), where (**a**) represents emissions of exports and (**b**) represents emissions of imports.

4.2. Emissions Characteristics of China's Net Exports

As the "factory of the world", China has always been an export-oriented country, meaning that its carbon emissions from exports are higher than those from imports from other countries. Based on this, we calculate China's net emissions from international trade based on the previous results. As shown in Figure 5, China exports a large amount of carbon emissions in industries with high energy consumption, such as nonferrous metals, minerals, coal, and petroleum gas production. Excluding a few major sectors, the net emissions of other sectors are much lower. This situation is related to the long-term economic growth mode of the Chinese government. The government has invested heavily in energy-intensive industries to drive the rapid growth of the country's GDP. However, this situation is currently improving. With the adjustment of national strategies, environmental governance has been given equal importance to economic growth. Green sustainable development and the ensuing energy consumption revolution both reflect the Chinese government's determination to adapt to climate change. China is determined to start from multiple angles to resolve the contradiction between trade development and emissions growth.

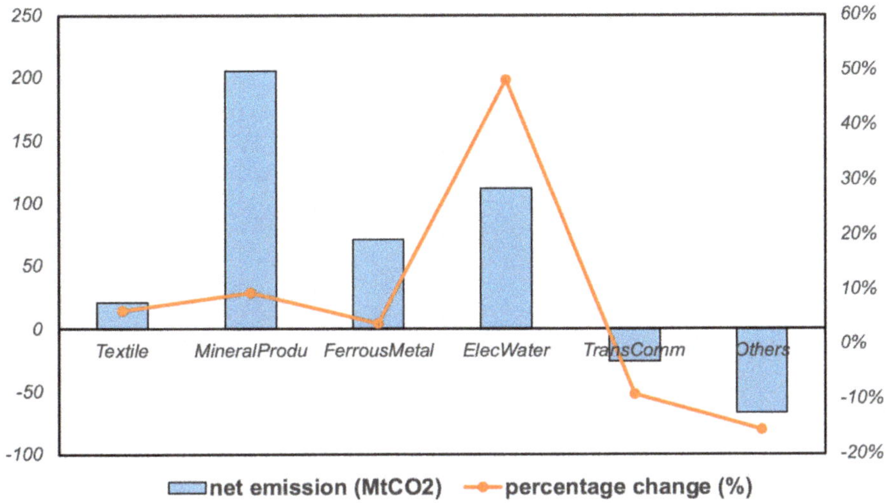

Figure 5. China's net export emissions and percentage change.

4.3. The Impact of the Trade Conflict on Climate Change

The most direct impact of human activities on climate change occurs through increases or decreases in carbon dioxide emissions. Based on the above results, China's carbon emissions in international trade have not changed much in the short term after the start of the trade conflict. Although import emissions have decreased, emissions from exports, the main component of China's trade, have not decreased but rather have increased. However, from a long-term perspective, the impact of the trade conflict on China's adaptation to climate change is likely to be more pronounced.

First, the trade conflict between China and the United States has had an impact not only on trade but also on the social economy of the two countries. The Chinese economy is in a "new normal" phase, the transitional stage from extensive growth based on scale and speed to intensive growth based on quality and efficiency. The negative impact of the trade conflict on China's economy is bound to delay its progress. As mentioned above, China's main exports in international trade come from the massive output of its energy-intensive industries. To ensure the steady development of the domestic economy and eliminate the negative effects of the trade conflict, government investment in these industries is not likely to change significantly.

Second, China has recently put forward a goal of achieving carbon neutrality by 2060. This plan is closely related to China's abundant wind power, hydropower, geothermal, and other new energy potential. China's abundant natural resources make it possible to achieve this goal. However, the new energy industry has a great demand for technology and equipment. China's current level of technology is not adequate to support independent achievement of its objectives. The import of technology and equipment is thus vital to the development of new energy. The trade conflict between China and the United States is set to have an impact on China's imports in and slow down the development of its domestic new energy industry, such that more effort will be required to achieve green development goals such as carbon neutrality.

Finally, as the economy develops, China's energy dependence will increase. Although China has a large amount of low-cost coal resources, considering the concept of green development, coal energy must gradually be replaced. On the other hand, China will have a higher degree of dependence on oil and natural gas, which are not abundant in the country, and thus will face much external uncertainty. In the international energy market, the United States will become a major oil and gas exporter in the market with the realization of its shale gas revolution and energy independence strategy. In the face of China's massive natural gas demand, energy cooperation between China and the United States may offer a new opportunity to improve the trade imbalance between the two countries.

5. Conclusions

The trade conflict between China and the United States has had an impact on China's import and export markets, which in turn has affected the carbon emissions contained in China's imports and exports in international trade and will affect China's response to climate change. In China–US trade, the trade volume of goods subject to tariffs has been greatly reduced, while in other sectors, import emissions have increased and export emissions have decreased. For the global market, China's export emissions to the rest of the world have increased slightly, while import emissions have decreased slightly. The trade conflict will cause China's net export emissions to continue to increase, with the change concentrated in energy-intensive industries.

At the same time, it can be seen that although the trade share between China and the United States is not large in comparison with the world total, some of the sectors involved in the trade war are the main sectors involved in trade between the two countries, and they all contribute a large share to China's total trade volume. The sharp decline in trade in these sectors will also have impacts and raise opportunities in China's inland markets. On the other hand, the trade conflict will affect China's social economy from other angles in the long run as well as some of China's strategies to adapt to climate change. Whether through a negative impact on the domestic economy or restricted imports of technology and equipment, the trade conflict will slow down the development of China's new energy industry. The energy trade may provide an opportunity to solve the problem of the trade imbalance between the two countries.

This paper still has many shortcomings, especially in terms of data. On the one hand, due to the difficulty of obtaining data from all countries, we assume that the regional emission intensity is consistent, and there will be considerable uncertainty; on the other hand, we have also simplified the additional levy departments when it comes to tariff plans, due to the GTAP model. It is not easy to completely match the actual situation. We selected key departments to impose tariffs and simulate.

Author Contributions: Conceptualization, J.C. and W.D.; methodology, F.Y.; software, F.Y.; validation, F.Y. and Z.W.; formal analysis, Z.W.; resources, J.C.; data curation, F.Y.; writing—original draft preparation, F.Y.; writing—review and editing, J.C.; visualization, F.Y.; supervision, J.C.; project administration, J.C. All authors have read and agreed to the published version of the manuscript.

Funding: This research was funded by the National Key Research and Development Program of China (2018YFC1509003, 2016YFA0602703) and the National Natural Science Foundation of China (42075167).

Data Availability Statement: Not applicable.

Acknowledgments: The authors sincerely thank State Key Laboratory of Earth Surface Processes and Resource Ecology for supporting this research.

Conflicts of Interest: The authors declare that they have no known competing financial interests or personal relationships that could have appeared to influence the work reported in this paper.

Appendix A

Table A1. Category of countries.

Region Abbreviations	Comprising	Description
China	chn	China
USA	usa	USA
Oceania	aus, nzl, xoc	Oceania
EastAsia	hkg, jpn, kor, mng, twn, xea, brn	East Asia (Except China)
SEAsia	knm, idn, lao, mys, phl, sgp, tha, vnm, xse	Southeast Asia
Namerica	can, mex, xna	North America
LatinAmer	arg, bol, bra, chl, col, ecu, pry, per, ury, ven, xsm, cri, gtm, hnd, nic, pan, slv, xca, dom, jam, pri, tto, xcb	Latin Amercia
EU_28	aut, bel, bgr, hrv, cyp, cze, dnk, est, fin, fra, deu, grc, hun, irl, ita, lva, ltu, lux, mlt, nld, pol, prt, rou, svk, svn, esp, swe, gbr	European Union 28
MENA	bhr, irm, isr, jor, kwt, omn, qat, sau, tur, are, xws, egy, mar, tun, xnf	Middle East and North Africa
SSA	ben, bfa, cmr, civ, gha, gin, nga, sen, tgo, xwf, xcf, xac, eth, ken, mdg, mwi, mus, moz, rwa, tza, uga, zmb, zwe, xec, bwa, nam, zaf, xsc	Sub-Saharan Africa
RestofWorld	che, nor, xef, alb, blr, rus, ukr, xee, xer, kaz, kgz, tjk, xsu, arm, zae, geo, xtw	Rest of World

Table A2. Category of sectors.

Sectors Reclassified	Sectors in GTAP	Sectors in China	Category
GrainsFesFis	Grain, Fes, Fis	Farming, Forestry, Animal Husbandry, Fishery, and Water Conservancy	Primary industry
Coal	Coal	Coal Mining and Dressing	Energy production
OilGas	Oil, Gas	Petroleum and Natural Gas Extraction	Energy production
OtherMineral	Mineral	Ferrous Metals Mining and Dressing, Nonferrous Metals Mining and Dressing, Nonmetal Minerals Mining and Dressing, Other Minerals Mining and Dressing	Energy production
ProcFood	Food Production	Food Processing, Food Production	Light industry
BeveragesTob	Beverage production, Tobacco Production	Beverage Production, Tobacco Processing	Light industry
Textile	Textile	Textile Industry	Light industry
Wearing	Wearing	Garments and Other Fiber Products	Light industry
LeatherProd	Leather Production	Leather, Furs, Down, and Related Products	Light industry
WoodProduct	Wood Production	Logging and Transport of Wood and Bamboo, Timber Processing, Bamboo, Cane, Palm Fiber and Straw Products	Light industry
PaperProduct	Paper Production	Papermaking and Paper Products	Light industry
Transport	Transport Equipment	Transportation Equipment	Light industry
MetalProduct	Metal Production	Metal Products	Heavy industry
OthLightMnfc	Light Manufacture	Furniture Manufacturing, Printing and Record Medium Reproduction, Cultural, Educational and Sports Articles	Light industry
PetroleumCoa	Petroleum, Coal production	Petroleum Processing and Coking, Raw Chemical Materials, and Chemical Products	Energy production
ChemicalPro	Chemical Production	Chemical Fiber	Heavy industry
BasicPharmac	Basic Pharmacy	Medical and Pharmaceutical Products	Light industry
RubberPlasti	RubberPlastic	Rubber Products, Plastic Products	Heavy industry
MineralProdu	Mineral Production	Nonmetal Mineral Products	Heavy industry
FerrousMetal	Ferrous Metal Production	Smelting and Pressing of Ferrous Metals	Heavy industry
OtherMetal	Other Metal Production	Smelting and Pressing of Nonferrous Metals	Heavy industry
ElectronicEq	Electronic Equipment	Electric Equipment and Machinery	Electric Equipment and Machinery

Table A2. *Cont.*

Sectors Reclassified	Sectors in GTAP	Sectors in China	Category
ElectricalEq	Electrical Equipment	Electronic and Telecommunications Equipment, Instruments, Meters, Cultural and Office Machinery	Electric Equipment and Machinery
OthHeavyMnfc	Other Heavy Manufacture	Ordinary Machinery, Equipment for Special Purposes, Instruments, Meters, Cultural and Office Machinery, Other Manufacturing Industry	Heavy industry
ElecWater	Electricity, Water	Production and Supply of Electric Power, Steam and Hot Water, Production and Supply of Tap Water	Energy production
GasManufactu	Gas Manufacture	Production and Supply of Gas	Energy production
Constructio	Construction	Construction	Construction
TransComm	Trans Commerce	Transportation, Storage, Post and Telecommunication Services	Services industry
OthServices	Other Services	Wholesale, Retail Trade and Catering Services, Others	Services industry

Table A3. Emission factors of each type of fuels.

No.	Fuels in China's Energy Statistics	Fuels in This Study	$NCV_i \times CC_i$ (*t C*/10^4 ton)
1	Raw coal	Raw coal	5.5272
2	Cleaned coal	Cleaned coal	6.8432
3	Other washed coal	Other washed coal	3.948
4	Briquettes	Briquette	4.7376
5	Gangue Coke	Coke	8.7864
6	Coke oven gas	Coke over gas	34.5989
7	Blast furnace gas Converter gas Other gas	Other gas	17.8367
8	Other coking products	Other coking products	7.686
9	Crude Oil	Crude oil	8.6344
10	Gasoline	Gasoline	8.316
11	Kerosene	Kerosene	8.624
12	Diesel oil	Diesel oil	8.686
13	Fuel oil	Fuel oil	9.073
14	Naphtha Lubricants Paraffin White spirit Bitumen asphalt Petroleum coke Other petroleum products	Other petroleum products	8.772
15	Liquefied petroleum gas (LPG)	LPG	9.4
16	Refinery gas	Refinery gas	8.686
17	Nature gas	Nature gas	59.5948

There are 26 kinds of fossil fuels in China's energy statistics system. Because the quality of some fuels is similar to that of other fuels, this paper combines these fuels into 17 types. Among the 17 types of fossil fuels, raw coal, crude oil, and natural gas are the main energy sources, and the other 14 fuels are classified as secondary energy sources.

Appendix B

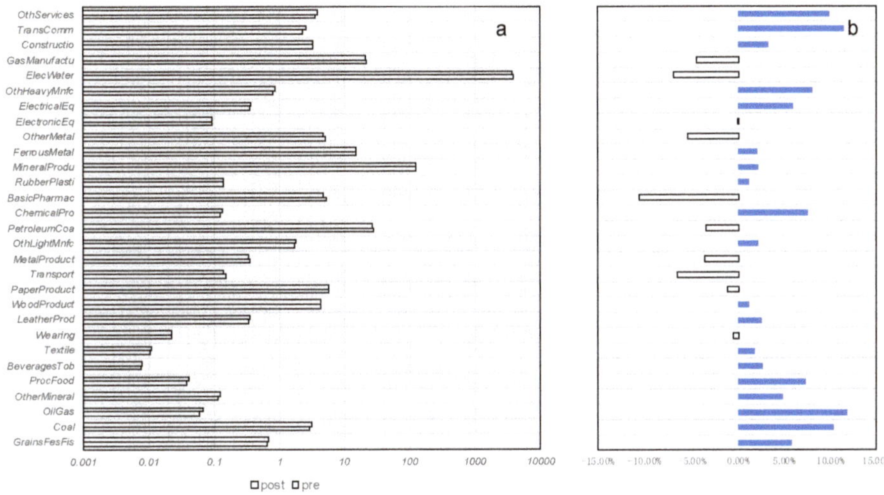

Figure A1. Changes in the world's carbon emissions from China's exports, where (**a**) represents the change in carbon emissions ($MtCO_2$) and (**b**) represents the percentage change.

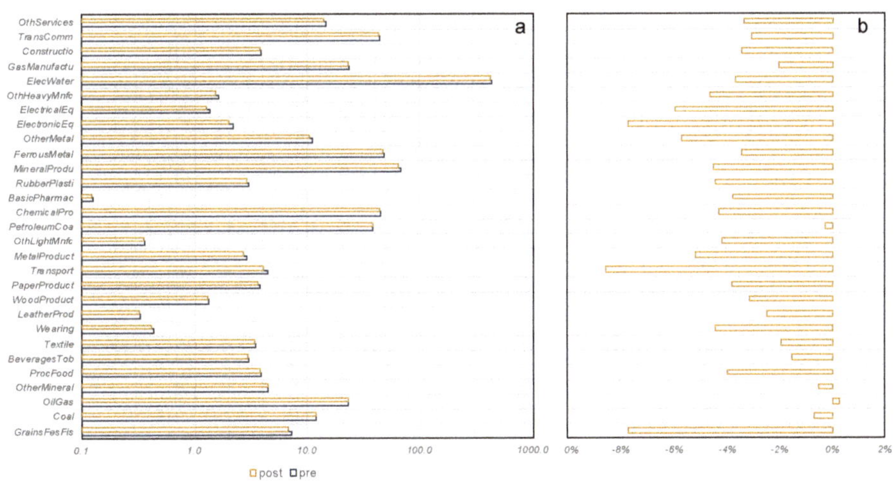

Figure A2. Changes in the world's carbon emissions from China's imports, where (**a**) represents the change in carbon emissions ($MtCO_2$) and (**b**) represents the percentage change.

References

1. World Bank. *Global Economic Prospects*; World Bank: Washington, DC, USA, 2020. Available online: https://openknowledge.worldbank.org/handle/10986/33748 (accessed on 20 June 2021).
2. Qiu, Y.; Chen, X.; Shi, W. Impacts of social and economic factors on the transmission of coronavirus disease 2019 (COVID-19) in China. *J. Popul. Econ.* **2020**, *33*, 1127–1172. [CrossRef]
3. Fong, M.W.; Gao, H.; Wong, J.Y.; Xiao, J.; Shiu, E.Y.; Ryu, S.; Cowling, B.J. Nonpharmaceutical Measures for Pandemic Influenza in Nonhealthcare Settings—Social Distancing Measures. *Emerg. Infect. Dis.* **2020**, *26*, 976. [CrossRef]
4. United States Inflation Rate. Available online: https://tradingeconomics.com/united-states/inflation-cpi (accessed on 21 June 2021).

5. Fauzi, M.A.; Paiman, N. COVID-19 pandemic in Southeast Asia: Intervention and mitigation efforts. *Asian Educ. Dev. Stud.* **2020**, *10*, 176–184. [CrossRef]
6. MoHFW | Home. Available online: https://www.mohfw.gov.in (accessed on 21 June 2021).
7. Office of the United States Trade Representative. USTR Issues Tariffs on Chinese Products in Response to Unfair Trade Practices. 2018. Available online: https://ustr.gov/about-us/policy-offices/press-office/press-releases/2018/june/ustr-issues-tariffs-chinese-products (accessed on 5 September 2020).
8. Customs Tariff Commission of the State Council. Proclamation by Customs Tariff Commission of the State Council about Imposing Tariffs on 60 Billions Dollars Imports Originating from the United States. (in Chinese). 2018. Available online: http://gss.mof.gov.cn/zhengwuxinxi/zhengcefabu/201809/t20180918_3022592.html (accessed on 5 September 2020).
9. World Trade Organization (WTO). Panels Established to Review US Steel and Aluminum Tariffs, Countermeasures on US Imports. 2018. Available online: https://www.wto.org/english/news_e/news18_e/dsb_19nov18_e.html (accessed on 10 September 2020).
10. Morrison, W.M. *China-US Trade Issues*; Congressional Research Service: Washington, DC, USA, 2018.
11. Lester, S. *Trade Policy under the Biden Administration*; Free Trade Bulletin No.76; Cato Institute: Washington, DC, USA, 2020. Available online: https://www.cato.org/sites/cato.org/files/2020-06/ftb-76.pdf (accessed on 21 June 2021).
12. Immergluck, D. Housing Policy Recommendations for the Biden/Harris Administration. *Hous. Policy Debate* **2021**, 1–4. [CrossRef]
13. Raupach, M.R.; Marland, G.; Ciais, P.; Le Quéré, C.; Canadell, J.G.; Klepper, G.; Field, C.B. Global and regional drivers of accelerating CO_2 emissions. *Proc. Natl. Acad. Sci. USA* **2007**, *104*, 10288–10293. [CrossRef] [PubMed]
14. Le Quéré, C.; Andres, R.J.; Boden, T.A.; Conway, T.; Houghton, R.A.; House, J.I.; Marland, G.; Peters, G.P.; van der Werf, G.R.; Ahlström, A.; et al. The Global Carbon Budget 1959–2011. *Earth Syst. Sci. Data* **2013**, *5*, 165–185. [CrossRef]
15. Weber, C.L.; Peters, G.; Guan, D.; Hubacek, K. The contribution of Chinese exports to climate change. *Energy Policy* **2008**, *36*, 3572–3577. [CrossRef]
16. Jakob, M.; Marschinski, R. Interpreting trade-related CO_2 emission transfers. *Nat. Clim. Chang.* **2012**, *3*, 19–23. [CrossRef]
17. Lin, B.; Sun, C. Evaluating carbon dioxide emissions in international trade of China. *Energy Policy* **2010**, *38*, 613–621. [CrossRef]
18. Ying, C.; Jiahua, P.; LaiHui, X. Energy Embodied in Goods in International Trade of China: Calculation and Policy Implications. *Chin. J. Popul. Resour. Environ.* **2011**, *9*, 16–32. [CrossRef]
19. Wiebe, K.S.; Bruckner, M.; Giljum, S.; Lutz, C.; Polzin, C. Carbon and materials embodied in the international trade of emerging economies: A multiregional input-output assessment of trends between 1995 and 2005. *J. Ind. Ecol.* **2012**, *16*, 636–646. [CrossRef]
20. Jiang, X.; Zhang, Q.; Zhao, H.; Geng, G.; Peng, L.; Guan, D.; Kan, H.; Huo, H.; Lin, J.; Brauer, M.; et al. Revealing the Hidden Health Costs Embodied in Chinese Exports. *Environ. Sci. Technol.* **2015**, *49*, 4381–4388. [CrossRef]
21. Liu, Z.; Guan, D.; Wei, W.; Davis, S.J.; Ciais, P.; Bai, J.; Peng, S.; Zhang, Q.; Vogt-Schilb, A.; Marland, G.; et al. Reduced carbon emission estimates from fossil fuel combustion and cement production in China. *Nature* **2015**, *524*, 335–338. [CrossRef] [PubMed]
22. Davis, S.J.; Ken, C. Consumption-based accounting of CO_2 emissions. *Proc. Natl Acad. Sci. USA* **2010**, *107*, 5687–5692. [CrossRef] [PubMed]
23. Zhang, Y. Carbon Contents of the Chinese Trade and Their Determinants: An Analysis Based on Non-Competitive (Import) Input-Output Tables. *China Econ. Q.* **2010**, *4*, 1290–1310.
24. Huimin, L.; Ye, Q. Carbon Embodied in International Trade of China and Its Emission Responsibility. *Chin. J. Popul. Resour. Environ.* **2010**, *8*, 24–31. [CrossRef]
25. Yan, Y.; Zhao, Z. CO_2 Emissions Embodied in China's International Trade: A Perspective of Allocating International Responsibilities. *J. Int. Trade* **2012**, *1*, 131–142.
26. Nakano, S.; Okamura, A.; Sakurai, N.; Suzuki, M.; Tojo, Y.; Yamano, N. *The Measurement of CO_2 Embodiments in International Trade: Evidence from the Harmonised Input-Output and Bilateral Trade Database*; OECD: Paris, France, 2009.
27. Wiedmann, T. A review of recent multi-region input–output models used for consumption-based emission and resource accounting. *Ecol. Econ.* **2009**, *69*, 211–222. [CrossRef]
28. Li, J.; Chandio, A.A.; Liu, Y. Trade Impacts on Embodied Carbon Emissions—Evidence from the Bilateral Trade between China and Germany. *Int. J. Environ. Res. Public Health* **2020**, *17*, 5076. [CrossRef]
29. Su, B.; Ang, B. Input–output analysis of CO_2 emissions embodied in trade: Competitive versus non-competitive imports. *Energy Policy* **2013**, *56*, 83–87. [CrossRef]
30. Sun, C.; Ding, D.; Yang, M. Estimating the complete CO_2 emissions and the carbon intensity in India: From the carbon transfer perspective. *Energy Policy* **2017**, *109*, 418–427. [CrossRef]
31. Feng, K.; Davis, S.; Sun, L.; Li, X.; Guan, D.; Liu, W.; Liu, Z.; Hubacek, K. Outsourcing CO_2 within China. *Proc. Natl. Acad. Sci. USA* **2013**, *110*, 11654–11659. [CrossRef] [PubMed]
32. Liu, Z.; Davis, S.; Feng, K.; Hubacek, K.; Liang, S.; Anadon, L.D.; Chen, B.; Liu, J.; Yan, J.; Guan, D. Targeted opportunities to address the climate–trade dilemma in China. *Nat. Clim. Chang.* **2016**, *6*, 201–206. [CrossRef]
33. Hertel, T.W. *Global Trade Analysis: Modeling and Applications*; Cambridge University Press: Cambridge, MA, USA, 1997.
34. GTAP v10 Data Base. 2019. Available online: https://www.gtap.agecon.purdue.edu/about/project.asp (accessed on 21 June 2021).
35. Walmsley, T.L.; Dimaranan, B.V.; McDougall, R.A. A Baseline Scenario for the Dynamic GTAP Model. *Dyn. Model. Appl. Glob. Econ. Anal.* **2012**, *2012*, 136–157. [CrossRef]
36. NBS. *China Energy Statistical Yearbook*; China Statistics Press: Beijing, China, 2019.

37. NBS. *China Statistical Yearbook 2001–2019*; China Statistics Press: Beijing, China, 2019.
38. NBS. National Data. National Bureau of Statistics. 2019. Available online: http://data.stats.gov.cn/index.htm (accessed on 21 June 2021).
39. P. R. China National Development and Reform Commission (NDRC). *Guidelines for Provincial Greenhouse Gas Inventories*; NDRC: Beijing, China, 2011. (In Chinese)
40. P. R. China National Development and Reform Commission (NDRC). *Energy Production and Consumption Revolution Strategy (2016–2030)*; NDRC: Beijing, China, 2017. (In Chinese)
41. Change, I. *2006 IPCC Guidelines for National Greenhouse Gas Inventories*; Institute for Global Environmental Strategies: Hayama, Japan, 2006.
42. Leontief, W. Environmental Repercussions and the Economic Structure: An Input-Output Approach. *Rev. Econ. Stat.* **1970**, *52*, 262–271. [CrossRef]

sustainability

MDPI

Article

The Effects of Land-Use Change/Conversion on Trade-Offs of Ecosystem Services in Three Precipitation Zones

Qiang Feng [1], Siyan Dong [2,*] and Baoling Duan [1]

[1] College of Resources and Environment, Shanxi University of Finance and Economics, Taiyuan 030006, China; fengqiang921@163.com (Q.F.); sxnddbl@163.com (B.D.)

[2] National Climate Center, Laboratory for Climate Studies, China Meteorological Administration, Beijing 100081, China

* Correspondence: dongsy@cma.gov.cn; Tel.: +86-01058991567

Abstract: Revealing the spatial differentiation of ecosystem service (ES) trade-offs and their responses to land-use change along precipitation gradients are important issues in the Loess Plateau of China. We selected three watersheds called Dianshi (300 mm < MAP (mean annual precipitation) < 400 mm), Ansai (400 mm < MAP < 500 mm), and Linzhen (500 mm < MAP < 600 mm). A new ES trade-off quantification index was proposed, and quantile regression, piecewise linear regression, and redundancy analysis were used. The results were as follows. (1) Carbon sequestration (TC) and soil conservation (SEC) increased, but water yield (WY) decreased in the three watersheds from 2000 to 2018. (2) The effect of forests on trade-offs was positive in three watersheds, the main effect of shrubs was also positive, but the negative effect appeared in the TC-WY trade-off in Ansai. Grassland exacerbated trade-offs in Dianshi, whereas it reduced trade-offs in Ansai and Linzhen. These effects exhibited respective trends with the quantile in the three watersheds. (3) There were threshold values that trade-offs responded to land-use changes, and we could design land-use conversion types to balance ESs. In general, the water consumption of grass cannot be ignored in Dianshi; shrubs and grass are suitable vegetation types, and forests need to be restricted in Ansai; more forests and shrubs can be supported in Linzen due to higher precipitation, but the current proportions of forests and shrubs are too high. Our research contributes to a better understanding of the response mechanisms of ES trade-offs to land-use changes.

Keywords: ecosystem services trade-offs; land-use change; soil conservation; carbon storage; water yield; precipitation gradient; Loess Plateau

Citation: Feng, Q.; Dong, S.; Duan, B. The Effects of Land-Use Change/Conversion on Trade-Offs of Ecosystem Services in Three Precipitation Zones. *Sustainability* **2021**, *13*, 13306. https://doi.org/10.3390/su132313306

Academic Editors: Xiaodong Yan, Jia Yang and Shaofei Jin

Received: 19 September 2021
Accepted: 22 November 2021
Published: 01 December 2021

1. Introduction

Ecosystem services (ESs) are defined as the benefits that humans derive from natural ecosystems directly and indirectly [1,2]. The relationships among ESs are often identified as trade-offs and synergies. A trade-off is defined as one ES increasing at the expense of another, and a synergy is a situation in which multiple ESs increase or decrease synchronously [3,4]. How to balance multiple ESs is still a major challenge in ecosystem management. The trade-off analysis of ESs provides an integrative and dialectical approach to understand ES relationships, and based on trade-off analysis, land management decisions can be made to realize maximal and sustainable ES supply [5–8].

Precipitation is the key factor controlling many ecological processes, especially in water-stressed regions [9,10]. Previous studies have found that primary production and carbon sequestration [11,12], biodiversity [13], plant cover and growth [14,15], soil-water carrying capacity [16], nutrient recycling and storage [17–19], elemental stoichiometry [20], and hydrologic processes [21,22] change along precipitation gradients. Therefore, precipitation gradients should be an important factor in ecosystem management. There is an obvious precipitation gradient in the Loess Plateau of China that provides an appropriate geographical area for studying regional differences on a landscape scale.

The vegetation cover increased observably by the Grain-for-Green Program (GFGP) launched in 1999. As a result, ecosystem services are undergoing significant change, whereby soil conservation services and the carbon sequestration of local ecosystems are enhanced, but water yield and soil moisture are decreased. These changes threaten regional water resource security and revegetation sustainability, especially in arid and semiarid regions. Therefore, ES trade-offs on the Loess Plateau have attracted the attention of scholars and local governments. A series of studies have been carried out on trade-off and synergy identification, the spatial distribution of ES trade-offs, drivers for trade-offs, ES optimization models, and land-use management in view of trade-offs [5,23–27]. A key conclusion of previous studies is that ES trade-offs and water scarcity are caused by excessive revegetation (land-use conversion) in arid areas [25,27]. However, only a few studies focused on ES trade-offs along the precipitation gradient and only performed trade-off analyses of soil moisture, aboveground carbon, soil organic carbon, total nitrogen, and plant diversity based on field sampling and investigation along the precipitation gradient [28,29]. Thus, the spatial differentiation of ES (carbon sequestration, soil conservation, water yield) trade-offs and the response of the trade-offs to land-use change at the watershed scale along the precipitation gradient have not been clarified, and the existing theoretical basis is insufficient for land management and decision making across the precipitation gradient on the Loess Plateau.

We selected three watersheds called Dianshi (300 mm < MAP (mean annual precipitation from 2000 to 2018) < 400 mm), Ansai (400 mm < MAP < 500 mm), and Linzhen (500 mm < MAP < 600 mm) along the precipitation gradient in the central Loess Plateau. We used the InVEST (Integrated Valuation of Ecosystem Services and Trade-offs) model to calculate carbon sequestration, soil conservation, and water yield in 2000 and 2018. We proposed a new indicator to quantify ES trade-off intensity. The objectives of this study were to (1) reveal the spatial differentiation of land-use conversion and ES trade-offs in various precipitation regions, (2) reveal the effects of land-use conversion on ES trade-offs at different intensity levels (trade-off intensity under different quantiles) and determine the threshold values at which trade-offs respond to land-use conversion, and (3) develop recommendations for land-use planning for three precipitation regions in the central Loess Plateau.

2. Materials and Methods

2.1. Study Area

The Loess Plateau of China is located in the middle reaches of the Yellow River basin, where there is an obvious precipitation gradient. The mean annual precipitation (MAP) gradually decreases from 700 mm in the southeast to 200 mm in the northwest, and the precipitation contours are nearly parallel in the central Loess Plateau. The vegetation types change with precipitation from dry steppe to forest-steppe and deciduous broad-leaf forest. Three independent catchments along the precipitation gradient in the central Loess Plateau were selected (Figure 1). They are controlled by the "Dianshi", "Ansai", and "Linzhen" hydrometric stations, where runoff and sediment are observed. The Dianshi (300 mm < MAP < 400 mm), Ansai (400 mm < MAP < 500 mm), and Linzhen (500 mm < MAP < 600 mm) watersheds are located in three precipitation zones, as illustrated in Figure 1. The study area is characterized by an arid and semiarid continental monsoon climate zone and has typical loess geomorphic landforms and eroded terrains. The soils are mainly derived from loess, a fine silt soil that is weakly resistant to erosion, and this region is considered the most eroded in the world. The soil layer of the Loess Plateau is deep, the groundwater level is mostly between 30 and 80 m, and it hardly participates in the water cycle process of the soil–vegetation–atmosphere transfer system [30]. The natural vegetation was destroyed, and considerable secondary vegetation was planted. The main forest species include *Robinia pseudoacacia*, *Pinus tabulaeformis*, and *Platycladus orientalis*, and the shrubs include *Caragana korshinskii* and *Hippophae rhamnoides*. Grassland communities are mainly composed of *Artemisia gmelinii*, *Lespedeza davurica*, and *Stipa bungeana*. Local people cultivate millet, maize, and broom corn millet in croplands. The percentage of the population dependent on agriculture (including crop farming, forestry, animal husbandry and fishery) decreased from 72.3% in 2000 to 57.9% in 2015 [31].

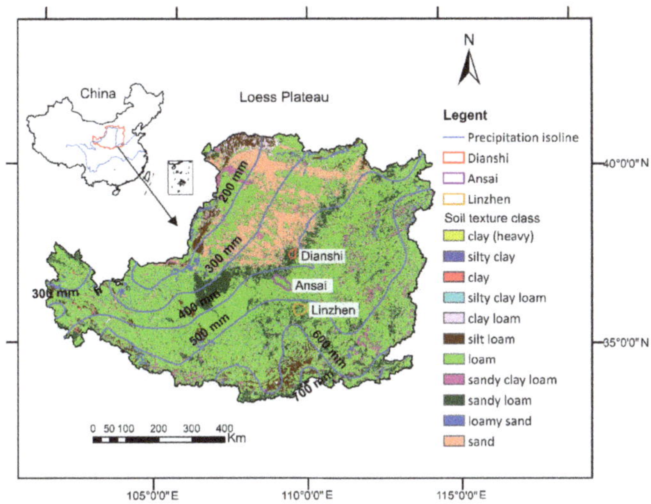

Figure 1. The location of the study area (soil texture class was obtained from the soil map of China in the Harmonized World Soil Database).

2.2. Data Sources

We downloaded Landsat images from the USGS [32] and generated land-use maps (30 m × 30 m) by supervised classification. We obtained meteorological data from the China Meteorological Data Service Center [33]. We obtained a DEM (30 m × 30 m) from the ASTER Global Digital Elevation Model provided by the Geospatial Data Cloud, Computer Network Information Center, Chinese Academy of Sciences [34]. The soil data were obtained from the Soil Map of China in the Harmonized World Soil Database [35].

2.3. Assessment of ESs and Land-Use Changes

The Hydrology Tool of ArcGIS 10 was used to divide the Dianshi, Ansai, and Linzhen watersheds into 240, 817, and 543 subwatersheds, respectively. We calculated and analyzed ESs at the subwatershed level in this study.

2.3.1. Soil Conservation (SEC)

Soil conservation was assessed by the "sediment delivery ratio" (*SDR*) model of InVEST3.3.2. First, the average amount of annual soil loss with existing vegetation was calculated by the revised universal soil loss equation (RUSLE) [36]. Then, the *SDR* was used to calculate the soil loss actually reaching the watershed outlet, and the *SDR* was directly calculated from the conductivity index using the sigmoid function [37]:

$$SLA_x = R_x \cdot K_x \cdot LS_x \cdot C_x \cdot P_x SDR_i \quad (1)$$

where SLA_x is the annual actual soil loss that reaches the watershed outlet on pixel x; R_x is the rainfall erosivity factor on pixel x; K_x is the soil erodibility factor; LS_x is the topographic factor; C_x is the cover and management factor; and P_x is the engineering measures factor. SDR_i is the sediment delivery ratio on pixel x.

The potential soil loss reaching the watershed outlet (SLP_x) was the condition without vegetation coverage and engineering measures (C_x = 1, P_x = 1):

$$SLP_x = R_x \cdot K_x \cdot LS_x SDR_i \quad (2)$$

Finally, the actual value of the soil conservation on pixel x (SC_x) was calculated as the difference between the SLA_x and SLP_x:

$$SC_x = R_x \cdot K_x \cdot LS_x (1 - \cdot C_x \cdot P_x) SDR_i \quad (3)$$

The model accuracy was evaluated by the sediment loading data at the outlet of the watershed, and the relative error between the calculated value and measured value was 3.5%, 2.3%, and 2.8% in Dianshi, Ansai, and Linzhen, respectively.

2.3.2. Water Yield (WY)

The water yield was assessed from the "Water Yield" submodel of InVEST based on the Budyko curve and annual average precipitation. The annual water yield Y_x on pixel x is calculated as follows:

$$Y_x = (1 - AET_x / P_x) \cdot P_x \quad (4)$$

where AET_x is the annual actual evapotranspiration on pixel x and P_x is the annual precipitation on pixel x. The calculations of AET_x are core technology in WY estimation, which can be found in the InVEST User's Guide [38]. The overall process is as follows: For forest, shrub, grassland, and cropland, AET_x was computed by an expression of the Budyko curve proposed by Fu [39] and Zhang et al. [40]; for construction land and water bodies, AET_x was directly computed from the reference evapotranspiration and has an upper limit defined by the precipitation.

2.3.3. Carbon Sequestration (TC)

Carbon storage directly depends on the carbon content of the four major carbon pools in the ecosystem, namely, aboveground biomass, underground biomass, soil carbon, and dead matter. The carbon model of InVEST can evaluate the total carbon storage by summing the four carbon pools according to the land-use maps. The data for the four major carbon pools were obtained by our field survey [41].

2.3.4. Calculation of Land-Use Changes

The ESs and land-use changes were defined as the values of the final stage minus those of the initial stage. The land use transfer matrix in a watershed was obtained by "Raster Calculator", and the area of one land-use type transfer to another in a small watershed was calculated by the "TabulateArea" of ArcGIS 10.

2.4. Calculation of the Trade-Offs between ESs

The methods for quantifying ES trade-offs are still inadequate. The correlation coefficient [42] and root mean square error [5,29,43] are often used to calculate trade-off values. However, the correlation coefficient usually needs time series data, and the root mean square error method only reflects static trade-offs at a single given time. Thus, we proposed one indicator for quantifying ES trade-offs on the premise that trade-offs have been identified between two ESs. This indicator is based on the idea that trade-off intensity is determined by the degree of relative waxing and waning between ESs, and it is calculated as follows:

If trade-off relation can be identified between ecosystem service A and B:

$$TB_{AB} = \frac{1}{2}\left(\sqrt{\left(\frac{ESA_{T2} - ESA_{T1}}{ESA_{T1}}\right)^2} + \sqrt{\left(\frac{ESB_{T2} - ESB_{T1}}{ESB_{T1}}\right)^2}\right) \times 100\% \quad (5)$$

If synergy relation can be identified between ecosystem service A and B:

$$TB_{AB} = -\frac{1}{2}\left(\sqrt{\left(\frac{ESA_{T2} - ESA_{T1}}{ESA_{T1}}\right)^2} + \sqrt{\left(\frac{ESB_{T2} - ESB_{T1}}{ESB_{T1}}\right)^2}\right) \times 100\% \quad (6)$$

where TR_{AB} is the trade-off/synergy value between ecosystem service A and ecosystem service B; ESA_{T1} and ESA_{T2} correspond to ecosystem service A at times T1 and T2, respectively (time T1 is earlier than T2); and ESB_{T1} and ESB_{T2} correspond to ecosystem service B at times T1 and T2, respectively.

First, the ES data of the subwatersheds were prepared for the three watersheds. Second, the data in which the change direction of ES_A and ES_B was inverse (the increase in one coincides with the decrease in the other), indicating a trade-off between the two ESs, were selected, and TR_{AB} was calculated by these data. Finally, the data in which ES_A and ES_B increased or decreased simultaneously, indicating synergy between the two ESs, were selected, and TR_{AB} was calculated by these data to represent synergy intensity.

2.5. Statistical Analyses

Most ES trade-off and land-use change data did not conform to a normal distribution and exhibited heteroscedasticity, so robust statistical methods were employed. Spearman correlation analysis and quantile regression do not require the homogeneity of variance assumption, and they are robust to outliers [44,45] and have been widely used in macroecology. Thus, they were used to reveal the relationship between ES trade-offs and land-use changes. Quantile regression estimates a portion (certain quantiles reflecting various levels of trade-off intensity) of the response variable instead of estimating the mean of the response variable as in ordinary least squares regressions. Thus, quantile regression provides a much more complete picture of the land-use changes influencing ES trade-offs. In addition, piecewise linear regression was used to identify the thresholds of ES trade-off responses to land-use changes. In contrast to simple linear regression, piecewise linear regression explores a more detailed trend of the relationship between variables [46,47]. Redundancy analysis (RDA) was applied to reveal the effects of land-use conversion (e.g., cropland conversion to forest and grassland conversion to shrub) on ES trade-offs. A Monte Carlo permutation test based on 499 random permutations was conducted to test the significance of the marginal and conditional effects [48].

Spearman correlation analysis and quantile regression were conducted using Stata15.1. Piecewise linear regression was performed with the segmented package in the statistical software R. RDA was performed using CANOCO5.0.

3. Results and Discussion

3.1. Temporal and Spatial Variations in ESs along the Precipitation Gradient

3.1.1. Land-Use Transformation along the Precipitation Gradient

As illustrated in Table 1, cropland and grassland were the major land-use types and covered 98.7% of the Dianshi watershed in 2000, whereas the grassland, cropland, and shrub types became dominant and covered 93.7% of the area in 2018. From 2000 to 2018, cropland was mainly transformed into grassland, and grassland was mainly transformed into shrubs and cropland. Therefore, "planting grass" was the primary vegetation rehabilitation method (grassland reached 61.2% of the total area), while "afforestation" was secondary in Dianshi (forest and shrub accounted for only 17.6% of the total area).

The major land-use types were grassland and cropland (96.8%) in the Ansai watershed in 2000, whereas forest and grassland (75.8%) became the major types in 2018. The main land-use transformation characteristics in Ansai were that cropland was mainly converted to forest and grassland, grassland was mainly converted to forest and shrubland, and forest and grassland had equal areas in 2018 (approximately 37%).

Table 1. Land-use transformation matrix from 2000 to 2018 (%).

		FoL in 2018	ShL in 2018	GrA in 2018	CrO in 2018	CoL in 2018	WaB in 2018	Total in 2000
Dianshi	FoL in 2000	0.11	0.09	0.41	0.04	0.01	0.00	0.66
	ShL in 2000	0.12	0.06	0.08	0.01	0.00	0.00	0.27
	GrA in 2000	1.18	6.41	28.03	5.01	0.62	0.23	41.47
	CrO in 2000	2.92	6.72	32.55	14.07	0.91	0.03	57.20
	CoL in 2000	0.00	0.01	0.03	0.02	0.08	0.00	0.14
	WaB in 2000	0.00	0.01	0.12	0.04	0.00	0.09	0.27
	Total in 2018	4.34	13.29	61.22	19.19	1.62	0.35	
	Change from 2000 to 2018	3.68	13.02	19.74	−38.01	1.48	0.08	
Ansai	FoL in 2000	0.96	0.12	0.27	0.10	0.05	0.01	1.51
	ShL in 2000	0.59	0.21	0.54	0.09	0.03	0.00	1.46
	GrA in 2000	20.01	6.97	22.89	3.63	1.07	0.25	54.82
	CrO in 2000	17.36	5.46	13.09	4.82	1.26	0.04	42.03
	CoL in 2000	0.04	0.00	0.01	0.04	0.07	0.00	0.16
	WaB in 2000	0.01	0.00	0.01	0.00	0.00	0.00	0.03
	Total in 2018	38.96	12.77	36.82	8.67	2.47	0.30	
	Change from 2000 to 2018	37.45	11.31	−18.00	−33.35	2.31	0.27	
Linzhen	FoL in 2000	2.37	1.75	0.50	0.30	0.05	0.00	4.97
	ShL in 2000	23.35	12.44	7.14	4.34	0.89	0.01	48.18
	GrA in 2000	10.10	9.55	2.97	1.91	0.40	0.08	25.02
	CrO in 2000	4.75	3.43	5.45	6.78	1.07	0.14	21.61
	CoL in 2000	0.00	0.02	0.01	0.03	0.03	0.00	0.09
	WaB in 2000	0.00	0.01	0.01	0.09	0.00	0.02	0.12
	Total in 2018	40.58	27.20	16.08	13.45	2.44	0.25	
	Change from 2000 to 2018	35.61	−20.98	−8.94	−8.16	2.35	0.13	

FoL: forestland, ShL: shrubland, GrA: grassland, CrO: cropland, CoL: construction land, WaB: water body. An example to explain the meaning of the transformation matrix: Figure 6.41 in the second column of the third row means that 6.41% of grassland in 2000 was changed to shrubland in 2018.

The dominant land-use types were shrub, grassland, and cropland (94.8%) in the Linzhen watershed in 2000, whereas forest, shrub, and grassland (83.9%) became the dominant types in 2018. From 2000 to 2018, shrubs were mainly converted to forest, grassland was mainly converted to forest and shrubs, and cropland was mainly converted to grassland, forest, and shrubs. In brief, forested land was preferentially selected for revegetation in Linzhen.

The soil organic matter content was 0.76%, 1.18%, and 1.33% in Dianshi, Ansai, and Linzhen, respectively, before the GFGP, according to data from the Second National Soil Survey of China. The soil carbon sequestration rates were 0.43, 0.51, and 0.21 Mg $ha^{-2}\cdot year^{-1}$ by revegetation, respectively [49]. Therefore, land-use transformation improved vegetation and soil conditions, which meant that local soil erosion would be reduced, and the agricultural production environment and farmers' livelihoods would become increasingly better.

3.1.2. Changes in ESs from 2000 to 2018 along the Precipitation Gradient

As illustrated in Figure 2, TC and SEC generally increased in the three watersheds from 2000 to 2018 (most ΔTC and ΔSEC were positive values), whereas WY decreased (most ΔWY were negative values). This phenomenon indicated that TC and SEC increased at the cost of decreasing WY. Therefore, a synergistic relationship can be identified between TC and SEC as well as trade-offs between the two ESs and WY. Similar conclusions were found by some studies on the Loess Plateau of China [5,50]. The amount of variation in the ESs changed with the precipitation gradient. With regard to the mean value of ΔESs, the increments of TC were approximative in the Ansai and Linzhen watersheds and were larger than those in Dianshi. The increment of SEC exhibited the trend of Linzhen > Ansai > Dianshi. The decrease in WY exhibited the trend of Dianshi > Ansai > Linzhen. These findings indicate that although TC and SEC were enhanced to some extent by the

GFGP in Dianshi, where rainfall was the smallest, this gain was offset by the significant decrease in WY. However, there were some small watersheds where the change trends of TC and WY were opposite, especially in the Linzhen and Ansai watersheds. The reason for this is that construction land increased in these small watersheds, which reduced TC and increased WY.

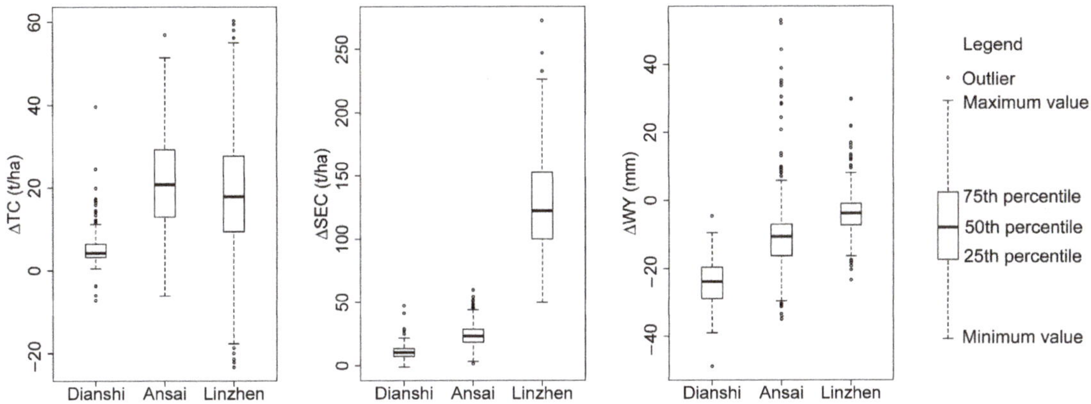

Figure 2. Boxplot of ES changes in three watersheds (ΔTC, ΔSEC, and ΔWY represent ES changes from 2000–2018).

3.1.3. The Correlation between Land-Use and ESs Change

The ΔTC value was significantly positively and negatively correlated with the change in forest and grassland, respectively, in the three watersheds (Table 2). ΔTC was significantly negatively correlated with the change in cropland in Dianshi and Ansai only. The correlations between ΔSEC and ΔForest and ΔShrub were significantly positive in Ansai and Linzhen, whereas the correlations were significantly negative with grassland change. ΔWY was significantly positively and negatively correlated with the changes in cropland and shrubs, respectively, in the three watersheds. In general, increasing forest was beneficial to TC and SEC and was not conducive to WY, whereas the influence of cropland showed the opposite trend [25,27,51]. Somewhat differently, shrubs were beneficial to TC, and grasslands were not conducive to WY in Dianshi, but the influences of shrubs and grasslands were opposite in Ansai, and this influence was not significant in Linzhen. Land-use change characteristics and rainfall differences in the three watersheds caused these phenomena.

Table 2. Spearman correlation analysis between land use and ES change.

ΔESs	Watershed	ΔForest	ΔShrub	ΔGrassland	ΔCropland
ΔTC	Dianshi	0.730 **	0.402 **	−0.291 **	−0.410 **
	Ansai	0.922 **	−0.167 **	−0.723 **	−0.262 **
	Linzhen	0.891 **	−0.002	−0.445 **	−0.02
ΔSEC	Dianshi	−0.014	0.087	0.035	−0.157 *
	Ansai	0.299 **	0.076*	−0.169 **	−0.330 **
	Linzhen	0.196 **	0.237 **	−0.351 **	0.063
ΔWY	Dianshi	−0.006	−0.145 *	−0.527 **	0.852 **
	Ansai	−0.530 **	−0.099 **	0.412 **	0.203 **
	Linzhen	−0.082	−0.276 **	−0.023	0.917 **

The asterisks * and ** indicate $p < 0.05$ and $p < 0.01$ respectively.

3.2. ESs Trade-Offs along the Precipitation Gradient

3.2.1. Comparing ES Trade-Offs in Three Precipitation Regions

As illustrated in Figure 3, the variation tendencies in the two types of trade-offs along the precipitation gradient were not consistent. The mean value of the TC-WY trade-off exhibited the trend of Ansai > Linzhen > Dianshi, and the mean value of the SEC-WY trade-off exhibited the trend of Linzhen > Dianshi > Ansai. Additionally, the variability of the ES trade-offs was large, and the variability increased with the rainfall gradient. Small watersheds with negative trade-off values appeared, indicating that synergic relationships could be identified among ESs in some small watersheds.

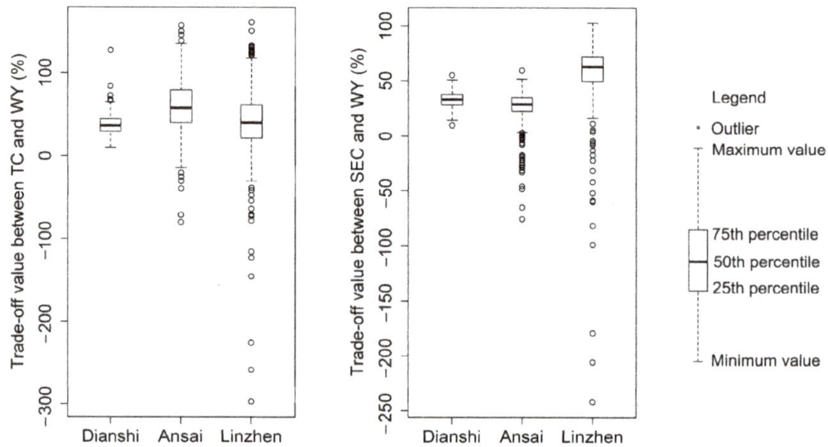

Figure 3. Boxplot of ESs trade-offs in three watersheds.

3.2.2. The Spatial Distribution of ESs Trade-Offs

As illustrated in Figure 4, the spatial distributions of the TC–WY trade-off and SEC–WY trade-off were similar in the Dianshi watershed; the high trade-off value regions were distributed in the northern watershed (lower reaches), whereas the low value regions were dispersedly distributed in the central section of the watershed. The trade-off intensity is determined by the degree of relative waxing and waning between ESs. The land-use change trends were that forest increased and cropland decreased in Dianshi, and such land-use transfers usually enhanced TC and SEC and reduced WY [23,25,27,51]; as a result, trade-off intensity increased. Using map overlay analysis between land-use and ES trade-offs, we found that the high trade-off-value areas overlapped with the areas where the increments of forest and grassland and the decrement of cropland were high. Therefore, land-use change is the direct reason for the ES trade-off.

The TC–WY and SEC–WY trade-offs decreased gradually from the southeast (lower reaches) to the northwest (upper reaches) in the Ansai watershed, which was consistent with the spatial distribution of the forest increment and in the decrements in grassland and cropland. In addition, there were small watersheds with negative trade-off values (two ESs both increased and decreased) near the main valley, indicating a synergistic relationship between the two ESs. The main reason for this was that forest and construction land slightly increased near the valley due to the flat terrain and convenient transportation; thus, the three ESs increased synchronously.

Except for some high trade-off areas assembling in the southwest, other high and low trade-off areas were fragmented and interlaced in the Linzhen watershed. The high trade-off areas coincided with the areas where grassland and cropland decreased and forests increased, whereas the low trade-off areas mainly coincided with the areas where grassland and forests increased slightly, cropland increased moderately, and shrubland decreased.

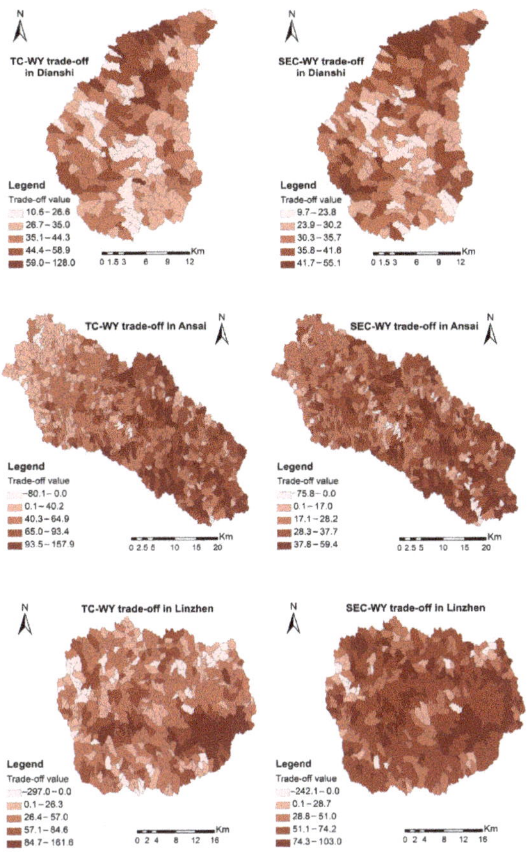

Figure 4. The spatial distribution of ES trade-offs in the three watersheds.

3.3. *The Effects of Land-Use Changes on ESs Trade-Offs*

3.3.1. The Effects of Land-Use Changes on ES Trade-Offs in Different Quantiles

Quantile regression is a method of estimating the conditional quantiles of a response variable distribution in a linear model that provides a more complete view of possible relationships between variables in ecological processes [45]. Forested land had significant positive effects (enhancing the trade-off) on TC–WY trade-offs in the three watersheds, and this positive effect exhibited the trend of Dianshi > Ansai > Linzhen (Table 3). The positive effect (regression coefficient) fluctuated by approximately 1.4 in various quantiles in the Dianshi watershed. This positive effect increased with the increase in quantiles, and it was the highest in the 80th and 90th quantiles in the Ansai and Dianshi watersheds, respectively. Forested land also had positive effects on the SEC–WY trade-off in the three watersheds, and this positive effect decreased along the precipitation gradient until the positive effect was not significant (the regression coefficients were only significant in the 30th–60th quartiles in the Linzhen watershed where the precipitation was the highest). The positive effect was highest in the 10th quantile; then, it declined with the quantile in the Dianshi watershed, and it was the highest in the 60th and 40th quantiles in the Ansai and Linzhen watersheds, respectively. Therefore, although the forest cover was low in the Dianshi watershed, the response of the trade-offs to forest cover was the strongest in this watershed, especially at a low trade-off level. This response was weaker as precipitation increased, possibly because higher rainfall can compensate for the negative effect of forest cover on water yield.

Table 3. Quantile regression between land-use changes and ES trade-offs.

Land-Use	Quantile	TC–WY Trade-Offs			SEC–WY Trade-Offs		
		Dianshi	Ansai	Linzhen	Dianshi	Ansai	Linzhen
Forest	10th	1.485 **	1.061 **	0.614 **	0.411 **	0.168 **	0.051
	20th	1.366 **	1.099 **	0.658 **	0.271 **	0.168 **	0.089
	30th	1.378 **	1.173 **	0.648 **	0.296 **	0.225 **	0.126 **
	40th	1.395 **	1.186 **	0.695 **	0.316 **	0.257 **	0.135 **
	50th	1.453 **	1.213 **	0.724 **	0.263 **	0.254 **	0.092 **
	60th	1.374 **	1.286 **	0.81 **	0.256 **	0.287 **	0.07 **
	70th	1.411 **	1.316 **	0.816 **	0.307 **	0.284 **	0.045
	80th	1.378 **	1.322 **	0.96 **	0.273 **	0.286 **	0.032
	90th	1.309 **	1.296 **	1.125 **	0.232 *	0.258 **	0.004
Shrub	10th	0.801 **	0.231 *	0.016	0.362 **	0.282 **	0.083 **
	20th	0.687 **	0.041	0.024	0.3 **	0.211 **	0.078 **
	30th	0.668 **	−0.149 *	0.017	0.273 **	0.213 **	0.086 **
	40th	0.668 **	−0.281 **	0.066	0.264 **	0.173 **	0.079 **
	50th	0.644 **	−0.313 **	0.082 *	0.238 **	0.183 **	0.057 **
	60th	0.6 **	−0.327 *	0.109 *	0.186 **	0.183 **	0.055 **
	70th	0.552 **	−0.204	0.151 **	0.162 **	0.125 **	0.056 **
	80th	0.455 **	−0.179	0.235 **	0.151	0.12 *	0.085 **
	90th	0.284	0.009	0.376 **	0.185	0.183 **	0.095 **
Grassland	10th	0.282 **	−0.716 **	−0.227 **	0.441 **	−0.09 **	−0.12 **
	20th	0.221 **	−0.86 **	−0.287 **	0.382 **	−0.118 **	−0.072 **
	30th	0.194 *	−0.942 **	−0.295 **	0.324 **	−0.171 **	−0.046
	40th	0.081	−1.004 **	−0.308 **	0.286 **	−0.199 **	−0.046
	50th	0.009	−1.038 **	−0.343 **	0.251 **	−0.222 **	−0.034
	60th	−0.04	−1.075 **	−0.331 **	0.175 **	−0.228 **	−0.024
	70th	−0.152	−1.125 **	−0.381 **	0.139 *	−0.268 **	−0.017
	80th	−0.155	−1.152 **	−0.454 **	0.112 *	−0.265 **	−0.014
	90th	−0.158	−1.185 **	−0.511 **	0.125 *	−0.297 **	−0.035 *
Cropland	10th	−0.837 **	−0.561 **	−0.493 **	−0.637 **	−0.256 **	−0.726 **
	20th	−0.742 **	−0.719 **	−0.344 **	−0.59 **	−0.308 **	−0.667 **
	30th	−0.717 **	−0.789 **	−0.302 **	−0.613 **	−0.35 **	−0.664 **
	40th	−0.738 **	−0.887 **	−0.317 **	−0.621 **	−0.353 **	−0.644 **
	50th	−0.753 **	−0.955 **	−0.317 **	−0.609 **	−0.362 **	−0.564 **
	60th	−0.735 **	−0.983 **	−0.281 *	−0.581 **	−0.403 **	−0.574 **
	70th	−0.724 **	−1.047 **	−0.17	−0.553 **	−0.403 **	−0.551 **
	80th	−0.791 **	−1.084 **	−0.114	−0.538 **	−0.373 **	−0.536 **
	90th	−0.866 **	−1.207 **	−0.164	−0.476 **	−0.361 **	−0.569 **

The asterisks * and ** indicate $p < 0.05$ and $p < 0.01$ respectively.

Shrubland had significant positive effects on the TC–WY trade-off, except for the 90th quantile in the Dianshi watershed, and the positive effects were larger in the Dianshi watershed than in the Ansai and Linzhen watersheds, denoting that the TC–WY trade-off in the Dianshi watershed is most sensitive to changes in shrubland. The positive effects decreased as the quantile increased in Dianshi. The effects of shrubs on the TC–WY trade-off presented a "U" pattern as the quantile increased in Ansai (shrubs intensified the trade-off first and then reduced the trade-off). The effects on the TC–WY trade-offs were enhanced as the quantile increased in Linzhen. Shrubs had significant positive effects on SEC–WY trade-offs in the three watersheds, exhibiting the trend of Dianshi > Ansai > Linzhen. The positive effects declined first and then rose as the quantile increased in the three watersheds. For low- and medium-rainfall regions, shrubland had a stronger influence on the SEC–WY trade-offs at low trade-off levels, and the influence was weaker at high trade-off levels. For high rainfall regions, the influence was stronger at high trade-off levels and weaker at moderate trade-off levels.

Grassland only had significant positive effects on TC–WY trade-offs in the 10th–30th quantiles in the Dianshi watershed, denoting that increasing grassland still exacerbated trade-offs at low trade-off levels, but the influence of grassland disappeared at moderate and high trade-off levels. In contrast, grassland significantly inhibited TC–WY trade-offs at all trade-off levels in the Ansai and Dianshi watersheds. The inhibitory effect of grassland was strongest at high trade-off levels and stronger in Ansai than in Linzhen. Grassland had significant positive effects on the SEC–WY trade-offs in the Dianshi watershed, and the effects decreased as the quantile increased. Whereas grassland significantly inhibited SEC–WY trade-offs in the Ansai watershed, the inhibitory effects were the strongest at high trade-off levels. The inhibitory effects were weaker in Linzhen than in Ansai, and the regression coefficients were statistically significant only in the 10th, 20th, and 90th quantiles. The results indicate that the water consumption of grassland could exacerbate trade-offs in low precipitation regions, especially at low trade-off levels. However, grassland could inhibit trade-offs in medium-precipitation and high-precipitation regions, so grassland could be arranged in high trade-off regions.

Cropland had inhibitory effects on the TC–WY trade-offs in the three watersheds. The regression coefficients fluctuated by approximately 0.77 in Dianshi, and the inhibitory effects were the highest at high trade-off levels in Ansai and low trade-off levels in Linzhen. Cropland also had inhibitory effects on the SEC–WY trade-offs in the three watersheds, exhibiting the general trend of Linzhen > Dianshi > Ansai. The inhibitory effects were highest at low trade-off levels in Dianshi and Linzhen and were highest at moderate trade-off levels in Ansai.

3.3.2. The Threshold Values at Which ES Trade-Offs Respond to Land-Use Changes

The influences of independent variables of different ranges on dependent variables can be revealed by piecewise linear regression, and we can use piecewise linear regression to identify the inflection point of trade-off responses to land-use changes. As illustrated in Figure 5, the intensive effect (slope of piecewise functions) of forest cover on trade-offs was higher when the increment of forested land exceeded 34.0% and 18.6% in Ansai and Linzhen, respectively, but this intensive effect was almost unchanged across the total range of the independent variable in Dianshi. Shrubland reduced the trade-offs when the shrub increased less than 8.2%, 20.1%, and −45.4% in Dianshi, Ansai, and Linzhen, respectively, whereas shrubland exacerbated trade-offs when the increment exceeded these thresholds. This phenomenon meant that the "golden mean" and limited increase in shrubs were good strategies. TC and SEC were very small, and WY was large (trade-off was intense) when there were few shrubs. In these circumstances, planting shrubs would enhance TC and SEC with low water consumption, and the trade-off would be reduced. However, the sustained planting of shrubs would reverse the relative size of ESs and strengthen the trade-off due to water deficiency. Therefore, although the water consumption of shrubs was less than that of forests, the water consumption of shrubs still cannot be ignored. Grassland reduced the trade-offs when the grassland increased less than 16.5% in Dianshi, but grassland exacerbated trade-offs when the increment exceeded the threshold, which was caused by very low rainfall in Dianshi. Generally, it is more reasonable to establish grassland in arid regions [5,52] because the water consumption of grassland is usually lower and grassland can provide higher WY and maintain other ESs at a relatively high level [5,25,27,53]. However, too much grassland may exceed the soil-water carrying capacity for vegetation, especially in low rainfall areas such as the Dianshi watershed. Grassland reduced trade-offs in Ansai, and the inhibitory action was relatively stable across the total range of grassland change. Grassland reduced trade-offs in the initial stage of grassland change in the Linzhen watershed, whereas grassland slightly enhanced trade-offs when the increment of grassland exceeded −16.9%. The comprehensive analysis of the threshold value, segmental slope, and land-use proportions of the initial period

(2000) showed that to control ES trade-offs, forest needed to be limited, especially in the Dianshi and Ansai watersheds, and shrubland proportions needed to be controlled at 8.5% and 21.6% in Dianshi and Ansai, respectively, whereas grassland proportions needed to be controlled at 58.0% in Dianshi. In addition, Wang et al. [29] also calculated the trade-offs between soil moisture and soil organic carbon, total nitrogen, and species richness at a single time point in the Loess Plateau of China, and the inflection point of the trade-offs responding to the precipitation gradient was identified.

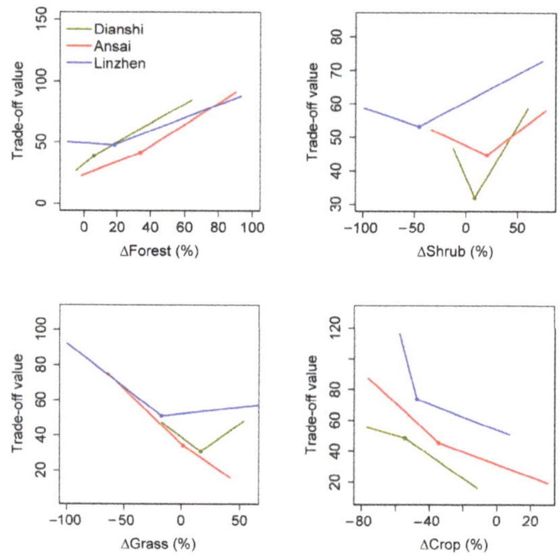

Figure 5. The threshold value at which ES trade-offs respond to land-use changes (trade-off value of ordinate label was the mean value of TC–WY and SEC–WY trade-offs).

3.3.3. The Effects of Land-Use Transformation on ESs Trade-Offs

The conversion of cropland and grassland to forests and shrubland was the main reason for the trade-off increase in the three watersheds, but which land-use transformations could reduce the trade-offs were different in the three watersheds (Table 4). Grassland conversion to cropland, cropland remaining unchanged, and grassland and cropland conversion to construction land could reduce the trade-offs in the Dianshi watershed. Grassland remaining unchanged, cropland conversion to grassland and construction land, and grassland conversion to construction land could reduce the trade-offs in the Ansai watershed. Forest and shrubland conversion to cropland could reduce the trade-offs in the Linzhen watershed. Therefore, it is necessary to ensure a certain proportion of cropland in low precipitation regions as well as certain grassland in medium precipitation regions, and forested land and shrubland need to be reduced appropriately in the high precipitation region.

3.4. Recommendations of ES Regulation for Various Precipitation Regions

Forests are propitious to carbon sequestration, soil and water conservation, climatic regulation, and atmospheric purification, but they are not propitious to water yield, especially in arid regions [11,50,54,55]. However, the function of cropland is contrary to that of forested land [25,27]. As a land-use type, grassland is a compromise [5,27]. Thus, regulating the proportions of several land use types can realize the balance of various ESs and reduce their trade-offs. For the low precipitation region (Dianshi watershed), the balancing of trade-offs could be achieved by restricting forest, shrub, and grassland, increasing cropland properly in low trade-off areas, while keeping most of the current cropland unchanged, converting grassland to cropland reasonably, and converting grassland

and cropland to construction land moderately in high trade-off areas. For the medium precipitation region (Ansai watershed), restricting shrubs in low trade-off areas, restricting forests, and increasing shrubs, grasslands, and croplands appropriately in high trade-off areas could increase the balance in trade-offs, and specific land-use transformations such as keeping current grasslands unchanged, converting cropland to grassland and construction land, and slightly converting grassland to construction land could also improve the balance. For high precipitation regions (Linzhen watershed), increasing cropland in low trade-off areas, restricting forested land and shrubland, and increasing grassland and cropland in high trade-off areas could help achieve balance. The main land-use transformation tactic is the conversion of forest and shrubs to cropland. In addition, some useful measures can be implemented in the whole study area, such as enhancing the quality of the current vegetation, improving the community structure, increasing biodiversity, tending forests and shrubs, rational grazing in grassland, employing conservation tillage, developing water-saving agriculture, and applying forest-crop intercropping.

Table 4. Marginal (Marg, %) and conditional effects (Cond, %) of land-use transformation (LUT) on ES trade-offs (mean value of TC–WY and SEC–WY trade-offs).

Dianshi Watershed				Ansai Watershed				Linzhen Watershed			
LUT	Marg	LUT	Cond	LUT	Marg	LUT	Cond	LUT	Marg	LUT	Cond
$L_{CrO\text{-}FoL}$	43.1	$L_{CRO\text{-}FOL}$	43.1	$L_{CRO\text{-}FOL}$	45.9	$L_{CRO\text{-}FOL}$	45.9	$L_{ShL\text{-}C}$	43.4	$L_{ShL\text{-}C}$	43.4
$L_{GrA\text{-}FoL}$	36.2	$L_{GrA\text{-}C}$	22.6	$L_{GrA\text{-}FoL}$	43.7	$L_{CrO\text{-}C}$	19.7	$L_{FoL\text{-}C}$	35.3	$L_{FoL\text{-}C}$	14.5
$L_{GrA\text{-}C}$	34.3	$L_{CrO\text{-}ShL}$	9.6	$L_{GrA\text{-}G}$	20.3	$L_{GrA\text{-}FoL}$	7.5	$L_{GrA\text{-}FoL}$	22.6	$L_{GrA\text{-}FoL}$	10.8
$L_{CrO\text{-}C}$	28.2	$L_{CrO\text{-}GrA}$	7.1	$L_{CrO\text{-}G}$	19.1	$L_{GrA\text{-}ShL}$	7.3	$L_{CRO\text{-}FOL}$	6.6	$L_{CRO\text{-}FOL}$	11.3
$L_{CrO\text{-}ShL}$	12.6			$L_{CrO\text{-}C}$	12.1	$L_{GrA\text{-}C}$	5.5	$L_{GrA\text{-}ShL}$	6.6		
$L_{GrA\text{-}C}$	7.9			$L_{GrA\text{-}C}$	9.9						
$L_{CrO\text{-}C}$	6.3			$L_{ShL\text{-}G}$	5.9						
$L_{GrA\text{-}G}$	5.0										

FoL: forestland, ShL: shrubland, GrA: grassland, CrO: cropland, CoL: construction land, WaB: water body. $L_{CrO\text{-}FoL}$ represents the land-use type change from cropland to forestland. Land-use transformation with gray shadow represents negative effects on trade-offs, and others represent positive effects.

3.5. The Limitation of the Methods and Results

The InVEST model was used to calculate ESs in this study. Depending on many advantages, such as simplicity and convenience, flexible operation, and strong spatial expression ability of output results, the InVEST model is widely applied to ecosystem management and decision making. However, the principles of calculations are simplified for many ESs. For example, errors in the empirical parameters of the revised universal soil loss equation (RUSLE) will therefore have a large effect on SEC predictions; WY calculation is based on annual averages, which neglect extremes; TC is calculated by the summation of the carbon content of the four major carbon pools, and ecological processes are ignored. Other important issues are that biophysical parameters are too dependent on land-use types, but land-use maps cannot reveal the differences within the same land-use type. Despite this, the InVEST model is a powerful tool to map and value ecosystem services at watershed and regional scales.

ESs are affected by land use, climate, soil, topography, and many other factors. Land use influences carbon sequestration [56] and soil conservation directly [57], and it also influences water yield by hydrologic processes [58]. Similarly, precipitation is the key factor for carbon sequestration [59], soil conservation [60], and water yield [61]. Soil properties affect plant growth [62], soil erosion [63], and hydrologic processes [64], and they are also important factors for these three ESs. In this study, the effects of land use on ESs trade-offs were analyzed in three precipitation regions, and other factors, such as soil, were not introduced, which would limit the results and corresponding management recommendations. Nevertheless, vegetation and soil exhibit zonal distributions across precipitation gradients in the Loess Plateau of China [14]. The current geographical landscape is formed by the collaborative evolution of many natural geographical factors. The regional differentiation of ESs across precipitation gradients implied differentiation across other environmental gradients. Therefore, we only employ land use and precipitation to study ESs trade-offs. Although there are certain limitations, the conclusions are reliable, and it is propitious to apply the results in practice.

4. Conclusions

From 2000 to 2018, TC and SEC were enhanced, but WY decreased in the three watersheds, and a trade-off relation can be identified between these two ESs and WY. The effects of forests on the trade-offs were positive in the three watersheds, and the main effects exhibited decreasing, increasing, and increasing trends as trade-off intensity increased in Dianshi, Ansai, and Linzhen, respectively. The positive effect of shrubs on TC–WY trade-offs exhibited a decreasing trend, a "U" pattern, and an increasing trend as the trade-off intensity increased in Dianshi, Ansai, and Linzhen, respectively; these effects on the SEC–WY trade-offs first declined and then rose in the three watersheds. Grassland exacerbated trade-offs in Dianshi, whereas it reduced trade-offs in Ansai and Linzhen. Regulating land-use proportions can realize the balance of three ESs and reduce their trade-offs. For Dianshi, we can restrict forest, shrub, and grassland and increase cropland properly. For Ansai, we can restrict shrubs in low trade-off areas, and we can also restrict forests and increase shrubs, grasslands, and croplands appropriately in high trade-off areas. For Linzhen, we can increase cropland in low trade-off areas, and we can restrict forest and shrub but increase grassland and cropland in high trade-off areas.

Author Contributions: Methodology, writing—original draft preparation, Q.F.; writing—review and editing, supervision, S.D.; investigation and data curation, B.D. All authors have read and agreed to the published version of the manuscript.

Funding: This research was funded by the National Natural Science Foundation of China (No. 41501201) and the Fundamental Research Program of Shanxi Province (No. 20210302123481).

Institutional Review Board Statement: Not applicable.

Informed Consent Statement: Not applicable.

Data Availability Statement: The data presented in this study are available on request from the corresponding author.

Conflicts of Interest: The authors declare no conflict of interest.

References

1. Costanza, R.; D'Arge, R.; Groot, R.D.; Farber, S.; Grasso, M.; Hannon, B. The value of the world's ecosystem services and natural capital. *Nature* **1997**, *25*, 3–15. [CrossRef]
2. Gentry, R.R.; Alleway, H.K.; Bishop, M.J. Exploring the potential for marine aquaculture to contribute to ecosystem services. *Rev. Aquac.* **2020**, *12*, 499–512. [CrossRef]
3. Millennium Ecosystem Assessment. *Ecosystem and Human Well-Being*; Island Press: Washington, DC, USA, 2005; pp. 137–142.
4. Bennett, E.M.; Peterson, G.D.; Gordon, L.J. Understanding relationships among multiple ecosystem services. *Ecol. Lett.* **2009**, *12*, 1394. [CrossRef]
5. Wu, X.; Wang, S.; Fu, B.; Liu, Y.; Zhu, Y. Land use optimization based on ecosystem service assessment: A case study in the Yanhe watershed. *Land Use Policy* **2018**, *72*, 303–312. [CrossRef]
6. Karimi, J.D.; Corstanje, R.; Harris, J.A. Understanding the importance of landscape configuration on ecosystem service bundles at a high resolution in urban landscapes in the UK. *Landsc. Ecol.* **2021**, *36*, 2007–2024. [CrossRef]
7. Lu, N.; Liu, L.; Yu, D.; Fu, B. Navigating trade-offs in the social-ecological systems. *Curr. Opin. Environ. Sustain.* **2011**, *48*, 77–84. [CrossRef]
8. Agol, D.; Reid, H.; Crick, F. Ecosystem-based adaptation in Lake Victoria Basin; synergies and trade-offs. *R. Soc. Open Sci.* **2021**, *8*, 201847. [CrossRef] [PubMed]
9. Knapp, A.K.; Fay, P.A.; Blair, J.M. Rainfall variability, carbon cycling, and plant species diversity in a mesic grassland. *Science* **2002**, *298*, 2201–2205. [CrossRef] [PubMed]
10. Meier, I.C.; Leuschner, C. Nutrient dynamics along a precipitation gradient in European beech forests. *Biogeochemistry* **2014**, *120*, 51–69. [CrossRef]

11. Feng, X.; Fu, B.; Lü, N.; Zeng, Y.; Wu, B. How ecological restoration alters ecosystem services: An analysis of carbon sequestration in China's Loess Plateau. *Sci. Rep.* **2012**, *3*, 2846. [CrossRef] [PubMed]
12. Li, T.; Ren, B.; Wang, D. Spatial variation in the storages and age-related dynamics of forest carbon sequestration in different climate zones-evidence from black locust plantations on the Loess Plateau of China. *PLoS ONE* **2015**, *10*, e0121862. [CrossRef]
13. Zhang, J.; Xu, B.; Li, M. Diversity of communities dominated by *Glycyrrhiza uralensis*, an endangered medicinal plant species, along a precipitation gradient in China. *Bot. Stud.* **2011**, *52*, 493–501.
14. Zhang, Y.; Huang, M.; Lian, J. Spatial distributions of optimal plant coverage for the dominant tree and shrub species along a precipitation gradient on the central Loess Plateau. *Agric. For. Meteorol.* **2015**, *206*, 69–84. [CrossRef]
15. Shi, S.; Li, Z.; Wang, H.; Wu, X.; Wang, S.; Wang, X. Comparative analysis of annual rings of perennial forbs in the Loess Plateau, China. *Dendrochronologia.* **2016**, *38*, 82–89. [CrossRef]
16. Jia, X.; Shao, M.; Yu, D. Spatial variations in soil-water carrying capacity of three typical revegetation species on the Loess Plateau, China. *Agric. Ecosyst. Environ.* **2019**, *273*, 25–35. [CrossRef]
17. Cao, Y.; Li, Y.; Chen, Y. Non-structural carbon, nitrogen, and phosphorus between black locust and Chinese pine plantations along a precipitation gradient on the Loess Plateau, China. *Trees* **2018**, *32*, 835–846. [CrossRef]
18. Du, C.; Gao, Y. Opposite patterns of soil organic and inorganic carbon along a climate gradient in the alpine steppe of northern Tibetan Plateau. *Catena* **2020**, *186*, 104366. [CrossRef]
19. Zhang, X.; Song, Z.; Hao, Q. Storage of soil phytoliths and phytolith-occluded carbon along a precipitation gradient in grasslands of northern China. *Geoderma* **2020**, *364*, 114–200. [CrossRef]
20. Wang, X.; Lü, X.; Dijkstra, F.; Zhang, H.; Han, X. Changes of plant N: P stoichiometry across a 3000-km aridity transect in grasslands of northern China. *Plant Soil* **2019**, *443*, 107–119. [CrossRef]
21. Li, S.; Liang, W.; Zhang, W.; Liu, Q. Response of soil moisture to hydro-meteorological variables under different precipitation gradients in the yellow river basin. *Water Resour. Manag.* **2016**, *30*, 1867–1884. [CrossRef]
22. Li, Z.; Coles, A.; Xiao, J. Groundwater and streamflow sources in China's Loess Plateau on catchment scale. *Catena* **2019**, *181*, 104075. [CrossRef]
23. Jia, X.; Fu, B.; Feng, X.; Hou, G.; Liu, Y.; Wang, X. The tradeoff and synergy between ecosystem services in the Grain-for-Green areas in Northern Shaanxi, China. *Ecol. Indic.* **2014**, *43*, 103–113. [CrossRef]
24. Zheng, Z.; Fu, B.; Hu, H.; Sun, G. A method to identify the variable ecosystem services relationship across time: A case study on Yanhe Basin, China. *Landsc. Ecol.* **2014**, *29*, 1689–1696. [CrossRef]
25. Feng, Q.; Zhao, W.W.; Fu, B.J.; Ding, J.Y.; Wang, S. Ecosystem service trade-offs and their influencing factors: A case study in the Loess Plateau of China. *Sci. Total Environ.* **2017**, *607–608*, 1250–1263. [CrossRef] [PubMed]
26. Hu, H.; Fu, B.; Lü, Y.; Zheng, Z. SAORES: A spatially explicit assessment and optimization tool for regional ecosystem services. *Landsc. Ecol.* **2014**, *30*, 547–560. [CrossRef]
27. Feng, Q.; Zhao, W.; Hu, X.; Liu, Y.; Daryanto, S.; Cherubini, F. Trading-off ecosystem services for better ecological restoration: A case study in the Loess Plateau of China. *J. Clean. Prod.* **2020**, *257*, 120469. [CrossRef]
28. Lu, N.; Fu, B.; Jin, T.; Chang, R. Trade-off analyses of multiple ecosystem services by plantations along a precipitation gradient across Loess Plateau landscapes. *Landsc. Ecol.* **2014**, *29*, 1697–1708. [CrossRef]
29. Wang, C.; Wang, S.; Fu, B.; Li, Z.; Wu, X.; Tang, Q. Precipitation gradient determines the tradeoff between soil moisture and soil organic carbon, total nitrogen, and species richness in the Loess Plateau, China. *Sci. Total Environ.* **2017**, *575*, 1538. [CrossRef]
30. Shao, M.; Wang, Y.; Jia, X. Ecological construction and soil desiccation on the Loess Plateau of China. *Bull. Chin. Acad. Sci.* **2015**, *30*, 257–264. (In Chinese)
31. Wu, X.; Wang, S.; Fu, B.; Feng, X.; Chen, Y. Socio-ecological changes on the loess plateau of China after grain to green program. *Sci. Total Environ.* **2019**, *678*, 565–573. [CrossRef] [PubMed]
32. Landsat Images. Available online: http://glovis.usgs.gov/ (accessed on 21 May 2020).
33. Meteorological Data. Available online: http://data.cma.cn/ (accessed on 10 April 2020).
34. Digital Elevation Model. Available online: http://www.gscloud.cn/ (accessed on 29 January 2020).
35. Harmonized World Soil Database. Available online: http://www.crensed.ac.cn/ (accessed on 18 June 2020).
36. Renard, K.; Foster, G.; Weesies, G.; McCool, D.; Yoder, D. *Predicting Soil Erosion by Water: A Guide to Conservation Planning with the Revised Universal Soil Loss Equation (RUSLE)*; USDA, Agricultural Handbook Number 703; U.S. Government Printing Office: Washington, DC, USA, 1997; pp. 1–18.
37. Vigiak, O.; Borselli, L.; Newham, L.; Mcinnes, J.; Roberts, A. Comparison of conceptual landscape metrics to define hillslope-scale sediment delivery ratio. *Geomorphology* **2012**, *138*, 74–88. [CrossRef]
38. Sharp, R.; Tallis, H.T.; Ricketts, T.; Guerry, A.D.; Wood, S.A.; Chaplin-Kramer, R. *InVEST + VERSION + User's Guide*; The Natural Capital Project, Stanford University, University of Minnesota, The Nature Conservancy, and World Wildlife Fund: Palo Alto, CA, USA, 2016; pp. 115–120.
39. Fu, B. On the calculation of the evaporation from land surface. *Sci. Atmos. Sin.* **1981**, *5*, 23–31.
40. Zhang, L.; Hickel, K.; Dawes, W.R.; Chiew, F.H.S.; Western, A.W.; Briggs, P.R. A rational function approach for estimating mean annual evapotranspiration. *Water Resour. Res.* **2004**, *40*, 89–97. [CrossRef]
41. Feng, Q. Ecosystem Services Trade-Offs in the Loess Hilly and Gully Region. Ph.D. Thesis, Beijing Normal University, Beijing, China, 2018. (In Chinese).

42. Li, B.; Chen, N.; Wang, Y.; Wang, W. Spatio-temporal quantification of the trade-offs and synergies among ecosystem services based on grid-cells: A case study of Guanzhong Basin, NW China. *Ecol. Indic.* **2018**, *94*, 246–253. [CrossRef]
43. Bradford, J.B.; D'Amato, A.W. Recognizing trade-offs in multi-objective land management. *Front. Ecol. Environ.* **2012**, *10*, 210–216. [CrossRef]
44. Xiao, C.; Ye, J.; Esteves, R.M. Using Spearman's correlation coefficients for exploratory data analysis on big dataset. *Concurr. Comput.* **2016**, *28*, 3866–3878. [CrossRef]
45. Cade, B.S.; Noon, B.R. A gentle introduction to quantile regression for ecologists. *Front. Ecol. Environ.* **2003**, *1*, 412–420. [CrossRef]
46. Toms, J.D.; Lesperance, M.L. Piecewise regression: A tool for identifying ecological thresholds. *Ecology* **2003**, *84*, 2034–2041. [CrossRef]
47. Peng, J.; Tian, L.; Liu, Y.; Zhao, M.; Hu, Y.; Wu, J. Ecosystem services response to urbanization in metropolitan areas: Thresholds identification. *Sci. Total Environ.* **2017**, *607–608*, 706–714. [CrossRef]
48. Šmilauer, P.; Lepš, J. *Multivariate Analysis of Ecological Data Using Canoco 5*; Cambridge University Press: New York, NY, USA, 2014; pp. 50–69.
49. Deng, L. *Responsing Mechanism of Ecosystem Carbon Sequestration Benefits to Vegetation Restoration on the Loess Plateau of China*; Northwest A&F University: Yangling, China, 2014. (In Chinese)
50. Lü, Y.; Fu, B.; Feng, X.; Zeng, Y.; Liu, Y.; Chang, R.; Sun, G.; Wu, B. A policy-driven large scale ecological restoration: Quantifying ecosystem services changes in the Loess Plateau of China. *PLoS ONE* **2012**, *7*, e31782. [CrossRef]
51. Yang, S.; Zhao, W.; Liu, Y. Influence of land use change on the ecosystem service trade-offs in the ecological restoration area: Dynamics and scenarios in the Yanhe watershed, China. *Sci. Total Environ.* **2018**, *644*, 556–566. [CrossRef] [PubMed]
52. Chisholm, R.A. Trade-offs between ecosystem services: Water and carbon in a biodiversity hotspot. *Ecol. Econ.* **2010**, *69*, 1973–1987. [CrossRef]
53. Mark, A.F.; Dickinson, K.J.M. Maximizing water yield with indigenous non-forest vegetation: A New Zealand perspective. *Front Ecol. Environ.* **2008**, *6*, 25–34. [CrossRef]
54. Fu, B.J.; Liu, Y.; Lü, Y.H.; He, C.S.; Zeng, Y.; Wu, B.F. Assessing the soil erosion control service of ecosystems change in the Loess Plateau of China. *Ecol. Complex.* **2011**, *8*, 284–293. [CrossRef]
55. Wang, S.; Fu, B. Trade-offs between forest ecosystem services. *For. Policy. Econ.* **2013**, *26*, 145–146. [CrossRef]
56. Zhang, F.; Xu, N.; Wang, C.; Wu, F.; Chu, X. Effects of land use and land cover change on carbon sequestration and adaptive management in Shanghai, China. *Phys. Chem. Earth.* **2020**, *120*, 102948. [CrossRef]
57. Napoli, M.; Altobelli, F.; Orlandini, S. Effect of land set up systems on soil losses. *Ital. J. Agron.* **2020**, *15*, 306–314. [CrossRef]
58. Khorchani, M.; Nadal-Romero, E.; Lasanta, T.; Tague, C. Carbon sequestration and water yield tradeoffs following restoration of abandoned agricultural lands in Mediterranean mountains. *Environ. Res.* **2021**, *10*, 112203. [CrossRef] [PubMed]
59. Hein, C.J.; Usman, M.; Eglinton, T.I.; Haghipour, N.; Galy, V.V. Millennial-scale hydroclimate control of tropical soil carbon storage. *Nature* **2020**, *581*, 63–66. [CrossRef] [PubMed]
60. Fang, H. Responses of runoff and soil loss to rainfall regimes and soil conservation measures on cultivated slopes in a hilly region of Northern China. *Int. J. Environ. Res. Public Health* **2021**, *18*, 2102. [CrossRef] [PubMed]
61. Lian, X.H.; Qi, Y.; Wang, H.W.; Zhang, J.; Yang, R. Assessing changes of water yield in Qinghai Lake watershed of China. *Water* **2020**, *12*, 11. [CrossRef]
62. Afonso, S.; Arrobas, M.; Rodrigues, M.A. Soil and plant analyses to diagnose Hop fields irregular growth. *J. Soil Sci. Plant Nutr.* **2020**, *20*, 1999–2013. [CrossRef]
63. Hao, H.X.; Qin, J.H.; Sun, Z.X.; Guo, Z.L.; Wang, J.G. Erosion-reducing effects of plant roots during concentrated flow under contrasting textured soils. *Catena* **2021**, *203*, 105378. [CrossRef]
64. Bonetti, S.; Wei, Z.W.; Or, D. A framework for quantifying hydrologic effects of soil structure across scales. *Commun. Earth Environ.* **2021**, *2*, 107. [CrossRef]

sustainability

MDPI

Article

Human Resource Allocation in the State-Owned Forest Farm of China for the Changing Climate

Xiaofang Deng [1], Junkui Li [1,*], Lijuan Su [1], Shan Zhao [1] and Shaofei Jin [2,*]

1 National Academy of Forestry and Grassland Administration of China, Beijing 102600, China
2 Department of Geography, Minjiang University, Fuzhou 350108, China
* Correspondence: lgyxshqk@163.com (J.L.); jinsf@tea.ac.cn (S.J.); Tel.: +86-188-5010-2506 (S.J.)

Abstract: Global climate change has become a hot topic in today's international political, economic, environmental and diplomatic arenas. China has implemented a series of strategies, measures and actions to cope with climate change, which has promoted industrial transformation and human resource adjustment in China's state-owned forest areas. However, little is known about the role of current human resource allocation in adaptation to climate change in the state-owned forest farm of China. To address these gaps, this study calculated the current situation of human resource structure and the contribution rate of three industries to the allocation of human resources and the evaluation model of coordinated fitness to the climate changes in key state-owned forest farms. The results show that: (1) The current situation of talent in key state-owned forest areas shows a shortage of total amount, a shortage of high-level and highly educated talents, and aging of talents. (2) The coefficient of structural deviation increased and the coefficient of structural-change synergy kept decreasing, indicating that the coordination between human resource allocation and industrial structure in key state-owned forest areas nowadays is only at the intermediate level of synergistic fitness. The paper highlights the trained-professional human resource and the industrial structure changes in the context of climate change as the main limited factors for the key state-owned forest farms of China. Increasing the education investment for climate change and the economic income for the employees are suggested to be promoted for policy makers in future.

Keywords: state-owned forest farms; human resource allocation; industrial structure; coordination and adaptation; personal structure

Citation: Deng, X.; Li, J.; Su, L.; Zhao, S.; Jin, S. Human Resource Allocation in the State-Owned Forest Farm of China for the Changing Climate. *Sustainability* **2022**, *14*, 9667. https://doi.org/10.3390/su14159667

Academic Editor: Hyo Sun Jung

Received: 17 June 2022
Accepted: 4 August 2022
Published: 5 August 2022

1. Introduction

The crisis brought about by climate change is a daily occurrence, and the Copenhagen Global Climate Change Summit in Denmark began to generate unprecedented global attention towards ecological protection and global warming. Since the 21st century, China has been very sensitive to global climate change and has gradually changed its strategy of trading the environment for economic growth, and has begun to establish a red line for ecological protection. The 17th National Congress of the Communist Party of China (CPC) included the Scientific Outlook on Development in the Party Constitution, the 18th National Congress of the CPC included the Scientific Outlook on Development in the Party's guiding ideology, and the 19th National Congress of the CPC held in 2017 made significant innovations in both theoretical thinking and practical initiatives for the construction of ecological civilization, including "Harmonious coexistence between human beings and nature" in the new era of adherence to and development of the country, all of which reflect the importance China attaches to environmental protection and sustainable development and its role as a great nation. Northeast and Inner Mongolia state-owned forest areas are an important part of China's forestry construction and development, the largest forest reserve resource cultivation base and timber and forest by-product supply base, and play an important role in the construction of the forestry ecological system, as well as an irreplaceable ecological barrier for the country. Additionally, in the context of global warming, in order to further protect the ecological environment, China launched the implementation of the natural forest resources protection project for state-owned forest areas in 2000, and started to implement the policy of completely stopping commercial logging of natural forests in key state-owned forest areas in Northeast China and Inner Mongolia in 2015, which brings great challenges to the development of forestry economy in China's state-owned forest areas, but also brings important opportunities for industrial transformation. Green water and green mountains are golden mountains, and protecting ecology is the process of protecting natural values and value-added natural capital, and protecting the potential and backbone of economic and social development [1]. In the context of global warming, the development strategy of ecological civilization has put forward higher requirements for the development of state-owned forest areas, promoting the change in the forestry development mode of state-owned forest areas from timber-production-oriented to ecological-restoration-and-construction-oriented, and from using forest resources for economic benefits to protecting forests and providing ecological services [2]. The change in the development model represents a comprehensive transformation of the social, economic, and ecological fields in state-owned forest areas, and will inevitably have a significant impact on the human resource structure of state-owned forest areas.

In the context of global climate change, key state-owned forest farms in China have been influenced by relevant national policies, and industrial transformation has been more drastic and rapid. In this process, especially after the implementation of the natural forest protection project, the interaction and coordination between human resources allocation and industrial structure is of great significance in promoting the development of forestry economic transformation, but this issue has not yet received extensive attention. Therefore, we make assumptions: (1). Climate change has an impact on the personal structure of key state-owned forest areas. (2). Climate change has an impact on the industrial structure of key state-owned forest areas. (3). Under the background of climate change, the allocation of human resources and the coordination and adaptability of industrial structure in key state-owned forest areas have changed.

The aims of this study are to (1) explore the current situation of the team of state-owned forest farms; (2) discover the coordination of human resource allocation and industrial structure. Following these aims, this study can support scientific evidence for the future transformation and development of state-owned forest farms in team construction. Deeply, this study can put forward countermeasures and suggestions to strengthen team construction within a larger and more reasonable framework to provide strong support for the reform and development of state-owned forest farms and forestry and grassland industry.

This paper is organized as follows. The Introduction section introduces the research background, literature review and theoretical analysis. The Materials and Methods section explains the research design and data processing. The Results section describes the empirical results and analysis. The last section offers discussion, conclusions and policy recommendations.

1.1. Literature Review

Human resources are an important subject of social and economic construction and development, and many scholars have studied the importance of human resources: Schultz believes that human capital is a form of capital that expresses the indicators including mental and physical labor condensed on the workers themselves in the form of quantity [3–5]; Becker believes that human capital is the cost of investment in human resources. Additionally, this investment is mainly in the form of money to influence the monetary income and psychological activities of the talent [6,7]. As the most basic production factor in the industrial sector, the reasonable allocation of forest workers will not only affect the effective performance of social productivity and the optimization and upgrading of industrial structure, but also affect the smooth transformation and development of the state-owned forest economy. The study of human resource allocation and industrial structure has been a key concern in economics, and British economists put forward the "Allotment-Clark theorem", which holds that with the increase in productivity and per capita national income level, labor force employment is gradually transferred from primary industry to secondary and tertiary industries until tertiary industry employment takes absolute advantage [8]. Some scholars further explored the law of change in industrial structure and verified the theorem based on the time series data of 57 countries with different levels of development around the world [9], after which American economists also put forward the theory that employment structure transformation lags behind industrial structure transformation [10]. With the development of China's economy, the problem of rational allocation of human resources and industrial structure has gradually emerged, and more scholars have studied the relationship between economic growth and employment, the relationship between three industrial structures and employment in three industries, the impact of industrial structure upgrading on employment structure, the relationship between tertiary industry development and employment, and the interaction mechanism between talent structure and industrial structure in China [11–13], but state-owned forest areas as the key area of reform in the context of climate change. However, as a key area of reform in China under the context of climate change, not much attention has been paid to the human resource allocation problem in state-owned forest areas in the existing relevant studies, and they mainly focus on the development of micro subjects and their measurement, and labor migration [9,14–22].

1.2. Theoretical Contribution

According to the review for the existing theoretical research, the main contribution of this study is to extend the talent predicament theory. To adapt to climate change, state-owned forest farms in China have undertaken substantial activities to relieve the negative effect of global warming. The talent predicament occurred because the professional employees were needed when implementing these activities [23,24]. Therefore, this study can connect the talent predicament theory with climate change. The second contribution is to extend the study area of the climate change, from the physical theory to the human resource theory. As one of the consequences of climate change, state-owned forest farms in China are facing industrial transformation to better deal with the climate change; the human resource is also coevolutionary to these changes. Therefore, in the context of global warming and industrial transformation of state-owned forest areas in China, it is of great practical significance to study the coordinated allocation of industrial structure and human resources in state-owned forest areas to promote the development of economic and social transformation of state-owned forest areas.

2. Materials and Methods

In this section, two sub-sections were designed to describe the detailed calculation processes to address the aims concerned in this study. The first part is to introduce the data sources from where the data were obtained. The second part is to draw the entire models.

2.1. Data Sources

Data on the output value and number of employed persons in each industry from 1998 to 2018 were obtained from the *China Forestry and Grassland Statistical Yearbook* (1998–2018). As the *China Forestry and Grassland Statistical Yearbook* no longer provides statistics on the number of people employed in each industry after 2018, only specific data for the period 1998–2018 will be analyzed.

This study conducted a questionnaire survey from October to December 2021 using paper and electronic questionnaires to employees at all levels in key state-owned forest areas in China; 800 documents were distributed and 731 valid questionnaires were received. In general, the workforce in key state-owned forest farms is characterized by an obvious aging trend, a low level of education, unreasonable job title levels, diverse professional backgrounds and generally low incomes, etc. Various types of workforces have not had many opportunities to participate in training and learning in recent years, and problems such as insufficient updating of knowledge, more serious brain drain and the need to strengthen their ability to perform their duties are also prominent.

2.2. Data Processing Methods

In this sub-section, the data model of the industrial structure and human resource industry structure and the construction of a coordinated suitability evaluation model are described, respectively.

2.2.1. Industrial Structure and Human Resource Industry Structure

This study explores the regional industrial structure in terms of the share of each industry in the regional economy in the total regional output, and measures its human resource industrial structure in terms of the share of the population employed in each industry in the total regional employment [18].

2.2.2. Coordinated Suitability Evaluation Model Construction

The evaluation model of the degree of coordination and appropriateness mainly constructs the structural deviation coefficient, structure departure degree coefficient and structural-change synergy coefficient to quantitatively evaluate the degree of synergy and appropriateness between the existing human resources allocation and industrial structure in key state-owned forest farms. The quantitative evaluation of the degree of synergistic appropriateness between the existing human resource allocation and industrial structure in key state-owned forest farms is calculated as follows [19,20]:

The structural deviation coefficient is a measure of the match with human resources at the level of a single industry. When the structural deviation coefficient is greater than zero, it indicates that the growth in output value of the industry is ahead of the growth in employment of people, indicating that the industry can absorb more people into employment. When the structural deviation coefficient is less than zero, it means that the growth of output value of the industry lags behind the growth of employment, indicating that there is already a certain number of hidden unemployed people in the industry, and this excessive number of people should be squeezed out from the industry. The closer the structural deviation coefficient is to zero, the more the industrial structure of the industry tends to match the human resource structure. The formula for the structural deviation coefficient is:

$$D_i = P_i / W_i - 1 \tag{1}$$

where D_i is the deviation coefficient of the structure of industry i; P_i is the ratio of the output value of industry i to the total output value of the forestry industry in that year (%); W_i is the ratio of the number of persons in industry i to the total number of employed persons at the end of the year (%).

The structure departure degree coefficient assesses the extent to which the industrial structure and human resources move in tandem at the overall regional level. The structure departure degree coefficient represents the difference between the distribution of human resources allocation and industrial structure in terms of "mean", while the structural deviation function measures the suitability of human resources and industrial structure between different regions in terms of "variance". The larger the actual measured value of the total variance function (L_{P-W}), the greater the variance between the human resource allocation and industrial structure of the region, and the less synchronous the development; conversely, the smaller the actual measured value (closer to zero), the more synchronous the development of the human resource structure and industrial structure of the region, and the closer to reasonable. The formula is:

$$L_{P-W} = \sum_{i=1}^{3} |P_i - W_i| \tag{2}$$

The structural-change coefficient describes the dynamic adaptation of the regional human resources due to changes in the industrial structure. The mathematical nature of the structural-change coefficient (C_{P-W}) shows that $0 \leq C_{P-W} \leq 1$. The closer the actual measured value of C_{P-W} is to 1, the better the synergistic change property between the 2 structures in the region, i.e., the more sensitive the change in human resources is to the change in industrial structure. The equation for the structural-change synergy function is

$$C_{P-W} = \frac{\sum_{i=1}^{3} P_i W_i}{\sqrt{\sum_{i=1}^{3} P_i{}^2 \sum_{i=1}^{3} W_i{}^2}} \quad (3)$$

using the grading standard of coordination fitness formulated, i.e., coordination fitness grade. It refers to dividing the range of coordination fitness into several continuous intervals; that is, dividing the coordination degree from 0 to 1 into 10 level intervals, each interval represents a coordination level, and each level represents a kind of coordination state, so as to form a continuous ladder (Table 1).

Table 1. Harmonized suitability grading scale.

C_{P-W} in the Interval	Degree of Coordination and Adaptation
Excellent coordination and adaptation	1.00 to 0.90
Good coordination and adaptation	0.89 to 0.80
Intermediate coordination adaptation	0.79 to 0.70
Primary coordination adaptation	0.69 to 0.60
Critical coordination adaptation	0.59 to 0.50
On the verge of maladjustment	0.49 to 0.40
Mild maladjustment	0.39 to 0.30
Moderate adaptation detuning	0.29 to 0.20
Severe maladjustment	0.19 to 0.10
Extremely well adapted to detuning	0.09 to 0.00

A linear regression analysis was used to investigate the trends in industrial structure, human resource structure, structural deviation coefficient, structure departure degree coefficient and structural-change coefficient in key state-owned forest farms during 1998–2018. The S-W test showed that all indicators were approximately normally distributed except for the structural deviation coefficients of secondary and tertiary industries, and data that did not conform to normal distribution were processed by taking the natural logarithm e.

3. Results

In this section, three sub-sections were displayed to answer, for the state-owned forest farms, (1) the current situation of the personal structure; (2) changes in industrial structure and human resource allocation; and (3) changes in human resource allocation and industrial structure coordination and adaptation, respectively.

3.1. Analysis of the Current Situation of the Personal Structure in State-Owned Forest Farms

The sample survey workers returned 731 questionnaires. In terms of academic structure, among the 731 people surveyed, 17 had postgraduate degrees, accounting for 2.33% of the total; 262 had undergraduate degrees, accounting for 35.84% of the total; 241 had college (secondary) degrees, accounting for 32.97% of the total; and 211 had high school degrees or below, accounting for 28.86% of the total (Figure 1). In terms of the structure of titles, among the 731 people surveyed, 143 had senior titles, accounting for 19.56% of the total; 205 had intermediate titles, accounting for 28.04% of the total; 154 had junior titles, accounting for 21.07% of the total; and 229 had no titles, accounting for 31.33% of the total (Figure 2). In terms of the gender and age structure of the trained human resources, among the 731 people surveyed, there were 497 men, accounting for 67.99% of the total; and 234 women, accounting for 32.01%. There were 92 young trained human resources under 35 years old, accounting for 12.59% of the total; 437 trained human resources aged 36–50, accounting for 59.78% of the total; and 202 people aged 50 or above, accounting for 27.63% of the total (Figure 3).

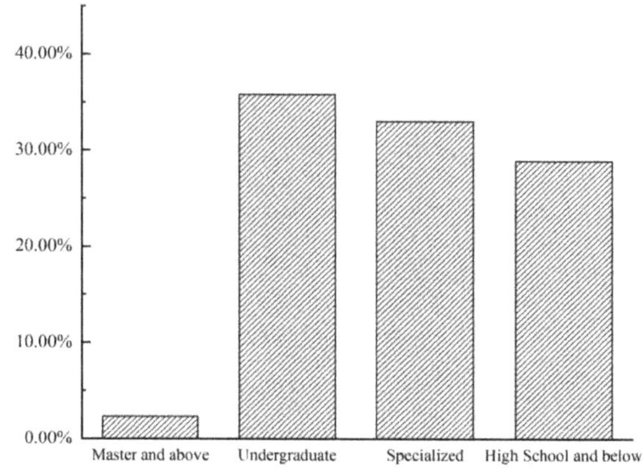

Figure 1. Educational structure of trained human resources in key state-owned forest farms.

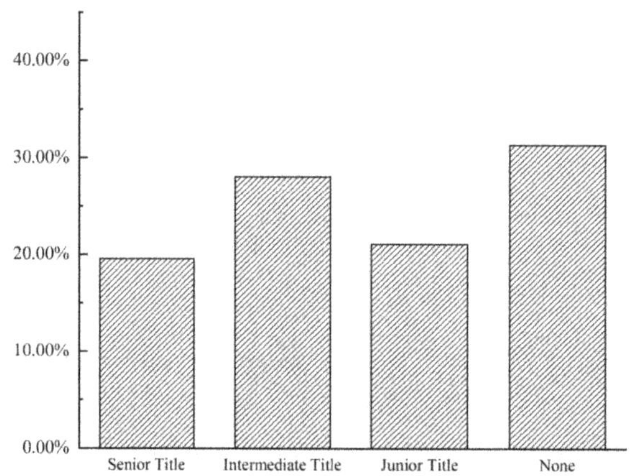

Figure 2. Structure of trained human resources titles in key state-owned forest farms.

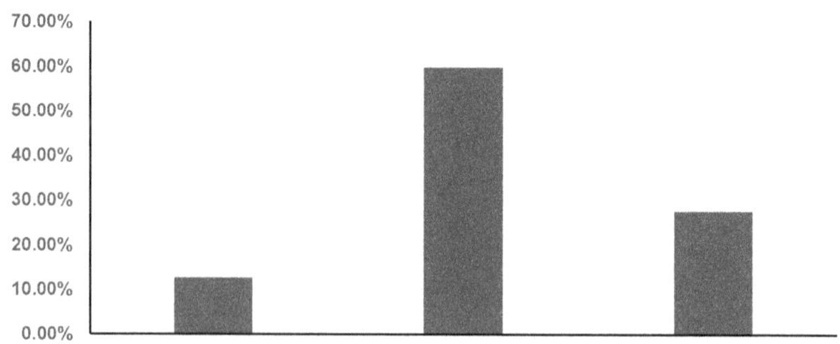

Figure 3. Age distribution of employees in key state-owned forest farms.

The survey results found that there is a lack of participation in education and training activities for cadres and workers in state-owned forest farms to update their knowledge and improve their quality and ability. The investigation shows that 31.33% of the trained human resources participating in the survey have not participated in any form of education and training activities in the past three years, and only 23.12% of them have participated in more than three training activities. The investigation shows that 67.72% of the respondents have a sense of career crisis, of which 15.18% of them had a very strong sense of occupational crisis, and only about 32% did not feel occupational crisis or did not think about the issue of occupational crisis; when faced with the question of whether they were confident in completing their work tasks, only 24.49% gave a positive answer, and nearly 80% said they could not very well or were not sure whether they could complete their work tasks.

According to the survey results, 74.56% of the respondents believe that the current human resources situation in state-owned forest farms is serious or very serious. As for the current talent environment and talent policy, nearly half of the respondents thought that the talent environment and policy were relatively good or very good. In terms of income, only about 12% of the state forestry employees surveyed were satisfied with their current income, and more than half were less than satisfied, with nearly 16% being very dissatisfied with their income.

3.2. Changes in Industrial Structure and Human Resource Allocation in State Forestry from 1998 to 2018

There are differences in the trends in changes in the three industries in terms of industrial structure (Figure 4): the proportion of primary and secondary industries in the total output value of key state-owned forest farms in the 20-year period both show fluctuating changes until 2010, reaching maximum values in 2002 (42.83%) and 2007 (30.84%), respectively, while both start to decline rapidly around 2010, reaching minimum values (39.10% and 22.63%). Linear regression analysis showed that the primary and secondary industries in key state-owned forest farms showed a significant downward trend between 1998 and 2018 ($p < 0.05$). The proportion of the output value of the tertiary forestry industry in the total forestry output value increased year by year, especially after 2015, and gradually exceeded the proportion of the output value of the secondary industry, and was roughly the same as that of the primary industry (Figure 4), with the smallest proportion occurring in 1998 (12.53%) and the largest in 2018 (38.28%). Linear regression analysis showed that the tertiary industry in key state forest areas showed a significant upward trend ($p < 0.05$) during 1998–2018.

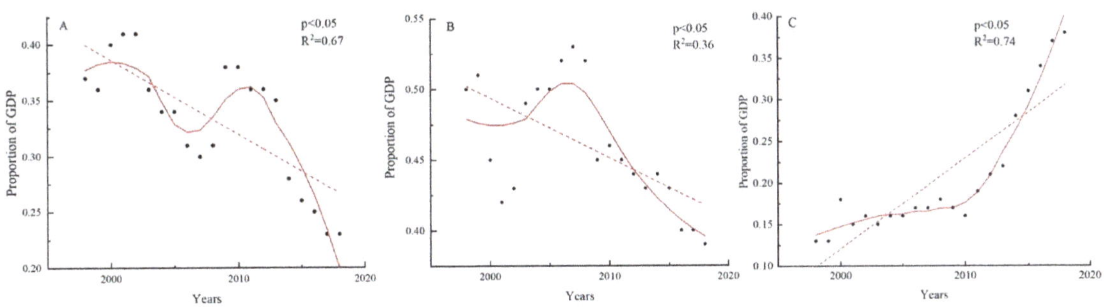

Figure 4. Changes in the structure of output value of three industries in 20a key state-owned forest farms. Note: (**A–C**) refer to primary industry, secondary industry and tertiary industry, respectively. The dot represents the observed data; the solid line represents the locally weighted scatterplot smoothing line; the dot line represents the linear regression.

In Figure 5, the allocation of human resources in the three industries in key state-owned forest farms shows the following characteristics: (1) The proportion of employment in the primary industry fluctuated continuously between 1998 and 2018, without showing a significant trend of change, reaching a maximum value of 95.16% in 2008 and a minimum value of 90.32% in 2005, respectively, with an average value of 92.06%, which is significantly higher than the proportion of employment in the primary industry in that period. The average value was 92.06%, which was significantly higher than the proportion of the primary industry in that period, and higher than the proportion of employment in the secondary and tertiary industries in key state-owned forest farms in the same period. (2) Linear regression analysis shows that the proportion of employment in the secondary industry in key state-owned forest farms showed a significant downward trend during the period 1998–2018 ($p < 0.05$), with the proportion of employment in the secondary industry decreasing from the highest value of 7.87% in 1998 to 3.15% in 2010 from 1998 to 2010, with an average decrease of 0.22 percentage points per year, and from 2011 to 2015 produced some fluctuations and showed a downward trend again after 2016. (3) The share of employment in the tertiary industry in key state-owned forest farms showed a steady upward trend from 1998 to 2018 ($p < 0.05$), similar to the trend in the share of the tertiary industry in GDP, rising from 0.08% in 1998 to 4.21% in 2018, with an average annual increase of 0.21 percentage points, which indicates that the tertiary industry has a strong characteristic of absorbing labor.

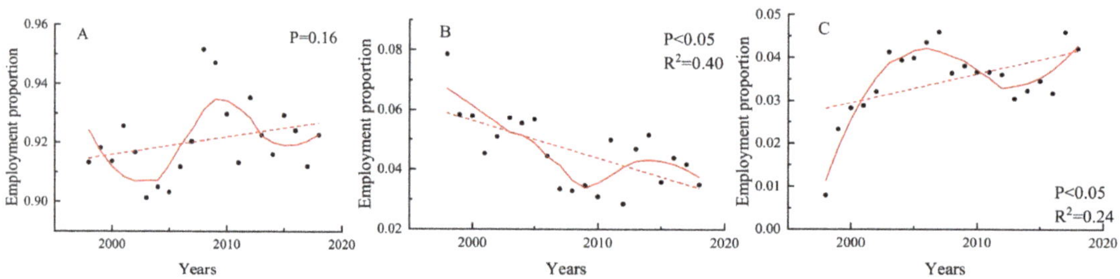

Figure 5. Changes in human resource allocation in the three industries in 20a key state-owned forest. Note: (**A**–**C**) refer to primary industry, secondary industry and tertiary industry, respectively. The dot represents the observed data; the solid line represents the locally weighted scatterplot smoothing line; the dot line represents the linear regression.

3.3. *Changes in Human Resource Allocation and Industrial Structure Coordination and Adaptation in State-Owned Forest Farms*

The deviation coefficient of primary industry structure (D1) in key state-owned forest areas was negative from 1998 to 2018 (Figure 6), and it showed fluctuations between 1998 and 2007, while it rapidly decreased after 2008, showing a significant downward trend overall ($p < 0.05$). The structural deviation coefficient (D1) of the primary sector reached a minimum value of −0.58 in 2018, indicating that the negative deviation of the primary sector from the structural equilibrium reached 0.58%, i.e., at least 0.58% of the employees in the primary sector urgently need to shift to other industrial sectors. The coefficient of deviation from the structure of the secondary industry in key state-owned forestry areas was positive over the 20-year period, with a mean value of 8.03, and experienced a trend of increasing and then decreasing: it increased each year between 1998 and 2009, with the minimum value (3.75) occurring in 1998 and the maximum in 2009 (10.18), while rapidly declining after 2010 and stabilizing after 2015, with the structural deviation coefficient in 2018 (5.42) basically the same as in 1998, which also indicates that there is still a small shortage of forestry secondary industry personnel. The structural deviation coefficient of the tertiary forestry industry in key state-owned forest farms gradually decreased between 1998 and 2009, then rapidly increased after 2010, and generally showed an upward trend between 1998 and 2018 ($p < 0.05$), reaching a maximum value of 9.78 in 2018, which indicates that there is still a large human resource gap in the tertiary forestry industry in 2018, with a strong ability to absorb surplus personnel.

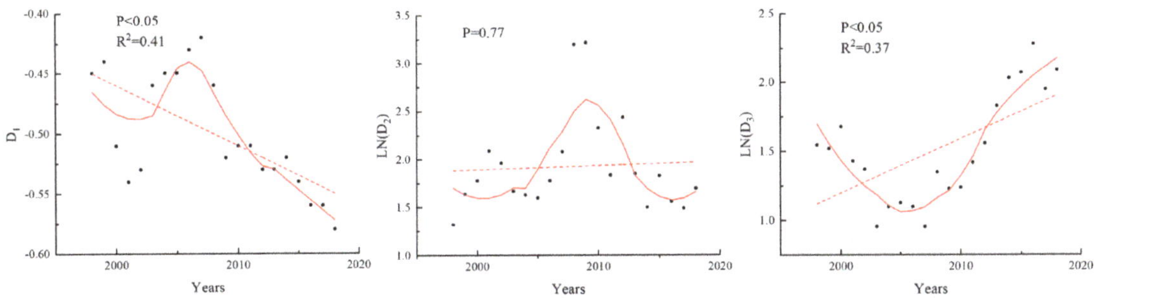

Figure 6. Coefficient of deviation of the structure of the three industries in 20a key state-owned forest farms. Note: D1, D2 and D3 represent the primary, secondary and tertiary industries of state-owned forest areas, respectively. The dot represents the observed data; the solid line represents the locally weighted scatterplot smoothing line; the dot line represents the linear regression.

From Figure 7, it can be seen that the overall industrial structure deviation $L_{P\text{-}W}$ in key state-owned forest farms is relatively small, maintaining a mean value of 0.93, with greater volatility between 1998–2010, while showing a significant upward trend overall between 1998–2018 ($p < 0.05$). While the synergy coefficient of structural change in key state-owned forest farms $C_{P\text{-}W}$ increased and then decreased during the 20a period, reaching a maximum value (0.86) in 2007 and rapidly decreasing after 2010, reaching a minimum value of 0.70 in 2018, showing a significant downward trend overall ($p < 0.05$), the synergy suitability grading scale shows that the human resource allocation in state-owned forest farms increased from 1998. This may be closely related to the reform of state-owned forest farms and the promotion of natural forest protection projects.

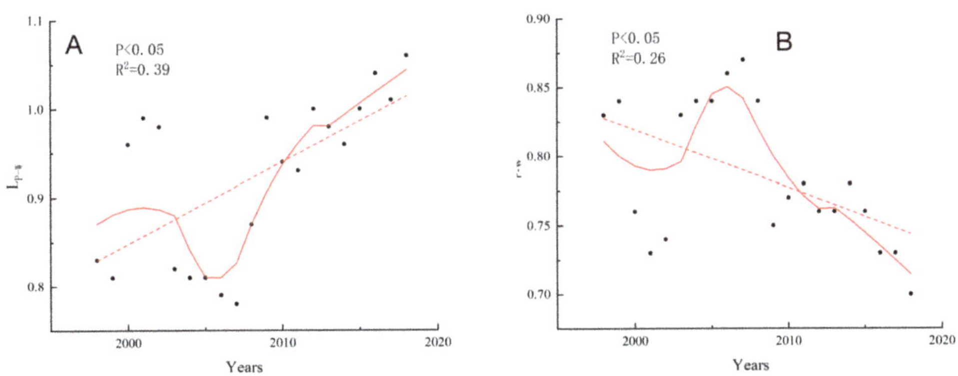

Figure 7. 20a structure departure degree coefficient (**A**) and synergy coefficients (**B**) for structural change in key state forest areas. The dot represents the observed data; the solid line represents the locally weighted scatterplot smoothing line; the dot line represents the linear regression.

4. Discussion

Through the analysis of the current situation of the structure of the people in the key state-owned forest farms, the following points can be found: (1) Trained human resources are still in short supply. There is a lack of high-level innovative scientific and technological trained human resources, and there are few leading forestry-trained human resources with high visibility and greater influence, showing a trend of youthfulness. There is a shortage of urgent specialists in emerging fields such as wetland protection, forestry to cope with climate change, and forest rights transfer assessment. There is a large gap between the supply and demand of practical and skilled trained human resources in forest management, forest breeding and cultivation, forest tourism, landscape engineering, wood processing and special industries. The proportion of biased trained human resources is large, and the number of composite trained human resources is too small. There is a relative lack of trained human resources in the western region, and a general shortage of trained human resources in grassroots units and small and medium-sized enterprises. Some practitioners have low professionalism. Forestry system grassroots workers do not have many opportunities to receive education and training, and the speed of knowledge updating lags behind. Through continuing education and learning, employees' cultural literacy has improved, but most of the professions studied are non-forestry professions; there is the problem of "learning not using". In cadres of workers newly entered into the forestry system and a large number of new forestry construction and production practitioners, professional literacy needs to be improved. There is an urgent need to supplement the learning of forestry knowledge and skills [23,24]. Grassroots units to attract the workforce are not strong. Forestry is a tough industry with a long growth cycle and slow output, in the fierce competition for workforce, placing it in a disadvantageous position. In particular, forestry and grassroots units are remote, have difficult conditions, are closed information, and suffer from economic underdevelopment and a low level of treatment; the attractiveness of trained workforce is not enough, there is a long-standing shortage of trained human resources into the problem, and the best workforce cannot be stabilized. In one study [25], there was insufficient investment in the development of talent resources for forestry enterprises and institutions for staff education, and training funds did not reach the actual proportion of national regulations. Special funds for the development of the forestry workforce were not included in the stable financial budget; workforce-training funds for forestry key projects and major scientific research projects were not guaranteed, resulting in the training and introduction of high-level forestry human resources. There is an urgent shortage of trained human resources, forestry workforce education, and training infrastructure and weak infrastructure for the education and training of forestry-trained human resources; in addition, and the construction of a trained-human-resources service system is relatively lagging behind. Zhi et al. [26] conducted a study on the current situation of human resources in state-owned forest farms in China, which include state-owned forest areas. The results indicate that state-owned forest farms tend to suffer from a lack of human resources, aging employees, and a shortage of managerial and technical employees, similar to the findings of this study [27–29].

The degree of coordination between human resource allocation and industrial structure determines the effective development of regional social productivity and the optimization and upgrading of industrial structure [30]. This study is based on data related to industry and employment in key state-owned forest farms during 1998–2018. Based on the data related to industry and employment in key state-owned forest farms during 1998–2018, this study calculates the "structural deviation coefficient", "structure departure degree coefficient" and "structural change synergy coefficient" to establish a model for evaluating the coordination and appropriateness of industrial structure and human resource allocation in each year. The results are as follows: the coefficient of structural deviation of the primary industry in key state-owned forest farms was negative from 1998 to 2018, and showed a rapid decline after 2008, while the coefficients of structural deviation of the secondary and tertiary industries were all positive, and the coefficient of structural deviation of the tertiary industry decreased from 1998 to 2007. The coefficient of structural deviation of the tertiary industry declined from 1998 to 2007, but increased rapidly after 2008, showing a significant upward trend overall (Figure 6). The above may indicate that the primary forestry industry in key state-owned forest areas does not have the ability to absorb more labor, and the number of existing workers is still high, while there is a large employment gap in the tertiary forestry industry, and the gap has gradually expanded after 2008. The coefficient of structural deviation and the coefficient of structural-change synergy in key state-owned forest farms showed significant upward and downward trends during 1998–2018, respectively, but both of them fluctuated significantly during the period 1998–2007, and showed a more obvious increase and decrease after 2008. In recent years, the coefficient of structural change in state-owned forest farms has been hovering between 0.7 and 0.8, while the coefficient of structural deviation has been increasing (Figure 7). Although the theoretical level indicates that the development of human resources and industrial structure in key state-owned forest farms is more synchronous, in terms of the volume of the economy and the actual situation, there are no obvious leading industries in key state-owned forest farms today, and the scale of high-value-added industries is small. The changes in the industrial structure, human resource allocation and the suitability of the two in key state-owned forest farms over the past 20 years are due to two reasons: firstly, the adjustment of the industrial structure due to the change in national industrial policy during the period; and secondly, the lag in the transformation of the workforce compared to the adjustment of industries. The second is the lag in labor force conversion relative to industrial restructuring. Li and Wang [27] analyzed the coordination measures of human resource allocation and industrial structure in key state-owned forest areas in China from 2007 to 2015, and the results showed that the coordination of human resource allocation and industrial structure in state-owned forest areas was at the primary synergistic adaptation stage, similar to the findings of this study.

With global climate change and the strengthening of China's attention to environmental protection, China carried out a series of reforms in key state-owned forest areas from 1998 to 2018, and the Natural Forest Resource Protection Project is one of the most important reform measures. As the world's first super ecological project focusing on the protection of natural forests, the Natural Forest Resources Protection Project included more timber restriction and logging suspension policies for key state-owned forest areas in Northeast China and Inner Mongolia, which also led to the industrial transformation of state-owned forest areas and gradually reduced timber production, which had a great impact on the industrial structure of state-owned forest areas during 1998–2007, leading to fluctuations and repetitions. Meanwhile, the secondary industry in state-owned forest areas began to develop high-value-added products and large-scale production in this phase, which also led to its GDP share reaching the highest value in this phase during 1998–2018 (Figure 4) [31]. With the acceleration of climate change, and in accordance with the principle of "common but differentiated responsibilities" established by the UNFCCC [32,33], although China is not currently obligated to reduce GHG emissions, as a responsible power, China is very concerned about the unique role and strategic position of forestry in addressing climate change. In 2007 and 2008, the State Council released the National Program for Addressing Climate Change in China (hereinafter referred to as the National Program) and the Policies and Actions for Addressing Climate Change in China, which explicitly included forestry among the six priority areas for climate change mitigation and the four priority areas for climate change adaptation in China. The 17th National Congress of the Communist Party of China (CPC) held in 2007 included the scientific concept of development in the party constitution, and strict forest harvesting limits were set in the 11th, 12th and 13th Five-Year Plans [33]. The above national policy adjustments led to an accelerated industrial transformation in key state-owned forest areas after 2008, with the GDP share of primary and secondary industries rapidly decreasing and the share of tertiary industries increasing (Figure 4). However, there is often a lag in the transformation of human resources allocation relative to changes in industrial structure (Figure 5) [34] Although the change in the industrial structure of key state-owned industries has caused changes in the allocation of human resources in the three industries, a cross-sectional comparison between industries reveals that, compared to the gradually declining marginal output levels per capita in the primary and secondary industries, there is still a large degree of room for improvement in the productivity and innovation development levels of the tertiary industry, and there is still a large human resource gap, and the industry has a large potential for allocating human resources in the future.

5. Research Limitation

This study theoretically contributed one new line between the climate change and personal structure in the state-owned forest farms, which are facing multiple challenge to adaptation to climate change. The behavior of the educated trained employees will play crucial roles in mitigation of climate change [35]. The main limitations of the study and future direction of research will be considered as follows: (1) The latest data were not collected. The time range of this study is from 1998 to 2018, and there is no relevant collection from 2019 to 2021, which can be collected and further studied in the future. (2) In addition to state-owned forest areas, there are also certain changes in the personal structure of state-owned forest farms in various regions of China under the background of climate change, but our study does not involve state-owned forest farms, which can be studied in the future. (3) Due to limited conditions, only 731 questionnaires were withdrawn. The number of questionnaires should be increased in future research.

6. Policy Recommendations

The current situation of trained human resources in key state-owned forest farms shows a shortage of the total number of trained human resources, a shortage of high-level and highly educated trained human resources, and the aging of trained human resources, which requires strengthening and improving the work of trained human resources in state-owned forest farms in terms of the concept and development direction of the workforce, the key workforce, the working system and mechanism, the education and training system, and the livelihood protection. During the period 1998–2018, the contribution rate of the primary industry in state-owned forest areas to the economy and to the allocation of human resources in key state-owned forest farms gradually declined. The contribution rate of the secondary industry in state-owned forest areas to the allocation of human resources gradually stabilized, and its ability to absorb surplus workers was limited. The tertiary industry developed most rapidly among the three industries, and its development efficiency showed a gradual improvement, with a greater potential to absorb surplus workers subsequently. The development of human resources and industrial structure in key state-owned forest farms has generally been relatively synchronous over the past 20 years, and the degree of coordination and appropriateness of the two can be maintained at an intermediate level or above, but the fluctuation and repetition of industrial structure and human resources allocation in each industry were more drastic between 1998 and 2007, and the coordination between human resources allocation and industrial structure in key state-owned forest farms has been decreasing in recent years since 2008, with 2018's coefficient of structural deviation and synergy coefficient of structural change both reaching their maximum values during the period 1998–2018, showing that the allocation of human resources in industrial sectors is not as reasonable and at a lower level. The reasons for this may be (1) the industrial restructuring in key state-owned forest farms due to the promotion of the CPC's natural forest protection project and the policy of banning logging and halting logging in forestry; (2) the large lag in labor force conversion itself compared to the rapid restructuring of industries driven by national policies.

The large-scale reduction in forestry production as well as the transformation and development process of key state-owned forest farms, human resources allocation, and industrial structure adjustment are the two key issues of concern in the transformation and development to achieve a reasonable and coordinated development of human resources allocation and industrial structure. In terms of the workforce, it should not be rushed, and the solution should be combined with the actual situation to explore and innovate a reform path suitable for the development of trained human resources in forest areas according to local conditions, which can be started from the following aspects: In terms of industrial structure, firstly, the construction and investment of the tertiary forestry industry should be strengthened to increase the ability of the tertiary industry to absorb surplus personnel. The tertiary industry in forestry is the main direction of comprehensive transformation and development in key state-owned forest farms; key state-owned forest farms can make full use of natural snow and ice resources in winter with skiing and fog sightseeing as the main content. The forest recreation industry can implement forest rafting in summer with forest resources as the main content throughout the forest area as the theme, and gradually expand and strengthen the tertiary industry chain to provide employment opportunities for surplus personnel. Secondly, the industry should actively promote the optimization and upgrading of the secondary industry; deepen the wood deep-processing industry chain; increase investment in science and technology; expand the scale of production; enrich the variety of forest products in order to achieve the purpose of driving the development of other production projects; change the situation of the primary industry and the raw material production industry to make a large contribution to employment; and enhance the ability of the secondary and tertiary forestry industries to absorb human resources.

7. Conclusions

This study analyzes the current situation of personal structure in key state-owned forest farms of China under the background of climate change, and empirically analyzes the coordination between human resource allocation and industrial structure in key state-owned forest farms within the background of climate change by establishing an evaluation model of the contribution and coordinated suitability of each industry to human resource allocation. The trained-professional human resources and the industrial structure changes in the context of climate change are the main limited factors for the key state-owned forest farms of China. Therefore, increasing the education investment for climate change and the economic income for the employees are suggested to be promoted for policy makers in future.

Author Contributions: Conceptualization, J.L. and S.J.; methodology, S.J.; software, X.D. and S.J.; validation, J.L. and S.J.; formal analysis, X.D. and L.S.; investigation, S.J. and S.Z.; resources, S.J.; data curation, S.J.; writing—original draft preparation, X.D., J.L., L.S., and S.Z.; writing—review and editing, X.D., J.L., L.S. and S.Z.; visualization, S.J.; supervision, J.L. and S.J.; project administration, J.L. and S.J. funding acquisition, S.J. All authors have read and agreed to the published version of the manuscript.

Funding: This research was funded by the start-up funding of Minjiang University (32304307), the Natural Science Foundation of Fujian Province, China (Grant No. 2022J011140), and Project of Central Leading Local "Fujian Mental Health Human-Computer Interaction Technology Research Center" (2020L3024).

Institutional Review Board Statement: Not applicable.

Informed Consent Statement: Informed consent was obtained from all subjects involved in the study.

Data Availability Statement: Not applicable.

Conflicts of Interest: The authors declare no conflict of interest.

References

1. Chen, K. Some thoughts on the transformation of forestry economic development in the context of global climate change. In Proceedings of the 12th Annual Conference of the Chinese Association for Science and Technology, Fuzhou, China, 1–3 November 2010.
2. Ding, S.; Liao, L.T.; Wen, Z.M. Analysis on Two Impelled-System Model in Structure of Forestry Industry. *Appl. Mech. Mater.* **2013**, *273*, 49–54. [CrossRef]
3. Theodore, S.; Wu, Z. *On Investment in Human Capital*; Beijing Institute of Economics Press: Beijing, China, 1990.
4. Theodore, S. *The Economic Value of Education*; Jilin People's Publishing House: Jilin, China, 1982.
5. Theodore, S. *Economic Growth and Agriculture*; People's University of China Press: Beijing, China, 2015.
6. Gary, B. *Human Capital*; (Chinese Edition); Machinery Industry Press: Beijing, China, 2016.
7. Gary, B. *Economic Analysis of Human Behavior*; (Chinese Edition); Gezhi Press: Shanghai, China, 2008.
8. Brown, A.J. The Conditions of Economic Progress. *Int. Aff.* **1952**, *28*, 79–80. [CrossRef]
9. Rostow, W.W.; Kuznets, S. Economic Growth of Nations. Total Output and Production Structure. *Political Sci. Q.* **1971**, *86*, 654. [CrossRef]
10. Helleiner, G.K.; Chenery, H.B.; Syrquin, M. Patterns of Development. *Afr. Econ. Hist.* **1976**, *56*, 1950–1970.

11. Zhang, Y.; Li, M. Study on the evaluation of the coordination and suitability of China's regional personal structure optimization and industrial structure upgrading. *China Soft Sci.* **2011**, *3*, 177–192.
12. Liu, L.H.; Cao, Y.K. Exploring the strategy of updating the survival value system of workers in state-owned forest farms. *Acad. Exch.* **2011**, *7*, 137–141.
13. Zhao, G. Research on the general rule of interaction between personal structure and industrial structure. *Bus. Res.* **2008**, *2*, 34–39.
14. Liang, W. *Talent Resource Development in State-Owned Forest Farms of Heilongjiang Province: Efficiency, Mobility and Countermeasures*; Northeast Forestry University: Harbin, China, 2010.
15. Zhu, H. Investigation and analysis of the frequency of talent flow in state-owned forest farms in Heilongjiang Province. *For. Econ. Issues* **2009**, *29*, 214–217.
16. Johnson, D.R.; Hoopes, D.G. Managerial cognition, sunk costs, and the evolution of industry structure. *Strateg. Manag. J.* **2003**, *24*, 1057–1068. [CrossRef]
17. Tullio, G. Long run implications of the increase in taxation and public debt for employment and economic growth in Europe. *Eur. Econ. Rev.* **1987**, *31*, 741–774. [CrossRef]
18. Maity, S.; Sinaha, A. Linkages between Economic Growth and Population Ageing with a Knowledge Spillover Effect. *J. Knowl. Econ.* **2020**, *12*, 1905–1924. [CrossRef]
19. Moss, H.E.; Rodd, J.M.; Stamatakis, E.A.; Bright, P.; Tyler, L.K. Anteromedial Temporal Cortex Supports Fine-grained Differentiation among Objects. *Cereb. Cortex* **2005**, *15*, 616–627. [CrossRef]
20. Jenkin, M.E.; Saunders, S.M.; Wagner, V.; Pilling, M.J. Protocol for the development of the Master Chemical Mechanism, MCM v3 (Part B): Tropospheric degradation of aromatic volatile organic compounds. *Atmos. Chem. Phys.* **2003**, *3*, 181–193. [CrossRef]
21. Liao, H.; Yang, L.; Dai, S.; Van Assche, A. Outward FDI, industrial structure upgrading and domestic employment: Empirical evidence from the Chinese economy and the belt and road initiative. *J. Asian Econ.* **2021**, *74*, 101303. [CrossRef]
22. Dastmalchian, A.; Blyton, P. Organizational Structure, Human Resource Practices and Industrial Relations. *Pers. Rev.* **1992**, *21*, 58–67. [CrossRef]
23. Chreif, M.; Farmanesh, P. Applying Green Human Resource Practices toward Sustainable Workplace: A Moderated Mediation Analysis. *Sustainability* **2022**, *14*, 9250. [CrossRef]
24. Freire, C.; Pieta, P. The Impact of Green Human Resource Management on Organizational Citizenship Behaviors: The Mediating Role of Organizational Identification and Job Satisfaction. *Sustainability* **2022**, *14*, 7557. [CrossRef]
25. Peng, F.; Altieri, B.; Hutchinson, T.; Harris, A.J.; McLean, D. Design for Social Innovation: A Systemic Design Approach in Creative Higher Education toward Sustainability. *Sustainability* **2022**, *14*, 8075. [CrossRef]
26. Zhi, Y. *Survey on the Current Situation of State-Owned Forestry Talents and Research on Development Countermeasures*; Beijing Forestry University: Harbin, China, 2017.
27. Li, J.; Wang, Y. Measuring the coordination between human resource allocation and industrial structure in key state-owned forest farms. *For. Econ. Issues* **2017**, *37*, 66–71.
28. Jin, W. An empirical study on the relationship between industrial structure and employment structure changes in Sichuan. *Math. Stat. Manag.* **2010**, *29*, 578–585.
29. Zhang, B.; Yang, B. Industrial structure, employment contribution rate and employment growth: Empirical evidence based on the tertiary industry in Yunnan Province. *J. Chongqing Univ. Commer. Ind.* **2007**, *1*, 80–84.
30. Qin, Q.; Changchun, L. A study on the coordination of industrial structure and employment structure in Chongqing. *Geogr. Res. Dev.* **2013**, *32*, 46–50.
31. Li, N.; Yuan, J. Forestry construction in China in the context of climate change. *Prot. For. Technol.* **2011**, *1*, 4–6.
32. Ott, H.E.; Sterk, W.; Watanabe, R. The Bali Roadmap—New Horizons for Global Climate Policy. *Clim. Policy* **2008**, *8*, 91–95. [CrossRef]
33. Zhang, J.Y.; Wang, G.Q.; Liu, J.F. Introduction to China's National Program to Address Climate Change. *China Water Resour.* **2008**, *3*, 72–74.
34. Duan, L.; Guan, P. Research on the Dynamic Equilibrium of Human Resource Structure Based on Markov Chain. In Proceedings of the 2008 4th International Conference on Wireless Communications, Networking and Mobile Computing, Dalian, China, 12–17 October 2008; IEEE: Piscataway, NJ, USA, 2008.
35. Omarova, L.; Jo, S.-J. Employee pro-environmental behavior: The impact of environmental transformational leadership and GHRM. *Sustainability* **2022**, *14*, 2046. [CrossRef]

sustainability

MDPI

Article

Population Exposure Changes to One Heat Wave and the Influencing Factors Using Mobile Phone Data—A Case Study of Zhuhai City, China

Junrong Li, Peng Guo *, Yanling Sun *, Zifei Liu, Xiakun Zhang and Xinrui Pei

School of Geography and Environmental Science, Tianjin Normal University, Tianjin 300387, China; lijunrong98@163.com (J.L.); Liu_zifei0801@163.com (Z.L.); zhang_xiakun@163.com (X.Z.); pei_xinrui@163.com (X.P.)

* Correspondence: guopgis@163.com (P.G.); flyling99@163.com (Y.S.)

Abstract: The frequent occurrence of extreme high temperature weather and heat waves has greatly affected human life. This paper analyzes population exposure and its influencing factors during a heat wave incident in Zhuhai from 6 to 12 September 2021 based on real-time mobile phone data and meteorological data. The results show that the most areas of Zhuhai are affected by high temperature during this heat wave incident. The hourly population exposure is directly proportional to hourly heat wave coverage. In terms of time dimension, the overall population exposure shows a trend of decreasing and then increasing. In terms of spatial dimensions, high population exposure is concentrated in areas such as primary and secondary schools, colleges and universities, office buildings, and residential areas. Low exposure is distributed in most of the mountainous areas along the southern coast. In addition, the leading factors that cause changes in population exposure in different periods of the heat wave cycle are different, which rely more on either climatic factors or population factors.

Keywords: heat wave; high temperature; population exposure; mobile phone data; impact factor; Zhuhai City

Citation: Li, J.; Guo, P.; Sun, Y.; Liu, Z.; Zhang, X.; Pei, X. Population Exposure Changes to One Heat Wave and the Influencing Factors Using Mobile Phone Data—A Case Study of Zhuhai City, China. *Sustainability* **2022**, *14*, 997. https://doi.org/10.3390/su14020997

Academic Editors: Xiaodong Yan, Jia Yang and Shaofei Jin

Received: 18 December 2021
Accepted: 14 January 2022
Published: 17 January 2022

1. Introduction

In October 2018, the United Nations Intergovernmental Panel on Climate Change (IPCC) stated in its special report "Global Warming of 1.5 °C" that human activities are estimated to have caused global warming to be about 1.0 °C higher than the pre-industrial era, with a possible range of 0.8 °C to 1.2 °C [1]. Extreme weather has occurred more frequently in recent years due to climate change. The "2021 Global Risk Report" indicates that in the next 10 years, incidents caused by extreme weather are one of the top risks to be faced by the world which poses a threat to human well-being [2]. Heatwaves, as one of the most typical extreme weather incidents, shows a potential increase in frequency, intensity, and duration under the effects of climate change [3], which not only causes irreversible destruction to natural environment but also a great impact on human life [4–6].

Heatwaves are generally defined as a long spell of hot days with extremely high temperature which puts a risk to human health, while most studies also take duration and intensity of heatwaves into account [7]. However, in China, a day with a maximum temperature ≥35 °C is generally defined as a high temperature incident, and a span of over three consecutive high temperature incidents is defined as heat wave [8]. Studies show that the frequent occurrence of extreme heat incidents has posed a threat to human health and also has a negative impact on people's daily life and social economic development [9]. Therefore, research on high temperature disasters has attracted widespread attention from government departments and the scientific community and has become a hot spot in climate change research [10–12].

Current research on high temperature disasters mainly focuses on high temperatures' temporal and spatial characteristics, including intensity, occurrence frequency, and duration, etc. [13–24]. Risk assessments [11,25–30] and vulnerability (sensitivity) assessments [31–49] of high temperature disasters are increasing, which becomes a strong support for urban high temperature risk prevention and community disaster prevention management. In addition, some scholars conduct research on urban structure to improve the climate change adaptability and urban sustainability [50,51]. It cannot be ignored that humans are the main victims of heat wave disasters. Studies have confirmed a significant relationship between high temperature and population morbidity and mortality [52,53]. Therefore, it is particularly important to identify the population exposure of a region under heat waves. At present, some scholars have studied the population exposure under heat waves based on climate and population scenario data [54–58]. However, earlier studies are mostly based on static statistical data from city administrative units which are not up to date and lack precision, thus, it is hard to reflect the near-real-time exposure risk of population in the affected areas and to measure risk and damage degree of disaster to population spatially and temporally [59]. With the development of big data, data such as mobile phone data and Weibo sign-in data with high accuracy, large coverage area, and high update frequency have provided richer data sources for research that requires temporal and spatial accuracy of the population.

Under these conditions, this paper takes a heat wave incident from 6 to 12 September 2021 in Zhuhai city as an example to discuss the spatial issues of population exposure under one heat wave incident from a geographical perspective. Meteorological monitoring data and mobile phone data were processed on a GIS software platform to obtain real-time temperature and population distributions. Meanwhile, the dynamic changes of population exposure were analyzed by using grid as the basic analysis unit, and the influencing factors of population exposure changes were further explored. This study not only shows areas that are affected more by the heat wave with finer real-time population distribution and temperature but can also be used to prevent and cope with heat waves and improve the urban environment.

2. Materials and Methods

2.1. Study Area

Zhuhai City is located in the southcentral part of Guangdong Province, China, with longitude from 113°03′ E to 114°19′ E and latitude from 21°48′ N to 22°27′ N. The inland area of Zhuhai City consists of hills, coasts, and plains of the Phoenix Mountain and Jiangjun Mountain, which faces Hong Kong across the sea to the east, connects with Macau in the south, Taishan City to the west, and Zhongshan City to the north. It has three administrative regions. As of the end of 2020, the land area of Zhuhai is 1736.45 km^2. Zhuhai City has a transitional oceanic climate between south subtropical and tropical, with obvious alternating winter and summer monsoons, high temperature throughout the year, with an average annual temperature of 23.8 °C, and an average annual rainfall of 1799.2 mm. From 1979 to 2000, the daily extreme maximum temperature occurred on 10 July 1980, with a temperature of 38.5 °C [60]. During the study period of this article, the highest maximum temperature of some weather stations in Zhuhai City was monitored as high as 39.5 °C. The "Communiqué of the Seventh National Census of Guangdong Province" pointed out that the 10-year growth rate of Zhuhai's permanent population ranks second in Guangdong Province. With the support of national policies and its own location advantages, Zhuhai has become an important city in Pearl River Delta and even the Guangdong–Hong Kong–Macao Greater Bay Area.

2.2. Data

2.2.1. Meteorological Data

The temperature data involved in this article come from the monitoring data released by the Zhuhai Meteorological Administration. There are 31 weather stations in the research area, which are more evenly distributed throughout Zhuhai City, the spatial distribution is shown in Figure 1. In early September 2021, due to the combined effects of the downdraft surrounding the strong tropical storm "Conson" and the super typhoon "Sando", along with subtropical high pressure, the weather in Zhuhai was hot, and the high temperature weather intensified, which led to the city's socioeconomic and population exposure risks to increase. Based on the definition of heat waves given by China Meteorological Administrations: A heat wave is a span of at least 3 consecutive days with highest temperature over 35 °C [8]. Meteorological monitoring and high temperature warning information issued by the Zhuhai Meteorological Administration in September 2021 are screened for heat wave incidents, and one from 6 to 12 September 2021 is finally selected as the study period. The average temperature of various meteorological stations in Zhuhai (Figure 2a) indicates that the average temperature of Zhuhai has obviously increased during the study period. Figure 2b shows the hourly maximum temperature and the number of stations reaching high temperatures of various meteorological stations in Zhuhai. We can see that the maximum temperature of meteorological stations in a day generally occurs between 12:00 and 18:00. Starting from 9 September, the number of high-temperature stations in the city has increased significantly; there are 14, 9, 18, and 27 high-temperature stations above 35 °C. Since the 11th, more than half of the meteorological stations have monitored continuous high temperature, and the highest temperature reached 38.81 °C. In order to better present the exposure of Zhuhai's population under heat waves, this article is based on the ordinary kriging method to conduct meteorological data. Spatial interpolation analysis is used to obtain a spatial and temporal distribution map of temperature in the entire area of Zhuhai, and then the hourly temperature data are rasterized and resampled to a 500 m × 500 m grid.

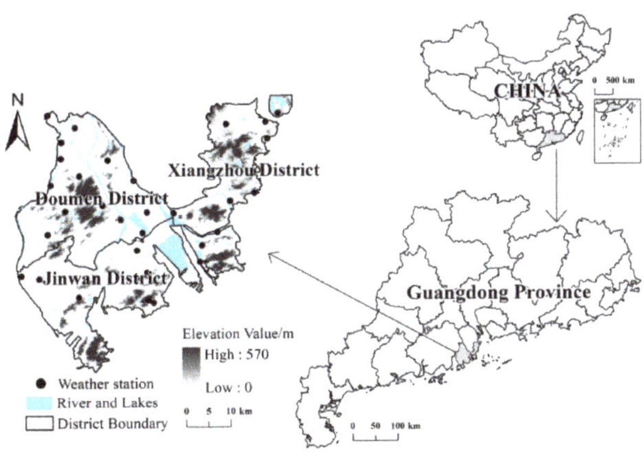

Figure 1. Location of Zhuhai in China and weather stations.

2.2.2. Population Data

The population data in this article involve both statistical population and mobile phone data, among which the permanent resident population comes from the "Zhuhai City Seventh National Census Bulletin" issued by the Zhuhai Bureau of Statistics in 2021, and the mobile phone data come from China Mobile operators, including all China Mobile users in Zhuhai in September 2021. The hourly population data are the anonymous mobile phone data, which are collected at the same frequency as temperature monitoring data. After the data are obtained, the geographic information system platform software ArcGIS is used to divide Zhuhai into 500 m × 500 m grids. According to the mobile phone data, the geographic coordinates of the base station to which the data belong will be gathered at the grid level, and each grid contains the population data within the corresponding geographic location.

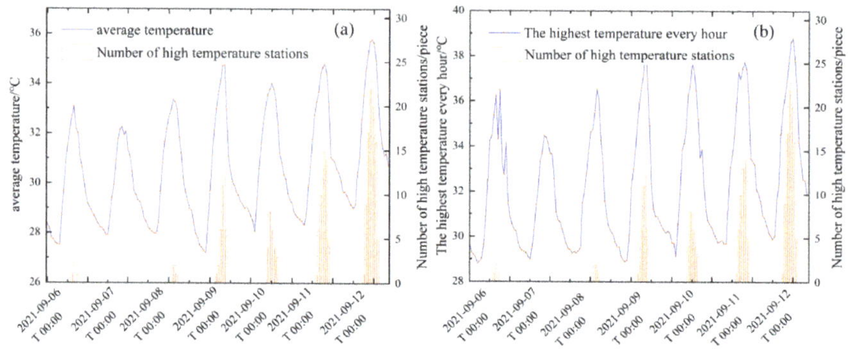

Figure 2. (**a**) Hourly average temperature and number of high temperature weather stations; (**b**) Hourly maximum temperature and number of high temperature weather stations.

2.3. Methods

2.3.1. Hourly Population Distribution Estimates

The mobile phone data can accurately locate spatial locations of the population and obtain real-time population data in a small scale, which makes it possible to achieve accurate small-scale research. However, there are still residents without mobile phones, one person can have multiple SIMs, and there are market share and data barriers between different providers. Raw mobile phone data cannot show real-time population distribution accurately without a trim. Therefore, this article draws on some scholars' processing methods of mobile phone data [59,61] in order to estimate the real-time population distribution by integrating mobile phone data and Zhuhai City's seventh census data. The first step is to calculate the mobile phone data population weight of each grid, then calculate the weighted statistical population and add it to the mobile phone population of each grid. Therefore, the result obtained is closer to the actual population distribution. The calculation steps are as follows:

$$pw_{ij} = \frac{p_{ij}}{\sum_{j=1}^{m} p_{ij}} \quad (1)$$

$$PT_{ij} = pw_{ij} \times p_{total} + p_{ij} \quad (2)$$

where pw_{ij} is the population weight, p_{ij} is the number of cell phone locations in the grid number j at the i time point, m represents the total number of grids, PT_{ij} is the total population in the grid number j at the i time point, and p_{total} is the seventh census population of Zhuhai City.

2.3.2. Population Exposure Estimation

This article defines the population exposure of heat waves as the number of people in the high temperature grid at 35 °C and above per hour. In order to study the population exposure changes during the heat wave cycle in Zhuhai, this paper calculated the population exposure during the study period. The specific calculation formula is as follows:

$$PE_{ij} = \frac{PT_{ij} \times T_{\geq 35\,^{\circ}\mathrm{C}}}{t} \quad (3)$$

where PE_{ij} is the population exposure of the grid number j at the i time point, PT_{ij} represents the population number in the grid number j at the i time point, $T_{\geq 35\,^{\circ}\mathrm{C}}$ represents all grids where the temperature is 35 °C and above, and t is Time (h).

2.3.3. Analysis of Factors Affecting the Change of Population Exposure under High Temperature

The population exposure of the heat wave is a function of the two factors of population and high temperature. By drawing on the contribution rate of different influencing factors to the population exposure under the heat wave calculated by other scholars [12], the results suggests that the population exposure of high temperature changes are mainly affected by high temperature weather, population number, and population distribution changes, which can be decomposed into the impact of climate factors (temperature changes, population numbers remain unchanged), population factors (temperature unchanged, population changes), and the combined effects of population and climate factors (both temperature and population change).

The calculation method of the contribution rate of different influencing factors in the population exposure changes under the heat wave is as follows:

$$\text{Contribution rate of climate factors}: \frac{PT_{ij} \times |\Delta T_{\geq 35\ ^\circ\mathrm{C}}|}{PE_{ij}} \times 100\% \tag{4}$$

$$\text{Contribution rate of population factor}: \frac{T_{\geq 35\ ^\circ\mathrm{C}} \times |\Delta PT_{ij}|}{PE_{ij}} \times 100\% \tag{5}$$

$$\text{The common contribution rate of population and climate factors}: \frac{|\Delta t_{\geq 35\ ^\circ\mathrm{C}}| \times |\Delta PT_{ij}|}{PE_{ij}} \times 100\% \tag{6}$$

In the formula, $PT_{ij} \times |\Delta T_{\geq 35\ ^\circ\mathrm{C}}|$ is the influence of the population factor; $T_{\geq 35\ ^\circ\mathrm{C}} \times |\Delta PT_{ij}|$ is the influence of the climate factor; $|\Delta T_{\geq 35\ ^\circ\mathrm{C}}| \times |\Delta PT_{ij}|$ is the combined effect of both the population and climate factors. About delta, it refers to a fluctuation value. For example, $\Delta T_{\geq 35\ ^\circ\mathrm{C}}$ refers to the temperature change on different days at the same time. For example, $\Delta T_{\geq 35\ ^\circ\mathrm{C}}$ refers to the change in temperature difference between 6 September at 14:00 and 7 September at 14:00. ΔPT_{ij} refers to the amount of population change on different days at the same time. For example, ΔPT_{ij} refers to the population change in the jth grid at the ith time point between 7 September and 6 September.

3. Results

3.1. *Temporal and Spatial Distribution Characteristics of Heat Waves*

According to the temperature distribution map of Zhuhai City (Figure 3), it shows that on 6 September and 8 September, some meteorological stations had short-term high temperatures at 13:00, 14:00, and 15:00. However, there is no significant high temperature phenomenon in the study area within the period. Throughout the entire heat wave incident, starting at 12:00 on 9 September, it gradually strengthened on 11 September and reached a peak on 12 September. Table 1 shows the statistical results of the evolution of percentage of heat wave coverage over time from 12:00 to 18:00 during the study period, which lasted four days (9 September–12 September), starting at 12:00 on 9 September. The temperature dropped on 10 September, and the range of the heat wave expanded rapidly from 12:00 on 11 September, with a growth rate of 36.64 times that of 12:00 on 9 September. The duration was also delayed by 2 h from the previous 2 days. As of 12 September, the large-scale heat wave radiation effect appeared 2 h earlier than the previous 3 days. The heat wave coverage reached 71.57%, and the growth rate was 1.94 times of 9 September at 14:00.

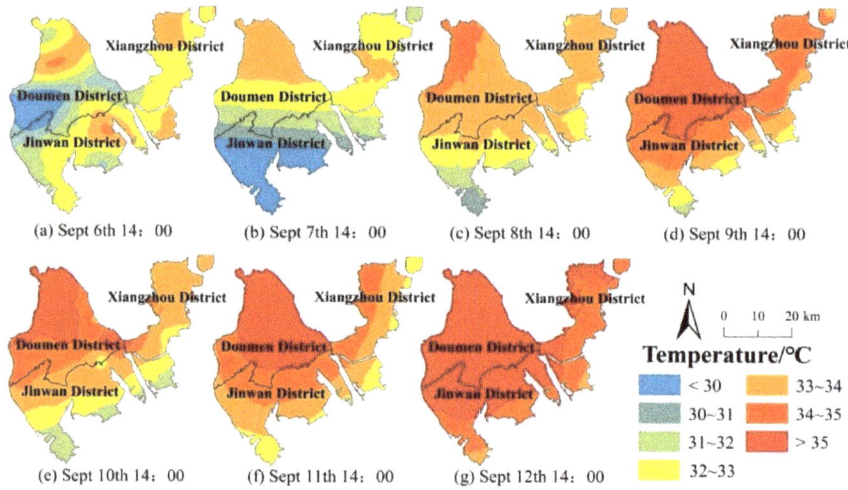

Figure 3. Spatial distribution of temperature at 14:00 during the study period.

Table 1. Percentage of area covered by heat wave (%).

	12:00	13:00	14:00	15:00	16:00	17:00	18:00
2021/9/6	0	0.55	0	1.51	0	0	0
2021/9/7	0	0	0	0	0	0	0
2021/9/8	0	0	0	0	0	0	0
2021/9/9	0.74	8.18	36.79	42.78	5.06	0	0
2021/9/10	0	6.43	18.79	16.54	7.95	0	0
2021/9/11	27.11	5.72	31.97	40.97	44.50	33.68	4.60
2021/9/12	16.74	56.13	71.57	69.23	62.01	41.57	7.38

According to the diurnal variation of temperature, the highest temperature in a day appears at 14:00. This article takes the spatial distribution of temperature at 14:00 during the heat wave cycle day as an example (Figure 3). It shows that the temperature trend in Zhuhai City is generally distributed in a north–south direction or a southeast–northwest direction. The high temperature area ($\geq$35 °C) in Zhuhai City has the characteristics of high in the north and low in the south, high in the northwest, and low in the southeast. The mountainous and coastal areas in the southeast of Jinwan District and east of Xiangzhou District are less affected by heat waves than other areas. Sustained high temperatures are mainly concentrated in most areas of Doumen District, the northern part of Jinwan District, and parts of western Xiangzhou District, where these areas have flatter terrain, higher population densities, and higher industrial densities. The dual effects of natural high temperature and man-made heat have caused the temperature in these areas to be higher than in other areas.

3.2. Temporal and Spatial Distribution Characteristics of Population Density

Based on the weighted calculation of hourly mobile phone data and statistical population data, the hourly population spatial distribution during the study period in Zhuhai City is obtained from 12:00 to 18:00. Figure 4 shows the population density distribution map at 14:00. It indicates that the high-value areas of population density are located in the central part of Xiangzhou District (the central city of Zhuhai City), and the low-value areas are scattered in the western area of Zhuhai City. The population spatial distribution of Zhuhai shows "less at both ends and more in the middle" in north–south direction; "more coastal areas in the east and less inland areas in the west" in east–west direction. The population density distribution of Doumen District and Jinwan District in the west is in a ring shape, while the Xiangzhou District in the east shows a clump distribution. Judging from the distribution trend, in the north–south direction, the population of Zhuhai City is concentrated at the central border of Doumen District and Jinwan District and the middle of Xiangzhou District. In the east–west direction, the population of Zhuhai City is mainly concentrated in the eastern coastal areas, and it is gradually decreasing to the western inland areas. During the study period, the population of Zhuhai City did not change much, and the areas with significant population increase were mainly scattered and dotted. From the daily hourly population density change rate during the study period (Figure 5), the hourly population density change rate in Zhuhai City has the largest fluctuation on 6 September (Monday), and the smallest fluctuation occurred on 12 September (Sunday). Obviously, there are significantly more commuters on weekdays than on weekends. In addition, we can also see that the population density change rate of non-heat wave cycle days (6 September–8 September) during the study period is significantly higher than that of the heat wave cycle days (9 September–12 September). During the 4 days of the continuous heat wave, the hourly population density change rate decreased day by day. The sudden increase on 11 September (Saturday) was related to the increase in crowd travel activities. The high heat wave reached its peak on 12 September. The weather caused a decrease in outdoor activities, which explains that the population density change rate is the lowest compared to previous days.

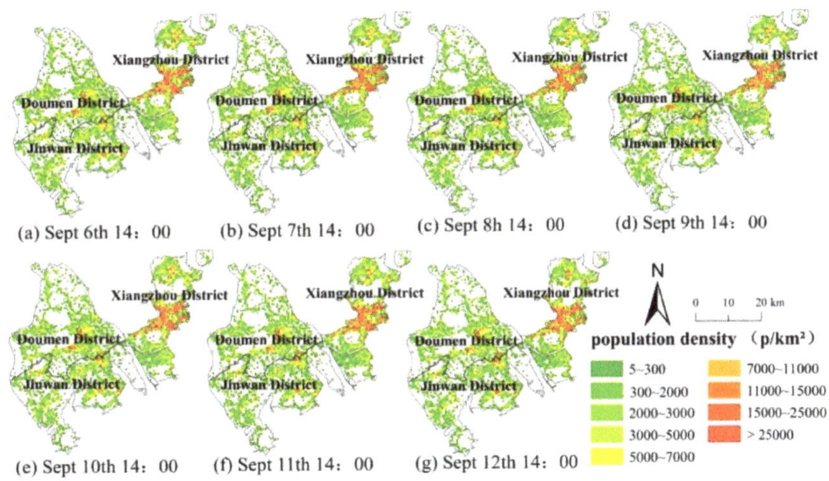

Figure 4. The spatial distribution of population density at 14:00 during the study period.

3.3. *Temporal and Spatial Characteristics of Population Exposure*

In order to clarify the temporal and spatial distribution of population exposure during the heat wave period, this article defines the area below 35 °C as the heat wave zero exposure area, and people in this environment are accordingly deemed not to be harmed; meanwhile, for part of the time period when there is no population distribution in the heat wave area, the population exposure is counted as 0, and finally, the heat wave area containing the population distribution is regarded as the population exposure area under high temperature, and the hourly population exposure percentage analysis is carried out from 12:00 to 18:00. The statistical results are shown in Table 2. By comparing Figure 6, it shows that the time-to-hour exposure of the dynamic population is consistent with the time-to-hour high temperature range. The strongest heat wave population exposure was on 12 September, and the population exposure percentage was as high as 55.99%, followed by 9 and 11 September. It indicates that during the heat wave cycle, the population exposure of the first two days (9 September and 10 September) was mainly concentrated from 13:00

to 15:00, it began to fade at 16:00, and it finally hit 0 at 17:00 and 18:00; starting from 11 September, the population exposure starts to increase, when the time has been delayed by 2 h compared to previous two days, and the number of people affected by high temperature weather has increased. Taking the heat wave population exposure at 14:00 during the heat wave period (9 September–12 September) as an example, the population exposure to high temperature in Zhuhai is 543,900, 149,200, 479,000, and 1.214 million, respectively. Due to temperature changes on 10 September, the population exposure has decreased compared to 9 September. Starting on 11 September, the population exposure has increased, increasing by 3.2 times and 8.13 times, respectively, compared to 10 September. After 10 September, the population exposure of Zhuhai City has increased significantly. It is necessary to pay attention to negative impacts of high temperature on human health and focus on strengthening the early warning and prevention of high temperature.

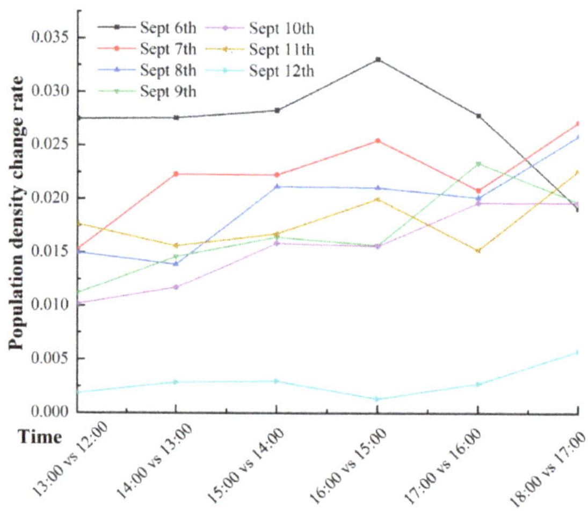

Figure 5. Hourly population density change rate during the study period.

Table 2. Hourly population exposure percentage during the study period (%).

	12:00	13:00	14:00	15:00	16:00	17:00	18:00
2021/9/6	0	0.1	0	0.19	0	0	0
2021/9/7	0	0	0	0	0	0	0
2021/9/8	0	0	0	0.17	0.22	0	0
2021/9/9	0.17	1.78	23.52	24.3	1.35	0	0
2021/9/10	0	3.06	6.39	4.12	1.36	0	0
2021/9/11	0	1.16	20.75	25.98	30.76	23.27	0.64
2021/9/12	10.61	55.99	52.64	42.4	37.63	22.43	1.49

Taking the spatial distribution of temperature and population density in Zhuhai at 14:00 during the study period as the background, combined with the spatial distribution of population exposure at the same time (Figure 6), it indicates that the persistent heat wave concentration area is always in Doumen District. The population in the north and west regions of the district is distributed in a circular pattern, with population exposure between 0~1000/person (km^2), which is at a relatively low level; Jingan Town and Baiteng Sub-districts in the east region with the most primary/secondary schools, colleges, office buildings, and residential areas have a population exposure rate higher than 5000 per person (km^2).

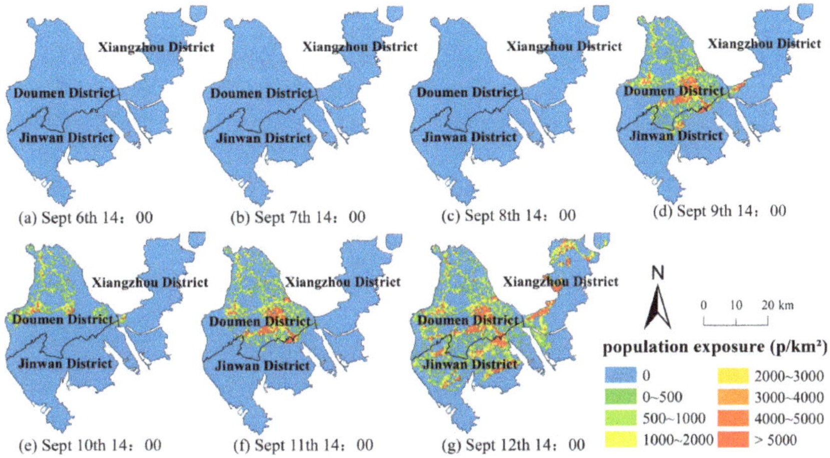

Figure 6. Spatial distribution of population exposure at 14:00 during the study period.

3.4. Analysis of Factors Affecting the Change of Population Exposure under Heat waves

Changes in population exposure under heat waves not only depend on climatic factors (temperature) but also on the scale of the population and its distribution. Looking into heat wave cycle (Table 3) on 10 September, compared to 9 September, the population exposure change on 10 September is mainly affected by population factors, while the population exposure changes in the last two days of the heat wave cycle are mainly affected by the combined influence of population factors and climate factors.

Table 3. Contribution rate of influencing factors of the change in population exposure throughout the day during the high temperature cycle in Zhuhai City.

Population Exposure Changes	Climate Factors (%)	Population Factors (%)	Climate and Population Factors (%)
Sept 10 vs. Sept 9	31.17	46.86	21.97
Sept 11 vs. Sept 10	10.69	27.50	61.82
Sept 12 vs. Sept 11	29.98	34.82	35.20

Considering that the population exposure of Zhuhai City under the heat wave incident mainly appeared at 13:00, 14:00, 15:00, and 16:00, only the factors affecting the changes in population exposure at these timestamps were analyzed (Figure 7). From the perspective of different periods of the heat wave cycle, the leading factors of population exposure changes are different. From 13:00 and 14:00, the changes in population exposure between 10 September and 9 September and 11 September and 10 September are most affected by the population factor. The change in population exposure between 12 September and 11 September is mainly affected by the combined effects of climatic factors and population factors. At other time points, the dominant factors affecting population exposure changes on each day of the heat wave cycle are different. Comparing the contribution rate of each factor in the heat wave cycle, it shows that the population factor's contribution rate of exposure changes gradually decreases, and its contribution rate gradually decreases from 66.16% and 77.69% to 3.13% and 17.19% between 10 September and 9 September and 11 September and 10 September, respectively. The contribution rate of climate factors to changes in population exposure has shown an increasing trend over time between 11 September and 10 September and between 12 September and 11 September. The comprehensive influence of climate factors and population factors has no obvious regularity in each period.

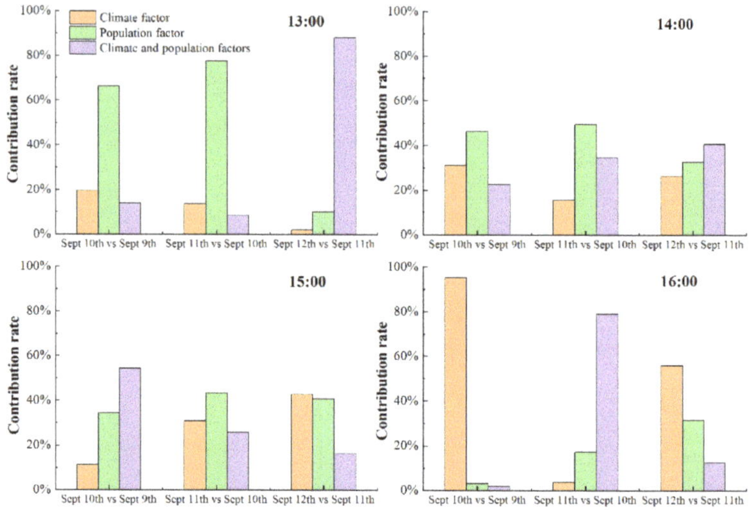

Figure 7. Contribution rate of influencing factors of population exposure changes at different time points in the heat wave cycle of Zhuhai City.

4. Summary and Discussion

This paper analyzes the population exposure situation in Zhuhai City under one heat wave incident from the perspective of geography. The real-time weather station data and mobile phone data present the temperature changes and population distribution in Zhuhai City more accurately, which helps to reveal the changes in population exposure under high temperature and to identify the hot spots of population exposure changes. Meanwhile, this article also analyzes the influencing factors of population exposure changes under high temperature, which has certain practical guiding significance for more effective high temperature warning and high temperature disaster risk prevention.

This paper finds that population exposure under heat waves is not only related to climate factors but also related to population factors. Areas with high population density and that are economically developed may not necessarily have high population exposure, and vice versa. In general, the population exposure of Zhuhai City has increased significantly during this high temperature cycle. Although it is not concentrated in the downtown area of Zhuhai, most of the other areas are affected by the high temperature. It is worth noting that the population exposure areas are mainly concentrated near schools, commercial buildings, industrial parks, and residential areas. The population in these areas is susceptible to high temperatures and has weak adaptability to high temperatures. In fact, indoor and outdoor population heat exposure is different [62–64], and the resident's age, gender, occupation, disease status, medical resources, economic level, and other factors will affect the degree of residents' response to high temperatures and heat waves. Especially, construction workers, couriers, and takeaway delivery workers who work under high temperature exposure are mobile sensitive groups that need urgent attention under heat waves [65,66]. Therefore, these areas should focus on strengthening the early warning and prevention of high temperature, and this article needs to make improvements in the above aspects.

However, the research in this article still has some shortcomings: First, in terms of data acquisition, it is difficult to obtain hourly meteorological data throughout the year. In addition, mobile phone data are confidential and are difficult to obtain. Therefore, this paper only analyzes the population exposure under one typical heat wave incident during the data collection period. As far as the heat wave incidents occurring in history, this is only an ordinary heat wave incident. In fact, a typical event in history is more convincing [67]. Secondly, in terms of methods, for the analysis of the influencing factors affecting population exposure under heat waves, this paper discusses the two aspects of climate factors and population factors at the macro-scale. In fact, urban structure, socioeconomic status, and urban ecological conditions at the micro-scale also have an impact on population exposure to heat waves, which is not explored in this paper [12,50,51]. Finally, the results of this paper only refer to the population exposure situation in Zhuhai City under one heat wave event, which can contribute to the disaster prevention and mitigation planning of Zhuhai City and are not universally representative. If more abundant data can be obtained in the future, we will improve the above shortcomings.

5. Conclusions

This paper selected Zhuhai City as the heat wave research area. Taking a heat wave incident in September 2021 as an example, based on Zhuhai city's hourly meteorological data, statistical population data, and hourly mobile phone data, the three types of data are spatialized to a 500 m × 500 m grid. Based on the high temperature distribution characteristics and the population weighting model, the qualitative and quantitative analysis of the population exposure situation by time under the influence of heat waves is carried out. At the same time, the influencing factors of population exposure changes were studied on this basis. The main conclusions are as follows:

(1) The results show a dynamicity in heat wave coverage which adjusts to temporal and spatial differences, and the coverage of the heat wave has an "increasing-decreasing-increasing" pattern over time. The heat wave was concentrated from 12:00 to 18:00, and from a spatial perspective, the area of heat wave coverage in west and north side of the city is bigger and lasts longer than in the east and south side regions.

(2) On the whole, the high-value areas of population density are located in central part of Xiangzhou District (the central city of Zhuhai City), and the low-value areas are scattered in the western area of Zhuhai City. The population spatial distribution of Zhuhai shows "less at both ends and more in the middle" in the north–south direction and "more coastal areas in the east and less inland areas in the west" in the east–west direction. The population density distribution of Doumen District and Jinwan District in the west is in a ring shape, while the Xiangzhou district in the east shows a clump distribution. Judging from the distribution trend, in the north–south direction, the population of Zhuhai City is concentrated at the central border of Doumen District and Jinwan District and the middle of Xiangzhou District. In the east–west direction, the population of Zhuhai City is mainly concentrated in the eastern coastal areas, and it is gradually decreasing to the western inland areas.

(3) The downtown area of Zhuhai (the central area of Xiangzhou District) is less affected by this heat wave incident, but most areas are affected by high temperature. The hourly population exposure is direct proportional to hourly heat wave coverage. In terms of the time dimension, the overall population exposure shows a trend of first decreasing and then increasing. In terms of spatial dimensions, high population exposure is concentrated in areas such as Jing'an Town and Baiteng Street in Doumen District. These areas are densely populated areas such as primary and secondary schools, colleges and universities, office buildings, and residential areas. Not only that, but high population exposure is also distributed in the area where industrial parks and commercial buildings are widely distributed in the east of Jinwan District. Low exposure is distributed in most of Xiangzhou District and the mountainous areas along the southern coast of Jinwan District due to topography.

(4) Looking into the heat wave cycle, compared to 9 September, the population exposure change on 10 September is mainly affected by population factors, but its contribution to population exposure changes gradually decreases over time. In other periods, the contribution rate of climate factors to changes in population exposure is gradually increasing, and the population exposure changes in the last two days of the heat wave cycle are mainly affected by the combined influence of population factors and climate factors. From the

perspective of different periods of the heat wave cycle, the leading factors of population exposure changes are different.

Author Contributions: Conceptualization, J.L., Y.S. and P.G.; data curation, J.L., Y.S. and Z.L.; formal analysis, J.L., P.G., X.Z. and X.P.; funding acquisition, Y.S.; investigation, P.G., Z.L., X.Z. and X.P.; methodology, J.L. and Y.S.; resources, P.G.; software, P.G. and Z.L.; supervision, Y.S.; validation, J.L.; visualization, X.Z.; writing—original draft, J.L.; writing—review and editing, J.L. and Y.S. All authors have read and agreed to the published version of the manuscript.

Funding: This research was funded by the National Natural Science Foundation of China: 41001022.

Institutional Review Board Statement: Not applicable.

Informed Consent Statement: Not applicable.

Data Availability Statement: Not applicable.

Acknowledgments: Special thanks to anonymous reviewers for their valuable comments. In addition, the authors gratefully acknowledge Jie Liu and Fanyang Qian for assisting with the data preparation and grammar revision.

Conflicts of Interest: The authors declare no conflict of interest. The funders had no role in the design of the study; in the collection, analyses, or interpretation of data; in the writing of the manuscript, or in the decision to publish the results.

References

1. IPCC. An IPCC Special Report on the Impacts of Global Warming of 1.5 °C above Pre-Industrial Levels and Related Global Greenhouse Gas Emission Pathways. Available online: https://www.ipcc.ch/sr15/ (accessed on 4 December 2021).
2. World Economic Forum. The Global Risks Report 2021. Available online: https://www.weforum.org/reports/the-global-risks-report-2021,2021-01-19 (accessed on 4 December 2021).
3. Meehl, G.; Tebaldi, C. More Intense, More Frequent, and Longer Lasting Heat Waves in the 21st Century. *Science* **2004**, *305*, 994–997. [CrossRef]
4. Bell, A.M.L. Heat Waves in the United States: Mortality Risk during Heat Waves and Effect Modification by Heat Wave Characteristics in 43 U.S. Communities. *Environ. Health Perspect.* **2011**, *119*, 210–218.
5. Im, E.S.; Pal, J.S.; Eltahir, E. Deadly heat waves projected in the densely populated agricultural regions of South Asia. *Sci. Adv.* **2017**, *3*, e1603322. [CrossRef]
6. Xia, Y.; Li, Y.; Guan, D.B.; Tinoco, D.M.; Xia, J.J.; Yan, Z.W.; Yang, J.; Liu, Q.Y.; Huo, H. Assessment of the economic impacts of heat waves: A case study of Nanjing, China. *J. Clean. Prod.* **2017**, *171*, 811–819. [CrossRef]
7. Xue, Q.; Xie, M.M.; Guo, Q.; Wang, Y.N.; Wu, R.R.; Liu, Q. Research progress on urban heat wave vulnerability assessment: A geographical perspective. *Prog. Geogr.* **2020**, *39*, 685–694. [CrossRef]
8. *QX/T 595-2021*; Meteorological Industry Standard of the People's Republic of China. Meteorological Press: Beijing, China, 2021.
9. Gao, J.H.; Sun, Y.Z.; Liu, Q.Y.; Zhou, M.G.; Lu, Y.G.; Li, L.P. Impact of extreme high temperature on mortality and regional level definition of heat wave: A multi-city study in China. *Sci. Total Environ.* **2015**, *505*, 535–544. [CrossRef] [PubMed]
10. Liu, G.L.; Zhang, L.C.; He, B.; Jin, X.; Zhang, Q.; Razafindrabe, B.; You, H.L. Temporal changes in extreme high temperature, heat waves and relevant disasters in Nanjing metropolitan region, China. *Nat. Hazards* **2015**, *76*, 1415–1430. [CrossRef]
11. Jedlovec, G.; Crane, D.; Quattrochi, D. Urban Heat Wave Hazard and Risk Assessment. *Results Phys.* **2017**, *7*, 4294–4295. [CrossRef]
12. Huang, D.P.; Zhang, L.; Gao, G.; Sun, S. Changes in population exposure to high temperature under a future scenario in China and its influencing factors. *Acta Geogr. Sin.* **2018**, *28*, 1371–1384.
13. Ding, T.; Qian, W.; Yan, Z. Changes in hot days and heat waves in China during 1961–2007. *Int. J. Climatol.* **2009**, *30*, 1452–1462. [CrossRef]
14. Habeeb, D.; Vargo, J.; Stone, B. Rising heat wave trends in large US cities. *Nat. Hazards* **2015**, *76*, 1651–1665. [CrossRef]
15. Schoetter, R.; Cattiaux, J.; Douville, H. Changes of western European heat wave characteristics projected by the CMIP5 ensemble. *Clim. Dyn.* **2015**, *45*, 1601–1616. [CrossRef]
16. Wang, W.W.; Zhou, W.; Li, X.Z.; Wang, X.; Wang, D.X. Synoptic-scale characteristics and atmospheric controls of summer heat waves in China. *Clim. Dyn.* **2016**, *46*, 2923–2941. [CrossRef]
17. Shin, J.; Olson, R.; An, S.I. Projected Heat Wave Characteristics over the Korean Peninsula During the Twenty-First Century. *Asia-Pac. J. Atmos. Sci.* **2018**, *54*, 53–61. [CrossRef]
18. Naveena, N.; Satyanarayana, G.C.; Rao, K.K.; Umakanth, N.; Srinivas, D. Heat wave characteristics over India during ENSO events. *J. Earth. Syst. Sci.* **2021**, *130*, 166. [CrossRef]
19. Naveena, N.; Satyanarayana, G.C.; Rao, K.K.; Umakanth, N.; Srinivas, D. Changes in daily climate extremes in China and their connection to the large scale atmospheric circulation during 1961–2003. *Clim. Dyn. Obs. Theor. Comput. Res. Clim. Syst.* **2011**, *36*, 2399–2417.

20. Trenberth, K.E.; Fasullo, J.T. Climate extremes and climate change: The Russian heat wave and other climate extremes of 2010. *J. Geophys. Res.-Atmos.* **2012**, *117*, 127–135. [CrossRef]
21. Min, S.K.; Son, S.W.; Seo, K.H.; Kug, J.S.; An, S.I.; Choi, Y.S.; Jeong, J.H.; Kim, B.M.; Kim, J.W.; Kim, Y.H.; et al. Changes in weather and climate extremes over Korea and possible causes: A review. *Asia-Pac. J. Atmos. Sci.* **2015**, *51*, 103–121. [CrossRef]
22. Luterbacher, J.; Dietrich, D.; Xoplaki, E.; Grosjean, M.; Wanner, H. European Seasonal and Annual Temperature Variability, Trends, and Extremes Since 1500. *Science* **2004**, *303*, 1499–1503. [CrossRef]
23. Shi, J.; Tang, X.; Cui, L. Climatic characteristics of high temperature in East China during 1961–2005. *J. Geogr. Sci.* **2008**, *18*, 283–294. [CrossRef]
24. Kim, O.Y.; Wang, B.; Shin, S.H. How do weather characteristics change in a warming climate? *Clim. Dyn.* **2013**, *41*, 3261–3281. [CrossRef]
25. Blumberg, G.; DPhil, S.M. Assessing the Potential Impact of Heat Waves in Cities: Implications for Hazard Preparation and Planning. *Procedia Econ. Financ.* **2014**, *18*, 727–735. [CrossRef]
26. Chen, K.; Bi, J.; Chen, J.; Chen, X.D.; Huang, L.; Zhou, L. Influence of heat wave definitions to the added effect of heat waves on daily mortality in Nanjing, China. *Sci. Total Environ.* **2015**, *506*, 18–25. [CrossRef] [PubMed]
27. Zuo, J.; Pullen, S.; Palmer, J.; Bennetts, H.; Chileshe, N.; Ma, T. Impacts of heat waves and corresponding measures: A review. *J. Clean. Prod.* **2015**, *92*, 1–12. [CrossRef]
28. Zaidi, R.Z.; Pelling, M. Institutionally configured risk: Assessing urban resilience and disaster risk reduction to heat wave risk in London. *Urban Stud.* **2015**, *52*, 1218–1233. [CrossRef]
29. Huang, X.; Li, Y.; Guo, Y.; Zheng, D.; Qi, M. Assessing Urban Risk to Extreme Heat in China. *Sustainability* **2020**, *12*, 2750. [CrossRef]
30. Lee, S.H.; Kang, J.E.; Park, C.S.; Yoon, D.K.; Yoon, S. Multi-risk assessment of heat waves under intensifying climate change using Bayesian Networks. *Int. J. Disast. Risk Reduct.* **2020**, *50*, 101704. [CrossRef]
31. Eid, C.E.; O'Neill, M.S.; Gronlund, C.J.; Brines, S.J.; Brown, D.G.; Diez-Roux, A.V.; Schwartz, J. Mapping Community Determinants of Heat Vulnerability. *Environ. Health Perspect.* **2009**, *117*, 1730–1736.
32. Wolf, J.; Adger, W.N.; Lorenzoni, I.; Abrahamson, V.; Raine, R. Social capital, individual responses to heat waves and climate change adaptation: An empirical study of two UK cities. *Glob. Environ. Chang.* **2010**, *20*, 44–52. [CrossRef]
33. Depietri, Y.; Welle, T.; Renaud, F.G. Social vulnerability assessment of the Cologne urban area (Germany) to heat waves: Links to ecosystem services. *Int. J. Disast. Risk Reduct.* **2013**, *6*, 98–117. [CrossRef]
34. Benmarhnia, T.; Deguen, S.; Kaufman, J.S.; Smargiassi, A. Vulnerability to Heat-related Mortality A Systematic Review, Meta-analysis, and Meta-regression Analysis. *Epidemiology* **2015**, *26*, 781–793. [CrossRef] [PubMed]
35. Xie, P.; Wang, Y.L.; Liu, Y.X.; Peng, J. Incorporating social vulnerability to assess population health risk due to heat stress in China. *Acta Geogr. Sinica* **2015**, *70*, 1041–1051.
36. Azhar, G.; Saha, S.; Ganguly, P.; Mavalankar, D.; Madrigano, J. Heat Wave Vulnerability Mapping for India. *Int. J. Environ. Res. Public Health* **2017**, *14*, 357. [CrossRef] [PubMed]
37. Shui, W.; Chen, Z.C.; Deng, J.M.; Li, Y.J.; Wang, Q.F.; Wang, W.L.; Chen, Y.P. Evaluation of urban high temperature vulnerability of coupling adaptability in Fuzhou, China. *Acta Geogr. Sin.* **2017**, *72*, 830–849.
38. Reischl, C.; Rauter, R.; Posch, A. Urban vulnerability and adaptation to heatwaves: A case study of Graz (Austria). *Clim. Policy* **2018**, *18*, 63–75. [CrossRef]
39. Semenza, J.C.; Rubin, C.H.; Falter, K.H.; Selanikio, J.D.; Flanders, W.D.; Howe, H.L.; Wilhelm, J.L. Heat-related deaths during the July 1995 heat wave in Chicago. *N. Engl. J. Med.* **1996**, *335*, 84–90. [CrossRef]
40. O'Neill, M.S.; Zanobetti, A.; Schwartz, J. Modifiers of the Temperature and Mortality Association in Seven US Cities. *Am. J. Epidemiol.* **2003**, *157*, 1074–1082. [CrossRef] [PubMed]
41. Hajat, S.; Armstrong, B.; Baccini, M.; Biggeri, A.; Bisanti, L.; Russo, A.; Paldy, A.; Menne, B.; Kosatsky, T. Impact of High Temperatures on Mortality: Is There an Added Heat Wave Effect? *Epidemiology* **2006**, *17*, 632–638. [CrossRef]
42. Kim, D.W.; Deo, R.C.; Chung, J.H.; Lee, J.S. Projection of heat wave mortality related to climate change in Korea. *Nat. Hazards* **2016**, *80*, 623–637. [CrossRef]
43. Lee, W.K.; Lee, H.A.; Park, H. Modifying Effect of Heat Waves on the Relationship between Temperature and Mortality. *J. Korean Med. Sci.* **2016**, *31*, 702–708. [CrossRef]
44. Van, S.Y.; Ntarladima, A.M.; Grobbee, R.; Karssenberg, D.; Vaartjes, I. Sex differences in mortality after heat waves: Are elderly women at higher risk? *Int. Arch. Occup. Environ. Health* **2019**, *92*, 37–48.
45. Zhang, M.S.; Wang, H.; Jin, W.; Van Dijk, M.P. Assessing heat wave vulnerability in Beijing and its districts, using a three dimensional model. *Int. J. Glob. Warm.* **2019**, *17*, 297–314. [CrossRef]
46. Canoui, P.F.; Cadot, E.; Spira, A. Excess deaths during the August 2003 heat wave in Paris, France. *Rev. Dépidémiologie Et De St. Publique* **2006**, *54*, 127–135. [CrossRef]
47. Fouillet, A.; Rey, G.; Laurent, F.; Pavillon, G.; Bellec, S.; Guihenneuc, J.C.; Clavel, J.; Jougla, E.; Hemon, D. Excess mortality related to the August 2003 heat wave in France. *Int. Arch. Occup. Environ. Health* **2006**, *80*, 16–24. [CrossRef] [PubMed]
48. Tran, K.V.; Azhar, G.S.; Nair, R.; Knowlton, K.; Jaiswal, A.; Sheffield, P.; Mavalankar, D.; Hess, J. A Cross-Sectional, Randomized Cluster Sample Survey of Household Vulnerability to Extreme Heat among Slum Dwellers in Ahmedabad, India. *Int. J. Environ. Res. Public Health* **2013**, *10*, 2515–2543. [CrossRef]

49. China's national assessment report on extreme weather and climate events and disaster risk management and adaptation is released. *China Emerg. Manag.* **2015**, *3*, 63.
50. Kazak, J. The Use of a Decision Support System for Sustainable Urbanization and Thermal Comfort in Adaptation to Climate Change Actions-The Case of the Wrocław Larger Urban Zone (Poland). *Sustainability* **2018**, *10*, 1083. [CrossRef]
51. Peng, J.; Dan, Y.Z.; Qiao, R.L.; Liu, Y.X.; Dong, J.Q.; Wu, J.S. How to quantify the cooling effect of urban parks? linking maximum and accumulation perspectives. *Remote Sens. Environ.* **2021**, *252*, 112135. [CrossRef]
52. Zhang, L.Y.; Zhang, Z.; Ye, T.; Zhou, M.G.; Wang, C.Z.; Yin, P.; Hou, B. Mortality effects of heat waves vary by age and area: A multi-area study in China. *Environ. Health* **2018**, *17*, 54. [CrossRef]
53. Mcgeehin, M.A.; Mirabelli, M. The potential impacts of climate variability and change on temperature-related morbidity and mortality in the United States. *Environ. Health Perspect.* **2001**, *109*, 185–189.
54. Jones, B.; O'Neill, B.C.; McDaniel, L.; McGinnis, S.; Mearns, L.O.; Tebaldi, C. Future population exposure to US heat extremes. *Nat. Clim. Chang.* **2015**, *5*, 652–655. [CrossRef]
55. Zhang, L.; Huang, D.P.; Yang, B.Y. Future population exposure to high temperature in China under RCP4.5 scenario. *Geogr. Res.* **2016**, *35*, 2238–2248.
56. Hu, L.; Wilhelmi, O.V.; Uejio, C. Heat exposure in cities: Combining the dynamics of temperature and population. *Sci. Total Environ.* **2019**, *655*, 1–12. [CrossRef]
57. Li, L.; Zha, Y. Population exposure to extreme heat in China: Frequency, intensity, duration and temporal trends. *Sustain. Cities Soc.* **2020**, *60*, 102282. [CrossRef]
58. Jones, B.; Tebaldi, C.; O'Neill, B.C.; Oleson, K.; Gao, J. Avoiding population exposure to heat-related extremes: Demographic change vs climate change. *Clim. Chang.* **2018**, *146*, 423–437. [CrossRef]
59. Xie, C.; Huang, B.; Liu, X.Q.; Zhou, T.; Wang, Y. Population exposure to heatwaves in Shenzhen based on mobile phone location data. *Prog. Geogr. Sci.* **2020**, *39*, 231–242. [CrossRef]
60. The Lowest Temperature in Zhuhai's History. Available online: https://www.lishixinzhi.com/miwen/1217303.html (accessed on 5 January 2022).
61. Chen, B.; Song, Y.M.; Jiang, T.T.; Chen, Z.Y.; Huang, B.; Xu, B. Real-Time Estimation of Population Exposure to PM2.5 Using Mobile- and Station-Based Big Data. *Int. J. Environ. Res. Public Health* **2018**, *15*, 573. [CrossRef]
62. Loughnan, M.; Carroll, M.; Tapper, N.J. The relationship between housing and heat wave resilience in older people. *Int. J. Biometeorol.* **2015**, *59*, 1291–1298. [CrossRef]
63. Zuurbier, M.; van Loenhout, J.A.F.; le Grand, A.; Greven, F.; Duijm, F.; Hoek, G. Street temperature and building characteristics as determinants of indoor heat exposure. *Sci. Total Environ.* **2020**, *766*, 144376. [CrossRef]
64. White-Newsome, J.L.; Sanchez, B.N.; Jolliet, O.; Zhang, Z.Z.; Parker, E.A.; Dvonch, J.T.; O'Neill, M.S. Climate change and health: Indoor heat exposure in vulnerable populations. *Environ. Res.* **2012**, *112*, 20–27. [CrossRef]
65. Al-Bouwarthan, M.; Quinn, M.M.; Kriebel, D.; Wegman, D.H. Assessment of Heat Stress Exposure among Construction Workers in the Hot Desert Climate of Saudi Arabia. *Ann. Work. Expo. Health* **2019**, *63*, 505–520. [CrossRef] [PubMed]
66. Calkins, M.M.; Bonauto, D.; Hajat, A.; Lieblich, M.; Seixas, N.; Sheppard, L.; Spector, J.T. A Case-Crossover Study of Heat Exposure and Injury Risk among Outdoor Construction Workers in Washington State. *Scand. J. Work. Environ. Health.* **2019**, *45*, 588–599. [CrossRef] [PubMed]
67. Robinea, J.M.; Cheung, S.L.K.; Roy, S.L.; Van Oyen, H.; Griffiths, C.; Michel, J.P.; Herrmannd, F.R. Death toll exceeded 70,000 in Europe during the summer of 2003. *Comptes Rendus Biol.* **2008**, *331*, 171–178. [CrossRef] [PubMed]

sustainability

MDPI

Article

Impact of Extreme Heatwaves on Population Exposure in China Due to Additional Warming

Leibin Wang [1], Robert V. Rohli [2,3], Qigen Lin [4], Shaofei Jin [5,*] and Xiaodong Yan [6,*]

1 Postdoctoral Research Station of Geography, Hebei Technology Innovation Center for Remote Sensing Identification of Environmental Change, School of Geographic Sciences, Hebei Normal University, Shijiazhuang 050024, China
2 Department of Oceanography & Coastal Sciences, College of the Coast & Environment, Louisiana State University, Baton Rouge, LA 70803, USA
3 Coastal Studies Institute, Louisiana State University, Baton Rouge, LA 70803, USA
4 Institute for Disaster Risk Management, School of Geographical Sciences, Nanjing University of Information Science & Technology, Nanjing 210044, China
5 Department of Geography, MinJiang University, Fuzhou 350108, China
6 State Key Laboratory of Earth Surface Processes and Resource Ecology, Faculty of Geographical Science, Beijing Normal University, Beijing 100875, China
* Correspondence: jinsf@tea.ac.cn (S.J.); yxd@bnu.edu.cn (X.Y.)

Abstract: Extreme heatwaves are among the most important climate-related disasters affecting public health. Assessing heatwave-related population exposures under different warming scenarios is critical for climate change adaptation. Here, the Coupled Model Intercomparison Project phase 6 (CMIP6) multi-model ensemble output results are applied over several warming periods in the Intergovernmental Panel on Climate Change AR6 report, to estimate China's future heatwave population exposure under 1.5 °C and 2.0 °C warming scenarios. Our results show a significant increase in projected future annual heatwave days (HD) under both scenarios. With an additional temperature increase of 0.5 °C to 2.0 °C of warming, by mid-century an additional 20.15 percent increase in annual HD would occur, over 1.5 °C warming. If the climate warmed from 1.5 °C to 2.0 °C by mid-century, population exposure would increase by an additional 40.6 percent. Among the three influencing elements that cause the changes in population exposure related to heatwaves in China–climate, population, and interaction (e.g., as urbanization affects population redistribution)–climate plays the dominant role in different warming scenarios (relative contribution exceeds 70 percent). Therefore, considering the future heat risks, humanity benefits from a 0.5 °C reduction in warming, particularly in eastern China. This conclusion may provide helpful insights for developing mitigation strategies for climate change.

Keywords: heatwaves; population exposure change; global warming; 1.5 °C warming scenario; 2.0 °C warming scenario

Citation: Wang, L.; Rohli, R.V.; Lin, Q.; Jin, S.; Yan, X. Impact of Extreme Heatwaves on Population Exposure in China Due to Additional Warming. *Sustainability* **2022**, *14*, 11458. https://doi.org/10.3390/su141811458

Academic Editor: Samuel Asumadu-Sarkodie

Received: 28 June 2022
Accepted: 6 September 2022
Published: 13 September 2022

1. Introduction

Meteorological measurements and general circulation model (GCM) simulations yield strong evidence for warming globally over the last several decades. Continued warming will likely cause increases in the frequency of extreme climatic and weather occurrences, challenging human systems. For instance, occurrences such as the 2013 heatwave over China, the 2011–2014 persistent drought over California, and the devastating 2013 flood in India resulted in significant large economic losses and left many people wounded or homeless [1]. Climate extremes include a broad range of physical effects that have been recorded extensively across the globe, including a significant danger to natural resources, the environment, and the social economy [2].

As is revealed by the newest climate change scenario design—called 'shared socio-economic pathway' (SSP)—identifying vulnerabilities in climate change risks is crucial for the public and governments [3]. In general, vulnerability is characterized as a result of a society's exposure and sensitivity to hazards, as well as its ability to adapt. Therefore, assessing future climate change risks, particularly the vulnerability of population populations to a certain disasters, is critical in developing social risk management measures [4].

Global-warming-affected extreme events that alter risks include many aspects, such as heatwaves, droughts, hurricanes, floods, mudslides, and many other hazards [5–11]. Heatwaves represent one of the most serious meteorological disasters impacting public health, and their frequency and severity are on the rise, with a tendency to worsen [12–14]. For example, the 1995 Chicago heatwave claimed the lives of 739 people [15]. At least 70,000 people died in the 2003 heatwave in central Europe, in addition to the catastrophic socioeconomic effects [16,17]. In 2010, a catastrophic heatwave in eastern Europe and western Russia killed over 55,000 people in Russia alone [18]. This was followed by a severe drought, which reduced food production by 25 percent and cost the local economy almost USD 15 billion [19]. In 2019, two heatwaves in western Europe caused deaths that exceeded normal rates by 50 percent, with temperatures in the Netherlands, Belgium, France, Germany, and the United Kingdom setting new highs in meteorological history [20]. In China, heatstroke (HS) during the summer has also become a serious public health issue in recent years [1]. The 2013 record-breaking high-temperature event in Shanghai caused about 160 deaths in the Pudong New District [21].

China is the most populated country in the world, and the overwhelming majority of the population is concentrated in the heatwave-prone eastern monsoon region [22], where air temperatures will continue to rise under current emissions scenarios. Population exposure to extreme drought is projected to increase by 17% if the global temperature increase is limited to 2.0 °C rather than 1.5 °C in the future [23], increasing the threat to the living conditions of most people living in China [24,25]. Like other disasters, heatwave-related disasters are impactful due to two major factors–the exposure to the phenomenon and the consequence of its occurrence. On the one hand, climate change scenarios including the magnitude of warming affect the hazard exposure [26,27]; while on the other hand, consequence is influenced by population and the exposure and socioeconomic vulnerability of that population [28,29]. Thus, heatwave risk can be described as a function of climate and population [28], both of which change over time and geographic area [4]. According to Jones et al., climate change is a stronger determinant of exposure than demographic change [30]. The determinants of population exposure, in general, are mostly determined by geographical disparities, taking into consideration regional population policies. The overall change in exposure could also result predominantly from the interactive effect of both factors, such as in the case of climate migrations across areas such as Africa [31,32].

Previous studies on heatwaves in China have mostly concentrated on the traits of intense heatwave episodes and their fluctuations [33–40]. Recent work has also provided a clearer understanding of the dangers of heatwaves in future temperature increase scenarios. Owing to recognition of the importance of public health management techniques, population exposure to intense heatwaves has received more attention recently [41]. Evidence suggests that climate change may be a major driver of changes in population exposure in China [42,43], and some studies show that the interaction between climate and population determines the population exposure to future extreme heatwaves in China [44]. However, there is currently little understanding of the relative contributions of heatwaves and demographic changes on population exposure changes on a regional scale. It is important to fill this gap as a critical next step for risk assessment of extreme heatwaves.

This study explores changes in heatwave and population exposure under two temperature increasing scenarios of 1.5 °C and 2.0 °C in China based on a daily maximum and minimum surface air temperature (T_{max} and T_{min}) dataset from CMIP6 and gridded daily datasets over China, along with population data that consider local population policies, from Oak Ridge National Laboratory's LandScan program and the Shared Socioeconomic Pathways (SSPs) from Tsinghua University. In addition, differences in population exposure under the different warming scenarios are analyzed, with explanations of possible reasons for the differences. It is important to separately examine the differences in impact of the one-half degree of difference in increases from 1.5 °C to 2.0 °C, because the Intergovernmental Panel on Climate Change (IPCC) [45] recognizes these two thresholds, with the former being deemed likely to occur—with high confidence—between 2030 and 2052 if current rates are extrapolated. The 2 °C threshold is selected because IPCC (2018) notes, with medium confidence, that it would result in an additional 420 million people being frequently exposed to extreme heatwaves beyond those exposed to 1.5 °C of warming.

2. Materials and Methods

2.1. Dataset

For comparison and bias correction, daily T_{max} and T_{min} across China from a high-resolution (0.25° × 0.25°) gridded daily dataset, namely CN05.1, generated from more than 2400 stations affiliated with China Meteorological Administration [46] are taken into considered. Daily T_{max} and T_{min} output from 20 CMIP6 model simulations are employed to project the spatial distribution of future heatwaves under 1.5 °C and 2.0 °C warming scenarios. Information about these CMIP6 models is provided in Table S1. The historical scenario used is the "all-forcing" simulations (1995–2014). Projections of GCMs under three scenarios based on the combination of the shared socioeconomic pathways (SSPs) [47] and the representative concentration pathways (RCPs) [48]—including SSP1–2.6, SSP2–4.5, and SSP5–8.5—are included. The time frame of 1.5 °C and 2.0 °C warming globally is shown in Table 1, which is consistent with the IPCC AR6 [49]. Output datasets from the CMIP6 simulations are downscaled into 0.25° × 0.25° grid cells using the bilinear interpolation method via xarray (version 2022.3.0) [50].

Table 1. Earliest 20-year averaging period that displays 1.5 °C and 2.0 °C of global warming, by SSP threshold. The change is displayed in °C relative to the 1850–1900 reference period for the selected time periods.

SSPs / Warming Scenarios	SSP1-2.6	SSP2-4.5	SSP5-8.5
1.5 °C	2023–2042	2021–2040	2018–2037
2.0 °C	post–2100	2043–2062	2032–2051

The historical population data come from LandScan, which contains the global population for the period 2000–2020 and is widely used in natural and human-caused disaster risk assessments [51]. The projected future population data come from the gridded dataset of Tsinghua University from 2010 to 2100 [52]. Both historical and future population data have a resolution of 30 arc-seconds (~1 km) and are up-scaled to 0.25° × 0.25° using the grid area scaling method. The population SSPs scenarios included in this dataset are consistent with CMIP6, including data from three scenarios: SSP1–2.6, SSP2–4.5, and SSP5–8.5.

2.2. Definition of Present, 1.5 °C, and 2.0 °C Warming

In this study, the historical baseline period of 1995–2014 is used, based on the recommendation of IPCC AR6. The historical climate or historical scenario in this work refers to the climate of present. The selected time periods of 1.5 °C and 2.0 °C of warming relative to 1850–1900 for 20-year averaged global surface air temperature (GSAT) changes are based on multiple lines of evidence (the time span of each SSP as shown in Table 1). Therefore, the meteorological and population data for 1.5 °C and 2.0 °C warming are averaged using multi-model ensembles in the corresponding warming scenarios, respectively. Historical climate and population data (1995–2014) are compared with the projected data for 1.5 °C and 2.0 °C of warming. Because the population of LandScan is as early as 2000, the historical population is represented by 16 years of data (2000–2014).

2.3. Quantile Mapping Bias Correction

Before using GCM data, it is critical to correct for biases in the GCM simulations against observed data. After downscaling the climate data into 0.25° × 0.25° grid cells, a quartile mapping technique is used to bias correct the historical simulation and future projection of the GCM models, using cumulative density function matching methods [53]. The key to quantile deviation correction is to establish a transfer function, which generally includes a theoretical probability distribution function [54–56] and empirical probability function [57], using theoretical cumulative distribution function and cumulative distribution function of observed data, respectively. In comparison, the empirical probability function has wider utility because it does not require any assumptions about the statistical distribution of the original data [58]. For each grid cell, given month, and variable (T_{max} and T_{min}), the transform function is fitted by the qq-plot of GCM historical simulations and observations [58]. Then, bias correction using transfer functions for historical simulation and warming scenarios results. In this work, the quantile mapping bias correction of the GCM output is implemented with xclim [59].

2.4. Heatwave Events and Population Exposure

Early research on heatwaves only considered whether the highest temperature or lowest temperature on a single day exceeded a certain threshold. However, exceedance thresholds that correspond optimally to population exposure are likely to be affected by antecedent and subsequent temperature [60,61]. More recently, compound hot extremes have considered both daily T_{max} and T_{min} based on a bivariate definition framework [62–64]. Recently, studies on heatwaves in the Northern Hemisphere have used relative temperature thresholds and diurnal temperature range (DTR) to identify heatwaves. This method has been shown to offer improvements by defining heatwave events to be defined more strictly and requiring both T_{max} and T_{min} to exceed the 95th quantile [65].

In this study, a heatwave is considered to occur if all of the following criteria are met: (1) daily $T_{max} \geq$ 95th percentile of observed daily T_{max} in May, June, July, August, and September during 1985–2014); (2) daily $T_{max} \geq 35$ °C [66]; (3) daily DTR in those same months ≤ 10 °C, to ensure that uncomfortable conditions persisted throughout the day and in recognition that both sensible and latent heating are key factors in the warming and stress effects [65,67]. The number of days spent experiencing a heatwave in a year is referred to as heatwave days (*HD*). When a heatwave event straddles two calendar years, it is counted in the year in which the heatwave began. Expressing "risk" as the product of hazard intensity and hazard exposure [68] population exposure (*E*) is defined as the product of *HD* and the total population (*Pop*) [69] at the grid point, and the unit is person-days, shown as

$$E = HD \times Pop \tag{1}$$

2.5. Amplified Impacts and Relative Contributions

In this study, heatwave effect is defined as E of compound heatwave events. Li et al. [70] and Jones et al. [28] showed that the absolute amplified impact (AI_a) due to an 0.5 °C additional warming is defined as

$$AI_a = E_{2.0} - E_{1.5} \tag{2}$$

where $E_{1.5}$ and $E_{2.0}$ represent the population exposure under the 1.5 and 2.0 °C warming scenarios, respectively. Relative amplified impact (AI_r) is defined as the ratio of the additional exposure to the baseline exposure, thereby indicating the proportion of the population that will be additionally exposed to heatwaves by increasing the delta temperature relative to the baseline temperature rise scenario. The equation is

$$AI_r = \frac{E_{2.0} - E_{1.5}}{E_{1.5}} \times 100\% \tag{3}$$

According to the definition of population exposure, the increases in population exposure relative to historical simulations can be calculated as

$$\begin{aligned}\Delta E &= E_f - E_h \\ &= HD_f \times Pop_f - HD_h \times Pop_h \\ &= \Delta HD \times Pop_h + \Delta Pop \times HD_h + \Delta HD \times \Delta Pop\end{aligned} \tag{4}$$

where ΔE is relative to the historical baseline period, E_h and E_f depict population exposure under historical simulation and future global warming scenarios, respectively; HD_f and HD_h represent HD under future global warming scenarios and historical simulation, respectively; Pop_f and Pop_h correspond to the population numbers under the future global warming scenarios and historical simulation, respectively; and ΔHD and ΔPop represent the changes of HD and population in the future global warming scenarios and historical simulations, respectively. Furthermore, $\Delta HD \times Pop_h$ represents the change in population exposure due to climate change, $\Delta Pop \times HD_h$ represents the change in population exposure due to population change, and $\Delta HD \times \Delta Pop$ represents the change in population exposure due to interaction of climate and population change. The consistency of sigsn between these three elements and ΔE represents the contribution to the change in exposure, respectively. Consistency in sign signifies a positive effect on the change of population exposure; otherwise, it is a negative effect.

3. Results

3.1. Changes in Heatwave Events

To evaluate the ability of GCM to simulate heatwaves, the spatial distribution of HD in China is analyzed based on CN05.1 and multi-model ensemble of GCM outputs. Figure 1a,b shows the annual number of HD from 1995 to 2014 based on observational data and GCM simulations. The present number of HD is generally 3–10 days per year, and the HD simulated by the GCM is between 3 and 11 days per year. Heatwaves mainly occur in the central and eastern regions of China and the Sichuan Basin, including: the Yangtze River Basin (YRB) region, the Chongqing-Chengdu (CC) region, and the Pearl River Basin (PRB) region (blue box in Figure 1). The consistency between the historical simulation of the model and the CN05.1 observation data is high, which indicates that a multi-model ensemble of GCM can better simulate the spatial distribution of the heatwave hotspots. The simulated error of annual HD is less than 0.2 days, the regional average (areas with heatwaves) annual HD of the model simulation and CN05.1 are 4.45 and 4.63 days, respectively. As shown in Figure 1c, in places with more HD, the simulation error of HD is less, and vice versa. In general, the multi-model ensemble can predict the number of HD in China accurately, with the exception of slight overestimation in a few places.

Not surprisingly, compared with the present simulation, the HD under the 1.5 °C and 2.0 °C global warming scenarios show greater HD frequency, and the regional average annual HD increases from 4.45 days to 5.36 and 6.44 days, respectively, as shown in Figure 1d,f. The 0.5 °C of additional warming in the 2.0 °C warmer future would lead to an increase in HD by approximately 20.15 percent when compared to that under the 1.5 °C warmer climate. The annual HD at the core of the heatwave region increased by 1 and 1.8 times. Heatwave-free areas under the 1.5 °C scenario—such as Northeast China, Xinjiang, and Inner Mongolia—are expected to experience heatwaves in the 2.0 °C scenario, but the annual HD generally do not exceed 5 days in the 2.0 °C scenario. Figure 1c,g show the spatial variation of HD in the historical simulation for the 1.5 °C and 2.0 °C global warming scenarios, and more than 95 percent of the regions where heatwaves have occurred show the characteristics of increased HD, with annual HD frequency increases of 2.68 and 4.22 days, respectively. Except for newly added heatwave areas, the spatial characteristics of HD are consistent with historical periods. The HD in the Yangtze and Pearl River Basin regions showed a more obvious increasing trend under the 2.0 °C warming scenario than in the 1.5 °C warming scenario. These areas will have high population concentration in the future (as shown in Figure S1), indicating that the population may face more severe heatwave exposure risks.

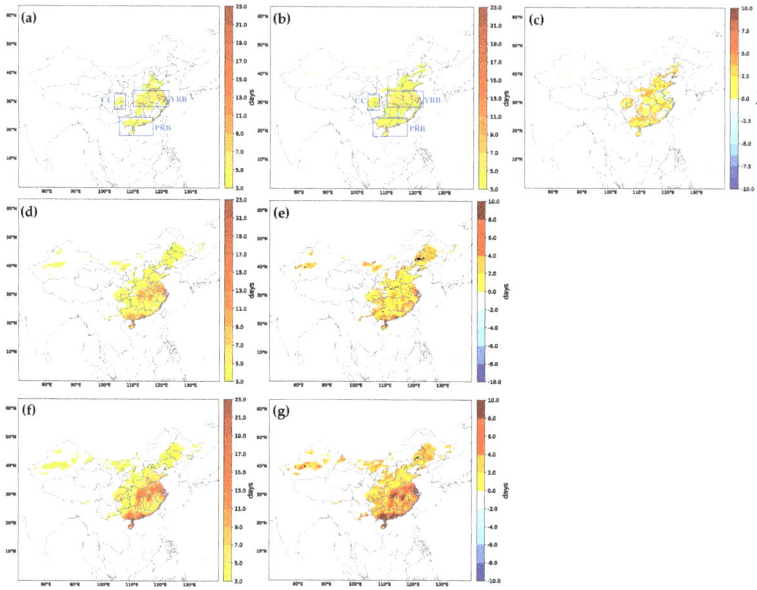

Figure 1. Spatial distribution of heatwave days (HD) per year. (**a**,**b**,**d**,**f**) HD at observed, historical simulation, 1.5 °C, and 2.0 °C warming scenarios, respectively. (**c**) Difference of HD between historical simulations and observed. (**e**,**g**) Difference in HD relative to the historical heatwave for the 1.5 °C and 2.0 °C warming scenarios, respectively; black dots indicate an insignificant difference according to the one-sided *t*-test at the 0.05 significance level.

3.2. Present Distribution Patterns and Future Changes in Population Exposure

Population exposure is an important parameter for revealing the risk of future heatwaves. Figure 2 shows the spatial distribution of E under the present, 1.5 °C, and 2.0 °C warming scenarios based on GCM models. During the present period, increased E to heatwaves occurred mainly in central and eastern China and the Sichuan Basin. Under the 1.5 °C and 2.0 °C warming scenarios, the annual E increased from 4.09 billion person-days in the present period to 7.98 billion and 11.22 billion person-days, respectively. The per capita annual HD frequency increased from 4.72 days to 7.0 and 9.89 days. Despite the expansion of E areas, the spatial distribution of E centers is relatively concentrated. To a certain extent, the spatial variation of E is determined by both climate change and population redistribution.

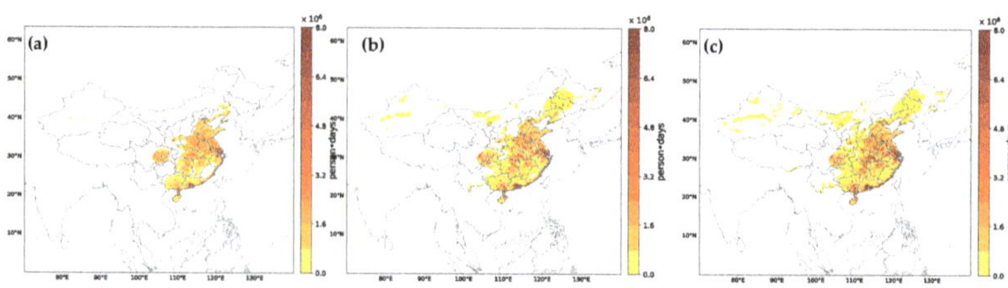

Figure 2. Spatial distribution of population exposure. From left to right: present (**a**), 1.5 °C warming (**b**), and 2.0 °C warming (**c**).

To focus on the changes of population exposure, as shown in Figure 3, the spatial variation characteristics of E are explored. The newly added heatwave regions—such as Northeast China, Xinjiang, and Inner Mongolia—show increased E, which is mainly caused by the increase in the frequency of heatwave events under climate warming. In other regions, E I ncreased or decreased differently, depending on the redistribution of population and heatwave occurrence. Compared with the 1.5 °C warming scenario, E under the 2.0 °C warming showed a steady increase in most areas of central and eastern China, and the annual E increased by 3.24 million person-days. This means that if the climate continues to warm by 0.5 °C without emission reduction measures, E related to heatwaves will increase by an additional 3.24 million person-days.

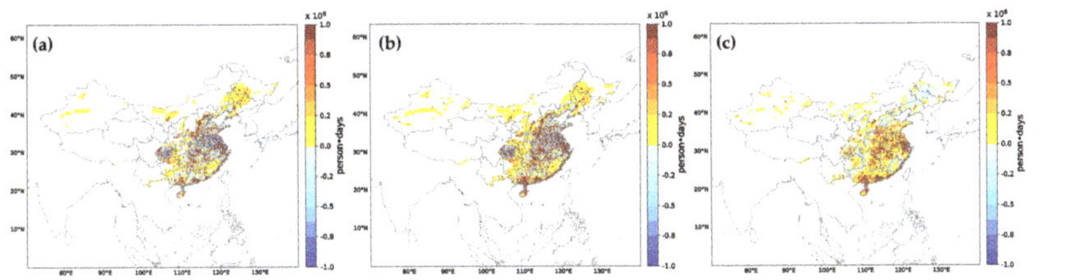

Figure 3. Differences in heatwave-related population exposure during different periods. From left to right: 1.5 °C–present (**a**), 2.0 °C–present (**b**), and 2.0 °C–1.5 °C (**c**).

3.3. Amplified Impacts and Relative Contributions of Climate and Population

As analyzed in Section 3.2, 0.5 °C of additional warming has an absolute amplified impact of 3.24 million person-days on E. This means that the relative risk of E (AI_r) under 2.0 °C warming would increase by 40.6 percent for a 1.5 °C warming scenario. Relative to present, 1.5 °C and 2.0 °C warming increase the impact by 95.1 and 174.3 percent, respectively. Although it is known that climate and population redistribution are the main contributors to increased E by heatwaves, the magnitude of their respective contributions is unknown. Here, changes in E are decomposed into the effects of climate ($\Delta HD \times Pop_{his}$), population ($\Delta Pop \times HD_{his}$), and interactions ($\Delta HD \times \Delta Pop$), based on Equation (4).

Figure 4 shows the contribution of these three effects to the change in E to heatwaves in China under the 1.5 °C and 2.0 °C climate warming scenarios, respectively. Under the 1.5 °C climate warming scenario, the main factor causing the change in E is climate, which accounts for 71.99 percent of the change in E to heatwaves. The second factor is population, which accounts for 16.91 percent of the total changes. The contribution of interaction was the smallest, at about 11.09 percent. Under the 2.0 °C warming scenario, the contribution of climate is still larger (80.34 percent), followed by interaction (12.42 percent), and population (7.24 percent). This suggests that climate plays a more important role in changes of E with further increases in climate warming. In addition, population dynamics are also one of the key factors affecting population exposure changes. Although urbanization has led to the migration of China's population centers to cities, the future population will begin to decline slowly after reaching a peak (1.45 billion) around 2030, as a result of many factors, including birth, death, migration, education level, and population policies under each SSP. The 1.5 °C warming scenario period coincides with the peak of China's total population, while the 2.0 °C warming scenario period occurs during the stage after the population peaks (as shown in Figure S2). This may partially explain why the population contribution under the 2.0 °C warming scenario is smaller than in the 1.5 °C warming scenario. The decomposition of E changes from the 1.5 °C to 2.0 °C warming scenarios also illustrates this problem, with the population contributing negatively to the total change.

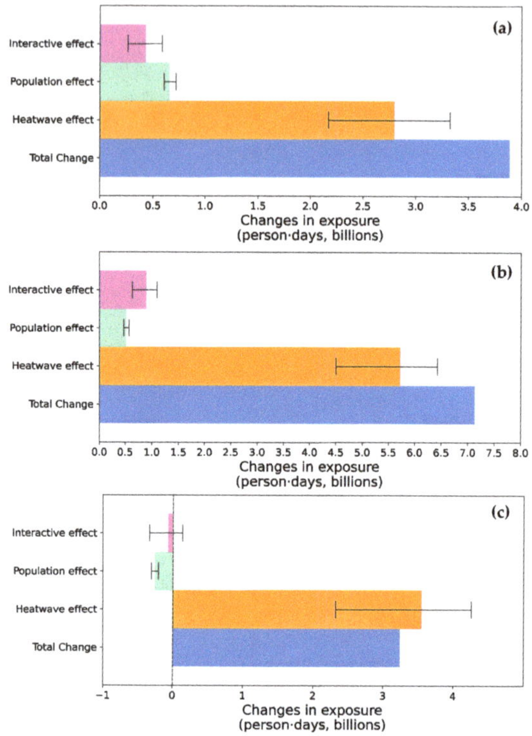

Figure 4. Contribution of three effects to the change in population exposure. From left to right: 1.5 °C–present (**a**), 2.0 °C–present (**b**), and 2.0 °C–1.5 °C scenarios (**c**).

The relative contributions of the three elements to the overall change in population exposure to heatwaves are further analyzed at the regional scale. Figure 5 shows the spatial distribution of the relative contributions of climate, population, and interaction to changes in E under the warming scenarios. Climate and population show positive contributions in most regions, with climate contributing substantially. The proportion of positive and negative contributions for the interaction component is similar. For the 1.5 °C warming scenario, the climate elements in the newly exposed areas—the Yangtze and Pearl River Basins and the eastern Sichuan Basin—play a positive role in E. The magnitude and area of this effect are further expanded under 2.0 °C warming scenario. The negative impact area of population on E has an increasing trend. While both positive and negative contributions of the interaction were characterized by an increase, the increase in the overall positive effect was greater. In addition, we can find that population and interaction are negative in most regions, while the contribution of climate is positive in the total change in population exposure from 1.5 to 2.0 °C warming. It is important to note that at 0.25° × 0.25° resolution, the relative contribution of individual elements can exceed 100 percent because of the extreme drastic changes in population and HD at the individual grid scale. However, the sum of the relative contributions of climate, population, and interaction at each grid point equals 100 percent.

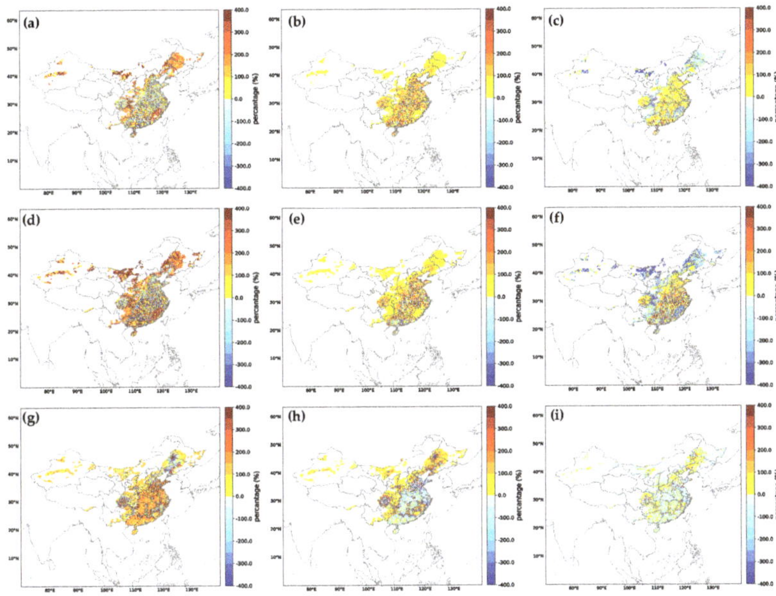

Figure 5. Spatial distribution of contributions (climate, population, and interaction, from left to right) in population exposure changes at 1.5 °C–present (**a**–**c**), 2.0 °C–present (**d**–**f**), and 2.0 °C–1.5 °C (**g**–**i**).

4. Discussion

This study assessed heatwave risk via E—under 1.5 and 2.0 °C warming scenarios—based on results from the latest CMIP multi-model ensemble, through changes in heatwave and regional population changes. Instead of using the Coupled Model Intercomparison Project phase 5 (CMIP5) data in most previous studies, we used the CMIP6 model output data, which shows improvements in simulation of climate indices in China, especially the daily maximum and minimum temperatures [8,25]. Results suggest that if nothing occurs to curtail continued global warming, the risk of regional heatwaves in China will increase by 2–2.8 times by mid-century. The spatial distribution of heatwave is consistent with the results of previous studies [40,42,62]. Of these, climate change was the dominant factor (>70 percent) in increasing exposure to heatwave risk.

These results are only preliminary conclusions. The main limitations include that the heatwave risk in this study mainly considers E, whereas future work should consider differences in vulnerability of different populations (including age, gender, income, education level, etc.) to quantify the assessment of heatwave exposure risk more accurately. In addition, the spatial redistribution of population is critical to the estimation of E relative to heatwave. However, urbanization is also an important factor affecting the future redistribution of China's population. Therefore, analyzing changes in E from urban and rural perspectives is conducive to improved scientific understanding of heatwave risk changes. Recent studies have suggested that the wet bulb temperature is closer to the human apparent temperature, which can better reflect the intensity of the heatwave. Therefore, compound heatwaves should be considered more carefully in future studies.

5. Conclusions

In this study, the latest CMIP6 multi-model ensemble results were used to assess changes in frequency of HD at 1.5 °C and 2.0 °C global warming levels compared to present. On this basis, the changes in E to heatwaves were assessed based on population projection data, and a relative contribution decomposition analysis was performed on the changes in E. Results showed a significant increase in the frequency of HD at two different warming levels. The assessment highlighted that 1.5 °C and 2.0 °C warming increased E by 95.1 and 174.3 percent compared to present. An additional 0.5 °C temperature rise would significantly affect changes in E with about 40.6 percent more E. In addition, the climate element played a significant positive leading role (>70 percent) in total changes of E to heatwaves. Of the additional 0.5 °C warming, climate change positively contributes to the increase in the risk of E, with 109.69 percent of total E. Population and interaction tend to decrease the risk of E with relative contributions of −7.71 and −1.98 percent, respectively, in total change of E. Therefore, considering future heat risks, humanity benefits from a 0.5 °C reduction in warming, particularly in eastern China. This conclusion provides useful insights for advancing climate change adaptation.

Supplementary Materials: The following supporting information can be downloaded at: https://www.mdpi.com/article/10.3390/su141811458/s1, Figure S1: Spatial distribution of population at 0.25° × 0.25° grid cells. (a) is LansScan at present. (b) is Tsinghua population at 1.5 °C warming scenarios. (c) is Tsinghua population at 2.0 °C warming scenarios. Only grid cells with population over 500 are plotted; Figure S2: Dynamics of projected total population in China. Table S1: Table S1. Overview of twenty CMIP6 models used in this study.

Author Contributions: Conceptualization, L.W. and X.Y.; methodology, L.W. and S.J.; software, L.W.; validation, S.J. and X.Y.; formal analysis, L.W. and X.Y.; investigation, L.W., Q.L. and S.J.; resources, L.W.; data curation, L.W. and S.J.; writing—original draft preparation, L.W. and R.V.R.; writing—review and editing, L.W., R.V.R., Q.L. and S.J.; visualization, L.W. and S.J; supervision, X.Y.; funding acquisition, L.W. and S.J. All authors have read and agreed to the published version of the manuscript.

Funding: This research was funded by Science Foundation of Hebei Normal University, grant number L2018B23. This research was funded by Fujian Province Forestry Science and Technology Research Project, grant number 2022FKJ02. This research was funded by Fujian Mental Health Human-Computer Interaction Technology Research Center, grant number 2020L3024.

Institutional Review Board Statement: Not applicable.

Informed Consent Statement: Not applicable.

Data Availability Statement: The data presented in this study are available on request from the corresponding author.

Acknowledgments: This research paper was conducted at School of Geographical Sciences, Hebei Normal University. Thanks to the LandScan program at Oak Ridge National Laboratory (ORNL) for providing the present population dataset, to Tsinghua University for providing the population projection dataset, to Laboratory for Climate Studies, China Meteorological Administration for providing a gridded daily observation dataset over China, and to Lawrence Livermore National Laboratory (LLNL) for providing the CMIP6 dataset.

Conflicts of Interest: The authors declare no conflict of interest.

References

1. Chen, H.; Sun, J. Significant Increase of the Global Population Exposure to Increased Precipitation Extremes in the Future. *Earths Future* **2021**, *9*, e2020EF001941. [CrossRef]
2. IPCC. *Summary for Policymakers. Climate Change 2021: The Physical Science Basis. Contribution of Working Group I to the Sixth Assessment Report of the Intergovernmental Panel on Climate Change*; Cambridge University Press: Cambridge, UK; New York, NY, USA, 2021; pp. 3–32.
3. van Vuuren, D.P.; Kriegler, E.; O'Neill, B.C.; Ebi, K.L.; Riahi, K.; Carter, T.R.; Edmonds, J.; Hallegatte, S.; Kram, T.; Mathur, R.; et al. A New Scenario Framework for Climate Change Research: Scenario Matrix Architecture. *Clim. Chang.* **2014**, *122*, 373–386. [CrossRef]
4. KC, S.; Lutz, W. The Human Core of the Shared Socioeconomic Pathways: Population Scenarios by Age, Sex and Level of Education for All Countries to 2100. *Glob. Environ. Chang.* **2017**, *42*, 181–192. [CrossRef] [PubMed]
5. Chen, H.; Sun, J. Changes in Drought Characteristics over China Using the Standardized Precipitation Evapotranspiration Index. *J. Clim.* **2015**, *28*, 5430–5447. [CrossRef]
6. Chen, J.; Liu, Y.; Pan, T.; Liu, Y.; Sun, F.; Ge, Q. Population Exposure to Droughts in China under the 1.5 °C Global Warming Target. *Earth Syst. Dyn.* **2018**, *9*, 1097–1106. [CrossRef]

7. Dunn, R.J.H.; Alexander, L.V.; Donat, M.G.; Zhang, X.; Bador, M.; Herold, N.; Lippmann, T.; Allan, R.; Aguilar, E.; Barry, A.A.; et al. Development of an Updated Global Land In Situ-Based Data Set of Temperature and Precipitation Extremes: HadEX3. *J. Geophys. Res. Atmos.* **2020**, *125*, e2019JD032263. [CrossRef]
8. Zhu, H.; Jiang, Z.; Li, J.; Li, W.; Sun, C.; Li, L. Does CMIP6 Inspire More Confidence in Simulating Climate Extremes over China? *Adv. Atmos. Sci.* **2020**, *37*, 1119–1132. [CrossRef]
9. Sanderson, B.M.; Xu, Y.; Tebaldi, C.; Wehner, M.; O'Neill, B.; Jahn, A.; Pendergrass, A.G.; Lehner, F.; Strand, W.G.; Lin, L.; et al. Community Climate Simulations to Assess Avoided Impacts in 1.5 and 2 °C Futures. *Earth Syst. Dyn.* **2017**, *8*, 827–847. [CrossRef]
10. Qu, X.; Huang, G. Global Monsoon Changes under the Paris Agreement Temperature Goals in CESM1(CAM5). *Adv. Atmos. Sci.* **2019**, *36*, 279–291. [CrossRef]
11. Good, P.; Booth, B.B.B.; Chadwick, R.; Hawkins, E.; Jonko, A.; Lowe, J.A. Large Differences in Regional Precipitation Change between a First and Second 2 K of Global Warming. *Nat. Commun.* **2016**, *7*, 13667. [CrossRef]
12. Coffel, E.D.; Horton, R.M.; de Sherbinin, A. Temperature and Humidity Based Projections of a Rapid Rise in Global Heat Stress Exposure during the 21st Century. *Environ. Res. Lett.* **2018**, *13*, 014001. [CrossRef]
13. Matthews, T.K.R.; Wilby, R.L.; Murphy, C. Communicating the Deadly Consequences of Global Warming for Human Heat Stress. *Proc. Natl. Acad. Sci. USA* **2017**, *114*, 3861–3866. [CrossRef]
14. Ranasinghe, R.; Ruane, A.C.; Vautard, R.; Arnell, N.; Coppola, E.; Cruz, F.A.; Dessai, S.; Islam, A.S.; Rahimi, M.; Ruiz Carrascal, D.; et al. *Climate Change Information for Regional Impact and for Risk Assessment. Climate Change 2021: The Physical Science Basis. Contribution of Working Group I to the Sixth Assessment Report of the Intergovernmental Panel on Climate Change*; Cambridge University Press: Cambridge, UK; New York, NY, USA, 2021; pp. 1767–1926.
15. Dematte, J.E. Near-Fatal Heat Stroke during the 1995 Heat Wave in Chicago. *Ann. Intern. Med.* **1998**, *129*, 173. [CrossRef]
16. Christidis, N.; Jones, G.S.; Stott, P.A. Dramatically Increasing Chance of Extremely Hot Summers since the 2003 European Heatwave. *Nat. Clim. Chang.* **2015**, *5*, 46–50. [CrossRef]
17. Robine, J.-M.; Cheung, S.L.K.; Le Roy, S.; Van Oyen, H.; Griffiths, C.; Michel, J.-P.; Herrmann, F.R. Death Toll Exceeded 70,000 in Europe during the Summer of 2003. *C. R. Biol.* **2008**, *331*, 171–178. [CrossRef]
18. Katsafados, P.; Papadopoulos, A.; Varlas, G.; Papadopoulou, E.; Mavromatidis, E. Seasonal Predictability of the 2010 Russian Heat Wave. *Nat. Hazards Earth Syst. Sci.* **2014**, *14*, 1531–1542. [CrossRef]
19. Barriopedro, D.; Fischer, E.M.; Luterbacher, J.; Trigo, R.M.; García-Herrera, R. The Hot Summer of 2010: Redrawing the Temperature Record Map of Europe. *Science* **2011**, *332*, 220–224. [CrossRef]
20. Vautard, R.; van Aalst, M.; Boucher, O.; Drouin, A.; Haustein, K.; Kreienkamp, F.; van Oldenborgh, G.J.; Otto, F.E.L.; Ribes, A.; Robin, Y.; et al. Human Contribution to the Record-Breaking June and July 2019 Heatwaves in Western Europe. *Environ. Res. Lett.* **2020**, *15*, 094077. [CrossRef]
21. Sun, Q.; Miao, C.; AghaKouchak, A.; Duan, Q. Unraveling Anthropogenic Influence on the Changing Risk of Heat Waves in China. *Geophys. Res. Lett.* **2017**, *44*, 5078–5085. [CrossRef]
22. Tong, J.; Jing, Z.; Cheng, J.; Lige, C.; Yanjun, W.; Hemin, S.; Anqian, W.; Jinlong, H.; Buda, S.; Run, W. National and Provincial Population Projected to 2100 under the Shared Socioeconomic Pathways in China. *Adv. Clim. Chang. Res.* **2017**, *13*, 128.
23. Chen, H.; Sun, J. Increased Population Exposure to Extreme Droughts in China Due to 0.5 °C of Additional Warming. *Environ. Res. Lett.* **2019**, *14*, 064011. [CrossRef]
24. Tang, S.; Qiao, S.; Feng, T.; Jia, Z.; Zang, N.; Feng, G. Predictability of the Mid-Summer Surface Air Temperature over the Yangtze River Valley in the National Centers for Environmental Prediction Climate Forecast System. *Int. J. Climatol.* **2021**, *41*, 811–829. [CrossRef]
25. Zhu, H.; Jiang, Z.; Li, L. Projection of Climate Extremes in China, an Incremental Exercise from CMIP5 to CMIP6. *Sci. Bull.* **2021**, *66*, 2528–2537. [CrossRef]
26. Rohat, G.; Flacke, J.; Dosio, A.; Dao, H.; Maarseveen, M. Projections of Human Exposure to Dangerous Heat in African Cities Under Multiple Socioeconomic and Climate Scenarios. *Earths Future* **2019**, *7*, 528–546. [CrossRef]
27. Russo, S.; Sillmann, J.; Sippel, S.; Barcikowska, M.J.; Ghisetti, C.; Smid, M.; O'Neill, B. Half a Degree and Rapid Socioeconomic Development Matter for Heatwave Risk. *Nat. Commun.* **2019**, *10*, 136. [CrossRef] [PubMed]
28. Jones, B.; O'Neill, B.C.; McDaniel, L.; McGinnis, S.; Mearns, L.O.; Tebaldi, C. Future Population Exposure to US Heat Extremes. *Nat. Clim. Chang.* **2015**, *5*, 652–655. [CrossRef]
29. Liao, X.; Xu, W.; Zhang, J.; Li, Y.; Tian, Y. Global Exposure to Rainstorms and the Contribution Rates of Climate Change and Population Change. *Sci. Total Environ.* **2019**, *663*, 644–653. [CrossRef]
30. Jones, B.; Tebaldi, C.; O'Neill, B.C.; Oleson, K.; Gao, J. Avoiding Population Exposure to Heat-Related Extremes: Demographic Change vs Climate Change. *Clim. Chang.* **2018**, *146*, 423–437. [CrossRef]
31. Liu, Z.; Anderson, B.; Yan, K.; Dong, W.; Liao, H.; Shi, P. Global and Regional Changes in Exposure to Extreme Heat and the Relative Contributions of Climate and Population Change. *Sci. Rep.* **2017**, *7*, 43909. [CrossRef]
32. Rohat, G.; Flacke, J.; Dosio, A.; Pedde, S.; Dao, H.; van Maarseveen, M. Influence of Changes in Socioeconomic and Climatic Conditions on Future Heat-Related Health Challenges in Europe. *Glob. Planet. Chang.* **2019**, *172*, 45–59. [CrossRef]
33. Ding, T.; Qian, W. Geographical Patterns and Temporal Variations of Regional Dry and Wet Heatwave Events in China during 1960–2008. *Adv. Atmos. Sci.* **2011**, *28*, 322–337. [CrossRef]

34. Kong, Q.; Guerreiro, S.B.; Blenkinsop, S.; Li, X.-F.; Fowler, H.J. Increases in Summertime Concurrent Drought and Heatwave in Eastern China. *Weather Clim. Extrem.* **2020**, *28*, 100242. [CrossRef]
35. Wang, P.; Tang, J.; Wang, S.; Dong, X.; Fang, J. Regional Heatwaves in China: A Cluster Analysis. *Clim. Dyn.* **2018**, *50*, 1901–1917. [CrossRef]
36. Wang, W.; Zhou, W.; Li, X.; Wang, X.; Wang, D. Synoptic-Scale Characteristics and Atmospheric Controls of Summer Heat Waves in China. *Clim. Dyn.* **2016**, *46*, 2923–2941. [CrossRef]
37. Wang, Y.; Ren, F.; Zhang, X. Spatial and Temporal Variations of Regional High Temperature Events in China. *Int. J. Climatol.* **2014**, *34*, 3054–3065. [CrossRef]
38. Wei, K.; Chen, W. An Abrupt Increase in the Summer High Temperature Extreme Days across China in the Mid-1990s. *Adv. Atmos. Sci.* **2011**, *28*, 1023–1029. [CrossRef]
39. Zhan, L.-F.; Wang, Y.; Sun, H.; Zhai, J.; Zhan, M. Study on the Change Characteristics of and Population Exposure to Heatwave Events on the North China Plain. *Adv. Meteorol.* **2019**, *2019*, 7069195. [CrossRef]
40. Dong, S.; Zhou, B.; Hou, M.; Li, R.; Xu, Y.; Yu, L.; Zhang, Y. Projected Risk of Extreme Heat in China Based on CMIP5 Models. *Adv. Clim. Chang. Res.* **2014**, *10*, 365–369. [CrossRef]
41. Zhang, G.; Zeng, G.; Liang, X.-Z.; Huang, C. Increasing Heat Risk in China's Urban Agglomerations. *Environ. Res. Lett.* **2021**, *16*, 064073. [CrossRef]
42. Chen, H.; He, W.; Sun, J.; Chen, L. Increases of Extreme Heat-Humidity Days Endanger Future Populations Living in China. *Environ. Res. Lett.* **2022**, *17*, 064013. [CrossRef]
43. Wu, X.; Hao, Z.; Tang, Q.; Zhang, X.; Feng, S.; Hao, F. Population Exposure to Compound Dry and Hot Events in China under 1.5 and 2 °C Global Warming. *Int. J. Climatol.* **2021**, *41*, 5766–5775. [CrossRef]
44. Ma, F.; Yuan, X. Impact of Climate and Population Changes on the Increasing Exposure to Summertime Compound Hot Extremes. *Sci. Total Environ.* **2021**, *772*, 145004. [CrossRef] [PubMed]
45. IPCC. *Global Warming of 1.5 °C: IPCC Special Report on Impacts of Global Warming of 1.5 °C above Pre-Industrial Levels in Context of Strengthening Response to Climate Change, Sustainable Development, and Efforts to Eradicate Poverty*, 1st ed.; Cambridge University Press: Cambridge, UK, 2022; ISBN 978-1-00-915794-0.
46. Wu, J.; Gao, X. A Gridded Daily Observation Dataset over China Region and Comparison with the Other Datasets. *Chin. J. Geophys. Chin.* **2013**, *56*, 1102–1111. [CrossRef]
47. O'Neill, B.C.; Tebaldi, C.; van Vuuren, D.P.; Eyring, V.; Friedlingstein, P.; Hurtt, G.; Knutti, R.; Kriegler, E.; Lamarque, J.-F.; Lowe, J.; et al. The Scenario Model Intercomparison Project (ScenarioMIP) for CMIP6. *Geosci. Model Dev.* **2016**, *9*, 3461–3482. [CrossRef]
48. van Vuuren, D.P.; Edmonds, J.; Kainuma, M.; Riahi, K.; Thomson, A.; Hibbard, K.; Hurtt, G.C.; Kram, T.; Krey, V.; Lamarque, J.-F.; et al. The Representative Concentration Pathways: An Overview. *Clim. Chang.* **2011**, *109*, 5–31. [CrossRef]
49. IPCC. *Climate Change 2021: The Physical Science Basis. The Working Group I Contribution to the Sixth Assessment Report*; Cambridge University Press: Cambridge, UK; New York, NY, USA, 2021; *in press*.
50. Hoyer, S.; Hamman, J. Xarray: N-D Labeled Arrays and Datasets in Python. *J. Open Res. Softw.* **2017**, *5*, 10. [CrossRef]
51. Rose, A.; McKee, J.; Sims, K.; Bright, E.; Reith, A.; Urban, M. *LandScan Global 2020*; Oak Ridge National Laboratory: Oak Ridge, TN, USA, 2021.
52. Chen, Y.; Guo, F.; Wang, J.; Cai, W.; Wang, C.; Wang, K. Provincial and Gridded Population Projection for China under Shared Socioeconomic Pathways from 2010 to 2100. *Sci. Data* **2020**, *7*, 83. [CrossRef]
53. Li, H.; Sheffield, J.; Wood, E.F. Bias Correction of Monthly Precipitation and Temperature Fields from Intergovernmental Panel on Climate Change AR4 Models Using Equidistant Quantile Matching. *J. Geophys. Res.* **2010**, *115*, D10101. [CrossRef]
54. Chen, J.; Brissette, F.P.; Chaumont, D.; Braun, M. Finding Appropriate Bias Correction Methods in Downscaling Precipitation for Hydrologic Impact Studies over North America. *Water Resour. Res.* **2013**, *49*, 4187–4205. [CrossRef]
55. Gudmundsson, L.; Bremnes, J.B.; Haugen, J.E.; Engen-Skaugen, T. Technical Note: Downscaling RCM Precipitation to the Station Scale Using Statistical Transformations—A Comparison of Methods. *Hydrol. Earth Syst. Sci.* **2012**, *16*, 3383–3390. [CrossRef]
56. Piani, C.; Weedon, G.; Best, M.; Gomes, S.; Viterbo, P.; Hagemann, S.; Haerter, J. Statistical Bias Correction of Global Simulated Daily Precipitation and Temperature for the Application of Hydrological Models. *J. Hydrol.* **2010**, *395*, 199–215. [CrossRef]
57. Themeßl, M.J.; Gobiet, A.; Heinrich, G. Empirical-Statistical Downscaling and Error Correction of Regional Climate Models and Its Impact on the Climate Change Signal. *Clim. Chang.* **2012**, *112*, 449–468. [CrossRef]
58. Boé, J.; Terray, L.; Habets, F.; Martin, E. Statistical and Dynamical Downscaling of the Seine Basin Climate for Hydro-Meteorological Studies. *Int. J. Climatol.* **2007**, *27*, 1643–1655. [CrossRef]
59. Logan, T.; Aoun, A.; Bourgault, P.; Huard, D.; Lavoie, J.; Rondeau-Genesse, G.; Smith, T.J.; Alegre, R.; Barnes, C.; Biner, S.; et al. *Ouranosinc/Xclim: V0.37.0*; CERN: Meyrin, Switzerland, 2022.
60. Nairn, J.; Fawcett, R. The Excess Heat Factor: A Metric for Heatwave Intensity and Its Use in Classifying Heatwave Severity. *Int. J. Environ. Res. Public Health* **2014**, *12*, 227–253. [CrossRef]
61. Ma, F.; Yuan, X.; Jiao, Y.; Ji, P. Unprecedented Europe Heat in June–July 2019: Risk in the Historical and Future Context. *Geophys. Res. Lett.* **2020**, *47*, e2020GL087809. [CrossRef]
62. Chen, Y.; Zhai, P. Revisiting Summertime Hot Extremes in China during 1961–2015: Overlooked Compound Extremes and Significant Changes. *Geophys. Res. Lett.* **2017**, *44*, 5096–5103. [CrossRef]

63. Li, Y.; Ding, Y.; Liu, Y. Mechanisms for Regional Compound Hot Extremes in the Mid-lower Reaches of the Yangtze River. *Int. J. Climatol.* **2021**, *41*, 1292–1304. [CrossRef]
64. Purich, A.; Cowan, T.; Cai, W.; van Rensch, P.; Uotila, P.; Pezza, A.; Boschat, G.; Perkins, S. Atmospheric and Oceanic Conditions Associated with Southern Australian Heat Waves: A CMIP5 Analysis. *J. Clim.* **2014**, *27*, 7807–7829. [CrossRef]
65. Qiu, W.; Yan, X. The Trend of Heatwave Events in the Northern Hemisphere. *Phys. Chem. Earth Parts ABC* **2020**, *116*, 102855. [CrossRef]
66. Rohli, R.V.; Keim, B.D. Spatial and Temporal Characteristics of Extreme-High-Summer-Temperature Events in the South-Central United States. *Phys. Geogr.* **1994**, *15*, 310–324. [CrossRef]
67. Han, L.; Yu, X.; Xu, Y.; Deng, X.; Yang, L.; Li, Z.; Lv, D.; Xiao, M. Enhanced Summertime Surface Warming Effects of Long-term Urbanization in a Humid Urban Agglomeration in China. *J. Geophys. Res. Atmos.* **2021**, *126*, e2021JD03500. [CrossRef]
68. Cardona, O.D.; Van Aalst, M.K.; Birkmann, J.; Fordham, M.; Mc Gregor, G.; Rosa, P.; Pulwarty, R.S.; Schipper, E.L.F.; Sinh, B.T.; Décamps, H.; et al. Determinants of Risk: Exposure and Vulnerability. In *Managing the Risks of Extreme Events and Disasters to Advance Climate Change Adaptation: Special Report of the Intergovernmental Panel on Climate Change*; Cambridge University Press: Cambridge, UK, 2012; pp. 65–108.
69. IPCC. *Managing the Risks of Extreme Events and Disasters to Advance Climate Change Adaptation: Special Report of the Intergovernmental Panel on Climate Change*; Cambridge University Press: Cambridge, UK, 2012; ISBN 978-1-107-02506-6.
70. Li, D.; Zhou, T.; Zou, L.; Zhang, W.; Zhang, L. Extreme High-Temperature Events Over East Asia in 1.5 °C and 2 °C Warmer Futures: Analysis of NCAR CESM Low-Warming Experiments. *Geophys. Res. Lett.* **2018**, *45*, 1541–1550. [CrossRef]

sustainability

MDPI

Article

Impact of Abnormal Climatic Events on the CPUE of Yellowfin Tuna Fishing in the Central and Western Pacific

Weifeng Zhou [1,2], Huijuan Hu [1,3], Wei Fan [2,*] and Shaofei Jin [4,*]

1 Key Laboratory of Fishery Resources Remote Sensing Information Technology, Chinese Academy of Fishery Sciences, Shanghai 200090, China; zhouwf@ecsf.ac.cn (W.Z.); huhuijuan2020631@163.com (H.H.)
2 East China Sea Fishery Research Institute, Chinese Academy of Fishery Sciences, Shanghai 200090, China
3 College of Marine Sciences, Shanghai Ocean University, Shanghai 201306, China
4 Department of Geography, Minjiang University, Fuzhou 350108, China
* Correspondence: fanwee@126.com (W.F.); jinsf@tea.ac.cn (S.J.); Tel.: +86-21-65680117 (W.F.)

Abstract: To explore the impact of climate change on fishery resources, the temporal and spatial characteristics of the thermocline in the main yellowfin tuna purse-seine fishing grounds in the western and central Pacific Ocean during La Niña and El Niño years were studied using the 2008–2017 Argo grid data (BOA_Argo) and the log data of commercial fishing vessels. A generalized additive model (GAM) was used to analyze the variables affecting yellowfin tuna fishing grounds. The results showed that in La Niña years, the catch per unit effort (CPUE) moved westward as the high-value zone of the upper boundary contracted westward to 145° E, and in the El Niño years this moved eastward to 165° E. Compared with normal years, the upper boundary depth difference of the thermocline on the east and west sides of the equatorial Pacific was larger in La Niña years, and the upper boundary depth of 80–130 m shifted westward. The thermocline strength was generally weaker in the west and stronger in the east. The thermocline had two band-like distribution structures with an axis at 15° N and 15° S. The CPUE was distributed from 120 m to 200 m. The CPUE distribution was dense when the temperature range of the upper boundary of the thermocline was 27.5–29.5 °C, and the intensity was 0.08–0.13 $°C \cdot m^{-1}$. The upper-boundary temperature had the greatest impact on the CPUE. The eastward shift of the CPUE during El Niño and the westward shift during La Niña were associated with the optimal thermocline parameter values. The factor of year had a fluctuating effect on the CPUE, and the influence of the La Niña years was greater. The areas with high abundance were 5° N–5° S and 150° E–175° E. The results showed that the changes in the thermocline caused by abnormal climate events significantly affected the CPUE.

Keywords: central and western Pacific; thermocline; yellowfin tuna; CPUE; El Niño; La Niña; GAM model

Citation: Zhou, W.; Hu, H.; Fan, W.; Jin, S. Impact of Abnormal Climatic Events on the CPUE of Yellowfin Tuna Fishing in the Central and Western Pacific. *Sustainability* **2022**, *14*, 1217. https://doi.org/10.3390/su14031217

Academic Editor: Tim Gray

Received: 24 November 2021
Accepted: 18 January 2022
Published: 21 January 2022

1. Introduction

Marine fisheries, especially tropical marine fisheries, provide high-quality protein for human diets and make a significant contribution to human and societal wellbeing. Tuna fisheries are one of the four most highly valued fisheries worldwide [1]. Tuna fisheries had 5.2 million MT volume and USD 11.7 billion in landed value in 2018 [2]. Fleets using purse seines are one of two primary harvest strategies in global tuna production [3]. Tuna-fishing license fees to operate in the exclusive economic zones of several Pacific Island countries and territories provide 30–90% of all (non-aid) government revenue, such as Kiribati, Nauru, and Tokelau [4]. However, fishery resources and tuna capture are highly correlated with the marine environment, and are increasingly threatened by various physical and biogeochemical responses to climate change [5]. The fishery resources in the entire western and central Pacific Ocean (WCPO) support major industrial tuna fisheries and a variety of small-scale coastal fisheries. The total annual average catch from the Parties to the Nauru Agreement (PNA) purse-seine fishery contributed more than 50% of the recent (2014–2018) average tuna catch from the WCPO and equates to almost 30% of the total global tuna supply [6]. An increasing number of fishery habitat studies have focused on the impact of marine surface environmental factors on the distribution of fishery resources. However, there has been less research on the characteristics of the subsurface marine environment under abnormal climate conditions and the impact on fishery resources. Climate change threatens tropical marine fisheries [7]. The large-scale effects of abnormal climate events on the marine environment have a significant impact on fishery resources and fishing ground distribution. El Niño events are a prominent feature of climate variability, with global climatic impacts [8]. The El Niño–Southern Oscillation (ENSO) is considered to be one of the major climate events impacting tuna fisheries [9]. Research by Hampton et al. [10,11] has shown that skipjack tuna is very sensitive to temperature changes, and the tuna distribution shifts with the occurrence of ENSOs. The thermocline is a layer between the upper warm water and the lower cold water where the temperature drops sharply. The thermocline is one of the most important physical phenomena in the ocean. As the strongest signal of inter-annual global climate change, ENSO's occurrence, development, and extinction are closely related to the tropical Pacific thermocline. During La Niña events, most fleets prefer to fish in the western part of the region. The converse occurs during El Niño episodes. Changes in the depth of the thermocline have also been demonstrated to explain the variability in purse-seine catch rates [12]. The Oceanic Niño Index (ONI) was calculated from the moving average of sea surface temperature anomalies (SSTAs) for three consecutive months in the El Niño 3.4 area. According to the definitions of El Niño and La Niña events by the National Oceanic and Atmospheric Administration (NOAA) of the United States, an El Niño event is considered to have occurred when the ONI is greater than +0.5 °C for five consecutive months, whereas a La Niña event is considered to have occurred when the ONI is less than −0.5 °C for five consecutive months. According to the definitions of La Niña and El Niño events established by the NOAA, the ONI can be used to characterize El Niño and La Niña events as well as environmental conditions [13]. For example, the ONI and the Southern Annular Mode (SAM) were used as proxy indices of environmental conditions affecting penguins and krill, respectively [14], whereas Kuo-Wei Lan used ONI as a climatic index to explore the relationship between climate change and grey mullet (*Mugil cephalus* L.) in the Taiwan Strait [15].

How the thermocline changes under abnormal climate events and its relationship with changes in fishery resources is well worth studying. In this paper, an overlay map of thermocline parameters, including the upper boundary temperature, upper boundary depth, thermocline thickness, and thermocline strength, was drawn with the catch per unit effort (CPUE) distribution of yellowfin tuna under the abnormal climate mode. The distribution patterns of the parameters in La Niña and El Niño years and the effects on the spatial distribution of yellowfin tuna were analyzed. In addition, a generalized additive model (GAM) was used to analyze the impact of each variable on the CPUE of yellowfin tuna. The results can provide additional thermocline distribution information and serve as a reference for tuna production in this area.

2. Materials and Methods

2.1. Data Sources

This study area was located at 130° E–130° W and 25° N–25° S in the WCPO. The fishery data were obtained from the monthly fishing log data of tuna purse-seine vessels in the WCPO from January 2008 to July 2017. Data included the name of the production vessel, production date, longitude and latitude of the operation location, operation net times, species type, and yield.

To better reflect the ENSO signal, the temperature and salinity data were obtained from the Argo grid data (BOA_Argo) set [16], provided by the China Argo Real-time Data Center. This data, with 1° × 1° horizontally and 58 standard layers with unequal distance from 0–1975 m vertically, can provide more explicit distribution information of surface temperature and salinity data compared with other common datasets [17]. In this study, we compiled the data periods consistent with the fishing log data.

The climate event data were calculated using the sea surface temperature in the El Niño 3.4 area (https://origin.cpc.ncep.noaa.gov/products/analysis_monitoring/ensostuff/ONI_v5.php (accessed on 11 January 2022)) (Table 1).

Table 1. Definition of El Niño and La Niña events.

ONI	Type of Event	ONI	Type of Event
$0.5 \leq ONI \leq 0.9$	Weak El Niño event, WE	$-0.9 \leq ONI \leq -0.5$	Weak La Niña event, WL
$1.0 \leq ONI \leq 1.4$	Moderate El Niño event, ME	$-1.4 \leq ONI \leq -1.0$	Moderate La Niña event, ML
$1.5 \leq ONI \leq 1.9$	Strong El Niño event, SE	$-1.9 \leq ONI \leq -1.5$	Strong La Niña event, SL
$ONI \geq 2.0$	Very strong El Niño event, VSE	$ONI \leq -2.0$	Very strong La Niña event, VSL

2.2. Data Analysis

First, to match the spatio-temporal resolution of environmental variables, the monthly CPUE was converted to a $1^\circ \times 1^\circ$ grid. The equation is as follows:

$$CPUE_{ymij} = \frac{\text{Catch}_{\text{ymij}}}{\text{Effort}_{\text{ymij}}} \tag{1}$$

where y is the year, m is the month, i is the longitude, and j is the latitude. For the mth month, $CPUE_{ymij}$ is the catch per unit effort (t/net) at the ith longitude and the jth latitude, $\text{Catch}_{\text{ymij}}$ is the total catch at the ith longitude and the jth latitude, and $\text{Effort}_{\text{ymij}}$ is the fishing effort, i.e., the cumulative total operation net time, at the ith longitude and the jth latitude.

Because the BOA_Argo datasets included 58 standard layers with unequal vertical distances in the range of 0–1975 m, the Akima [18] interpolation method was used to interpolate the temperature data of Argo buoy profiles with an uneven distribution to regular depth layers at 2 m intervals. The thermocline discrimination method [19] was used to calculate the gradient of the temperature profile ($\Delta T/\Delta H$). In short, if the vertical gradient of a certain section in a temperature profile was greater than or equal to the standard of the lowest limit value of the deep-water thermocline (0.05 $^\circ C \cdot m^{-1}$), then the section was determined to be the thermocline. The depths of the upper and lower boundary points of the section were the upper and lower boundary depths of the thermocline, respectively. Finally, the width of the section was the thermocline thickness. The entire vertical temperature gradient was defined as the thermocline strength. In total, six characteristic parameters of the thermocline were extracted: the temperature and depth of the upper boundary, the temperature and depth of the lower boundary, and the thermocline strength and thickness. In this study, the characteristic parameters of the thermocline from 2008 to 2017 were divided into groups by month and displayed as filled isolines.

2.3. Generalized Additive Model

The GAM can simulate the nonlinear relationship between the response variable and multiple explanatory variables and has been widely used in fishery management [20–22]. In this study, a GAM was used to analyze the effects of spatial and temporal variables and environmental variables on the CPUE. The temporal variables were the year and month, and the spatial variables were the longitude and latitude. The environmental variables were the ONI and the six extracted parameters. The Akaike information criterion (AIC) value method was used to select the best model. The smaller the AIC value, the higher the fitting degree of the model [23]. The CPUE was log transformed prior to analysis [24,25]. The GAM equation is as follows:

$$\log(CPUE + 1) = \text{s}(year) + \text{s}(month) + \text{s}(lon) + \text{s}(lat) + \text{s}(\text{environmental variable}) \tag{2}$$

where *year* represents the year, *month* represents the month, *lon* represents the longitude, *lat* represents the latitude, and environmental variable represents an environmental variable.

2.4. Data Processing and Analysis

Excel was used to store the fishing log data of tuna purse-seine vessels in the WCPO and to calculate the CPUE. MATLAB R2016a software was used to read the BOA_Argo data and to calculate and extract the six parameters of thermocline upper boundary temperature and depth, lower boundary temperature and depth, and thermocline thickness and strength, and to visually display the superposition of the subsurface SST vertical parameters and CPUE. The GAM was constructed using the mgcv library of the R programming software [26].

3. Results

The year of occurrence of an El Niño or La Niña event is defined as an El Niño year or a La Niña year. Therefore, 2009, 2014, and 2015 were El Niño years, whereas 2008, 2010, 2011, 2016, and 2017 were La Niña years (Table 2) [27].

Table 2. ONI from 2008 to 2017 *.

Type of ENSO	Year	ONI Month											
		1	2	3	4	5	6	7	8	9	10	11	12
WL	2008	**−1.6**	**−1.4**	**−1.2**	**−0.9**	**−0.8**	**−0.5**	−0.4	−0.3	−0.3	−0.4	**−0.6**	**−0.7**
ME	2009	**−0.8**	**−0.7**	**−0.5**	−0.2	0.1	0.4	**0.5**	**0.5**	**0.7**	**1**	**1.3**	**1.6**
SL	2010	**1.5**	**1.3**	**0.9**	0.4	−0.1	**−0.6**	**−1**	**−1.4**	**−1.6**	**−1.7**	**−1.7**	**−1.6**
ML	2011	**−1.4**	**−1.1**	**−0.8**	**−0.6**	**−0.5**	−0.4	**−0.5**	**−0.7**	**−0.9**	**−1.1**	**−1.1**	**−1**
NORMAL	2012	**−0.8**	**−0.6**	**−0.5**	−0.4	−0.2	0.1	0.3	0.3	0.3	0.2	0	−0.2
NORMAL	2013	−0.4	−0.3	−0.2	−0.2	−0.3	−0.3	−0.4	−0.4	−0.3	−0.2	−0.2	−0.3
WE	2014	−0.4	−0.4	−0.2	0.1	0.3	0.2	0.1	0	0.2	0.4	**0.6**	**0.7**
VSE	2015	**0.6**	**0.6**	**0.6**	**0.8**	**1**	**1.2**	**1.5**	**1.8**	**2.1**	**2.4**	**2.5**	**2.6**
WL	2016	**2.5**	**2.2**	**1.7**	**1**	**0.5**	0	−0.3	**−0.6**	**−0.7**	**−0.7**	**−0.7**	**−0.6**
WL	2017	−0.3	−0.1	0.1	0.3	0.4	0.4	0.2	−0.1	−0.4	**−0.7**	**−0.9**	**−1**
WE	2018	**−0.9**	**−0.8**	**−0.6**	−0.4	−0.1	0.1	0.1	0.2	0.4	**0.7**	**0.9**	**0.8**

* WE stands for weak El Niño event, WL stands for weak La Niña event, ME stands for moderate El Niño event, ML stands for moderate La Niña event, SE stands for strong El Niño event, SL stands for strong La Niña event, VSE stands for very strong El Niño event, and VSL stands for very strong La Niña event. The bold numbers represent abnormal climatic events.

3.1. Temporal and Spatial Variation of Upper Boundary Temperature and Depth of the Thermocline with Catch

The ENSO had a significant impact on the spatial distribution of the yellowfin tuna purse-seine fishing grounds in the WCPO. The superposition of thermocline parameters in a typical La Niña year (2010), normal year (2013), and El Niño year (2015) and the CPUE of yellowfin tuna were analyzed. The results (Figure 1) show that in the La Niña year, the high-value zone of 28–29 °C of the upper boundary temperature of the thermocline moved westward. The CPUE moved westward to the west of 170° E as the high-value zone of the upper boundary temperature contracted westward to 145° E. In the El Niño year, the high-value zone of 28–29 °C of the upper boundary temperature of the thermocline expanded eastward. The CPUE moved eastward to the east of 165° E with the eastward expansion of the high-value zone of the upper boundary temperature to 173° W. Yellowfin tuna is a warm-water fish species whose habitat and spawning area need to be above a certain water temperature. Therefore, the upper boundary temperature of the thermocline will affect the spatial distribution of yellowfin tuna [25].

In the normal year (2013), the upper boundary depth of the thermocline in the equatorial Pacific was deeper in the west and shallower in the east. In the typical La Niña year (2010) compared with the normal year (2013), the upper boundary depth difference of the thermocline on the east and west sides of the equatorial WCPO was larger. The upper boundary depth of 80–130 m was shifted westward, which may have been due to the enhancement of Walker circulation and southeast trade winds in the La Niña year. In the El Niño year (2015) compared with the normal year, the difference of the upper boundary depth of the thermocline on the east and west sides of the equatorial WCPO in the El Niño year was smaller. The upper boundary depth of 80–130 m moved eastward.

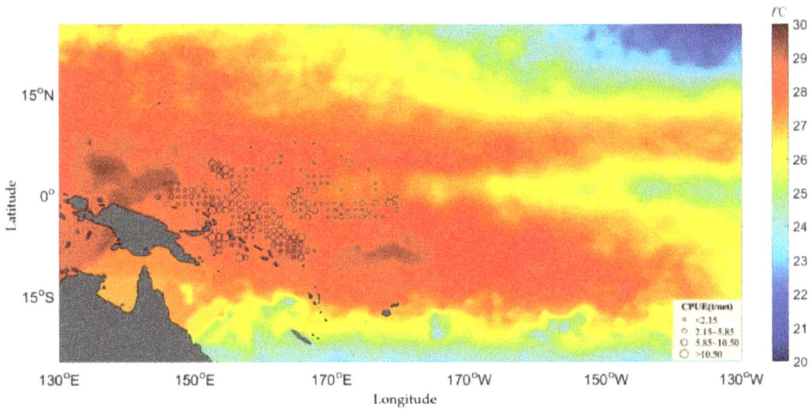

(a)

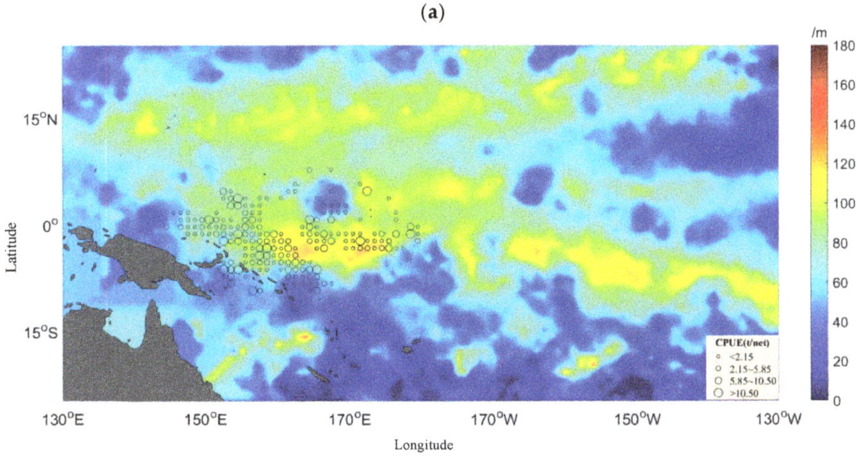

(b)

Figure 1. *Cont.*

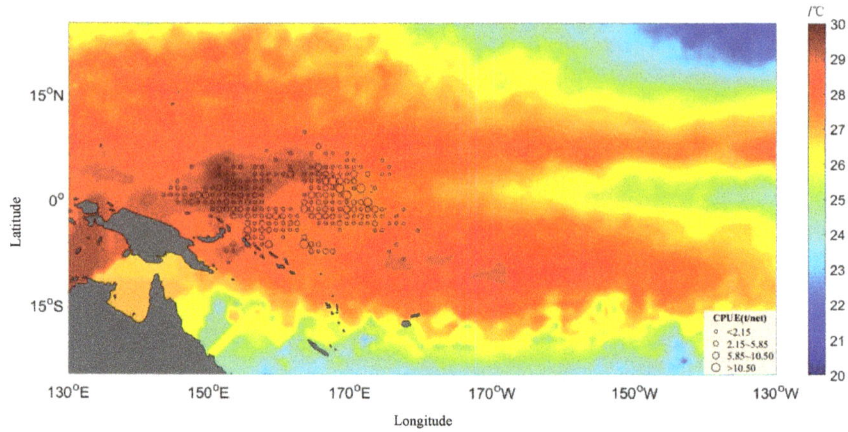

(**c**)

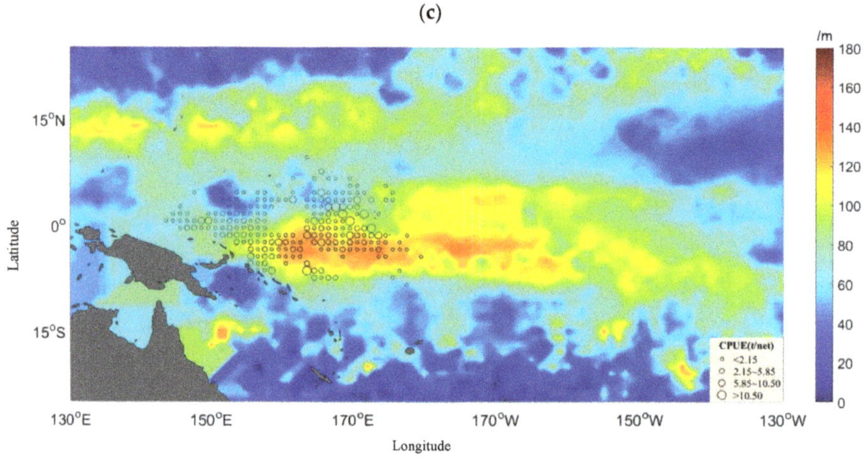

(**d**)

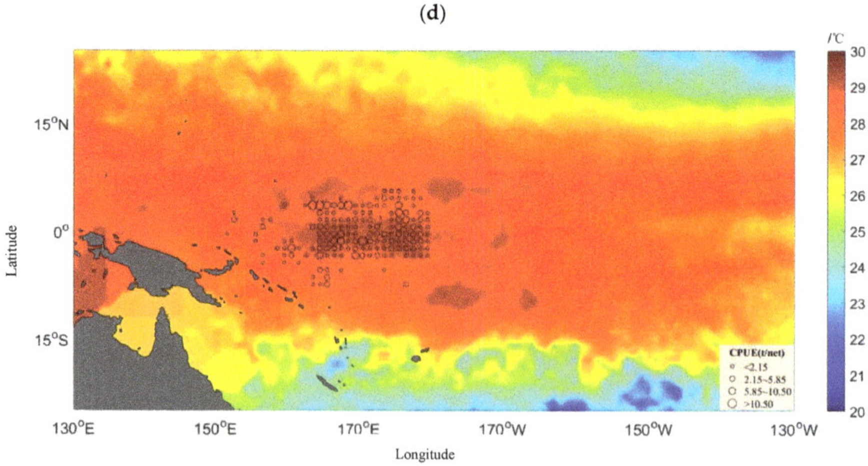

(**e**)

Figure 1. *Cont.*

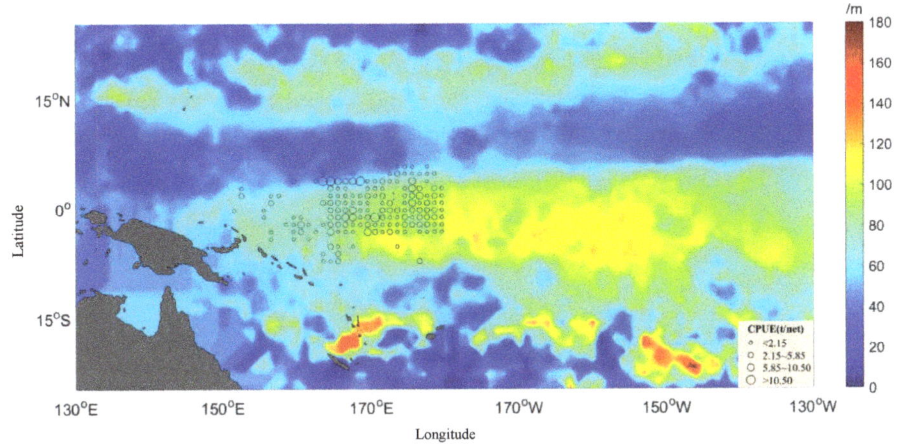

(f)

Figure 1. Overlay distribution of the upper temperature of the thermocline, the upper depth of the thermocline, and the catch per unit effort (CPUE) of yellowfin tuna. (**a**) The upper temperature of the thermocline in a La Niña year (2010); (**b**) the upper depth of the thermocline in a La Niña year (2010); (**c**) the upper temperature of the thermocline in a normal year (2013); (**d**) the upper depth of the thermocline in a normal year (2013); (**e**) the upper temperature of the thermocline in an El Niño year (2015); (**f**) the upper depth of the thermocline in an El Niño year (2015).

3.2. Temporal and Spatial Variation of Thermocline Strength and Thickness with Catch

As shown in Figure 2, the thermocline strength was weaker in the west and stronger in the east. The thermocline strength in the La Niña year was greater than that in the El Niño year in the area of 180° W of the equatorial WCPO. Conversely, the thermocline strength in the La Niña year was less than that in the El Niño year in the area of the 180° E region. On the whole, the thermocline thickness was greater in the west and lesser in the east. There was a thick band-like structure on each side, with an axis of 15° N and 15° S. The CPUE was mainly distributed in the thickness of 120–200 m. Figure 3 illustrates how the maximum CPUE value of yellowfin tuna was shifted westward in longitude and southward in latitude in the La Niña year, and shifted eastward in longitude and northward in latitude in the El Niño year.

The annual output was high in La Niña years such as 2008, 2010, 2011, and 2016 (Figure 4). The main operation area was located in the center of the WCPO warm pool in La Niña years. The SST was very suitable for the growth of yellowfin tuna. Therefore, both the catch and resource abundance were high in La Niña years. The annual production was low in El Niño years such as 2009 and 2015. In El Niño years, the western Pacific thermocline became shallower, whereas the eastern and central Pacific thermocline became deeper. In La Niña years, however, the western Pacific thermocline became deeper, whereas the eastern and central Pacific thermocline became shallower. Hence, the thermocline variation would have caused changes in the yellowfin tuna central fishing grounds. The variation patterns between annual catch and CPUE were consistent.

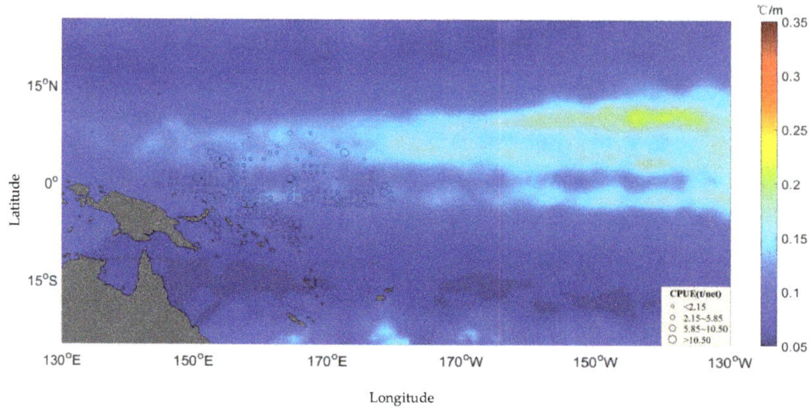

(**a**)

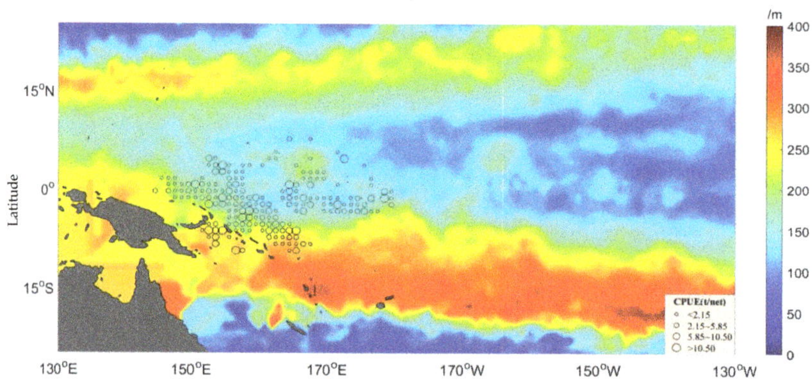

(**b**)

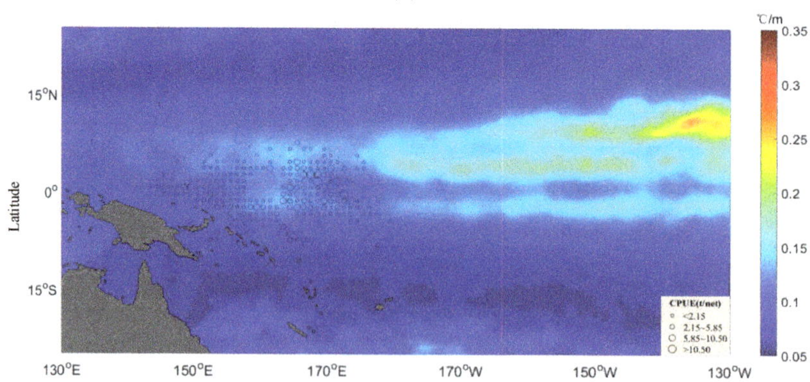

(**c**)

Figure 2. *Cont.*

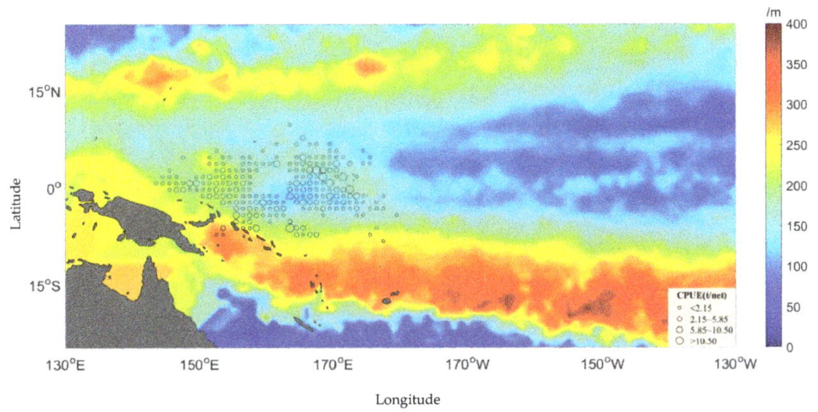

(**d**)

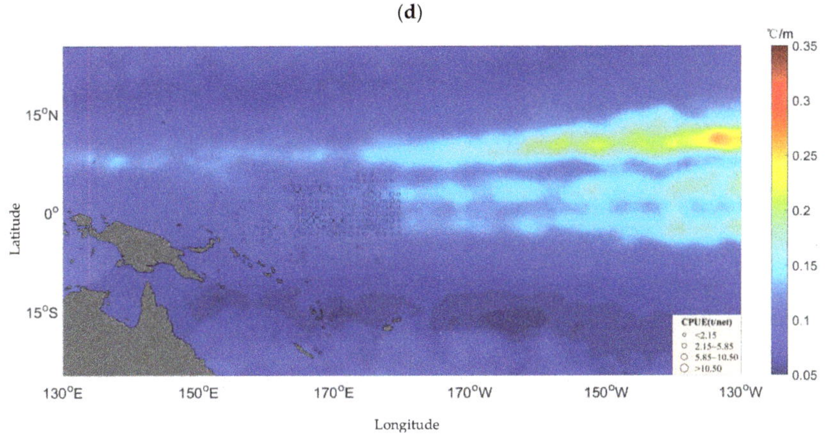

(**e**)

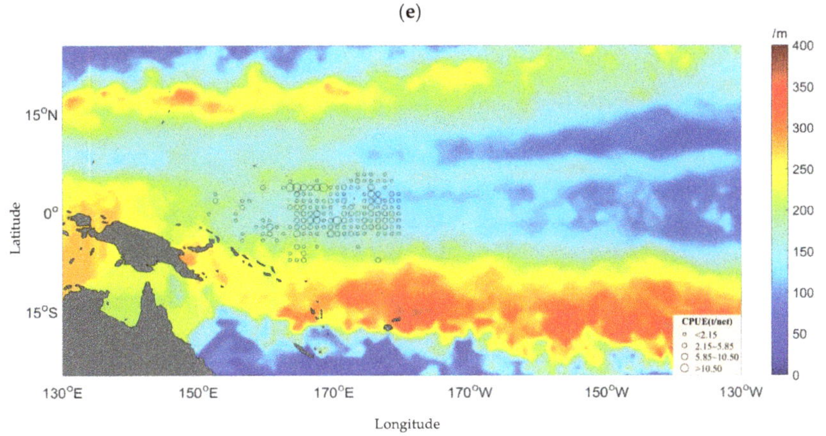

(**f**)

Figure 2. Overlay distribution of the strength and thickness of the thermocline and the catch per unit effort (CPUE) of yellowfin tuna. (**a**) The strength of the thermocline in a La Niña year (2010); (**b**) the thickness of the thermocline in a La Niña year (2010); (**c**) the strength of the thermocline in a normal year (2013); (**d**) the thickness of the thermocline in a normal year (2013); (**e**) the strength of the thermocline in an El Niño year (2015); (**f**) the thickness of the thermocline in an El Niño year (2015).

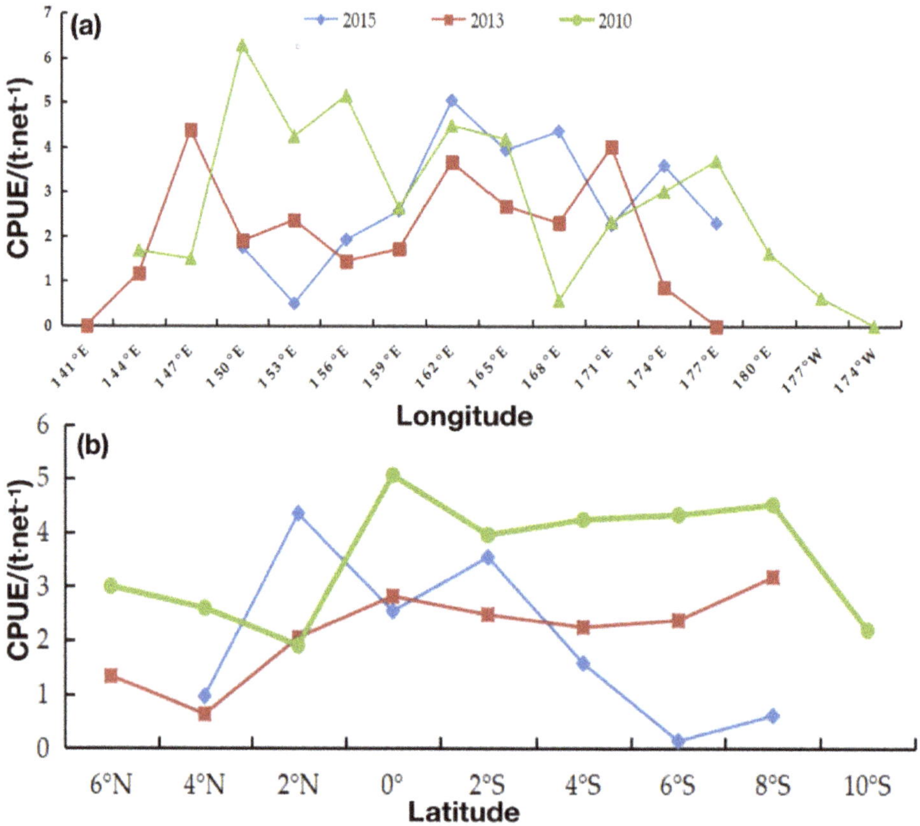

Figure 3. Distribution of catch per unit effort (CPUE) with longitude (**a**) and latitude (**b**) for yellowfin tuna in the central and western Pacific.

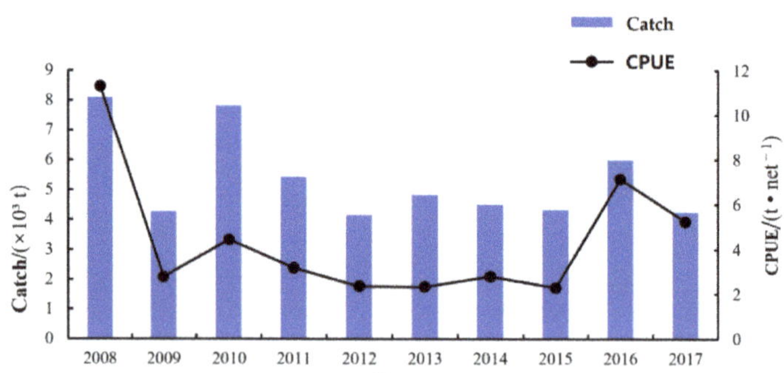

Figure 4. Annual catch and catch per unit effort (CPUE) statistics of yellowfin tuna in the central and western Pacific.

3.3. Analysis Results of the GAM

The GAM was used to analyze the impact of environmental variables on the temporal and spatial variation of yellowfin tuna catch. The results show that the AIC value of the model decreased with the number of factors. The final model retained all the input variables. Meanwhile, the AIC value of the model was the lowest, with a CPUE total variance interpretation of 21.1%, as illustrated in Table 3. In the model, the contribution of variables explaining the change in CPUE indicated the degree of impact on CPUE. The contribution of the time factor was 15.4%, that of the spatial factor was 2.5%, and that of the environmental factors was 3.2% (Table 4). The year was the most important factor, followed by the ONI, month, latitude, longitude, and upper boundary temperature and depth of thermocline, whereas thermocline strength and thickness had little impact on the GAM.

Table 3. Statistical characteristics of GAM models.

Formulae	AIC	Deviance/%	R2adj
Log(CPUE + 1) = s(y)	11,712.65	13.7	0.135
Log(CPUE + 1) = s(y) + s(m)	11,637.99	15.4	0.151
Log(CPUE + 1) = s(y) + s(m) + s(lat)	11,570.29	16.8	0.164
Log(CPUE + 1) = s(y) + s(m) + s(lat) + s(lon)	11,521.25	17.9	0.174
Log(CPUE + 1) = s(y) + s(m) + s(lat) + s(lon) + s(upt)	11,485.4	18.6	0.181
Log(CPUE + 1) = s(y) + s(m) + s(lat) + s(lon) + s(upt) + s(dh)	11,472.67	18.9	0.184
Log(CPUE + 1) = s(y) + s(m) + s(lat) + s(lon) + s(upt) + s(dh) + s(uph)	11,468.99	19.1	0.186
Log(CPUE + 1) = s(y) + s(m) + s(lat) + s(lon) + s(upt) + s(dh) + s(uph) + s(intensity)	11,458.37	19.2	0.187
Log(CPUE + 1) = s(y) + s(m) + s(lat) + s(lon) + s(upt) + s(dh) + s(uph) + s(intensity) + s(ONI)	11,361.8	21.1	0.205

Note: y—year; m—month; lat—latitude; lon—longitude; upt—upper temperature of the thermocline; uph—upper depth of the thermocline; dh—thickness of the thermocline; intensity—strength of the thermocline; ONI—Oceanic Niño Index; GAM—generalized additive models.

Table 4. Test values of generalized additive models (GAM).

Variable	edf	Ref. df	F	P	Contribution Rate (%)
Year	8.046	8.757	75.019	<0.001	13.7
Month	6.394	7.550	6.030	2.72×10^{-7}	1.7
Latitude	4.757	5.809	12.212	5.51×10^{-13}	1.4
Longitude	5.768	6.783	10.550	3.27×10^{-12}	1.1
Upper temperature	1.000	1.000	9.311	0.002 29	0.7
Thickness of thermocline	1.000	1.000	1.052	0.305 02	0.3
Upper depth of thermocline	1.000	1.000	0.287	0.592 13	0.2
Intensity	1.000	1.001	9.390	0.002 19	0.1
ONI	8.109	8.750	14.491	<0.001	1.9

The effects of spatial and temporal variables on the CPUE were nonlinear, whereas the effects of the upper boundary temperature and depth of the thermocline and thermocline thickness and strength on the CPUE among environmental variables were linear (Figure 5). As shown in Figure 5, (1) the purse-seine CPUE of yellowfin tuna decreased from 2008 to 2009, barely fluctuated from 2009 to 2010, decreased from 2011 to 2012, and then increased by 2016. Hence, the impact of the year on the CPUE fluctuated. The confidence intervals were narrow in 2008, 2011, and 2013, indicating that those La Niña years had a strong impact on the CPUE. (2) The CPUE of yellowfin tuna changed little in different months. It remained unchanged from January to April, increased slightly by June, and then began to fluctuate. (3) The impact curve of latitude on the CPUE presented a dome shape. The CPUE value was the largest in the 0° equatorial region. The 95% confidence interval was narrow, and the confidence level was high. The CPUE increased with latitude from 10° S–0° and decreased with latitude from 0–5° N, showing that the CPUE increased gradually as the latitude approached the equator. (4) The CPUE increased with the increase in longitude from 140–150° E. However, the confidence interval was large, and the confidence level was low. The CPUE fluctuated and the confidence interval was narrow at 150–175° E, indicating that the meridional space close to the CPUE was 150–175° E. Moreover, the confidence level dropped with the increase in longitude and the expansion of the confidence interval in the east of 175° E.

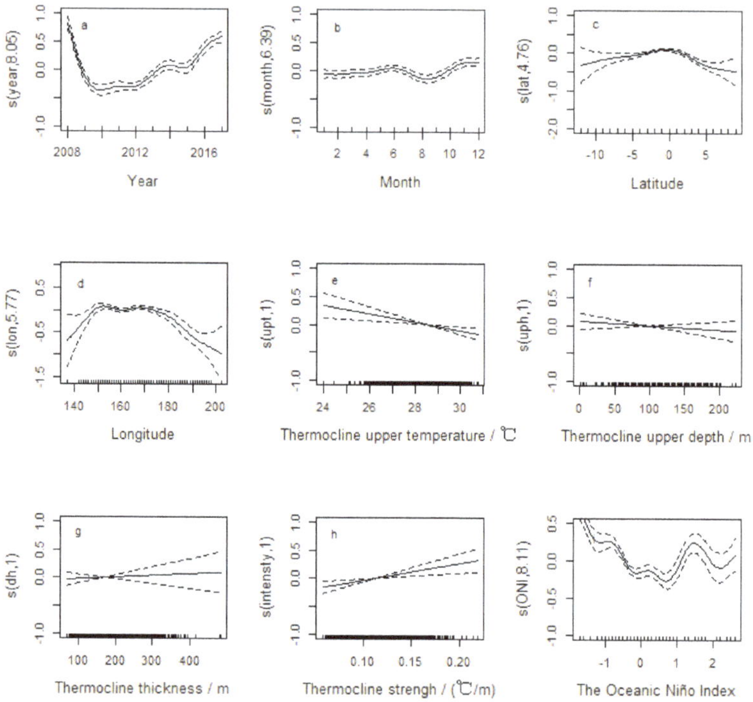

Figure 5. Effects of predictor variables: (**a**) year, (**b**) month, (**c**) Latitude, (**d**) Longitude, (**e**) thermocline upper temperature, (**f**) thermocline upper depth, (**g**) Thermocline thickness, (**h**) thermocline strength, and (i) the Oceanic Niño Index, derived from the generalized additive model (GAM) on catch per unit effort (CPUE).

Among the environmental variables, the relationship between the upper boundary temperature of the thermocline and the CPUE showed that the upper boundary temperature of the thermocline in the yellowfin tuna fishing grounds was between 27 and 30 °C, with an optimal upper boundary temperature range of 27.5–29.5 °C. The upper boundary depth of the thermocline was between 50 and 150 m. The optimal upper boundary depth range was 80–120 m. The thermocline thickness of the fishing grounds was between 100 and 200 m. The CPUE increased with the thickness of the thermocline. However, the confidence interval increased, and the confidence level decreased. The thermocline strength of fishing grounds was between 0.08 and 0.13 $°C{\cdot}m^{-1}$ and was positively correlated with the CPUE. The CPUE increased with the increase in ONI, and 0.3 was the maximum value. ONI in the range of −1–0.6 was closely related to CPUE.

4. Discussion

Temperature is a key environmental factor affecting fish activities [28]. It can directly and indirectly affect fish life activities, such as spawning [29], embryonic development [30,31], survival rate [32], feeding metabolism [33], migration [34], and habitat distribution [35]. Yellowfin tuna is a warm-water fish species that needs a certain water temperature to inhabit and spawn. The temporal and spatial variation of thermocline characteristic parameters of yellowfin tuna fishing grounds in the WCPO in El Niño and La Niña years shows that the CPUE in a La Niña year moved westward to 170° E as the high-value zone of the upper boundary temperature contracted westward to 145° E. In an El Niño year, the CPUE moved eastward with the eastward expansion of the high-value zone of the upper boundary temperature, which generally moved eastward to the east of 165° E. The reason may be that the upper boundary temperature of the thermocline affects the tuna spatial distribution. The high-value zone of 27–29.5 °C of the upper boundary temperature in El Niño years expanded eastward, whereas that in La Niña years moved westward. Furthermore, changes in the vertical structure of the water temperature profile also affect the horizontal spatial distribution and fishing mode for tuna [36]. To summarize, the CPUE distribution of fishing grounds in El Niño and La Niña years change with the eastward expansion or westward shift, respectively, of the upper boundary temperature of the thermocline, which makes new areas suitable for fishing grounds. Thermocline changes in the tropical Pacific are caused by SSTAs, which consist of enhanced deep atmospheric convection and westerly wind anomalies in the central Pacific. During all El Niño events, weaker equatorial trade winds produce an eastward shift of the western Pacific warm pool, and a deeper thermocline response to this wind anomaly in turn reinforces the initial warming. Sea-level anomalies in this area exhibit a zonal seesaw indicative of a deeper (shallower) thermocline in the eastern (western) equatorial Pacific [37,38]. The strong temperature gradient of the thermocline is a physical barrier for skipjack and juvenile yellowfin and bigeye tuna, whereas adult yellowfin and bigeye tuna can dive below the thermocline to chase mesopelagic prey. Therefore, changes in the vertical thermal structure of the ocean associated with ENSO can potentially impact the catchability of tuna species by different fishing gears. Purse seiners targeting surface tuna use the top of the thermocline as a lower barrier to trap tuna schools [39].

In order to accurately analyze the relationship between the abnormal climate phenomena and the abundance and spatial-temporal distribution of yellowfin tuna resources in the WCPO, this study explored the impacts of environmental variables on the temporal and spatial variation in yellowfin tuna catch by using a generalized additive model (GAM) and analyzing the optimal environmental parameters of the thermocline where the fishing grounds were located. The results showed that the contribution of the time factor was 15.4%, that of the spatial factor was 2.5%, and that of the environmental factors was 3.2%. The upper boundary temperature of the thermocline in yellowfin tuna fishing grounds was mostly between 27 and 30 °C. The optimal upper boundary temperature range was 27.5–29.5 °C. The upper boundary temperature of the thermocline had the greatest impact among the subsurface environmental factors.

According to the analysis results of spatial superposition and the GAM, the thermocline strength in the WCPO was weaker in the west and stronger in the east. The 180° W region of the equatorial WCPO was the main fishing area in which the thermocline strength in La Niña years was greater than that in El Niño years. Conversely, the thermocline strength in La Niña years was less than that in El Niño years in the 180° E region. Moreover, the CPUE was closely related to the thermocline in the strength range of 0.08–0.13 $°C \cdot m^{-1}$.

Regarding the upper boundary depth of the thermocline, the thermocline was thicker in the west and thinner in the east. There was a thick band-like structure on each side with an axis spanning 15° N and 15° S. The CPUE was distributed in the range of 120–200 m. The width of the thermocline in the WCPO in La Niña years was lesser than in El Niño years. GAM analysis showed that the upper boundary depth of the thermocline where the fishing grounds were located was between 50 and 150 m. In addition, the optimal upper boundary depth of the thermocline where the fishing grounds were located ranged from 80 to 120 m, which was consistent with the 70–109 m suitable upper boundary depth of the thermocline for yellowfin tuna in the WCPO [40]. Compared with the normal years, the upper boundary depth difference of the thermocline on the east and west sides of the equatorial WCPO was larger in La Niña years, and the upper boundary depth value of 80–130 m was shifted westward. Compared with normal years, the upper boundary depth difference of the thermocline on the east and west sides of the equatorial WCPO decreased in El Niño years. The upper boundary depth value of 80–130 m moved eastward.

Studies [41,42] have suggested that the center of gravity of the catch in the WCPO moves eastward and southward in El Niño years, and moves slightly westward and northward in La Niña years. This is consistent with the conclusion of this study for longitude.

The slight difference in latitude may be the result of different research objects. The research object of this paper was yellowfin tuna, whereas other studies in the literature [28] considered skipjack tuna. Moreover, the time series of the research data were inconsistent. The data used in this article are relatively updated data, which may also have led to differences. According to the statistical results of the catch in this paper, the yield in La Niña years was higher than that in El Niño years. Chen et al. [43] found that the yield was higher in La Niña years and lower in El Niño years when studying the effects of El Niño and La Niña on skipjack abundance in the WCPO. Deary et al. [44] also found that the production of yellowfin tuna in the central Pacific increased significantly in La Niña years. This may be because the upper boundary depth of the thermocline in the WCPO in La Niña years is deeper than in El Niño years, and the thermocline strength is higher than in El Niño years. In La Niña years, the suitable vertical habitat space of yellowfin tuna in this area is compressed, which is conducive to surface fishing and high catch. In contrast, the catch is lower during El Niño years [45,46]. In addition, the vertical structure of the thermocline has little impact on the fishing ground due to purse-seine operation. However, the thermocline changes caused by abnormal climate events have significant impacts on the CPUE. This study provides a reference for purse-seine tuna production in the tropical WCPO in abnormal climate years and for the study of the relationship between the temporal and spatial distribution of tuna and the thermocline.

Author Contributions: Conceptualization, W.Z., H.H. and S.J.; methodology, W.Z.; software, H.H.; validation, W.Z., H.H. and S.J.; formal analysis, W.Z.; investigation, W.Z., H.H. and S.J.; resources, W.F.; data curation, W.Z. and H.H.; writing—original draft preparation, W.Z. and S.J.; writing—review and editing, W.Z., H.H., S.J. and W.F.; visualization, H.H.; supervision, W.F.; project administration, W.F.; funding acquisition, W.F. and S.J. All authors have read and agreed to the published version of the manuscript.

Funding: This work was financially supported by the National Key Research and Development Project of China (2019YFD0901405) and the startup funding of Minjiang University (32304307).

Institutional Review Board Statement: Not applicable.

Conflicts of Interest: The authors declare no conflict of interest.

References

1. FAO. *The State of Food and Agriculture 2020. Overcoming Water Challenges in Agriculture*; FAO: Rome, Italy, 2020.
2. McKinney, R.; Gibbon, J.; Wozniak, E.; Galland, G. *Netting Billions 2020: A Global Tuna Valuation*; The Pew Charitable Trusts: Philadelphia, PA, USA, 2020.
3. McCluney, J.K.; Anderson, C.M.; Anderson, J.L. The fishery performance indicators for global tuna fisheries. *Nat. Commun.* **2019**, *10*, 1641. [CrossRef] [PubMed]
4. FFA Tuna Development Indicators 2016. Available online: https://www.ffa.int/system/files/FFATunaDevelopmentIndicatorsBrochure.pdf (accessed on 11 January 2022).
5. Lehodey, P.; Senina, I.; Calmettes, B.; Hampton, J.; Nicol, S. Modelling the impact of climate change on Pacific skipjack tuna population and fisheries. *Clim. Chang.* **2013**, *119*, 95–109. [CrossRef]
6. Clark, S.; Bell, J.; Adams, T.; Allain, V.; Aqorau, T.; Hanich, Q.; Jaiteh, V.; Lehodey, P.; Pilling, G.; Senina, I.; et al. The Parties to the Nauru Agreement (PNA) 'Vessel Day Scheme': A cooperative fishery management mechanism assisting member countries to adapt to climate variability and change. In *Fisheries and Aquaculture Technical Paper 667. Adaptive Management of Fisheries in Response to Climate Change*; FAO: Rome, Italy, 2021; pp. 209–224.
7. Asch, R.G.; Cheung, W.W.L.; Reygondeau, G. Future marine ecosystem drivers, biodiversity, and fisheries maximum catch potential in Pacific Island countries and territories under climate change. *Mar. Policy* **2018**, *88*, 285–294. [CrossRef]
8. Lam, V.W.Y.; Allison, E.H.; Bell, J.D.; Blythe, J.; Cheung, W.W.L.; Frölicher, T.L.; Gasalla, M.A.; Sumaila, U.R. Climate change, tropical fisheries and prospects for sustainable development. *Nat. Rev. Earth Environ.* **2020**, *1*, 440–454. [CrossRef]
9. Cai, W.J.; Borlace, S.; Lengaigne, M.; Rensch, P.V.; Collins, M.; Vecchi, G.; Timmermann, A.; Santoso, A.; McPhaden, M.J.; Wu, L.X.; et al. Increasing frequency of extreme El Niño events due to greenhouse warming. *Nat. Clim. Chang.* **2014**, *4*, 111–116. [CrossRef]
10. Lehodey, P.; Bertignac, M.; Hampton, J.; Lewis, A.; Picaut, J. El Niño Southern Oscillation and tuna in the western Pacific. *Nature* **1997**, *389*, 715–718. [CrossRef]
11. Hampton, J.; Lewis, A.; Williams, P. *The Western and Central Pacific Tuna Fishery: Overview and Status of Stocks*; Oceanic Fisheries Programme; Secretariat of the Pacific Community: Nouméa, New Caledonia, 1999; Volume 39.

12. Hampton, J. Estimates of tag-reporting and tag-shedding rates in a large-scale tuna tagging experiment in the western tropical Pacific Ocean. *Oceanogr. Lit. Rev.* **1997**, *11*, 1346.
13. Climate Variability. Oceanic Niño Index. Available online: https://www.climate.gov/news-features/understanding-climate/climate-variability-oceanic-ni%C3%B1o-index (accessed on 11 January 2022).
14. Watters, G.M.; Hinke, J.T.; Reiss, C.S. Long-term observations from Antarctica demonstrate that mismatched scales of fisheries management and predator-prey interaction lead to erroneous conclusions about precaution. *Sci. Rep.* **2020**, *10*, 2314. [CrossRef]
15. Lan, K.W.; Lee, M.A.; Zhang, C.I.; Wang, P.Y.; Wu, L.J.; Lee, K.T. Effects of climate variability and climate change on the fishing conditions for grey mullet (Mugil cephalus L.) in the Taiwan Strait. *Clim. Chang.* **2014**, *126*, 189–202. [CrossRef]
16. Li, H.; Xu, F.; Zhou, W.; Wang, D.; Wright, J.S.; Liu, Z.; Lin, Y. Development of a global gridded Argo data set with B arnes successive corrections. *Geophys. Res. Ocean.* **2017**, *122*, 866–889. [CrossRef]
17. Liu, S.L.; Liu, Z.D.; Li, H.; Li, Z.Q.; Wu, X.F.; Sun, C.H.; Xu, J.P. Manual of Global Ocean Argo gridded data set. *Geophys. Res. Ocean.* **2020**, *122*, 14.
18. Akima, H.A. New Method of Interpolation and Smooth Curve Fitting Based on Local Procedures. *J. ACM (JACM)* **1970**, *17*, 589–602. [CrossRef]
19. Zhou, Y.X.; Li, B.L.; Zhang, Y.J.; Ba, N.C. World oceanic thermocline characteristics in winter and summer. *Mar. Sci. Bull.* **2002**, *21*, 16–22.
20. Guisan, A.; Edwards, J.T.C.; Hastie, T. Generalized linear and generalized additive models in studies of species distributions: Setting the scene. *Ecol. Model.* **2002**, *157*, 89–100. [CrossRef]
21. Mainuddin, M.; Ssiton, K.; Saiton, S.I. Albacore fishing ground in relation to oceanographic conditions in the western North Pacific Ocean using remotely sensed satellite data. *Fish Oceanogr.* **2008**, *17*, 61–73. [CrossRef]
22. Briand, K.; Molony, B.; Lehodey, P. A study on the variability of albacore (Thunnus alalunga) longline catch rates in the southwest Pacific Ocean. *Fish Oceanogr.* **2011**, *20*, 517–529. [CrossRef]
23. Wu, S.N.; Chen, X.J.; Liu, Z.N. Establishment of forecasting model of the abundance index for chub mackerel (Scomber japonicus) in the northwest Pacific Ocean based on GAM. *Acta Oceanol. Sin.* **2019**, *41*, 36–42.
24. Yu, J.; Hu, Q.; Tang, D.; Pimao, C. Environmental effects on the spatiotemporal variability of purpleback flying squid in Xisha-Zhongsha waters, South China Sea. *Mar. Ecol. Prog. Ser.* **2019**, *623*, 25–37. [CrossRef]
25. Yang, S.L.; Zhang, B.B.; Tang, B.J.; Hua, C.J.; Zhang, S.M.; Fang, X.M.; Dai, Y.; Feng, C.L. Influence of vertical structure of the water temperature on bigeye tuna longline catch rates in the tropical Atlantic Ocean. *Fish Sci. China* **2017**, *4*, 875–883.
26. Wood, S.N. *Generalized Additive Models: An Introduction with R*, 2nd ed.; Chapman & Hall/CRC: Boca Raton, FL, USA, 2017; p. 476.
27. Historical El Niño/La Niña Episodes (1950-Present). Available online: https://origin.cpc.ncep.noaa.gov/products/precip/CWlink/MJO/enso.shtml (accessed on 11 January 2022).
28. Dai, D.N.; Liu, H.S.; Dai, X.J.; Tian, S.Q. The relationship between ENSO and spatio-temporal distribution of CPUE of yellowfin tuna (Thunnus albacares) by purse seine in the Eastern Pacific Ocean. *Shanghai Ocean Univ.* **2011**, *20*, 571–578.
29. Kurita, Y.; Fujinami, Y.; Amano, M. The effect of temperature on the duration of spawning markers—Migratory-nucleus and hydrated oocytes and postovulatory follicles—In the multiple-batch spawner Japanese flounder (Paralichthys olivaceus). *Fish. Bull.* **2011**, *109*, 79–89.
30. Arenzon, A.; Lemos, C.A.; Bohrer, M. The influence of temperature on the embryonic development of the annual fish cynopoecilus melanotaenia (cyprinodontiformes, rivulidae). *Braz. J. Biol.* **2002**, *62*, 743. [CrossRef]
31. Schirone, R.C.; Gross, L. Effect of temperature on early embryological development of the zebra fish, Brachydanio rerio. *J. Exp. Zool. Part A Ecol. Integr. Physiol.* **1968**, *169*, 43–52. [CrossRef]
32. Emilie, R.D.; Alain, P.; Daniel, D.C.; Pascal, F.; Fabrice, T.; Rummer, J.L. Strong effects of temperature on the early life stages of a cold stenothermal fish species, brown trout (Salmo trutta L.). *PLoS ONE* **2016**, *11*, e0155487.
33. Jobling, M. The influences of feeding on the metabolic rate of fishes: A short review. *J. Fish Biol.* **1981**, *18*, 385–400. [CrossRef]
34. Jansen, T.; Gislason, H. Temperature affects the timing of spawning and migration of north sea mackerel. *Cont. Shelf Res.* **2011**, *31*, 64–72. [CrossRef]
35. Freitas, C.; David, V.R.; Moland, E.; Olsen, E.M. Sea temperature effects on depth use and habitat selection in a marine fish community. *J. Anim. Ecol.* **2021**, *90*, 1787–1800. [CrossRef]
36. Yang, S.L.; Zhang, B.B.; Zhang, H.; Zhang, S.M.; Wu, Y.M.; Zhou, W.F.; Feng, C.L. A Review: Vertical Swimming and Distribution of Yellowfin Tuna Thunnus alalunga. *Fish. Sci.* **2019**, *38*, 119–126.
37. Barnston, A.G.; Tippett, M.K.; L'Heureux, M.L.; Li, S.; DeWitt, D.G. Skill of real-time seasonal ENSO model predictions during 2002–11: Is our capability increasing? *Bull. Am. Meteorol. Soc.* **2012**, *93*, 631–651. [CrossRef]
38. Kirtman, B.P.; Min, D.; Infanti, J.M.; Kinter, J.L., III; Paolino, D.A.; Zhang, Q.; van den Dool, H.; Saha, S.; Mendez, M.P.; Becker, E.; et al. The North American Multimodel Ensemble: Phase-1 Seasonal-to-Interannual Prediction; Phase-2 toward Developing Intraseasonal Prediction. *Bull. Am. Meteorol. Soc.* **2014**, *95*, 585–601. [CrossRef]
39. Lehodey, P.; Bertrand, A.; Hobday, A.J.; Kiyofuji, H.; McClatchie, S.; Menkès, C.E.; Pilling, G.; Polovina, J.; Tommasi, D. ENSO Impact on Marine Fisheries and Ecosystems. In *El Niño Southern Oscillation in a Changing Climate*; McPhaden, M.J., Santoso, A., Cai, W., Eds.; AGU and John Wiley & Sons: Washington, DC, USA; New York, NY, USA, 2020; pp. 429–451.

40. Yang, S.L.; Zhang, B.B.; Jin, S.F.; Fan, W. Relationship between the temporal-spatial distribution of longline fishing grounds of yellowfin tuna (Thunnus albacares) and the thermocline characteristics in the Western and Central Pacific Ocean. *Acta Oceanol. Sin.* **2015**, *37*, 78–87.
41. Shen, J.H.; Chen, X.D.; Cui, X.S. Analysis on spatial-temporal distribution of skipjack tuna catches by purse seine in the Western and Central Pacific Ocean. *Mar. Fish.* **2006**, *1*, 13–19.
42. Guo, A.; Chen, X.J. The relationship between ENSO and tuna purse-seine resource abundance and fishing grounds distribution in the Western and Centra1 Pacific Ocean. *Mar. Fish.* **2005**, *27*, 338–342.
43. Chen, Y.Y.; Chen, X.J. Influence of El Nino/La Nina on the abundance index of skipjack in the Western and Central Pacific Ocean. *Shanghai Ocean Univ.* **2017**, *26*, 113–120. [CrossRef]
44. Deary, A.L.; Moret, F.S.; Engels, M.; Zettler, E.; Jaroslow, G.; Sancho, Y.G. Influence of Central Pacific Oceanographic Conditions on the Potential Vertical Habitat of Four Tropical Tuna Species1. *Pac. Sci.* **2015**, *69*, 461–476. [CrossRef]
45. Prince, E.D.; Goodyear, C.P. Hypoxia-based habitat compression of tropical pelagic fishes. *Fish Oceanogr.* **2006**, *15*, 451–464. [CrossRef]
46. Prince, E.D.; Luo, J.; Phillip, G.C.; Hoolihan, J.P.; Snodgrass, D.; Orbesen, E.S.; Serafy, J.E.; Ortiz, M.; Schirripa, M.J. Ocean scale hypoxia-based habitat compression of Atlantic istiophorid billfishes. *Fish Oceanogr.* **2010**, *19*, 448–462. [CrossRef]

sustainability

MDPI

Article

Analysis of the Temporal and Spatial Distribution of Extreme Climate Indices in Central China

Yan Li [1], Junfang Zhao [2,*], Rui Miao [3,*], Yan Huang [1], Xiaoqing Fan [1], Xiaoqing Liu [1], Xueqi Wang [1], Ye Wang [1,4] and Yuyang Shen [1,4]

1 CMA Public Meteorological Service Centre, Beijing 100081, China; liyan81_cma@163.com (Y.L.); dandanhy123@163.com (Y.H.); xqfan@sina.com (X.F.); 18801383698@163.com (X.L.); xueqiwang95@163.com (X.W.); wangyeye09@163.com (Y.W.); shenyuyang_00@163.com (Y.S.)
2 State Key Laboratory of Severe Weather, Chinese Academy of Meteorological Sciences, Beijing 100081, China
3 Tianjin Meteorological Service Center, Tianjin 300074, China
4 Huafeng Meteorological Media Group Co., Ltd., Beijing 100081, China
* Correspondence: zhaojf@cma.gov.cn (J.Z.); miaorui_mr@163.com (R.M.); Tel.: +86-10-58995896 (J.Z.)

Abstract: Using the daily precipitation and temperature data of 101 meteorological stations in four provinces of central China (Henan, Hubei, Hunan, Jiangxi) from 1988 to 2017, we analyzed the temporal and spatial dynamics and periodicity of nine extreme climate indices in central China, using the predefined methods for analyzing extreme climate events, such as a M-K test, a linear trend analysis, and a wavelet analysis. The extreme climate characteristics and changes in central China in the past 30 years were revealed. The results showed that the CSDI was significantly reduced linearly at a rate of −0.19 d/10a, and the WSDI and TXx increased significantly at rates of 0.25 d/10a and 0.30 °C/10a, respectively. The CDD decreased significantly at a rate of −1.67 d/10a. The duration of extreme low-temperature and drought events in central China showed a gradual shortening, while the duration of extreme high-temperature events and the high-temperature values increased. The results of the abrupt climate change test showed that some extreme climate indices in central China had significant abrupt climate changes after 2000. Analyzing the cyclicality of each index, it was determined that the extreme climate index in central China had a significant cyclical change every 2–4 years, and the change was more notable after 2000. Analyzing the spatial distribution of the extreme climate indices, it was determined that Jiangxi had the longest duration of all high-temperature events, and was the largest and longest of events of extreme precipitation. It was also determined that the Jiangxi region was at greater risk of extreme climate events in central China. The results of this study can provide a scientific basis for climate change trends, local disaster prevention, and mitigation management in central China.

Keywords: extreme climate indices; temporal and spatial dynamics; linear trend; climate abrupt change; central China

Citation: Li, Y.; Zhao, J.; Miao, R.; Huang, Y.; Fan, X.; Liu, X.; Wang, X.; Wang, Y.; Shen, Y. Analysis of the Temporal and Spatial Distribution of Extreme Climate Indices in Central China. *Sustainability* **2022**, *14*, 2329. https://doi.org/10.3390/su14042329

Academic Editor: Xiaodong Yan

Received: 9 December 2021
Accepted: 14 February 2022
Published: 18 February 2022

1. Introduction

The Sixth Assessment Report (AR6) of the Intergovernmental Panel on Climate Change (IPCC) reported that the average global surface temperature has risen by approximately 1 °C compared to 1850–1900, and forecasted that the global average temperature rise is expected to reach or exceed 1.5 °C in the next 20 years [1,2]. This report predicts that, because of this increase in temperature, climate change will intensify over the next few decades. When the global temperature rises by 1.5 °C, the frequency of heat waves will increase, the length of the warm season will be extended, and the cold season will be shortened; when the global temperature rises by 2 °C, the high temperature extremes will increase. These changes will cause the critical tolerance thresholds for agriculture and human health to be reached more frequently. Issues caused by climate change are not only temperature-based, but will bring additional combinations of changes to different regions. Further warming will cause changes such as shifts in dryness, humidity, wind, snow, and ice [1,2]. In China, extreme events, such as regionally persistent heavy rainfall, high temperatures and heat waves, and persistent drought, have attracted wide attention due to their potential regional hazards.

Central China is a key area for the transient between China's east–west and north–south borders and is a hub for land and water transportation. This region has a developed agriculture and a foundation for light and heavy industries [3]. Central China has a large north–south span with a complex weather system. Extreme precipitation events can trigger floods in central China and can have a significant impact on the industrial and agricultural production, including the local population [4]. In particular, low-probability, high-intensity, long-lasting rainstorms can lead to frequent large-scale severe floods, and can often cause a major loss of life and property [5,6]. Therefore, to ensure the sustainable development of the economy in central China, it is of great significance to study the development and evolution of extreme climate events in central China.

Research on climate extremes mainly focuses on the extreme value of various climactic events, the analysis of the changing trend of the extreme climate index, the frequency, intensity and change in major precipitation and temperature events, and the research on some major extreme climate events as they occur. Many have conducted an analysis on the impact of extreme climate events on agriculture, ecology, energy, etc. [7–14].

When determining the extreme climate index, many scholars choose suitable indices to discuss the changing characteristics and trends of the climate events, based on the extreme climate index published by the World Meteorological Organization (WMO). For example, Klein Tank and Können used the extreme climate index to study the trends of daily extreme climate and precipitation in Europe [15]. Studies by some scholars have shown that, in the past half-century, the difference between the maximum and minimum temperature has been significantly reduced [16]. Gallant analyzed the trend of the extreme climate index changes in the United States, Europe, and Australia from 1950 to 2012, to study the impact of climate change on those countries [17]. Pita-Díaz and Ortega-Gaucin used the climate change index to analyze changes in temperature and precipitation extremes in Zacatecas, Mexico [18]. Lucas et al. used daily high-resolution grid data to analyze 22 extreme climate indices from 1980 to 2016 in the XRB, located in the Amazon region of Brazil. The research results show that, in recent decades, both the day and night have become warmer [19]. The rainfall in the central and southern parts of the XRB shows a downwards trend, while the rainfall in the northern part of the basin shows an increasing trend [19]. Some scholars have found that, in the past 100 years, there has been a clear trend of humidification in Central Asia [20]. A variety of papers analyzed extreme climate indices in many regions of the world, including Asia, America, and Africa [21–24].

China's research on extreme climates is mainly focused on the above five aspects. In terms of extreme climate indices research, Shi et al. analyzed the extreme continuous diurnal trends of temperature and precipitation in China from 1961 to 2015. The results show that China's duration of cold periods (CSDI) and duration of warm periods (WSDI) are 0.9 days and 0.1 days per decade, respectively [25]. The CSDI has dropped significantly, while the WSDI has increased significantly, at a rate of 0.8 days per decade [25]. He et al. used an independent and comprehensive climate index to quantitatively assess extreme climate events in Guangdong [26]. Many scholars have conducted detailed analyses on the distribution and changing trends of multiple extreme climate indices in the region [27–30]; Zhai and Liu pointed out that the frequency of heavy rainfall events in the middle and lower reaches of the Yangtze River has increased, while droughts in the northeast, north, and southwest regions are clearer, and high temperatures and heat waves occur more frequently in the eastern region [31]. Liu and Xu pointed out that the total amount of precipitation in the southwestern region decreased slightly, but the maximum daily precipitation and precipitation intensity increased significantly [32]. Ju et al. showed that the highest and lowest temperature in East China are both increasing [33].

However, few have considered the entirety of central China as the area of research when studying the regional trends and impact of extreme climate indices. This article aims to analyze the extreme climate characteristics of the four provinces in central China by nine selected extreme climate indices, and quantify the variation patterns and temporal-spatial distribution characteristics of temperature and precipitation extreme events of central China from 1988 to 2017.This work aims to also provide a scientific basis and reference for decision-making in disaster prevention and mitigation in central China under the background of future climate change.

2. Materials and Methods

The daily precipitation series, daily maximum and minimum temperature series of 101 meteorological stations (Figure 1) in central China (Henan, Hubei, Hunan, Jiangxi) from 1988 to 2017 (30 years) were used. The data came from the National Meteorological Information Center. We conducted a quality control process on the data to eliminate the impact of site relocation or missing records. For example, Songshan in Henan Province was eliminated due to the lack of years recorded.

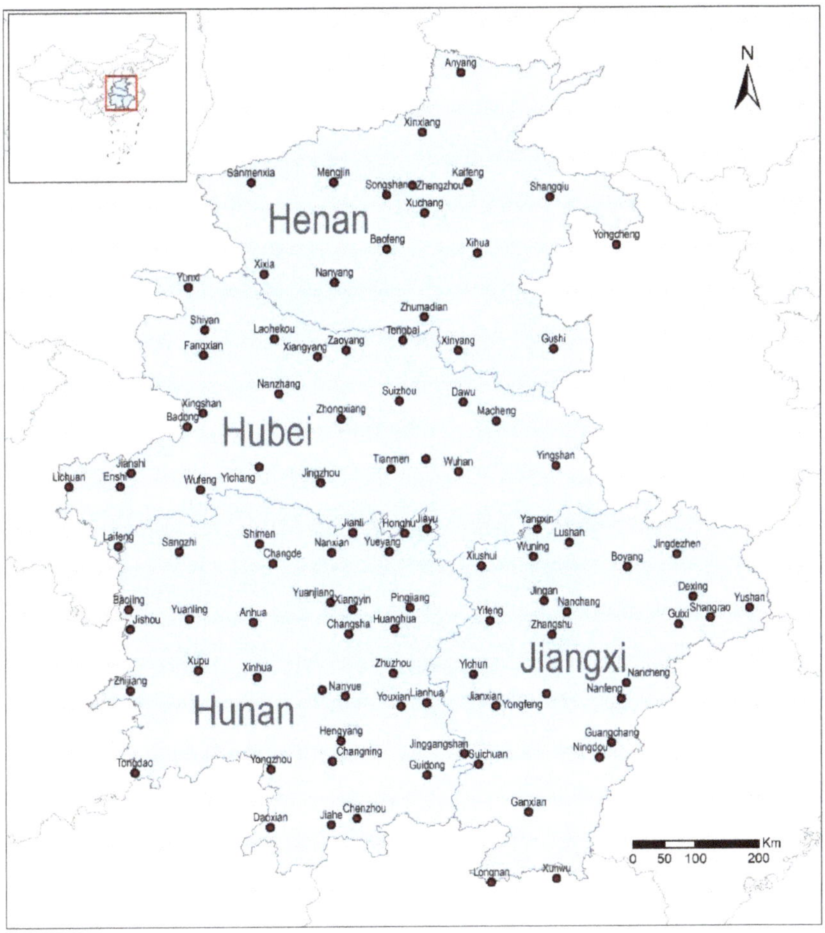

Figure 1. Distribution map of meteorological observation stations in central China.

The purpose of this study was to quantify the changes in extreme events of temperature and precipitation in the four provinces of central China, from 1988 to 2017. Considering the impact of continuous extreme climate events, we selected four of the extreme climate indices published by the WMO, which are WSDI, CSDI, CDD, and CWD. Considering the temperature and precipitation extremes, the five indices of ID, FD, TNn, TXx, and R95 were selected. By studying the temporal changes and spatial distribution of these nine extreme climate indices, the characteristics of the extreme climate events in the four provinces of central China were analyzed. The definitions of the indices are shown in Table 1. Since the daily maximum air temperature (TXx) had a greater impact in the summer, and the daily minimum air temperature (TNn) had a greater impact in the winter, TXx in the summer and TNn in the winter were emphatically analyzed.

Table 1. Climate indices derived from daily rainfall data and maximum and minimum temperatures, with definitions and units.

Indices	Name	Definition	Unit
ID	Icy days	Annual count of days with Tmax < 0 °C	d
FD	Frost days	Annual count of days with Tmin < 0 °C	d
WSDI	Duration of warm periods	Annual count of days with at least six consecutive days in which Tmax > 90 percentile	d
CSDI	Duration of cold periods	Annual count of days with at least six consecutive days in which Tmin > 90 percentile	d
TNn	Lowest Tmin	Lowest annual value of daily Tmin	°C
TXx	Highest Tmax	Highest annual value of daily Tmax	°C
R95	Very wet days	Total annual rainfall when PRCP > 95 percentile	mm
CDD	Consecutive dry days	Maximum length of drought, or maximum number of consecutive days with PRCP < 1 mm	d
CWD	Consecutive wet days	Maximum length of wet period, or maximum number of consecutive days with PRCP > 1 mm	d

To analyze the characteristics of the extreme climate index, a linear trend analysis method was used to discuss the variation trend of each element. The Mann–Kendall test method was used for the climate mutation testing of the extreme climate indices. This method was not limited to the distribution of samples, and could also exclude a few interferences of outliers [10]. It was suitable for meteorological non-normally distributed data, and the calculation was convenient. Climate mutation referred to the change of statistical characteristics of the climate state, which was manifested as the discontinuity of climate change. It was a phenomenon that the climate changes abruptly from one stable state to another. Among several detection methods of climate mutation, the Mann–Kendall method was very confident in the detection of mean mutation. It is a nonparametric statistical test method. Its advantage is that it does not need samples to follow a certain distribution, nor is it disturbed by a few outliers, and the calculation is relatively simple. Therefore, the Mann–Kendall test method is an effective tool recommended by WMO (World Meteorological Organization) for analyzing abrupt climate changes and related climatological research.

To analyze the temporal variation characteristics of each extreme climate index, the wavelet analysis method was used to study its periodicity. The wavelet analysis method introduced the window Fourier transform, which can obtain the frequency characteristics of physical quantities, and can also present the variation of the period with time, which is one of the important methods of period analysis.

To understand and master the cyclical change characteristics of the extreme climate indices in central China, wavelet analyses were conducted on the nine extreme climate indices, the significant oscillation periods, and the oscillation strength changes of the main periods at different times were obtained. First, the original time series of the extreme climate indices was standardized to eliminate the influence of units on the data analysis. Subsequently, the 95% confidence interval of the wavelet and the period range that passed the 0.05 significance level test were obtained by analyzing the wavelet power spectrum and the real part of the wavelet coefficients, to understand the characteristics of the main period change with time. By drawing the full spectrum of the wavelet, the changes of the main period of the extreme climate index and the main period passing the significance test were obtained.

3. Results

3.1. Spatial Distribution of the Extreme Temperature Index

From the spatial distribution map of the six extreme temperature indices in central China (Figure 2), it can be seen that the extreme temperature indices have clear spatial distribution characteristics, and are closely related to latitude and topography.

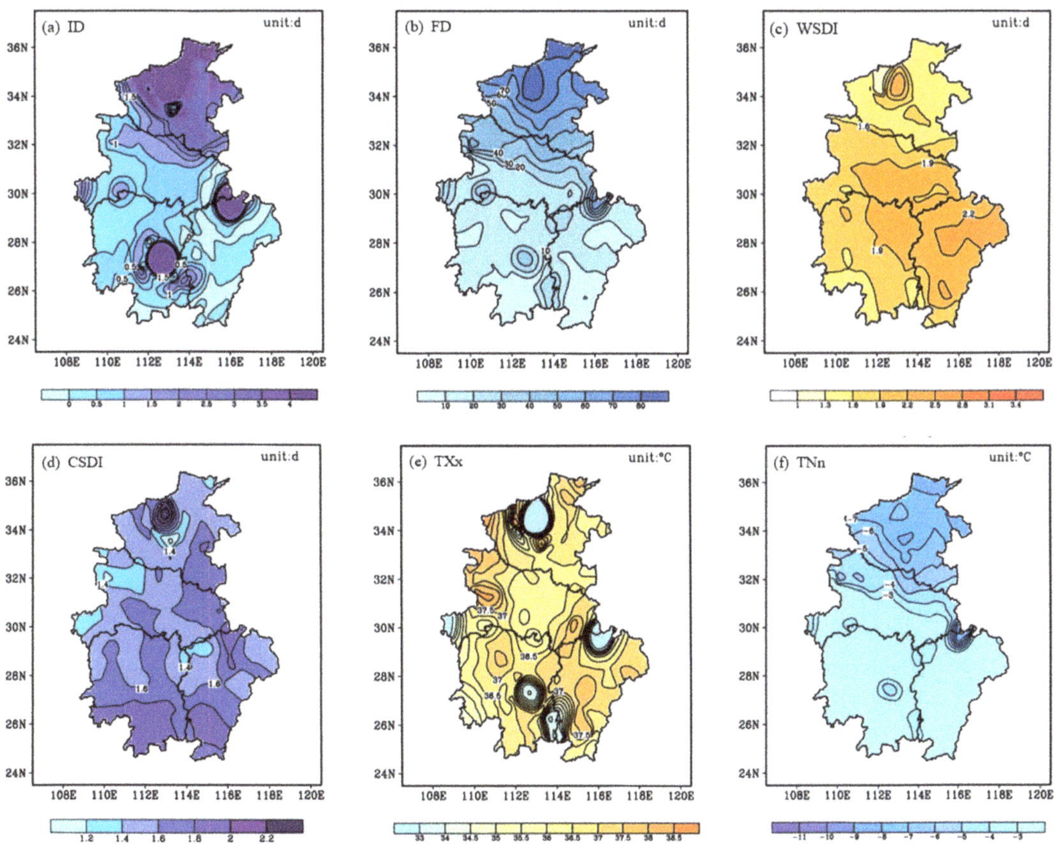

Figure 2. Spatial distribution of extreme temperature indices in central China 1988–2017.

The cold series indices include icing days (ID), frost days (FD), duration of cold periods, and the lowest minimum temperature (TNn). The spatial distribution map shows that the number of icing days (ID) and the number of frost days (FD) both gradually decrease from north to south. The large value areas are concentrated in Henan and northern Hubei, while southern Hubei, Hunan, and Jiangxi are the low-value areas in central China. In addition, Lushan Mountain in northern Jiangxi and Nanyue Mountain in southeastern Hunan have significantly more freezing days and frost days than surrounding areas, due to their high altitudes.

The duration of cold periods was shorter in the north and longer in the south. The area with the highest values were located in Hunan and southwestern Jiangxi. In contrast, Henan and Hubei had fewer cold duration days, and the distribution in Hubei was shorter in the west and longer in the east. The lowest minimum temperature (TNn) in central China gradually decreased from south to north. The average low-temperature center in winter in central China was mainly located in Henan. Hunan and Jiangxi were the areas with high winter temperatures in central China; generally higher than −3°C. However, affected by the mountainous topography, the TNn indices of Nanyue in Hunan and Lushan in Jiangxi reached −8.0 °C and −8.8 °C, respectively.

The warm series indices include the duration of warm periods and highest maximum temperature. The duration of the warm periods was longer in the south and shorter in the north. Most of the Jiangxi Province, central and eastern Hubei Province and eastern Hunan Province were high-value areas located in central China, and the occurrence of warm persistent events was generally more than double; the high-value areas gradually increased from west to east, indicating that the high-temperature events were persistent in these areas, making the regions prone to drought. Among them, Jiangxi had the highest number of persistently warm days. The number of warm sustained days in Hunan was generally characterized by a decreasing distribution from northeast to southwest, which are mostly related to the geomorphological characteristics of the mountains on the east, south, and west sides of Hunan.

The average value of the TXx index in central China in summer was 36.5 °C, and the high-value areas were mainly concentrated in Henan and western Hubei. Among them, Xingshan in Hubei had the highest value of the summer daily maximum temperature in central China, reaching 38.9 °C. The TXx index in Hubei showed the distribution characteristics of low in the middle and high in the east and west. The high-value area of TXx index in Hunan was located in the northern region, and the high-value area of TXx index in Jiangxi was located in the northwest and central regions of Jiangxi, both exceeding 37 °C.

3.2. Time Change Trend Analysis of the Extreme Temperature Index

The Mann–Kendall method (Figure 3) was used to test the trend of each extreme temperature index in central China in combination with trend analysis (see Table 2 for test statistics). The results showed that the trends of each extreme climate index were different. Combined with Figure 3 and Table 2, if the UF_K value is greater than 0, it indicates that the sequence shows an upward trend, and if it is less than 0, it indicates a downward trend. When UF_K and UB_K exceed the critical value, it indicates that the upward or downward trend is significant. The time that UF_K and UB_K curves intersect is the time when the mutation starts.

From the overall trend, only the CSDI index in the cold series index (ID, FD, CSDI, TNn) had a significant change, and the rate of change decreases with −0.19n/10a, indicating that the number of cold duration days had a tendency to decrease annually, with the duration of low temperature events becoming shorter. The warm series indices (WSDI, TXx) showed a significant increasing trend, WSDI increased with a linear trend of 0.25 d/10 a, and TXx increased with a linear trend of 0.30 °C/10a, indicating that there were more continuous warm events in central China, and the high value of the hottest temperature increases significantly. The duration of extremely high temperature events and the intensity of the events increased.

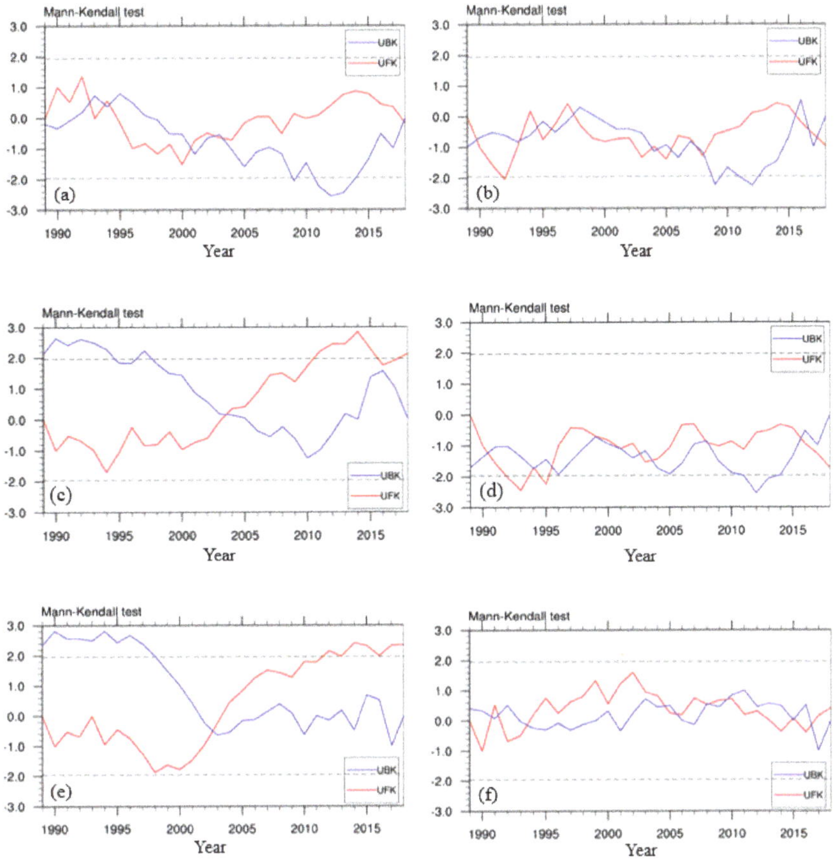

Figure 3. The Mann–Kendall test of the extreme temperature index in central China from 1988 to 2017: (**a**) Icing days (ID); (**b**) Frost days (FD); (**c**) Duration of warm periods (WSDI); (**d**) Duration of cold periods (CSDI); (**e**) Highest Tmax (TXx); (**f**) Lowest Tmin (TNn).

Table 2. Linear trends and abrupt climate changes in extreme temperature indices in central China in 1988–2017.

Extreme Temperature Index	Linear Trend Value	The Year of the Mutation	Trend Values before and after Climate Mutations	
			Before	After
ID	0.11 d/10 a	2004	−0.29 d/10 a	−1.22 d/10 a
FD	−1.35 d/10 a	2008	−0.29 d/10 a	−1.86 d/10 a *
WSDI	0.25 d/10 a *	2004	0.07 d/10 a	−0.1 d/10 a
CSDI	−0.19 d/10 a *		/	
TXx (Summer)	0.30 °C/10 a *	2003	−0.14 °C/10 a	−0.006 °C/10 a
TNn (Winter)	0.14 °C/10 a		/	

Note: * means value passed the 90% significance test, indicating that the linear trend is significant.

According to the results of the Mann–Kendall mutation test, there were climate mutations in several extreme temperature indices, as shown by the mutation of both the ID index and WSDI index in 2004, the mutation of the FD index in 2008, and the mutation of the TXx index in 2003. The results of the segmented trend analysis of the extreme climate index show that, after 2008, the frost index had a significant decrease, indicating that the winter and spring in central China will be warmer after 2008, and the number of days with minimum daily temperature below 0 °C will be decreased.

As seen from the above, the warm series indices (WSDI, TXx) have a significant linear trend, and abrupt climate change exists. The averages of the values before and after the mutation were further calculated, and the results showed that the warm series index average in central China increased significantly after the mutation. Specifically, the average durations of the warm periods were 1.6 times more than in 1988–2003, and 2.1 times more than in 2004–2017. The TXx in central China was 36.1 °C in 1988–2002, and 36.8 °C in 2003–2017. In addition, the maximum WSDI (3d) and TXx index (38 °C) appeared in 2013. This may be related to the sustained, large-scale, high-temperature event in China during the mid-summer of 2013. Some studies suggest that the sustained high temperature event in 2013 may be affected by the continued strength of the western Pacific subtropical high and the westward ridge point.

3.3. Periodic Analysis of the Extreme Temperature Index

By standardizing the original time series of the extreme temperature index and analyzing the wavelet power spectrum (as shown in Figure 4), the wavelet 95% confidence interval (the area inside the red solid line on the left of Figure 4) and the period range that passed the 0.05 significance level test (the area surrounded by the green dotted line on the left of Figure 4) were obtained. Considering the boundary effect of wavelet analysis, in the 30 years from 1988 to 2017, the confidence level of the wavelet analysis results for the 2–8 year cycle from 1990 to 2015 is greater than 95%.

This is done by plotting the full spectrum of the wavelet (the colored part in the left image of Figure 4) to obtain the main change period of the extreme temperature index (the peaks shown in the blue solid line on the right image of Figure 4) and the main period that passes the significance test (as shown in the right image of Figure 4, the crest of the blue solid line on the right of the red dotted line). Through the analysis of the real part of the wavelet coefficient (as shown in the black isoline in the left figure of Figure 4), the characteristics of the main change period with time and the stage of significant change were understood. The results of the wavelet analysis show that different extreme temperature indices show different oscillation periods, and there are some differences in the main periods of separate time periods.

In the cold series index of central China, the number of icing days (ID) shows a cycle, which is most notable from 2002 to 2013, and is dominated by a cycle of 2–3 years and sometimes four years (Figure 4a). Frost days show a significant cycle of 3–4 years, which was most notable in 1990–1998 and 2002–2015. The period from 1990 to 1998 was dominated by 3–4 years and from 2002 to 2015 by a 2–4 years cycle (Figure 4b). The number of cold duration days showed a significant period of 2–8 years, which was most notable in 1991–1996 and 2000–2015. Among them, the period of 1991–1996 was mainly approximately 2–4 years and 6–8 years, and the periods of 2000–2015 were mainly approximately 3–4 years and 6–8 years (Figure 4d). The average daily minimum temperature in the winter has a cycle of 2–4 years, and the period is most notable in 1991–1995 and 1996–2013. Among them, the cycle of 1991–1995 is dominated by approximately two years, and 1996–2013 is approximately two and four years (Figure 4f).

In central China, the warm duration days showed a cycle of 3–4 years, which was notable from 1990 to 2000, and from 2009 to 2015. The above two stages are mainly based on a cycle of approximately 3–4 years (Figure 4c). The extremely high value of the average daily maximum temperature during the summer shows a cycle of 2–4 years, which is most notable in 1991–1996 and 2008–2015. The above two phases are in a cycle of approximately 2–4 years (Figure 4e).

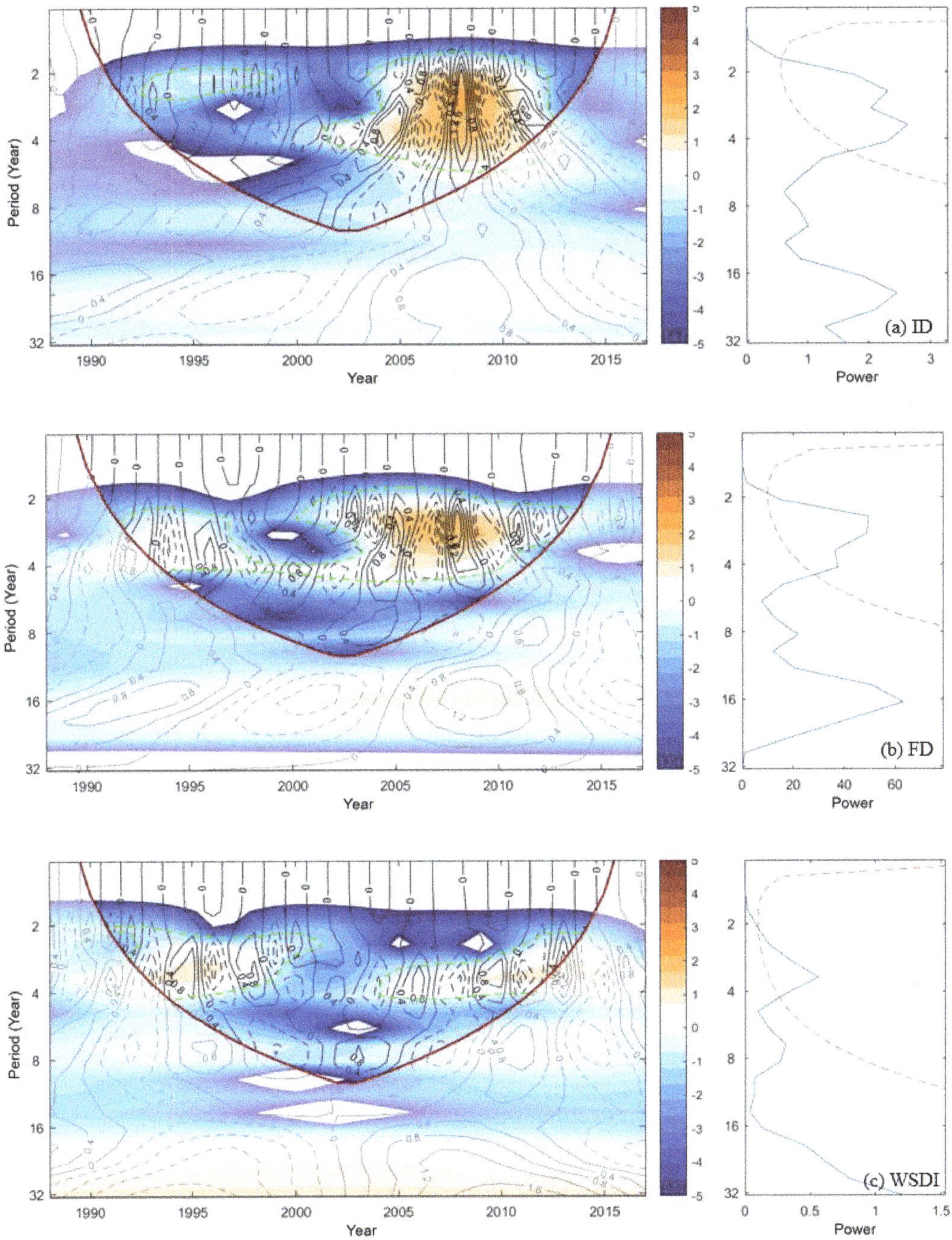

Figure 4. *Cont.*

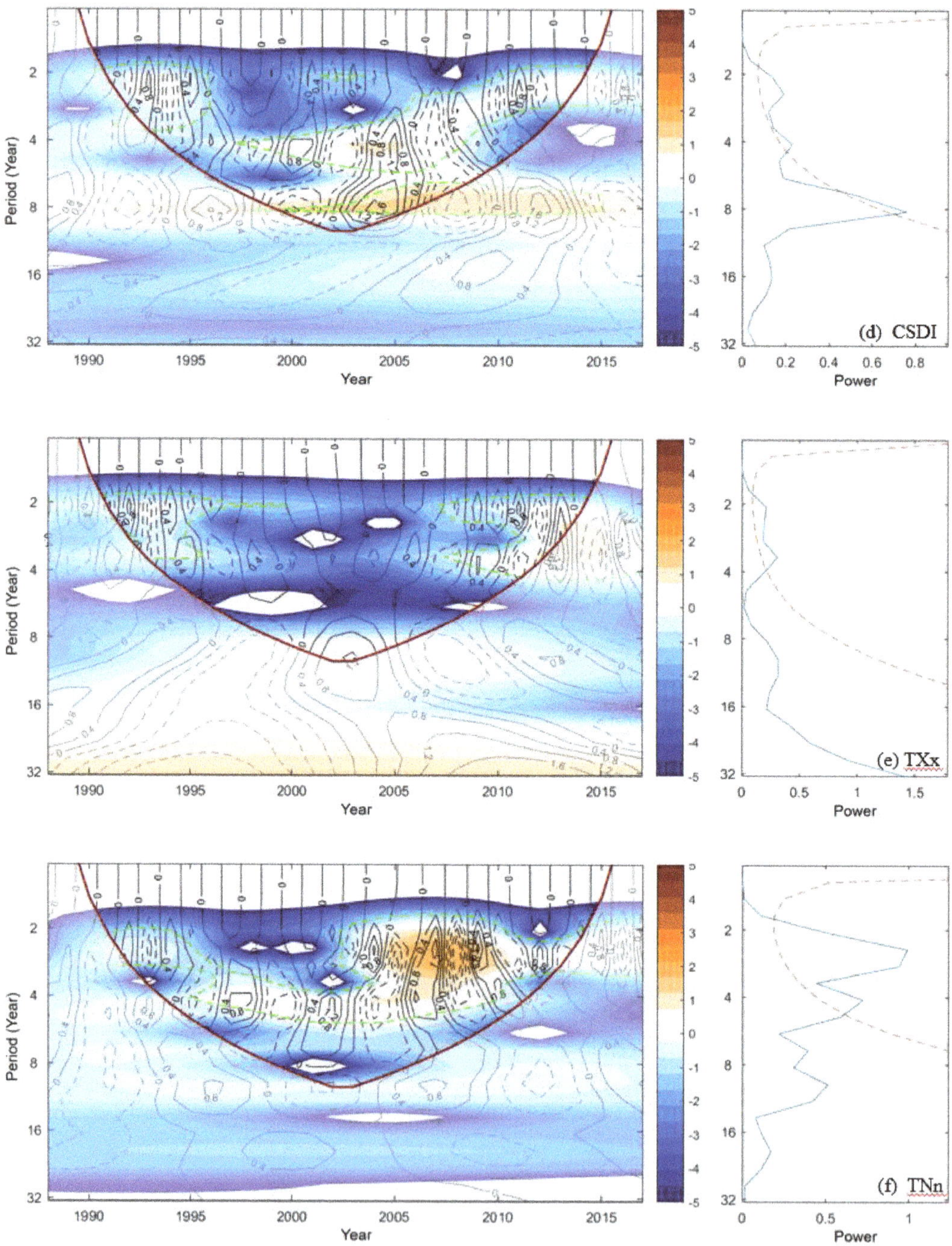

Figure 4. Wavelet power spectrum, real part of the wavelet coefficient, and full spectrum of the extreme temperature indices in central China in 1988–2017: (**a**) Icing days (ID); (**b**) Frost days (FD); (**c**) Duration of warm periods (WSDI); (**d**) Duration of cold periods (CSDI); (**e**) Highest Tmax (TXx); (**f**) Lowest Tmin (TNn)).

In conclusion, with the exception of CSDI, the types of extreme temperature indices all have significant cycles of 2–4 or 3–4 years, and the CSDI has a 6–8 years cycle, in addition to the 2–4 years cycle. The main periods of ID and FD reflecting the extreme value of the low temperature were both notable in 2002–2013. The WSDI and TXx (highest Tmax) represent extremely high temperatures, and their main cycle was significant from 2000 to 2015.

3.4. Spatial Distribution of the Extreme Precipitation Index

Figure 5 shows the distribution of extreme precipitation indices in central China. The extreme precipitation indices include extreme precipitation (R95), consecutive dry days, and consecutive wet days.

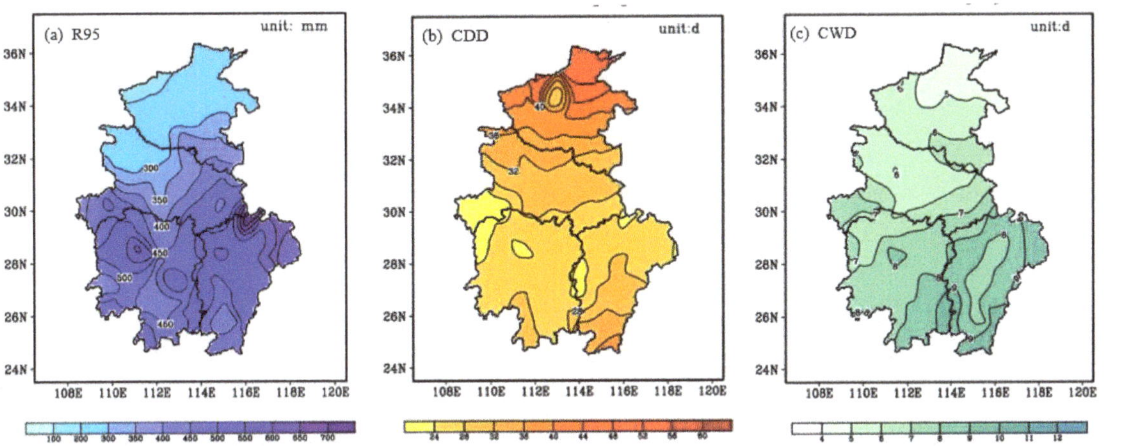

Figure 5. Spatial distribution map of the extreme precipitation index in central China from 1988 to 2017.

The extreme precipitation and the number of consecutive wet days generally showed the distribution of a gradual increase from north to south and from west to east, and the distribution of high-value areas to low-value areas in the region was clear, with an average value of 431.6 mm for the region. Jiangxi was the area with the highest extreme precipitation in central China, and the extreme values were concentrated in the north and east, generally exceeding 500 mm. At the same time, Jiangxi was also the region with the longest consecutive wet days in central China, generally higher than eight days. Jiangxi had a higher probability of extreme consecutive heavy precipitation events, and was more prone to flood disasters. Western Henan and north-western Hubei were the areas with the lowest extreme precipitation in central China, which was below 300 mm; Sanmenxia, in Henan Province, had the lowest extreme precipitation (161.6 mm) in central China. Meanwhile, Henan was also the region with the shortest consecutive wet days in central China.

3.5. Time Change Trend Analysis of the Extreme Precipitation Index

Additionally, using the Mann–Kendall method (Figure 6), combined with trend analysis, the trend change test was performed on the three extreme precipitation indices in central China (see Table 3 for test statistics).

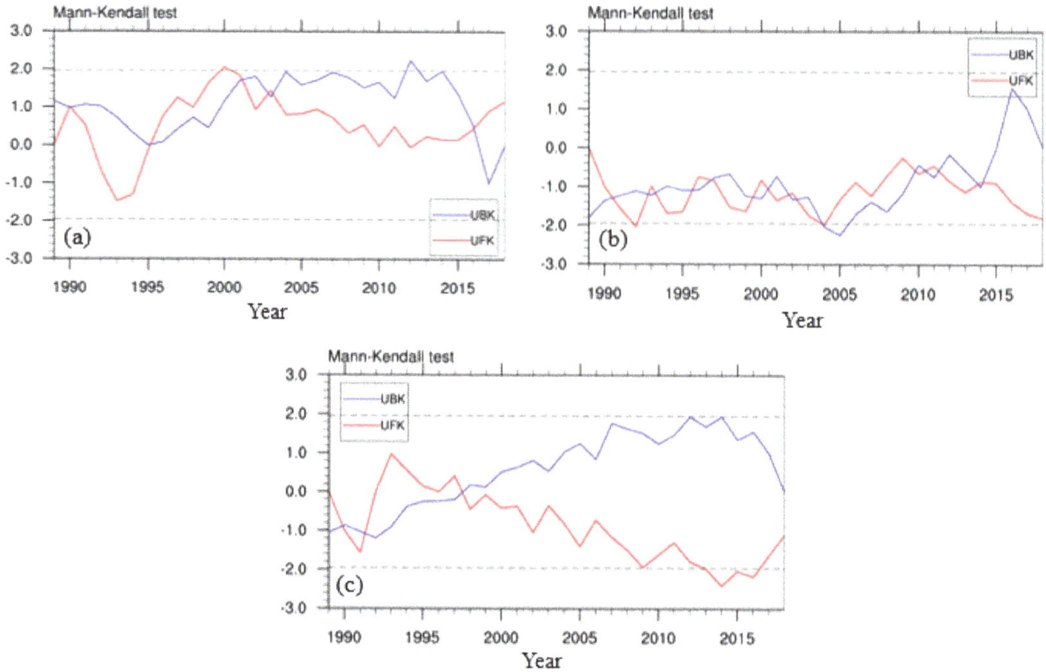

Figure 6. The Mann–Kendall test of the extreme precipitation index in central China from 1988 to 2017: (**a**) Very wet days (R95); (**b**) Consecutive dry days (CDD); (**c**) Consecutive wet days (CWD).

Table 3. Linear trends and abrupt climate changes of extreme precipitation indices in central China from 1988 to 2017.

Extreme Precipitation Index	Linear Trend Value	The Year of the Mutation	Trend Values before and after Climate Mutations		
			Before		After
R95	14.4 mm/10 a	1995 2004	108.4 mm/10 a	−143.3 mm/10 a	72.4 mm/10 a
CDD	−1.67 d/10a *		/		
CWD	−0.23 d/10 a	1998	−0.4 d/10 a		0.003 d/10 a

(Note: * means passing the 90% significance test, indicating that the linear trend is significant.).

The extreme precipitation (R95) in central China had a climate mutation (Figure 6a). The mutation years were 1995 and 2004. The trend analysis of the three periods before and after the mutations showed that the R95 index had increasing, decreasing, and increasing trends in 1988–1995, 1996–2003, and 2004–2017, respectively. The maximum and minimum values in the study period both appeared in the 1996–2003 period, which were 1998 (615.6 mm) and 2001 (295.0 mm), respectively.

The consecutive dry days in central China showed a significant decreasing trend, with a linear trend value of −1.67 d/10a, but there was no clear climate mutation (Figure 6b), indicating that the duration of the extreme drought events in central China was gradually shortened.

The overall trend of consecutive wet days in central China was not significant, but there was a sudden climate mutation (Figure 6c) in 1998. The maximum value of the CWD index appeared in 1992 (10 days), and the minimum value appeared in 2013 (5.5 days) in the pre-mutation and post-mutation periods.

3.6. Periodic Analysis of the Extreme Precipitation Index

The wavelet analysis method was also used to analyze the periodicity of the extreme precipitation index (Figure 7). The results showed that the extreme precipitation (R95) in central China showed a 2–4 years cycle, which was more notable in 1994–2005 and 2007–2015. Specifically, the 1994–2005 period was mainly 2–4 years, and the 2007–2015 period was mainly 2 years (Figure 7a).

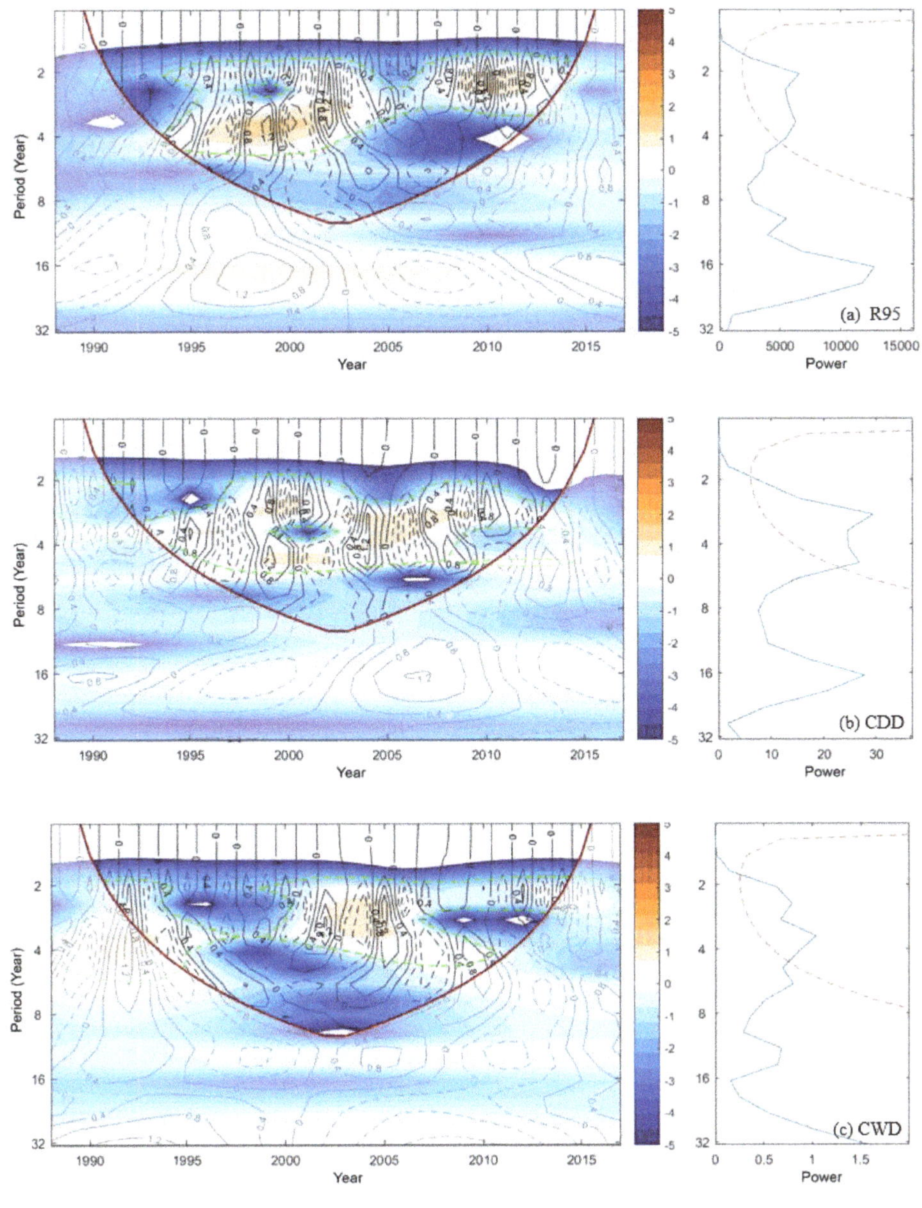

Figure 7. Wavelet power spectrum, real part of wavelet coefficients, and the full spectrum of the extreme precipitation indices in the four provinces of central China from 1988 to 2017: (**a**) Very wet days (R95); (**b**) Consecutive dry days (CDD); (**c**) Consecutive wet days (CWD).

The number of consecutive dry days in central China showed a significant period of 2–6 years, which was more significant from 1993 to 2014, with periods of approximately two years and 4–6 years (Figure 7b).

The number of consecutive wet days in central China had a significant period of 2–4 years, which was more significant in 1990–1996 and 2000–2015 (Figure 7c).

In conclusion, the significant periods of extreme precipitation, consecutive dry days, and consecutive wet days in central China are not consistent. However, from the early 1990s to approximately 2015, these three indices had significant periods in most years, with a 2–4 years cycle dominating.

4. Discussion

This article analyzes the temporal and spatial variation characteristics of extreme climate indices in central China, which can help us to understand the changing regularity of severe weather in central China. Many studies have found that the evolution of extreme climate events is often closely related to the abnormal atmospheric circulation situation [34–36]. For example, the maximum value of ID index and FD index appeared in 2008, which was mainly due to the obvious abnormality of the upper-level jet in East Asia at the beginning of 2008, compared with normal years. The stronger East Asian upper-level jet means that the intensity of cold air in the north is stronger, and the high-level divergence area on the anticyclone side of the jet inlet area provides the occurrence and maintenance of low-temperature events [37,38]. As a result, there has been long-lasting and strong low-temperature rain and snow freezing weather in many places, including central China, which made the ID index and FD index of central China in 2008 abnormally high.

The inter-annual variation of the WSDI index is large, and the average value after 2004 is larger than that in 1998–2004, which is considered to be closely related to global warming. The maximum value appeared in 2013. In addition, the WSDI index in 2010 was also abnormally high. The abnormal WSDI index in these two years was related to the abnormally strong western Pacific subtropical high. The subtropical high air prevails in the subtropical high control area, and the subtropical air heating and clear sky radiative heating lead to the appearance of high-temperature weather. While the western Pacific subtropical high had a larger range and stronger intensity in the abnormal years (such as 2010 and 2013), and stably extended westward to most of central China, the high temperature weather is persistent in this area [39].

Affected by the super El Niño in 1997–1998, the subtropical high in the Northwest Pacific became more stable and stronger. The water vapor along the edge of the subtropical high was continuously transported to the middle and lower reaches of the Yangtze River, resulting in the emergence of severe floods, which appeared with an abnormally high R95 index in 1998 [40]. The maximum value of the CDD index appeared in 1988. Some scholars believe that this is related to the large-scale drought caused by the strong La Niña event in 1988–1989 [41]. It is mainly related to the weak subtropical high and weak water vapor transport caused by the La Niña event.

According to the cycle analysis results of six extreme temperature indices and three extreme precipitation indices, it can be extrapolated that the low-temperature events in central China will be more significant around 2022, 2025, and 2028; the drought events in central China around 2025 will be more significant, so disaster prevention and mitigation projects are suggested to be prepared in advance around these years.

5. Conclusions

Based on the above analysis, the spatial and temporal variation characteristics of extreme climate indices in central China can be obtained.

Spatially, the WSDI presented a distribution characteristic of more in the south and less in the north, where the highest value was mainly in Jiangxi. The extreme precipitation index generally showed the distribution characteristics increasing from north to south and from west to east. Jiangxi is the area with the largest extreme precipitation and the longest continuous precipitation days in central China. It has a greater probability of heavy rainfall events, and is more likely to be flooded. Combined with the analysis of extreme temperature indices and precipitation indices, it is found that the risk of extreme climate events in Jiangxi is greater.

The results of time evolution characteristics and the Mann–Kendall test show that the CSDI in the cold series index had a significant decreasing trend. The ID index and the FD index showed a significant decreasing trend after the mutation, indicating that the winter and spring in central China are gradually warmer. Both the warm series indices (WSDI and TXx) showed a significant increasing trend, and the mean values of the WSDI index and the TXx index increased significantly after the mutation, indicating that the duration of extreme high temperature events in central China became longer, and the intensity of extreme high temperature increased.

The results of the wavelet analysis show that the extreme climate index had significant cyclical change characteristics, mainly 2–4 years of periodic oscillation, and different indices have certain differences during the main period in separate time periods. Most of the extreme climate indices have significant periodic changes after 2000.

This study can provide a theoretical basis for the study of the occurrence regularity of extreme disastrous weather and climate events in central China. The next step is to carry out research on the impacts of extreme climate events on urban lifelines such as agriculture, transportation, water supply, power supply, and other different industries. In addition, more disaster mechanism studies should be carried out based on different disaster types. This will provide more scientific bases for the assessment of climate change and its impacts on stable grain production, energy supply, and urban safety management.

Author Contributions: J.Z. and R.M. contributed to the study conception and design. Material Preparation and analysis were performed by Y.H.; X.F. and Y.L. Data collection and processing were performed by X.L.; X.W.; Y.W. and Y.S. The first draft of the manuscript was written by Y.L. All authors commented on subsequent versions of the manuscript. All authors have read and agreed to the published version of the manuscript.

Funding: This research was funded by the Fund Project of CMA Public Meteorological Service Centre (YW20180044) and Research Project of China Yangtze Power Co., Ltd. (241802002).

Data Availability Statement: Not applicable.

Acknowledgments: We thank the reviewers for their help in improving this manuscript.

Conflicts of Interest: The authors declare no conflict of interest.

References

1. Zhou, B.T.; Qian, J.; Changes of weather and climate extremes in the IPCC sixth assessment report. *Clim. Chang. Res.* **2021**, 1–7. Available online: https://kns.cnki.net/kcms/detail/11.5368.P.20210929.1740.006.html (accessed on 1 November 2021).
2. Arias, P.A. Technical Summary. In *Climate Change 2021: The Physical Science Basis*; Cambridge University Press: Cambridge, UK, 2021.
3. Ren, Y.J.; Wan, S.Q.; Xiao, Y.; Liu, M.; Sun, S.L. Simulative evaluation and projection of air temperatures over central China in the SRES scenarios. *Acta Meteorol. Sina.* **2012**, *70*, 1098–1106. (In Chinese) [CrossRef]
4. Zhong, M.; Che, Q.; Zhang, M.M.; Dong, L.P.; Zhang, P.P.; Chen, X. Characteristics of extreme precipitation weather situation and physical quantity anomaly in central China. *Meteor Mon.* **2020**, *46*, 503–516. (In Chinese)
5. Xiao, Y.; Li, Z.X.; Zhang, X.P.; Zhou, J.J. Research on extreme precipitation event characteristics of Xiangjiang river basin in recent 48 years. *Resour. Environ. Yangtze Basin* **2010**, *19*, 1356–1362. (In Chinese)
6. Ke, D.; Guan, Z.Y. Precipitation and circulation patterns of daily precipitation extremes over central China during boreal summer. *Acta Meteorol. Sina.* **2014**, *72*, 478–493. (In Chinese) [CrossRef]
7. Yan, Z.; Jones, P.D. *Trends of Extreme Temperatures in Europe and China Based on Daily Observations*; Springer: Dordrecht, The Netherlands, 2002; pp. 355–392. [CrossRef]
8. Zhong, Y.H.; Yang, M.Z.; Yuan, C.X. Temporal and spatial characteristics of summer extreme precipitation in Eastern China and possible causalities. *J. Geosci. Environ. Prot.* **2020**, *8*, 36–46. [CrossRef]
9. Harrison, M.T.; Cullen, B.R.; Rawnsley, R.P. Modelling the sensitivity of agricultural systems to climate change and extreme climatic events. *Agric. Syst.* **2016**, *148*, 135–148. [CrossRef]
10. Liu, X.F.; Zhu, X.F.; Pan, Y.Z.; Bai, J.J.; Li, S.S. Performance of different drought indices for agriculture drought in the North China Plain. *J. Arid Land.* **2018**, *10*, 507–516. [CrossRef]

11. Perera, A.T.D.; Nik, V.M.; Chen, D.; Scartezzini, J.L.; Hong, T.Z. Quantifying the impacts of climate change and extreme climate events on energy systems. *Nat. Energy* **2020**, *5*, 150–159. [CrossRef]
12. Cheng, Q.P.; Zhong, F.L.; Wang, P. Potential linkages of extreme climate events with vegetation and large-scale circulation indices in an endorheic river basin in northwest China. *Atmos. Res.* **2020**, *247*, 105256. [CrossRef]
13. Wang, A.F.; Chen, S.H.; Tang, X.L. The effects of extreme weather events on crops in Guangxi from 1978 to 2017. *Chin. Agric. Sci. Bull.* **2021**, *37*, 106–115. (In Chinese)
14. Ren, J.Y.; Tong, S.Q.; Bao, Y.H.; Guo, X.M. Changes of extreme climate and its effect on net primary productivity in Inner Mongolia. *Chin. J. Ecol.* **2021**, *40*, 2410–2420. (In Chinese) [CrossRef]
15. Klein Tank, A.M.G.; Können, G.P. Trends in indices of daily temperature and precipitation extremes in Europe, 1946–1999. *J. Clim.* **2003**, *16*, 3665–3680. [CrossRef]
16. Frich, P.L.; Alexander, L.; Della-Marta, P.; Gleason, B.; Peterson, T.C. Observed coherent changes in climatic extremes during 2nd half of the 20th century. *Clim. Res.* **2002**, *19*, 193–212. [CrossRef]
17. Gallant, A.J.E.; Karoly, D.J.; Gleason, K.L. Consistent trends in a modified climate extremes index in the United States, Europe and Australia. *J. Clim.* **2014**, *27*, 1379–1394. [CrossRef]
18. Pita-Díaz, O.; Ortega-Gaucin, D. Analysis of anomalies and trends of climate change indices in Zacatecas, Mexico. *Climate* **2020**, *8*, 55. [CrossRef]
19. Lucas, E.; Souza, F.; Silva, F.; Júnior, R.; Pinto, D.; Silva, V. Trends in climate extreme indices assessed in the Xingu river basin-Brazilian Amazon. *Weather Clim. Extremes* **2021**, *4*, 100306. [CrossRef]
20. Liu, Y.R.; Li, Y.P.; Yang, X.; Huang, G.H.; Li, Y.F. Development of an integrated multivariate trend-frequency analysis method: Spatial-temporal characteristics of climate extremes under global warming for Central Asia. *Environ. Res.* **2021**, *195*, 110859. [CrossRef] [PubMed]
21. Uddin, M.J.; Hu, J.; Islam, A.; Eibek, K.U.; Nasrin, Z.M. A comprehensive statistical assessment of drought indices to monitor drought status in Bangladesh. *Arabian J. Geosci.* **2020**, *13*, 323. [CrossRef]
22. Naderi, M. Extreme climate events under global warming in northern Fars Province, southern Iran. *Theor. Appl. Climatol.* **2020**, *142*, 1–23. [CrossRef]
23. Marigi, S.N.; Njogu, A.K.; Githungo, W.N. Trends of extreme temperature and rainfall indices for arid and semi-arid lands of south eastern Kenya. *J. Geosci. Environ. Prot.* **2016**, *04*, 158–171. [CrossRef]
24. Bennett, K.E.; Walsh, J.E. Spatial and temporal changes in indices of extreme precipitation and temperature for Alaska. *Int. J. Climatol.* **2015**, *35*, 1434–1452. [CrossRef]
25. Shi, J.; Cui, L.L.; Wen, K.M.; Tian, Z.; Wei, P.P.; Zhang, B.W. Trends in the consecutive days of temperature and precipitation extremes in China during 1961–2015. *Environ. Res.* **2018**, *161*, 381–391. [CrossRef] [PubMed]
26. He, Y.H.; Lin, K.R.; Tang, G.P.; Chen, X.H.; Guo, S.L.; Gui, F.L. Quantifying the changing properties of climate extremes in Guangdong Province using individual and integrated climate indices. *Int. J. Climatol.* **2016**, *37*, 781–792. [CrossRef]
27. Cai, W.X.; Lu, W.H.; Yu, G.Q.; Liu, J.F. Time trend analysis of continuous extreme precipitation events in Songhua River Basin from 1951 to 2017. *Sci. Technol. Eng.* **2021**, *21*, 3887–3893. (In Chinese)
28. Qi, Q.H.; Cai, R.S.; Guo, H.X. The climatic variations of temperature extremes in the Eastern of China. *Sci. Geogr. Sin.* **2019**, *39*, 1340–1350. (In Chinese) [CrossRef]
29. Jiao, Y.; Li, Y.Z.; Tian, X.S.; Zhou, Y. Analysis on the changes of extreme temperature and precipitation events in Ji'nan and its surrounding areas. *Meteorol. Environ. Sci.* **2021**, *44*, 43–52. (In Chinese) [CrossRef]
30. Li, C.X.; Tian, Q.H. Attribution of extreme precipitation in the lower reaches of the Yangtze river during may 2016. *Environ. Res. Lett.* **2018**, *13*, 014015. [CrossRef]
31. Zhai, P.M.; Liu, J. Extreme weather/climate events and disaster prevention and mitigation under global warming background. *Strateg. Study CAE* **2012**, *14*, 55–63. (In Chinese) [CrossRef]
32. Liu, L.; Xu, Z.X. Spatiotemporal distribution of the extreme climate indices in the five southwestern Provinces of China. *Resour. Environ. Yangtze Basin* **2014**, *23*, 294–301. (In Chinese) [CrossRef]
33. Ju, L.L.; Shi, J.; Zhang, M. A study of extreme temperature changes in East China from 1961 to 2015. *Desert Oasis Meteo.* **2020**, *1*, 112–121. (In Chinese) [CrossRef]
34. Zhang, Z.W.; Wang, S.G.; Shang, K.Z.; Chen, L.; Wang, H.B.; Chen, X.Y.; Wang, M.Z. Analysis of the atmospheric circulation anomalies of large-scale extreme high temperature events over central China in recent 50 years. *J. Lanzhou Univ. Nat. Sci.* **2011**, *47*, 50–55. (In Chinese) [CrossRef]
35. Zhang, M. Variation Features of Summer High-Temperature and its Influence Process in China. Master's Thesis, Nanjing University of Information Science and Technology, Nanjing, China, 2011. (In Chinese).
36. Huang, Z.Y. Climatic Characteristics and Influencing Factor of Persistent High-Temperature Events in Hunan. Master Thesis, Lanzhou University, Lanzhou, China, 2016. (In Chinese).
37. Zhang, C.Y.; Zhang, Y.C. The characteristics of East Asian jet stream in severe snow storm and freezing rain processes over Southern China in early 2008. *J. Trop. Meteorol.* **2013**, *29*, 306–314. (In Chinese) [CrossRef]
38. Qian, X. Analyses on Cold-wet Spells in China and its Correspondingly Anomalous Large Scale Circulation. Master Thesis, Nanjing University of Information Science and Technology, Nanjing, China, 2014. (In Chinese).

39. Tang, K.; Jin, R.H.; Peng, X.Y.; Niu, R.Y. The analysis of causes about extremely high temperature in the summer of 2013 year in the Southern Region. *J. Chengdu Univ. Inf. Technol.* **2014**, *29*, 652–659. (In Chinese) [CrossRef]
40. Xu, C.W.; Wang, W.; Ma, J.S.; Yang, X. The characteristics causes of formation and climatic impact of the 1997–1998 El Nino event. *Donghai Mar. Sci.* **2004**, *22*, 1–8. (In Chinese)
41. Chen, M.C.; Huang, Y.R. Causes of the distinctive evolution patterns between the typical and the 1988 to 1989 La Niña events. *Trans. Atmos. Sci.* **2021**, *44*, 428–440. (In Chinese) [CrossRef]

sustainability

MDPI

Article

Assessment of Seasonal Variability of Extreme Temperature in Mainland China under Climate Change

Weixiong Yan [1,2], Junfang Zhao [2,*], Jianping Li [1,*] and Yunxia Wang [1]

[1] Key Laboratory for Meteorological Disaster Monitoring and Early Warning and Risk Management of Characteristic Agriculture in Arid Regions, Ningxia Key Laboratory for Meteorological Sciences, Ningxia Institute of Meteorological Sciences, Yinchuan 750003, China; laobencau@163.com (W.Y.); wyxia1995@nwafu.edu.cn (Y.W.)

[2] State Key Laboratory of Severe Weather, Chinese Academy of Meteorological Sciences, Beijing 100081, China

* Correspondence: zhaojf@cma.gov.cn (J.Z.); lijp_111@163.com (J.L.); Tel.: +86-130-0198-1203 (J.Z.); +86-138-9519-3106 (J.L.)

Abstract: Some studies have suggested that variations in the seasonal cycle of temperature and season onset could affect the efficiency in the use of radiation by plants, which would then affect yield. However, the study of the temporal variation in extreme climatic variables is not sufficient in China. Using seasonal trend analysis (STA), this article evaluates the distribution of extreme temperature seasonality trends in mainland China, describes the trends in the seasonal cycle, and detects changes in extreme temperature characterized by the number of hot days (HD) and frost days (FD), the frequency of warm days (TX90p), cold days (TX10p), warm nights (TN90p), and cold nights (TN10p). The results show a statistically significant positive trend in the annual average amplitudes of extreme temperatures. The amplitude and phase of the annual cycle experience less variation than that of the annual average amplitude for extreme temperatures. The phase of the annual cycle in maximum temperature mainly shows a significant negative trend, accounting for approximately 30% of the total area of China, which is distributed across the regions except for northeast and southwest. The amplitude of the annual cycle indicates that the minimum temperature underwent slightly greater variation than the maximum temperature, and its distribution has a spatial characteristic that is almost bounded by the 400 mm isohyet, increasing in the northwest and decreasing in the southeast. In terms of the extreme air temperature indices, HD, TX90p, and TN90p show an increasing trend, FD, TX10p, and TN10p show a decreasing trend. They are statistically significant ($p < 0.05$). This number of days also suggests that temperature has increased over mainland China in the past 42 years.

Keywords: STA; China; warm days; cold days; warm nights; cold nights; hot days; frost days

Citation: Yan, W.; Zhao, J.; Li, J.; Wang, Y. Assessment of Seasonal Variability of Extreme Temperature in Mainland China under Climate Change. *Sustainability* **2021**, *13*, 12462. https://doi.org/10.3390/su132212462

Academic Editors: Xiaodong Yan, Jia Yang and Shaofei Jin

Received: 21 September 2021
Accepted: 3 November 2021
Published: 11 November 2021

1. Introduction

The global surface temperature was 1.09 °C higher in last the last decade than 1850–1900, which was more likely not higher than for any multi-century average during the Holocene [1]. The frequency and intensity of hot extremes have increased and those of cold extremes have decreased on the global scale since 1950 [1]. Agriculture is one of the most directly affected sectors by global climate change, especially crop production and food security [2]. China is the largest food-producing country in the world. Since 2003, grain output has increased continuously. In 2020, the sown area of grain reached 1.17 million × 10^8 hm^2 and a total yield of 6.69 × 10^8 t [3]. Chinese food production plays an important role in its own country and even food security in the world. China is located in eastern Eurasia, mostly at middle and high latitudes (Figure 1). This is a sensitive and significant area for global climate change. Climate change is likely to have a significant impact on global food production, and Chinese food production is also faced with the uncertainty caused by climate change and the risk of yield reduction caused by extreme climate [4–7]. In the past half-century, the yield of most major crops in the world has increased significantly, mainly due to irrigation, chemical input, and the extensive use of modern crop varieties [8]. However, an increasing number of studies have shown that there are two significant differences between the positive and negative effects of climate warming on crop growth and yield [9–13], and the results depend on the study areas, crops, and methods.

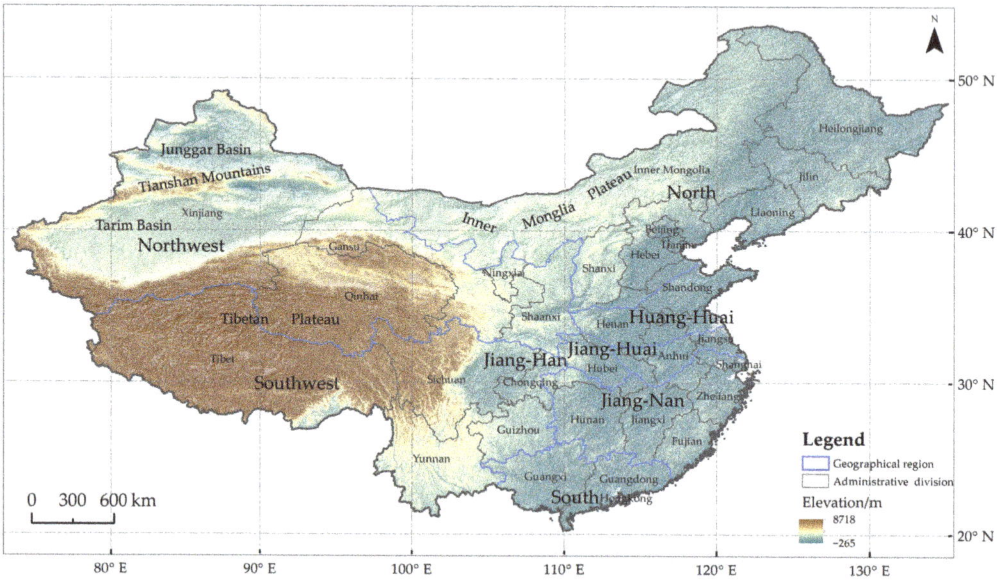

Figure 1. Map of mainland China including geographic divisions and provinces.

In the past 100 years, the trend of temperature increase in China has been higher than the global average [14]. In the last 50 years, surface air temperature has increased by nearly 1.40 °C with a change rate of 0.25 °C (10 years)$^{-1}$, indicating that the trend of climate warming has accelerated [14,15]. From 1961 to 2018, the start date of the average growing season in China advanced, the end date was delayed, the length was extended, and the advance of the start date had a greater impact on the length extension. In addition, the advance of the start date and the increase in the length of the average growing season in China are mainly due to warming in spring. The trend in the growing season length in China follows the Northern Hemisphere trend [16]. The extreme air temperatures also show an increasing trend. During 1961–2014, the temperatures of the hottest day and coldest night in China showed a rapid upward trend, which were 0.17 °C·10a^{-1} and 0.52 °C·10a^{-1}, respectively [17]. From 1961 to 2018, cold days (TX10p) and cold nights (TN10p) in China had a decreasing trend, and the decreasing trend of TN10p was significantly greater than that of TX10p. Warm days (TX90) and warm nights (TN90) had an increasing trend, and the increasing trend of TN90p was significantly greater than that of TX90p [18]. The trend of extreme air temperature was consistent with that on the global scale [1]. Extreme high temperatures and low temperatures can damage crop tissues and organs, delay crop growth and development, or hinder flowering and fruiting, resulting in lower yields [19]. Long-term global warming has changed the distribution of temperature changes and extremely high temperatures have become more common in some places, such as southwest China [20–22]. The same is true for agricultural regions, where the probability of crop exposure to maximum temperatures increases at the critical stage of reproductive growth [23]. According to the latest studies, the contribution rate of climate warming to the yield of spring maize in northeast China from 1981 to 2009 was 29.7%, and maximum temperatures above 30 °C caused a 14.1% yield reduction. The increase in high temperature during the vegetative period was the main reason for the yield reduction [24]. Many studies have also focused on the impact of temperature changes in different periods on crop yield. Some scholars have collected relevant literature worldwide and used meta-analysis to conclude that 0–5 °C warming during the reproductive period had significant negative effects on wheat yield and its components [25,26]. The decline in wheat yield is different in different climatic regions, and the negative effect of warming at night is greater than that during the daytime [25]. As far as China is concerned, wheat yield increases significantly in the monsoon region but decreases significantly in the temperate continental climate region. The winter wheat yield has increased significantly with the increase in night temperature in the monsoon region [26]. Another study indicated that warming up to +3 °C has increased winter yield by 5.8% per °C (change rate of yield/average of yield) while reducing spring wheat yield by 16.1% per °C [27].

The annual variability of temperature, precipitation, and plant phenology usually has seasonal cycles. With the growth of global climate studies, monitoring these seasonal trends as a means to detect the response of the Earth system to global change has sparked great interest [28–30]. Previous studies regarding the impact of climate warming on crop yield have more or less been related to the temporal dynamics of temperature [24–27]. Most existing studies have used long-series data to analyze annual and seasonal changes or the length of the growing season [14–18], but analysis of the seasonal trend of temperature is not sufficient, especially the changes in the time of occurrence for extreme air temperature. The purpose of this paper is to select several extreme air temperature indices from the ETCCDI (Expert Team on Climate Change Detection and Indices) [31], which are related to grain production. We analyze the spatial pattern and change in extreme air temperature seasonal trends in mainland China over the last 42 years and detect the changes in extreme temperature events. These results are expected to provide help for studying the long-term impact trend of climate change on food production.

2. Data and Methodology

2.1. Data

In recent years, with the development of automatic observation technology, the number of meteorological observation stations in China has increased greatly, which improves the spatial density and frequency of observations and partially meets the needs of land-atmosphere processes and weather climate analysis. However, at the beginning of the layout, automatic observation stations were mostly placed in sections with stable communication, convenient maintenance, and clear purpose, which have high relevance, but cannot provide uniform distribution and long-time series; therefore, it is difficult to meet the needs of long-term climate trend analysis at present. Reanalysis data are a set of gridded and long-series meteorological datasets based on data assimilation technology that integrate multisource observation data and numerical simulation results. This could compensate for the uneven spatial and temporal distribution of in-situ observations. At present, the main reanalysis datasets include a series of products (ERA5, ERA15, ERA40, and ERA-Interim) from the European Centre for Medium Term Weather Forecasts (ECMWF) [32], NCEP/NCAR reanalysis I (R1) jointly developed by National Centers for Environmental Prediction (NCEP) and National Center of Atmospheric Research (NCAR) [33], NCEP/DOE reanalysis II (R2) jointly developed by the Department of Energy (DOE) [34], Japanese 25 year reanalysis (JRA-25) and 55 year reanalysis (JRA-55) [35,36], NASA's Modern-Era Retrospective analysis for Research and Applications (MERRA and MERRA-2), etc. [37,38]. In May 2021, the China Meteorological Administration (CMA) released China's first-generation global/land surface reanalysis product (CRA). The product reproduces the global three-dimensional atmospheric conditions from the ground to a 55 km height since 1979, with a temporal resolution of 6 hours and a spatial resolution of 30 km. The quality of the product is generally equivalent to that of international third-generation global reanalysis products [39] (http://data.cma.cn/analysis/cra40, accessed on 5 November 2021).

ERA5 is the fifth generation ECMWF atmospheric reanalysis of the global climate covering the period from January 1950 to present. ERA5 combines vast amounts of historical observations into global estimates using advanced modeling and data assimilation systems. ERA5 provides hourly estimates of a large number of atmospheric, land, and oceanic climate variables. The data cover the Earth on a 30 km grid and resolve the atmosphere using 137 levels from the surface up to a height of 80 km [40]. Although ECMWF recently formed the global ERA5 dataset since 1950, the one from 1950 to 1978 is not the final version. There are some evaluation and application studies on the specific elements of the datasets at home and abroad which show that the quality of the datasets is significantly improved compared with the previous version [32,41,42]. Therefore, the hourly 2 m temperature data from ERA5 during 1979–2020 are used to form daily and monthly extreme air temperatures, annual frost days (FD), and hot days (HD) in mainland China. The ground spatial resolution of the reanalysis data is further improved to 0.25° × 0.25°.

2.2. Methodology

2.2.1. Seasonal Trend Analysis of Extreme Air Temperature

The seasonal trends of monthly maximum temperature (Txmax), monthly mean maximum temperature (Txmean), monthly minimum temperature (Tnmin), and monthly mean minimum temperature (Tnmean) were examined by seasonal trend analysis (STA) in mainland China. Seasonal trend analysis was initially applied to the trend analysis of image time series [43], and some scholars have used the methodology to analyze the minimum temperature over the La Plata River Basin in South America [44]. Due to the influence of solar radiation, atmospheric circulation, and other factors, meteorological elements change over time and can be considered a kind of fluctuation. For the time series of meteorological elements, we can regard it as the superposition of many harmonic waves [45]. A given time series y_t and taking $t = 0$ as origin, y_t can be decomposed into sine signals as Equation (1):

$$y_t = A_0 + \sum_{n=1}^{n=T/2} A_n \sin(\frac{2\pi n t}{T} + \phi_n) \tag{1}$$

in which A_0 is the arithmetic average value of the original series, and the other terms in the right hand side represent the nth harmonics over T. A_n are amplitudes, and ϕ_n are phase angles (from 0 to 360°). t is time, and T is the temporal length of the series. The time series of any meteorological element is limited, and the maximum number of harmonics can be decomposed into half of the length of the series. Although a long-term series of meteorological elements contains a variety of time scale changes, the annual cycle is the most important; therefore, the first two harmonics can be used to simulate the original series [43]. Therefore, $n = 2$.

First, harmonic analysis of the temperature series was carried out, including five characteristic parameters, namely, the annual average amplitude (A0), the amplitude and phase of the annual cycle (A1 and F1), the amplitude and phase of the semi-annual cycle (A2 and F2) (Table 1) [21,43]. Here, A0 actually represents the annual average, A1 is the annual temperature range, F1 indicates shifts in time, where a value of 30° corresponds to 1 month approximately, and A2 and F2 are not clear, which can be regarded as the shape factor of the annual curve [44]. Second, once the five harmonic parameters were obtained for each year using Equation (1), their Theil–Sen median slopes were estimated. This slope was then used to characterize their trend. The significance of the statistics was evaluated using the non-parametric Mann–Kendall test ($p < 0.05$) [46]. Theil–Sen median slope estimation is a robust non-parametric statistical method that is insensitive to outliers and is very effective against reflecting the trend of time series data [47,48]. Finally, the trend of these parameters can be visualized. Since there are as many as 3^5 combinations of these parameters, it is impossible to summarize all five seasonal curve shape parameters in a single image. It is generally found that the three amplitude images contain the largest amount of information and that rendering trends in A0, A1, and A2 provide an effective composition. A companion phase trend is created by rendering trends in A0, F1, and F2. According to the classes of combination, the region of interest can be selected to draw the fitting curve of the beginning and ending years of climate elements, and the seasonal trend and change of elements can be better understood by combining with the image [30,48].

Table 1. Five characteristic parameters of harmonic analysis.

ID	Name	Definition/Meaning
A0	annual average amplitude	the arithmetic average value of the original series/annual average temperature
A1	amplitude of annual cycle	difference between maximum and minimum of the 1st harmonic/annual temperature range
A2	amplitude of semi-annual cycle	difference between maximum and minimum of the 2nd harmonic/can be regarded as the shape factor of annual curve
F1	phase of annual cycle	start phase angle of the 1st harmonic/indicate the time when the sine waves reaches a peak
F2	phase of semi-annual cycle	start phase angle of the 2nd harmonic/can be regarded as the shape factor of annual curve

The advantages of employing ERA5 and STA are (1) obtaining the spatial variation of extreme temperature, unlike the uneven station data; (2) obtaining five parameters of time series, that is, the change trend of temperature, the change of temperature range, and time change of maximum value can be understood at the same time; and (3) after the five parameters of the same grid point are superimposed and visualized, we can understand the temperature change pattern (different parameter combinations, different change patterns).

MATLAB and ArcMap were used for data processing and plotting, respectively.

2.2.2. Extreme Air Temperature Indices

We use five extreme temperature indices defined by the ETCCDI [31] and hot days (HD) that characterize extreme temperature [49,50]. Table 2 has listed the details of TN10p, TN90p, TX10p, TX90P, FD, and HD. The 90th and 10th percentiles of daily maximum/minimum temperature are calculated for a 5-day window centered on each calendar day in the base 1991–2020 period. In the last part of the paper, we evaluate the spatial distribution of the trend in extreme air temperature indices trends and examine the temporal evolution of the regional averages of these indices using linear trend rates. Linear estimation is a trend analysis method commonly used in climate analysis. In order to be able to compare with existing results and those of our article, this method was used here.

Table 2. List of extreme air temperature indices.

ID	Name	Definition	Units
HD	hot days	Annual count of days when TX ≥ 35 °C	days
FD	frost days	Annual count of days when TN < 0 °C	days
TN10p	cold nights	Percentage of days when TN < 10th percentile [1]	%
TN90p	warm night	Percentage of days when TN > 90th percentile	%
TX10p	cold days	Percentage of days when TX < 10th percentile [1]	%
TX90p	warm days	Percentage of days when TX > 90th percentile	%

[1] TN/TX represent daily minimum air temperature/maximum air temperature.

3. Results

3.1. Seasonal Trends in Temperature

3.1.1. Maximum Temperature

Figure 2 shows the spatial distribution of statistically significant trends in the annual average amplitude (A0), amplitudes of the annual (A1) and semi-annual cycles (A2), and phases of the annual (F1) and semi-annual cycles (F2) for monthly maximum temperature (Txmax) and monthly average maximum temperature (Txmean). Complementarily, a summary of areas with significant trends is presented in Figure 3.

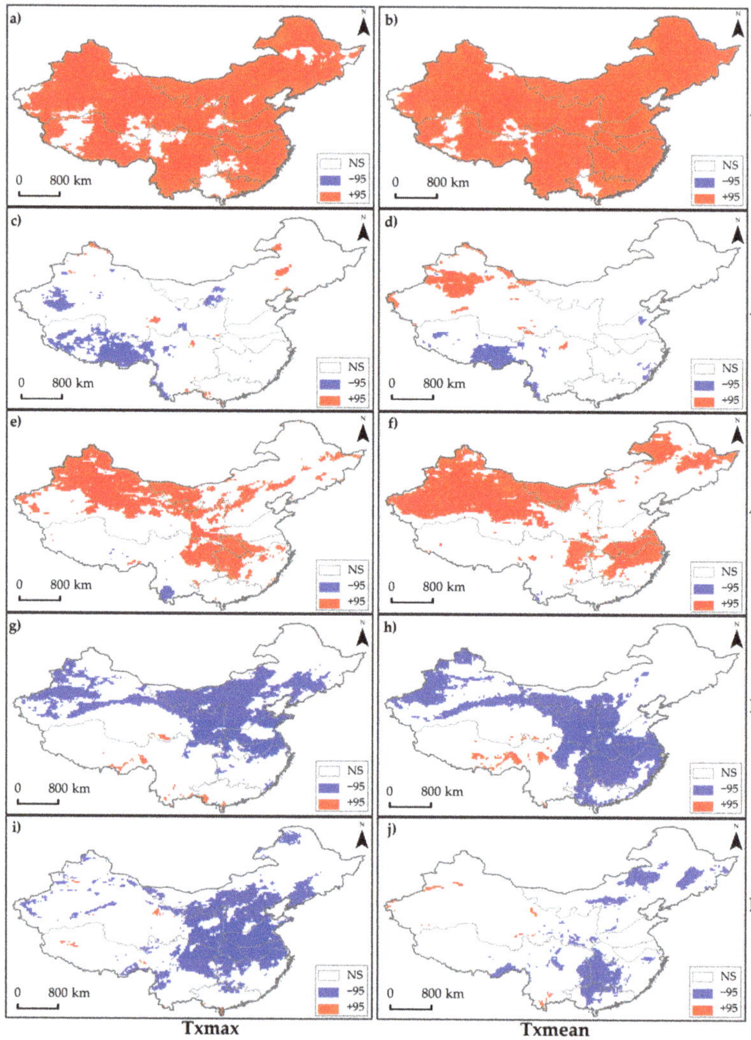

Figure 2. Areas in mainland China with trends of annual average amplitude (A0), the amplitude (A1) and phase (F1) of annual cycle, amplitude (A2) and phase (F2) of semi-annual cycle of Txmax (**a**,**c**,**e**,**g**,**i**) and Txmean (**b**,**d**,**f**,**h**,**j**). NS denotes no trend. −95 and +95 represent negative and positive trends, respectively, with a significant confidence level of 95% (same in Figure 7).

The amplitude variation in the maximum temperature was mainly positive, and the phase variation was negative over mainland China, as shown in Figure 2. During 1979–2020, the maximum temperatures in most parts of mainland China had a significant increase and there was no significant decrease. The A0 of Txmax and Txmean increased significantly in 84.5% and 93.2% of areas, respectively. The regions without significant change mainly occurred on the Tibetan Plateau and south China (Figures 2a,b and 3). Studies have revealed that the south is one of the regions with the weakest warming trend in China, and the warming trend on the Tibetan Plateau ranks first among the eight major regions in China [51]. In addition, the daily maximum temperature of the Tibetan Plateau from 1961 to 2015 had a warming trend [52]. The areas where Txmax and Txmean significantly changed in A1 accounted for 10.0% and 9.3% of the total area of mainland

China, respectively. The A1 of Txmax decreased significantly, mainly on the Tibetan Plateau and Tarim Basin. The area where the A1 of Txmean decreased significantly was slightly smaller than the area where it increased significantly. They appeared on the southeastern Tibetan Plateau, Tianshan Mountains, and Junggar Basin, respectively (Figures 2c,d and 3). Since the amplitude of the nonzero semi-annual is not easy to interpret, it may be related to the difference in the semi-annual period or annual curve shape in the seasonal curve [44]. Therefore, this paper only provides the results without analysis. Figure 2e,f show that the A2 of the maximum temperatures was approximately 1/3, showing a significant positive trend and mainly distributed in the northwestern region and Yangtze River Basin.

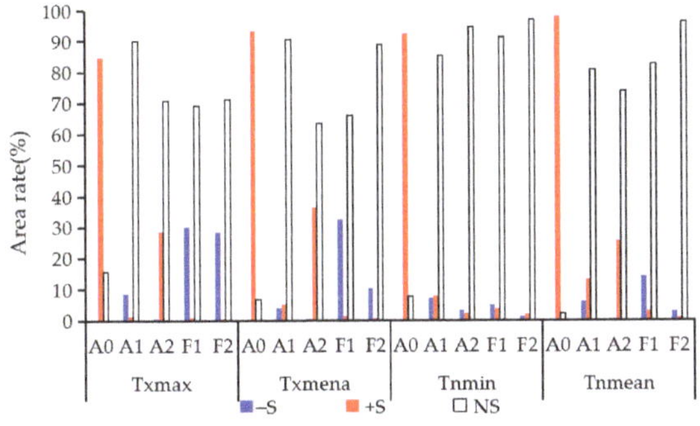

Figure 3. Area rate of mainland China with significant trends in the five parameters. +S/−S denotes positive/negative significant trends, respectively, with a significant confidence level of 95%.

F1 of the maximum temperature in mainland China, approximately 30% of the area, had a significant negative trend. Among these areas, Txmax mainly occurred in northwestern China and north of the Yangtze River, while Txmean had a pattern of 'shrinking in the north and expanding in the south', the area with a significant decrease in north China decreased, and the area with a significant decrease to the south of the Yangtze River increased (Figure 2g,h). F1 reflects the time when the sine waves reached a peak, which indicates that the time when the maximum temperature appeared in the above areas in the last 42 years was delayed. Some studies suggest that the phase change was related to a variety of mechanisms, but the influence of the change in thermal mass was greater [53]. Thermal mass on land is largely modulated by soil moisture. If soil moisture decreases, it will produce a positive phase shift [53]. Because of the lack of long-term and spatial high-resolution soil moisture datasets, it is very difficult to find conclusions supporting the above soil moisture and temperature changes from the existing studies on soil moisture changes in mainland China. The significant trend of F2 was also dominated by a negative trend, with 28.3% of Txmax decreasing significantly, mainly in the Yangtze River Basin and north China. The area where Txmean decreased significantly was approximately 1/3 of Txmax (Figure 2i,j).

Figure 4 shows the trends in monthly maximum temperature (Txmax) and monthly average maximum temperature (Txmean) during 1979–2020 (Figure 4a,b) together with the trends in A0 (Figure 4c,d). We can see that Txmax and Txmean in mainland China had a positive trend and the spatial distribution and magnitudes of the trend were very consistent with their A0. They had a significant linear relationship with a coefficient of determination of 0.92. This indicates that A0 from the seasonal trend analysis method, as a representative index of annual average temperature, is also suitable for the analysis of interannual temperature. From the perspective of spatial distribution, both Txmax and Txmean had a strong warming trend on the northeastern edge of the Tibetan Plateau, eastern coast, and Inner Mongolian Plateau. In addition, combined with Figures 2 and 4, the trend rates of the regions where the maximum temperature change was not significant were also small.

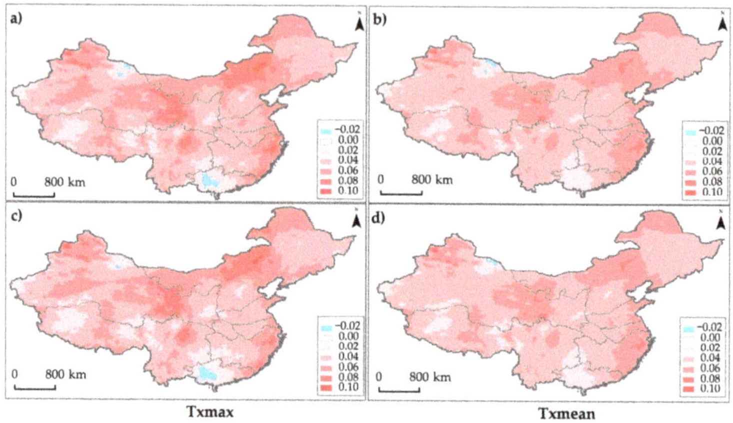

Figure 4. Theil–Sen trend (TS slope (**a**,**b**)) and linear trend for annual average amplitude (A0) of Txmax and Txmean (A0 slope (**c**,**d**)) during 1979–2020.

The five parameters of seasonal trends together represent the temporal dynamics of climate factors, up to 243 combinations. There may be one or several combinations with obvious advantages over mainland China. Therefore, the paper selects the first three significant combinations with the largest area from these combinations to examine the main classes and spatial distribution of the seasonal trend of each temperature element.

Figure 5 shows the first three classes of significant changes in Txmax and Txmean, which were characterized by a significant increase dominated by A0. The seasonal trend of Txmax was very distinct in mainland China, and only 9.6% of the areas did not change significantly. There were 63 significant change combinations and the first three accounted for 46.9% of the total area. A total of 28.7% (+0000, red) had a significant increase in A0 and no significant increase in other parameters. The combination is mainly distributed in the northeastern, southeastern coastal, and the Tibetan Plateau, indicating that while the annual average maximum temperature in these regions is increasing, the range of the annual maximum temperature and its occurrence time have not changed significantly.

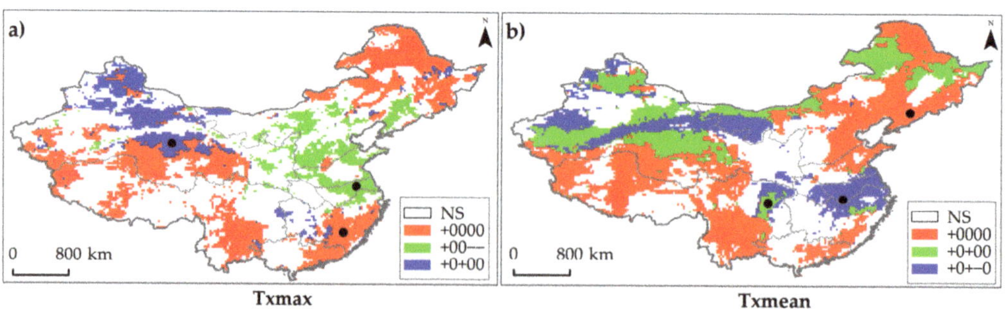

Figure 5. The combination of the top three areas of significant seasonal trends in Txmax and Txmean. Significant positive trends are marked with a "+" sign, significant negative trends with a "−" sign, and no-trends with a "0". Read from left to right, the symbols indicate the significance of trends in A0, A1, A2, F1, and F2. Black spots are randomly selected representative points of the three classes. For example, "+0000" has a significant positive trend in A0 and no significant trend in the other parameters.

The second combination accounted for 9.6%, with A0 increasing significantly, while F1 and F1 decreased significantly (+00−−, green), which was mainly distributed in the eastern northwest China, northern Huang-Huai, and Jiang-Huai regions, indicating that the extreme maximum temperature in these regions generally increased and that the time was delayed. Both A0 and A2 also increased by 9.1% (+0+00, blue) and were mainly distributed in northwest China and the northern margin of the Tibetan Plateau (Figure 5a).

There were 51 combinations with significant seasonal variations in Txmean in mainland China, accounting for 96.3%. The first three classes were when A0 increased significantly (+0000), A0 and A2 increased significantly (+0+00), A0 and A2 increased significantly, and F1 decreased significantly (+0+−0), accounting for 34.8%, 16.2%, and 13.0% of the total area of mainland China, respectively. The first class was distributed mainly in northeast China, north China, the Tibetan Plateau, and the southeastern coast. The second was mainly in the northeast and northwest and the third appeared in the middle and lower reaches of the Yangtze River and northwest (Figure 5b). It can be concluded that in some regions of Huang-Huai and Jiang-Huai, the maximum temperature not only had a significant upward trend but also the time at which its maximum value appeared was significantly delayed. These two places are one of the main grain-producing regions in China, which provides some ideas for follow-up studies on the effect of temperature increases on grain yield.

A grid was randomly selected from the first three classes of Txmax and Txmean, and the monthly dynamics of the start year (1979, black curve) and end year (2020, red curve) were fitted (Figure 6). We can see that the shapes of the curves are different mainly due to different parameter combinations. Even if the same class is different due to locations and elements, the seasonal trend of the same class of curves is consistent, regardless of their shapes. For example, when A0 increased significantly (+0000), the overall value in 2020 was higher than that in 1979. When F1 decreased significantly, the peak time was obviously delayed (+00−−, +0+−0). However, the curve of a significant increase only in A0 of Txmax seems to have significantly delayed in 2020, but the statistical test is not significant, which should be related to the large difference in the time of the maximum at this grid.

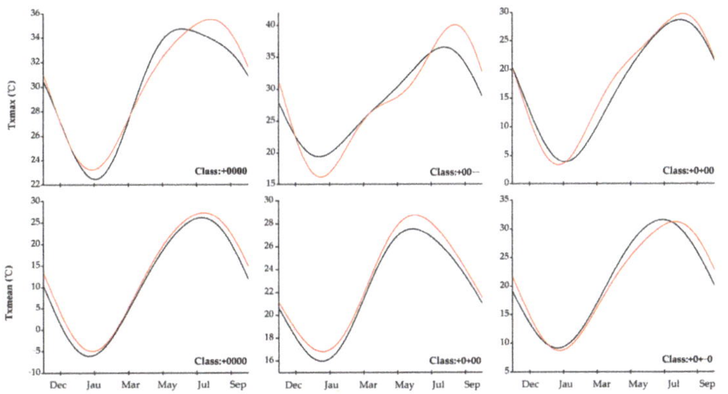

Figure 6. Examples of seasonal trends representative of each of the three major trend classes in Txmax and Txmean. The black curve represents the seasonal curve at the beginning of the series while the red curve represents the seasonal curve at the end of the series. The difference between the two indicates the change resulting from the trend.

3.1.2. Minimum Temperature

Similar to the maximum temperature, the minimum temperature also had a significant increase in mainland China (Figure 7). The A0 of monthly minimum temperature (Tnmin) and monthly average minimum temperature (Tnmean) increased significantly in 92.4% and 97.9% of areas, respectively. From this point of view, the warming of the minimum temperature was larger than that of the maximum temperature, which is consistent with existing studies [54,55], but the time variation of its minimum value was slightly smaller than that of the maximum temperature (Figure 7g,h). The regions where A0 of the minimum temperature did not change significantly were scattered on the Tibetan Plateau, northwest China, and northeast China (Figure 7a,b). In contrast, the minimum temperature in the mid-lower reaches of the Yellow River, Yangtze River Basin, Jiang-Nan, south, and eastern southwest China showed a positive trend in the last 42 years. The above regions are major agricultural areas in China. The regions where A1 of Tnmin and Tnmean increased and decreased significantly were bounded by the 400 mm isohyet in mainland China, i.e., the temperate continental and plateau mountain climatic areas mainly increased, while the monsoon climatic areas mainly decreased (Figure 4c,d). Comparing the spatial distribution of the parameters of maximum temperature and minimum temperature, we can find that A0, A1, A2, and F1 of maximum temperature, A0, A1, F1, and F2 of minimum temperature, and their trend of Txmax/Tnmin and Txmean/Tnmean had similar spatial patterns. However, there was an exception. A2 of Tnmin showed no change in most regions, while Tnmean showed a significant increasing trend in northern China (Figure 7e,f).

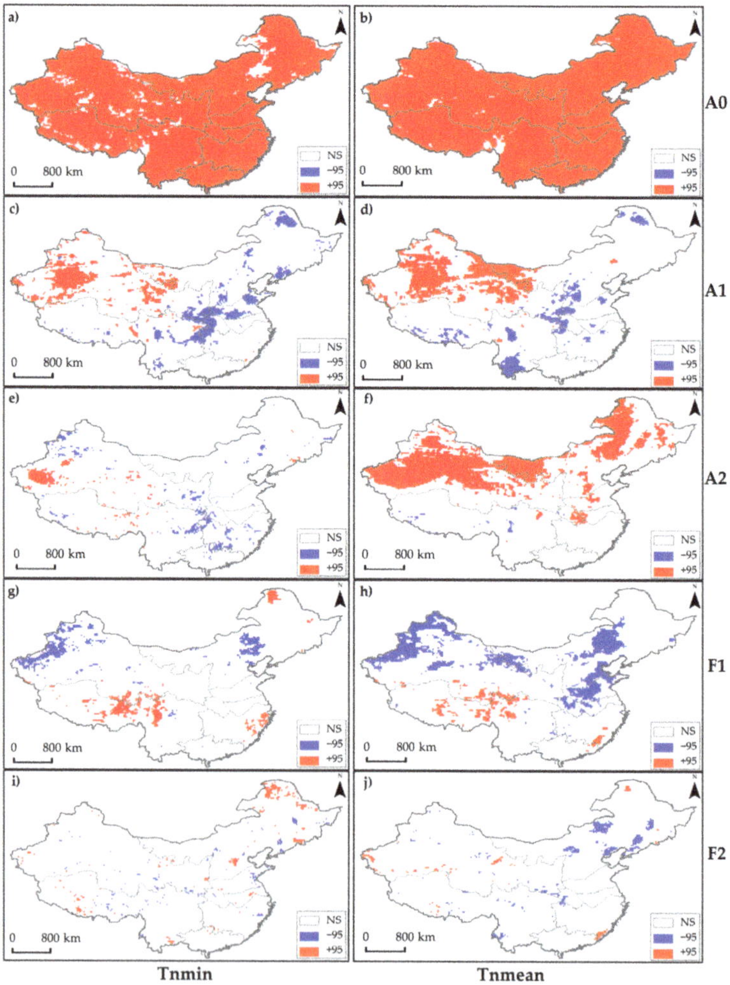

Figure 7. Areas in mainland China with trends of annual average amplitude (A0), the amplitude (A1) and phase (F1) of annual cycle, amplitude (A2) and phase (F2) of semi-annual cycle of Tnmin (**a**,**c**,**e**,**g**,**i**) and Tnmean (**b**,**d**,**f**,**h**,**j**).

F1 of Tnmin had a significant change in 8.6% of the area and the area with a significant increase was slightly smaller than that with a significant decrease (Figure 7g). F1 of Tnmean decreased significantly in 14.4% of the area and increased significantly in 3.1% of the area (Figure 7h). The area where F2 changed significantly was further reduced, accounting for 3.2% and 3.7%, respectively (Figure 7i,j).

In the last 42 years, the spatial distribution and magnitude of the trend of minimum temperature were also similar to the trend of their A0, and linear regression determination coefficients were 0.96 and 0.94, respectively (Figure 8a–d). The warming trends of Tnmin and Tnmean were generally higher in the north and lower in the south. The warming trend rate for most parts of the north was 0.04–0.08 $^{\circ}C{\cdot}a^{-1}$, and that for the south was not more than 0.04 $^{\circ}C{\cdot}a^{-1}$. The spatial patterns of the warming trends of Tnmin and Tnmean were also similar, and the warming trend rate of the former was higher than that of the latter (Figure 8).

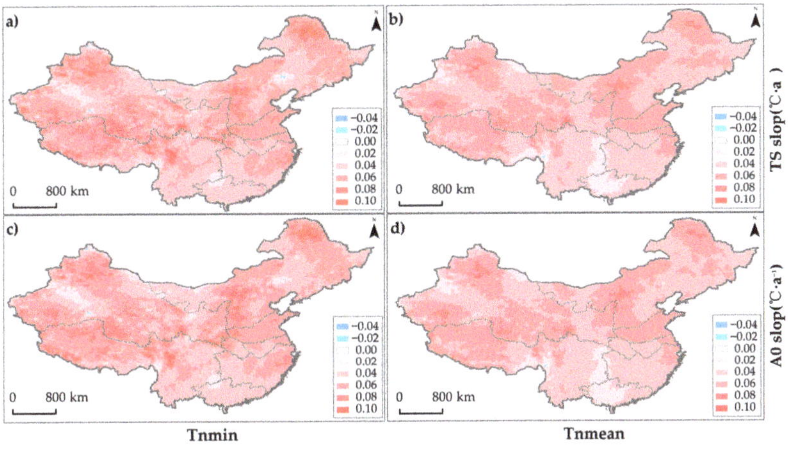

Figure 8. Theil–Sen trend (TS slope (**a**,**b**)) and linear trend for annual average amplitude (A0) of Tnmin and Tnmean (A0 slope (**c**,**d**)).

The seasonal trend of Tnmin was very distinct in mainland China and all were mainly characterized by significant changes in amplitude, which was somewhat different from the maximum temperature. A total of 94.7% of the areas had significant changes, including 57 combinations. The first three classes of significant change accounted for 78.1% of mainland China, and the first class (+0000) was the most distinct, accounting for 67.6% of the total area. The A0 and A1 classes increased (++000) and the A0 increase and A1 decrease (+−000) accounted for 5.3% and 5.1%, respectively, and they appeared in the western and central regions, respectively (Figures 9a and 10).

The first three classes with significant seasonal trends in Tnmean were also dominated by amplitude, and all had increased significantly. A0 increased significantly (+0000), and both A0 and A2 increased significantly (+0+00). The three amplitudes all increased significantly (+++00), accounting for 52.2%, 12.6%, and 5.8% of mainland China, respectively, and the latter two mainly appeared in northwestern China (Figures 9b and 10).

From the three fitting curves, we can see the seasonal trends of Tnmin and Tnmean. Because these three classes are amplitude combinations, the phase change was no-trend at the beginning year and the end year; that is, the time of peak appearance was no different (Figure 10).

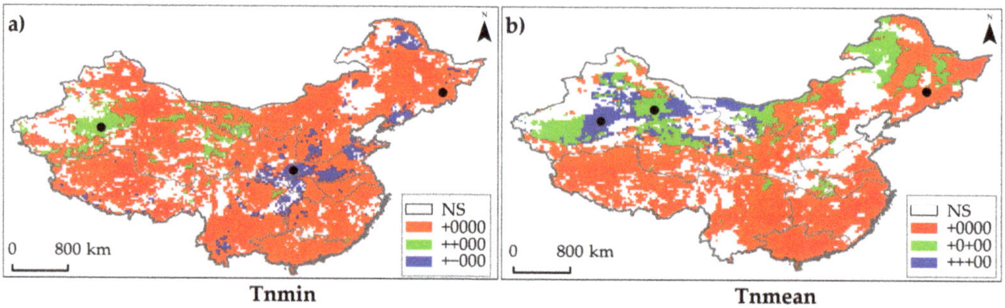

Figure 9. The combination of the top three areas of significant seasonal trends in Tnmin and Tnmean. The meaning of these symbols is the same as that in Figure 5.

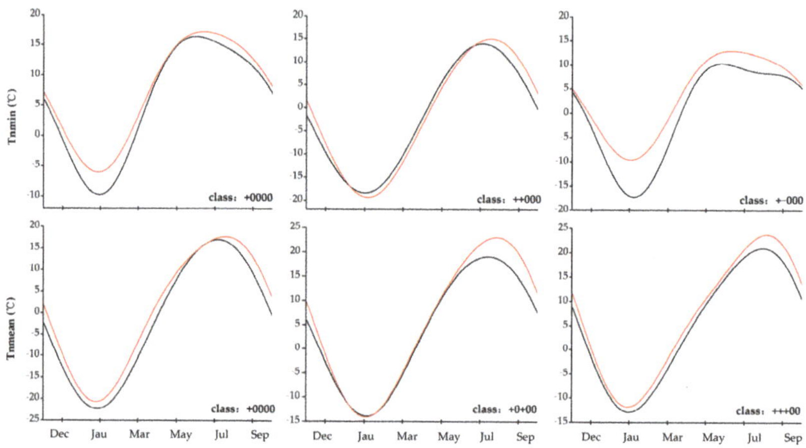

Figure 10. Examples of seasonal trends representative of each of the three major trend classes in Tnmin and Tnmean.

3.2. Change of Extreme Air Temperature Days

3.2.1. Hot Days, Cold Days, and Warm Days

During 1979–2020, hot days (HD) increased significantly in the eastern and northwestern regions (22.7%), with trend rates of 0.2~0.6 d·a^{-1} and 0.6~0.8 d·a^{-1} in some regions of the lower reaches of the Yangtze River. The area of significant decrease was small (5.3%), mainly in the northeast (Inner Mongolia and parts of Liaoning), and their trend rate was not less than −0.4 d·a^{-1} (Figure 11a,b). HD in most parts of the north, northwest and Tibetan Plateau had no significant change, which is related to the fact that there were few or no temperatures higher than 35 °C.

TX10p decreased significantly in eastern Chain, eastern northwest China, and most of the Tibetan Plateau. Most other regions had a decreasing but not significant trend. The trend rate of the significant reduction was −0.4–−0.2%d·a^{-1}. In western Xinjiang, there was an increasing trend and a significant increase in the Tianshan region, but the trend rate did not exceed 0.2% d·a^{-1} (Figure 11c,d). TX90p had a significant increase in most regions, and the trend was not significant in adjacent areas of the south, southwest and Jiang-Nan, eastern northwest, Tibetan Plateau, and most of the Tarim Basin. In terms of the spatial distribution of the increasing trend rate, most regions were less than 0.4% d·a^{-1}, and the increasing trend was slightly prominent in the eastern northwestern and southern southwestern regions (Figure 11e,f).

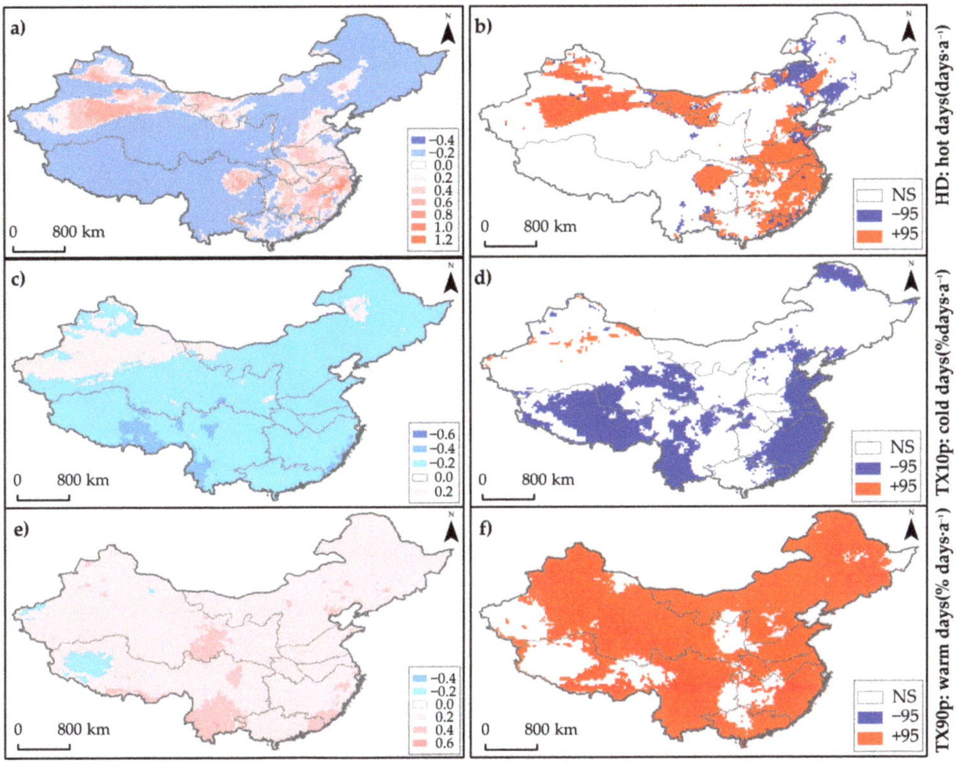

Figure 11. HD (**a**,**b**), TX10p (**c**,**d**), and TX90p (**e**,**f**). Left for trend rate and right for statistically significance.

The regional average of HD, TX10p, and TX90p show that HD and TX90p had an obvious increasing trend, while TX10p had a decreasing trend (Figure S1). The average HD does not seem to have been high in mainland China. This is mainly due to the vast territory of China, with large differences from east to south and from north to south. Some regions in the northeast and Tibetan Plateau do not experience daily temperatures above 35 °C, while the southeast and Turpan Basin may have temperatures as high as 40 days, which further indicates that HDs in warm regions have increased significantly. From the regional trend rate, HDs and TX10p were not as large as TX90p, which was mainly related to the increase in TX90p in most of mainland China.

3.2.2. Frost Days, Cold Nights, and Warm Nights

Frost days (FD) mainly occur in winter, early spring, and late autumn in China. HD is closely related to latitude and altitude. For example, some regions on the Tibetan Plateau have frost year-round, while most regions in south China have frost-free days for approximately 350 days out of the year. During 1979–2020, the change in FD was not significant in most of the area south of the Yangtze River, and the decreasing trend north of the Yangtze River was significant. In addition to the change in FD, this distribution pattern may also have been related to a few HD in the south. In the regions where FD decreased significantly, the trend rate was mostly -0.2–0.8 d·a^{-1}, and it could reach -0.8–1.0 d·a^{-1} in some regions on the Loess Plateau and Tibetan Plateau (Figure 12a,b).

TN10p decreased significantly in most regions, but they were not significant in the northwestern and northeastern regions. In western Inner Mongolia and the Tarim Basin, there were no trends. The trend rate of a significant decrease in most regions did not exceed 0.2% d·a^{-1} (Figure 12c,d). Warm nights (TN90p) had a significant increasing trend in most regions, but the increase was not significant only in some regions of southern Jiang-Nan and western Jiang-Han and east of southwest China. In terms of the spatial distribution of trend rates, TN90p increased more significantly in western than in eastern China (Figure 12e,f).

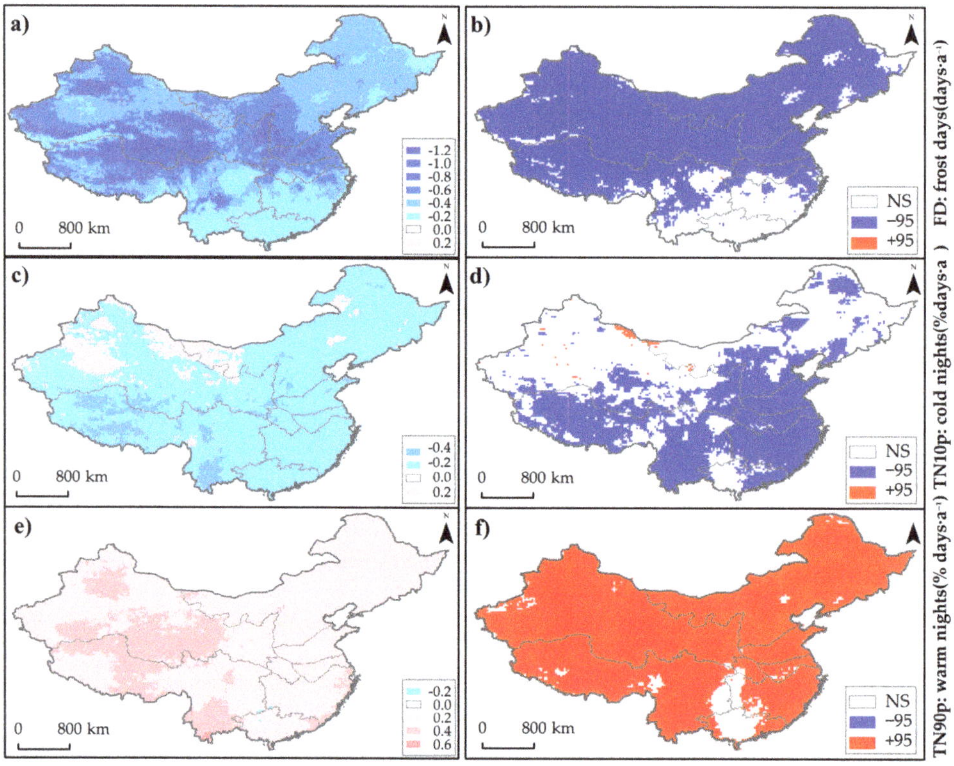

Figure 12. FD (**a**,**b**), TN10p (**c**,**d**), and TN90p (**e**,**f**). Left indicates trend rate and right for significance.

After the regional average in mainland China, both FDs and TN10p had a decreasing trend, and TN90p had an increasing trend. From the absolute value of the trend rate, TN10p was smaller than TN90p, which shows that the warmer minimum temperature increased more distinctly (Figure S2).

The present studies on extreme temperature changes in mainland China show that although the most extreme high temperatures were increasing and extreme low temperatures were decreasing, there were certain differences between regions and magnitudes. It has been reported that days of extreme temperatures at some observatories in mainland China do not conform to a normal distribution [56–59]. Therefore, this difference may be related to the methods and data.

4. Discussion

The reanalysis dataset can provide grid data with uniform spatial distribution and its surface temperature has high reliability [39,60,61]. The extreme temperatures are better captured by the ERA5 [60]. Based on the ERA5, we analyzed the seasonal trend of monthly extreme maximum and minimum temperature in mainland China during 1979–2020 and the time evolution of extreme air temperature days in this paper, hoping to provide help for studying the long-term impact of climate change on grain production.

Under the indisputable fact of climate warming, whether it is a single observation site, reanalyzed data, or model predictions, it was found that extreme warm events have increased significantly, while extreme cold events have decreased significantly in China [1,62–67]. This trend is consistent with the changing characteristics of global temperature extremes [62]. In recent years, some studies have attempted to interpret and assess the effects of extreme climate events on crop yield [19,68–70]. Due to the lack of long-term trials or yield observation data, China's research on this aspect is more based on crop simulation models or statistical yield, or short-term experiments [17,19,62,63,68,70]. The results show that an increased heat ETS (extreme temperature stress) and a decreased cold ETS would be expected for most areas of China during 2020–2049, and the spatial variability of rice yield loss will be greater [63]. Single rice in northeast China and early rice in south China will be under severe cold stress, while single rice in the middle and lower reaches of the Yangtze River and late rice in south China will be under severe heat stress [63]. During 1986–2015, the multi-year average extreme high temperature days increasing every additional day could result in a 226.62 $kg{\cdot}hm^{-2}$ multi-year average maize yield reducing in the main summer maize cultivating area of China [68]. The number of extreme high temperature days has shown an increasing trend during 2021–2050 under RCP4.5 and RCP8.5, which could result in maize yield decreased by 9.2% and 27.3%, respectively [68]. Based on analysis of more than 20,000 historical maize trials, Lobell et al. concluded that each degree day spent above 30 °C reduced the yield by 1% under optimal rain-fed conditions and by 1.7% under drought conditions [24,69]. The conclusion proves the effect of extreme high temperature with experimental data.

A study has pointed out that the yield of maize and rice will decrease with the increase of temperature in the mean growing season [70]. Conversely, the maize and rice yield would increase by approximately 6.947% and 2.885% with a 1 $MJ{\cdot}m^{-2}$ increase in the mean growing season downward shortwave solar radiation, respectively [70]. Furthermore, radiation is greatly affected by cloud cover, which is very important for agricultural production, crop distribution, and animal migration. With climate change, global cloudiness has also changed significantly [71–74]. Limited by space, this article only analyzed the seasonal trend of extreme temperature in mainland China and did not assess the impact of the trend on the crop yield, especially rice, maize, and wheat. Past studies have focused mainly on the number of days of extreme temperature changes and have paid less attention to changes in the monthly or annual range of extreme temperatures, as well as changes in the time of occurrence [49,65,67,75–78]. For these reasons, the result has certain significance for future studies on the impact of extreme climate on agricultural production. In addition, the effects of extreme temperatures, moisture conditions, and cloud cover on crop yields are comprehensive, and future research should focus on their compound and dynamic effects.

5. Conclusions

Monthly maximum temperature (Txmax) and monthly mean maximum temperature (Txmean) had the same annual cycle change with a significant positive trend. The five characteristic parameters for annual and semi-annual cycles in Txmax and Txmean were similar in spatial distribution, indicating that Txmax and Txmean in most parts of mainland China had analogously seasonal variations. The trend in annual average amplitude (A0) was largest, and the annual amplitude (A1) was smallest. The areas with significant changes in annual phase (F1) and semi-annual phase (F2) mainly decreased, which indicates that the time of maximum temperature in these regions had a delaying trend. The area with a higher trend rate in Txmax was larger than that in Txmean, but the trend rate of both was less than 0.06 $°C{\cdot}a^{-1}$ in most regions, showing a strong warming trend on the northeastern edge of the Tibetan Plateau, eastern coast, and Inner Mongolian Plateau. The maximum temperature changed significantly to over 90% of mainland China and the changes in A0 were dominant.

The A0 values of monthly minimum temperature (Tnmin) and monthly mean minimum temperature (Tnmean) in most of mainland China also had a significant warming trend; the trend rate was 0.02–0.08 °C·a^{-1}, which was higher in the north than in the south, and the Tibetan Plateau was especially prominent. The Tnmin warming trend was higher than that of Tnmean. The significant change area of A1 was significantly smaller than that of A0, dispersing on both sides of the 400 mm isohyet; that is, the northwestern area mainly increased and the southeastern area mainly decreased. Different from the maximum temperature, the area where F1 with the minimum temperature changes significantly decreased and the areas with increasing trends increased. The change in A0 in minimum temperature was also dominant and its proportion was higher than maximum temperature, which shows that the trend in minimum temperature in mainland China was more distinct than maximum temperature.

In recent years, the number of heat waves has increased [75–78]. HD has increased significantly in the eastern and northwestern regions (significantly increased areas account for 22.7% of mainland China). However, there was no significant change in those areas where HD may have occurred in south China, east of southwest China, south of north China, and northwest China. TX90p had a significant increase in most regions, while TX10p had a significant decrease on the eastern Tibetan Plateau, most of the Tibetan Plateau, and eastern northwest China. In particular, FD decreased significantly on the Tibetan Plateau. TN10p decreased significantly in most regions but did not change significantly in the northwest and northeast. TN90p increased significantly in most regions.

Supplementary Materials: The following are available online at https://www.mdpi.com/article/10.3390/su132212462/s1, Figure S1: Temporal evolution of regional averaged HD, TX90p, and TX10p and Figure S2: Temporal evolution of regional averaged FD, TN90p, and TN10p.

Author Contributions: J.Z. and J.L. contributed to the study conception and design. Material preparation, data collection, and analysis were performed by W.Y. and Y.W. The first draft of the manuscript was written by W.Y. and all authors commented on subsequent versions of the manuscript. All authors have read and agreed to the published version of the manuscript.

Funding: This research was funded by Foundation Key Research and Development Program of Ningxia Hui Autonomous Region, 2020BBF03009; Key Research and Development Program of Ningxia Hui Autonomous Region, 2020BBF03024; and Natural Science Foundation of Ningxia Hui Autonomous Region, 2020AAC03467.

Institutional Review Board Statement: Not applicable.

Informed Consent Statement: Not applicable.

Data Availability Statement: ERA5 daily temperature data used in this paper are available at ECMWF website (https://www.ecmwf.int/en/forecasts/datasets/reanalysis-datasets/era5, accessed on 4 November 2021).

Acknowledgments: We thank the reviewers for their help in improving this manuscript.

Conflicts of Interest: Authors declare no conflict of interest.

References

1. Arias, P.A. Coauthors. Technical Summary. In *Climate Change 2021: The Physical Science Basis*; Cambridge University Press: Cambridge, UK, 2021.
2. Guo, J. Advances in impacts of climate change on agricultural production in China. *J. App. Meteor. Sci.* **2015**, *26*, 1–11. (In Chinese) [CrossRef]
3. National Bureau of Statistics of China. *Statistical Communiqué of the People's Republic of China on the 2020 National Economic and Social Development*; China Statistics Press: Beijing, China, 2020; pp. 50–51. (In Chinese)
4. Lobell, D.B.; Schlenker, W.; Costa-Roberts. Climate Trends and Global Crop Production since 1980. *Science* **2011**, *333*, 616–620. [CrossRef]
5. Lv, Z.; Zhu, Y.; Liu, X.; Ye, H.; Tian, Y.; Li, F. Climate change impacts on regional rice production in China. *Clim. Chang.* **2018**, *147*, 523–537. [CrossRef]
6. Anderson, W.B.; Seager, R.; Baethgen, W.; Cane, M.; You, L. Synchronous crop failures and climate-forced production variability. *Sci. Adv.* **2019**, *5*, 1976. [CrossRef] [PubMed]

7. Khalili, P.; Masud, B.; Qian, B.; Mezbahuddin, S.; Dyck, M.; Faramarzi, M. Non-stationary response of rain-fed spring wheat yield to future climate change in northern latitudes. *Sci. Total Environ.* **2021**, *772*, 145474. [CrossRef]
8. Lobell, D.B.; Gourdji, S.M. The Influence of Climate Change on Global Crop Productivity. *Plant. Phys.* **2012**, *160*, 1686–1697. [CrossRef] [PubMed]
9. Goyal, M.K. Understanding the climate change impact on crop yield over Eastern Himalayan Region: Ascertaining GCM and scenario uncertainty. *Theor. Appl. Climatol.* **2020**, *142*, 467–482. [CrossRef]
10. Zhao, C.; Liu, B.; Piao, S.; Wang, X.; Lobell, D.B.; Huang, Y.; Huang, M.; Yao, Y.; Bassu, S.; Ciais, P.; et al. Temperature increase reduces global yields of major crops in four independent estimates. *Proc. Natl. Acad. Sci. USA* **2017**, *114*, 9326–9331. [CrossRef]
11. Liu, Y.; Li, N.; Zhang, Z.; Huang, C.; Chen, X.; Wang, F. The central trend in crop yields under climate change in China: A systematic review. *Sci. Total Environ.* **2020**, *704*, 135355. [CrossRef] [PubMed]
12. Liu, W.; Ye, T.; Shi, P. Decreasing wheat yield stability on the North China Plain: Relative contributions from climate change in mean and variability. *Int. J. Clim.* **2021**, *41*, 2820–2833. [CrossRef]
13. Zhang, Y.; Shou, W.; Maucieri, C.; Lin, F. Rainfall increasing offsets the negative effects of nighttime warming on GHGs and wheat yield in North China Plain. *Sci. Rep.* **2021**, *11*, 6505. [CrossRef] [PubMed]
14. Ding, Y.; Wang, H. Newly acquired knowledge on the scientific issues related to climate change over the recent 100 years in China. *Chin. Sci. Bull.* **2016**, *61*, 1029–1041. (In Chinese) [CrossRef]
15. Ren, G.; Zhao, Z.; Zheng, J.; Wu, T.; Tang, G.; Xu, Y. Recent Progress in studies of climate change in China. *Adv. Atoms Sci.* **2012**, *29*, 958–977. [CrossRef]
16. Wu, B.; Jiang, D.; Wang, X. Changes in the Growing Season Across China during 1961–2018. *Chin. J. Atmos. Sci.* **2021**, *45*, 424–434. (In Chinese) [CrossRef]
17. Chen, H.; Sun, J. Anthropogenic influence has increased climate extreme occurrence over China. *Sci. Bull.* **2021**, *66*, 749–752. [CrossRef]
18. Kong, F. The change of extreme cold and warm events in China from 1961 to 2018 and their spatial differences. *Water Resour. Hydropower Eng.* **2020**, *51*, 34–44. (In Chinese) [CrossRef]
19. Wei, X.; Wang, P.; Zhang, Z.; Chen, Y.; Song, X.; Shuai, J.; Shi, P.; Tao, F. Assessing the Impact of Climate Change on Crop Yield Based on Three-interval Temperature Theory: A Case Study of Maize in Heilongjiang Province (1960–2009). *J. Nat. Resour.* **2015**, *30*, 470–479. (In Chinese) [CrossRef]
20. Vogel, M.M.; Zscheischler, J.; Wartenburger, R.; Dee, D.; Seneviratne, S.R. Concurrent 2018 Hot Extremes Across Northern Hemisphere Due to Human-Induced Climate Change. *Earths Future* **2019**, *7*, 692–703. [CrossRef] [PubMed]
21. Avotniece, Z.; Klavins, M.; Rodinovs, V. Changes of Extreme Climate Events in Latvia. *Environ. Clim. Technol.* **2012**, *9*, 4–11. [CrossRef]
22. Xue, Y.; Chen, Q.; Zhang, J.; Huang, P. Trends in extreme high temperature at different altitudes of Southwest China during 1961–2014. *Atmos. Ocean. Sci. Lett.* **2020**, *13*, 417–425. [CrossRef]
23. Gourdji, S.M.; Sibley, A.M.; Lobell, D.B. Global crop exposure to critical high temperatures in the reproductive period: Historical trends and future projections. *Environ. Res. Lett.* **2013**, *8*, 024041. [CrossRef]
24. Zhang, Y.; Zhao, Y.; Sun, Q. Increasing maize yields in Northeast China are more closely associated with changes in crop timing than with climate warming. *Environ. Res. Lett.* **2021**, *16*, 054052. [CrossRef]
25. Gao, M.; Tang, L.; Wu, Z.; Zhang, X.; Sun, Z.; Sun, N.; Gao, Y.; Zhang, C. Response of wheat yield components in global climatic zones to the warming during reproductive period. *J. Plant Nutr. Fertil.* **2019**, *25*, 1247–1256. (In Chinese) [CrossRef]
26. Gao, M.; Zhang, X.; Sun, Z.; Sun, N.; Li, S.; Gao, Y.; Zhang, C. Wheat yield and growing period in response to field warming in different climatic zones in China. *Sci. Agric. Sin.* **2018**, *51*, 386–400. (In Chinese) [CrossRef]
27. He, D.; Fang, S.; Liang, H.; Wang, E.; Wu, D. Contrasting yield responses of winter and spring wheat to temperature rise in China. *Environ. Res. Lett.* **2020**, *15*, 124038. [CrossRef]
28. Sparks, T.; Menzel, A. Observed changes in seasons: An overview. *Int. J. Climatol.* **2020**, *22*, 1715–1725. [CrossRef]
29. Kucharik, C.J.; Barford, C.; Maayar, E.; Wofsy, S.C.; Monson, R.K.; Baldocchi, D. A multiyear evaluation of a dynamic global model at three AmeriFlux forest sites: Vegetation structure, phenology, soil temperature, and CO_2 and H_2O vapor exchange. *Ecol. Modell.* **2006**, *196*, 1–31. [CrossRef]
30. Eastman, J.R.; Sangermano, F.; Machado, E.; Rogan, J.; Anyamba, A. Global Trends in seasonality of normalized difference vegetation index (IDVI), 1982–2011. *Remote Sens.* **2013**, *5*, 4799–4818. [CrossRef]
31. Karl, T.R.; Nicholls, N.; Ghazi, A. Clivar/GCOS/WMO Workshop on Indices and Indicators for Climate Extremes Workshop Summary. *Clim. Chang.* **1999**, *42*, 3–7. [CrossRef]
32. Zhao, B.; Zhang, B.; Shi, C.; Liu, J.; Jiang, L. Comparison of the Global Energy Cycle between Chinese Reanalysis Interim and ECMWF Reanalysis. *J. Meteorol. Res.* **2019**, *33*, 563–575. [CrossRef]
33. Kalney, E. The NCEP/NCAR 40-year reanalysis project. *Bull. Am. Meteorol. Soc.* **1996**, *74*, 789–799. [CrossRef]
34. Kanamitsu, M.; Ebisuzaki, W.; Woollen, J.; Yang, K.; Hnilo, J.; Fiorino, M.; Potter, G.-L. NCEP-DOE AMIP-II Reanalysis (R-2). *Bull. Am. Meteorol. Soc.* **2020**, *74*, 1631–1643. [CrossRef]
35. Onogi, K.; Tsutsui, J.; Koide, H.; Sakamoto, M.; Kobayashi, S.; Hatsushika, H.; Matsumoto, T.; Yamazaki, N.; Kamahori, H.; Takahashi, K.; et al. The JRA-25 Reanalysis. *J. Meteor. Soc. Jpn.* **2007**, *85*, 369–432. [CrossRef]

36. Kobayashi, S.; Ota, Y.; Harada, Y.; Ebita, A.; Moriya, M.; Onoda, H.; Onogi, K.; Kamahori, H.; Kobayashi, C.; Endo, H.; et al. The JRA-55 Reanalysis: General specifications and basic characteristics. *J. Meteor. Soc. Jpn.* **2015**, *93*, 5–48. [CrossRef]
37. Rienecker, M.M.; Suarez, M.J.; Gelaro, R.; Todling, R.; Bacmeister, J.; Liu, E.; Bosilovich, M.G.; Schubert, S.D.; Takacs, L.; Kim, G.K.; et al. MERRA: NASA's Modern-Era Retrospective Analysis for Research and Applications. *J. Clim.* **2011**, *24*, 3624–3648. [CrossRef]
38. Gelaro, R.; McCarty, W.; Suárez, M.J.; Todling, R.; Molod, A.; Takacs, L.; Randles, C.A.; Darmenov, A.; Bosilovich, M.G.; Reichle, R.; et al. The Modern-Era Retrospective Analysis for Research and Applications, Version 2 (MERRA-2). *J. Clim.* **2017**, *30*, 5419–5454. [CrossRef] [PubMed]
39. Zhang, J.; Zhao, T.; Li, Z.; Li, C.; Li, Z.; Ying, K.; Shi, C.; Jiang, L.; Zhang, W. Evaluation of Surface Relative Humidity in China from the CRA-40 and Current Reanalyses. *Adv. Atmos. Sci.* **2021**, *38*, 1958–1976. [CrossRef]
40. Hersbach, H.; Bell, B.; Berrisford, P.; Hirahara, S.; Horányi, A.; Muñoz-Sabater, J.; Nicolas, J.; Peubey, C.; Radu, R.; Schepers, D.; et al. The ERA5 global reanalysis. *Q. J. R. Meteor. Soc.* **2020**, *146*, 1999–2049. [CrossRef]
41. Hu, G.J.; Zhao, L.; Wu, X.; LI, R.; Wu, T.; Su, Y.; Hao, J. Evaluation of reanalysis air temperature products in permafrost regions on the -Tibetan Plateau. *Theor. Appl. Climatol.* **2019**, *138*, 1457–1470. [CrossRef]
42. Wang, X.; Prigent, C. Comparisons of Diurnal Variations of Land Surface Temperatures from Numerical Weather Prediction Analyses, Infrared Satellite Estimates and In Situ Measurements. *Remote Sens.* **2020**, *12*, 583. [CrossRef]
43. Eastman, J.R.; Sangermano, F.; Ghimire, B.; Ghimire, H. Seasonal trend analysis of image time series. *Int. J. Remote Sens.* **2009**, *30*, 2721–2726. [CrossRef]
44. Cogliati, M.G.; Müller, G.; Lovino, M.A. Seasonal trend analysis of minimum air temperature in La Plata river basin. *Theor. Appl. Climatol.* **2021**, *144*, 25–37. [CrossRef]
45. Duan, D. Harmonic analysis and its application to weather forecast. *J. Cnuwrep* **1986**, *1*, 72–84. (In Chinese)
46. Hamed, K. Exact distribution of the Mann-Kendall trend test statistic for persistent data. *J. Hydrol.* **2009**, *365*, 86–94. [CrossRef]
47. Sen, P.K. Estimates of the Regression Coefficient Based on Kendall's Tau. *J. Am. Stat. Assoc.* **1968**, *63*, 1379–1389. [CrossRef]
48. Neeti, N.; Eastman, J.R. A contextual Mann-Kendall approach for the assessment of trend significance in image time series. *Trans. GIS* **2011**, *15*, 599–611. [CrossRef]
49. Zhai, P.; Pan, X. Trends in temperature extremes during 1951–1999 in China. *Geophys. Res. Lett.* **2003**, *30*, 1913. [CrossRef]
50. Lin, Z.; Yu, Z.; Zhang, H.; Wu, C. Quantifying the attribution of model bias in simulating summer hot days in China with IAP AGCM 4.1. *Atmos. Ocean. Lett.* **2016**, *9*, 436–442. [CrossRef]
51. Editing Commission of the Third National Report on Climate Change of China. In *The Third National Report on Climate Change*; Science Press: Beijing, China, 2015; pp. 20–21.
52. Jin, Q.; Yang, J.; Chen, H.; He, Q.; Tang, F. Comprehensive analysis of temperature variations over Tibet Plateau from 1961 to 2015. *J. Lanzhou Univ. Nat. Sci.* **2020**, *56*, 755–764. (In Chinese) [CrossRef]
53. Stine, A.; Huybers, P.; Fung, I. Change in the phase of the annual cycle of surface temperature. *Nature* **2009**, *457*, 435–440. [CrossRef] [PubMed]
54. Wang, L.; Liu, B.; Henderson, M.; Shen, X.; Yang, L. Warming across decades and deciles: Minimum and maximum daily temperatures in China, 1955–2014. *Int. J. Climatol.* **2018**, *38*, 2325–2332. [CrossRef]
55. Wu, X.; Wang, Z.; Zhou, X.; Lai, C.; Chen, X. Trends in temperature extremes over nine integrated agricultural regions in China, 1961–2011. *Theor. Appl. Climatol.* **2017**, *129*, 1279–1294. [CrossRef]
56. Qian, C.; Zhang, X.; Li, Z. Linear trends in temperature extremes in China, with an emphasis on non-Gaussian and serially dependent characteristics. *Clim. Dyn.* **2019**, *53*, 533–550. [CrossRef]
57. Shen, X.; Liu, B.; Lu, X.; Fan, G. Spatial and temporal changes in daily temperature extremes in China during 1960–2011. *Theor. Appl. Climatol.* **2017**, *130*, 933–943. [CrossRef]
58. Zhang, J.; Qian, C. Linear Trends in Occurrence of High Temperature and Heat Waves in China for the 1960–2018 Period: Method and Analysis Results. *Clim. Environ. Res.* **2020**, *25*, 225–239. (In Chinese) [CrossRef]
59. Xing, P.; Yang, R.; Du, W.; Dang, B.; Xuan, C.; Xiong, F. Spatiotemporal variation of high temperature day and heat wave in North China during 1961–2017. *Sci. Geogr. Sin.* **2020**, *40*, 1365–1376. (In Chinese) [CrossRef]
60. Reda, K.W.; Liu, X.; Tang, Q.; Gebremicael, T.G. Evaluation of Global Gridded Precipitation and Temperature Datasets against Gauged Observations over the Upper Tekeze River Basin, Ethiopia. *J. Meteorol. Res.* **2021**, *35*, 673–689. [CrossRef]
61. Zheng, Y. A Preliminary Analysis on the Applicability of ERA5 Reanalysis Data in Guangdong Province. *Meteorol. Environ. Res.* **2020**, *11*, 37–42, 48. [CrossRef]
62. Zhou, G.; He, Q.B. *Assessment Report of Climate Hazard of Agricultural Production in China*, 1st ed.; Social Sciences Academic Press: Beijing, China, 2020; pp. 17–20.
63. Zhang, Z.; Chen, Y.; Wang, C.; Wang, P.; Tao, F. Future extreme temperature and its impact on rice yield in China. *Int. J. Climatol.* **2017**, *37*, 4814–4827. [CrossRef]
64. Schleussner, C.F.; Lissner, T.K.; Fischer, E.M.; Wohland, J.; Perrette, M.; Golly, A.; Rogelj, J.; Childers, K.; Schewe, J.; Frieler, K.; et al. Differential climate impacts for policy-relevant limits to global warming: The case of 1.5 °C and 2 °C. *Earth Syst. Dynam.* **2016**, *6*, 2447–2505. [CrossRef]
65. Guo, X.; Huang, J.; Luo, Y.; Zhao, Z.; Xu, Y. Projection of heat waves over China for eight different global warming targets using 12 CMIP5 models. *Theor. Appl. Climatol.* **2017**, *128*, 507–522. [CrossRef]

66. Xiao, S.; Zhang, Z.; Chen, Y.; Wang, P.; Xiang, M.; Shi, P.; Tao, F. Spatiotemporal changes of global extreme temperature events (ETEs) since 1981 and the meteorological causes. *Nat. Hazards.* **2014**, *70*, 975–994. [CrossRef]
67. Zhang, G.; Zeng, G.; Yang, X.; Jiang, Z. Future Changes in Extreme High Temperature over China at 1.5 °C–5 °C Global Warming Based on CMIP6 Simulations. *Adv. Atmos. Sci.* **2021**, *38*, 256–267. [CrossRef]
68. Zhang, Q.; Yang, Z. Impact of Extreme Heat on Corn Yield in Main Summer Corn Cultivating Area of China at Present and Under Future Climate Change. *Int. J. Plant. Prod.* **2019**, *13*, 267–274. [CrossRef]
69. Lobell, D.B.; Bänziger, M.; Magorokosho, C.; Vivek, B. Nonlinear heat effects on African maize as evidenced by historical yield trials. *Nat. Clim. Chang.* **2011**, *1*, 42–45. [CrossRef]
70. Chen, X.; Wang, L.; Niu, Z.; Zhang, M.; Li, C.; Li, J. The effects of projected climate change and extreme climate on maize and rice in the Yangtze River Basin, China. *Agric. Forest Meteorol.* **2020**, *282*, 107867. [CrossRef]
71. Bai, Y.; Wang, J.; Zhang, B.; Zhang, Z.; Liang, J. Comparing the impact of cloudiness on carbon dioxide exchange in a grassland and a maize cropland in northwestern China. *Ecol. Res.* **2012**, *27*, 615–623. [CrossRef]
72. Valjarević, A.; Morar, C.; Živković, J.; Niemets, L.; Kićović, D.; Golijanin, J.; Gocić, M.; Bursać, N.M.; Stričević, L.; Žiberna, I.; et al. Long Term Monitoring and Connection between Topography and Cloud Cover Distribution in Serbia. *Atmosphere* **2021**, *12*, 964. [CrossRef]
73. Jones, P.A.; Henderson-Sellers, A. Historical Records of Cloudiness and Sunshine in Australia. *J. Clim.* **1992**, *5*, 260–270. [CrossRef]
74. Kaiser, D.P. Decreasing cloudiness over China: An updated analysis examining additional variables. *Geophys. Res. Lett.* **2000**, *27*, 2193–2196. [CrossRef]
75. Bucchignani, E.; Zollo, A.L.; Cattaneo, L.; Montesarchio, M.; Mercogliano, P. Extreme weather events over China: Assessment of COSMO-CLM simulations and future scenarios. *Int. J. Climatol.* **2017**, *37*, 1578–1594. [CrossRef]
76. Xie, W.; Zhou, B.; You, Q.; Zhang, Y.; Ullah, S. Observed changes in heat waves with different severities in China during 1961–2015. *Theor. Appl. Climatol.* **2020**, *141*, 1529–1540. [CrossRef]
77. Xu, F.; Chan, T.; Ming, L. Different changes in dry and humid heat waves over China. *Int. J. Climatol.* **2020**, *41*, 1369–1382. [CrossRef]
78. Wang, J.; Yan, Z. Rapid rises in the magnitude and risk of extreme regional heat wave events in China. *Weather Clim. Extrem.* **2021**, *34*, 100379. [CrossRef]

sustainability

MDPI

Article

High-Resolution Regional Climate Modeling and Projection of Heatwave Events over the Yangtze River Basin

Zhibo Gao and Xiaodong Yan *

State Key Laboratory of Earth Surface Processes and Resource Ecology, Faculty of Geographical Science, Beijing Normal University, Beijing 100875, China; gaozhibo@mail.bnu.edu.cn
* Correspondence: yxd@bnu.edu.cn

Abstract: Heatwave events (HWEs) have strong impacts on human health, ecosystems, and sustainable social development. Using a gridded observation dataset and a high-resolution regional climate model (RCM), this study analyzed the characteristics of HWEs over the Yangtze River Basin (YRB) in eastern China during the historical period and projected the changes in HWEs over the YRB in the future. The daily maximum temperature (Tmax), long-lived ($\geq$6 days) HWEs, and total ($\geq$3 days) HWEs in the YRB all showed an obvious upward trend from 1981 to 2018, while the increase in short-lived ($\geq$3 days and <6 days) HWEs was relatively moderate overall. The RCM of the Weather Research and Forecasting (WRF) model can simulate the characteristics of Tmax and HWEs in the historical period very well, and the projection results showed that Tmax, total HWEs, and long-lived HWEs will all increase obviously in both the SSP245 and SSP585 scenarios. Short-lived HWEs will also increase rapidly under SSP585, but they will rise slowly overall under SSP245. The changes in HWEs had distinct regional differences, and the intensity and coverage area of HWEs were greater under SSP585 overall. In the future, the increase in HWEs over the YRB region is likely to be associated with the enhancement of the western-Pacific subtropical high (WPSH) and South-Asian high (SAH), and this enhancement was also greater under SSP585. The results from the high-resolution simulation of the RCM can provide an important reference for disaster prevention and mitigation in the future.

Keywords: WRF model; projection; short-lived heatwave event; long-lived heatwave event; Yangtze River Basin

Citation: Gao, Z.; Yan, X. High-Resolution Regional Climate Modeling and Projection of Heatwave Events over the Yangtze River Basin. *Sustainability* **2022**, *14*, 1141. https://doi.org/10.3390/su14031141

Academic Editor: Amir Mosavi

Received: 28 December 2021
Accepted: 17 January 2022
Published: 20 January 2022

Publisher's Note: MDPI stays neutral with regard to jurisdictional claims in published maps and institutional affiliations.

1. Introduction

The Intergovernmental Panel on Climate Change (IPCC) in its 6th Assessment Report (AR6) reported that the global mean surface temperature in the first two decades of the 21st century (2001–2020) was 0.99 °C higher than in 1850–1900 [1]. Under the background of global warming, the occurrence of extreme weather and climate events, such as heatwave events (HWE), have increased significantly, which has led to a large impact on the social economy, natural ecosystems, and human health [2–5]. For example, during summer 2003, long-lasting HWEs occurred in Europe, which resulted in more than 4000 deaths in 18 countries and an estimated USD 13 billion in property damage [6,7]. Unusually, HWEs also persisted in the Yangtze River Basin (YRB) of eastern China during summer 2013, resulting in persistent drought that affected over nine provinces with a population of more than half a billion [5,8]. More importantly, the results of a number of numerical simulations suggest that HWEs in eastern China will still increase significantly in the future in the context of global warming [5,9,10].

Reliable weather and climate forecasts of HWEs are essential for risk mitigation and early-warning systems [11]. Most global climate models (GCMs), with resolutions from 100 km to 200 km, are generally good in capturing large-scale circulation [12–14], but they are not efficient enough to represent the fine-scale processes of the atmosphere, as well as the terrain and land-use distribution, which makes it difficult to accurately characterize regional HWEs [15–18]. Regional climate models (RCMs), with refined grid spacing and solid-model physics, can address the abovementioned problems associated with GCMs very well [19–21]. Several studies have indicated that RCMs are believed to provide more added value in their simulations with respect to GCMs [22–25]. Based on the above facts, it is important and meaningful to use RCMs to study the mechanisms and variation of HWEs in a certain area [26–28].

The YRB is one of the core regions for the occurrence of HWEs in China [29]. A large number of studies have focused on how HWEs in this region are generated from the perspective of large-scale circulation anomalies [29,30], variability in the sea-surface temperature (SST) [31], and anomalous boundary characteristics in observational diagnostics [32]. Li, Xiao, and Zhao [30] suggested that the occurrence of HWEs is strongly associated with the joint effects of the western-Pacific subtropical high (WPSH) and the South-Asian high (SAH). In addition, Teng and Branstator [33] pointed out that the East-Asian jet stream (EAJS) can act as a waveguide to adjust the intensity and position of the WPSH through quasi-zonal teleconnections, which further influence the HWEs in the YRB region. It is also indicated that the strong SST anomaly over the mid-North Atlantic connects to the WPSH and East-Asian upper-level westerlies via the teleconnection wave train and can further contribute to the variability of HWEs over the YRB region [34]. However, previous studies on HWEs over the YRB region have mostly focused on large-scale circulation characteristics and have lacked descriptions of small-scale information, such as the spatial distribution of HWEs and their regional differences in future climate scenarios. Considering the large population, rich production, and large economic volume in the YRB, it is necessary to conduct a high-resolution simulation study of HWEs in this region to provide a reference for sustainable development, disaster prevention, and early warning systems. Therefore, in this study, we attempted to answer three key questions: (1) What are the temporal and spatial characteristics of HWEs in the YRB during the historical period? (2) Do high-resolution RCMs have the ability to simulate the characteristics of HWEs very well? (3) How will HWEs change in the future?

The rest of the article is structured as follows. The following section describes the data, methods, and experimental design for the simulation of HWEs over the YRB region. Section 3 presents the analytical results. Finally, the discussion and conclusion are presented in Sections 4 and 5, respectively.

2. Materials and Methods

2.1. Model and Experimental Design

The Weather Research and Forecasting (WRF) model version 4.0 [35] was employed in this study. The simulation domain is centered at 30° N, 113.70° E, with a horizontal resolution of 40 km, which is more refined than the first phase of the Coordinated Regional Climate Downscaling Experiment (CORDEX-I). Figure 1 shows the simulation domain and the terrain height within it, and the domain covers all of the YRB region.

Most existing literature has used a particular GCM as the initial conditions and boundary conditions (ICs and BCs) for the RCM, which causes the simulation results of the RCM to have a large uncertainty because dynamic downscaling simulations are often degraded by biases in large-scale forcing [36]. Based on this, a bias-corrected GCM dataset [36] based on 18 models from the Coupled Model Intercomparison Project Phase 6 (CMIP6) was used as the ICs and BCs to drive the WRF model in this study. Previous studies have showed that this dataset is of better quality than individual CMIP6 models [36]. The simulation was divided into two periods. One is the reference period from 1995 to 2014, and the selection of this period was recommended by CMIP6. The other is the future period from 2015 to 2054. The simulation in the future period includes two scenarios: SSP245 and SSP585. SSP245 represents a moderate socioeconomic-development path with medium-low radiation forcing, while SSP585 represents the combined scenario of a high energy-intensive, socioeconomic developmental path with strong radiative forcing. The model starts to integrate one year in advance, and the first year of simulation was used as a spin-up time and was not involved in the analysis.

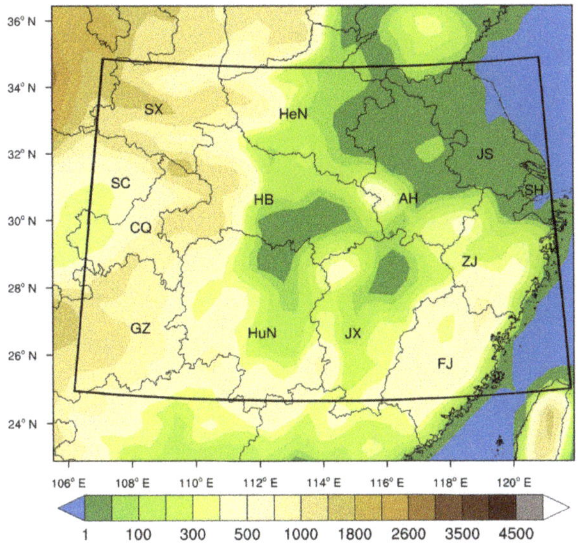

Figure 1. The topography (m; shading) of the simulation domain. The black quadrilateral represents the YRB region, and the meanings of the abbreviations in the figure are as follows: SX, Shaanxi Province; HeN, Henan Province; AH, Anhui Province; JS, Jiangsu Province; SC, Sichuan Province; CQ, Chongqing Province; HB, Hubei Province; ZJ, Zhejiang Province; SH, Shanghai City; GZ, Guizhou Province; HuN, Hunan Province; JX, Jiangxi Province; and FJ, Fujian Province.

The physics options used for WRF include the single-moment six-class microphysics scheme [37], the new Kain–Fritsch cumulus parameterization [38], the Noah land surface scheme [39], and the NCAR Community Atmosphere Model (CAM) shortwave and longwave radiation scheme [40].

2.2. Definition of HWEs

For the definition of HWE, the China Meteorological Administration (CMA) usually defines HWE as a daily maximum surface air temperature at 2 m (Tmax) greater than 35 °C and a duration of more than three days [41,42]. Based on this, in this study, if the Tmax at a certain grid point meets the above conditions, it is considered that a HWE has occurred at that grid point. Furthermore, for the entire YRB region, if the number of grid points that meet the above conditions exceeds one-eighth of the total number of grid points, it is considered that the entire YRB region experienced a HWE. If a HWE occurs, HWDs are the total participating days of the HWE. Similar definition methods have also been used in previous studies on regional HWEs [30,41]. Furthermore, to distinguish HWEs of different intensities, we defined HWEs with a duration of 3–5 days as short-lived HWEs and HWEs with a duration of more than 5 days as long-lived HWEs.

2.3. Gridded Observation Dataset

A set of gridded observations, CN05.1 (CN051) [43], were used to analyze the characteristics of historical HWEs in the YRB and to verify the reliability of WRF for the simulation of HWEs in this region. This dataset was produced by the National Climate Center of CMA from over 2400 weather stations in China and has been widely used as a reference for the validation of models [44,45]. It provides Tmax, daily minimum and mean surface air temperature at 2 m, and daily total precipitation from 1961 to 2018. Previous studies have indicated that the Tmax over the YRB region exhibits an obvious shift at approximately 1980 and then increased significantly [46]. Therefore, in terms of historical periods, this study mainly focused on the characteristics of HWEs from 1981 to 2018.

2.4. Methods

2.4.1. Regression Analysis

In order to explore the linear trends of Tmax and HWEs, linear regression was carried out. It has been a common method to determine the long-term linear trend of a certain meteorological variable [47,48]:

$$T^r = T_0^r + s(t - t_0) \tag{1}$$

where T^r and T_0^r are the variables for years t and t_0, respectively. The s represents the linear trend of the variable. A positive (negative) value indicates an increasing (decreasing) trend of the variable.

In addition, linear regression was also used to study the relationship between geopotential height (y) and HWEs (x):

$$y = ax + b \tag{2}$$

where a represents the regression coefficient (RC) and b is the intercept. A positive RC indicates that as the predictor variable increases, the response variable also increases and vice versa. The absolute value of RC can indicate the relationship strength. The larger the number, the stronger the relationship. The results of the linear regression were tested for significance using the t-test at the 90% confidence level.

2.4.2. Evaluation of the Model Performance

The correlation coefficient (R) was used to indicate the strength and direction of the relationship between the WRF model (x) and observation (y) during the 1995–2014 reference period:

$$R = \frac{\sum_{i=1}^{n}(x_i - x_m)(y_i - y_m)}{\sqrt{\sum_{i=1}^{n}(x_i - x_m)^2} \cdot \sqrt{\sum_{i=1}^{n}(y_i - y_m)^2}} \tag{3}$$

where n represents the number of years, and x_m and y_m are the mean values of x and y, respectively. The value of R ranges from -1 to 1, and the $-$ and + signs indicate negative and positive linear correlations, respectively. When the R value is positive, a larger absolute R value indicates that the model is closer to the observation. Significance levels of the R are estimated according to the two-tailed Student t-test.

3. Results

3.1. Characteristics of HWEs in the Historical Period

Because Tmax is an essential indicator for HWEs, we first gave attention to its changes. The summer mean Tmax over the YRB increased slowly, and it exhibited an obvious interdecadal shift at approximately 2000, which became a positive anomaly (Figure 2a). Correspondingly, the total number of HWEs also showed a slow rise and shifts in 2000 (Figure 2b). It is also worth mentioning that the YRB experienced persistent HWEs in 2013, and the Tmax of many weather stations set a new record. Interestingly, although the total number of HWEs is increasing; the short-lived HWEs have not risen so much, except for more occurrences in approximately 2005; and there was no obvious upward trend in other years (Figure 2c). However, it should be mentioned that the number of long-lived HWEs increased obviously during the study period, which is the main reason for the increase in the total number of HWEs (Figure 2d). Similar characteristics can be found for the variation in HWDs between the two classes of HWEs (Figure 2e,f). Therefore, HWEs with long durations were the main type of HWEs from 1981 to 2018, and they dominated the increase in total HWEs in the YRB region.

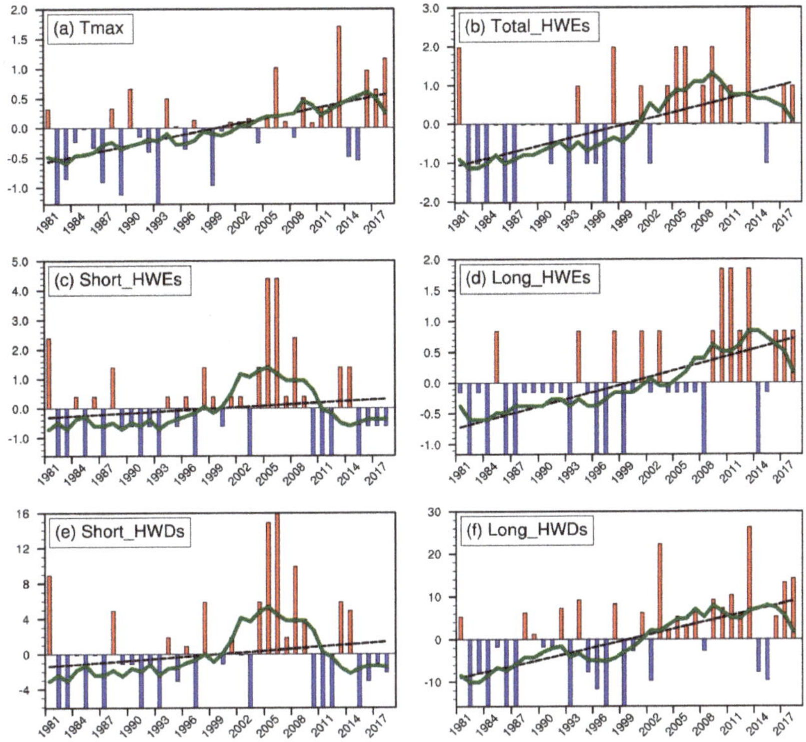

Figure 2. (**a**) Anomalies of mean daily maximum tempeature at 2 m (°C) and anomalies in the number of (**b**) total HWEs (events), (**c**) short-lived HWEs (events), (**d**) long-lived HWEs (events), (**e**) short-lived HWDs (days), and (**f**) long-lived HWDs (days) during summer over the YRB region. The black dashed lines denote the linear trends, and the green solid lines represent the nine-year running average.

The above results are based on regional averages. In order to clearly understand the characteristics of HWEs in different areas, we further analyzed the spatial distribution of the long-term trends in Tmax and the number of HWEs. The region with the greatest increasing trend of Tmax was located in the western YRB, including Shaanxi, Sichuan, and Chongqing provinces (Figure 3a). In addition, the trends were also obvious in eastern coastal areas (Jiangsu and Shanghai). For the trends of total HWEs, the western YRB also had the greatest value, followed by northern and eastern areas (Figure 3b). It should be mentioned that the increasing trend of total HWEs in the western and southern YRB was mainly caused by the rise of long-lived HWEs (Figure 3d,f), while the increase in HWEs in the northern and eastern YRB was mainly caused by the rise of short-lived HWEs (Figure 3c,e). Therefore, the increasement in the local short-lived HWEs was still not negligible.

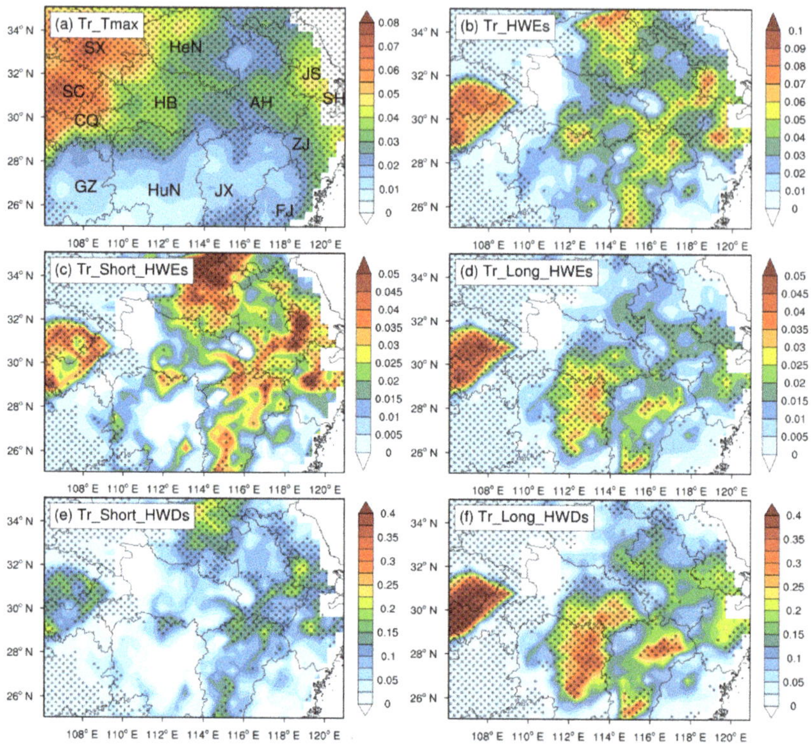

Figure 3. Spatial distribution of the linear trend of summer mean (**a**) daily maximum temperature (°C/a), (**b**) total HWEs (events/a), (**c**) short-lived HWEs (events/a), (**d**) long-lived HWEs (events/a), (**e**) short-lived HWDs (days/a), and (**f**) long-lived HWDs (days/a) over the YRB region from 1981 to 2018. The dots indicate that the 90% confidence level was exceeded.

3.2. Projection of HWEs

Before using the RCM to make projections of HWEs in the YRB region, we first analyzed the simulation results in reference periods to verify the model's ability to simulate HWEs. Because Tmax is the most important indicator for HWEs, we mainly focused on the prediction skill of Tmax. The model can simulate the spatial distribution of the average daily Tmax in summer very well (Figure 4a,b), although there is a certain cold bias in the northern YRB region. From the perspective of time evolution, the model can also simulate the interannual variation in the regional average Tmax very well (Figure 4c). The R between the model and observations reached 0.45, exceeding the 95% confidence level. In addition, the model also has a certain ability to simulate the number of HWDs in different years. The R reached 0.39 and passed the 90% confidence level (Figure 4d). Therefore, the model has a certain ability to simulate the HWEs in the YRB region, which indicates that it can be used to predict the future variation in HWEs in this region.

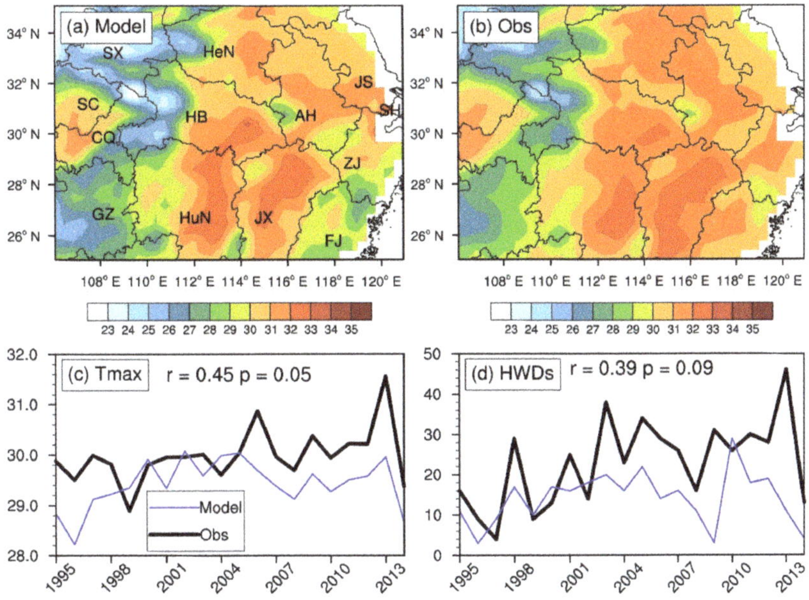

Figure 4. Spatial distributions of the average Tmax (°C) in summer from 1995 to 2014 for (**a**) model and (**b**) observation, (**c**) time series of the spatially averaged daily Tmax (°C) in the YRB during summer, and (**d**) time series of the total HWDs (days) in the YRB during summer.

In the future, the average Tmax in summer over the YRB will rise obviously (Figure 5a). The upward trend of SSP585 will be greater than that of SSP245. Relative to the average of 1995–2014 reference period, the temperature of SSP585 will rise by more than 1.8 °C in the middle of the century. The total number of HWEs will also rise under the impact of increasing temperatures, especially in the SSP585 scenario, which will increase by 1.8 events per year by mid-century (Figure 5b). It should be mentioned that there is a great difference in the changes of the short-lived HWEs under the two scenarios (Figure 5c). For SSP245, short-lived HWEs first increase and then decrease, with a relatively flat trend. However, for SSP585, although there is also a slight downward trend after 2024, the overall upward trend of short-lived HWEs is particularly obvious, which will increase by 1.5 events per year in the middle of the century. As for the long-lived HWEs, they will increase obviously in both future scenarios (Figure 5d), which may cause large challenges for future disaster prevention and mitigation work. It can also be found that after 2030, the increase in total HWEs under the SSP245 scenario is mainly caused by the rise in long-lived HWEs. In addition, the spatial coverages of each HWE will also increase in the future (Figure S1 in Supplementary Material). More areas will be affected by HWEs under SSP585. The above results show that adopting a moderate economic-development path can effectively alleviate the impact of HWEs in the future.

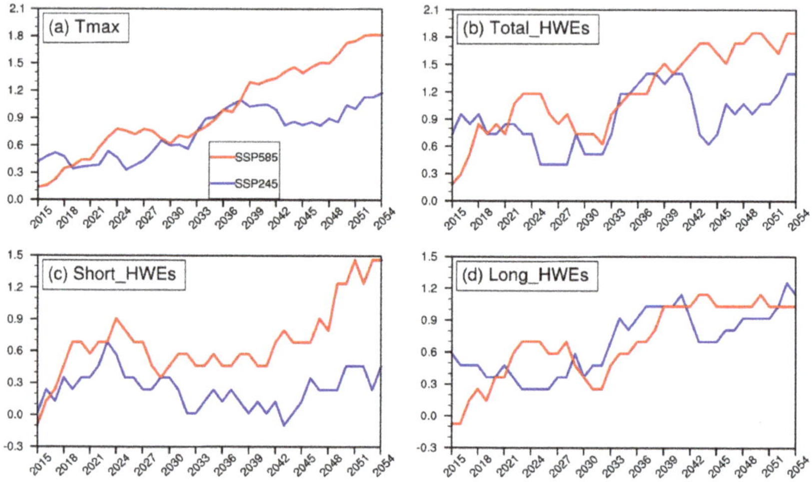

Figure 5. The anomalies in (**a**) Tmax (°C), (**b**) total HWEs (events), (**c**) short-lived HWEs (events), and (**d**) long-lived HWEs (events) relative to the 1995–2014 reference period averaged over the YRB region. The time series were smoothed with a 9-year running mean filter.

To thoroughly understand the regional differences in the changes in HWEs, we then analyzed the trends in spatial distribution. For Tmax, the northern region warms faster than the southern region, and the warming trend in SSP585 is greater, with the maximum value located in the northeastern part of the YRB (Figure 6a,e). Correspondingly, the occurrence of HWEs in the YRB also increases consistently across the region, with a greater increase in SSP585, especially in the central and eastern regions, including Henan, Hubei, Hunan, Jiangxi, Jiangsu, and Anhui provinces (Figure 6b,f). It is noteworthy that for short-lived HWEs, the rising trend in the northern region (including Henan and northern Anhui) is evident in the SSP245 scenario, and it is the main reason for the rising number of total HWEs in this region in the future. In contrast, for the southern region (including Jiangxi and Hubei), the number of short-lived HWEs is likely to decrease in the future (Figure 6c). It should be mentioned that the future trend of short-lived HWEs in the southern region is consistent with the historical period. At the same time, there will be a significant increase in long-lived HWEs in this region in the future (Figure 6d). For SSP585, both short-lived HWEs and long-lived HWEs will increase significantly in the central and eastern regions (Figure 6g,h), which once again proves the importance of controlling greenhouse-gas emissions. Furthermore, we also analyzed the differences in climatology of Tmax and HWEs from 2035 to 2054 relative to the 1995 to 2014 reference period (Figure 7). Under SSP245, it can also be found that the Tmax will rise more obviously in the northern YRB, and the numbers of HWEs will increase more significantly in the northern and southern YRB, with the northern region dominated by the rise of short-lived HWEs, while the southern region is dominated by the increase in long-lived HWEs. The changes in SSP585 will be more obvious, and it should be mentioned that short-lived HWEs will also rise in the southern YRB under SSP585.

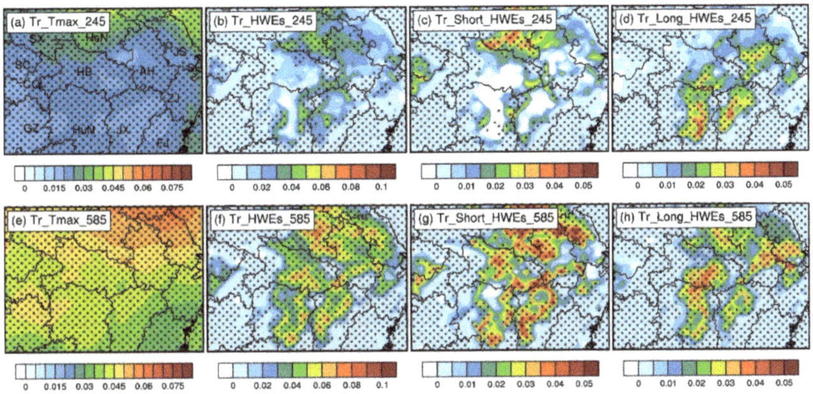

Figure 6. Spatial distribution of the trend of summer mean (**a**), (**e**) daily maximum temperature (°C/a), (**b**), (**f**) total HWEs (events/a), (**c**), (**g**) short-lived HWEs (events/a), and (**d**), (**h**) long-lived HWEs (events/a) over the YRB region from 2015–2054 under (**a–d**) SSP245 and (**e–h**) SSP585. The dots indicate that the 90% confidence level is exceeded.

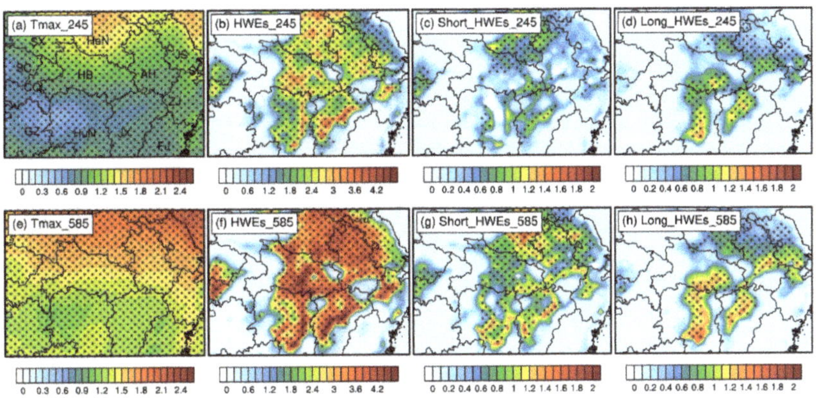

Figure 7. The changes in (**a**), (**c**) Tmax (°C); (**b**), (**f**) HWEs (events); (**c**), (**g**) short-lived HWEs (events); and (**d**), (**h**) long-lived HWEs (events) for (**a–d**) SSP245 and (**e–h**) SSP585 over the 2035–2055 period relative to the 1995–2014 reference period. The dots indicate that the 90% confidence level is exceeded.

The above results indicate that the temperature in the YRB region will be higher, and HWEs will be more frequent in the future compared to the reference period. Previous studies have proven that the WPSH and SAH play an important role in increasing Tmax and HWEs. Therefore, it is important to understand the changes in geopotential height in the YRB region in the future. Since geopotential heights at 500 hPa and 200 hPa levels have a good indication for these two systems, we mainly focused on analyzing the changes in geopotential heights at these two levels. The geopotential heights of these two levels in the YRB will increase in the future relative to the reference period, and the increasing trend will be more obvious under the SSP585 scenario (Figure 8). The results may indicate that the WPSH and SAH will both intensify in the future. To obtain a deep understanding of the relationship between the enhancement of these two systems and the HWEs, Figure 9 shows the regressions of geopotential height anomalies against the time series of HWEs. Under SSP245, the increase in short-lived HWEs is mainly related to the strengthening of the WPSH, and the impact of the SAH was not significant. The increase in long-lived HWEs is related to the combined effect of the WPSH and SAH. These results are consistent with the conclusions of previous studies based on the analysis of HWEs in historical periods [30]. The main physical process is that when the SAH intensifies, an anomalous anticyclone occurs at the 200 hPa level over the YRB, and a downward advection of vorticity by anomalous sinking motions exists. Consequently, the negative vorticity anomaly is formed at 500 hPa, which can lead to a strong westward extension of the WPSH and the occurrence of long-lived HWEs [49]. However, under the SSP585 scenario, the SAH will increase more obviously, which will result in a wider and stronger influence of the WPSH, and both short-lived HWEs and long-lived HWEs will be affected. The above results once again emphasize the importance of controlling greenhouse-gas emissions.

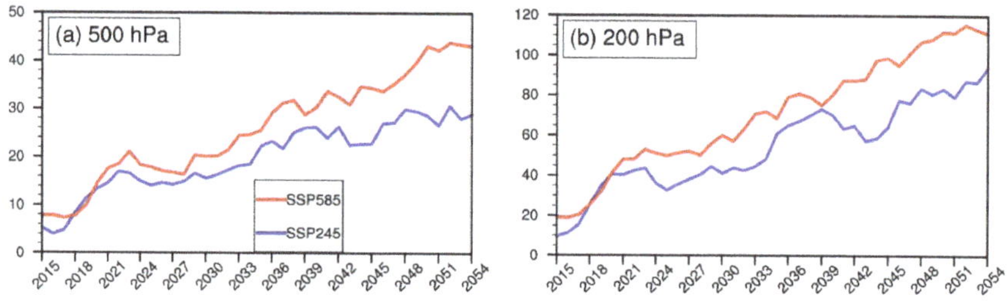

Figure 8. The anomalies of geopotential height (gpm) of (**a**) 500 hPa and (**b**) 200 hPa relative to the 1995–2014 reference period averaged over the YRB region. The time series are smoothed with a 5-year running mean filter.

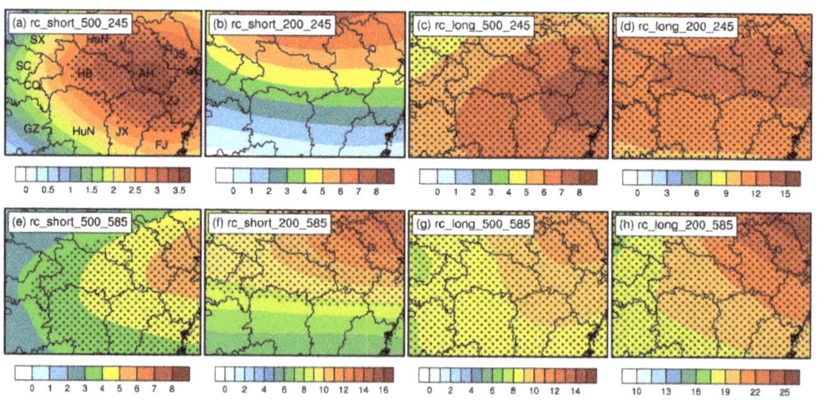

Figure 9. Regressions of geopotential height anomalies (gpm) of (**a**,**c**,**e**,**g**) 500 hPa and (**b**,**d**,**f**,**h**) 200 hPa against the time series of (**a**,**b**,**e**,**f**) short-lived HWEs and (**c**,**d**,**g**,**h**) long-lived HWEs. Time series of short-lived HWEs and long-lived HWEs were standardized before the calculation. The dots indicate that the 90% confidence level is exceeded.

4. Discussion

This work carried out high-resolution simulations on HWEs in both reference and future periods over the YRB region. Different from the GCM with coarser resolution, the high-resolution simulation from the RCM provides more details on the spatial and temporal variations in HWEs in the YRB. The results also provide valuable information to policy-makers in different regions and help to achieve regional sustainable development.

However, our research still has some limitations. First, due to the limitation of computing resources, our simulation area was relatively small, which made it impossible for us to analyze the changes in SST and large-scale atmospheric circulation in the future. Previous studies have indicated that SST and global atmospheric teleconnection also have an important impact on regional HWEs [11,29,31]. For example, the strong SST anomaly can connect the WPSH via the teleconnection wave train and can further contribute to Tmax variability [34]. Based on this fact, our future work will also focus on the simulation results of some GCMs. Although these models cannot provide the small-scale information as RCMs, they can provide more large-scale signals. Therefore, it is meaningful to use the optimal GCMs to explore the relationship between the large-scale factors and the short-lived and long-lived HWEs over the YRB region. Second, the land-use distribution does not change during our simulation, which makes the model unable to reflect the information of human activities in time. In the near future, we will also use the WRF model to study the contribution of land-use changes to the variations in short-lived HWEs and long-lived HWEs in the YRB region.

5. Conclusions

In the context of global warming, HWEs are becoming increasingly frequent, and obvious differences exist between different regions. The YRB region, an important economic center of China, is one of the regions where HWEs are more frequent and severe. Coarse-resolution models (such as GCMs) cannot resolve the detailed characteristics of HWEs. Based on this, we used a high-resolution RCM (WRF) and observation dataset (CN051) to analyze the characteristics of HWEs in the historical period and project the changes in HWEs in the YRB in the future. The main findings are as follows.

(1). During the 1981–2018 historical period, the Tmax, long-lived HWEs, and total HWEs in the YRB area all showed an obvious upward trend, and the turning point occurred in approximately 2000. Although the upward trend of short-lived HWEs was relatively small overall, there was a significant increase in certain regions, such as Henan, Sichuan, and the eastern coastal region. At the same time, the high incidence of long-lived HWEs was mainly concentrated in Hubei, Hunan, and eastern Sichuan provinces.

(2). Through the high-resolution simulation during the reference period, it was found that the WRF model can simulate the daily Tmax very well, which provides reliability for the projections. The projection results show that Tmax, long-lived HWEs, and total HWEs will increase obviously in both scenarios, and the upward trend of SSP585 is even greater. Short-lived HWEs will also increase under the SSP585 scenario, but they are relatively stable overall under SSP245. For SSP245, in the northern region, the daily Tmax will rise faster, and short-lived HWEs will increase, while the long-lived HWEs will rise significantly in the southern YRB. However, both short-lived and long-lived HWEs will increase in each subregion of the YRB under the SSP585 scenario.

(3). In both future scenarios, the geopotential heights at 500 hPa and 200 hPa over the YRB will increase, which may imply that both the WPSH and SAH will be enhanced, and this is more pronounced in the SSP585 scenario. Changes in the WPSH and SAH have an important impact on the HWEs of the YRB. The long-lived HWEs are caused by the joint effect of the WPSH and SAH, and this joint effect will also affect the short-lived HWEs under the SSP585 scenario. As mentioned in the fourth part, the detailed physical mechanism of the impact of geopotential height on HWEs still needs further study in the future.

Supplementary Materials: The following supporting information can be downloaded at: https://www.mdpi.com/article/10.3390/su14031141/s1, Figure S1: The anomalies of the number of grids covered by (a) total HWEs (grids), (b) short-lived HWEs (grids), and (c) long-lived HWEs (grids) relative to the 1995–2014 reference period over the YRB region. The time series are smoothed with a 9-year running mean filter.

Author Contributions: Conceptualization, X.Y.; visualization, Z.G.; writing—original draft, Z.G.; writing—review and editing, X.Y. All authors have read and agreed to the published version of the manuscript.

Funding: This research was funded by the National Key Research and Development Program of China, with grant numbers 2019YFA0606904 and 2018YFC509003.

Institutional Review Board Statement: Not applicable.

Informed Consent Statement: Not applicable.

Data Availability Statement: The coupled Model Intercomparison Project Phase 6 (CMIP6) dataset is available at https://www.scidb.cn/en/detail?dataSetId=791587189614968832&dataSetType=personal (accessed on 27 December 2021).

Acknowledgments: The authors appreciate Sichang Liu, Zhongfeng Xu, and Na Li at the Institute of Atmospheric Physics, Chinese Academy of Sciences (IAP, CAS), for their helpful discussions. The authors also thank Duoying Ji and Yan Guo in Beijing Normal University for both fruitful discussions and helping with computing issues. This research was supported by the Super Computing Center of Beijing Normal University, and the user name was gaozhibo.

Conflicts of Interest: The authors declare no conflict of interest.

References

1. Masson-Delmotte, V.; Zhai, P.; Pirani, A.; Connors, S.L.; Péan, C.; Berger, S.; Caud, N.; Chen, Y.; Goldfarb, L.; Gomis, M.I.; et al. (Eds.) *IPCC Climate Change 2021: The Physical Science Basis. Contribution of Working Group I to the Sixth Assessment Report of the Intergovernmental Panel on Climate Change*; Cambridge University Press: Cambridge, UK, 2021.
2. Zhang, G.; Zeng, G.; Yang, X.; Jiang, Z. Future changes in extreme high temperature over China at 1.5 °C–5 °C global warming based on CMIP6 simulations. *Adv. Atmos. Sci.* **2021**, *38*, 253–267. [CrossRef]
3. King, M.J.; Reeder, M.J. Extreme heat events from an object viewpoint with application to south-east Australia. *Int. J. Climatol.* **2021**, *41*, 2693–2709. [CrossRef]
4. Li, D.; Zhou, T.; Zou, L.; Zhang, W.; Zhang, L. Extreme high-temperature events over East Asia in 1.5 °C and 2 °C warmer futures: Analysis of NCAR CESM low-warming experiments. *Geophys. Res. Lett.* **2018**, *45*, 1541–1550. [CrossRef]
5. Sun, Y.; Zhang, X.; Zwiers, F.W.; Song, L.; Wan, H.; Hu, T.; Yin, H.; Ren, G. Rapid increase in the risk of extreme summer heat in Eastern China. *Nat. Clim. Change* **2014**, *4*, 1082–1085. [CrossRef]
6. Kinter, J.L.; Fennessy, M.J. Climatic feedbacks during the 2003 European heat wave. *J. Clim.* **2011**, *24*, 5953–5967.
7. Joughin, I.; Abdalati, W.; Fahnestock, M. Large fluctuations in speed on Greenland's Jakobshavn Isbrae glacier. *Nature* **2004**, *432*, 608–610. [CrossRef]
8. Wang, J.; Yan, Z.; Quan, X.-W.; Feng, J. Urban warming in the 2013 summer heat wave in eastern China. *Clim. Dyn.* **2016**, *48*, 3015–3033. [CrossRef]
9. Kong, Q.; Guerreiro, S.B.; Blenkinsop, S.; Li, X.-F.; Fowler, H.J. Increases in summertime concurrent drought and heatwave in Eastern China. *Weather Clim. Extremes* **2020**, *28*, 100242. [CrossRef]
10. Chen, H.; Sun, J. Projected changes in climate extremes in China in a 1.5 °C warmer world. *Int. J. Clim.* **2018**, *38*, 3607–3617. [CrossRef]
11. Yang, J.; Zhu, T.; Gao, M.; Lin, H.; Wang, B.; Bao, Q. Late-July barrier for subseasonal forecast of summer daily maximum temperature over Yangtze River Basin. *Geophys. Res. Lett.* **2018**, *45*, 12610–12615. [CrossRef]
12. Torres-Alavez, J.A.; Giorgi, F.; Kucharski, F.; Coppola, E.; Castro-García, L. ENSO teleconnections in an ensemble of CORDEX-CORE regional simulations. *Clim. Dyn.* **2021**, *57*, 1445–1461. [CrossRef]
13. Lin, L.; Gettelman, A.; Xu, Y.; Wu, C.; Wang, Z.; Rosenbloom, N.; Bates, S.C.; Dong, W. CAM6 simulation of mean and extreme precipitation over Asia: Sensitivity to upgraded physical parameterizations and higher horizontal resolution. *Geosci. Model Dev.* **2019**, *12*, 3773–3793. [CrossRef]
14. Anand, A.; Bhowmick, M.; Mishra, S.K.; Sahany, S.; Reddy Telukuntla, C.V. Response of Indian monsoon to increase of resolution in NCAR-CAM5. *Atmos. Res.* **2019**, *221*, 12–26. [CrossRef]
15. Yan, Y.; Tang, J.; Wang, S.; Niu, X.; Le, W. Uncertainty of land surface model and land use data on WRF model simulations over China. *Clim. Dyn.* **2021**, *57*, 1833–1851. [CrossRef]
16. Gao, Z.; Zhu, J.; Guo, Y.; Yan, X.; Wang, X.; Li, H.; Li, S. Comparison of different land-surface perturbation methods in short-range ensemble forecasts. *Atmos. Ocean. Sci. Lett.* **2021**, *14*, 100045. [CrossRef]
17. Kong, X.; Wang, A.; Bi, X.; Li, X.; Zhang, H. Effects of horizontal resolution on hourly precipitation in AGCM simulations. *J. Hydrometeorol.* **2020**, *21*, 643–670. [CrossRef]
18. Liang, X.-Z.; Sun, C.; Zheng, X.; Dai, Y.; Xu, M.; Choi, H.I.; Ling, T.; Qiao, F.; Kong, X.; Bi, X.; et al. CWRF performance at downscaling China climate characteristics. *Clim. Dyn.* **2018**, *52*, 2159–2184. [CrossRef]
19. Zhu, X.; Wei, Z.; Dong, W.; Ji, Z.; Wen, X.; Zheng, Z.; Yan, D.; Chen, D. Dynamical downscaling simulation and projection for mean and extreme temperature and precipitation over central Asia. *Clim. Dyn.* **2020**, *54*, 3279–3306. [CrossRef]
20. Tian, L.; Jin, J.; Wu, P.; Niu, G.-y.; Zhao, C. High-resolution simulations of mean and extreme precipitation with WRF for the soil-erosive Loess Plateau. *Clim. Dyn.* **2020**, *54*, 3489–3506. [CrossRef]

21. Cardoso, R.M.; Soares, P.M.M.; Miranda, P.M.A.; Belo-Pereira, M. WRF high resolution simulation of Iberian mean and extreme precipitation climate. *Int. J. Clim.* **2013**, *33*, 2591–2608. [CrossRef]
22. Yun, Y.; Liu, C.; Luo, Y.; Gao, W. Warm-season mesoscale convective systems over eastern China: Convection-permitting climate model simulation and observation. *Clim. Dyn.* **2021**, *57*, 3599–3617. [CrossRef]
23. Gao, Z.; Zhu, J.; Guo, Y.; Luo, N.; Fu, Y.; Wang, T. Impact of land surface processes on a record-Breaking rainfall event on May 06–07, 2017, in Guangzhou, China. *J. Geophys. Res. Atmos.* **2021**, *126*, e2020JD032997. [CrossRef]
24. Yun, Y.; Liu, C.; Luo, Y.; Liang, X.; Huang, L.; Chen, F.; Rasmmusen, R. Convection-permitting regional climate simulation of warm-season precipitation over Eastern China. *Clim. Dyn.* **2019**, *54*, 1469–1489. [CrossRef]
25. Guo, Z.; Fang, J.; Sun, X.; Yang, Y.; Tang, J. Sensitivity of summer precipitation simulation to microphysics parameterization over Eastern China: Convection-permitting regional climate simulation. *J. Geophys. Res. Atmos.* **2019**, *124*, 9183–9204. [CrossRef]
26. Kong, X.; Wang, A.; Bi, X.; Wang, D. Assessment of temperature extremes in China using RegCM4 and WRF. *Adv. Atmos. Sci.* **2019**, *36*, 363–377. [CrossRef]
27. Fita, L.; Polcher, J.; Giannaros, T.M.; Lorenz, T.; Milovac, J.; Sofiadis, G.; Katragkou, E.; Bastin, S. CORDEX-WRF v1.3: Development of a module for the weather research and forecasting (WRF) model to support the CORDEX community. *Geosci. Model Dev.* **2019**, *12*, 1029–1066. [CrossRef]
28. Yu, E.; Sun, J.; Chen, H.; Xiang, W. Evaluation of a high-resolution historical simulation over China: Climatology and extremes. *Clim. Dyn.* **2014**, *45*, 2013–2031. [CrossRef]
29. Qi, X.; Yang, J.; Gao, M.; Yang, H.; Liu, H. Roles of the tropical/extratropical intraseasonal oscillations on generating the heat wave over Yangtze River valley: A numerical study. *J. Geophys. Res. Atmos.* **2019**, *124*, 3110–3123. [CrossRef]
30. Li, N.; Xiao, Z.; Zhao, L. A recent increase in long-lived heatwaves in China under the joint influence of South Asia and Western North Pacific subtropical highs. *J. Clim.* **2021**, *34*, 7167–7179. [CrossRef]
31. Yang, K.; Zhang, J.; Wu, L.; Wei, J. Prediction of summer hot extremes over the middle and lower reaches of the Yangtze River valley. *Clim. Dyn.* **2018**, *52*, 2943–2957. [CrossRef]
32. Gao, M.; Wang, B.; Yang, J.; Dong, W. Are peak summer sultry heat wave days over the Yangtze–Huaihe River Basin predictable? *J. Clim.* **2018**, *31*, 2185–2196. [CrossRef]
33. Teng, H.; Branstator, G. Tropospheric waveguide teleconnections and their seasonality. *J. Atmos. Sci.* **2017**, *74*, 1513–1532.
34. Sun, J. Record-breaking SST over mid-North Atlantic and extreme high temperature over the Jianghuai–Jiangnan region of China in 2013. *Chin. Sci. Bull.* **2014**, *59*, 3465–3470. [CrossRef]
35. Skamarock, C.; Klemp, B.; Dudhia, J.; Gill, O.; Liu, Z.; Berner, J.; Wang, W.; Powers, G.; Duda, G.; Barker, D.M.; et al. *A Description of the Advanced Research WRF Model Version 4.1 (No. NCAR/TN-556+STR)*; University Corporation for Atmospheric Research: Boulder, CO, USA, 2019.
36. Xu, Z.; Han, Y.; Tam, C.-Y.; Yang, Z.-L.; Fu, C. Bias-corrected CMIP6 global dataset for dynamical downscaling of the historical and future climate (1979–2100). *Sci. Data* **2021**, *8*, 293. [CrossRef] [PubMed]
37. Hong, S.Y.; Lim, J.J. The WRF single-moment 6-class microphysics scheme (WSM6). *Asia-Pac. J. Atmos. Sci.* **2006**, *42*, 129–151.
38. Kain, J.S. The Kain–Fritsch convective parameterization: An update. *J. Appl. Meteorol.* **2004**, *43*, 170–181. [CrossRef]
39. Niu, G.-Y.; Yang, Z.-L.; Mitchell, K.E.; Chen, F.; Ek, M.B.; Barlage, M.; Kumar, A.; Manning, K.; Niyogi, D.; Rosero, E.; et al. The community Noah land surface model with multiparameterization options (Noah-MP): 1. Model description and evaluation with local-scale measurements. *J. Geophys. Res.* **2011**, *116*, D12109. [CrossRef]
40. Collins, W.; Rasch, P.; Boville, B.; McCaa, J.; Williamson, D.; Kiehl, J.; Briegleb, B.; Bitz, C.; Lin, S.-J.; Zhang, M.; et al. *Description of the NCAR Community Atmosphere Model (CAM 3.0) (No. NCAR/TN-464+STR)*; University Corporation for Atmospheric Research: Boulder, CO, USA, 2004.
41. Wang, P.; Tang, J.; Sun, X.; Wang, S.; Wu, J.; Dong, X.; Fang, J. Heat waves in China: Definitions, leading patterns, and connections to large-scale atmospheric circulation and SSTs. *J. Geophys. Res. Atmos.* **2017**, *122*, 10679–10699. [CrossRef]
42. Liu, L.; Sun, L.; Liao, Y.; Zhu, Y.; Zou, X.; Wang, Y.; Yan, J. Development and application of national prediction system for extreme high temperature. *Meteorol. Mon.* **2008**, *10*.
43. Wu, J.; Gao, X. A gridded daily observation dataset over China region and comparison with the other datasets. *Chin. J. Geophys.* **2013**, *56*, 1102–1111.
44. Gao, X.-J.; Shi, Y.; Giorgi, F. Comparison of convective parameterizations in RegCM4 experiments over China with CLM as the land surface model. *Atmos. Ocean. Sci. Lett.* **2016**, *9*, 246–254. [CrossRef]
45. Yu, E.; Wang, H.; Gao, Y.; Sun, J. Impacts of cumulus convective parameterization schemes on summer monsoon precipitation simulation over China. *Acta Meteorol. Sin.* **2011**, *25*, 581–592. [CrossRef]
46. Yuan, Z.; Yin, J.; Wei, M.; Yuan, Y. Spatio-temporal variations in the temperature and precipitation extremes in Yangtze River Basin, China during 1961–2020. *Atmosphere* **2021**, *12*, 1423. [CrossRef]
47. Song, S.; Yan, X. Changes in the frequency of extreme cooling events in winter over China and their relationship with Arctic oscillation. *Sustainability* **2021**, *13*, 11491. [CrossRef]
48. Huang, Y.; Zhang, X.; Zhang, D.; Zhang, L.; Zhang, W.; Ren, C.; Pan, T.; Chu, Z.; Chen, Y. Spatial-temporal characteristics of Arctic summer climate comfort level in the context of regional tourism resources from 1979 to 2019. *Sustainability* **2021**, *13*, 13056. [CrossRef]
49. Wei, W.; Zhang, R.; Wen, M.; Yang, S.; Li, W. Dynamic effect of the South Asian high on the interannual zonal extension of the western North Pacific subtropical high. *Int. J. Clim.* **2019**, *39*, 5367–5379. [CrossRef]

sustainability

MDPI

Article

The Study on Compound Drought and Heatwave Events in China Using Complex Networks

Kaiwen Li [1,2], Ming Wang [1,3,*] and Kai Liu [1,3]

1 Academy of Disaster Reduction and Emergency Management, Beijing Normal University, Beijing 100875, China; 202031051092@mail.bnu.edu.cn (K.L.); liukai@bnu.edu.cn (K.L.)
2 Faculty of Geographical Science, Beijing Normal University, Beijing 100875, China
3 The School of National Safety and Emergency Management, Beijing Normal University, Beijing 100875, China
* Correspondence: wangming@bnu.edu.cn; Tel./Fax: + 86-10-5880-6497

Abstract: Compound extreme events can severely impact water security, food security, and social and economic development. Compared with single-hazard events, compound extreme events cause greater losses. Therefore, understanding the spatial and temporal variations in compound extreme events is important to prevent the risks they cause. Only a few studies have analyzed the spatial and temporal relations of compound extreme events from the perspective of a complex network. In this study, we define compound drought and heatwave events (CDHEs) using the monthly scale standard precipitation index (SPI), and the definition of a heatwave is based on daily maximum temperature. We evaluate the spatial and temporal variations in CDHEs in China from 1961 to 2018 and discuss the impact of maximum temperature and precipitation changes on the annual frequency and annual magnitude trends of CDHEs. Furthermore, a synchronization strength network is established using the event synchronization method, and the proposed synchronization strength index (SSI) is used to divide the network into eight communities to identify the propagation extent of CDHEs, where each community represents a region with high synchronization strength. Finally, we explore the impact of summer Atlantic multidecadal oscillation (AMO) and Pacific decadal oscillation (PDO) on CDHEs in different communities. The results show that, at a national scale, the mean frequency of CDHEs takes on a non-significant decreasing trend, and the mean magnitude of CDHEs takes on a non-significant increasing trend. The significant trends in the annual frequency and annual magnitude of CDHEs are attributed to maximum temperature and precipitation changes. AMO positively modulates the mean frequency and mean magnitude of CDHEs within community 1 and 2, and negatively modulates the mean magnitude of CDHEs within community 3. PDO negatively modulates the mean frequency and mean magnitude of CDHEs within community 4. AMO and PDO jointly modulate the mean magnitude of CDHEs within community 6 and 8. Overall, this study provides a new understanding of CDHEs to mitigate their severe effects.

Keywords: compound drought and heatwave events; complex network; event synchronization; atmospheric circulation patterns

Citation: Li, K.; Wang, M.; Liu, K. The Study on Compound Drought and Heatwave Events in China Using Complex Networks. *Sustainability* **2021**, *13*, 12774. https://doi.org/10.3390/su132212774

Academic Editors: Xiaodong Yan, Jia Yang and Shaofei Jin

Received: 28 October 2021
Accepted: 16 November 2021
Published: 18 November 2021

1. Introduction

Global warming has led to an increase in drought and heatwave events [1,2]. The conjoined extreme events of droughts and heatwaves are considered as compound drought and heatwave events (CDHEs) [3–5], which have severe impacts on socioeconomic development and the environment. For example, severe drought and heatwaves in the 2003 European summer resulted in around 40,000 deaths [6]. Drought and heat anomalies in the central United States has caused at least USD 30 billion in damages [7]. Therefore, it is essential to study the spatial and temporal variations in, and spatial influences of, CDHEs to mitigate their effects.

There is no unified standard for the definition of CDHEs, because drought events and heatwave events correspond to a variety of definitions. Drought events are usually defined by various indicators, such as the standardized precipitation index (SPI) [8,9], the standardized precipitation evapotranspiration index [10,11], and the Palmer drought severity index [12,13]. Heatwave events are mainly defined by relative thresholds (percentile thresholds), absolute thresholds, and durations [14,15]. In recent studies, the definition of CDHEs mainly refers to the combination of drought and heatwave event definitions mentioned above [4,16–18]. There are also studies that directly describe the properties of CDHEs by corresponding metrics. For instance, Wu et al. [19] proposed a dry–hot magnitude index (DHMI) to characterize the magnitude of CDHEs. Their results show that high-magnitude CDHEs mainly occur in northeastern and southwestern China. Wu et al. [20] derived a standardized compound event indicator and a standardized dry-hot index to evaluate the severity of changes in CDHEs in the warm season. Their results show that the severity of CDHEs increases significantly in most parts of China, and temperature is the dominant driving factor compared with precipitation. The variation in various properties (frequency, magnitude, severity, etc.) of CDHEs, and the potential linkage of such variation to atmospheric circulation patterns, has also been investigated. Hao et al. [16] applied a logistic regression model to examine the relation between the occurrence of compound events and El Niño–Southern Oscillation (ENSO) represented by NINO34. The results show that ENSO plays an important role in the occurrence of CDHEs during the warm season in the northern parts of South America, southern Africa, southeastern Asia, and Australia. In China, there has been a significant increase in the frequency and spatial extent of compound dry-warm events, which is partly related to atmospheric circulation patterns [21]. Mukherjee et al. [22] used a Poisson generalized linear model to analyze the relation between the frequency of seasonal CDHEs and the warm and cold phases of ENSO, Pacific Decadal Oscillation (PDO), and North Atlantic Oscillation (NAO). The results show that ENSO is strongly related with CDHEs over the southern hemisphere, PDO influences the frequency of CDHEs over western North America during the boreal summer, while NAO is weakly related with CDHEs. The composite analysis, correlation analysis, and logistic regression model have all been used to investigate the influence of various atmospheric circulation patterns on CDHEs in China. These methods found that the Atlantic Multidecadal Oscillation (AMO) affects CDHEs for about 18.52% of mainland China, while NAO and PDO, respectively, account for around 14.64% and 12.96%, and ENSO affects CDHEs by about 5.27% [23]. In general, previous studies analyzed the spatial and temporal variations in properties of CDHEs using various definitions or metrics, revealing that atmospheric circulation patterns have a profound impact on CDHEs.

The complex systems theory applied to the study of extreme events has received increased attention. Boers et al. introduced network divergence to directed networks to predict extreme rainfall events in the central Andes [24]. Complex networks have been used to reveal global patterns of extreme rainfall teleconnections, and Rossby waves were shown to be the physical mechanism of these teleconnections [25]. Konapala and Mishra [26] applied the complex network approach to study the spatial and temporal evolution of droughts in the continental USA. They found that drought events propagate differently at different thresholds associated with their initiation. However, complex network approaches have rarely been applied to the study of CDHEs.

The objective of this study is to evaluate the spatiotemporal variations and propagation extent of CDHEs for the period of 1961–2018 in China, based on monthly SPI, daily maximum temperature, and a complex network approach. The correlations between AMO, PDO, and the frequency and magnitude of CDHEs are employed to explore the possible physical mechanism of the formation of CDHEs. The results of this study will enhance our understanding of compound extreme events and provide suggestions for decisionmakers to mitigate the negative impacts of these events in China.

2. Data and Definition

2.1. Data

Our study is based on monthly precipitation data and daily maximum temperature data from the CN05.1 dataset [27]. The time span of precipitation and maximum temperature data is from 1961 to 2018, the resolution is 0.25° × 0.25°, and there are 163 grid points in the north–south direction and 283 grid points in the east–west direction, where the number of grid points within China is 15,247 (precipitation and maximum temperature data for Taiwan Province are missing).

AMO and PDO are major large-scale circulation patterns that influence the climate of East Asia. The AMO index is defined as the average anomalies of sea surface temperature (SST) in the north Atlantic [28]. Its positive (negative) phase usually indicates that SST is warmer (cooler) than the average SST across the north Atlantic Ocean. The PDO index is defined as the leading principal component of the north Pacific (NP) monthly SST anomalies [29]. Its positive (negative) phase corresponds to cool (warm) SST in the central and western NP. In this study, we found that more than 83.5% of CDHEs occur during June, July, and August (JJA) in China (as shown in Figure 1); thus, we focused on the impact of the AMO and PDO indexes in JJA on CDHEs, where the temporal series of the AMO and PDO indexes in JJA are derived from the China National Climate Center (http://cmdp.ncc-cma.net, accessed on 18 November 2021), and shown in Figure 2.

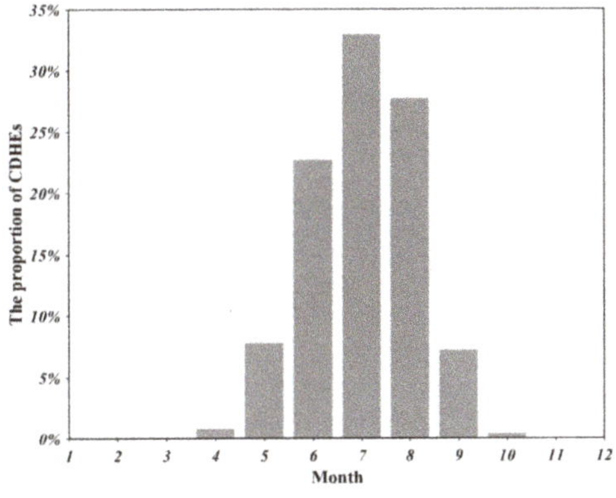

Figure 1. The proportion of CDHEs in different months for the period of 1961–2018.

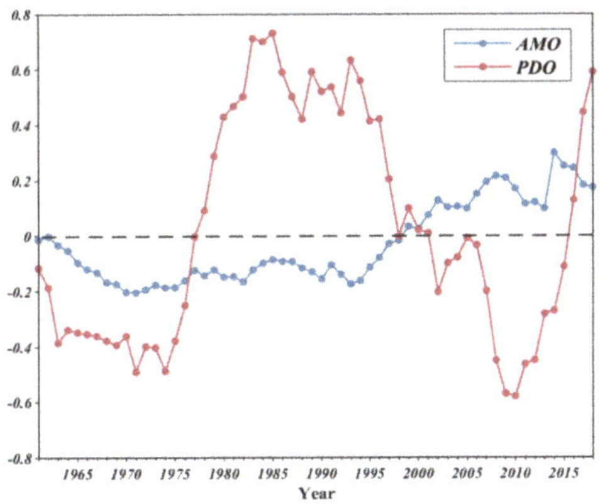

Figure 2. Temporal series of AMO (blue) and PDO (red) index during JJA for the period of 1961–2018 after 9 year moving average.

2.2. *Defining Compound Drought and Heatwave Events*

CDHEs are defined as extreme disaster events of drought in conjunction with a heatwave. The influences of drought events are usually long-lasting, so the monthly SPI is calculated to characterize the degree of drought in a given month [11], and −1 is selected as the threshold of drought conditions. According to the World Meteorological Organization's recommended criteria for heatwaves [30], we define a heatwave as an event with a daily maximum temperature (T_{max}) ≥ 32 °C and a duration of ≥3 consecutive days. As shown in Figure 3, the occurrence of a CDHE can be defined as a binary variable, where 1 is for occurrences and 0 is for non-occurrences. Δt represents the duration of a CDHE, and $\sum \Delta T$ represents the accumulative temperatures of a CDHE, which are the sum of the daily maximum temperature exceeding the threshold for the duration of a CDHE. The SPI of the month in which the CDHE is located, and its accumulative temperatures are used for the subsequent calculation of DHMI. More details are described in Section 3.1.

Compared with previous studies, we focused not only on the interannual variation in the frequency of CDHEs, but also on the interannual variation in the magnitude of CDHEs, where the magnitude of CDHEs is represented by DHMI.

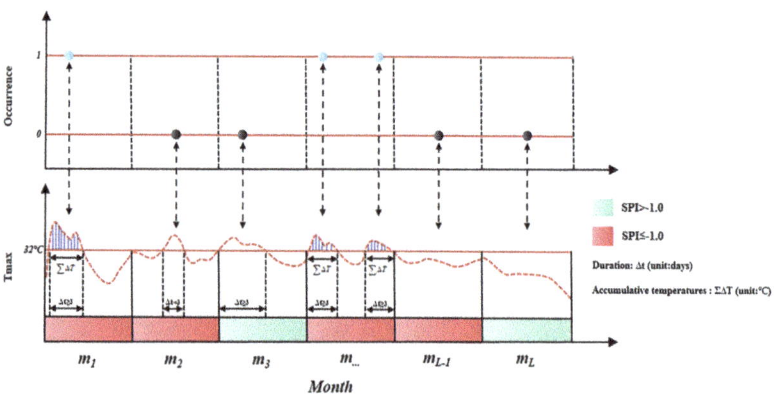

Figure 3. The schematic diagram of CDHEs. L is the total number of months for the period of 1961–2018. The top y-axis represents the occurrence of CDHEs, 1 for occurrences and 0 for non-occurrences. The bottom y-axis represents the maximum temperature. Δt represents the duration of a CDHE and $\sum \Delta T$ represents the accumulative temperatures of a CDHE.

3. Methodology

3.1. *The Drought–Heatwave Magnitude Index*

The magnitude of a CDHE is jointly determined by drought degree and heatwave conditions; then, the DHMI of a CDHE can be defined as [19]:

$$\text{DHMI} = \text{P}(\sum \Delta T) \times (\Delta SPI) \quad (1)$$

where $\sum \Delta T$ represents the accumulative temperatures of a CDHE, $\text{P}(\sum \Delta T)$ is obtained using a nonparametric method to estimate the cumulative density function (CDF) of $\sum \Delta T$, transforming $\sum \Delta T$ to a non-exceedance probability ranging from 0 to 1. A larger $\sum \Delta T$ value corresponds to a higher probability of $\text{P}(\sum \Delta T)$ [31]. The degree of drought is represented by monthly SPI. ΔSPI proposed by the runs theory [32,33] is the absolute value of the difference between the SPI for the month in which the CDHE occurs and the drought threshold. Therefore, a larger ΔSPI indicates a more severe drought. DHMI can also be treated as a weighted value of ΔSPI, so its maximum value is the maximum value of ΔSPI.

3.2. Event Synchronization

The method of event synchronization is used to reveal the global pattern of extreme rainfall teleconnections [25] and predict extreme floods [24]. In this study, grid i and grid j are selected to describe the definition of synchronized events. We suppose that, for grid i, a CDHE occurs at a moment t_m^i; for grid j, a CDHE occurs at a moment t_n^j, where $m \in [1, \text{M}]$, and $n \in [1, \text{N}]$. M and N denote the total number of CDHEs at grids i and j, respectively [34]. CDHEs occurring on consecutive days are counted as single events, and the occurrence time of each CDHE is determined as the time corresponding to the first day when the event occurs. A dynamical delay $\tau_{m,n}^{i,j}$ is introduced to decide whether a pair of events occurs at t_m^i, and t_n^j is counted as a synchronized event. Its definition is as follows:

$$\tau_{m,n}^{i,\,j} = min\left(\frac{\left\{t_{m+1}^i - t_m^i, t_m^i - t_{m-1}^i, t_{n+1}^j - t_n^j, t_n^j - t_{n-1}^j\right\}}{2}\right) \quad (2)$$

Furthermore, we introduce a maximum delay ($\tau_{max} = 2$ days) to constrain the formation of CDHEs [24]. $f(i/j)$ is used to define the number of times a CDHE shortly occurs in grid i after it occurs in grid j, i.e:

$$f(i/j) = \sum_{m=1}^{M}\sum_{n=1}^{N} S_{ij} \quad (3)$$

with

$$S_{ij} = \begin{cases} 1 & if\ 0 < t_m^i - t_n^j \le \min\left(\tau_{m,n}^{i,\,j}, \tau_{max}\right) \\ 1/2 & if\ t_m^i = t_n^j \\ 0 & else \end{cases} \quad (4)$$

and analogously for $f(j/i)$. Q_{ij} denotes the synchronization strength between grids i and j. Its definition is as follows

$$Q_{ij} = \frac{f(i/j) + f(j/i)}{\sqrt{M \times N}} \quad (5)$$

where Q_{ij} is normalized to $Q_{ij} \in [0, 1]$. There is $Q_{ij} = 1$ if CDHEs are fully synchronized between grids i and j.

3.3. Network Construction

In this study, the synchronization strength network is established through traversing the synchronization strength for all possible pairs of grid points, where the grid points are defined as nodes, and the synchronization strength between grid points is the weight of the connected edges in the network. To eliminate minor edges, we prune the network using Q_{95}, corresponding to the 95th percentile of all synchronization strengths. Take grids i and j as examples: there is a connected edge between grids i and j if $Q_{ij} \ge Q_{95}$. Finally, we extract some edges with higher weights through the abovementioned principle. The total number of edges in the synchronization strength network is 2.496 million, which also indicates that the network has a large scale.

3.4. Complex Network Metrics

The establishment of the synchronization strength network provides a possibility to identify the propagation extent of CDHEs. Complex network metrics aim to quantitatively describe the topological characteristics of nodes and edges in the network.

The degree centrality is a measurement to evaluate the local importance of nodes in the network, and the degree value of node i is defined as:

$$k_i = \sum_{j=1}^{\theta} A_{ij} \quad (6)$$

with

$$A_{ij} = \begin{cases} 1, & if\ node\ i\ and\ j\ is\ connected \\ 0, & otherwise \end{cases} \tag{7}$$

where k_i is the degree value of node i, θ is the total number of nodes, A_{ij} is the adjacency matrix, and $A_{ij} = 1$ if node i and node j are connected; otherwise, $A_{ij} = 0$. A higher degree value for a node indicates that the node is connected to more nodes that are strongly synchronized with its CDHEs.

The average distance of nodes is introduced to describe the average status of the geographic distance of all edges for a node, and its mathematical expression is given as:

$$AD_i = \frac{\Sigma(D_{ij} \times A_{ij})}{k_i} \tag{8}$$

where AD_i is the average distance of node i and D_{ij} is the geographic distance between node i and node j. The average distance of nodes is concerned with whether a single node has the possibility of connecting to distant nodes.

For node i, the synchronization strength index SSI_i is proposed to quantify the ability of node i to form a region with high synchronization strength. In general, the regions with high synchronization strength that are formed should satisfy three conditions: the degree of nodes in the region is larger, the average distance between these nodes and its neighboring nodes is shorter, and the synchronization strength between neighboring nodes is stronger. Thus, the definition of SSI_i is given as:

$$SSI_i = normalized\left(\frac{k_i \times CC_i}{AD_i}\right) \tag{9}$$

with

$$CC_i = \frac{2 \times E_{existed}}{k_i \times (k_i - 1)} \tag{10}$$

where CC_i is the clustering coefficient of node i, $E_{existed}$ is the number of existing edges between neighboring nodes of node i, and CC_i is used to measure the clustering degree of node i in the network [35,36]. A node with a larger SSI is more likely to form an initial region with high synchronization strength, and the occurrence of CDHEs in the region is more synchronous.

3.5. Community Detection

Community detection is a method for discovering closely related components in a network, and it is utilized to identify the propagation extent of CDHEs. The synchronization strength between the nodes within the community is stronger; conversely, the synchronization strength is weaker between different communities. The existing algorithms of community detection mainly include the fast Newman and GN algorithms for unweighted networks [37,38], and the spin glass algorithm, random walk algorithm, and label propagation algorithm for weighted networks [39,40]. The spin glass algorithm, developed from the theory of spin glass in physics, is suitable for weighted networks with a larger scale, and is used to divide communities in the synchronization strength network. The mechanism of this algorithm can be briefly described as the connection or disconnection between nodes, and the community structure of the network is interpreted as the spin configuration that minimizes the energy of spin glass [41]. The communities derived from the spin glass algorithm are cohesive subgraphs.

For community detection, the modularity Mod is introduced to compare the division quality under different numbers of communities. The definition of modularity Mod is given as [42]:

$$Mod = \frac{1}{2e_w} \sum_{i,j} \left[w_{ij} - \frac{w_i * w_j}{2e_w} \right] \delta(C_i, C_j) \tag{11}$$

where *Mod* is the global weighted modularity, $Mod \in [0, 1]$. e_w is the sum of the weights of all edges, w_{ij} is the weight of the edge between node i and node j, w_i is the sum of the weights of edges connected to node i, and analogously for w_j. A higher *Mod* indicates that the nodes within each community are more closely related and more divergent between the nodes of different communities. C_i and C_j are the communities of node i and node j, respectively. If node i and node j belong to the same community, $C_i = C_j$, $\delta(C_i, C_j) = 1$; otherwise, $\delta(C_i, C_j) = 0$.

4. Results and Discussions

The annual frequency and annual magnitude of CDHEs are examined in this study. The annual frequency of CDHEs is defined as the total number of heatwaves in drought months of each year, while the annual magnitude of CDHEs is calculated as the mean DHMI of all CDHEs occurring in the same year. In Section 4.1, we first analyze the linear trend in the mean frequency and mean magnitude of CDHEs over all grids during 1961–2018. For spatial variation, we calculate the mean annual frequency and mean annual magnitude of CDHEs, and then analyze the trends in the annual frequency and annual magnitude of CDHEs in each grid based on the linear fitting method for the period of 1961–2018. In Section 4.2, the synchronization strength index (SSI) is proposed to initially identify the number of communities and the spin glass algorithm is used to derive the propagation extent of CDHEs. We also explore the influence of AMO and PDO on the variations in the frequency and magnitude of CDHEs in each community.

4.1. Spatiotemporal Variation Analysis of CDHEs

From a national perspective, as shown in Figure 4, the trend in the mean frequency of CDHEs over all grids is -9.645×10^{-4} yr^{-1} with $p = 0.401$ (Figure 4a), and the trend in the mean magnitude of CDHEs over all grids is 7.991×10^{-6} yr^{-1} with $p = 0.959$ (Figure 4b); these values indicate a non-significant ($p > 0.05$) decrease in the mean frequency of CDHEs, as well as a non-significant increase in the mean magnitude of CDHEs.

We further investigated the spatial distribution of the characteristics of CDHEs between different regions, the mean annual frequency and mean annual magnitude of CDHEs in each grid are presented in Figure 5a,b; we found that the grids where CDHEs had not occurred from 1961 to 2018 were mainly located at high altitudes of the Chinese mainland (as shown in Figure 5c). Because high-altitude areas are relatively cold, air temperature struggles to surpass the threshold temperature corresponding to the definition of heatwaves, which is not conducive to the formation of CDHEs. The spatial distribution of the mean annual frequency of CDHEs is shown in Figure 5a. It can be seen that the mean annual frequency of CDHEs in Northeast China (NEC) and the middle-eastern region of North China (NC) is less than 1, while the mean annual frequency in the western region of Northwest China (NWC), the western region of NC, the eastern region of Southwest China (SWC), most regions of central China (CC), East China (EC), and South China (SC) is more than 1; the mean annual frequency of CDHEs in southern Guangxi province and northern Hainan province exceeds 2, and the maximum mean annual frequency is 2.6. As shown in Figure 5b, we found that the spatial distribution pattern of the mean annual magnitude of CDHEs in the northern China (including NC, NEC, and NWC) is relatively similar to that of the mean annual frequency, and the mean annual magnitude of CDHEs reaches a maximum (0.27) in the eastern region of Xinjiang province. The mean annual magnitude exceeds 0.2 in the western Chongqing province, southern Guangxi and Jiangxi provinces, and northern Hainan province.

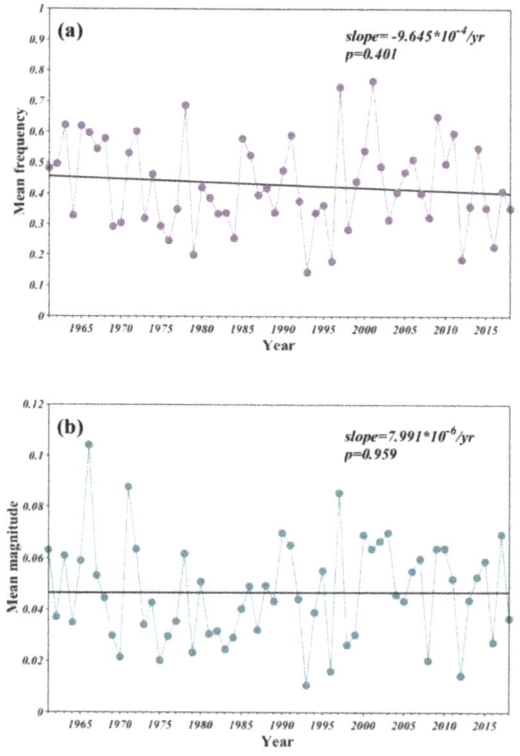

Figure 4. Temporal series of the mean frequency (**a**) and mean magnitude (**b**) of CDHEs over all grids for the period of 1961–2018 in China. Black solid lines represent the results of linear fitting.

The joint analysis of the mean annual frequency and mean annual magnitude of CDHEs shows that the high mean annual frequency and high mean annual magnitude of CDHEs in the western region of NWC and NC may be attributed to the fact that the corresponding grid points are located in the desert, where the continuous high temperature and precipitation deficit are more conducive to the formation of CDHEs, which also makes the magnitude of CDHEs higher. The reason for the low mean annual frequency and mean annual magnitude in NEC and the middle-eastern region of NC is the low maximum temperature in the region [43,44], which inhibits the formation of CDHEs and also makes the accumulative temperatures of CDHEs lower than those of CDHEs in other regions. Except for the northern China, the spatial distribution patterns of mean annual frequency and mean annual magnitude of CDHEs are dissimilar. At low latitudes, the number of drought months probably determines the mean annual frequency of CDHEs due to high maximum temperatures. The accumulative temperatures of CDHEs and the drought degree of the month in which CDHEs occur together determine the mean annual magnitude of CDHEs.

Trends in the annual frequency and annual magnitude of CDHEs in each grid are shown in Figure 6a,b. There is a significant increase in the annual frequency of CDHEs in NC, as well as in some parts of NWC. This may be due to the effect of unchanged precipitation and increased maximum temperature during JJA in these regions, as shown in Figure 6c,d. The significant increase in maximum temperature is more likely to lead to heatwaves. Therefore, the significant increase in the annual frequency of CDHEs in these regions can be attributed to the rise in maximum temperature. The regions where the annual frequency of CDHEs decreases significantly are mainly located in the western region of NWC, the northern region of CC, the western region of EC, and the southern region of SC. In the western region of NWC, which is located in the desert and has an environment characterized by a high temperature, there is potential for more heatwaves. A significant increase in precipitation in this region makes it possible to convert from drought to non-drought months and may lead to a significant decrease in the annual frequency of CDHEs in this region. The possible mechanism for this in the southern region of SC is similar to that of the western region of NWC. The reason for the significant decrease in the annual frequency of CDHEs in the southern region of SC is a significant increase in precipitation. In the northern region of CC and the western region of EC, unchanged precipitation and a decreased trend in maximum temperature are generally consistent with the results of previous studies [20,45], and likely inhibit the formation of CDHEs in these regions.

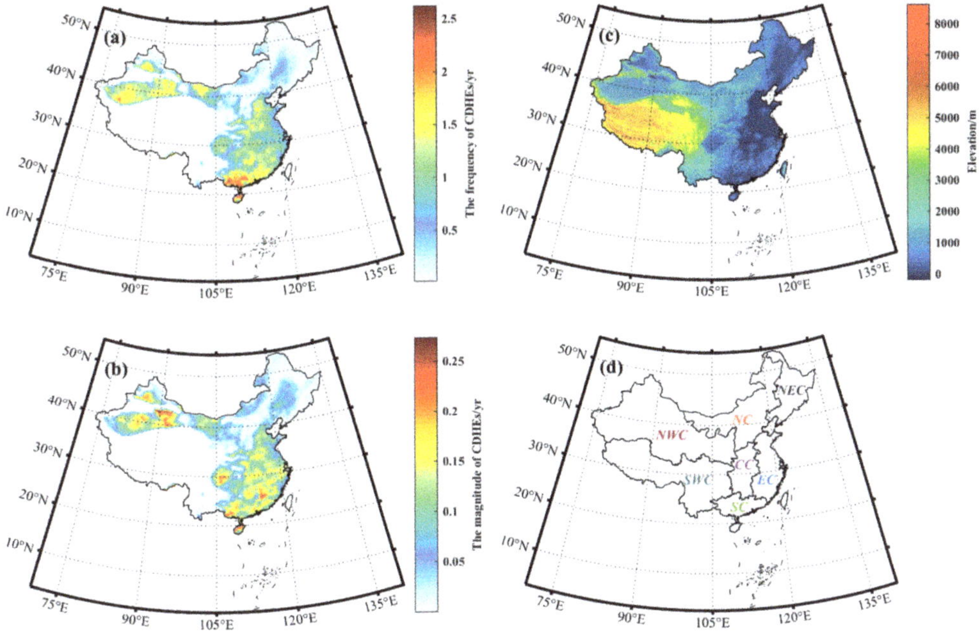

Figure 5. The spatial distribution of the mean annual frequency (**a**), mean annual magnitude (**b**) of CDHEs for the period of 1961–2018, elevation (**c**) and seven regions (**d**) in China.

According to the definition of the annual magnitude of CDHEs, the annual magnitude trend is likely affected by the individual or combined effect of variations in precipitation and temperature. For instance, with precipitation unchanged, the annual magnitude of CDHEs is expected to increase with the increase in maximum temperature. In some parts of NWC and NC, the annual magnitude shows a significantly increased trend. The precipitation decreases significantly, and the maximum temperature shows a significant increase in the eastern region of NC, which may cause an increase in ΔSPI and accumulative temperatures, resulting in a significant increase in the annual magnitude of CDHEs in this region. Except for the eastern region of NC, the interannual precipitation is unchanged, and the significantly increased maximum temperature dominates the significant increase in the annual magnitude of CDHEs. The regions showing a significant decrease in the annual magnitude of CDHEs are mainly located in the western region of NWC, CC, and the eastern region of SWC, where the western region of NWC and the eastern region of SWC show a significant increase in precipitation and a non-significant increase in maximum temperature; as a result, a significant increase in precipitation probably becomes the main factor that leads to a significant decrease in the annual magnitude. Precipitation in the northern and central regions of CC shows a non-significant increase, but the maximum temperature in the former shows a non-significant decrease; the combined effect of maximum temperature and precipitation may contribute to the significant decrease in the annual magnitude of CDHEs, while the maximum temperature in the latter shows a non-significant increase; thus, the main reason for the significant decrease in the annual magnitude of CDHEs is that the change in the annual magnitude in this region is more sensitive to changes in precipitation than it is to maximum temperature.

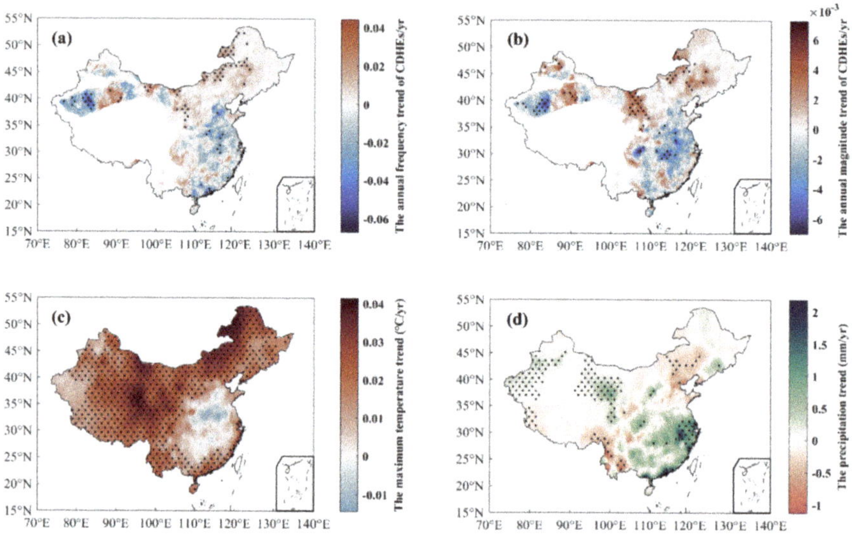

Figure 6. Trend analysis of the annual frequency (**a**) and annual magnitude (**b**) of CDHEs, and the maximum temperature (**c**) and precipitation (**d**) during JJA for the period of 1961–2018 in China. Black dots indicate statistical significance at a 0.05 significance level.

4.2. The Propagation Extent and Driving Factors of CDHEs

In this study, we use the method of event synchronization to establish the synchronization strength network, which is applied to reveal the synchronism of the occurrence time of CDHEs among nodes. A higher synchronization strength between a pair of nodes indicates that there are more CDHEs occurring at similar time points, and it also implies that the total number of CDHEs between nodes is similar. If a region exists in which the relations among nodes are relatively close, the occurrence time of CDHEs in the past may be similar in this region, and CDHEs are more likely to propagate within the region. In order to find the abovementioned regions, we propose an SSI based on the degree value, as well as the average distance and clustering coefficient of nodes in complex networks. The SSI of nodes is shown in Figure 7.

We found six regions that are formed by some nodes with high SSI ($SSI \in [0.6, 1.0]$). According to the definition of SSI, these six regions are initially identified as core regions where CDHEs can easily propagate. However, this method does not provide a clear boundary of the propagation extent of CDHEs. The spin glass algorithm, a nonoverlapping algorithm of community detection, is used to detect closely related components in the network. In comparison with the connected edges between communities, the connected edges between nodes within a community are denser and have higher weights; thus, each community can be considered as the propagation extent of CDHEs. In this study, the number of communities is initially determined to be four to eight, with reference to the number of core regions. Furthermore, the modularity Mod is introduced to judge the quality of community detection under the different numbers of communities. The modularity of the number of communities from four to eight is shown in Table 1.

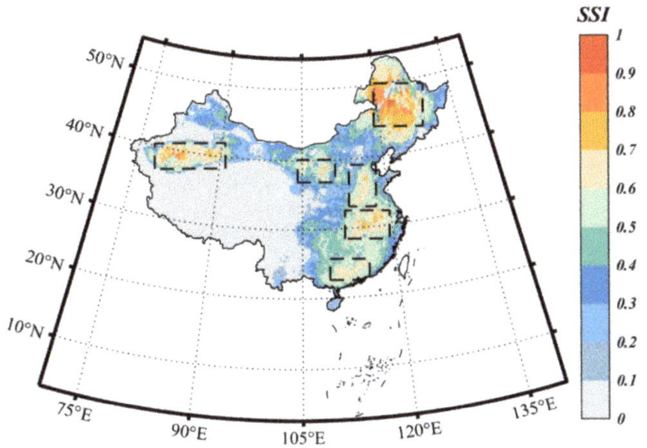

Figure 7. The synchronization strength index (SSI) of nodes. Black dashed box indicates regions where the $SSI \in [0.6, 1.0]$.

Table 1. The modularity under the different numbers of communities.

The Number of Communities	4	5	6	7	8
The modularity *Mod*	0.667	0.680	0.687	0.691	0.688

We found that the highest value of *Mod* occurs when the number of communities is seven. With this number of communities, the northern regions of NWC and southern regions of SWC are classified as same community. However, spatial verification shows that this teleconnection does not exist; thus, we chose eight, corresponding to the second highest *Mod*, as the number of communities for subsequent community detection. Meanwhile, in order to ensure the local optimum of community detection, we also calculated the modularity when the number of communities was 9, 10 and 11, and found that the modularity decreases as the number of communities increases. The results of community detection with eight communities are shown in Figure 8. It can be seen that community 1, 2, 3, 5, and 7 contain six core regions, indicating that the proposed SSI is indicative for the initial determination of the number of communities in the spin glass algorithm, and the spin glass algorithm further discovers the closely related components in the synchronization strength network and provides clear boundaries. Each community can provide the propagation extent of CDHEs, and the total number of CDHEs is relatively close between nodes within a community.

AMO and PDO have been proven to influence the formation of CDHEs in China [23,28]. In this study, we correlate a temporal series of the mean frequency and mean magnitude within different communities with a temporal series of AMO and PDO indexes during JJA, aiming to explore the driving factors of CDHEs in different regions of China. The results of the correlation analysis are presented in Tables 2 and 3. As shown in Tables 2 and 3, the variations in the mean frequency and mean magnitude of CDHEs within community 1 and 2 are significantly and positively correlated with the AMO index, implying that the AMO pattern may modulate variations in the frequency and magnitude of CDHEs within community 1 and 2, and that the frequency and magnitude of CDHEs within community 2 are more positively modulated by the AMO pattern. In community 7 and 8, which are also located in the northern China, variations in the mean frequency and mean magnitude of CDHEs are not significantly correlated with individual patterns, but variations in the mean magnitude of CDHEs in community 8 are significantly correlated with the compounding effect of AMO and PDO; the results of the standardized regression coefficients for individual patterns show that AMO plays a dominant role in positively modulating variations in the mean magnitude of CDHEs in community 8. Similarly, variations in the mean magnitude of CDHEs in community 6, located in the southern regions of SWC, are significantly correlated with the compounding effect of AMO and PDO, and AMO also dominates in positively modulating variations in the mean magnitude of community 6. Previous studies [23] have shown that the warm AMO phase tends to increase the frequency of CDHEs in northern China and the southern regions of SWC, and AMO is significantly and positively correlated with temperature in most of these regions. When AMO is in the warm phase, the warming of the northern China and the southern regions of SWC may lead to a high magnitude of CDHEs; this also explains the positive modulation of AMO in the frequency and magnitude of CDHEs in communities located in the northern China and the southern regions of SWC. During the warm phase of AMO, the western North Atlantic releases more heat flux from the ocean into the atmosphere, which triggers two types of stationary wave trains that propagate eastward. One part is an arching wave train that propagates from the western North Atlantic toward the polar region and even farther into northeastern Asia (the positive polar-Eurasian pattern), and the other part is a Rossby wave train zonally propagating from the western North Atlantic to East Asia (the negative Silk Road pattern). These wave trains form a downward motion in most of China, providing favorable conditions for high temperatures and precipitation deficits; when AMO is in the cold phase, the reverse applies [46]. Community 3 is located in the central-eastern region of China, where the correlations between the mean frequency of CDHEs and individual patterns are negative and none of them are significant, but the mean frequency of CDHEs is significantly and positively correlated with the compounding effect of AMO and PDO, which indicates that AMO and PDO jointly influence the variations in the frequency of CDHEs in this region. According to the results of the standardized regression coefficients, the contribution of AMO to the variations in the mean frequency of CDHEs is greater than that of PDO. For the variation in the magnitude of CDHEs within community 3, the AMO pattern is significantly negatively correlated with the variation in the mean magnitude of CDHEs. AMO has a significant positive correlation with precipitation and a non-significant positive correlation with temperature in the central-eastern region of China. The increase (decrease) in precipitation may inhibit (promote) the formation of CDHEs and decrease (increase) the magnitude of CDHEs. Therefore, AMO may negatively modulate variations in the frequency and magnitude of CDHEs, which is consistent with the results in this study.

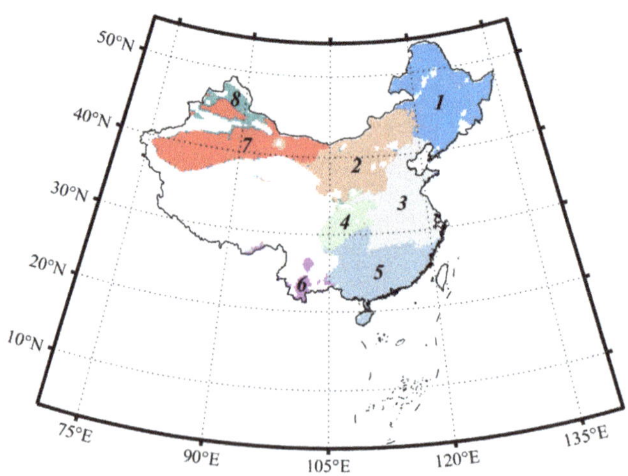

Figure 8. The spatial distribution of communities when the number of communities is 8.

Table 2. The single and multiple correlation coefficient between temporal series of AMO, PDO index during JJA and the mean frequency of CDHEs in 8 communities.

Teleconnection Pattern \ Community	1	2	3	4	5	6	7	8
AMO	0.27 *	0.40 **	−0.23	0.05	−0.06	0.15	0.07	0.13
PDO	−0.03	−0.19	−0.16	−0.26 *	−0.07	0.06	−0.07	−0.10
AMO and PDO	0.27 *	0.41 **	0.32 *	0.26 *	0.11	0.18	0.09	0.15

Note: the significance level of correlation $p < 0.01$ is denoted as **, $0.01 < p < 0.05$ is denoted as *.

Table 3. The single and multiple correlation coefficient between temporal series of AMO, PDO index during JJA and the mean magnitude of CDHEs in 8 communities.

Teleconnection Pattern \ Community	1	2	3	4	5	6	7	8
AMO	0.28 *	0.48 **	−0.27 *	0.06	−0.01	0.10	0.04	0.25
PDO	−0.04	−0.20	−0.17	−0.27 *	−0.09	0.23	0.05	0.06
AMO and PDO	0.28 *	0.49 **	0.36 **	0.27 *	0.09	0.28 *	0.07	0.28 *

Note: the significance level of correlation $p < 0.01$ is denoted as **, $0.01 < p < 0.05$ is denoted as *.

The variations in the mean frequency and mean magnitude of CDHEs within community 4 are significantly negatively correlated with the PDO pattern, which may be attributed to the significant negative correlation between PDO and temperature in the region. The potential mechanism of this correlation can be explained by the EAP teleconnection wave train [29]. The variations in the mean frequency and mean magnitude of CDHEs within community 5 do not significantly respond to AMO or PDO patterns. A reason for this may be that temperature and precipitation within community 5 are not sensitive to changes in AMO and PDO patterns [23].

5. Conclusions

In this study, we define CDHEs using daily maximum temperature and a monthly scale SPI index, and focus on the spatial and temporal variations in the annual frequency and annual magnitude of CDHEs during 1961–2018; this analysis shows that, on a national scale, the mean frequency and mean magnitude of CDHEs, respectively, show a non-significant decrease and a non-significant increase. In China, CDHEs rarely occur in areas with high altitude or low temperature. The spatial distribution of the mean annual frequency and mean annual magnitude of CDHEs is more consistent in northern China, while spatial distribution is divergent elsewhere. The trends in the annual frequency and annual magnitude of CDHEs within each grid point are also discussed, and significant increases or decreases in the annual frequency and annual magnitude can be attributed to the individual or combined effect of variations in precipitation and maximum temperature.

We further obtained the propagation extent of CDHEs through the constructed synchronization strength network combined with the proposed synchronization strength index, and discussed the correlation between the mean frequency and mean magnitude of CDHEs, AMO, and PDO within each community to explore possible physical mechanisms. AMO, through two types of teleconnection wave trains, positively modulates the mean frequency and mean magnitude of CDHEs within community 1 and 2, and negatively modulates the mean magnitude of CDHEs within community 3. PDO negatively modulates the mean frequency and mean magnitude of CDHEs within community 4. AMO and PDO jointly modulate the mean magnitude of CDHEs within community 6 and 8. CDHEs have a serious impact on the development of society, and there is minimal research regarding the spatial propagation direction of CDHEs, these research have positive implications for forecasting and emergency management of compound disaster events. In future research, we will study the spatial propagation of CDHEs from the perspective of complex networks.

Author Contributions: Conceptualization, all authors; methodology, K.L. (Kaiwen Li); data collection, K.L. (Kaiwen Li); first draft, K.L. (Kaiwen Li); draft review and editing, M.W., K.L. (Kai Liu); formal analysis, K.L. (Kaiwen Li); supervision and validation, K.L. (Kaiwen Li), M.W.; investigation, all authors; Funding Acquisition, M.W., K.L. (Kai Liu). All authors have read and agreed to the published version of the manuscript.

Funding: The research for this article was supported by the National Key Research and Development Plan (2017YFC1502902).

Institutional Review Board Statement: Not applicable.

Informed Consent Statement: Not applicable.

Data Availability Statement: CN05.1 dataset used in this paper is available at the China Meteorological Data Service Center website (http://data.cma.cn/, accessed on 18 November 2021).

Acknowledgments: The authors appreciate Tian Xie for his contribution to the language editing of this article.

Conflicts of Interest: The authors declare no conflict of interest.

References

1. Zhou, P.; Liu, Z. Likelihood of concurrent climate extremes and variations over China. *Environ. Res. Lett.* **2018**, *13*, 094023. [CrossRef]
2. Kirono, D.G.C.; Hennessy, K.J.; Grose, M.R. Increasing risk of months with low rainfall and high temperature in southeast Australia for the past 150 years. *Clim. Risk Manag.* **2017**, *16*, 10–21. [CrossRef]
3. Seneviratne, S.; Nicholls, N.; Easterling, D.; Goodess, C.; Kanae, S.; Kossin, J.; Luo, Y.; Marengo, J.; McInnes, K.; Rahimi, M.; et al. *Changes in Climate Extremes and Their Impacts on the Natural Physical Environment: An Overview of the IPCC SREX Report*; Cambridge University Press: Cambridge, UK; New York, NY, USA, 2012; ISBN 9781139177245.
4. Chen, L.; Chen, X.; Cheng, L.; Zhou, P.; Liu, Z. Compound hot droughts over China: Identification, risk patterns and variations. *Atmos. Res.* **2019**, *227*, 210–219. [CrossRef]
5. Ye, L.; Shi, K.; Xin, Z.; Wang, C.; Zhang, C. Compound Droughts and Heat Waves in China. *Sustainability* **2019**, *11*, 3270. [CrossRef]
6. García-Herrera, R.; Díaz, J.; Trigo, R.M.; Luterbacher, J.; Fischer, E.M. A Review of the European Summer Heat Wave of 2003. *Crit. Rev. Environ. Sci. Technol.* **2010**, *40*, 267–306. [CrossRef]
7. Rippey, B.R. The US drought of 2012. *Weather Clim. Extrem.* **2015**, *10*, 57–64. [CrossRef]
8. McKee, T.B.; Nolan, J.; Kleist, J. The relationship of drought frequency and duration to time scales. In Proceedings of the 8th Conference on Applied Climatology, Anaheim, CA, USA, 17–22 January 1993.
9. Yuan, X.; Jian, J.; Jiang, G. Spatiotemporal Variation of Precipitation Regime in China from 1961 to 2014 from the Standardized Precipitation Index. *ISPRS Int. J. Geo-Inf.* **2016**, *5*, 194. [CrossRef]
10. Vicente-Serrano, S.M.; Beguería, S.; López-Moreno, J.I. A Multiscalar Drought Index Sensitive to Global Warming: The Standardized Precipitation Evapotranspiration Index. *J. Clim.* **2010**, *23*, 1696–1718. [CrossRef]
11. Liu, C.; Yang, C.; Yang, Q.; Wang, J. Spatiotemporal drought analysis by the standardized precipitation index (SPI) and standardized precipitation evapotranspiration index (SPEI) in Sichuan Province, China. *Sci. Rep.* **2021**, *11*, 1280. [CrossRef]
12. Wells, N.; Goddard, S.; Hayes, M.J. A self-calibrating Palmer Drought Severity Index. *J. Clim.* **2004**, *17*, 34–46. [CrossRef]
13. Zhao, H.; Gao, G.; An, W.; Zou, X.; Li, H.; Hou, M. Timescale differences between SC-PDSI and SPEI for drought monitoring in China. *Phys. Chem. Earth* **2017**, *102*, 48–58. [CrossRef]
14. Xu, Z.; Fitzgerald, G.; Guo, Y.; Jalaludin, B.; Tong, S. Impact of heatwave on mortality under different heatwave definitions: A systematic review and meta-analysis. *Environ. Int.* **2016**, *89–90*, 193–203. [CrossRef]
15. Tong, S.; Kan, H. Heatwaves: What is in a definition? *Maturitas* **2011**, *69*, 5–6. [CrossRef]
16. Hao, Z.; Hao, F.; Singh, V.P.; Zhang, X. Quantifying the relationship between compound dry and hot events and El Niño–southern Oscillation (ENSO) at the global scale. *J. Hydrol.* **2018**, *567*, 332–338. [CrossRef]
17. Olmo, M.; Bettolli, M.L.; Rusticucci, M. Atmospheric circulation influence on temperature and precipitation individual and compound daily extreme events: Spatial variability and trends over southern South America. *Weather. Clim. Extremes* **2020**, *29*, 100267. [CrossRef]
18. Geirinhas, J.L.; Russo, A.; Libonati, R.; Sousa, P.M.; Miralles, D.G.; Trigo, R.M. Recent increasing frequency of compound summer drought and heatwaves in Southeast Brazil. *Environ. Res. Lett.* **2021**, *16*, 034036. [CrossRef]
19. Wu, X.; Hao, Z.; Hao, F.; Singh, V.P.; Zhang, X. Dry-hot magnitude index: A joint indicator for compound event analysis. *Environ. Res. Lett.* **2019**, *14*, 064017. [CrossRef]
20. Wu, X.; Hao, Z.; Zhang, X.; Li, C.; Hao, F. Evaluation of severity changes of compound dry and hot events in China based on a multivariate multi-index approach. *J. Hydrol.* **2020**, *583*, 124580. [CrossRef]
21. Wu, X.; Hao, Z.; Hao, F.; Zhang, X. Variations of compound precipitation and temperature extremes in China during 1961–2014. *Sci. Total. Environ.* **2019**, *663*, 731–737. [CrossRef]

22. Mukherjee, S.; Ashfaq, M.; Mishra, A.K. Compound Drought and Heatwaves at a Global Scale: The Role of Natural Climate Variability-Associated Synoptic Patterns and Land-Surface Energy Budget Anomalies. *J. Geophys. Res. Atmos.* **2020**, *125*, e2019JD031943. [CrossRef]
23. Wu, X.; Hao, Z.; Hao, F.; Zhang, X.; Singh, V.P.; Sun, C. Influence of Large-Scale Circulation Patterns on Compound Dry and Hot Events in China. *J. Geophys. Res. Atmos.* **2021**, *126*, 1–15. [CrossRef]
24. Boers, N.; Bookhagen, B.; Barbosa, H.D.M.J.; Marwan, N.; Kurths, J.; Marengo, J.A. Prediction of extreme floods in the eastern Central Andes based on a complex networks approach. *Nat. Commun.* **2014**, *5*, 5199. [CrossRef]
25. Boers, N.; Goswami, B.; Rheinwalt, A.; Bookhagen, B.; Hoskins, B.; Kurths, J. Complex networks reveal global pattern of extreme-rainfall teleconnections. *Nature* **2019**, *566*, 373–377. [CrossRef]
26. Konapala, G.; Mishra, A. Review of complex networks application in hydroclimatic extremes with an implementation to characterize spatio-temporal drought propagation in continental USA. *J. Hydrol.* **2017**, *555*, 600–620. [CrossRef]
27. Wu, J.; Gao, X.J. A gridded daily observation dataset over China region and comparison with the other datasets. *Chin. J. Geophys.* **2013**, *56*, 1102–1111. [CrossRef]
28. Li, H.; He, S.; Gao, Y.; Chen, H.; Wang, H. North Atlantic Modulation of Interdecadal Variations in Hot Drought Events over Northeastern China. *J. Clim.* **2020**, *33*, 4315–4332. [CrossRef]
29. Zhang, G.; Zeng, G.; Li, C.; Yang, X. Impact of PDO and AMO on interdecadal variability in extreme high temperatures in North China over the most recent 40-year period. *Clim. Dyn.* **2020**, *54*, 3003–3020. [CrossRef]
30. Zuo, J.; Pullen, S.; Palmer, J.; Bennetts, H.; Chileshe, N.; Ma, T. Impacts of heat waves and corresponding measures: A review. *J. Clean. Prod.* **2015**, *92*, 1–12. [CrossRef]
31. Russo, S.; Dosio, A.; Graversen, R.G.; Sillmann, J.; Carrao, H.; Dunbar, M.B.; Singleton, A.; Montagna, P.; Barbola, P.; Vogt, J.V. Magnitude of extreme heat waves in present climate and their projection in a warming world. *J. Geophys. Res. Atmos.* **2014**, *119*, 12–500. [CrossRef]
32. Yevjevich, V.; Ica Yevjevich, V. An Objective Approach to Definitions and Investigations of Continental Hydrologic Drought s an Objective Approach to Definitions and Investigations of Continental Hydrologic Droughts. Ph.D. Thesis, Colorado State University, Fort Collins, CO, USA, 1967.
33. Nam, W.-H.; Hayes, M.J.; Svoboda, M.D.; Tadesse, T.; Wilhite, D.A. Drought hazard assessment in the context of climate change for South Korea. *Agric. Water Manag.* **2015**, *160*, 106–117. [CrossRef]
34. Quiroga, R.Q.; Kreuz, T.; Grassberger, P. Event synchronization: A simple and fast method to measure synchronicity and time delay patterns. *Phys. Rev. E* **2002**, *66*, 041904. [CrossRef] [PubMed]
35. Li, Y.; Shang, Y.; Yang, Y. Clustering coefficients of large networks. *Inf. Sci.* **2017**, *382–383*, 350–358. [CrossRef]
36. Masuda, N.; Sakaki, M.; Ezaki, T.; Watanabe, T. Clustering Coefficients for Correlation Networks. *Front. Aging Neurosci.* **2018**, *12*, 7. [CrossRef] [PubMed]
37. Newman, M.E.J.; Girvan, M. Finding and evaluating community structure in networks. *Phys. Rev. E* **2004**, *69*, 026113. [CrossRef]
38. Girvan, M.; Newman, M.E.J. Community structure in social and biological networks. *Proc. Natl. Acad. Sci. USA* **2002**, *99*, 7821–7826. [CrossRef]
39. Pons, P.; Latapy, M. Computing Communities in Large Networks Using Random Walks. *J. Graph Algorithms Appl.* **2006**, *10*, 191–218. [CrossRef]
40. Zhu, X.; Ghahramani, Z. Learning from labeled and unlabeled data with label propagation. *Sch. Comput. Sci. Carnegie Mellon Univ. Tech. Rep. C* **2002**, *3175*, 237–244. [CrossRef]
41. Reichardt, J.; Bornholdt, S. Statistical mechanics of community detection. *Phys. Rev. E* **2006**, *74*, 016110. [CrossRef]
42. Haq, N.F.; Moradi, M.; Wang, Z.J. Community structure detection from networks with weighted modularity. *Pattern Recognit. Lett.* **2019**, *122*, 14–22. [CrossRef]
43. Ayantobo, O.O.; Li, Y.; Song, S.; Yao, N. Spatial comparability of drought characteristics and related return periods in mainland China over 1961–2013. *J. Hydrol.* **2017**, *550*, 549–567. [CrossRef]
44. Yao, N.; Li, Y.; Lei, T.; Peng, L. Drought evolution, severity and trends in mainland China over 1961–2013. *Sci. Total Environ.* **2018**, *616–617*, 73–89. [CrossRef]
45. Cao, L.; Zhu, Y.; Tang, G.; Yuan, F.; Yan, Z. Climatic warming in China according to a homogenized data set from 2419 stations. *Int. J. Clim.* **2016**, *36*, 4384–4392. [CrossRef]
46. Zhu, Y.; Wang, T.; Ma, J. Influence of internal decadal variability on the summer rainfall in Eastern China as simulated by CCSM4. *Adv. Atmospheric Sci.* **2016**, *33*, 706–714. [CrossRef]

sustainability

MDPI

Article

Spatial-Temporal Variation of Snow Black Carbon Concentration in Snow Cover in Northeast China from 2001 to 2016 Based on Remote Sensing

Yanjiao Zheng [1,2], Lijuan Zhang [1,*], Wenliang Li [3], Fan Zhang [1,4] and Xinyue Zhong [5]

1 Information Service in Cold Regions, Harbin Normal University, Harbin 150025, China; zhengyj@nuist.edu.cn (Y.Z.); beibei1923@163.com (F.Z.)
2 Collaborative Innovation Center on Forecast and Evaluation of Meteorological Disasters, Nanjing University of Information Science and Technology, Nanjing 210044, China
3 Department of Geography, Environment and Sustainability, The University of North Carolina at Greensboro, Greensboro, NC 27412, USA; w_li3@uncg.edu
4 College of Tourism Management, Sanya University, Sanya 572000, China
5 Key Laboratory of Remote Sensing of Gansu Province, Northwest Institute of Eco-Environment and Resources, Chinese Academy of Sciences (CAS), Lanzhou 730000, China; xyzhong@lzb.ac.cn
* Correspondence: zlj19650205@163.com

Abstract: The amount of black carbon (BC) on snow surface can significantly reduce snow surface albedo in the visible-light range and change the surface radiative forcing effect. Therefore, it is key to study regional and global climate changes to understand the BC concentration on snow. In this study, we simulated the BC concentration on the surface snow of northeast China using an asymptotic radiative transfer model. From 2001 to 2016, the BC concentration showed no significant increase, with an average increase of 82.104 ng/g compared with that in the early 21st century. The concentration of BC in December was the largest (1344.588 ng/g) and decreased in January and February (1248.619 ng/g and 983.635 ng/g, respectively). The high black carbon content centers were concentrated in the eastern and central regions with dense populations and concentrated industries, with a concentration above 1200 ng/g, while the BC concentration in the southwest region with less human activities was the lowest (below 850 ng/g), which indicates that human activities played an important role in snow BC pollution. Notably, Heilongjiang province has the highest concentration, which may be related to its atmospheric stability in winter. These findings suggest that the BC pollution in northeast China has been aggravated from 2001 to 2016. It is estimated that the snow surface albedo will decrease by 16.448% due to the BC pollution of snow in northeast China. The problem of radiative forcing caused by black carbon to snow reflectivity cannot be ignored.

Keywords: snow cover; black carbon concentration; radiative forcing; northeast China

Citation: Zheng, Y.; Zhang, L.; Li, W.; Zhang, F.; Zhong, X. Spatial-Temporal Variation of Snow Black Carbon Concentration in Snow Cover in Northeast China from 2001 to 2016 Based on Remote Sensing. *Sustainability* **2022**, *14*, 959. https://doi.org/10.3390/su14020959

Academic Editors: Xiaodong Yan, Jia Yang, Shaofei Jin and Luigi Aldieri

Received: 23 September 2021
Accepted: 10 January 2022
Published: 15 January 2022

1. Introduction

Snow cover is widely distributed on land surfaces, with a high albedo, strong radiation, and high insulation, which directly affects the land surface temperature, air temperature, surface albedo, soil moisture, and then affects the radiation balance of the earth-atmosphere system. Thus, snow cover is an important factor in the climate system [1]. The surface albedo of pure fresh snow is very high, generally between 60% and 95% [2]. When black carbon (BC) aerosols deposit into the surface snow through dry and wet deposition, even a trace of BC will greatly reduce the albedo of the snow surface, thereby increasing the absorption of solar radiation on the snow surface [3]. Its heating effect will accelerate snow melting and reduce the snow retention time and area [4,5], which, in turn, leads to regional climate change and even affects global climate change [3,6–8]. The fifth report of the IPCC indicates that the global average surface temperature has increased by approximately 0.85 °C in the past 100 years, and the annual average surface temperature has shown an increasing trend in most regions of the world, especially in winter and spring in the mid-high latitudes of the northern hemisphere [9]. In October 2018, the IPCC issued a special report on global warming of 1.5 °C [10]. However, studies have confirmed that the spatial variation of global temperature is the fastest at high latitudes [11–13]. The impact of human activities on global climate change can be attributed to two factors: (1) change in the chemical composition of the Earth's atmosphere through greenhouse gas emissions, and the "umbrella effect" caused by the excessive increase of aerosols, such as sulfates, and (2) land use and land-cover change (LUCC) [14,15]. Previous studies have shown that on a regional or smaller scale, the impact of LUCC on climate is even greater than the contribution of greenhouse gases due to changes in surface energy [16,17]. The most critical surface parameter of LUCC affecting regional climate is to change the surface reflectance; pure fresh snow has a high surface albedo, which is a key climate parameter that significantly affects surface energy balance in the Earth system [18], however, when BC aerosols enter the surface snow and ice through dry and wet deposition, even a small amount of BCwill greatly reduce the albedo of the snow surface, thus increasing the absorption of solar radiation on the snow surface [19], and its heating effect will accelerate snow melting and reduce the snow retention time and area [20], which leads to regional climate change and even affects global climate change [21]. Therefore, as the main emission material of human activities, BC has greatly changed the surface reflectance and has become an important pollutant that changes regional and even global climate. To assess the impact of BC on climate change, it is most important to clarify the concentration of BC on regional snow, and then accurately evaluate its impact on snow and ice albedo and its radiation forcing effect [22].

Owing to the influence of the local environment, human activities, source emissions, and atmospheric circulation, the distribution of BC aerosols in the global snow concentration have strong regional and spatial differences [22,23]. The study of BC concentration in ice and snow began in the 1980s in the South and North Poles [24–26], and then gradually expanded to the middle and low latitudes, achieving numerous results [27–31]. Published data show that BC concentrations in glaciers, ice sheets, and snow cover range from 0.015 to more than 1000 ng/g globally. Low BC values, which can be pronounced, have been found in Antarctica; while high values, due to BC emissions and transport, have been observed in mid-latitude snow and ice (Alps, North America, and the Tibetan Plateau) [32]. Doherty et al. [33] indicated that the BC concentration in the snow in the Arctic ranges from a few ng/g to dozens of ng/g, and is about 5 ng/g in the Arctic; 21–34 ng/g in western and eastern Russia; The BC concentration in the snow cover in high-latitude regions are all low, such as Scandinavia, Svalbard, and Fram Strait are 88 ng/g, 7–42 ng/g, and 9 ng/g, respectively [34]. The concentration of BC in snow in most parts of North America ranges from 10 to tens of ng/g. In central and western North America, the concentration of BC ranges from 5 to 111 ng/g and has a strong latitude zonality. But up to now, the study of BCconcentration in snow is mainly based on laboratory testing methods based on field sampling samples. The biggest advantage of this method is that the results are more accurate, but it requires a lot of manpower, material resources, and the sampling points are obtained. The quantity is limited, and it is difficult to effectively reflect the distribution characteristics of snow BCconcentration in large-scale areas. Compared with a laboratory test, remote-sensing technology has the advantages of wide coverage and high resolution, it is not limited by climate and geographical conditions, and it objectively and truly reflects the information of land features [35], which has been applied more and more in recent years. Some scholars have proposed a method for simulating the BCcontent in snow with a model based on remote sensing data. For example, Flanner et al. [36] used the SNICAR model to simulate the global average snow surface BCconcentration in 1998 and 2001; M. Namazi et al. [37] used the CanAM4.2 model to simulate the BCcontent of snow in the northern hemisphere. The results of the two studies both show that the BCcontent in the snow in northeast China is the highest in the world, but they only revealed the black carbon content in the snow in the northeast in a certain year or two. Huang et al. [38] collected

29 snow samples in northeast China from January to February of 2010 and January of 2014. The results of two field snow samples showed that the BC concentration in snow cover in northeast China was 40–1600 ng/g and 50–3700 ng/g, respectively. Negi and Kokhanovsky [39,40] developed the ART model based on the SNICAR model, and realized the inversion of snow particle size and albedo. Wei [41] used the ATR model to invert the BC concentration of snow cover in China from 2000 to 2018; It can be seen that the ART theory has a wide range of applications for snow cover research. It can not only be used to retrieve albedo, but also to retrieve snow particle size and pollutant content [42]. Northeast China is the second largest stable snowpack area in China, and it is also an important heavy industrial base in China. Coal, petroleum, and biomass fuels are the main energy sources consumed and the combustion mode is backward. BC aerosols enter the snow through a dry and wet deposition. Based on the asymptotic radiative transfer (ART) model and remote sensing image data, this study simulates the snow carbon concentration in northeast China from 2001 to 2016 and analyzes its spatiotemporal variation characteristics, which provides a basis for studying global snow carbon concentration and its radiative forcing.

2. Materials and Methods

2.1. Study Area

The study area is located in the northeast of China, between 37°–53° N and 115°–135° E, which includes Heilongjiang province, Jilin province, and Liaoning province (Figure 1). This area is the region with the widest range of stable snow cover and the largest average snow depth. The snow cover period here ranges from south to north for 30 to 190 days [43], and the annual cumulative snow depth is 498 cm [44]. It is distinguished from other snow areas by its large annual average snow reserves and the most obvious inter-annual variability [45]. Simultaneously, the northeast region is a famous heavy industry base of China, and its energy consumption is based on fossil fuels, such as coal and petroleum. In the vast rural areas it still relies on biofuels and its combustion methods are relatively backward. In addition, the use of motor vehicles in urban areas continues to increase. Therefore, BC aerosols are an important group of atmospheric aerosols in northeast China.

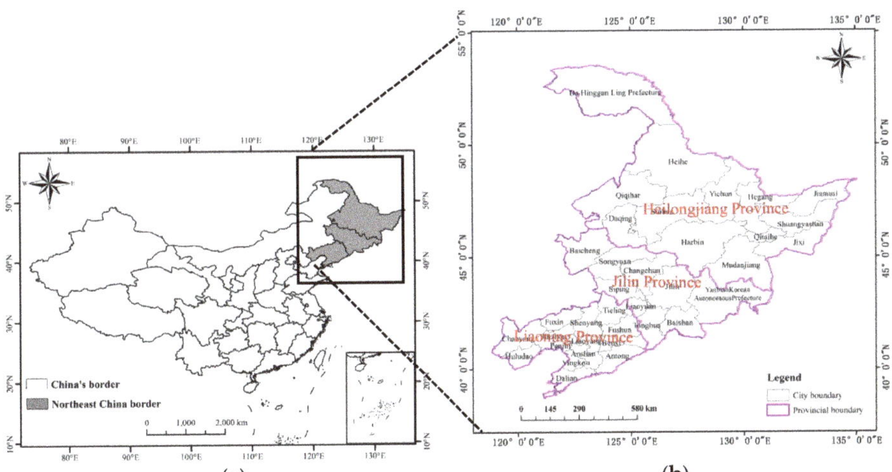

Figure 1. Location of study areas. (**a**) study area in northeast China; (**b**) three provinces comprising study area (Heilongjiang, Jilin, and Liaoning) ranging in latitude from 37° N to 53° N and in longitude from 115° E to 135° E.

2.2. Data Sets

2.2.1. Remote Sensing Data

The visible reflectance, near-infrared reflectance, solar zenith angle, observational zenith angle, and relative azimuth angle were the main parameters driving the ART model. The above parameters were obtained from the MOD09GA (V6) dataset, which was registered and downloaded from the U.S. National Aeronautics and Space Administration (NASA) data website (https://modis.gsfc.nasa.gov/) (accessed on 28 February 2019). The track numbers were h25v03, h26v03, h26v04, h27v04, and h27v05. The timescale was from January 2001 to December 2016. The data format is HDF. The dataset has been corrected by atmospheric bidirectional reflection and solar altitude angle.

The ground reflectance in the visible-light band is the MOD09GA (V6) dataset in band 2 and that in the near-infrared band is the MOD09GA (V6) dataset in band 3. The spatial resolution of the ground reflectance data in the two bands is 500 m. The solar zenith angle and observation zenith angle are calculated using the corresponding solar zenith angle and observation zenith angle in the MOD09GA (V6) dataset, and the relative azimuth is calculated by using the solar azimuth angle and observation azimuth in the data set. The spatial resolution of the four angles is 1000 m. To match the reflectance data with the angle data in spatial scale, it is necessary to resample the angle data of 1000 m to 500 m. The data pre-processing process of MOD09GA (V6) includes data-format conversion, resampling, re-projection, splicing, and clipping.

To study the influencing factors of the distribution and variation trend of snow BC concentration in northeast China, the spatial distribution data of China's elevation (DEM) released by the resource and environment data cloud official website (http://www.resdc.cn/data.aspx?DATAID=284) (accessed on 10 May 2019), the average wind-speed data released by the scientific data center in cold and arid regions (http://westdc.westgis.ac.cn/data/7a35329c-c53f-4267-aa07-e0037d913a21) (accessed on 10 May 2019), and the night light time data of the DMSP satellite (https://ngdc.noaa.gov/eog/dmsp/downloadV4composites.html) (accessed on 12 May 2019) were used to discuss the effects of topographic factors, population density, energy factors, and meteorological conditions on the distribution and variation trend of snow BC concentration in northeast China.

2.2.2. Field Data

To validate the model, twice field samples were taken from January to February 2010 and January 2014, totaling 28 points, including 19 in 2010 and 9 in 2014 (Figure 2). The specific sampling process was the following: clean stainless-steel utensils were used to dig out snow pits and make a vertical profile in one direction. When sampling, according to the order of "left" and "right", two snow samples of 0–5 cm cover in vertical section were collected and placed in Whirlpak sampling bags, sealed, and put into an incubator to ensure that they remained in a solid-state until the next processing step in the temporary laboratory. In the temporary laboratory, samples of the collected snow were scooped into clean glass beakers with clean stainless-steel containers and quickly melted in a microwave oven. After measuring the volume of the melted samples, the melted snow water was filtered through a 0.4-mm Nuclepore filter by a manual pump immediately to separate BC and other water-insoluble substances [46], and then an ISSW spectrophotometer was used to analyze the BC and other light-absorbing substances in the filter samples [25,47]. Since the BC concentration was estimated based on the Ångström absorption index of the sample on the filter, the BC concentration of the sample given was a range, not a definite value [48]. For example, the uncertainty range of the BC concentration of a snow sample in Changchun City in early February 2010 was 250–650 ng/g.

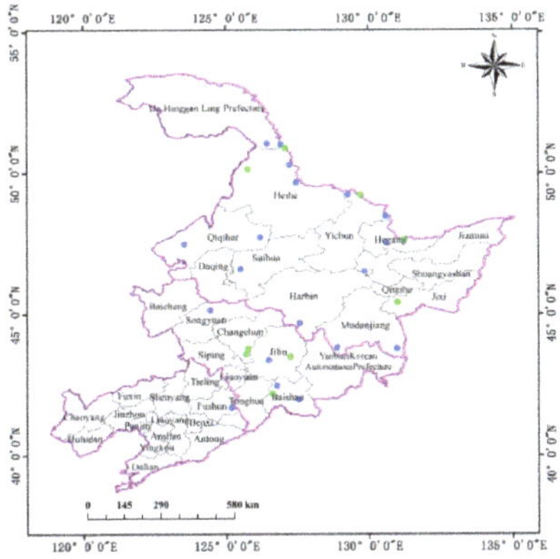

Figure 2. Locations of the 28 sampling sites (blue is the sampling point in 2010 and green is the sampling point in 2014), ranging in latitude from 42° to 51° N and in longitude from 120° to 131° E.

2.3. Methods

2.3.1. Snow Recognition Method

Snow pixel judgment was the basic data for simulating snow BC concentration. In this study, we used the SNOWMAP algorithm to identify snow-covered pixels, which was proposed by Hall et al. [49], and subsequently improved for the production of MODIS snow cover products [50] and has been validated and applied worldwide [51,52].

The core concentration of the SNOWMAP algorithm is normalized difference snow index (NDSI), which uses the spectrum characteristic of the high reflectivity of snow in the visible light band and low reflectance in the near-infrared band to distinguish between clouds and snow. The calculation method is

$$NDSI = \frac{R(b4) - R(b6)}{R(b4) + R(b6)} \tag{1}$$

where, $R(b4)$ and $R(b6)$ representrepresents the surface reflectance of band 4 (545–565 nm) and band 6 (1628–1652 nm) in the MOD09GA data, respectively.

Hao et al. [53] found that the NDSI threshold was 0.4. The NDSI of water is also greater than 0.4, so we eliminated the influence of water on snow pixel based on the characteristic that the reflectivity of snow in the second waveband was greater than, or equal to, 0.11, and the reflectivity of water in this waveband was not greater than 0.11 [50]. Therefore, the discriminant formulas of snow pixels in this study were

$$\begin{cases} NDSI \geq 0.4 \\ R(b2) > 0.1 \end{cases} \tag{2}$$

where, $R(b2)$ represents the surface reflectance of band 2 (841–876 nm).

In this paper, the snow pixels identified by the SNOWMAP algorithm were marked as 1, and the other non-snow pixels were marked as 0. A snow cover map with a spatial resolution of 500 m × 500 m was generated.

2.3.2. ART Model

The ART model is a simplified radiative transfer model developed by Kokhanovsky et al. [54,55]. The basic principle is to regard snow particles as irregular particles, consider the effect of pollutants on the absorption characteristics of the snow surface, the optical characteristics of a single particle are calculated using the principles of geometrical optics, and then the asymptotic analysis solution of the radiative transfer model is obtained using the asymptotic analysis method. The model driving parameters mainly include visible-light reflectance, near-infrared reflectance, solar zenith angle, viewing zenith angle, and relative azimuth angle. Snow characteristics, such as the grain effective size (α_{ef}) and BC concentration (c), can be estimated on regional and global scales.

The formula for simulating the BC concentration of each grid in snow based on the ART model is:

$$R_1 = R_0 \exp\left[-\frac{4f(\mu, \mu_0, \varphi)}{\sqrt{3(1-g)}}\sqrt{\frac{2}{3}B\alpha_{s,1}c\alpha_{ef}}\right] \tag{3}$$

$$R_2 = R_0 \exp\left[-\frac{4f(\mu, \mu_0, \varphi)}{\sqrt{3(1-g)}}\sqrt{\alpha_{i,2}\beta K\alpha_{ef} + \frac{2}{3}B\alpha_{s,2}c\alpha_{ef}}\right] \tag{4}$$

There are 15 parameters in the equations (Equation (3) and (4)), and the physical meaning and units are listed in Table 1. The snow BC concentration in this study is relative snow concentration.

Table 1. Parameters used in ART model.

Parameter	Physical Meaning	Unit/Value
R_1	Surface reflectance of band 3 in MODIS dataset	
R_2	Surface reflectance of band 2 in MODIS dataset	
R_0	Reflection function of the semi-infinite snow layer	
$f(\mu, \mu_0, \phi)$	Determined by escape function and semi-infinite space sno layer reflectivity function	
g	Asymmetry parameter	0.76
B	Constant	0.84
$\alpha_{s,i}$	Absorption coefficient of BC	
c	Relative BC concentration	ng/g
α_{ef}	Snow grain size	mm
$\alpha_{i,2}$	Absorption coefficient of ice	
β	Probability of photon absorption	0.47
K	Constant	2.63
ϑ_0	Solar zenith angle	degree
ϑ	Viewing zenith angle	degree
φ	Relative azimuth angle	degree

R_1, R_2, ϑ_0, and ϑ in equations (Equations (3) and (4)) can be downloaded from the MODIS data. Other parameters can be obtained using the following equations.

1. Reflection function of the semi-infinite snow layer R_0

   R_0 was calculated by the following equation (Equation (2)) [56,57]

$$R_0(\mu, \mu_0, \varphi) = \frac{A + B(\mu+\mu_0) + C\mu\mu_0 + P(\theta)}{4(\mu+\mu_0)} \tag{5}$$

The parameters in Equation (5) are listed in Table 2.

Table 2. Parameters needed to calculate R_0.

Parameter	Mean	Value	Parameter	Mean	Value
A	constant	1.247	μ	The cosine of viewing zenith angle	[-]
B	constant	1.186	μ_0	The cosine of solar zenith angle	[-]
C	constant	5.157	$P(\theta)$	The phase function	[-]

The phase function $P(\theta)$ does not depend on the wavelength, therefore the function R_0 (μ, μ_0, ϕ) is the same in the entire spectrum. The phase function $P(\theta)$ is calculated as:

$$P(\theta) = 11.1\exp(-0.087\theta) + 1.1\exp(-0.014\theta) \quad (6)$$

where, θ is defined as $\theta = a\cos(-\mu\mu_0 + ss_0\cos\varphi)$, $s = \sin(\vartheta)$, $s_0 = \sin(\vartheta_0)$, $\mu_0 = \cos(\vartheta_0)$, and $\mu = \cos(\vartheta)$.

2. Parameter $f(\mu, \mu_0, \phi)$

$f(\mu, \mu_0, \phi)$ is determined by the escape function and the semi-infinite space snow reflectance function, and its calculation formula is as follows:

$$f(\mu, \mu_0, \varphi) = \frac{u(\mu_0)u(\mu)}{R_0(\mu, \mu_0, \varphi)} \quad (7)$$

where, $u(\mu_0)$ is called the escape function, which is solved according to the following empirical formulas:

$$u(\mu_0) = \frac{3}{7}(1 + 2\mu_0),\ u(\mu) = \frac{3}{7}(1 + 2\mu) \quad (8)$$

3. Absorption coefficient of ice crystal and BC

In the visible light band (channel 1), the absorption of light by the snow layer is mainly caused by pollutants (mainly BC); in the near-infrared band (channel 2), the absorption of light by the snow layer is mainly caused by ice crystals and BC.

The formula of the absorption coefficient of BC is

$$\alpha_{s,i} = 4\prod \chi_s(\lambda)/\lambda_i \quad (9)$$

where, $i = 1,2$, indicating the first and second channels. Since the refractive index of BC is independent of wavelength, $\chi_s(\lambda)$ is a constant of 0.46. λ is the wavelength; that is, $\lambda_1 = 460$ nm and $\lambda_2 = 865$ nm.

The expression of the light absorption coefficient of ice crystals is

$$\alpha_{ice,2} = 4\prod \chi_{ice}(\lambda)/\lambda \quad (10)$$

where, $\alpha_{ice,2}$ is the ice absorption coefficient at the wavelength λ_2; $\chi_{ice}(\lambda)$ is the imaginary part of the refractive index of ice, which is a function of wavelength. Here, we use $\chi_{ice}(\lambda) = 2.5 \times 10^{-9}$.

4. Relative azimuth angle ϕ

ϕ is calculated with Equation (11):

$$\varphi = saa - vaa \quad (11)$$

where *saa* is the solar azimuth angle and *vaa* is the viewing azimuth angle.

Entering all the above parameters into the ART model, we simulated the BC concentration of each snow pixel with a spatial resolution of 500 × 500 m.

Based on the above data and formula, we analyzed the BC concentration of snow cover in northeast China, and the research framework is shown as follows (Figure 3):

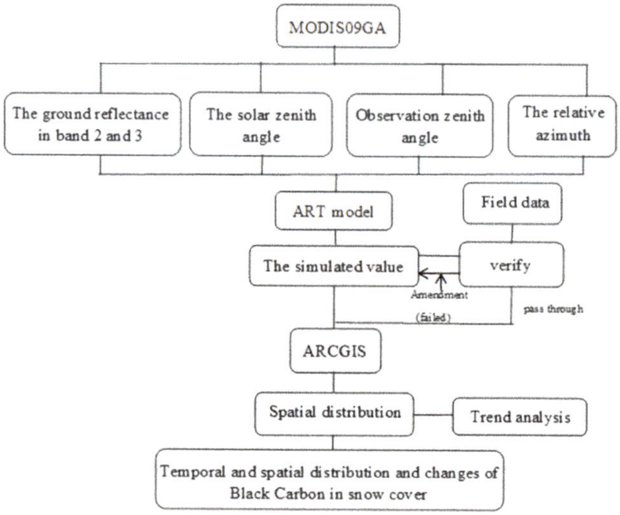

Figure 3. The framework of the study.

3. Results

3.1. Model Verification

We verified the accuracy of the simulated values by comparing the simulated snow BC concentration with the measured values. Since the measured value was an uncertain range, as long as the simulated value was distributed within the range of the measured value we could consider the simulation result credible. We also used the literature comparison method to verify the simulated value. The field sampling measured datum were used to verify the simulation values of the same period (Figure 4). There are significant disagreements in the measured and modeled snow BC concentrations at individual sites, but over the region as a whole there is no apparent systematic bias in the modeled vs. measured concentrations. Zhao et al. [58] simulated the BC concentration of snow cover in northeast China in 2010, and the simulated value was above 1000 ng/g, similar to the results of this study (1197.468 ng/g). For areas north of 48° N, the simulated value was higher than the measured value. However, Flanner et al. [36] simulated the snow BC content in the high latitudes of northeast China, and the simulation result was above 1000 ng/g, which is similar to the simulation result of this study. Based on the above comparison, the ART model has a high ability to simulate the BC concentration of snow cover in northeast China, and the simulation results are credible.

3.2. Annual BC Concentration

3.2.1. Temporal Variation

From 2001 to 2016, the average annual BC concentration of snow cover in northeast China was between 1098.927 and 1257.300 ng/g, with an average value of 1197.468 ng/g. The maximum value appeared in 2016 and the minimum value appeared in 2002 (Figure 5). No significant ($p > 0.05$) increase in BC concentration was observed from 2001 to 2016, with an annual growth rate of 5.137 ng/g.

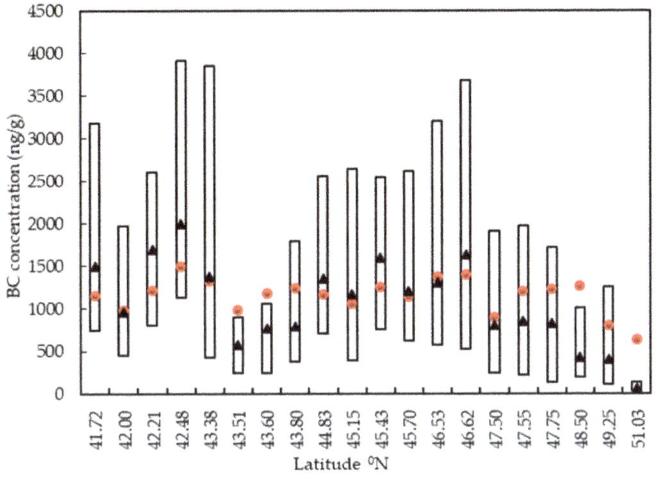

Figure 4. BC concentrations in the top snow layer from the campaign observations (black) and the corresponding ART simulations (red). The column plot of observations shows the minimum and maximum possible values from estimates; the triangle within the column shows the average of observations; and the circle within the column shows the model results.

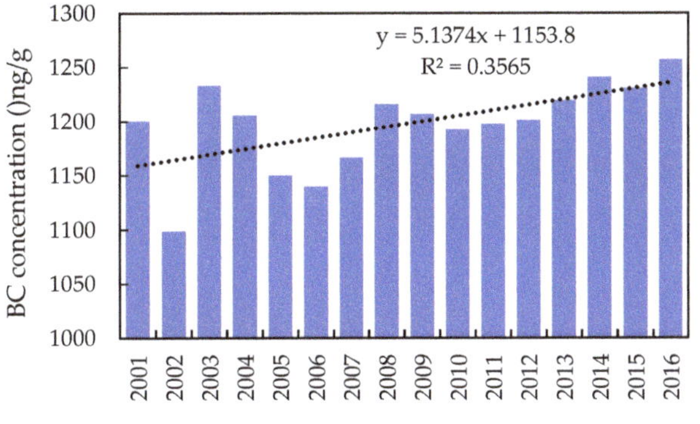

Figure 5. Average annual variability of BC concentration on snow cover in northeast China from 2001 to 2016.

3.2.2. Spatial Distribution and Variation

Figure 6 shows the spatial distribution of the average BC concentration over northeastern China during 2001 to 2016. There was generally a strong spatial difference in BC concentration that varied considerably over the 310–1561 ng/g range in the study area. Generally, there were two high-value (>1200 ng/g) regions of BC concentration in northeast China, both of which were located in Heilongjiang province. One is an industrial corridor composing Harbin, Daqing, Qiqihar, Suihua, and Heihe; and the other is a coal-forest industrial area composing Hegang, Jiamusi, and Shuangyashan. Conversely, the average BC concentration values were largely <850 ng/g in the mountainous areas of the study area, where human activity is minimal.

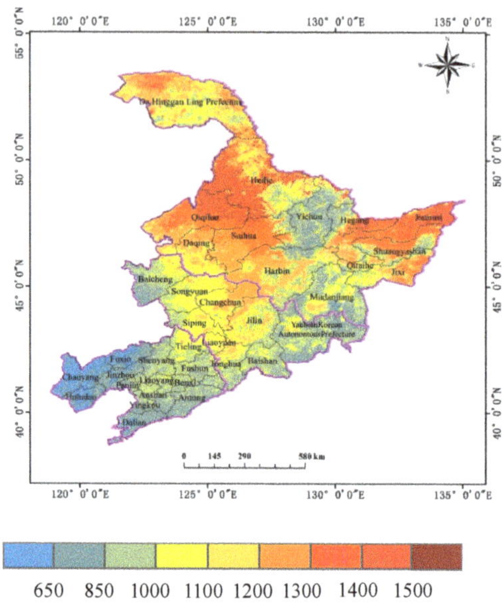

Figure 6. Spatial distributions of the averaged BC concentration over the study area during December to February 2001–2016.

From 2001 to 2016, the BC concentration in 81.805% of the study area showed an increasing trend (Figure 7). Approximately 6.975% of the study area showed a significant ($p < 0.05$) increase, which was mainly distributed in the northeast of the Daxinganling region, east of Heihe, north of Qiqihar, northeast of Harbin, north of Suihua, and southeast of Shuangyashan. In contrast, only 18.195% of the study area showed an insignificant ($p > 0.05$) decrease in BC concentration, which was mainly distributed in the eastern part of Heilongjiang Province and the central and eastern parts of Liaoning Province. Notably, by combining Figures 5 and 6, we found that the heavy snow BC polluted areas centered on the southern part of the Heihe had significantly worsened pollution. In addition, areas where snow cover was less polluted by BC, such as the Xiaoxing'an Mountains, Changbai Mountains, and southern Liaoning, were also becoming more polluted. These results indicate that the BC pollution degree of snow cover in northeast China increased overall from 2001 to 2016, especially in heavily polluted areas.

3.3. *Monthly Snow BC Concentration*

3.3.1. Temporal Variation

The average BC concentration was the highest in December (1344.588 ng/g), and it was gradually decreased in January (1248.619 ng) and February (983.635 ng/g), which was related to the cleaning effect of snowfall on the air. The BC concentration in December and January showed an increasing trend from 2001 to 2016, with annual increases of 7.087 ng/g and 6.556 ng/g, respectively (Figure 8); while in February, it showed a decreasing trend, with an annual reduction of 0.079 ng/g. The coefficients of variation of BC concentration from December to February were 5.601, 8.392, and 8.594, respectively, indicating that the inter-annual distribution of BC concentration in each month was relatively stable.

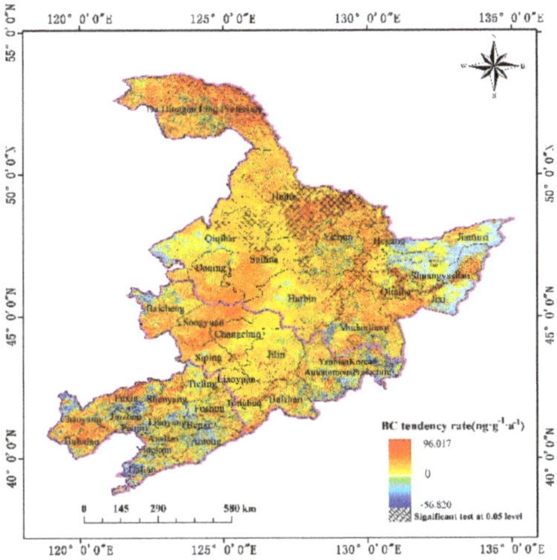

Figure 7. Trends of BC concentration in the top snow layer in northeast China from 2001 to 2016.

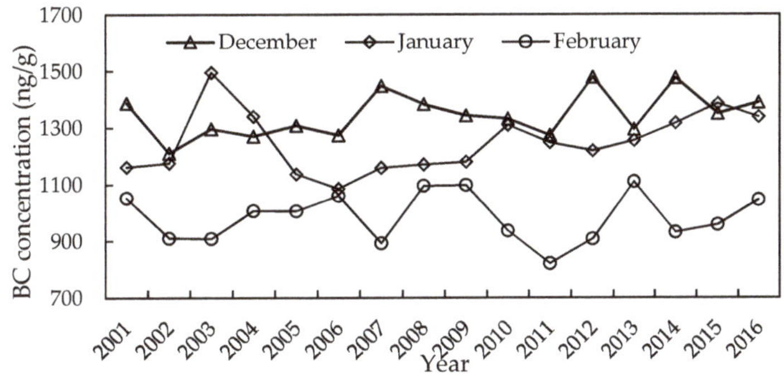

Figure 8. Average annual variability of BC concentration on snow cover in December, January, and February from 2001 to 2016.

3.3.2. Spatial Distribution and Variation

Figure 9 shows the spatial distribution of the average BC concentration in December, January, and February over northeastern China during 2001 to 2016. There was generally a strong spatial difference in BC concentration monthly that showed higher values in the north than in the south. In December, the BC content was greater than 850 ng/g, and there were three high-value centers (>1400 ng/g) in the spatial distribution, located in the Daxinganling region, northern Qiqihar and southern Heihe, and Hegang and Jiamusi regions. In January, the high-value center range shrank sharply, especially in northern Qiqihar and southern Heihe. By February, the BC content in the study area was drastically reduced, falling below 1200 ng/g. Areas with low BC concentration (<900 ng/g) were located in Yichun City, Huludao City, Chaoyang City, Fuxin City, and Dalian City.

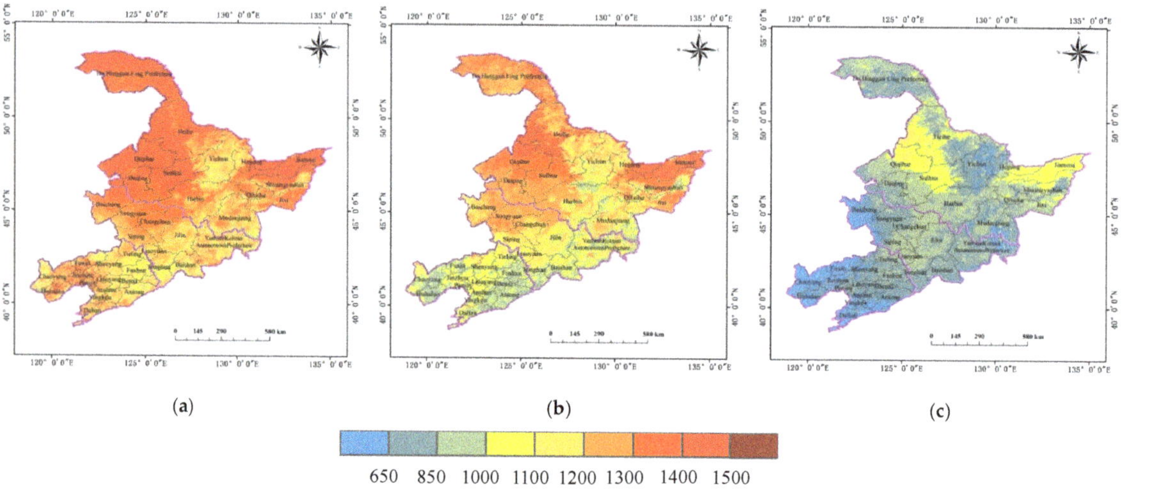

Figure 9. Monthly mean value of BC concentration on snow cover in the northeast of China from 2001 to 2016: (**a**) December, (**b**) January, and (**c**) February.

From 2001 to 2016, the increasing trend of BC content in December and January accounted for 82.519% and 82.789% of the total area, respectively (Figure 10). Among them, the BC concentration in central Heilongjiang Province and southern Liaoning increased significantly ($p < 0.05$), accounting for 9.922% and 10.603% of the total area, respectively, indicating that BC concentration in central Heilongjiang Province increased significantly in December and January. On the contrary, the area with a decreasing trend of BC content in February accounted for 60.110% of study area, and BC concentration decreased significantly ($p < 0.05$) in the southern Greater Xing'an Mountains and northern Heihe, accounting for 2.015% of the entire northeast area.

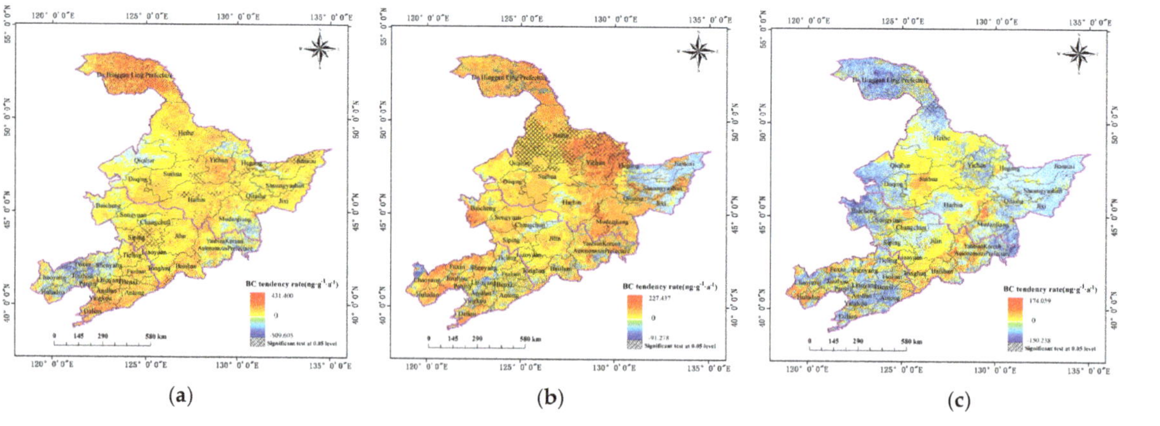

Figure 10. Monthly average snow BC concentration tendency rate of northeast China from 2001 to 2016: (**a**) December, (**b**) January, and (**c**) February.

3.4. Influencing Factors

3.4.1. Terrain

The BC on the snow is mainly the dry deposition of BC aerosols in the atmosphere, therefore the difference in the spatial distribution of BC concentration on the snow is mainly determined by the difference in the spatial distribution of the BC aerosols in the atmosphere. Lv et al. [59] found that elevation is the most important factor contributing to the concentration distribution of BC aerosols, followed by population density, and heavy industry scale. Owing to the sparse population of high-altitude areas, the lack of large-scale heavy industrial production, and the high coverage of surface vegetation in these areas, the concentration of BC aerosols is low. The northeast of China is surrounded by mountains on three sides (Figure 11), while the central and eastern regions are plains and terraces. Therefore, the concentration of BC aerosol in the central region is relatively high. This is consistent with the conclusion that the BC content of snow in the central plain area is higher than that in the surrounding mountainous areas. It shows that geomorphology types are one of the factors that affect the BC concentration in snow.

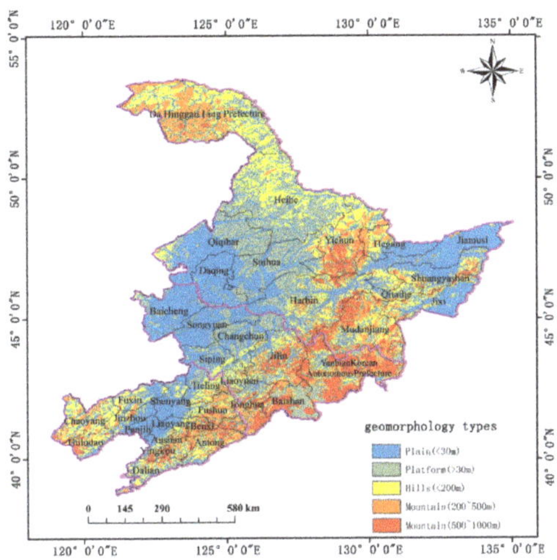

Figure 11. Geomorphology types of northeast China.

3.4.2. Energy Consumption

BC aerosols in the atmosphere mainly come from the incomplete combustion of fossil fuels and biomass fuels. Studies showed that population density and the scale of heavy industry are also important factors affecting the distribution of BC aerosol concentration [59]. The value of the night light time index can reflect the amount of power consumption, representing information such as city size, population density, and energy use [60,61]. Therefore, this study selects the spatial distribution of nighttime light index data in northeast China in 2001 and 2013 (Figure 12), which represents information such as population density and energy usage. The high values of night light hours in 2001 and 2013 were located in Shenyang, Changchun, Harbin, and Daqing, showing a clear zonal distribution. The Songnen Plain in the west and the Sanjiang Plain in the east are also areas with relatively high nighttime light index, suggesting that they are the areas with large cities, concentrated industrial production, and high population density in the three northeastern provinces. Comparing the spatial distribution of snow BC in northeast China in 2001 and 2013, it can be seen that the high value area of the nighttime light index in northeast China in 2001 and 2013 is basically consistent with the high value area of snow BC concentration. Comparing the spatial changes of the nighttime light index in 2013 and 2001, the areas with light index in western, central and eastern Heilongjiang Province have increased significantly. This is the main reason why BC concentration of winter snow in Heilongjiang province became the largest region in northeast China from 2001 to 2016, and BC concentration of winter snow in central and eastern Heilongjiang Province increased significantly. In addition, it can also be seen that the density and scale of cities in the southern part of northeast China have increased significantly. Therefore, the concentration of BC in the snow in the southern part of northeast China in 2013 showed an increasing trend in most areas compared with 2001. The above results indicate that population density, city size, and energy consumption are important factors affecting the concentration of BC in snow.

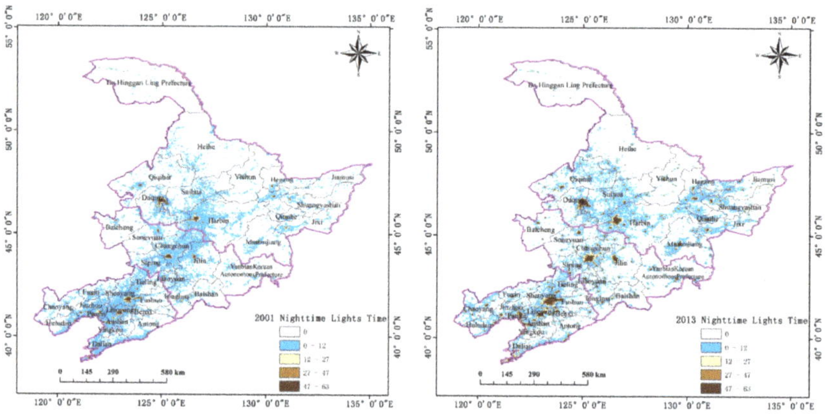

Figure 12. Spatial distributions of Nighttime Lights Time in northeastern China in 2001 and 2013.

3.4.3. Atmospheric Stability

The spatial distribution characteristics of BC concentration on snow cover in northeast China are not only affected by terrain and energy consumption, but also related to winter atmospheric stability. When the atmosphere stratification is in a stable state, the poor diffusion ability of the atmosphere makes it difficult for the pollutants to diffuse; otherwise, the pollutants are easy to diffuse [62]. In this study, the average wind speed was used to characterize the stratification stability of the near-surface atmosphere. The lower the wind speed, the more stable the atmospheric layer structure, and the more likely it is to form high-concentration pollution in local areas. Figure 13 shows the average winter wind speed from 2001 to 2016 (December, January, and February the following year). It can be seen from Figure 12 that the winter wind speed was generally low and the atmospheric stratification was relatively stable. The wind speed in the eastern region was relatively high, mainly due to the formation of canyon winds between the Xiaoxing'anling Mountains and Changbai Mountains. In particular, in the central and northern regions of Heilongjiang Province, the wind speed was the lowest, therefore the atmosphere stratification in this area was stable. In addition, the winter temperature was low and the mixed layer was not high, resulting in frequent occurrence of temperature inversions. These climatic conditions hinder the diffusion of BC aerosols. Moreover, these areas were surrounded by mountains on three sides, and the BC aerosols were more difficult to transport outwards and accumulate near the ground. They enter the snow surface through dry deposition, resulting in northern Qiqihar, southern Heihe, and northern Suihua being the high centers of BC in the northeast.

3.4.4. Land Use Type

Land use type can not only indicate the coverage of surface vegetation, but also a direct result of the impact of human activities on the surface. Comparing the land use types in northeast China in 2020 with the simulation of the spatial distribution of BC in northeast China from 2001 to 2016 (Figure 14), it can be seen that the areas where cultivated land and cities are consistent with high-value areas of BC in the snow, the correlation coefficient between urban area expansion and BC concentration increase is 0.584 ($p < 0.05$). It shows that human activities can also affect the concentration of BC in the snow by changing the land use type.

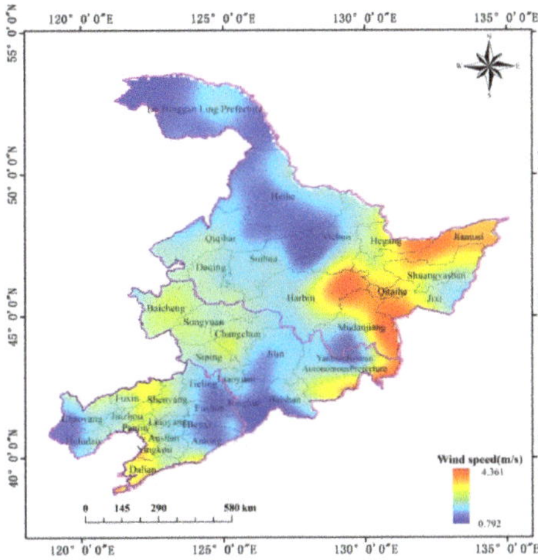

Figure 13. Average winter (December, January, and February the following year) wind speed in northeast China from 2001 to 2016.

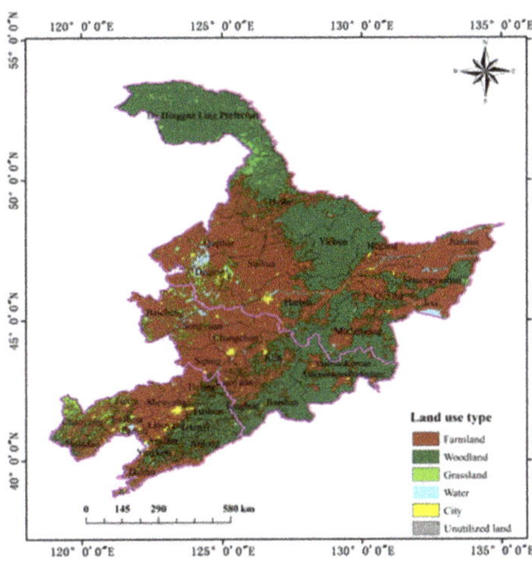

Figure 14. Spatial distributions of land use type in northeastern China in 2020.

In conclusion, the BC content of snow cover in northeast China is affected by both natural and artificial factors. Natural factors are mainly affected by terrain and stable atmosphere stability. Due to sparse population and no large-scale heavy industry production, the concentration level of BC aerosol in high-altitude areas is relatively low. When the atmosphere stratification is in a stable state, the poor diffusion ability of the atmosphere makes it difficult for the pollutants to diffuse, and the BC concentration increased. Human factors mainly include urban expansion, energy structure change, and land use change, among which urban density, urban size, and population density are important factors affecting the distribution of snow BC concentration. The BC concentration in snow cover is the highest in the region where farmland and city are located, suggesting that human activities affect the spatial distribution of BC content in snow cover by changing the surface type.

4. Discussion

This study used the ART model combined with remote-sensing data to simulate the BC concentration on snow cover in northeast China. Compared with the existing research results, this study not only analyzed the spatial and temporal distribution of BC concentration on snow cover in northeast China from 2001 to 2016, but also further analyzed the characteristics of snow BC concentration in northeast China during the year and inter-annually, and briefly analyzes its influencing factors.

In this paper, through simulations, a long-term series of snow BC concentrations in a certain area, and the spatial distribution and changes of snow BC concentrations with high spatial resolution year by year were obtained. Compared with previous studies, there are few similar studies, and it is also one of the main contributions of this paper. The research results are of great significance for snow albedo simulation research, improvement of regional climate model simulation accuracy, and snow pollution prevention and control. The results showed that the BC content of snow cover in northeast China did not increase significantly from 2001 to 2016, but increased by 52.299 ng/g from 2012 to 2016, compared with that from 2001 to 2005. The 93.25% of the northeast showed an insignificant trend, and the significant increase areas were mainly distributed in the northeast of the Daxinganling region, the east of Heihe City, the northern edge of Yichun City, and the southeast of Shuangyashan City.

Owing to the influence of local environment, human activities, source emissions, and atmospheric circulation, the global distribution of BC aerosol concentration is highly regional, therefore the BC concentration on snow is also significantly different in space. The results show that in the inaccessible Antarctic, the average BC concentration on snow is only 0.2 ng/g [26], while the BC concentration in the Arctic is higher than that of the Antarctic, with an average of 25 ng/g [24]. At the same time, the results of studies by Clarke and Noone [24], Warren and Clarke [26], Doherty et al. [25], Grenfell et al. [63,64], and McConnell et al. [65] show that the variation of snow BC concentration in the European Arctic are greater than in the Canadian Arctic and Arctic Oceans, and the reason may be related to the distance to the source of BC emissions. The snow BC concentration in the middle and low latitudes of mountainous and urban areas is much higher than that in the Antarctic and Arctic regions. For example, the snow BC concentration in the Alps in December 1992 was 22–302 ng/g [31], and the BC concentration of Olympus in the United States in 2012 was 250 ng/g [66]. The average BC concentration of glaciers in the Tibetan Plateau and Xinjiang from 2001 to 2004 was 63 ng/g [67]. The ice core in the monsoon climate zone of the Tibetan Plateau shows that BC concentration has increased rapidly since the 1980s, and the glaciers in the southeast of the Tibetan Plateau have accelerated to retreat [68], indicating that anthropogenic BC aerosol emissions in southeast Asia play an increasingly important role in glacial melting on the Tibetan Plateau. Mid-latitude urban areas have the highest snow BC concentration. For example, snow samples from southern New Mexico and northern Texas in the United States from 1982 to 1985 show that the BC concentration in urban areas is three times that of remote rural areas [28,29]; the range of snow BC concentration in the city of Detroit from 1984 to 1985 was 17–5700 ng/g [30]. In 2007–2009, the BC concentration in the surface snow reached 2740 ng/g in the Sapporo area of Japan [27]. In January and February 2010 and January 2014, the snow BC concentration of northeast China was 40–1600 ng/g and 50–3700 ng/g, respectively. At the same time, it can be seen from the two sampling results that the snow BC concentration in the northeast heavy industrial area is higher than that in the northern border area [38,48,69]. The above statements all reflect the impact of human activities on regional snow BC concentration.

Since the strong absorption of light by BC aerosols in the atmosphere was discovered in the 1980s, scientists have subsequently discovered that the albedo of visible light in snow depends largely on the concentration of BC on snow. Warren and Wiscombe [3] believe that trace amounts of light-absorbing substances can significantly reduce the snow surface albedo in the visible light band, but have little effect on the albedo in the near-infrared band. This study summarized previous measurements of simulated snow BC concentrations and corresponding reduced albedo to obtain a simple equation between the BC concentration and its reduced albedo (Figure 15). Using the equation given in Figure 15, the amount of albedo reduction caused by the BC concentration simulated in this study was roughly calculated (Table 3). The estimation results showed that the larger the snow BC concentration, the more the albedo decreases. These results are only theoretical estimates and require field measurements for future verification.

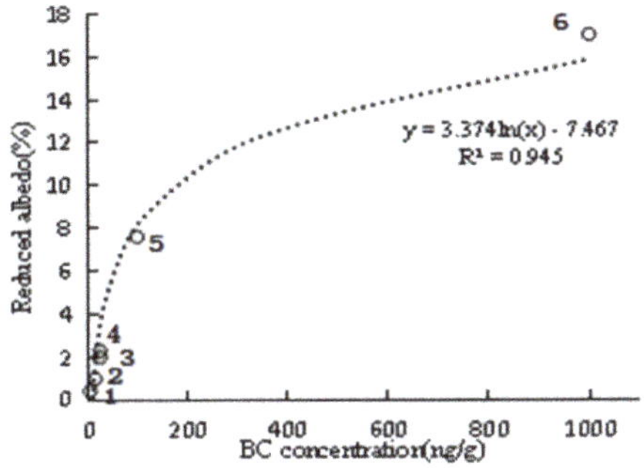

Figure 15. Logarithmic scatter plot for BC concentrations and reduced albedos (hollow circles) in snow and ice based on previous studies, and the simple linear relationship possibly existing between them. The data of the 6 points in the figure are from the literature: 1-Jacobson [8]; 2-Warren [3]; 3-Jacobson [8]; 4-Clarke and Noone [24]; 5-Jacobson [8]; 6-Flanner [7].

Table 3. BC concentration and its reduced albedo in the three northeastern provinces and capital cities.

	BC Concentration (ng/g)	Reduced Albedo (%)
Annual-mean value	1197.468	16.448
Heilongjiang Province	1103.617	16.172
Jilin Province	962.440	15.711
Liaoning Province	788.96	15.040
Harbin City	1051.787	16.010
Changchun City	1093.404	16.141
Shenyang City	864.628	15.349

In addition to BC, light-absorbing pollutants on snow cover include organic carbon (OC) and mineral dust (MD), accounting for 11.7% of total light absorption [69]. OC comes from the incomplete combustion of fossil fuels and biomass fuels, which are deposited on the snow surface after dry and wet deposition. MD mainly comes from the arid and semi-arid desert areas of Mongolia. Under the influence of winter monsoon, the dust in the desert area is lifted up and transported to the northeast region and deposited on the surface of the snow there. ART model does not have the ability to detect OC and MD concentration on snow. Therefore, in future research, OC and MD must be taken into account. Based on the optical characteristics of the three substances, BC, OC, and MD, the ART model parameters should be improved to make it more suitable for the characteristics of snow pollution in northeast China.

Compared with the previous research, this study proposed to increase the influence factors of land use type and reached the conclusion that cultivated land and snow BC high value area are highly consistent. But why is the content of BC in the snow in the area where the cultivated land is located so high? The mechanism needs to be further discussed. Because northeast China is one of the three largest black lands in the world, is there a high BC content in the simulation due to incomplete snow cover in the grid and part of the black lands? This also needs to be further revised through field investigations. In addition, this paper used the lighting index to indirectly analyze the impact of population density, energy consumption, and urban expansion on the concentration of BC in the snow. The relationship between the lighting index and population density, energy consumption, and city size and expansion need further analysis. The research conclusions of this study showed that the BC content of northeast China's snow cover increased by 52.299 ng/g from 2012 to 2016 compared with 2001 to 2005. In fact, there was a net loss of population in northeast China in the recent five years. Why is the BC content in snow cover increasing? This also needs to be combined with the actual data for further analysis.

The model only considers the situation that BC is deposited on the snow surface through dry sedimentation, but it does not consider the situation that BC particles dissolve in the snow particles. After BC is deposited in the snow, the snow gradually ages and deforms until the snow melts as time goes by. During this period, the BC particles migrate with the snow melt water, and their effective particle size and optical properties change. In the future, we can observe the change process of BC in snow and perform experimental simulation to establish the model mechanism, so as to more accurately simulate its influence on albedo.

5. Conclusions

(1) From 2001 to 2016, the average annual BC concentration in northeast China was between 1098.927 and 1257.300 ng/g, with an average value of 1197.468 ng/g. Industrial corridors composed of Harbin, Daqing, Qiqihar, Suihua, and Heihe and coal-forest industrial production areas composed of cities such as Hegang, Jiamusi, and Shuangyashan were high-value centers of snow BC concentration. The area with increased BC concentration accounted for 81.805% of the study area, of which 6.975% was a significant increase area, mainly distributed in the middle of Heilongjiang Province.

(2) The BC concentration of northeast China was the highest in December (1344.488 ng/g), followed by January and February from 2001 to 2016. From 2001 to 2016, the BC concentration of snow cover in December and January in northeast China showed an increasing trend, while the BC concentration in February showed a decreasing trend, with no significant increase or decrease trend in each month.

(3) The high vales of BC concentration in Shenyang, Changchun, Harbin, and Daqing are due to large population density and concentrated industrial production. In the central part of Heilongjiang province, due to the stability of the atmospheric stratum and the influence of topographical conditions, the pollutants were not easy to diffuse, causing Heilongjiang province to become the center of high BC concentration. The increase of cultivated land area, urban density and scale, and the expansion of population area are the main reasons for the increase of BC concentration in snow cover, and also the reasons for the high value of BC content in snow cover in Heilongjiang Province.

This study simulated the BC concentration of surface snow, analyzed the temporal and spatial distribution and variation characteristics of BC concentration, and analyzed the influencing factors of BC concentration of surface snow in northeast China from 2001 to 2016. Compared with previous studies, the main contribution of this study is to analyze the distribution characteristics of in snow cover with large scale and high resolution, and to obtain the variation characteristics of BC concentration in snow (2001–2016) in northeast China. The results of this study provide important basic data for simulating the albedo of snow cover in winter in northeast China and analyzing the influence of snow cover on the regional climate. The shortcomings of current studies lie in the need to further improve the spatial resolution of the simulation of snow BC concentration, refine the analysis of temporal and spatial distribution and variation factors of snow BC, and deepen the study of the influence of BC on snow albedo.

Author Contributions: Y.Z. designed the research framework and drafted the manuscript; L.Z. completed the manuscript and made major revision; W.L. was involved in the design and revision of the manuscript; F.Z. and X.Z. were involved in data acquisition and analysis. All authors have read and agreed to the published version of the manuscript.

Funding: This research was funded by the National Natural Science Foundation of China under grant number 41771067 and the Key projects of Natural Science Foundation of Heilongjiang Province under grant number ZD2020D002.

Institutional Review Board Statement: Not applicable.

Informed Consent Statement: Not applicable.

Data Availability Statement: The data presented in this study are available by request from the corresponding author.

Acknowledgments: We sincerely thank Wang Xin of Lanzhou University for providing snowfield sampling data for this study.

Conflicts of Interest: The authors declare no conflict of interest.

References

1. Wang, X.; Zhang, Q.; Guo, N.; Cai, D.H. Method for Discrimination Snow in Qilian Mountain Region Based on MODIS Data. *Arid Meteorol.* **2007**, *25*, 29–34. (In Chinese)
2. Lei, X.C.; Song, K.S.; Du, J.; Wu, Y.Q.; Wang, Y.D.; Tang, X.G.; Zeng, L.H. Analysis on the spectral reflectance response to snow contaminants in Northeast China. *J. Grad. Sch. Acad. Sin.* **2011**, *28*, 611–616. (In Chinese)
3. Warren, S.G.; Wiscombe, W.J. A Model for the Spectral Albedo of Snow. II: Snow Containing Atmospheric Aerosols. *J. Atmos. Sci.* **1980**, *37*, 2734–2745. [CrossRef]
4. Painter, T.H.; Barrett, A.P.; Landry, C.C.; Neff, J.C.; Cassidy, M.P.; Lawrence, C.R.; McBride, K.E.; Farmer, G.L. Impact of disturbed desert soils on duration of mountain snow cover. *Geophys. Res. Lett.* **2007**, *34*, 237–254. [CrossRef]
5. Ming, J.; Xiao, C.D.; Qin, D.H.; Helene, C. Climate Forcing of Black Carbon in Snow and Ice. *Adv. Clim. Chang. Res.* **2006**, *2*, 238–241.
6. Flanner, M.G.; Zender, C.S.; Hess, P.G.; Mahowald, N.M.; Painter, T.H.; Ramanathan, V.; Rasch, P.J. Springtime warming and reduced snow cover from carbonaceous. *Atmos. Chem. Phys.* **2009**, *9*, 2481–2497. [CrossRef]
7. Flanner, M.G.; Zender, C.S.; Randerson, J.T.; Rasch, P.J. Present-day climate forcing and response from black carbon in snow. *J. Geophys. Res.* **2007**, *112*, D11202. [CrossRef]
8. Jacobson, M.Z. Climate response of fossil fuel and biofuel soot, accounting for soot's feedback to snow and sea ice albedo and emissivity. *J. Geophys. Res. Atmos.* **2004**, *109*, D21201. [CrossRef]
9. Allen, S.K.; Plattner, G.-K.; Nauels, A.; Xia, Y.; Stocker, T.F. Climate Change 2013: The Physical Science Basis. An overview of the Working Group 1 contribution to the Fifth Assessment Report of the Intergovernmental Panel on Climate Change (IPCC). In Proceedings of the EGU General Assembly Conference Abstracts, Vienna, Austria, 27 April–1 May 2014.
10. Bongaarts, J. Intergovernmental Panel on Climate Change Special Report on Global Warming of 1.5 °C Switzerland: IPCC, 2018. *Popul. Dev. Rev.* **2019**, *45*, 251–252. [CrossRef]
11. Kim, S.J.; Choi, H.; Kim, B.; Park, S.; Shim, T.; Kim, J. Analysis of recent climate change over the Arctic using ERA-Interim reanalysis data. *Chin. J. Polar Res.* **2013**, *24*, 326–338.
12. Screen, J.A.; Deser, C.; Simmonds, I. Local and remote controls on observed Arctic warming. *Geophys. Res. Lett.* **2012**, *39*, 10709. [CrossRef]
13. Xu, W.; Li, Q.; Jones, P.; Wang, X.L.; Trewin, B.; Su, Y.; Chen, Z.; Zhai, P.; Wang, J.; Vincent, L. A new integrated and homogenized global monthly land surface air temperature dataset for the period since 1900. *Clim. Dyn.* **2017**, *50*, 2513–2536. [CrossRef]
14. Concepts, R. Fourth Assessment Report: Climate Change, 2007 (AR4). *PLoS ONE* **2014**, *6*, e16126.
15. Hua, W.; Chen, H.; Xing, L. Review of Land Use and Land Cover Change in China and Associated Climatic Effects. *Adv. Earth Sci.* **2014**, *29*, 1025–1036.
16. Sung, J.H.; Eum, H.-I.; Park, J.; Cho, J. Assessment of Climate Change Impacts on Extreme Precipitation Events: Applications of CMIP5 Climate Projections Statistically Downscaled over South Korea. *Adv. Meteorol.* **2018**, *2018*, 1–12. [CrossRef]
17. Forestieri, A.; Arnone, E.; Blenkinsop, S.; Candela, A.; Fowler, H.; Noto, L.V. The impact of climate change on extreme precipitation in Sicily, Italy. *Hydrol. Process.* **2018**, *32*, 332–348. [CrossRef]
18. He, C.; Liou, K.N.; Takano, Y.; Ling, P.Y.; Qi, Y.; Chen, F. Impact of Grain Shape and Multiple Black Carbon Internal Mixing on Snow Albedo: Parameterization and Radiative Effect Analysis. *J. Geophys. Res. Atmos.* **2018**, *123*, 1253–1268. [CrossRef]
19. Skiles, S.M.; Flanner, M.; Cook, J.M.; Dumont, M.; Painter, T.H. Radiative forcing by light-absorbing particles in snow. *Nat. Clim. Chang.* **2018**, *8*, 964–971. [CrossRef]
20. Dou, T.F.; Xiao, C.D. An overview of black carbon deposition and its radiative forcing over the Arctic. *Adv. Clim. Chang. Res.* **2016**, *7*, 115–122. [CrossRef]
21. Schmale, J.; Flanner, M.; Kang, S.; Sprenger, M.; Zhang, Q.; Guo, J.; Li, Y.; Schwikowski, S.; Farinotti, D. Modulation of snow reflectance and snowmelt from Central Asian glaciers by anthropogenic black carbon. *Sci. Rep.* **2017**, *7*, 40501. [CrossRef]
22. Liu, C.M.; Dang, S.Z.; Wang, Z.G.; Wang, S.G. Research Progress of Black Carbon in Snow and Ice. *South-North Water Transf. Water Sci. Technol.* **2012**, *10*, 44–51.

23. Liu, X.Q.; Wang, N.L.; Yao, T.D.; Xu, B.Q.; Zhao, H.B.; He, J.Q.; Guo, X.J. Carbonaceous aerosols in snow and ice in the Tibetan plateau. *Earth Sci. Front.* **2006**, *13*, 335–341.
24. Clarke, A.D.; Noone, K.J. Soot in the Arctic snowpack: A cause for perturbations in radiative transfer. *Atmos. Environ.* **1985**, *19*, 2045–2053. [CrossRef]
25. Doherty, S.J.; Warren, S.G.; Grenfell, T.C.; Clarke, A.D.; Brandt, R.E. Light-absorbing impurities in Arctic snow. *Atmos. Chem. Phys.* **2010**, *10*, 11647–11680. [CrossRef]
26. Warren, S.G.; Clarke, A.D. Soot in the Atmosphere and Snow Surface of Antarctica. *J. Geophys. Res. Atmos.* **1990**, *95*, 1811–1816. [CrossRef]
27. Aoki, T.; Kuchiki, K.; Niwano, M.; Kodama, Y.; Hosaka, M.; Tanaka, T. Physically based snow albedo model for calculating broadband albedos and the solar heating profile in snowpack for general circulation models. *J. Geophys. Res. Atmos.* **2011**, *116*, D11114. [CrossRef]
28. Chylek, P.; Johnson, B.; Damiano, P.A.; Taylor, K.C.; Clement, P. Biomass burning record and black carbon in the GISP2 Ice Core. *Geophys. Res. Lett.* **1995**, *22*, 89–92. [CrossRef]
29. Chylek, P.; Srivastava, V.; Cahenzli, L.; Pinnick, R.G.; Dod, R.L.; Novakov, T.; Cook, T.L.; Hinds, B.D. Aerosol and graphitic carbon content of snow. *J. Geophys. Res. Atmos.* **1987**, *92*, 9801. [CrossRef]
30. Dasch, J.M.; Cadle, S.H. Atmospheric Carbon Particles in the Detroit Urban Area: Wintertime Sources and Sinks. *Aerosol Sci. Technol.* **1989**, *10*, 236–248. [CrossRef]
31. Fily, M.; Bourdelles, B.; Dedieu, J.P.; Sergent, C. Comparison of in situ and Landsat Thematic Mapper derived snow grain characteristics in the alps. *Remote Sens. Environ.* **1997**, *59*, 452–460. [CrossRef]
32. Kang, S.C.; Zhang, Y.L.; Qian, Y.; Wang, H.L. A review of black carbon in snow and ice and its impact on the cryosphere. *Earth-Sci. Rev.* **2020**, *210*, 103346. [CrossRef]
33. Doherty, S.J.; Dang, C.; Hegg, D.A.; Zhang, R.D.; Warren, S.G. Black carbon and other light-absorbing particles in snow of central North America. *J. Geophys. Res.-Atmos.* **2014**, *119*, 12807–12831. [CrossRef]
34. Forsström, S.; Ström, J.; Pedersen, C.A.; Isaksson, E.; Gerland, S. Elemental carbon distribution in Svalbard snow. *J. Geophys. Res.* **2009**, *114*, D19112. [CrossRef]
35. Tang, X.G.; Liu, D.W.; Zhang, B.; Du, J.; Lei, X.C.; Zeng, L.H.; Wang, Y.D.; Song, K.S. Research on Hyperspectral Remote Sensing in Monitoring Snow Contamination Concentration. *Spectrosc. Spectr. Anal.* **2011**, *31*, 1318–1321.
36. Flanner, M.G.; Liu, X.; Zhou, C.; Penner, J.E.; Jiao, C. Enhanced solar energy absorption by internally-mixed black carbon in snow grains. *Atmos. Chem. Phys.* **2012**, *12*, 4699–4721. [CrossRef]
37. Namazi, M.; Salzen, K.V.; Cole, J.N.S. Simulation of black carbon in snow and its climate impact in the Canadian Global Climate Model. *Atmos. Chem. Phys.* **2015**, *15*, 18839–18882. [CrossRef]
38. Huang, J.; Fu, Q.; Zhang, W.; Wang, X.; Zhang, R.; Ye, H.; Warren, S.G. Dust and Black Carbon in Seasonal Snow Across Northern China. *Bull. Am. Meteorol. Soc.* **2011**, *92*, 175–181. [CrossRef]
39. Negi, H.S.; Kokhanovsky, A. Retrieval of snow albedo and grain size using reflectance measurements in Himalayan basin. *J. Cryosphere* **2011**, *5*, 203–217. [CrossRef]
40. Negi, H.S.; Kokhanovsky, A. Retrieval of snow grain size and albedo of western Himalayan snow cover using satellite data. *J. Cryosphere* **2011**, *5*, 831–847. [CrossRef]
41. Wei, Y.R. Remote Sensing Inversion and Spatiotemporal Variation of Black Carbon and Snow Grain Size in Typical Snow Regions of China Based on MODIS Data. Master Thesis, Lanzhou Jiaotong University, Lanzhou, China, June 2020.
42. Pan, H.Z.; Wang, J.; Li, H.Y. Retrieval Snow Albebo Using MODIS based on the ART Model. *J. Remote Sens. Technol. Appl.* **2015**, *30*, 1059–1065. [CrossRef]
43. Shi, Y.F. *Glaciers and the Environment in China: Present, Past and Future*, 1st ed.; Science Press: Beijing, China, 2000.
44. Chen, G.Y.; Li, D.L. Temporal-Spatial characteristics of Cumulative Snow Depth in Northeast China and Its vicinity. *Meteorol. Mon.* **2011**, *37*, 513–521.
45. Che, T.; Li, X. Spatial distribution and temporal variation of snow water resources in china during 1993–2002. *J. Glaciol. Geocryol.* **2005**, *27*, 64–67.
46. Ogren, J.A.; Charlson, R.J.; Groblicki, P.J. Determination of elemental carbon in rainwater. *Anal. Chem.* **1983**, *55*, 1569–1572. [CrossRef]
47. Grenfell, T.C.; Doherty, S.J.; Clarke, A.D.; Warren, S.G. Light absorption from particulate impurities in snow and ice determined by spectrophotometric analysis of filters. *Appl. Opt.* **2011**, *50*, 2037–2048. [CrossRef]
48. Xin, W.; Doherty, S.J.; Huang, J. Black carbon and other light-absorbing impurities in snow across Northern China. *J. Geophys. Res. Atmos.* **2013**, *118*, 1471–1492.
49. Hall, D.K.; Salomonson, V.V.; Riggs, G.A. Development of Methods for Mapping Global Snow Cover Using Moderate Resolution Imaging Spectroradiometer Data. *Remote Sens. Environ.* **1995**, *54*, 127–140. [CrossRef]
50. Hall, D.K.; Riggs, G.A.; Salomonson, V.V.; DiGirolamo, N.E.; Bayr, K.J. MODIS snow-cover products. *Remote Sens. Environ.* **2002**, *83*, 181–194. [CrossRef]
51. Li, Y.; Feng, X.Z.; Xiao, P.F.; Ye, N. Estimating per-pixel snow cover fraction from MODIS in typical area of Bayanbulak. *J. Nanjing Univ.* **2015**, *51*, 1022–1029.

52. Shao, D.H.; Li, H.Y.; Wang, J.; Hao, X.H.; Wang, R.K.; Ma, Y. Retrieval of Snow Albedo based on Multi-source Remote Sensing Data. *Remote Sens. Technol. Applic.* **2017**, *32*, 71–77.
53. Hao, X.H.; Wang, J.; Li, H.Y. Evaluation of the NDSI Threshold Value in Mapping Snow Cover of MODIS—A Case Study of Snow in the Middle Qilian Mountains. *J. Glac. Geocryol.* **2008**, *30*, 132–138.
54. Kokhanovsky, A.A.; Aoki, T.; Hachikubo, A.; Hori, M.; Zege, E.P. Reflective properties of natural snow: Approximate asymptotic theory versus in situ measurements. *IEEE Trans. Geosci. Remote Sens.* **2005**, *43*, 1529–1535. [CrossRef]
55. Kokhanovsky, A.A.; Zege, E.P. Scattering optics of snow. *Appl. Opt.* **2004**, *43*, 1589–1602. [CrossRef] [PubMed]
56. Kokhanovsky, A.; Rozanov, V.V.; Aoki, T.; Odermatt, D.; Brockmann, C.; Krüger, O.; Bouvet, M.; Drusch, M.; Hori, M. Sizing snow grains using backscattered solar light. *Int. J. Remote Sens.* **2011**, *32*, 6975–7008. [CrossRef]
57. Kokhanovsky, A.; Schreier, M. The determination of snow specific surface area, albedo and effective grain size using AATSR space-borne measurement. *Int, J Remote. Sens.* **2009**, *30*, 919–933. [CrossRef]
58. Zhao, C.; Hu, Z.; Qian, Y.; Ruby Leung, L.; Huang, J.; Huang, M.; Jin, J.; Flanner, M.G.; Zhang, R.; Wang, H.; et al. Simulating black carbon and dust and their radiative forcing in seasonal snow: A case study over North China with field campaign measurements. *Atmos. Chem. Phys.* **2014**, *14*, 11475–11491. [CrossRef]
59. Lv, X.; Xu, J.H.; Deng, K. Analysis of Temporal-Spatio Distribution of Annual Aerosol Optical Depth of Anhui Province. *Environ. Sci. Technol.* **2017**, *40*, 209–216.
60. Zhang, J.; Yuan, X.D.; Lin, H. The Extraction of Urban Built-up Areas by Integrating Night-time Light and POI Data—A Case Study of Kunming, China. *IEEE Access* **2021**, *9*, 22417–22429.
61. Li, J.F.; Pan, J.H. Spatial expansion of cities at county-level or above in Gansu Province from 1992 to 2012 based on DMSP nighttime light images. *J. Glaciol. Geocryol.* **2016**, *38*, 829–835.
62. Bi, X.Y.; Liu, F.; Chen, H.; Wu, D.; Deng, X.J.; Huang, J. Vertical Distribution Features of Atmospheric Stability over Beijing Area. *J. Trop. Meteorol.* **2003**, *19*, 173–179.
63. Grenfell, T.C.; Light, B.; Sturm, M. Spatial distribution and radiative effects of soot in the snow and sea ice during the SHEBA experiment. *J. Geophys. Res. Ocean.* **2002**, *107*, SHE 7-1–SHE 7-7. [CrossRef]
64. Grenfell, T.C.; Warren, S.G.; Radionov, V.F.; Makarov, V.N.; Zimov, S.A. Expeditions to the Russian Arctic to Survey Black Carbon in Snow. *Eos. Trans. Am. Geophys. Union* **2013**, *90*, 386–387. [CrossRef]
65. McConnell, J.R.; Edwards, R.; Kok, G.L.; Flanner, M.G.; Zender, C.S.; Saltzman, E.S.; Banta, J.R.; Pasteris, D.R.; Carter, M.M.; Kahl, J.D.W. 20th-Century Industrial Black Carbon Emissions Altered Arctic Climate Forcing. *Science* **2007**, *317*, 1381–1384. [CrossRef] [PubMed]
66. Kaspari, S.; McKenzie Skiles, S.; Delaney, I.; Dixon, D.; Painter, T.H. Accelerated glacier melt on Snow Dome, Mount Olympus, Washington, USA, due to deposition of black carbon and mineral dust from wildfire. *J. Geophys. Res. Atmos.* **2015**, *120*, 2793–2807. [CrossRef]
67. Ming, J.; Xiao, C.D.; Du, Z.C.; Flanner, M. Black Carbon in Snow/Ice of West China and Its Radiative Forcing. *Adv. Clim. Chang. Res.* **2009**, *5*, 328–335.
68. Wang, M.; Xu, B.Q.; Wu, G.J.; Yang, W. Carbonaceous Aerosols Recorded in Ice Core in Southeastern Tibetan Plateau. *Adv. Clim. Chang. Res.* **2010**, *6*, 175–180.
69. Xin, W.; Wei, P.; Yong, R.; Zhang, X.; Zhang, X.; Shi, J.; Jin, H.; Dai, M.; Chen, Q. Observations and model simulations of snow albedo reduction in seasonal snow due to insoluble light-Absorbing particles during 2014 Chinese survey. *Atmos. Chem. Phys.* **2017**, *17*, 2279–2296.

sustainability

MDPI

Article

Spatial Distribution of, and Variations in, Cold Regions in China from 1961 to 2019

Yumeng Wang [1], Jingyan Ma [2,*], Lijuan Zhang [1], Yutao Huang [1], Xihui Guo [1], Yiping Yang [1], Enbo Zhao [1], Yufeng Zhao [1], Yue Chu [1], Meiyi Jiang [1] and Nan Wang [1]

[1] Heilongjiang Province Key Laboratory of Geographical Environment Monitoring and Spatial Information Service in Cold Regions, Harbin Normal University, Harbin 150025, China; wangyumeng19980603@163.com (Y.W.); zlj19650205@163.com (L.Z.); huangyutao0128@163.com (Y.H.); 18235889220@163.com (X.G.); ep_yangyiping@163.com (Y.Y.); zeb030410@163.com (E.Z.); janekabesilsen@163.com (Y.Z.); cy254654@163.com (Y.C.); jiangmi199608@163.com (M.J.); wangnan199710@163.com (N.W.)

[2] Teaching Management and Teaching Office, Harbin Normal University, Harbin 150025, China

* Correspondence: 18686818800@163.com

Abstract: In this study, on the basis of the temperature data collected at 612 meteorological stations in China from 1961 to 2019, cold regions were defined using three indicators: an average temperature of <-3.0 °C during the coldest month; less than five months with an average temperature of >10 °C; and an annual average temperature of ≤ 5 °C. Spatial interpolation, spatial superposition, a trend analysis, and a spatial similarity analysis were used to obtain the spatial distribution of the cold regions in China from 1961 to 2019. Then, the areas of the cold regions and the spatial change characteristics were analyzed. The results reveal that the average area of the cold regions in China from 1961 to 2019 was about 427.70×10^4 km^2, accounting for about 44.5% of the total land area. The rate of change of the area of the cold regions from 1961 to 2019 was -14.272×10^4 km^2/10 a, exhibiting a very significant decreasing trend. On the basis of the changes between 1991–2019 and 1961–1990, the area of China's cold regions decreased by 49.32×10^4 km^2. The findings of this study provide references for studying changes in the natural environment due to climate change, as well as for studying changes on a global scale.

Keywords: cold region of China; spatiotemporal distribution; spatiotemporal variation; 1961–2019

Citation: Wang, Y.; Ma, J.; Zhang, L.; Huang, Y.; Guo, X.; Yang, Y.; Zhao, E.; Zhao, Y.; Chu, Y.; Jiang, M.; et al. Spatial Distribution of, and Variations in, Cold Regions in China from 1961 to 2019. *Sustainability* **2022**, *14*, 465. https://doi.org/10.3390/su14010465

Academic Editors: Xiaodong Yan, Jia Yang and Shaofei Jin

Received: 9 December 2021
Accepted: 29 December 2021
Published: 2 January 2022

1. Introduction

Cold regions are a valuable part of the earth system [1], and they are generally defined as areas with low temperatures [2] and with the presence of ice and snow for at least part of the year. Therefore, the ice and snow resources in cold regions are very rich and are very important to human production activities [3]. They are an essential and lasting factor that affects the natural and social systems on the Earth's surface [4]. Cold regions are very sensitive to climate change [5], and climate change has changed the distribution of the cold regions around the world [6–10]. The sixth report of the IPCC pointed out that the global annual average surface temperature has risen by 1.09 °C in the past 100 years, that the climate warming trend in China was much higher than the global average [11,12], and that the area and spatiotemporal distribution characteristics of China's cold regions are undergoing significant changes. However, at present, little attention has been paid to the temporal and spatial changes in the cold regions in China. Gaining an accurate understanding of the distribution of, and variations in, cold regions has important theoretical and practical significance for engineering projects in cold regions, industrial and agricultural production, as well as for the rational development and utilization of the water, ice, and snow resources in cold regions [13,14].

Different methods have been proposed for the classification of cold regions. Koppen et al. [15] was the first to propose a division index for cold regions. Two indices were used to classify the cold regions in Canada: (1) The average temperature of the coldest month is ≤ -3.0 °C; and (2) The number of months with an average monthly temperature of >10 °C is less than four. Gerdel et al. [16] suggested dividing the Canadian cold regions on the basis of the criterion of an annual average temperature of 0 °C and lower. However, Wilson et al. [17] reported that there were some problems with only considering the temperature factor and proposed the use of both temperature and precipitation for cold region division. Hamelin et al. [18] proposed 10 indicators to divide the Canadian cold regions. For the classification of the cold regions in China, Yang et al. [19] proposed climate indicators on the basis of the above studies. The indicators included: (1) The average temperature in the coldest month is below -3 °C; (2) The number of months with an average monthly temperature of above 10 °C is less than four; (3) The freezing period of rivers and lakes is more than 100 days, and more than 50% of the precipitation is solid precipitation; (4) The number of months with an average monthly temperature of >10 °C is no greater than five; (5) The average temperature in October and April is below 0 °C; (6) The annual average temperature does not exceed 5 °C; (7) The number of days with an average daily temperature of >10 °C is less than 150; (8) The accumulated temperature is 500–1000 °C; (9) The percentage of solid precipitation is greater than 30%; and (10) The average annual number of snow cover days is >30 days. They divided the cold regions in China on the basis of these 10 indicators [3]. Their results show that China's cold regions were mainly distributed in four main regions: (1) Gansu, Qinghai, and Xinjiang; (2) Tibet, Aba, and Ganzi, in western Sichuan, northern Yunnan, the Yulong Mountains, and the north part of the Gaoligong Mountains in Yunnan; (3) Northeast and northwest Heilongjiang; and (4) The northeastern part of Inner Mongolia, except for the desert areas in the Junggar Basin, the Tarim Basin, and the northern part of Heihe [3,19]. The data reveals that the cold regions accounted for about 43% of the land area of China. To consolidate the various indicators proposed by Yang et al. [19], Chen et al. [20] proposed three indicators: the average temperature in the coldest month is <-3.0 °C; the number of months with an average temperature of >10 °C is less than five; and the annual average temperature is ≤ 5 °C. Then, using these three indicators, Chen et al. [20] created a spatial distribution map of the cold regions in China from 1961 to 1998 based on observation data collected four times a day at 571 stations in China from 1961 to 1998. They reported that China's cold regions were mainly distributed in the Greater Khingan Mountains, in the Changbai Mountains, on the Sanjiang Plain in northeastern China, in the Hexi Corridor, in most of the mountainous areas in Xinjiang, and on the Qinghai-Tibetan Plateau. The cold regions covered an area of 417.4×10^4 km^2, which accounted for about 43.5% of the land area. Their results provide an important basis for studying China's permafrost, glaciers, stable seasonal snow, climatic divisions, and vegetation divisions, and these results are used to this day. However, 23 years have passed since Chen's study period ended (i.e., 1998). Since the beginning of the 21st century, the average annual temperature in China has risen by about 0.63 °C [21]. Therefore, it is time to revisit China's cold region divisions and to further analyze the spatial evolution characteristics of the cold regions within the context of global warming [22–24]. Currently, such studies are lacking.

In the present study, on the basis of the three indicators proposed by Chen et al. [20] and observation data collected at 612 meteorological stations in China from 1961–2019, the cold regions in China were identified, and the spatial distribution of the cold regions was analyzed in order to provide scientific references for development in the cold regions in China, and for studying the impacts of climate change in these cold regions.

2. Data and Methods

2.1. Data Sources

1. Temperature data: The indicators used in this study were all temperature indices. The temperature data were the monthly temperature values for 612 meteorological stations in China from January 1961 to December 2019, which were downloaded from the China Meteorological Data Network (http://data.cma.cn accessed on 21

September 2020). All the 612 stations have complete data availability and there are no gaps. The distribution of the meteorological stations is shown in Figure 1.

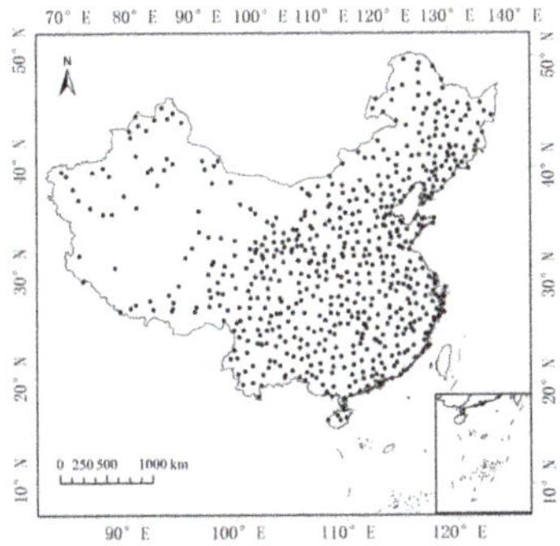

Figure 1. Distribution of meteorological stations in China.

2. Elevation data: The digital elevation model (DEM) data were obtained from the cloud geospatial database (http://www.gscloud.cn/ accessed on 20 September 2020), with a resolution of 1 km × 1 km.

2.2. *Methods*

2.2.1. Spatial Interpolation of Meteorological Data

The distribution of the meteorological stations in China is uneven. For example, in remote areas such as the Qinghai-Tibetan Plateau, there are only a limited number of stations. In order to obtain high-resolution meteorological data, spatial interpolation is often used. There are many available methods of interpolating meteorological data. Of these methods, the Kriging method is commonly used. The Kriging method performs interpolation using the distribution of the meteorological elements of nearby stations, on the basis of the principle of covariance. However, since the number of meteorological stations in cold regions is small and the spatial resolution is very low, the Kriging method is not suitable. Liu et al. [25] proposed an elevation-based spatial interpolation method, i.e., ANUSPLIN. In addition to ordinary spline independent variables, this method introduces linear covariate submodels, such as the relationship between the temperature and the elevation, precipitation, and coastline. The basic principle is to allow the introduction of multivariate and covariate linear submodels. The coefficients of the models can be automatically determined on the basis of the data. This method is able to smoothly process splines of more than two dimensions, and multiple influencing factors are introduced as covariates to perform the spatial interpolation of the meteorological elements. In addition, the temperature is affected by the altitude. Because only considering the relationship between the latitude and longitude and temperature compromises the accuracy, the terrain factor was introduced in this study as a covariate. The partial thin-plate smooth spline function of the elevation linear submodel was used for the spatial temperature interpolation. The calculation formula is as follows:

$$Z_i = f(x_i) + b^T y_i + e_i (i = 1, 2, \ldots, n) \tag{1}$$

where Z_i is the dependent variable at point *i*; x_i is a d-dimensional spline independent variable; f is the unknown smoothing function to be estimated; y_i is a p-dimensional independent covariate; b is the p-dimensional coefficient of y_i; e_i is the independent variable random error with an expected value of 0 and a variance of $w_i\sigma^2$; and w_i is the known local relative coefficient of variation, of which σ^2 is taken as the weight, and is the error variance, which is an unknown constant for all data points. When interpolating temperature, this paper uses a three-variable local thin-disk smooth spline function (longitude and latitude are independent variables, and altitude is a covariate), and the number of splines is set to 2.

2.2.2. Trend Analysis Method

In order to reflect the change trend of the cold area from 1961 to 2019, linear trend rate estimation was adopted:

$$\mathrm{y} = \mathrm{ax} + \mathrm{b} \tag{2}$$

In the formula, y is the area of the cold area; x is the year; a is the linear regression coefficient, reflecting the change trend of the area of the cold area, and a > 0 indicates that the area of the cold area is increasing, and a < 0 indicates that the area is decreasing; b is the intercept; and a × 10 is called the "climate tendency rate", and the unit is 10^4 km^2/10 a.

2.2.3. Spatial Similarity Analysis

The kappa coefficient is generally used to determine the degree of agreement or accuracy between two images, and its calculation formula is:

$$K = \frac{P_0 - P_e}{1 - P_e} \tag{3}$$

Among them, P_0 is the sum of the number of samples correctly classified in each category divided by the total number of samples, which is the overall classification accuracy. Assume that the numbers of real samples in each category are a_1, a_2, ..., a_c, and that the predicted numbers of samples in each category are b_1, b_2, ..., b_c, and that the total number of samples is n. Then:

$$P_e = \frac{a_1 \times b_1 + a_2 \times b_2 + \ldots + a_c \times b_c}{n \times n} \tag{4}$$

The kappa coefficient calculation results are −1–1, but usually the kappa falls between 0 and 1, which can be divided into five groups to indicate the different levels of consistency: 0.0–0.20, very low consistency; 0.21–0.40, general consistency; 0.41–0.60, moderate consistency; 0.61–0.80, high consistency; and 0.81~1 are almost identical [26].

2.2.4. Mann–Kendall Mutation Test

In addition to the trend analysis, the MK method can also be used to test for mutation. This method is very effective for verifying a change of state from a relatively stable state to another state. For a time series, x, with n sample sizes, construct an order column:

$$S_k = \sum_{i=1}^{k} r_i \ (k = 2, 3, \ldots, n) \tag{5}$$

where

$$r_i = \begin{cases} 1, \ x_i > x_j \\ 0, \ x_i \leq x_j \end{cases} \qquad (j = 1, 2, \ldots, i) \tag{6}$$

It can be seen that the rank sequence, S_k, is the cumulative number of times the value of i at the moment, i, is greater than the number of values at time, j. Under the assumption of the random independence of the time series, define statistics:

$$UF_k = \frac{[S_k - E(S_k)]}{\sqrt{Var(S_k)}} \quad (k = 1, 2, \ldots, n) \tag{7}$$

where $UF_1 = 0$; and $E(S_k)$ and $Var\ (S_k)$ are the mean and variance of the cumulative number, S_k, respectively. This value is calculated when x_1, x_2, ..., x_n are independent, and when they have the same continuous distribution as:

$$E(S_k) = \frac{n(n-1)}{4} \quad (8)$$

$$Var(S_k) = \frac{n(n-1)(2n+5)}{72} \quad (9)$$

UF_i is a standard normal distribution, which is a sequence calculated according to a time series (x) order ($x_1, x_2, \ldots, x_n$). Given a significance level, a, in comparison with the data in the known normal distribution table, and if $UF_i > Ua$, then significant changes exist in the trend. This method can also be applied to the inverse sequence of the time series, and the above procedure can be repeated by $x_n, x_{n-1}, \ldots, x_1$, thus making $UF_k = -UB_k$, $k = n$, $n - 1, \ldots$, and $UB = 0$. Given the significance level, α, the two curves of UF_k and UB_k and the significant horizontal line are plotted on the same graph. If the values of UF_k and UB_k are greater than 0, then the sequence shows an upward trend, and values below 0 indicate a downward trend. When the value exceeds the critical line, this indicates that the rising or falling trend is significant. The range beyond the critical line is defined as the time zone of mutation. If the UF_k and UB_k curves appear on an intersection point, and the intersection point is between the critical line, then the intersection point corresponds to the time the mutation begins. More detailed descriptions of this method are introduced in [27].

3. Results and Analysis

3.1. Time Series Changes in the Area of Cold Regions in China from 1961 to 2019

Once the data layers corresponding to the three indicators were obtained, the three layers were superimposed to extract the overlapping area, which is the spatial distribution map of the cold regions in China from 1961 to 2019. The area of the cold regions from 1961 to 2019 was calculated, and its variation with time was obtained (Figure 2a). It can be seen that the average area of the cold regions in China from 1961 to 2019 was 427.70×10^4 km^2. The largest area of the cold regions occurred in 1969 (485.92×10^4 km^2), and the smallest area occurred in 2007 (368.80×10^4 km^2). The coefficient of variation was 0.07, indicating that the interannual variation was about 7%.

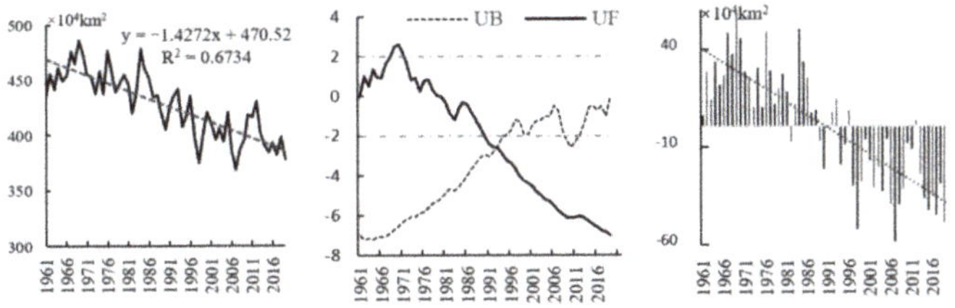

(**a**) (**b**) (**c**)

Figure 2. (**a**) Interannual changes; (**b**) MK test; and (**c**) Anomaly in the area of cold regions in China from 1961 to 2019.

The rate of change of the area of the cold regions in China from 1961 to 2019 was -14.272×10^4 km^2/10 a, suggesting a significant decreasing trend ($p < 0.01$). Between 1961 and 2019, the area of the cold regions decreased by 84.20×10^4 km^2. The Mann–Kendall (MK) test results (Figure 2b) reveal that there have been no significant sudden changes in the area of the cold regions since 1961. However, on the basis of the area anomaly map for 1961 to 2019 (Figure 2c), the area of the cold regions in China experienced a significant turning point around 1987. From 1961 to 1987, the average area of the cold regions was 453.64×10^4 km^2, and from 1988 to 2019, the area decreased to 405.81×10^4 km^2, i.e., a reduction of 47.83×10^4 km^2. The results of the analysis of variance revealed that there was a significant difference in the area of the cold regions in China between 1961–1987 and 1988–2019 ($p < 0.05$). Thus, the area of the cold regions in China entered a declining stage in 1987. Furthermore, the rate of change of the area of the cold regions from 1961 to

1987 was -5.697×10^4 km^2/10 a, indicating an insignificant decreasing trend; whereas the rate of change from 1988 to 2019 increased to -11.688×10^4 km^2/10 a, and this change was significant ($p < 0.01$). In conclusion, the area of the cold regions in China has not only entered a relatively low value period since 1987, but it has also decreased significantly.

3.2. Spatial Distribution of the Cold Regions in China from 1961 to 2019

Figure 3 shows the spatial distribution of the cold regions in China from 1961 to 2019, including northeastern China, northern China, northwestern China, and southwestern China. In order to analyze the spatial distribution of China's cold regions, the cold regions were analyzed according to the administrative divisions. The cold regions were distributed in 14 provinces and autonomous regions, including Heilongjiang, Jilin, Liaoning, the Inner Mongolia Autonomous Region, Hebei, Shanxi, Shaanxi, Gansu, Ningxia, Qinghai, Sichuan, Yunnan provinces, the Xinjiang Autonomous Region, and the Tibet Autonomous Region. If you superimpose the boundaries of each administrative region on the cold area spatial distribution map calculated in this paper, using the ArcGIS Tabulate Area function, you can get the cold area of each administrative area. The areas of the cold regions in the provinces and autonomous regions are listed in Table 1, from high to low. It can be seen that the Tibet Autonomous Region contained the largest area of cold regions (105.06×10^4 km^2), which was significantly larger than that of the other provinces and autonomous regions. Moreover, the areas of the cold regions in the Inner Mongolia Autonomous Region, the Xinjiang Autonomous Region, Qinghai, and Heilongjiang were also large, ranging from 54.09×10^4 km^2 to 75.04×10^4 km^2, which were significantly larger than those in Sichuan, Gansu, Jilin, and other provinces. By comparison, the cold regions in Hebei, Yunnan, Liaoning, the Ningxia Hui Autonomous Region, and Shaanxi were relatively small. Shaanxi had the smallest area of 0.32×10^4 km^2. The entire Tibet Autonomous Region was a cold region, except for the southern part of the Nyainqen Tanglha Mountains. In the Inner Mongolia Autonomous Region, the cold regions were mainly distributed in the northeastern part of the Inner Mongolia Plateau, and in the Yinshan, Langshan, Daqingshan, and Helan mountains. In the Xinjiang Autonomous Region, the cold regions were mainly located in the Bogda Mountains in the northern part of Hami, in the Tianshan Mountains, in the Bolokonu Mountains west of Urumqi, in the Harke Mountains in the southern part of the Tianshan Mountains, and in the Altai Mountains on the Sino-Mongolian border. All of the areas in Qinghai, except for the Qaidam Basin, were cold regions. The cold regions in Heilongjiang included the Greater Khingan Mountains, the Lesser Khingan Mountains, the Changbai Mountains, the Sanjiang Plain, and the Songnen Plain. The cold regions in Sichuan Province included the high mountains in northwestern Sichuan, such as the Daxue Mountains and the Qionglai Mountains. The cold regions in Gansu Province were mainly distributed in the Qilian Mountains in the west, in the Minshan Mountains in the Longnan area, in the northern mountains in the Hexi Corridor, in the Longzhong Plateau area, and in the Mazong Mountains in the northern part of Gansu Province. The cold regions in Jilin Province were mainly located in the Changbai Mountains in the eastern part of Jilin Province and the areas east of the Hada Mountains. The cold regions in Hebei were mainly distributed in Hengshan and Yanshan in the northwest. The cold regions in Shanxi Province were mainly distributed in the high mountains, including in the Taihang Mountains, the Lvliang Mountains, the Wutai Mountains, and the Heng Mountains. The cold regions in Yunnan Province were scattered in the southern part of the Hengduan Mountains in the northwest. The cold regions in the Ningxia Hui Autonomous Region were mainly distributed in the northern part of the Liupan Mountains. In Shaanxi, the cold regions were scattered in the middle of the Qinling Mountains.

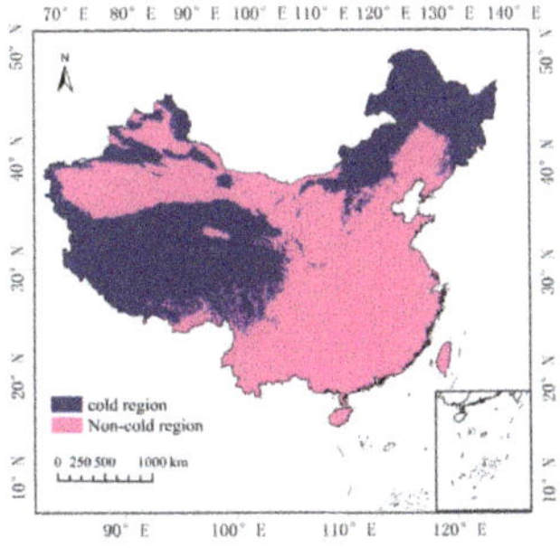

Figure 3. Spatial distribution of cold regions in China from 1961 to 2019.

Table 1. Area of cold regions in each administrative division from 1961 to 2019 ($\times 10^4$ km^2).

Administrative District	Area of Cold Region	Administrative District	Area of Cold Region	Administrative District	Area of Cold Region
Tibet Autonomous Region	105.06	Sichuan	19.85	Yunnan	1.10
Inner Mongolia Autonomous Region	75.04	Gansu	16.26	Liaoning	0.86
Xinjiang Autonomous Region	69.00	Jilin	11.82	Ningxia	0.50
Qinghai	66.82	Hebei	3.95	Shaanxi	0.32
Heilongjiang	54.09	Shanxi	3.03		

3.3. Spatial Variations in the Cold Regions in China from 1961 to 2019

The World Meteorological Organization (WMO) defines the climate reference value (i.e., climatic state) as the average value of a certain meteorological element for 30 years, which is considered to be able to represent the climate of a location [28]. Thus, the 30-year average value has been used as the reference climate state in scientific research on climate and climate change. In this study, the periods of 1961–1990 and 1991–2019 were defined as two climatic states. The statistics show that the areas of the cold regions in China in 1961–1990 and 1991–2019 were 453.18 $\times$ 10^4 km^2 and 403.86 $\times$ 10^4 km^2, respectively. It can be seen that, against the background of global warming, the area of China's cold regions has been decreasing (i.e., by 49.32 $\times$ 10^4 km^2). Table 2 shows the reduction in the areas of the cold regions in each administrative division. The results show that the areas of the cold regions in all 14 administrative regions decreased. Specifically, the reductions of 12.23 $\times$ 10^4 km^2, 10.62 $\times$ 10^4 km^2, and 8.09 $\times$ 10^4 km^2 in Inner Mongolia, Xinjiang, and Jilin Province, respectively, were relatively large. The areas of the cold regions in Yunnan, the Ningxia Hui Autonomous Region, and Shaanxi Province decreased slightly, by 0.44 $\times$ 10^4 km^2, 0.38 $\times$ 10^4 km^2, and 0.27 $\times$ 10^4 km^2, respectively.

Table 2. Reduction in areas of cold regions in each administrative division ($\times 10^4$ km^2).

Administrative District	Area of Cold Region	Administrative District	Area of Cold Region
Inner Mongolia Autonomous Region	12.23	Sichuan	1.25
Xinjiang Autonomous Region	10.62	Liaoning	1.21
Jilin	8.09	Heilongjiang	1.17
Qinghai	4.05	Hebei	1.12
Gansu	3.81	Yunnan	0.44
Shanxi	2.39	Ningxia	0.38
Tibet Autonomous Region	1.46	Shaanxi	0.27

On the basis of the spatial distribution maps of the cold regions in 1961–1990 (Figure 4a) and 1991–2019 (Figure 4b), the kappa values of the spatial distribution maps of the two climatic states were both 0.934, indicating excellent consistency. That is, the spatial distribution of China's cold regions did not change significantly, but there were some differences. Figure 4c shows the difference in the spatial distributions of the cold regions in China in 1991–2019 and 1961–1990. From Figure 4c, the cold regions that disappeared in the second 30-year period (1991–2019) were mainly concentrated in the central and northern parts of Jilin Province, in the central-northern parts of the Inner Mongolia Autonomous Region, in the Altai region in Xinjiang, and in the central part of the Qaidam Basin. Scattered reductions also occurred in the Hengshan, Wutai, and Yunzhong mountains in the northern part of Shanxi Province, in the southern part of the Qinling Mountains in Shaanxi Province, in the southeastern part of the Tibet Autonomous Region, in Beishan, on the Longzhong Plateau in Gansu Province, in the Qionglai Mountains and Jiajin Mountains in Sichuan Province, in the Longgang Mountains and Qianshan Mountains in Liaoning Province, in the Taihang Mountains in Hebei Province, in Southern Duanyunling in Yunnan Province, and in the northern Liupan Mountains in the Ningxia Hui Autonomous Region.

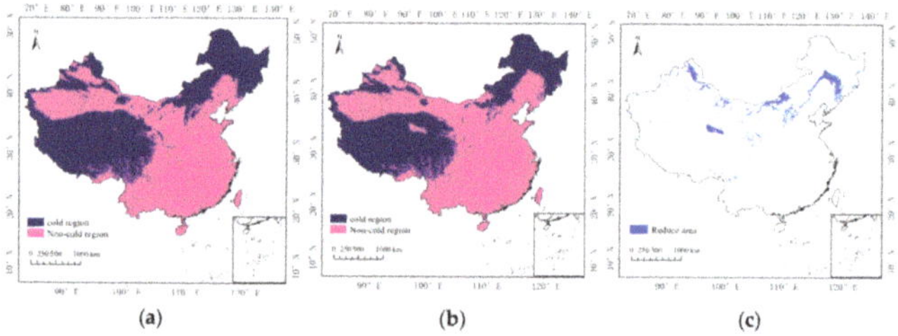

Figure 4. Spatial distributions of cold regions: (**a**) 1961–1990; (**b**) 1991–2019; and (**c**) Difference.

4. Discussion

Currently, the spatial distribution of the cold regions in China is primarily based on the results reported by Chen et al. [20]. However, they delineated the spatial distribution of China's cold regions from 1961 to 1998 without considering the variation characteristics of the cold regions. In this study, we used the observation data collected at meteorological stations in China from 1961 to 2019 to analyze the spatial distribution of China's cold regions, and we analyzed the time series changes in the area of the cold regions and the variations in their spatial distribution. The analysis of the impact of climate change on cold regions is of great scientific significance and practical value for development in the cold regions in China, and for the rational development and utilization of the water, ice, and snow resources in these cold regions.

The data period of Chen et al. [20] was from 1961 to 1998, and the data used was from 571 meteorological stations in China. In this study, the data period was from 1961 to 2019, and the data used was from 612 meteorological stations in China. In order to compare our results with those of Chen et al. [20], the cold region distribution map of Chen et al. [20] (Figure 5b) was vectorized, and the area of the distribution map was compared with that of this study (Figure 5a). The comparison results are shown in Figure 5c. The cold area obtained in this study was 427.70 × 10^4 km^2, accounting for 44.5% of the total area of China. The cold area obtained by Chen et al. [20] was 417 × 10^4 km^2, accounting for 43.5% of the total area. Thus, there was a difference of 10.3 × 10^4 km^2. As can be seen from Figure 5c, the results of this study reveal that the areas of the cold regions increased in the Qianshan and Longgang mountains in the northeastern part of Liaoning Province, in the Hengshan, Wutai, and Yunzhong mountains in Shanxi Province, in the Altay region in Xinjiang, in the Tianshan, Altun, Qilian, and Daxue Mountains, and in the Duanyun Mountains and the Mangkang Mountains on the southern Qinghai-Tibetan Plateau. In contrast, the areas of the cold regions decreased in the southern Greater Khingan Mountains, on the southern Songnen Plain, in Shanding Hural in the Inner Mongolia Autonomous Region, and in the Qaidam Basin in Qinghai. Thus, the cold regions identified in this study are different from those identified by Chen et al. [20] because these studies were based on different research periods.

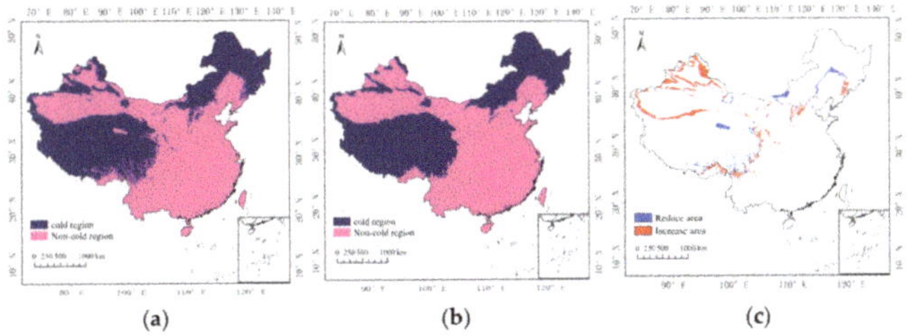

(a) (b) (c)

Figure 5. Comparison of spatial analysis results of cold regions in China: (**a**) 1961–2019; (**b**) The cold areas obtained by Chen et al.; and (**c**) Difference.

In order to compare our results with the research results of Chen Rensheng, this paper recalculated the distribution of cold regions in China from 1961 to 1998, as shown in the Figure 6 below. It can be seen that the spatial distribution map of China's cold regions calculated in this paper has higher spatial resolution than the research results of Chen Rensheng, so the statistical area should be more accurate. In addition, different spatial interpolation methods were used in the two studies, which will also make the statistical cold area different from the research results of Chen Rensheng. The main difference between the two results is due to the different time scales. More importantly, the research period of this paper is 1961–2019, which is 23 years longer than that of 1961–1998. With the extension of the research period, the spatial superposition area of the three indicators for dividing the cold region will also increase. Therefore, the area of China's cold region will increase from 1961 to 2019.

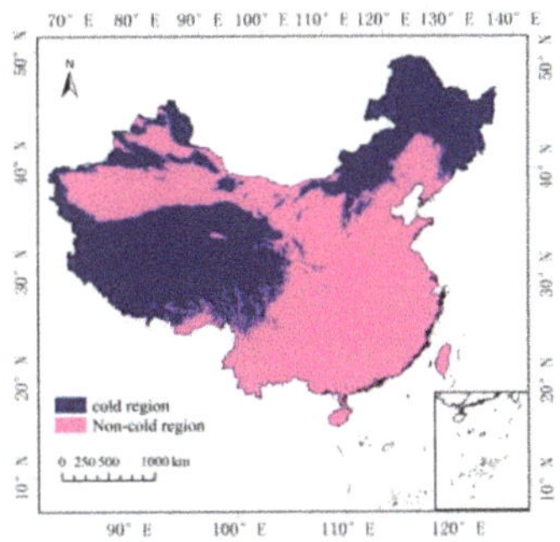

Figure 6. Spatial distribution of cold regions in China from 1961 to 1998.

On the basis of the area anomaly map and the results of the 5-year moving average, the year when the area anomaly became negative was 1987. Since 1987, the area of the cold regions has decreased significantly. According to the fifth assessment report of the International Panel on Climate Change (IPCC), the temperature increase became more significant after the 1980s [29]. Moreover, on the basis of the temperature data collected at 349 meteorological stations in China from 1953 to 2012, Liu et al. [30] found that the temperature increase in China during this period was significant, and that most of the years with abrupt changes were after 1986. Before the mid-1980s, the temperature in China fluctuated within a relatively small range. However, since then, the temperature has exhibited a significant upward trend. On the basis of the above discussion, it can be concluded that China's surface temperature underwent major changes in the mid-1980s.

Therefore, it is concluded that the temperature change in the mid-1980s was the main influencing factor of the change in the area of the cold regions in China.

By comparing the areas of the cold regions in China under two climatic states, it was found that the area of the cold regions in northeastern China decreased by 13.62×10^4 km^2, and that this was the region with the largest decrease. The results of many studies have shown that northeastern China is sensitive to climate change, and that it has experienced a significant temperature increase. For example, Chen et al. [31] studied climate change in China from 1951 to 1995 and they report that the temperature increases in China mainly occurred north of 35° N, with the largest temperature increase occurring in northern Heilongjiang. Haiying et al. [32] analyzed the spatial and temporal characteristics of the climate change in China from 1900 to 2000, and they found that the overall climate had been cooling since 1950, but that warming occurred in the northeastern, northern, and northwestern regions. The annual average temperature in northeastern China increased by about 1 °C during 1900–2000, especially from 1981 to 1998, when there was a steep jump. In addition, Liang et al. [33] analyzed the temperature characteristics in northern China from 1951 to 2014. They also found that the northeastern region had the largest rate of temperature increase, and this was also the region with the fastest temperature increase after the sudden change in temperature across the country. Therefore, the northeastern region had the largest decrease in the area of the cold regions.

The range of China's cold regions defined in this article is mainly distributed in northern Xinjiang, northeastern China, and on the Qinghai-Tibet Plateau, and this is the main distribution area of glaciers, frozen soil, and snow [34]. According to the results of this article, the area of China's cold regions is decreasing, and studies have shown that, with climate warming, glaciers shrink, the permafrost, as a whole, degenerates, snow cover shrinks, glaciers melt and shrink faster, and meltwater increases year by year. The changes in the temporal and spatial distributions of the water resources and water cycle processes caused by changes in glaciers will undoubtedly have a profound impact on the social and economic development of cold regions [35]. For example, they could lead to an increase in the glacier meltwater runoff, causing thermal melt slump, thermal melt subsidence, and other permafrost thermal melt disasters. As glacier retreat intensifies, the amount of meltwater increases, and glacier floods and glacial debris flow disasters increase with the increase in the glacier meltwater runoff [36]. Therefore, the reduction in the area of the cold region has a great impact on the ice layer, on ecosystems, and on human activities.

5. Conclusions

From 1961 to 2019, the area of the cold regions in China was about 427.70×10^4 km^2, accounting for about 44.5% of China's total land area. The rate of change in the area of the cold regions was -14.272×10^4 km^2/10 a, exhibiting a very significant decreasing trend. The area of the cold regions of China decreased significantly after 1987.

The cold regions in China were mainly distributed in the Greater Khingan Mountains, in the Changbai Mountains, on the Sanjiang Plain in northeastern China, in the central part of the Inner Mongolia Plateau, and in most of the mountainous areas in Xinjiang, and on the Qinghai-Tibetan Plateau. Among the 14 provinces and autonomous regions in China, the area of the cold regions in the Tibet Autonomous Region was the largest (105.06×10^4 km^2), and the area of the cold regions in Shaanxi Province was the smallest (0.32×10^4 km^2).

The areas of the cold regions in China during 1961–1990 and 1991–2019 were 453.18 $\times 10^4$ km^2 and 403.86×10^4 km^2, respectively. The difference between the two climatic states was 49.32×10^4 km^2. The largest decrease in the area of the cold regions occurred in the Inner Mongolia Autonomous Region (12.23×10^4 km^2). The difference in the areas of the cold regions during the two periods was mainly distributed in northeastern China, in Xinjiang, and in Inner Mongolia.

Author Contributions: Y.W. analyzed the data and drafted the manuscript; L.Z., J.M., and Y.H. completed the manuscript and made major revisions; X.G. and Y.Y. completed the data visualization;

E.Z. and Y.Z. downloaded data and searched references; Y.C., N.W., and M.J. checked and proofread the manuscript. All authors have read and agreed to the published version of the manuscript.

Funding: This research was funded by the National Natural Science Foundation of China (Grant No. 41771067, and supported by the Key Project of Natural Science Foundation of Heilongjiang Province, No. ZD2020D002.

Institutional Review Board Statement: Not applicable.

Informed Consent Statement: Not applicable.

Data Availability Statement: The data presented in this study are available on request from the corresponding author.

Conflicts of Interest: The authors declare no conflict of interest.

References

1. Shen, H. *Colde Regions Science and Marine Technology*; Eolss Publishers Company Limited: Oxford, UK, 2015.
2. Gelfan, A.N.; Motovilov, Y.G. Long-term Hydrological Forecasting in Cold Regions: Retrospect, Current Status and Prospect. *Geogr. Compass* **2009**, *3*, 1841–1864. [CrossRef]
3. Yang, Z.N. *Glacier Water Resources in China*; Gansu Science and Technology Press: Lanzhou, China, 1991; pp. 1–5.
4. Vaughan, D.G.; Comiso, J.C.; Allison, I.; Carrasco, J.; Kaser, G.; Kwok, R.; Mote, P.; Murray, T.; Paul, F.; Ren, J. Observations: Cryosphere. *Clim. Chang.* **2013**, *2103*, 317–382.
5. Hu, Z.Y.; Kuenzer, C.; Dietz, A.J.; Dech, S. The Potential of Earth Observation for the Analysis of Cold Region Land Surface Dynamics in Europe—A Review. *Remote Sens.* **2017**, *9*, 1067. [CrossRef]
6. Farquharson, L.M.; Romanovsky, V.E.; Cable, W.L.; Walker, D.A.; Kokelj, S.V.; Nicolsky, D. Climate change drives widespread and rapid thermokarst development in very cold permafrost in the Canadian High Arctic. *Geophys. Res. Lett.* **2019**, *46*, 6681–6689. [CrossRef]
7. Nitzbon, J.; Westermann, S.; Langer, M.; Martin, C.P.L.; Strauss, J.; Laboor, S.; Boike, J. Fast response of cold ice-rich permafrost in northeast Siberia to a warming climate. *Nat. Commun.* **2020**, *11*, 1–11. [CrossRef]
8. Ran, Y.; Li, X.; Cheng, G. Climate warming over the past half century has led to thermal degradation of permafrost on the Qinghai–Tibet Plateau. *Cryosphere* **2018**, *12*, 595–608. [CrossRef]
9. Gobiet, A.; Kotlarski, S.; Beniston, M.; Heinrich, G.; Rajczak, J.; Stoffel, M. 21st century climate change in the European Alps—A review. *Sci. Total Environ.* **2014**, *493*, 1138–1151. [CrossRef]
10. Maurer, J.M.; Schaefer, J.M.; Rupper, S.; Corley, A. Acceleration of ice loss across the Himalayas over the past 40 years. *Sci. Adv.* **2019**, *5*, eaav7266. [CrossRef]
11. IPCC. 2021: Summary for Policymakers. In *Climate Change 2021: The Physical Science Basis. Contribution of Working Group I to the Sixth Assessment Report of the Intergovernmental Panel on Climate Change*; Cambridge University Press: Cambridge, UK, 2021.
12. Yan, Z.W.; Ding, Y.H.; Zhai, P.M.; Song, L.C.; Cao, L.J.; Li, Z. Re-assessing climatic warming in China since the last century. *Acta Meteorol. Sin.* **2020**, *78*, 370–378. (In Chinese) [CrossRef]
13. Yang, Z.N.; Zeng, Q.Z. *Glacier Hydrology*; Chongqing Press: Chongqing, China, 2001; p. 48. (In Chinese)
14. Yang, Z.N. *Research on Cold Regions Hydrology in China*; Science Press: Bejing, China, 1997; pp. 50–56. (In Chinese)
15. Koppen, W. Das Geographische System der Klimate. In *Hand-Buch der Klimatologie*; Borntrager: Berlin, Germany, 1936; Volume 1, Part C.
16. Gerdel, R.W. *Characteristics of the Cold Regions*; Cold Regions Research and Engineering Laboratory, US Army Corps of Engineers: Hanover, NH, USA, 1969; p. 51.
17. Wilson, C. *Cold Regions Climatology*; Cold Regions Research and Engineering Laboratory, US Army Corps of Engineers: Hanover, NH, USA, 1967; p. 147.
18. Hamelin, L.E. *Canadian Nordicity: It's Your North Too*; Harvest House: Eugene, OR, USA, 1979; p. 373.
19. Yang, Z.; Liu, X.; Zeng, Q. *Hydrology in Cold Regions of China*; Science Press: Beijing, China, 2000; pp. 2–21. (In Chinese)
20. Chen, R.S.; Kang, E.S.; Wu, L.Z.; Yang, J.P.; Ji, X.B.; Zhang, Z.B. Cold Regions in China. *J. Glaciol. Geocryol.* **2005**, *27*, 469–475. (In Chinese) [CrossRef]
21. Su, B.D.; Sun, H.M.; Li, X.C.; Li, Z.J.; Zhang, J.P.; Wang, Y.J.; Huang, J.L.; Gao, M.N.; Jiang, T.; Si, L.L. Impact of climate change on terrestrial water cycle in China. *Trans. Atmos. Sci.* **2020**, *43*, 1096–1105. (In Chinese) [CrossRef]
22. Ding, Y.J.; Liu, S.Y.; Ye, B.S.; Zhao, L. Climatic Implications on Variations of Lakes in the Cold and Arid Regions of China during the Recent 50 Years. *J. Glaciol. Geocryol.* **2006**, *28*, 623–632. (In Chinese)
23. Li, B.Q.; Xiao, W.; Wang, Y.C.; Sun, Q.Y.; Liu, Z.Y. Recent research progress in the hydrological cycle model of cold regions. *J. Southwest Univ. Natl.* **2018**, *44*, 338–346. (In Chinese)
24. Wang, Z.C.; Zhang, Y.; Zhang, Y.; Yang, C.Y.; Yuan, K. Experimental Research on Intelligent Energy—Saving Ventilation System in Cold Regions. *Energy Conserv. Technol.* **2021**, *39*, 117–122. (In Chinese)

25. Liu, Z.H.; Li, L.T.; McVicar, T.R.; Van Niel, T.G.; Yang, Q.R.; Li, R. Introduction of the Professional Interpolation Software for Meteorology Data: ANUSPLIN. *Meteorol. Mon.* **2008**, 92–100. (In Chinese) [CrossRef]
26. Guo, Y.B.; Guo, W.; Qin, Y.C.; He, Q.; Zhang, X.J.; Wu, C. Consistency check based on Kappa coefficient and its software implementation. *Chin. J. Health Stat.* **2016**, *33*, 169–170. (In Chinese)
27. Mann, H.B. Nonparametric test against trend. *Econometrica* **1945**, *13*, 245–259. [CrossRef]
28. Lin, J.J.; Zhang, Q. Characteristics of China Climate States Change and Its Impact on the Analysis of Climate Change. *Plateau Meteorol.* **2015**, *34*, 1593–1600. (In Chinese) [CrossRef]
29. Shen, Y.P.; Wang, G.Y. Key findings and assessment results of IPCC WGI Fifth Assessment Report. *J. Glaciol. Geocryol.* **2013**, *35*, 1068–1076. (In Chinese) [CrossRef]
30. Liu, F.; Wu, L.M.; Chen, L.C.; Wang, K.; Ma, J.H.; Zhang, J.L. Trend and abrupt change of air temperature in China in recent 60 years based on GIS. *Agric. Sci. Technol. Inf.* **2015**, *6*, 74–78. (In Chinese) [CrossRef]
31. Chen, L.X.; Zhu, W.Q.; Wang, W.; Zhou, X.J.; Li, W.L. Study on climate change in China in recent 45 years. *Acta Meteorol. Sin.* **1998**, *56*, 2–16. (In Chinese) [CrossRef]
32. Hai, Y.; Gao, Z.Q. The analysis of temporal and spatial features of climate change in the last 100 years in China. *Bull. Sci. Technol.* **2010**, *26*, 58–62. (In Chinese) [CrossRef]
33. Liang, L.T.; Ma, L.; Liu, Y.X.; Sun, B.L.; Zhou, Y. Spatiotemporal variation of the temperature mutation and warming hiatus over northern China during 1951~2014. *China Environ. Sci.* **2018**, *38*, 1601–1615. (In Chinese) [CrossRef]
34. Kang, S.C.; Guo, W.Q.; Wu, T.H. Cryospheric changes and their impacts on water resources in the Belt and Road regions. *Adv. Earth Sci.* **2020**, *35*, 1–17. (In Chinese) [CrossRef]
35. Yao, T.D.; Qin, D.H.; Shen, Y.P.; Zhao, L.; Wang, N.L.; Lu, A.X. Cryospheric changes and their impacts on regional water cycle and ecological conditions in the Qinghai-Tibetan Plateau. *Chin. J. Nat.* **2013**, *35*, 179–186. (In Chinese)
36. Shen, Y.P.; Su, H.C.; Wang, G.Y. The responses of glaciers and snow cover to climate change in Xinjiang (II): Hazards effects. *J. Glaciol. Geocryol.* **2013**, *35*, 1355–1370. (In Chinese)

sustainability

MDPI

Article

Snow-Disaster Risk Zoning and Assessment in Heilongjiang Province

Hao Li, Wenshuang Xi, Lijuan Zhang * and Shuying Zang *

Heilongjiang Province Key Laboratory of Geographical Environment Monitoring and Spatial Information Service in Cold Regions, Harbin Normal University, Harbin 150025, China; killerlight9023@163.com (H.L.); xws202112@163.com (W.X.)
* Correspondence: zlj@hrbnu.edu.cn (L.Z.); zsy6311@163.com (S.Z.)

Abstract: Heilongjiang Province is located in the northeast region of China, with the country's highest latitude. It has long and cold winters, and a temperate monsoon climate. Its unique geographic location and climatic conditions make it the second largest stable snow-covered region in China. The winter snow period starts in October and ends in April of the following year. Therefore, the long-term accumulation of snow causes road obstructions and low-temperature frost damage, which seriously affects local economic development and human safety. This study adopts snow parameters (e.g., snow depth and snow-cover period), natural environmental factors (e.g., elevation and slope), and socioeconomic factors (e.g., gross domestic product and light index). On the basis of the disaster risk assessment theory, we constructed a disaster risk index from four aspects (i.e., disaster risk, susceptibility, vulnerability, and disaster prevention and mitigation). Then, we performed snow-disaster risk zoning and an assessment in Heilongjiang Province. The main findings are as follows: the snow-disaster risk in the northern and eastern regions of Heilongjiang Province was high; the central and northern regions were highly sensitive to disasters; the main urban areas were highly vulnerable; and the economically developed regions had strong disaster prevention and mitigation capabilities. Overall, the spatial distribution of the snow-disaster risk followed a decreasing trend from east to west. High-risk areas were distributed in the east and northwest (covering 34.3% of the entire Heilongjiang Province area); medium-risk areas were distributed in the north and center (accounting for 45.2% of the entire Heilongjiang Province area); and low-risk areas were concentrated in the west (constituting 20.5% of the entire Heilongjiang Province area).

Keywords: snow disaster; risk assessment; risk zoning; Heilongjiang Province

Citation: Li, H.; Xi, W.; Zhang, L.; Zang, S. Snow-Disaster Risk Zoning and Assessment in Heilongjiang Province. *Sustainability* **2021**, *13*, 14010. https://doi.org/10.3390/su132414010

Academic Editors: Jia Yang, Xiaodong Yan and Shaofei Jin

Received: 28 October 2021
Accepted: 14 December 2021
Published: 18 December 2021

1. Introduction

Since the 1990s, the focus of emergency management has gradually turned from emergency rescue and postevent recovery and reconstruction, to prepreventive preparations [1]. The United States, Britain, Germany, France, Japan, and other countries have promoted major disaster risk assessments. In 2004, the United Nations International Strategy for Disaster Reduction (UNISDR) and the United Nations Development Programme (UNDP) released two reports, respectively titled: "Living with Risk: A Global Review of Disaster Reduction Initiatives", and "Disaster risk reduction: a development concern". In March 2015, the "Sendai Framework for Disaster Risk Reduction 2015–2030", finally adopted by the Third World Conference on Disaster Risk Reduction, pointed out that two of the four priorities for disaster reduction are "Understanding disaster risk" and "Strengthening disaster risk governance to manage disaster risk" [2,3]. A snow disaster is large-scale snow, caused by heavy snowfall, which seriously affects the survival and health of humans and livestock. It is a meteorological disaster that can affect and damage traffic, communications, agriculture, and electricity [4]. In 1977, the snow disaster in Xilingol (Inner Mongolia) caused more than 70% of livestock deaths [5]. From 2000 to 2012, the agriculture sector of Liaoning Province suffered 20 large-scale disasters resulting from snowstorms [6]. In 2008, the 100-year snow disaster swept over half of China, and affected circuits, communications, the water supply, and heating, to different degrees. According to statistics, the snow disaster caused 129 deaths; the emergency resettlement of 1.66 million people; 485,000 house collapses; and 178 million mu of crop damage; and resulted in direct economic losses of CNY 151.65 billion. The heavy snow in Heilongjiang Province in 2007 caused 754 houses to collapse. Local snowstorms affected more than 900 people in Suifenhe City and other places, with direct economic losses of nearly CNY 100 million [7]. It is evident that snow disasters have had a huge impact on human society. Therefore, snow-disaster risk assessment provides an important theoretical basis for predisaster preparations, as well as for scientifically setting the meteorological disaster prevention standards for various regional infrastructures, which is more conducive to promoting the construction of resilient cities and villages, and to fundamentally improving the ability to resist snow disasters.

Snow disasters occur in various forms, such as snow melting leading to avalanches and ice floods, less (more) snow leading to black (white) disasters, and abnormal snowfall in windy weather leading to snowstorms. The existing studies on snow-disaster risk mainly focus on three aspects: (1) Studies on single snow-disaster risks. Some scholars have evaluated the avalanche risk. Schmitt et al. assessed the avalanche risk on the Alps on the basis of topographical parameters, such as slope, aspect and elevation, forestry-related variables, and rocks [8]. Seliverstov et al. [9] carried out avalanche risk zoning in Russia on the basis of the recurrence interval of avalanches (avalanche frequency), the percentage of the whole investigated territory that is occupied by avalanche-prone areas, the duration of the avalanche danger period, the probability of a person's stay in an avalanche-prone area for 1 day (24 h) and for 1 year, and the total population of the area and its density. Cappabianca et al. [10] presented an avalanche risk estimation procedure that combines a statistical analysis of the snowfall record, iterative simulations of avalanche dynamics, and empirically based vulnerability relations. Germain et al. [11] reconstructed past avalanche events in the north of the Gaspe Peninsula on the basis of tree rings and assessed the avalanche risk. In 2016, Germain also analyzed the avalanche risk from natural factors, the population, and environmental factors for avalanche disaster in northern Quebec, eastern Canada [12]. Some scholars have studied the risk of snowstorm disasters. Zhang et al. [13] calculated the probability of each level of the snowstorm information diffusion theory on the basis of the daily snow precipitation in 63 cities and counties in Heilongjiang Province, China, from 1961 to 2015, and established a hazard index model by using the snowstorm probability and the amount of snow precipitation. Next, a hazard assessment and the regionalization of snowstorms was performed for Heilongjiang Province from 1961 to 2015, and it was proposed that the high-risk areas increased by 30.7% from the 1960s to the 2010s, as opposed to the 38.9% reduction in the low-risk areas. Liu et al. [14] constructed the snowstorm disaster risk index on the basis of the frequency of snowstorms and assessed and regionalized the snowstorm risk in northeast China in the future (2020–2099). They showed that, under the RCP4.5 and RCP8.5 scenarios, the areas of the high-risk and light-risk regions would increase, while the areas of the low- and medium-risk regions would decrease. (2) Studies on the impact of snow disasters. Snow disasters have a major impact on agriculture, animal husbandry, and road transportation. Barbolini et al. [15], on the basis of avalanche accidents that occurred during outdoor winter activities over the Italian Alps in the last 20 years, proposed a vulnerability relation for people directly exposed to avalanches. Sinickas et al. [16] discussed the effects of the long-term changes in the avalanche occurrence rates in terms of consequences and vulnerability. Casteller et al. [17] report on the relationship between Nothofagus broadleaf forests and avalanche runout distances. Gao [18] constructed an agricultural risk estimation index and divided the risks of China's ice and snow on agriculture. Their results propose that the regions with high snow and freezing occurrences are located in northwestern China, and that the regions with the high-loss areas are located in the coastal and southeastern parts. Sa et al. [19] established a snow-cover index using passive microwave remote sensing data from 1978 to 2012, and evaluated the risk factors of disaster in the pastoral areas of the Inner Mongolia

Grasslands. Liu et al. [20] constructed the snow-disaster index and the vulnerability index by using the highway network, social and economic development data, meteorology data, and ArcGIS software of Guoluo Prefecture, Qinghai Province, combined with the analytic hierarchy process (AHP) and cluster analysis, to calculate the snow-road-disaster risk index of Guoluo Prefecture. Qi et al. [21] proposed the risk assessment index system of highway snow cover on the basis of a theoretical model and put forward a distribution of trunk highway snow-disaster risks in Shaanxi. Leone et al. [22] evaluated the impact of avalanches on road traffic in three Alpine departments (Alpes-de-Haute-Provence, Hautes-Alpes, and Alpes-Maritimes). Snow disasters can also amplify the effects of other disasters, such as earthquakes [23]. (3) Studies on the comprehensive risk regionalization of snow disasters. Wang et al. [24] further analyzed the formation mechanism of snow disasters (SDs), and constructed the integrated risk index (IRI) of a snow disaster from the aspects of historical disasters, snowfall events, disaster-formation environments, livestock overload, and livestock vulnerability and adaptability. The regions with high IRIs are mainly concentrated in the middle, east, and southwest of the Qinghai-Tibet Plateau, and appear as a contiguous risk belt from northeast to southwest. Liu et al. [25] conducted a comprehensive analysis of the 18 indexes of snow disaster on the Qinghai-Tibet Plateau, based on the hazard harmfulness data collected from historical records and the data collected from the entities affected by this hazard in 2010, and classified the snow-disaster hazards. Gao et al. [26], on the basis of the logistic regression of 33 snow-disaster events in Qinghai province, with the maximum snow depths, snow-cover days (SCDs), slopes, annual average temperatures, and per capita gross domestic product (GDP), constructed a potential risk assessment model to regionalize the snow-disaster risk in Qinghai Province. Park et al. [27] selected the pressure index (PI), the state index (SI), and the response index (RI) to assess the comprehensive risk of snow disaster in the metropolitan city of Ulsan. Insang [28] also assessed and analyzed the snow disaster risks of Daegu City, Ulsan City, Gyeongsangbuk Province, and Gangwon Province on the basis of the subindicators of three hazards, six exposures, four vulnerabilities, and five adaptive capacities.

In summary, scholars have conducted a lot of studies on snow-disaster risk and its impact. In addition to studies on the risk of a single snow disaster, studies on snow-disaster risk regionalization, from the perspectives of risk, sensitivity, vulnerability, and disaster prevention and mitigation, have gradually increased in recent years. However, the existing risk studies of snow disaster are mostly based on snow-cover data, and the risk indicators of snowstorms and wind-blown snow are seldom considered. Therefore, the risk of snow disaster in this paper comprehensively considers snow, snowstorms, and wind-blown snow. Compared with previous studies, the index selection has certain characteristics and is comprehensive. Heilongjiang Province is the highest-latitude area in China, belonging to the second largest snow-covered area in China. It is different from other snow-covered areas because of its large annual average snow reserves and an obvious interannual variability [29]. Studies have shown the increasing intensity of winter snowfall in Heilongjiang Province in recent years [30]. The years with abnormal snowfall have also increased, especially in the first decade of the 21st century. Extreme winter precipitation events in Heilongjiang Province have frequently occurred, and the precipitation values in 2002, 2003, 2009, and 2010 were over 50% more than those in normal years [31]. Heilongjiang Province is the region with the highest average annual wind speed in China [32]. The secondary disasters caused by wind and snow are greater [7], and the persistence of snow disasters is more significant than in other regions. Therefore, we selected Heilongjiang Province as the study area. According to the natural disaster risk theory model, we established a snow-disaster risk index according to the disaster risk, susceptibility, vulnerability, and the disaster prevention and mitigation capability. Combined with geographic information system (GIS) spatial analysis tools, the weighted comprehensive evaluation method, and the analytic hierarchy process, this study classifies the risk zoning of snow disasters in Heilongjiang Province. It provides a quantitative reference for Heilongjiang province from which to determine the snow-disaster prevention standard, make a disaster prevention plan, and assess the losses after a disaster. It is also useful for carrying out snow-disaster risk assessments and regionalization-related research.

2. Materials and Methodology

2.1. Study Area

Heilongjiang Province, located in northeastern China (121°11′–135°05′ E, 43°26′–53°33′ N), is the most northernly territory of the country (Figure 1). It borders Russia in the north, along Heilongjiang River. Characterized by a temperate continental monsoon climate, it has long cold winters and short summers. The annual average temperature ranges from −4 °C to 5 °C, from north to south of the study area. Its total annual precipitation is around 400–650 mm, which gradually decreases from east to west. The average annual wind speeds are about 2–4 m/s. Heilongjiang Province is an important grain production base in China, with commercial grain accounting for about one-tenth of the national total.

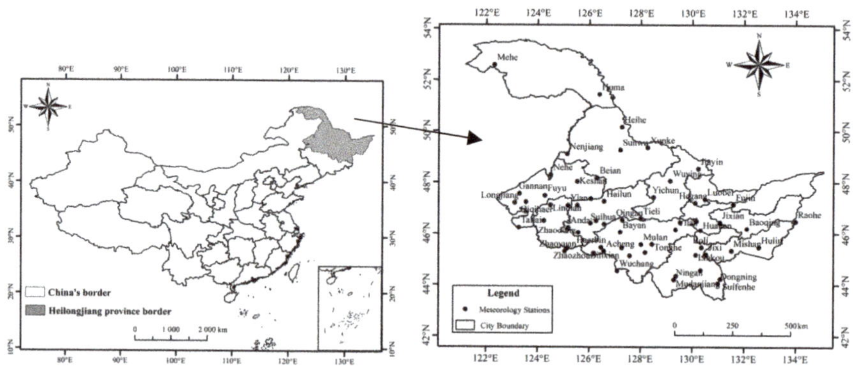

Figure 1. The study area and distribution of 62 meteorological stations in Heilongjiang Province, China.

2.2. Datasets

Precipitation and wind speed data: The daily meteorological data used in this study were provided by the Heilongjiang Meteorological Data Center. The earliest meteorological station in Heilongjiang Province was built in 1951, and 83 stations were established later. They have been divided into national standard meteorological stations and general meteorological stations. According to the continuity of data and the number of stations required for analysis, we selected 62 stations with continuous and complete observation data since 1961. The spatial distribution of the stations is shown in Figure 1. The basic data used in the study include the daily temperatures and precipitation data. The time scale is from 1983 to 2020. The study period is winter, which is defined as taking place from November of the current year to February of the second year.

Elevation, slope, and other data: The elevation, slope, and lighting data were generated by visual interpretation from the Resource and Environment Science and Data Center (www.resdc.cn. Accessed on 9 February 2021), with a spatial resolution of 90 m. The terrain standard deviation was calculated on the basis of the elevation.

Normalized difference vegetation index (NDVI) data: The NDVI data were downloaded from the National Aeronautics and Space Administration (NASA) website, from November of the current year to February of the following year, with a resolution of 1 km × 1 km. The time period is from 1996 to 2020, which is the winter average of the past five years.

Nighttime light index data: This study used national stable light images. Briefly, a stable light image is an annual raster image that calibrates the average nighttime light intensity. Each image included the lights of cities, towns, and the permanent light sources of other places (excluding the influence of occasional noises, such as moonlight clouds, firelight, and oil and gas burning). The pixel DN value of the image represents the average light intensity, and its range was 0–63. The nighttime light index data were downloaded from the Geographical Information Monitoring Cloud Platform (http://www.dsac.cn/DataProduct/Detail/201116. Accessed on 13 January 2021).

Socioeconomic factors: The socioeconomic factors (e.g., gross domestic product [GDP], per capita disposable income, and education level) used in this study were from the Heilongjiang Statistical Yearbook 1996–2020 (Accessed on 10 January 2021).

2.3. *Methods*

On the basis of the basic theory of disaster risk assessment, this paper constructs a theoretical model of a snow-disaster risk assessment. The trend analysis method is used to analyze the time characteristics of the risk factors; the analytic hierarchy process is used to determine the weights of the risk-influencing factors, and the comprehensive weighted evaluation method is used to form a risk index model.

2.3.1. Basic Theory of Disaster Risk Assessment

According to the theory of natural disaster risk formation, meteorological disaster risk is formed by four parts: risk, susceptibility (hazard-bearing object), vulnerability (hazard-bearing object), and the disaster prevention and mitigation capability. Each factor is composed of a series of subfactors. Its expression is:

$$\text{Disaster risk index} = f\,(\text{risk, susceptibility, vulnerability, disaster prevention and mitigation capability}) \quad (1)$$

The risk factors: All meteorological factors that may cause disasters can be called "meteorological hazards", and most of the meteorological hazards that exist in disaster-causing environments are abnormalities of certain natural phenomena, temporal and spatial laws, or natural substances, or a certain abnormality in the process of energy exchange. Generally, the greater the risk of meteorological factors, the greater the risk of meteorological disasters.

Susceptibility of the disaster-bearing object: The disaster-bearing object is the target of the hazard-causing factor and is the entity that bears the disaster. The susceptibility of the hazard-bearing object is the property shown by the hazard-bearing individual exposed to the hazard-causing agent and is the result of the interaction between the hazard-causing factor and the hazard-bearing object.

Vulnerability of disaster-bearing object: Disasters can only become disasters when they act on corresponding objects, namely, humans and their social and economic activities. Specifically, a disaster-bearing object refers to all objects that may be threatened by hazards in a given dangerous area, as well as to the degree of damage or loss caused by potential hazards, which comprehensively reflects the degree of loss of meteorological disasters.

The disaster prevention and mitigation capability refers to the various management measures and countermeasures that are used to prevent and mitigate meteorological disasters, including management capabilities, disaster-reduction investment, resource preparation, etc. With the proper management measures and strong management capabilities, the smaller the potential loss that may be suffered, and the smaller the risk of meteorological disasters.

On the basis of the above theory, we built a hierarchical analysis model for snow-disaster risk assessment (Figure 2). Please refer to Sections 3.1–3.4 for the selection basis of the specific indicators.

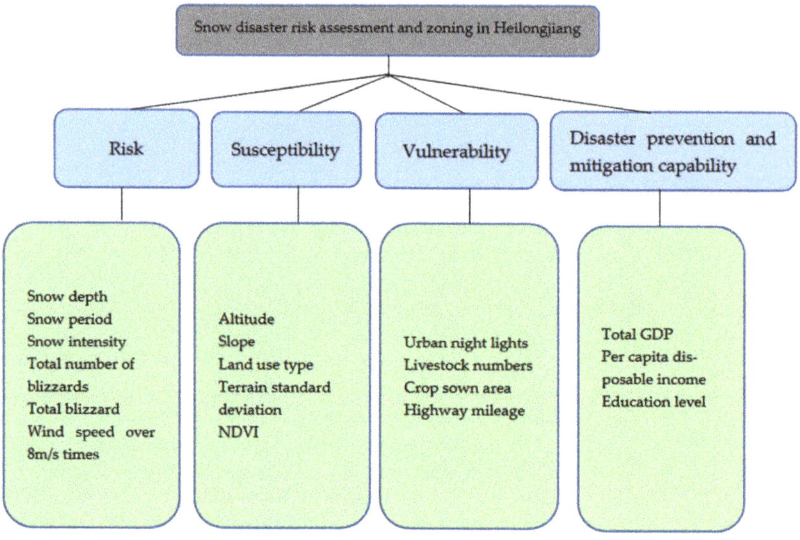

Figure 2. Hierarchical model of snow-disaster risk assessment.

2.3.2. Weighted Comprehensive Evaluation Method

The weighted comprehensive evaluation method is a method for solving the "bottom-up" indexes in the risk hierarchy analysis and the evaluation model [33]. This method comprehensively considers the degree of influence of each factor on the overall object and integrates the advantages and disadvantages of each specific index, using a numerical index to concentrate the data, indicating the pros and cons of the entire evaluation object. Therefore, this method is particularly suitable for the comprehensive analysis, evaluation, and optimization of technologies, strategies, or programs, and is currently one of the most commonly used calculation methods. Its expression is:

$$Y_{\rm i} = \sum_{j=1}^{m} \lambda_{ij} X_{ij} \quad i = 1,2,3,4; \quad j = 1,2, \ldots, m \tag{2}$$

In the formula, Y_i represents the disaster risk index, and i represents, respesctively, the risk, susceptibility, vulnerability, and disaster prevention and mitigation capabilities; X_{ij} is the factor that affects the risk, susceptibility, vulnerability, and disaster prevention and mitigation capability; and λ_{ij} is the weight value of the risk, susceptibility, vulnerability, and disaster prevention and mitigation capability ($0 \leq \lambda_j \leq 1$).

The expression for the comprehensive risk index of natural disasters is:

$$Y = \sum_{i=1}^{n} W_i Y_i \quad i = 1,2,3,4 \tag{3}$$

In the formula, Y represents the comprehensive disaster risk index; Y_i is the risk index, the susceptibility index, the vulnerability index, and the disaster prevention and mitigation capability index; and W_i is the weight value. The stronger the disaster prevention and mitigation ability, the smaller the comprehensive risk index and, thus, the "negative sign" is used.

The λ_j and W_i are determined by the analytic hierarchy process (see research method in Section 2.3.3 for details); each factor in the formula needs to be standardized because of different dimensions (see research method in Section 2.3.4 for details).

2.3.3. Analytic Hierarchy Process

The analytic hierarchy process (AHP) is a simple method for making decisions on some more complex and vague problems, especially for those problems that are difficult to fully quantitatively analyze [34]. This paper uses the operation principle of the analytic hierarchy process and uses the 1–9 scale method, provided by Saaty, to construct the judgment matrix for the pairwise relationships of the influence factors. The pairwise comparison of all of the influence factors determines the weight of each influence factor, which avoids the result error caused by the subjectivity of the expert. The qualitative comparison scale values between the two influencing factors are shown in Table 1 below.

Table 1. Scale of the AHP method.

Scale b_{ij}	Definition
1	The i factor is as important as the j factor.
3	The i factor is slightly more important than the j factor.
5	The i factor is more important than the j factor.
7	The i factor is much more important than the j factor.
9	The i factor is absolutely more important than the j factor.
2, 4, 6, 8	Between the noted levels.

Solve the maximum eigenvector value of the judgment matrix and its corresponding eigenvector by the sum-product method and check the consistency of the matrix (the following formula). After passing, solve it by the sum-product method.

$$CI = \frac{\lambda_{max} - n}{n-1} = \frac{-\sum_{i=1}^{n} \lambda_i}{n-1} \quad (4)$$

$$CR = \frac{CI}{RI} < 0.1 \quad (5)$$

In the formula, CI is the consistency index of the judgment matrix; $\lambda_{max}$is the largest characteristic root of the matrix; n is the order of the discrimination matrix; CR is the random consistency index of the judgment matrix; RI is the average random consistency index of the discrimination matrix. The RIs of the risk, susceptibility, vulnerability and the disaster prevention and mitigation capability are 1.24, 1.12, 0.9, and 0.58, respectively.

According to the analytic hierarchy process, taking risk as an example, the process of determining the weight of each factor is as follows:

The hazard factors include the snowfall intensity, snow depth, snow-period length, snowfall days, blizzard volume, blizzard number, and the days with wind speeds over 8 m/s. It is believed that, according to the degree of importance, the relative hazard factors of the blizzard amount are the most important, followed by the number of blizzards, the snowfall intensity, the snow depth, the snow-cover period, the snowfall days, and the wind speeds. According to the relative importance of each factor, the judgment matrix is constructed according to Table 1. The maximum characteristic root of the matrix, λ_{max}, is 6.242, CR = 0.038 < 0.10, and the weight result is valid. Therefore, the weights of the snow intensity, snow depth, snow period, the total blizzard volume, the total number of blizzards, and wind speeds over 8 m/s are 0.3848, 0.3054, 0.1648, 0.0718, 0.0468, and 0.0264, respectively.

In the same way, the weights of the susceptibility, vulnerability, disaster prevention and mitigation capability, and the comprehensive disaster risk factors are shown in Table 2.

Table 2. The weights of the risk, susceptibility, prevention and mitigation capacity, and comprehensive risk.

Sensitivity	**Elevation 0.4635**	**Slope 0.282**	**NDVI 0.1448**	**TSD 0.0727**	**Land Use Type 0.037**
Susceptibility	UNL 0.4742	Highway mileage 0.303	livestock numbers 0.154	crop sown area 0.0689	
Prevention and mitigation capacity	Total GDP 0.4885	Education level 0.1994	Per capita disposable income 0.3121		
Comprehensive risk	Risk 0.4627	Susceptibility 0.3272	Vulnerability 0.1357	Prevention and mitigation capacity 0.0744	

Note: TSD (terrain standard deviation); UNL (urban night lighting).

2.3.4. Standardization

In the process of zoning, because of the different dimensions of the selected factors, the magnitude of the difference is large. For example, the length of the snow-cover period is 150 days, and the average number of snowfalls is about 10 times per year. Therefore, when calculating the hazard-factor risk index, it needs to be normalized so that the value of each factor is between 0 and 1. When assessing the hazard risk, the hazard-bearing body susceptibility, the hazard-bearing body vulnerability, the disaster prevention and mitigation capability, the relationships between the selected influencing factors and the hazard-causing factor risk, the hazard-bearing body susceptibility, the hazard-bearing body vulnerability, and the disaster prevention and mitigation capability are different. Some are the greater the number of influencing factors, the greater the hazard risk, the greater the susceptibility of the hazard-bearing body, the greater the vulnerability of the hazard-bearing body, and the greater the ability of disaster prevention and mitigation, while some factors are the opposite. Therefore, in the evaluation process, the maximum value standard or the minimum value standardization are performed for ostentation, and an example of the formula is as follows: The greater the snowfall intensity, the greater the hazard risk of the hazard factor. Therefore, the maximum value of the snowfall intensity is standardized, and Formula (6) is selected; if the highway mileage is longer, the vulnerability is smaller and, thus, the highway mileage is minimized. For the value standardization, select Formula (7).

Maximum standardization:

$$X'_{max} = \frac{|X_{ij} - X_{min}|}{X_{max} - X_{min}} \tag{6}$$

Minimum standardization:

$$X'_{min} = \frac{|X_{max} - X_{ij}|}{X_{max} - X_{min}} \tag{7}$$

where X_{ij} is the index number of the j-th factor of the x factor; X'_{max} and X'_{min} are the dimensionalities of X_{ij}; and X_{max} and X_{min} are the minimum and maximum values in the index sequence, respectively.

2.3.5. Mann–Kendall Trend Test

In order to test the change trend of the meteorological elements, the Mann–Kendall trend test method is used to test the meteorological elements on the annual and seasonal scales [35]. The method assumes that the time data series $(x_1, x_2, \ldots, x_n)$ are independent, random, and uniformly distributed. For the sample sequence, the calculation equation for the test statistical variable, S, is:

$$S = \sum_{i=2}^{n} \sum_{j=1}^{i-1} sign(x_i - x_j) \tag{8}$$

when:

$$(x_i - x_j) = \left\{ \begin{array}{c} > 0 \\ = 0 \\ < 0 \end{array} \right\}, sign(x_i - x_j) = \left\{ \begin{array}{c} 1 \\ 0 \\ -1 \end{array} \right\} \quad (9)$$

$$\sigma s = \sqrt{\frac{n(n-1)(2n+5)}{18}} \quad (10)$$

when:

$$S\left\{ \begin{array}{c} > 0 \\ = 0 \\ < 0 \end{array} \right\}, Z = \left\{ \begin{array}{l} \frac{S-1}{\sigma s} \\ 0 \\ \frac{S+1}{\sigma s} \end{array} \right\} \quad (11)$$

where S is the test statistical variable of the normal distribution; x_i and x_j are two series of different distributions in the same sample, where $1 \leq j < I \leq n$; σ_S is the standard deviation; n is the total number of samples; and Z is the test value. If $Z > 0$, the tested time series has an upward trend, and if $Z < 0$, the tested time series has an upward and downward trend; the absolute value of Z is greater than or equal to 2.32, 1.64, and 1.28, which means they pass confidence. They are, respectively, 99%, 95%, and 90% significance test levels.

2.3.6. The Pettitt Test

The Pettitt test is a common tool to detect a single unknown mutation point, and it is also one of the most common nonparametric test methods [36]. It can be described as giving an observation data sequence, $X \mid x_t$, $t = 1, 2, \ldots, n$, where n is the sample size, if the sequence has a change point at $\tau(1 \leq \tau \leq n-1)$, defining the corresponding Pettitt statistics, U_T.

$$U\tau, n = \sum_{i=1}^{\tau} \sum_{j=\tau+1}^{n} sgn(x_i - x_j) = \left\{ \begin{array}{l} -1(x_i - x_j) < 0 \\ 0(x_i - x_j) = 0 \\ +1(x_i - x_j) > 0 \end{array} \right. \quad (12)$$

Let $K = \max(|U_{\tau,n}|)$, then the time, T, where K is located, is the possible change point position. The significance level of the corresponding change point is:

$$P = 2exp\left[-6K^2(n^3 + n^2)\right] \quad (13)$$

If $P \leq 0.5$, then time, T, is considered to be the position of the statistically significant change point.

3. Results

3.1. Snow-Disaster Risk Analysis

3.1.1. The Characteristics of the Risk Zoning Factors of Snow Disasters in Heilongjiang Province

In this paper, the snow depth, the length of the snow-cover period, the snowfall intensity, the total blizzard volume, the total number of blizzards, and the total days of wind speeds over 8 m/s were selected as the risk factors of snow disaster. The deeper the snow depth, the longer the snow-cover period, the stronger the snowfall intensity, the greater the blizzards, the greater the total number of blizzards, and the more days with wind speeds over 8 m/s, the greater the risk of snow disasters. Figure 3 shows the interannual variation characteristics of the snow depth, the snow-period length, the snow intensity, the total blizzard volume, the total number of blizzards, and the days with wind speeds over 8 m/s in Heilongjiang Province, from 1983 to 2020. It can be seen that the snow depth, snow intensity, total blizzard volume, and the total number of blizzards in Heilongjiang Province show a significant increase, and the M–K statistics are 2.869, 3.269, 1.968, 2.014, respectively. The length of the snow-cover period and the wind speeds showed a significant decrease, and the M–K statistical values were −3.223 and −6.43 (Table 3). It can be seen from the distribution of the box map (Figure 4) that the snow depth, the snow-period length, the snow intensity, the total blizzard volume, the total number of blizzards, and the days with average wind speeds over 8 m/s are 20.82 cm, 165.4 d, 1.24 cm/time, 142.0 mm, 9.9 times, and 92 d, respectively. The interannual changes in the wind speed, snow intensity, the number of blizzards, and the snow depth are relatively large, while the interannual changes in the snow depth and the blizzard volume are relatively small. It can be seen from the extreme value distribution characteristics that the maximum value of the snow intensity, the maximum value of the blizzard volume, and the number of blizzards are obviously larger than the average values, while the minimum value of the days with wind speeds over 8 m/s is obviously smaller than the average value. Using the Pettitt method to calculate the mutation year of each hazard factor, it was concluded that the snow depth, the snow-cover length, the snow intensity, and the days with wind speeds over 8 m/s days in Heilongjiang Province all have abrupt changes around the early 2000 s, and the catastrophes of the total snow cover and the total number of snowstorms occur in the middle and late parts of the 2000 s, respectively.

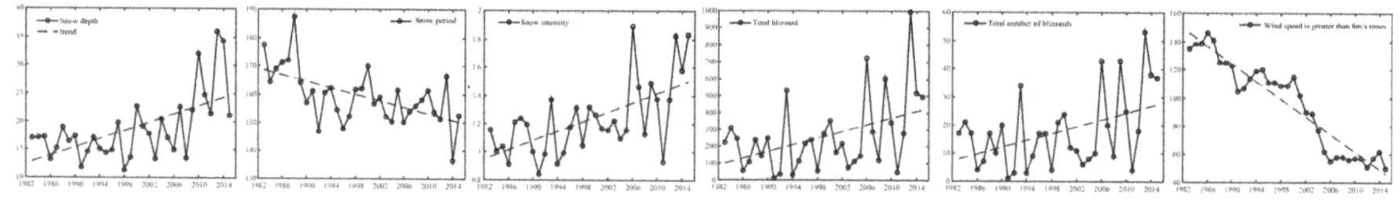

Figure 3. Time variation of risk factors.

Table 3. The M–K statistics and abrupt years of risk factors.

Disaster-Causing Factors	Snow Depth	Snow Period	Snow Intensity	Total Blizzard Volume	Total Number of Blizzards	Wind Speed Is Greater than 8 m/s Times
M–K statistics	2.867 **	−3.223 **	3.269 **	1.968 *	2.014 *	−6.43 **
Abrupt year	1999	2002	1998	2005	2008	2000

Note: "*", "**": significance at 0.05 and 0.01 levels, respectively.

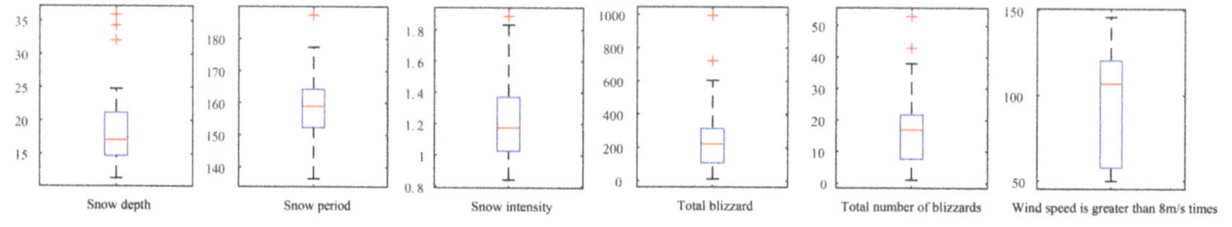

Figure 4. Boxplots of risk factors.

Figure 5 shows the spatial distribution of the snow depth, the snow-period length, the snow intensity, the total blizzard volume, the total number of blizzards, and the days with wind speeds over 8 m/s in Heilongjiang Province, from 1983 to 2020. It can be seen that the snow depths of Heilongjiang Province, from 1983 to 2015, are between 7.39 and 35.26 cm, and that the snow depths of 25–35.26 cm are mainly distributed in Mohe County in the northwest, northwest of Tahe County, northwest of Huzhong District, Xunke County, Jiayin County, north of Yichun City, north of Hegang City, and north of Luobei County; snow depths of 15–25 cm are mainly distributed in most cities and counties in central and southern Heilongjiang Province; snow depths of 7–15 cm are mainly distributed in the southwest of Heilongjiang Province, including Qiqihar cities and counties, Daqing, and the southwest of Suihua. The length of the snow period in Heilongjiang Province from 1983 to 2015 is 150–200.3 days. Snow periods of more than 180 days are mainly distributed in the Daxing'anling area, in the north of Heilongjiang Province; the areas of snow periods of more than 160 days are mainly distributed in central and northern Heilongjiang Province; the areas with a snow-cover period of less than 160 days are mainly distributed in the southern part of Heilongjiang Province. The length of the snow-cover period of the province generally increases from low latitudes to high latitudes. The snow intensity in Heilongjiang Province from 1983 to 2020 was from 1.23–2.10 cm/time. The areas with higher snow intensities are mainly located in the east of Hegang, Jiamusi, and in Shuangyashan in the east; the areas with moderate snow are mainly located in the east of Heilongjiang Province and Mohe County; and the areas with low snow intensity are mainly distributed in the west of Heilongjiang Province. The total blizzards in Heilongjiang Province from 1983 to 2020 were between 0 and 761.9 mm, and the blizzards above 200 mm were mainly distributed in Jiamusi, Hegang, Shuangyashan, Qitaihe, Jixi, and Mudanjiang, in eastern Heilongjiang Province; the areas with blizzards above 100 mm were mainly distributed in the Yichun, eastern Suihua, eastern Harbin, Mudanjiang, and Daxing'anling regions, in the central part of Heilongjiang Province; the areas with blizzards below 100 mm were mainly distributed in Heihe, Qiqihar, Daqing, Suihua, and Harbin, in the western part of Heilongjiang Province. The total number of blizzards in Heilongjiang Province from 1983 to 2020 was between 5 and 20 times. The total number of blizzards above 15 were mainly located in Jiamusi, Hegang, Shuangyashan, Qitaihe, Jixi, and Mudanjiang, in eastern Heilongjiang Province; the number of blizzards were 10 in the areas mainly distributed in Qiqihar, Daqing, Suihua, and Harbin, in the west of Heilongjiang Province. In Heilongjiang Province, from 1983 to 2020, the number of days with wind speeds over 8 m/s were 70–187.7 days, and the days with wind speeds over 8 m/s over 110 days were mainly distributed in Jiamusi, Hegang, Shuangyashan, Qitaihe, Jixi, and Mudanjiang, in eastern Heilongjiang Province. The areas with wind speeds> 8 m/s within 90–130 days are mainly distributed in the Qiqihar, Daqing, Heihe, and Daxing'anling areas, in western Heilongjiang Province; the areas with wind speeds over 8 m/s below 90 days are mainly distributed in Heihe, Suihua, and Iraq, in central Heilongjiang Province.

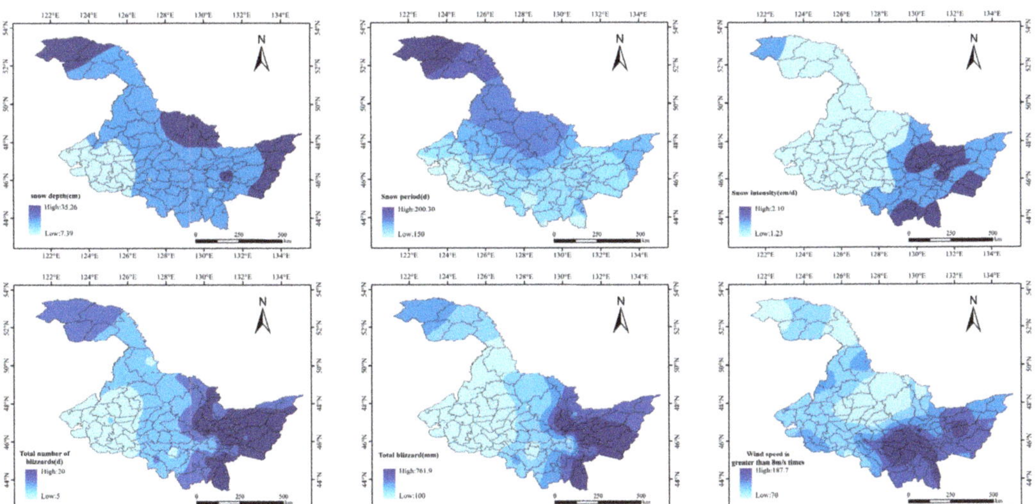

Figure 5. Distribution map of risk factors of snow disaster.

3.1.2. Results of the Risk Zoning of Snow Disasters in Heilongjiang Province

The spatial superpositions of the risk factors are used to obtain the disaster risk zoning results of snow disasters in Heilongjiang Province (Figure 6a). The spatial distribution of hazards is uneven, and there are high-value areas locally. The risk index is equally divided into three levels (as shown in Table 4), and the spatial distribution of the risk classification is presented in Figure 6b. The spatial distribution of the snow-disaster risk in Heilongjiang Province followed a decreasing trend from east to west. High-risk areas accounted for 46.3% of the entire Heilongjiang area, and were distributed mainly in the eastern region of Heilongjiang (Jiamusi, Yichun, Jixi, Qitaihe, Hegang, Shuangyashan, Xunke, and Mudanjiang) and the northwestern Daxing'anling area; the medium-risk areas covered 36.8% of the entire Heilongjiang area, and were distributed primarily in the central region of Heilongjiang Province (eastern Daxing'anling, Heihe, eastern Suihua, northern Mudanjiang, and eastern Harbin); and the low-risk areas accounted for 16.9% of the total area of Heilongjiang Province, and were distributed mostly in the western region of Heilongjiang Province (Qiqihar, Daqing, western Suihua, and western Harbin).

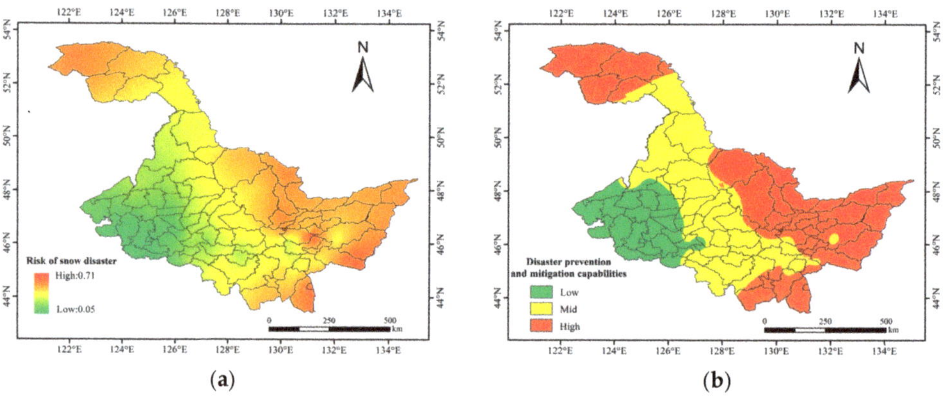

Figure 6. Risk map of snow disaster in Heilongjiang Province ((**a**): stretch, (**b**): classification).

Table 4. Risk classification criteria.

Index	Low	Mid	High
Risk	0.046–0.244	0.244–0.414	0.414–0.712
Susceptibility	0.014–0.129	0.129–0.245	0.245–0.667
Vulnerability	0–0.178	0.178–0.368	0.368–0.824
Prevention and mitigation capacity	0–0.178	0.178–0.280	0.280–0.969
Comprehensive Snow Disaster Risk	0.067–0.295	0.295–0.434	0.434–0.666

3.2. *Susceptibility Analysis of Snow Disaster*

3.2.1. The Characteristics of the Susceptibility Zoning Factors of Snow Disasters in Heilongjiang Province

Figure 7 shows the spatial distribution of the altitude, slope, land-use type, terrain standard deviation, and the NDVI in Heilongjiang Province. It can be seen that the altitude in Heilongjiang province is between 15 and 1651 m; the slope is between 0 and 57.4°; the terrain standard deviation is between 0 and 174.14; and the spatial distribution of the altitude, the slope, and the terrain standard deviation are roughly the same. The areas with high levels are mainly distributed in Huzhong District, Xinlin District, Yichun City, Hailin City, Muling City, and Dongning County in Heilongjiang Province. The land-use types in Heilongjiang Province are dominated by farmland and forest. The forest land is mainly distributed in the Greater Khingan Mountains, in the Lesser Khingan Mountains, and in Mudanjiang City in the south. The NDVI values corresponding to the forest area are also higher, and the highest is 0.83. Farmland is mainly distributed on the Songnen Plain, in the west of Heilongjiang Province, and on the Sanjiang Plain in the east.

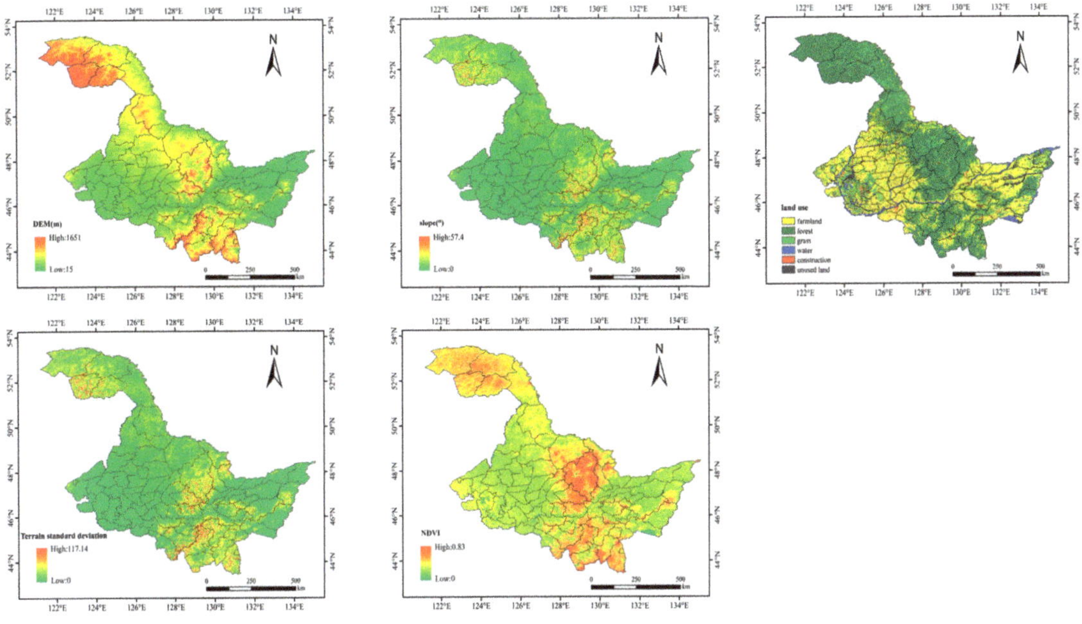

Figure 7. Heilongjiang Province altitude, slope, land-use type, terrain standard deviation, NDVI spatial distribution maps.

3.2.2. Results of the Susceptibility Zoning of Snow Disasters in Heilongjiang Province

The weight in Table 2 is used to perform the spatial superposition of the susceptibility factors to obtain the susceptibility zoning results of snow disasters in Heilongjiang Province (Figure 8a). The spatial distribution of the snow-disaster susceptibilities in Heilongjiang Province is high in the central and northwestern regions, and is low in the eastern and western regions. The susceptibility index is equally divided into three levels (as shown in Table 4), and the spatial distribution of the susceptibility classification is presented in Figure 8b. High susceptibilities accounted for 16.3% of the entire Heilongjiang area, and were distributed mainly in the northern region (southern Mohe County, Huzhong District, Xinlin District, Yichun City, Tieli City, and Mudanjiang City); medium susceptibilities covered 35.6% of the entire Heilongjiang area, and were distributed largely in the central region of Heilongjiang Province (Heihe City, Huma County, Tahe County, Jixi City, and Yichun); and low susceptibilities accounted for 48.1% of the total area of Heilongjiang Province, and were distributed mainly in the western (Qiqihar, Daqing, Suihua, and Harbin) and eastern (Jiamusi, Shuangyashan, and Jixi) regions of Heilongjiang Province. Heilongjiang Province is located in the Northeast Plain and includes the Songnen Plain and the Sanjiang Plain. Other regions with high susceptibilities were distributed in the Daxinganling, Xiaoxinganling, and Mudanjiang mountainous regions.

3.3. Vulnerability Analysis of Snow Disaster

3.3.1. The Characteristics of the Vulnerability Zoning Factors of Snow Disasters in Heilongjiang Province

Figure 9 shows the spatial distributions of urban lights, livestock numbers, crop-sown areas, and highway mileages in Heilongjiang. It can be seen that the urban night lights are between 0 and 63, and that the high values of urban night lights are scattered in prefecture-level cities and urban areas, while urban night lights in country-level cities are relatively low. The livestock numbers are from 498 to 665,511, and the areas with high levels are mainly distributed in Longjiang County, Shuangcheng City, and Zhaodong City, in the southwest of Heilongjiang Province. The crop-sown areas are from 214 to 342,188 km^2, and the areas with high levels are mainly distributed in Songnen Plain and Sanjiang Plain. The highway mileage is between 92.88 and 25,455, and the areas with high levels are distributed in Qiqihar city, Harbin city, Suihua city, and Heihe city, in Heilongjiang Province.

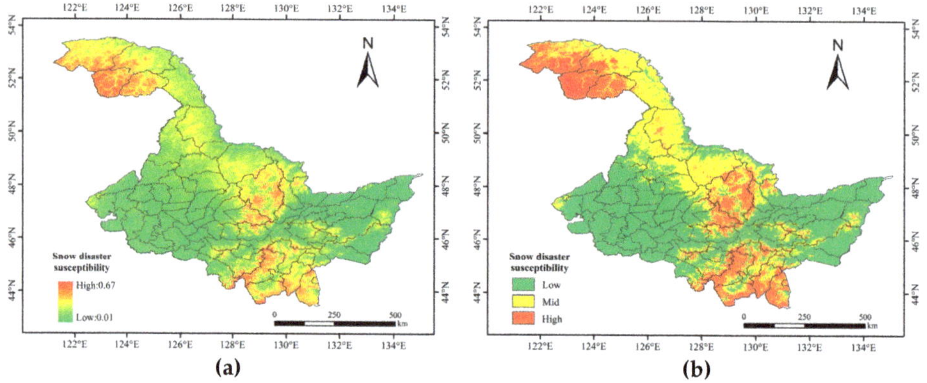

Figure 8. Map of snow-disaster susceptibilities in Heilongjiang Province ((**a**): stretch, (**b**): classification).

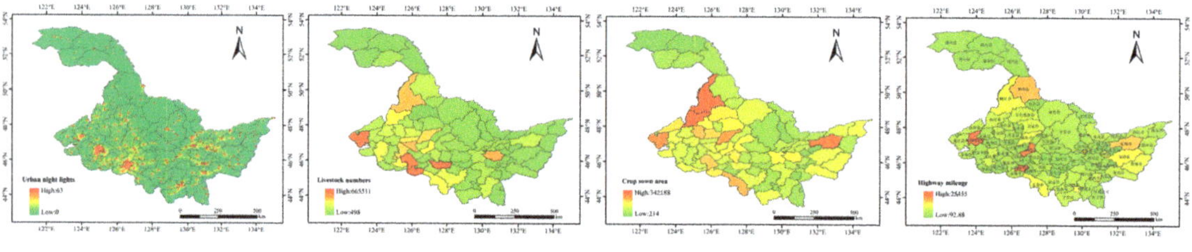

Figure 9. Spatial distributions of urban lights, livestock numbers, crop-sown areas, and highway mileages in Heilongjiang.

3.3.2. Results of the Vulnerability Zoning of Snow Disasters in Heilongjiang Province

The weight in Table 2 is used to perform the spatial superposition of the vulnerability factors to obtain the vulnerability zoning results of snow disasters in Heilongjiang Province (Figure 10a). The spatial distribution of the snow-disaster vulnerabilities in Heilongjiang Province is high in the western regions, and low in the central and northwestern regions. The vulnerability index is equally divided into three levels, and the spatial distribution of the vulnerability classification is presented in Figure 10b. High vulnerabilities accounted for 3.0% of the entire Heilongjiang area, and were distributed mainly in the economically developed main urban areas (Qiqihar City, Daqing City, Harbin City, and Suihua City), as well as other small parts of cities and counties; medium vulnerabilities constituted about 27.9% of the entire Heilongjiang area, and were distributed chiefly in Heihe City, Nenjiang County, Nehe City, Longjiang County, Zhaodong City, Wuchang City, Fujin City, Baoqing County, and other cities and counties of Heilongjiang Province; and low vulnerabilities accounted for 69.1% of the entire area of Heilongjiang Province, and were distributed mostly in the northern and central regions of Heilongjiang Province. As shown in the figure, high vulnerabilities were distributed mainly in regions with more urban lights and larger populations, which have been significantly affected by disasters.

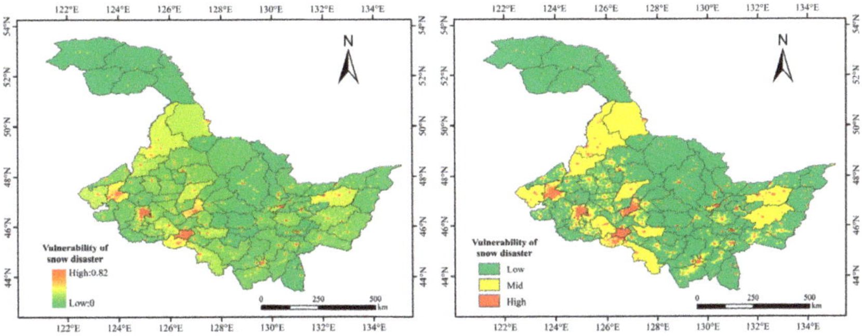

Figure 10. Vulnerability zoning map of snow disaster in Heilongjiang Province ((**a**): stretch, (**b**): classification).

3.4. Analysis of Snow-Disaster Prevention and Mitigation Capability

3.4.1. The Characteristics of the Prevention and Mitigation Capability Zoning Factors of Snow Disasters in Heilongjiang Province

Socioeconomic factors are the main indicators that affect the disaster prevention and mitigation capability. We selected the economic indicators of the Heilongjiang Province Statistical Yearbook, from 1996 to 2020 that each county has data for, as the factors of the disaster prevention and mitigation capability, and finally selected the total GDP, the per capita disposable income, and the education level. The higher the GDP, the stronger the region's economic strength, and the stronger its ability to fight disasters. The population size of each region is different, and the per capita disposable income further illustrates the level of economic development. Therefore, the per capita disposable income was selected as the disaster prevention and mitigation ability. The level of education represents the level of awareness of the impacts of, and the defenses against, disasters. Therefore, the higher the level of education, the stronger the ability to reduce disasters.

Figure 11 shows the spatial distributions of the total GDPs, per capita disposable incomes, and education levels in Heilongjiang. It can be seen that the total GDP is between 25,538 and 15,324,071, and that the high values of the total GDPs are mainly distributed in Daqing City, Qiqihar City, and Harbin City. The per capita disposable income is between 2514 and 14,850, and the areas with high levels are mainly distributed in Daqing City, Harbin City, Dongning County, and Hailin City. The education level is between 2300 and 73,770, and the areas with high levels are the same as those with high total GDPs.

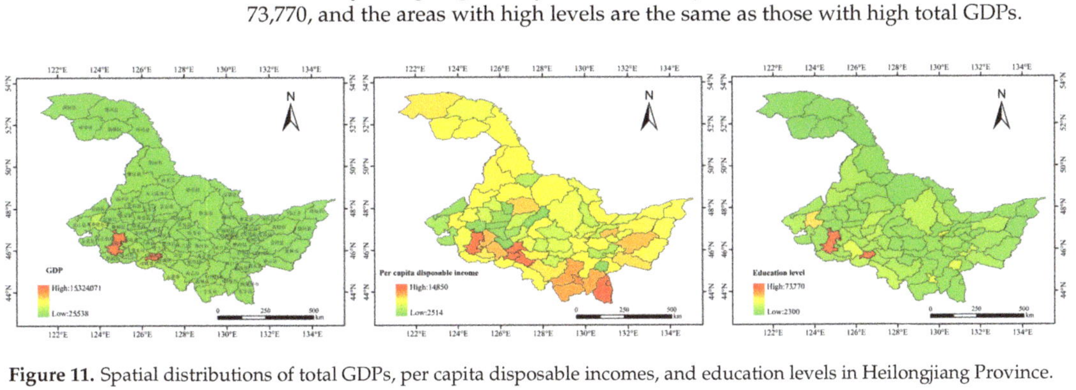

Figure 11. Spatial distributions of total GDPs, per capita disposable incomes, and education levels in Heilongjiang Province.

3.4.2. Results of the Prevention and Mitigation Capacity Zoning of Snow Disasters in Heilongjiang Province

The weight in Table 2 is used to perform the spatial superposition of the prevention and mitigation capacity factors to obtain the prevention and mitigation capacity zoning results of snow disaster in Heilongjiang Province (Figure 12a). The spatial distribution of the snow-disaster prevention and mitigation capacities in Heilongjiang Province is uneven and is high in the southern regions, and low in the northwestern regions. The prevention and mitigation capacity index is equally divided into three levels, and the spatial distribution of the prevention and mitigation capacity classification is presented in Figure 12b. High disaster prevention and mitigation capabilities accounted for 7.4% of the entire Heilongjiang area, and were distributed mainly in Qiqihar City, Daqing City, Zhaodong City, Hulan District, Harbin City, Acheng District, Shuangcheng City, Mudanjiang City, Jiamusi City, Suifenhe City, and Dongning County; middle disaster prevention and mitigation capabilities constituted 84.0% of the entire Heilongjiang area, and were distributed in most cities and counties in Heilongjiang Province; and low disaster prevention and mitigation capabilities covered 8.6% of the entire Heilongjiang Province area, and were distributed mainly in Sunwu County, Gannan County, Tailai County, Lindian County, Kedong County, Baiquan County, Mingshui County, Qinggang County, Wangkui County, Lanxi County, Suileng County, and Jixi City. The figure shows that the more developed the economy, the stronger the disaster prevention and mitigation capability.

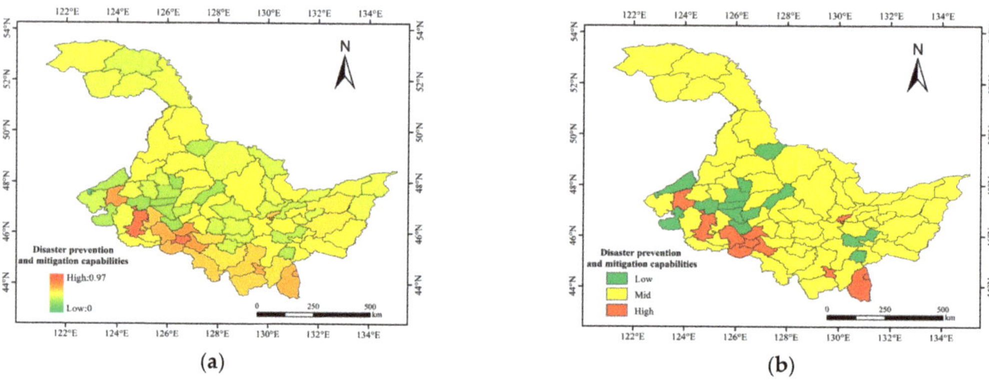

Figure 12. Map of Heilongjiang Province's snow-disaster prevention and mitigation capacities ((**a**): stretch, (**b**): classification).

3.5. *Comprehensive Snow-Disaster Risk Assessment and Zoning*

The risk factors, the susceptibility factors, the vulnerability factors, and the prevention and mitigation capacity factors are superimposed, according to the weights, and the comprehensive snow disaster risk index is obtained. The comprehensive risk assessment and zoning results for snow disasters in Heilongjiang Province are shown in Figure 13a. The comprehensive risk index is equally divided into three levels (Table 4), and the spatial distribution of the comprehensive risk classification is presented in Figure 13b.The spatial distribution of the comprehensive snow disaster risk in Heilongjiang Province followed a declining trend from east to west. High risks accounted for 34.3% of the entire Heilongjiang Province area (i.e., an area of 97,000 km^2), and were distributed mainly in Hulin City, the eastern part of Mishan City, Shuangyashan City, Qitaihe City, Suifenhe City, Dongning County, Muling City, Ning'an City, Hailin, Mudanjiang City, Yichun City, Jiayin County, Xunke, Hegang, Tangyuan, Jiamusi, Tieli, Mohe County, Huzhong District, Xinlin District, and western Tahe County; medium vulnerabilities covered 45.2% of the Heilongjiang area (i.e., an area of 214,000 km^2), largely distributed in eastern Tahe County, Huma County, Heihe City, Nenjiang County, Sunwu County, eastern Wudalianchi, Bei'an, Qing'an, Hailun, Suihua, Mulan, Tonghe, Yilan, Luobei County, Suibin County, Tongjiang City,

Fuyuan County, Fujin, Baoqing, Youyi, Jixian, Linkou, Jixi, Jidong, Shangzhi City, and Wuchang City; and low vulnerabilities accounted for 20.5% of the Heilongjiang Province area (i.e., an area of 162,000 km^2), and were distributed mostly in the southwestern region of Heilongjiang Province, including Qiqihar, Daqing, western Suihua, and western Harbin.

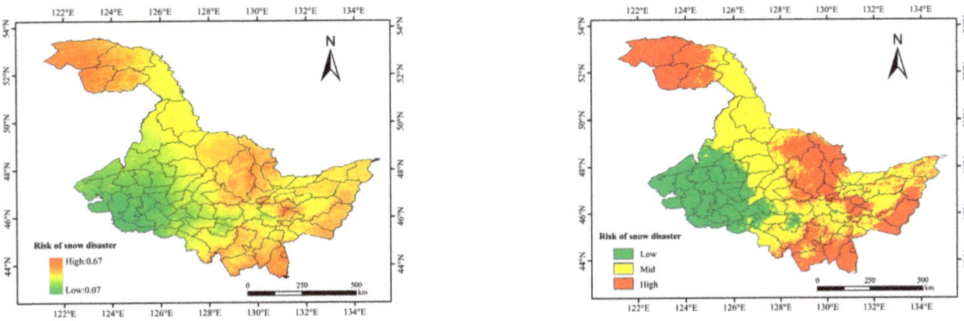

(a) (b)

Figure 13. Comprehensive risk zoning map of Heilongjiang snow disaster ((**a**): stretch, (**b**): classification).

4. Discussion

(1) This article evaluates and zones the snow-disaster risk in Heilongjiang Province from four aspects: risk, susceptibility, vulnerability, and the disaster prevention and mitigation capability. Compared to the snow-disaster risk research at home and abroad, the analysis content of this article is complete. The zoning indicators are relatively complete, which provides a research example for related research and has a certain reference effect. It should be noted that the constituent factors of risk, susceptibility, vulnerability, and the disaster prevention and mitigation capability are not necessarily fixed, and adjustments and replacements can be made without affecting the physical meaning when conducting research in different regions. Generally speaking, six index meteorological indicators, which included the snow depth, the snow-period length, snow days, total blizzard volume, the number of blizzards, and their scores were selected for risk, which are generally easier to obtain; the altitude and land-use types that constitute susceptibility are also easier to obtain. The standard deviations of the slope and terrain are calculated on the basis of altitude data. The NDVI data uses NDVI data; and the factors affecting vulnerability and the disaster reduction capabilities are mainly derived from statistical yearbooks. Since the indicators of the statistical yearbooks in different regions are different, the data in the statistical yearbooks can be replaced when conducting research. For example, the urban lighting index was used in this article because there is no spatial distribution data of populations in the statistical yearbooks. This article uses the MODIS lighting index to replace it.

(2) In the process of zoning, not only can we obtain a spatial distribution map of the different factors that affect not only the risk, susceptibility, vulnerability, and disaster prevention and mitigation capability, but also a zoning map of the risk, susceptibility, vulnerability, disaster prevention and mitigation capability, and the comprehensive disaster risk. Therefore, this method is conducive to the comprehensive analysis of disaster risk. For example, the Sanjiang Plain, in the east of Heilongjiang Province, is a high-risk area for snow disasters, compared with Figure 14, and most of the final comprehensive snow-disaster risk is in a medium-risk area. The main reason is the low susceptibility and the low vulnerability. The comparison and influencing factors can be further analyzed as follows: Although the snow depth, intensity, blizzard volume, and the number of blizzards are relatively large and strong here, it is a plain with a low altitude, small slopes, and small topographical fluctuations. This area is, especially, a primarily agricultural area, basically

consisting of arable land, and there are almost no crops in winter, so the comprehensive risk of snow disaster is low. Therefore, the results of the snow-disaster risk zoning in this paper provide theoretical support for the government and for disaster prevention and relief departments to formulate predisaster mitigation plans, and to help with postdisaster relief decision-making.

(3) Since the selection of zoning factors in this paper will be affected by human factors, on the one hand, and will be limited by statistical yearbook data, on the other, the uncertainty of the factor selection will affect the accuracy of the zoning results. Especially when zoning a large spatial area, it is difficult to obtain a unified socioeconomic index, so it will be restricted. It is recommended to use remote sensing products as much as possible, as with the NDVI and the urban lighting index in this article, which can achieve large-scale spatial risk zoning while ensuring high spatial resolution.

In addition, many meteorological disasters, such as droughts and floods, are related to precipitation data. The precipitation is affected by small-scale weather systems, as well as by the local topography, the underlying surface, and other factors. The spatial interpolation accuracy of precipitation will be affected. The precipitation data in the published precipitation reanalysis data also face accuracy problems. However, there is no good method for the spatial interpolation of precipitation, and research in this area needs to be strengthened.

(4) In the process of zoning in this paper, in addition to the factor selection, there were also problems in calculating the weight of each factor. Because there is no dependent variable, linear regression and other statistical methods cannot be used. The analytic hierarchy process method was used in this article to determine the weights, which are mainly based on the relative importance of each factor and are made by human judgment. Therefore, there are human factors that affect the impact. The results of the zoning will have a certain impact. The analytic hierarchy process is an analytical and discriminative decision-making method, proposed by the American scholar, Saaty, in 1997 [37]. However, although the analytic hierarchy process involves decision-making on data and information, it retains the judgment of the human perception of things and can be completed with less information. The advantage is judging the content of decision-making under the circumstances; however, the disadvantage is that the final result is very dependent on the decisionmaker's judgment of the objective conditions, which has a more obvious subjective intention. When the decisionmaker's perception of things is not accurate enough, it is easy for them to have bias. Therefore, some scholars have proposed methods, such as the multivariate instability index analysis method and the combination weighting method, for weight assignment. We will also conduct further related research.

(5) The current study, based on the meteorological data from 1983 to 2015, reveals the significant reduction in the snow-cover period in Heilongjiang Province over the past 33 years. This reduction is consistent with findings from an analysis of the spatiotemporal variation characteristics of snow in China from 1961 to 2012 [38]. The decrease in the snow-cover period, with interannual variations, has been caused by climate warming over the past 100 years [39]. In addition, the declining trend of the snow period with the latitude variation was similar to that identified in a previous study on the spatiotemporal distribution of snow cover in the northeast region [40]. The snowfall intensity in the east of Heilongjiang Province was higher than in the west, which was consistent with the research results of Huang et al. [30]. The total numbers of blizzards and the total amounts of snowfall in Heilongjiang Province were higher in the east and north, which is similar to the research findings of Zhang [13].

(6) In order to test the results of the snow-disaster risk zoning proposed in this paper, we used the snow-disaster occurrence data in Heilongjiang Province, recorded in the "Chinese Meteorological Disaster Dictionary-Heilongjiang Volume", to compare with this paper. According to the literature, snow disasters in Heilongjiang mostly occur in the eastern parts of the mountains. Traffic snow disasters are more serious, and there are frequent occurrences in Huanan, Boli, Linkou, Yilan, Tonghe, Jixi, Suifenhe, and other

places. The results are similar. At the same time, this article also counts the 2016–2020 snow-disaster loss data recorded by the Heilongjiang Provincial Civil Affairs Department, and the loss, according to the snow disaster, is shown in Figure 14. Compared with the results of the comprehensive risk zoning in this article, it is basically consistent. However, there are some differences in the Lesser Khingan area. The main reason is that the snow-disaster risk in the Lesser Khingan area threatens a high-value area. In the process of zoning in this article, the risk weight is higher, which leads to a higher overall risk. This result also reminds us that, when carrying out disaster risk zoning, we must appropriately increase the vulnerability weight of the disaster-bearing body. In addition, Zhang et al. [13] conducted research on the comprehensive risk of blizzards and found that the high-risk zone is located in the southeast, that the low-risk zone is located in the western region, and that the remaining areas are medium-risk zones, which are also the snow-disaster risks in this paper. These earlier studies provided references for the snow-disaster risk analysis used in this study.

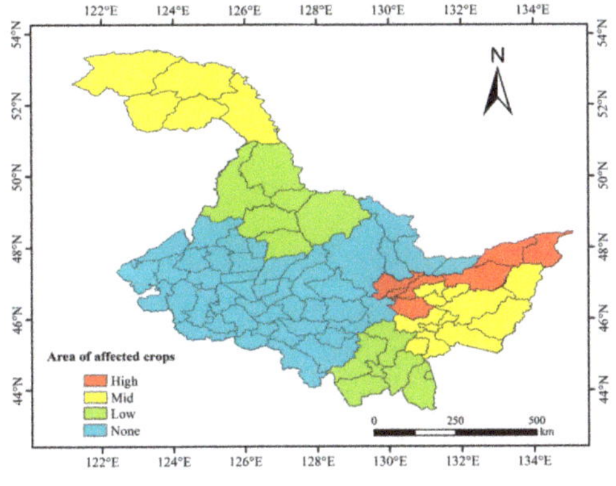

Figure 14. Snow-disaster loss data recorded by the Heilongjiang Provincial Civil Affairs Department.

5. Conclusions

On the basis of this research, the following conclusions can be made:

1. The snow depth, snowfall intensity, total number of snowstorms, and the total volume of snowstorms in Heilongjiang Province all followed an upward trend with interannual changes. The snow depth in the northeast was higher than that in the southwest. The total amount of snowfall, the total number of snowstorms, and the snowfall intensities in the east were all higher than those in the west. The snow-cover period and the number of days with wind speeds >8 m/s decreased with the interannual changes. The snow-cover period in the north was longer than that in the south. The number of days with wind speeds >8 m/s were greater than in the east, but relatively less than in the central area.

2. The snow-disaster risk was higher in eastern and northern Heilongjiang Province. Susceptibilities were higher for the central area and the northern Daxinganling area. Vulnerabilities were higher in the western part than in other areas and were distributed throughout various urban areas. The disaster prevention and mitigation capabilities of economically developed areas were stronger than those in less economically developed areas. The high-risk areas for snow disasters in Heilongjiang Province were located in the northern and eastern regions (e.g., Daxinganling, Yichun, and Mudanjiang), accounting for 34.3% of the total area of Heilongjiang Province. The medium-risk areas of snow disasters were located in the central region (e.g., Heihe, Suihua, Jiamusi, and eastern Harbin), accounting for 45.2% of the total area of Heilongjiang Province. The low-risk areas were located in the southwest of Harbin (e.g., Qiqihar, Daqing, western Suihua, and western Harbin), accounting for 20.5% of the entire Heilongjiang Province area.

Author Contributions: Writing—original draft preparation, H.L.; writing—review and editing, L.Z. and S.Z.; formal analysis, W.X. All authors have read and agreed to the published version of the manuscript.

Funding: This research was funded by the National Natural Science Foundation of China (No. U20A2082; No. 41771067), and was supported by the Natural Science Foundation of Heilongjiang Province of China (No. ZD2020D002).

Institutional Review Board Statement: Not applicable.

Informed Consent Statement: Not applicable.

Data Availability Statement: The data presented in this study are available on request from the corresponding author.

Acknowledgments: We gratefully acknowledge the assistance of Huang, Y.T. and Pan, T. in preparing software and photograph. We also thank Liu, Y.T. and Li, X.Z. for valuable discussion.

Conflicts of Interest: The authors declare no conflict of interest.

References

1. Zhong, K.B. International Experience and Enlightenment of Comprehensive Disaster Risk Assessment. *China Emerg. Manag.* **2021**, *5*, 78–81.
2. UN-ISDR. *Sendai Framework for Disaster Risk Reduction 2015–2030, International Strategy for Disaster Reduction*; United Nations Inter-Agency Secretariat: Geneva, Switherlands, 2015; Available online: www.unisdr.org (accessed on 14 October 2020).
3. Wang, P. Research on the zoning method of natural disasters based on GIS. *Front. Earth Sci.* **2001**, *8*, 2.
4. General Administration of Quality Supervision; Inspection and Quarantine of the People's Republic of China; Standardization Administration of the People's Republic of China. *GB/T 28921–2012 Natural Disaster Classification and Code*; China Standards Press: Beijing, China, 2013.
5. Gong, D.J.; Li, Z.J. Causes and mitigation measures of snowstorm disasters in Inner Mongolia. *Clim. Environ. Res.* **2001**, *1*, 132–138.
6. Chen, Y.Q.; Chen, Y.; Wu, M.L. Characteristic Analysis of Blizzard and Gale Disasters of Liaoning Facility Agriculture. *J. Anhui Agric. Sci.* **2013**, *41*, 1660–1661,1692.
7. Zhang, J.P. Research on the Spatial and Temporal Distribution and Prevention Technology of Highway Snowstorm Disasters in Heilongjiang Province. Master's Thesis, Chang'an University, Xi'an, China, 2012.
8. Schmitt, U.; Schardt, M.; Ninaus, J. Snow avalanche risk assessment supported by remote sensing Observing Our Environment from Space-New Solutions for a New Millennium. In Proceedings of the 21st EARSel Symposium, Paris, France, 14–16 May 2001; CRC Press: Boca Raton, FL, USA, 2021; pp. 14–16.
9. Seliverstov, Y.; Glazovskaya, T.; Shnyparkov, A.; Vilchek, Y.; Sergeeva, K.; Martynov, A. Assessment and mapping of snow avalanche risk in Russia. *Ann. Glaciol.* **2008**, *49*, 205–209. [CrossRef]
10. Cappabianca, F.; Barbolini, M.; Natale, L. Snow avalanche risk assessment and mapping: A new method based on a combination of statistical analysis, avalanche dynamics simulation and empirically-based vulnerability relations integrated in a GIS platform. *Cold Reg. Sci. Technol.* **2008**, *54*, 193–205. [CrossRef]
11. Germain, D.; Hétu, B.; Filion, L. *Tree-Ring Based Reconstruction of Past Snow Avalanche Events and Risk Assessment in Northern Gaspé Peninsula (Québec, Canada)*; Springer: Amsterdam, The Netherlands, 2010.
12. Germain, D. Snow avalanche hazard assessment and risk management in northern Quebec, eastern Canada. *Nat. Hazards* **2016**, *80*, 1303–1321. [CrossRef]
13. Zhang, F.; Zhang, L.J.; Huang, Y.T.; Wang, N.; Li, Y.S. Hazard assessment and regionalization of snowstorms in Heilongjiang Province from 1961 to 2015. *Phys. Chem. Earth Parts A/B/C* **2020**, *115*, 102833. [CrossRef]
14. Liu, D.; Zhang, L.; Jiang, S.; Shi, S.; Li, Y. Hazard prediction and risk regionalization of snowstorms in northeast China. *Phys. Chem. Earth Parts A/B/C* **2020**, *116*, 102832. [CrossRef]
15. Barbolini, M.; Cappabianca, F.; Sailer, R. Empirical estimate of vulnerability relations for use in snow avalanche risk assessment. In Proceedings of the 4th International Conference on Computer Simulation in Risk Analysis and Hazard Mitigation, Rhodes, Greece, 2004; Volume 77. Available online: https://www.witpress.com/elibrary/wit-transactions-on-ecology-and-the-environment/77/14328 (accessed on 14 October 2020).
16. Sinickas, A.; Jamieson, B.; Maes, M.A. Snow avalanches in western Canada: Investigating change in occurrence rates and implications for risk assessment and mitigation. *Struct. Infrastruct. Eng.* **2016**, *12*, 490–498. [CrossRef]
17. Casteller, A.; Häfelfinger, T.; Donoso, E.C.; Podvin, K.; Kulakowski, D.; Bebi, P. Assessing the interaction between mountain forests and snow avalanches at Nevados de Chillan, Chile and its implications for ecosystem-based disaster risk reduction. *Nat. Hazards Earth Syst. Sci.* **2018**, *18*, 1173–1186. [CrossRef]
18. Gao, J.M. Analysis and assessment of the risk of Snow and Freezing Disaster in China. *Int. J. Disaster Risk Reduct.* **2016**, *19*, 334–340. [CrossRef]

19. Sa, C.; Liu, G.; Wang, M.; Bao, Y. Study on Hazard Factor Assessment Caused by Snow Disaster in Pastoral Areas of Inner Mongolia Grassland based on Multi-source Remote Sensing Data. In Proceedings of the 7th Annual Meeting of Risk-Analysis-Council of China-Association-for-Disaster-Prevention (RAC), Changsha, China, 4 November 2016.
20. Liu, Z.; Hou, G.L.; Li, F. Preliminary Study of Risk Assessment of the Failure of Road Traffic Caused by Snow Disaster in Guoluo Prefecture. In *Medicine and Biopharmaceutical, Proceedings of the 2015 International Conference, London, UK, 8–12 June 2015*; World Scientific: Singapore, 2016; pp. 1420–1431.
21. Qi, H.L.; Tian, W.P.; Zhao, F. Risk assessment of snow disaster for trunk highway transportation in Shaanxi, China. *Nat. Hazards* **2017**, *85*, 523–536. [CrossRef]
22. Leone, F.; Colas, A.; Garcin, Y.; Eckert, N.; Jomelli, V.; Gherardi, M. The snow avalanches risk on Alpine roads network Assessment of impacts and mapping of accessibility loss. *Rev. Geogr. Alp. J. Alp. Res.* **2014**, 102–104.
23. Fujita, K.; Inoue, H.; Izumi, T.; Yamaguchi, S.; Sadakane, A.; Sunako, S.; Kouichi, N.; Walter, W.; Immerzeel, J.M.; Shea, R.B.; et al. Anomalous winter-snow-amplified earthquake-induced disaster of the 2015 Langtang avalanche in Nepal. *Nat. Hazards Earth Syst. Sci.* **2017**, *17*, 749–764. [CrossRef]
24. Wang, S.J.; Chen, S.Y.; Wei, Y.Q. Risk prevention and control strategies for the severely affected areas of snow disaster in the Three Rivers Source Region (TRSR), China. *Sci. Cold Arid. Reg.* **2019**, *11*, 248–252.
25. Liu, F.G.; Liu, F.; Mao, X.; Zhang, Y.; Chen, Q.; Liu, P.; Zhao, Z. Risk analysis of snow disaster in the pastoral areas of the Qinghai-Tibet Plateau. *J. Geogr. Sci.* **2014**, *24*, 411–426. [CrossRef]
26. Gao, J.L.; Huang, X.D.; Ma, X.F.; Feng, Q.S.; Liang, T.G.; Xie, H.J. Snow Disaster Early Warning in Pastoral Areas of Qinghai Province, China. *Remote. Sens.* **2017**, *9*, 475. [CrossRef]
27. Park, H.S.; Lee, S.J.; Yu, I.; Jeong, S.; Chung, G. Snow disaster risk index in the ulsan metropolitan city. *Korean Soc. Hazard Mitig.* **2014**, *14*, 97–106. [CrossRef]
28. Insang, Y. Methodology of an analysis of snow disaster risk for establishing climate change adaptation measures. *Korean Soc. Hazard Mitig.* **2020**, *20*, 351–364. [CrossRef]
29. Ju, T.; Li, X. Spatial Distribution and Temporal Variation of Snow Water Resources in China during 1993–2002. *J. Glaciol. Geocryol.* **2005**, *27*, 64–67.
30. Huang, Y.T. The Impact of Climate Warming on the Temporal and Spatial Changes of Snowfall in Heilongjiang Province. Master's Thesis, Harbin Normal University, Harbin, China, 2019.
31. Ban, J.; Sun, M.H. Causality analysis of winter precipitation anomalies in Heilongjiang province in 2012. *Heilongjiang Meteorol.* **2013**, *30*, 1–3,9.
32. Wang, N.; You, Q.L.; Liu, J.J. The long-term trend of surface wind speed in China from 1979 to 2014. *J. Nat. Resour.* **2019**, *34*, 1531–1542.
33. Zhang, L.J.; Chen, Y.; Yuan, N.W.A. Research on Weighted Combinatorial Evaluation Based on Reliability Analysis. *Manag. Rev.* **2012**, *24*, 170–176.
34. Beynon, M. An analysis of distrbutions of priority values from altemative comparison scales within AHP. *Eur. J. Oper. Res.* **2002**, *140*, 104–117. [CrossRef]
35. Mann, H.B. Non- parametric test against trend. *Econometrika* **1945**, *13*, 245–259. [CrossRef]
36. Pettitt, A.N. A Non-Parametric Approach to the Change-Point Problem. *Tournal R. Stat. Soc.* **1978**, *28*, 126–135. [CrossRef]
37. Saaty, T.L. A scaling method for priorities in hierarchical structures. *J. Math. Psychol.* **1977**, *15*, 234–281. [CrossRef]
38. Zhang, Z.F.; Shuang, X.; Liu, N.; Yu, Y.; Fan, S.H.; Feng, M.N. Snowfall change characteristics in China from 1961 to 2012. *Resour. Sci.* **2015**, *37*, 1765–1773.
39. Li, Y.; Wang, Y.; Chu, H.Y.; Tang, J.P. Climatic variability of near-surface wind energy resources on land in China and the impact of man-made changes on the underlying surface. *Chin. Sci. Bull.* **2008**, *21*, 2646–2653.
40. Yang, Q. Spatial and Temporal Distribution of Snow Cover and Simulation of Snowmelt Runoff in Northeast China. Ph.D. Thesis, Jilin University, Changchun, China, 2015.

sustainability

MDPI

Article

Integrated Assessments of Meteorological Hazards across the Qinghai-Tibet Plateau of China

Shao Sun [1], Qiang Zhang [2,3,4,*], Yuanxin Xu [5] and Ruyue Yuan [2,3]

1 Laboratory for Climate Studies, National Climate Center, China Meteorological Administration, Beijing 100081, China; sunshao@cma.gov.cn
2 Key Laboratory of Environmental Change and Natural Disasters, Ministry of Education, Beijing Normal University, Beijing 100875, China; yuanry@mail.bnu.edu.cn
3 Faculty of Geographical Science, Academy of Disaster Reduction and Emergency Management, Beijing Normal University, Beijing 100875, China
4 State Key Laboratory of Earth Surface Processes and Resource Ecology, Beijing Normal University, Beijing 100875, China
5 Key Laboratory of Drought-Flood Disasters in Plateau and Basin of Sichuan Province, Sichuan Climate Center, Chengdu 610072, China; xuyuanxin0710@gmail.com
* Correspondence: zhangq68@bnu.edu.cn

Citation: Sun, S.; Zhang, Q.; Xu, Y.; Yuan, R. Integrated Assessments of Meteorological Hazards across the Qinghai-Tibet Plateau of China. *Sustainability* **2021**, *13*, 10402. https://doi.org/10.3390/su131810402

Academic Editors: Xiaodong Yan, Jia Yang and Shaofei Jin

Received: 21 August 2021
Accepted: 10 September 2021
Published: 17 September 2021

Abstract: Recent decades have witnessed accelerated climate changes across the Qinghai-Tibet Plateau (QTP) and elevated socioeconomic exposure to meteorological hazards. The QTP is called the "the third pole", exerting remarkable impact on environmental changes in its surrounding regions. While few reports are available for addressing multi-hazard risks over the QTP, we develop an integrated indicator system involving multiple meteorological hazards, i.e., droughts, rainstorms, snowstorms and hailstorms, investigating the spatiotemporal patterns of major hazards over the QTP. The hazard zones of droughts and rainstorms are identified in the southern Gangdise Mountains, the South Tibet Valley, the eastern Nyenchen-Tanglha Mountains, the Hengduan Mountains and West Sichuan Basin. Snowstorm hazard zones distribute in the Himalayas, the Bayan Har Mountains and the central Nyenchen-Tanglha Mountains, while hailstorm hazard zones cluster in central part of the QTP. Since the 21st century, intensified rainstorms are detectable in the densely populated cities of Xining and Lhasa and their adjacent areas, while amplified droughts are observed in grain production areas of the South Tibet Valley and the Hengduan Mountains. Snowstorm hazards show large interannual variations and an increase in pastoral areas, although the overall trend is declining slightly. The frequency of hailstorms gradually decreases in human settlements due to thermal and landscape effects. Mapping meteorological hazards regionalization could help to understand climate risks in the QTP, and provide scientific reference for human adaptation to climate changes in highly sensitive areas.

Keywords: meteorological hazards; risk assessment; spatial pattern; population exposure; Qinghai-Tibet Plateau

1. Introduction

Due to drastic climate change and the expansion of socioeconomic exposure, the disaster-related losses due to meteorological hazards have been increasing rapidly in recent decades [1,2]. The Qinghai-Tibet Plateau (QTP) is known as the "third pole", a region highly sensitive to global warming. Over the past 50 years, warming magnitude of the QTP is twice as high as the global average [3,4]. A warming climate has changed the thermodynamic and kinetic conditions of the QTP [5], triggering an increase in the intensity, frequency and duration of meteorological hazards [6–8]. Snowstorms, along with low temperatures and strong winds, become one of the major hazards in the QTP that could have serious negative impacts on animal husbandry, industry, agriculture and transportation [9–12]. Climate projections indicate that snowstorms will continue to intensify with uncertainties and fluctuations throughout the decades to come [13,14]. Meteorological droughts directly affect the growth of plants, forests and crops in the QTP [15,16], causing decreased productivity and environmental degradation [17–19]. Rainstorms, as well as relevant geological hazards, pose a serious threat to human life, transportation and infrastructure, leading to serious casualties and traffic disruptions. Hailstorms are common disastrous hazards in the plateau and cause serious damages to buildings and crops [20]. Although the occurrence frequency of hailstorms decreases in the QTP [21,22], the overall intensity and economic losses are still increasing remarkably [23–25].

Previous studies indicate that single hazard assessment tends to underestimate the actual hazard intensity [26], which confirms the necessity and significance for multi-hazard risk assessment. In order to quantify the combined effects of the compound meteorological hazards in both space and time, a range of assessment methods have been developed, such as Bayesian networks and information diffusion theory [27,28], as well as some numerical and statistical analysis methods [29]. It should be noted that the standing studies about natural hazards over the QTP mainly focus on the mechanisms behind an individual hazard [30–33]. Therefore, an integrated assessment of multiple hazards and their potential risks are of great significance for disaster reduction and climate risk management in the coming decades.

In this study, we focus on the spatial patterns of major meteorological hazards in the QTP, i.e., droughts, rainstorms, snowstorms and hailstorms. By identifying the changes in the spatial extent of hazard zones, we analyze the historical evolution of the population exposure, and further estimate the potential trends of climate risks across the QTP. We attempt to address the following three scientific issues: (1) What is the spatial pattern of the major meteorological hazards across the QTP? (2) How do we evaluate the integrated intensity and socioeconomic impacts of meteorological hazards? (3) Where will be the hotspots of climate risks in the QTP?

2. Data and Methods

2.1. The Study Region

The QTP is known as the "third pole" and the highest plateau of the world, spanning 31 longitudes and 13 latitudes with a total area of 2.6 million km^2. The QTP could be subdivided into six geomorphic regions, i.e., the Northern Tibet Plateau, the South Tibet Valley, the Qaidam Basin, the Qilian Mountains, the Qinghai Plateau and the West Sichuan Basin, covering the entire Tibet province and parts of Qinghai, Xinjiang, Gansu, Sichuan and Yunnan Provinces (Figure 1). The population of the QTP is mainly distributed in the South Tibet Valley, the West Sichuan Plateau, and Xining city as well as its surrounding areas. With an average elevation of more than 4000 m, the QTP has a low population density and an undeveloped regional economy. Human activities are mainly in the valley areas and are highly sensitive to meteorological hazards. In the backdrop of climate warming, the frequency and intensity of meteorological hazards are rising over the QTP, posing serious threats to local people, primary industries and vulnerable infrastructures.

The historical disaster database covering the period from 2001 to 2019 is sourced from the China Meteorological Administration. The dataset contains about 4000 disaster records in the QTP at county level. These meteorological disasters have significant negative impacts on human well-being, causing a total of 36.8 million victims and direct economic losses of RMB 36.1 billion yuan. Droughts cause the largest affected population with an annual average of 783,000 people, accounting for 40.5% of the total victims. The population affected by hailstorms, rainstorms and snowstorms accounts for 21.2%, 19.4% and 18.9% of the total, respectively. In terms of direct economic losses, rainstorms cause the largest number, accounting for 38.7% of the total. The economic losses caused by droughts, snowstorms and hailstorms account for 29.4%, 16.2% and 15.7% of the total, respectively.

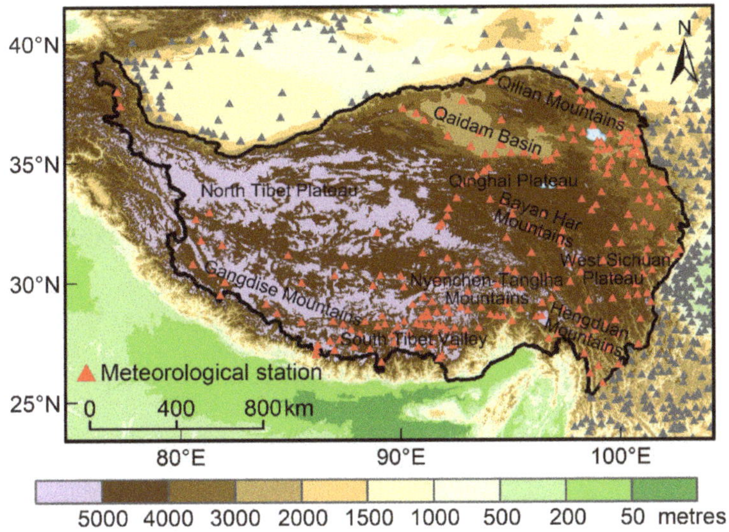

Figure 1. Topographic characteristics and meteorological station distribution in the QTP.

2.2. Data

The data applied in the research include meteorological observation data, geographic information data, meteorological disaster records and socioeconomic data. Meteorological data include daily temperature, precipitation, weather phenomena, snow depth, wind speed, and evaporation at 246 meteorological stations in QTP, covering the period from 2001–2019. The meteorological data are sourced from the China Meteorological Administration (CMA), available online: http://data.cma.cn/ (accessed on 21 August 2021).

Geographic information data of the QTP is sourced from the National Geomatics Center of China (NGCC). The dataset contains the basic geographic information of the QTP, including administrative boundaries, hydrographic network, land cover, and topographic data. Available online: http://www.ngcc.cn/ (accessed on 21 August 2021).

The meteorological disaster records are extracted from the China Meteorological Disaster Yearbook from 2001 to 2019 released by CMA, containing the records of droughts, rainstorms and floods, typhoon, hailstorms, chilling damage and their socioeconomic impacts.

The gridded datasets of population and GDP of China are released by the Institute of Geographic Sciences and Natural Resources Research, CAS, with a spatial resolution of 1 km. Available online: https://www.resdc.cn/ (accessed on 21 August 2021).

2.3. Methods

2.3.1. Indicator System of Hazard Assessments

Hazard assessments refer to a comprehensive analysis of the natural characteristics such as intensity, frequency and duration of each hazard. The evaluation steps include indicator system development, data normalization, weight coefficient determination, integrated assessment modeling, and geographic mapping.

An indicator system for hazard assessment in the QTP is designed based on the annual expected value of various indices (Table 1). The classification criteria of specific indicators are listed in Table 1. The assessment of meteorological droughts selects drought days, the longest duration of drought, and the accumulated intensity as indices. Rainstorm assessment selects rainstorm days (daily rainfall >20 mm), accumulated rainfall, and maximum 24-h rainfall as indices. Snowstorm assessment selects accumulated snowfall, snow days, and maximum snow depth as indices. Hailstorm assessment selects gale days and hail days as indices. The hazard levels are designated by referring to the spatial average and standard deviation of each index in QTP.

Table 1. The indicator system of meteorological hazards assessment over the QTP.

Hazard	Index	Hazard Level			
		High	Moderate	Low	Very Low
Drought	Drought Days (d)	≥60	50~60	40~50	<40
	Maximum Consecutive Drought Days (d)	≥35	30~35	25~30	<25
	Accumulated Intensity of Drought	≥70	60~70	50~60	<50
Rainstorm	Rainstorm Days (d)	≥4	2~4	1~2	<1
	Accumulated Rainfall (mm)	≥100	75~100	50~75	<50
	Maximum 24-h Rainfall (mm)	≥35	25~35	15~25	<15
Snowstorm	Accumulated Snowfall (mm)	≥120	70~120	20~70	<20
	Snow Cover Days (d)	≥30	20~30	10~20	<10
	Maximum Snow Depth (cm)	≥8	5~8	2~5	<2
Hailstorm	Gale Days (d)	≥80	40~80	10~40	<10
	Hail Days (d)	≥8	4~8	1~4	<1

For meteorological droughts, we apply a sliding 30-day SPI to identify drought days as well as its intensity in the growing season, as detailed in Section 2.3.2. For rainstorms, these indices are commonly applied in the daily operations of hazard monitoring. The reason why the threshold of a rainstorm day is defined as 20 mm is that daily precipitation of 20 mm could trigger geological hazards in mountain areas, according to the historical disaster records of the QTP. For snowstorms, the indices selected are major contributors to socioeconomic impacts, in which the snow cover days and maximum snow depth seriously threaten local transportation, livestock and infrastructures. For hailstorms, due to the lack of hailstone and lightning observation, the number of gale days and hail days are selected as two simple indicators to identify hazard zones.

2.3.2. Calculation of Meteorological Droughts

The precipitation in the QTP is subject to significant spatial heterogeneity. The annual precipitation in the northern part of the Kunlun Mountains is less than 100 mm, while in the Hengduan Mountains region is more than 800 mm. According to the climatic characteristics of the QTP, the region with annual precipitation less than 200 mm is defined as permanent arid region, which lacks the possibility of meteorological droughts. Additionally, the QTP has distinct dry and wet seasons, and the growing period of crops, forage grass and vegetation are concentrated in the rainy season. Hence, the hazard assessment of meteorological droughts mainly focuses on the period from April to October, and the annual expected drought days and other indices specifically refer to the rainy season.

The daily drought index is defined as the standardized precipitation index (SPI) of the last 30 days, which is calculated based on a sliding 30-day window of historical daily precipitation data. Daily SPI ≤ -0.5 has been set as the threshold of meteorological drought, and the indices of drought days, the longest consecutive drought days and accumulated drought intensity are calculated accordingly.

2.3.3. Calculation of Single Hazard Index

An assessment model of hazard index is designed based on the deviation of each index, so as to evaluate the intensity level of a meteorological hazard. The formula is shown in Equation (1).

$$\begin{cases} A = \sum_{i=1}^{n} Xi \\ \sigma = \sqrt{\frac{1}{n}\sum_{j=1}^{n}\left(A_j - \overline{A}\right)^2} \\ H = 5 + \frac{\left(A-\overline{A}\right)}{\sigma} \end{cases} \tag{1}$$

wherein, A stands for hazard value, Xi is the normalized value of each index. σ is the standard deviation of A. H stands for hazard index whose initial value is set as 5, and take the multiple of standard deviation (STD) as steps. For instance, H = 6.0 denotes 1 STD above the mean, while H = 4.5 denotes 0.5 STD less than the mean.

A hazard index above 6.0 is set to represent a high hazard level, a hazard index between 5.0 to 6 represents a moderate hazard level, a hazard index between 4.0 to 5.0 represents a low hazard level, and a hazard index below 4.0 represents a very low hazard level. In this way, the hazard level of rainstorms, droughts, snowstorms and hailstorms across the QTP are evaluated, respectively.

2.3.4. Calculation of Multi-Hazard Index

The historical affected population and economic losses of meteorological hazards are applied to determine the weight coefficients, then a multi-hazard index is developed on each hazard index and its weight coefficients, shown in Equation (2). The index indicates the integrated hazard level of four major meteorological hazards across the QTP. IH denotes multi-hazard index, H_D denotes drought hazard index, H_R denotes rainstorm hazard index, H_S denotes snowstorm hazard index, and H_W denotes hailstorm hazard index. The weight coefficients of a, b, c and d are determined by the percentage of total affected population and economic losses of each hazard in the QTP from 2001 to 2019 (shown in Section 2.1). For example, the total population affected by droughts accounts for 40.5% of all hazards, while the total economic losses accounts for 29.4% of all hazards. Then the average value of 35% is taken as the weight coefficient of drought hazard index. According to this rule, the weight coefficients of drought, rainstorm, snowstorm and hailstorm are 0.35, 0.29, 0.18 and 0.18, respectively.

$$IH = a \times H_D + b \times H_R + c \times H_S + d \times H_W \tag{2}$$

3. Results

3.1. Meteorological Droughts

Historical observations suggest that the drought hazard zones are mainly distributed in the South Tibet Valley, the Nianqing-Tanggula Mountains, the Hengduan Mountains, Western Sichuan Plateau and Songpan Plateau, among which the prefectures of Xigaze, Nyingchi, Diqing, Nujiang and Aba are expected to have drought days >60 d per year (Figure 2a). In terms of the longest consecutive drought days, the observations indicate consecutive drought days >30 d in the South Tibet Valley, Hengduan Mountain Area and Songpan Plateau at annual expected (Figure 2b). The accumulated drought intensity is the sum of the absolute value of the daily SPI index during drought days, and represents the total intensity of meteorological droughts. The areas with high drought intensity mainly distribute in the Nianqing-Tanggula Mountains, the Hengduan Mountains and SongPan Plateau (Figure 2c). Drought hazard index (DHI) is an integrated index based on the above three indices. The areas with DHI $\geq$ 6 (1 STD above the average) are identified as hazard zones, and totally account for 17.9% of the QTP and specifically include western Xigaze, eastern Nyingchi, southern Qamdo, Diqing, Nujiang and Aba (Figure 2d).

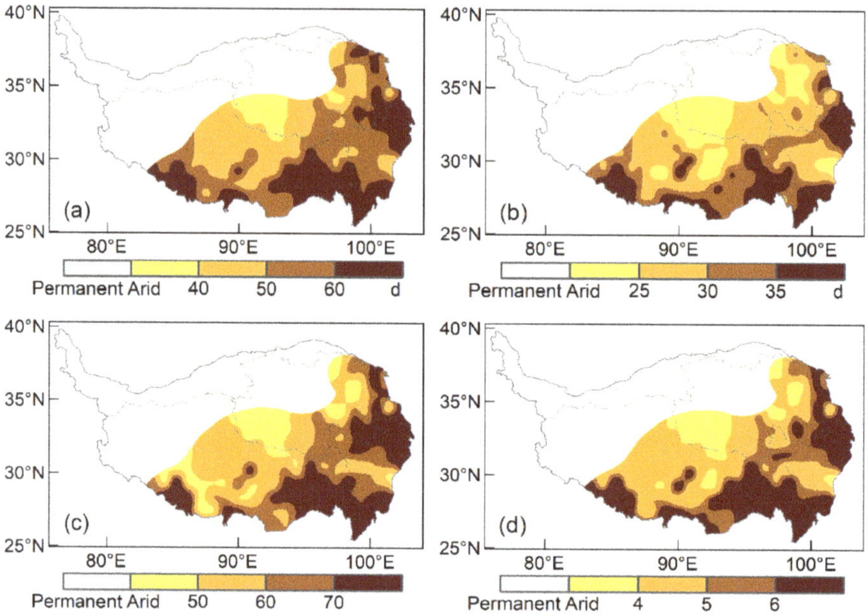

Figure 2. The spatial patterns of meteorological drought hazard across the QTP. (**a**) annual drought days; (**b**) the longest consecutive drought days; (**c**) annual accumulated drought intensity; (**d**) drought hazard index.

3.2. Rainstorms

The criterion for rainstorm events in China is generally defined by the China Meteorological Administration as a daily precipitation amount >50 mm. Due to the special geographic and climatic conditions in the QTP, a rainstorm with daily precipitation >20 mm could trigger natural disasters like flash floods and debris flows. Hence, a daily precipitation amount >20 mm is taken as the threshold to identify rainstorms in the QTP. A hazard assessment system of rainstorms in the QTP contains three indices, i.e., rainstorm days, accumulated rainstorm rainfall, and maximum daily precipitation, at annual expected value. Historical observations indicate that the spatial pattern of rainstorm days and accumulated rainstorm rainfall show a gradual decrease from southeastern to northwestern QTP. The Nyenchen-Tanglha Mountains and Western Sichuan Plateau are observed with rainstorm days >4 d and an accumulated rainstorm rainfall >100 mm at annual expected (Figure 3a,b). The spatial average of annual maximum daily precipitation in the QTP is 25 mm, and the areas above the average include the north Tibet plateau, the Gangdise Mountains, the South Tibet Valley, the Nyenchen-Tanglha Mountains, Hengduan mountains and western Sichuan plateau (Figure 3c). Therein, the maximum daily precipitation in the South Tibet Valley and the Hengduan Mountains exceeds 35 mm. Hazard assessment indicates that hazard zones (HI > 6) cover the prefectures of western Xigaze, Lhasa, Nyingchi, Nujiang, Diqing, Ganzi, Aba, Gannan, Huangnan and Xining (Figure 3d), totally accounting for 20.1% of the QTP.

3.3. Snowstorms

The QTP is an important livestock husbandry base in China, and also the largest region with snow cover in China. Snowstorms usually occur in autumn and winter, which are the most widespread and disastrous hazards in pastoral areas of the QTP and pose a serious threat to the production of local herders. The snowstorm hazard index is developed with three indices of accumulated snowfall, snow cover days, and maximum snow depth, at the annual expected value. The average of annual accumulated snowfall in the QTP is 67.3 mm, and the areas with heavy snowfall are mainly distributed in the Himalayas, Tanggula Mountains and Qinghai Plateau (Figure 4a). Therein, annual snowfall >120 mm is observed in the central Xigaze, southern Shannan, eastern Naqu, western Haixi, Yushu, Guoluo and southern Huangnan. The spatial average of annual snow cover days is 18.4 d in the QTP, and reaches more than 30 d in the Himalayas and Qinghai Plateau regions (Figure 4b). The spatial average of maximum snow depth in QTP is 6.5 mm, therein the snow depth reaches 6–10 cm in Bayankela and the Nyenchen-Tanglha Mountains, and exceeds 10 cm in the Himalayas (Figure 4c). The hazard assessment of snowstorms indicates that hazard zone (HI > 6) cover the prefectures of southern Ali, western Xigaze, southern Shannan, eastern Naqu, Yushu, Guoluo, southern Huangshan and northern Aba, totally accounting for 13.5% of the QTP (Figure 4d).

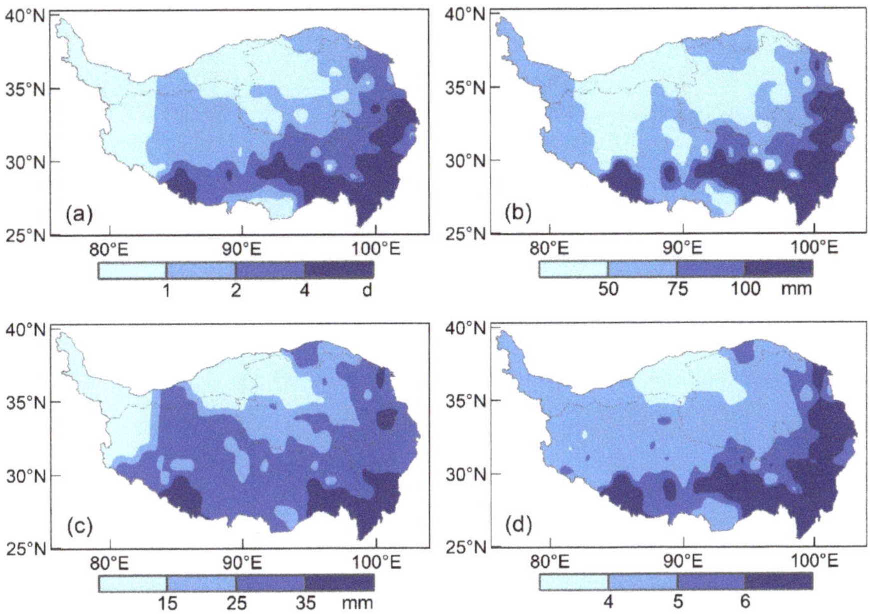

Figure 3. The spatial patterns of rainstorm hazard across the QTP. (**a**) annual rainstorm days; (**b**) annual accumulated rainstorm precipitation; (**c**) maximum daily precipitation; (**d**) rainstorm hazard index.

3.4. Hailstorms

Hailstorm refers to mesoscale or microscale catastrophic weather caused by the strong atmospheric convection, which mostly occurs in the mid-latitude mountainous areas and fragile ecological environment. Under the special thermal conditions induced by plateau topography, hailstorms become common hazards on the QTP in summer. The hailstorm hazard index is composed of annual expected gale days and annual expected hail days. Historical observations indicate that the average of annual gale days over the QTP is 43.1 d, and the areas above average are concentrated in the north Tibetan Plateau, Tanggula Mountains, Qinghai Plateau and the Gangdese Mountains. Wherein, the number of gale days in the prefectures of central Xigaze, Naqu, Haixi, and western Yushu exceeds 80 d (Figure 5a). Annual expected hail days in the QTP is 4.1 d at spatial average and show similar distribution as gale days. The prefectures of eastern Naqu, western Hernia, western Yushu and southern Guoluo have annual hail days > 8 d, as expected (Figure 5b). According to hailstorm hazard assessment, the hazard zones (HI > 6) account for 14.4% and are concentrated in the central part of the QTP, including the prefectures of Naqu, Haixi and western Yushu (Figure 5c).

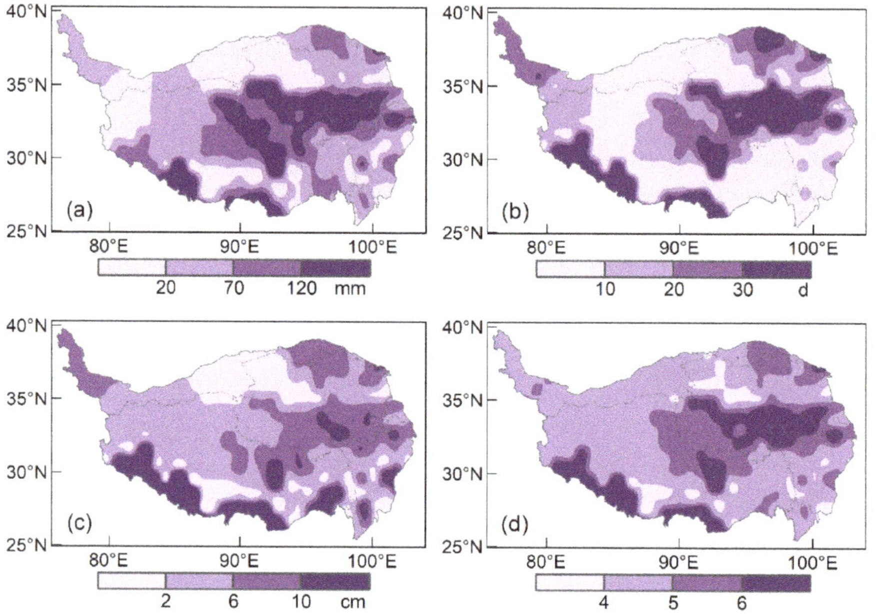

Figure 4. The spatial patterns of snowstorm hazard across the QTP. (**a**) annual accumulated snowfall; (**b**) annual snow cover days; (**c**) maximum snow depth; (**d**) snowstorm hazard index.

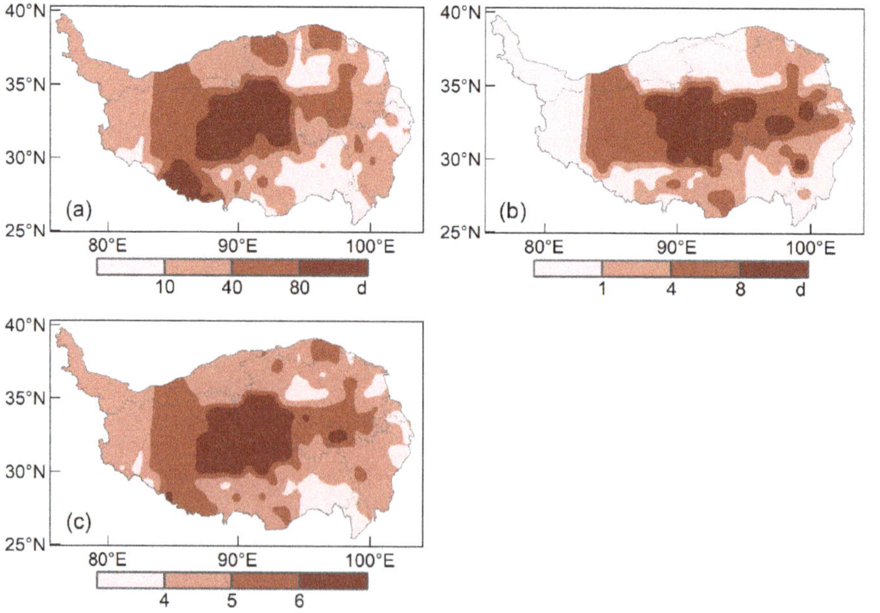

Figure 5. The spatial patterns of hailstorm hazard across the QTP. (**a**) annual gale days; (**b**) annual hail days; (**c**) hailstorm hazard index.

3.5. Multi-Hazard Assessments and Mapping

A multi-hazard index is developed to integrate the intensity of multiple meteorological hazards in the QTP, and the spatial pattern of multi-hazard levels is shown in Figure 6a, namely very low, low, moderate and high levels. The areas with high hazard level (HI > 6) account for 19.2% of the QTP and cover the prefectures of central Xigaze, southern Shannan, eastern Naqu, Nyingchi, Diqing, Nujiang, Liangshan, northwestern Aba, southern Huangnan, central Goluo, northern Ganzi and eastern Yushu (Figure 6a). Wherein, central Xigaze and southern Huangnan are severely affected by compound hazards of snowstorms, droughts and rainstorms. The Hengduan Mountains and Aba are affected by droughts and rainstorms, eastern Naqu is affected by snowstorms and hailstorms, central Guoluo and western Yushu are affected by snowstorms. All the areas mentioned above have been affected by high-level hazards in the past decades. The multi-hazard intensities in the northern and northwestern parts of QTP, including the Kunlun Mountains, Qaidam Basin, western Tibet Plateau and Ngari Plateau are relatively low (HI < 5). Most of these places belong to the classification of permanent arid region with no possibility of meteorological droughts, and have relatively low frequency of rainstorms, snowstorms and hailstorms as well.

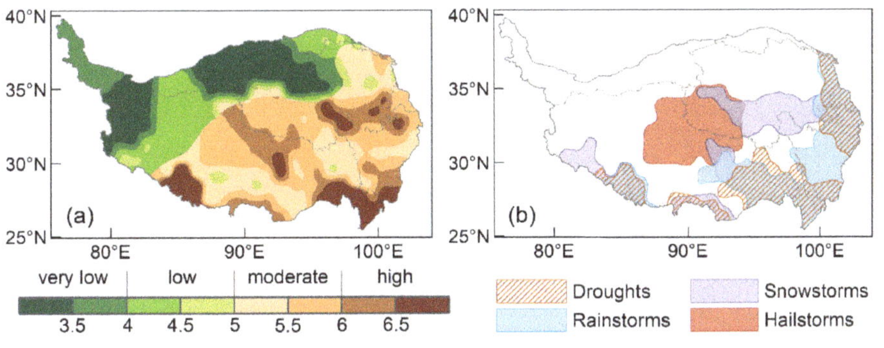

Figure 6. Mapping of (**a**) multi-hazard index and (**b**) meteorological hazards regionalization in the QTP.

Based on the identified hazard zones (HI > 6) of droughts, rainstorms, snowstorms and hailstorms in the QTP, the meteorological hazards regionalization in the QTP is mapped (Figure 6b). Drought hazard zones tend to overlay those of rainstorms due to the fact that rainstorm hazard zone is dominated by remarkable precipitation fluctuations and, hence, frequent occurrence of meteorological droughts. Three snowstorm hazard zones are identified in the map: the Himalayas Mountains, the Nyenchen-Tanglha Mountains and Qinghai Plateau. The snowfall amount, snow cover days and maximum snow depth in the hazard zones are significantly higher than the rest parts of the QTP. Hailstorm hazard zones distribute in the eastern Qiangtang Plateau and the Tanglha Mountains, and characterized by notable frequencies of gale and hail in the QTP.

4. Evolution of Population Exposure to Meteorological Hazards

Recent decades witness accelerated changes in global climate, and the spatial extent of meteorological hazards in the QTP also shows interannual variations. According to the indicator system for meteorological hazards assessment (Table 1), the hazard zones are identified when the grids in them meet the high-level criterion of all indices for a certain hazard. In this way, the spatial evolution of hazard zones for the period of 2001–2019 is analyzed. The population exposure to meteorological hazards is further adopted to reveal the historical changes in potential population risks of the QTP.

The maximum extent of drought hazard zones reaches 1.24 million km^2 in 2015, approximately half of the QTP, and the exposed population is estimated at 10.9 million people (Figure 7). The minimum extent of drought hazard zones is 0.23 million km^2 in 2018, accounting for 8.9% of the QTP, with 1.3 million people as the exposed population. Comparing the 2010s with the 2000s, it is found that the hazard zones expand by 5.4%, whereas the exposed population fell by 10%. The main reason is that the ascending trend of precipitation in the QTP weakens the frequency and intensity of meteorological droughts, especially in the densely populated areas of Western Sichuan Plateau, Songpan Plateau and the Bayan Har Mountains, and hence reduces the total affected population (Figure 8a). However, it is notable that the hazard zones are expanding in the South Tibet Valley and the Hengduan Mountains, posing a great threat to local agricultural and livestock industry. Hence, it is necessary to further improve the drought risk prevention in these areas.

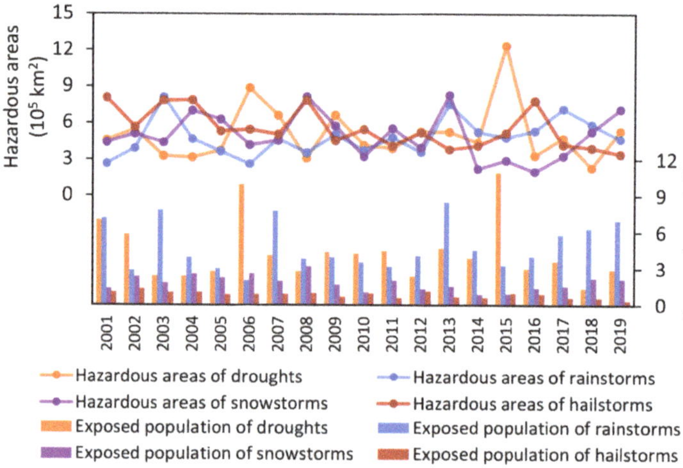

Figure 7. Temporal evolution in the spatial extent of hazard zones and exposed population of droughts, rainstorms, snowstorms and hailstorms for 2001–2019 in the QTP.

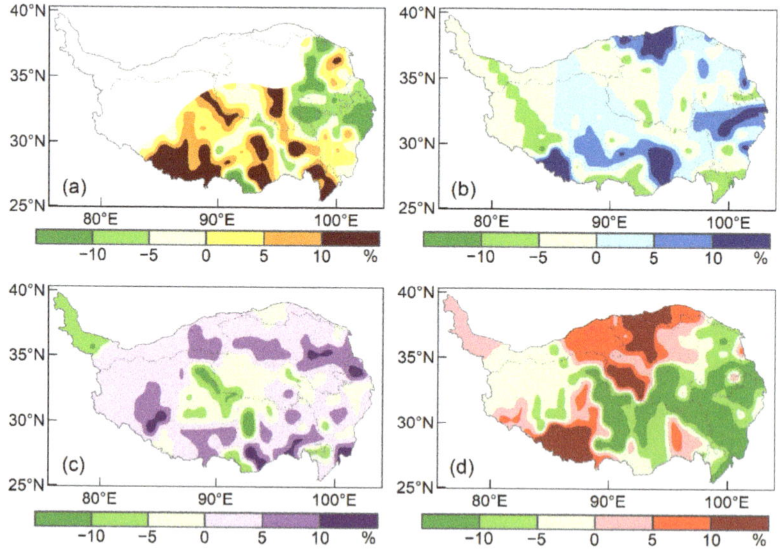

Figure 8. Spatial differences in (**a**) droughts hazard index, (**b**) rainstorms hazard index, (**c**) snowstorms hazard index, and (**d**) hailstorms hazard index, between 2010s and 2000s as percentages in the QTP.

In terms of rainstorms, the maximum extent of hazard zones reaches 0.81 million km^2 in 2003, whereas the largest exposed population is 8.5 million people in 2013 (Figure 7). The minimum hazard zones of rainstorms appears in 2006, at 0.26 million km^2, and the exposed population is about 2.0 million people. Comparing the second half to the first half period, we find that the hazard zones expands by 28.4% in 2010s, and the exposed population also increases by 13.1%. The spatial pattern indicates a notable increment of the rainstorms hazard index in the Qilian Mountains, the western Sichuan plateau and the northern Shannan (Figure 8b), resulting in the expansion of hazard zones. A particular concern is that rainstorm risk in Xining and Lhasa, two densely populated cities in the QTP, has increased significantly due to the combined effect of aggravated hazard and elevated exposure.

Snowstorms have always been one of the dominant meteorological hazards in the QTP. The maximum extent of snowstorm hazard zones reaches 0.83 million km^2 in 2013, whereas the maximum exposed population is nominated at 3.2 million people in 2008 (Figure 7). The minimum extent of hazard zones appears in 2016 at 0.20 million km^2, and the minimum exposed population is 0.8 million people in 2014. Comparing the second half to the first half period, we find that the hazard intensity reduces by 14.5% in 2010s, and the exposed population decreases by 26.0% as well. The spatial pattern indicates larger areas with intensified hazard in the QTP but are mainly distributed in low-hazard areas (Figure 8c). Moreover, two of the three snowstorm hazard zones, i.e., the Tanglha Mountains and Qinghai Plateau, decrease significantly in snowstorm hazard, resulting in an overall declining trend in exposed population. Nevertheless, from 2016 to 2019, the hazard zones of snowstorms are constantly expanding, suggesting the possibility of extreme events within the general trend. It is still necessary to strengthen the ability of monitoring and forecasting snow cover in pasturing areas, and improve the measures of forage storage and management, so as to reduce the impact of extreme snowstorms on livestock production in the future.

The maximum extent of hailstorm hazard zones reaches 0.81 million km^2 in 2001, whereas the maximum exposed population is nominated as 1.3 million people in 2002 (Figure 7). The minimum extent of hazard zones appears in 2019, at 0.34 million km^2, and the exposed population is about 0.4 million people. Comparing the second half to the first half period, we find that the hazard zones reduces by 26.1% in 2010s, and the exposed population decreases by 27.3% as well. The spatial pattern indicates that hailstorm hazard decreases in the Nyenchen-Tanglha Mountains, the Hengduan Mountains, western Sichuan plateau and the Bayan Har Mountains, whereas increases in Qaidam Basin, western Qinghai plateau and the Gangdise Mountains (Figure 8d). As for the climatological hazard zones of hailstorms, the hazard intensifies in western Qinghai plateau, whereas it weakens in the other parts, thus the exposed population of hailstorms in the QTP shows a downward trend.

5. Discussion and Conclusions

5.1. Discussion

Meteorological hazard research is highly dependent on long-term historical observations. At present, studies on meteorological hazards in the QTP are very limited due to the lack of observation data. Herein, we collect data from 246 meteorological stations in the QTP and attempt to carry out a multi-hazard assessment. It should be noted that the spatial distribution of meteorological stations is uneven, especially the limitation of observations in the northern Tibetan Plateau and the Kunlun Mountains increases uncertainty in hazard analysis. Future studies could study the applicability of remote sensing datasets in the QTP to improve the accuracy of hazard assessments.

The period from 2001 to 2019 is selected to analyze climate hazards over the QTP under the following two considerations. Firstly, climate variables such as temperature, precipitation and wind speed in the QTP are changing dramatically in recent decades due to accelerated global warming, thus the observations in the recent two decades are more likely to describe the current situation. Secondly, the weight coefficients in the assessment model are determined by historical disaster records. With the rapid growth of social wealth, the exposure decades ago is definitely different from the present, leading to low comparability of disaster losses over a long-time series. Therefore, the period of 2001–2019 is selected as a compromise between the robustness of climate conditions and the comparability in the disaster losses.

Current warning signals for meteorological hazard implemented by the China Meteorological Administration specifies the graded thresholds of multiple kinds of hazardous weather, which is applicable to the whole country. However, due to the special geographical conditions in the QTP, these thresholds could not be directly applied to the QTP. Hence, the relative thresholds are determined by standard deviation method in order to identify hazard zones in which the hazard intensity is at least 1 STD higher than the spatial mean of the QTP.

In comparison with previous studies on meteorological hazards, the hazard zone of droughts shows a similar pattern with existing works in the Tibet region [19,34]. Permanent arid zone is proposed for the first time in this paper to present the necessary conditions of meteorological droughts, which could provide reference for future research. The identification of rainstorm hazard zone has made certain progress in spatial resolution compared with previous studies [6]. We have also tried to use precipitation remote sensing datasets, but found many errors in plateau areas. This problem is expected to be solved as technology improves in the future. Studies on the snowfall over the QTP have been investigated by scholars in recent years [35]. On their basis, we further identify three hazard zones of snowstorms in the QTP, and reveal a downward trend in hazard intensity. As for hailstorms, there are few works that could be referenced in the QTP. However, a declining trend in hail days is also detected in another work by Zou et al. [23], thus we tend to believe that hailstorm risks would be moderated in a warming climate.

Monitoring changes in meteorological hazards and their effect on population in the QTP is exploratory work; it is an attempt aimed to reveal the evolutionary trends of socioeconomic exposure. Future studies may adopt climate models to simulate hazard intensities under radiative-forcing scenarios, and the population exposure could be estimated under socioeconomic scenarios. In this way, we could quantitatively assess the population risks of meteorological hazards in the future, and establish a solid scientific basis for meteorological risk management of the QTP.

5.2. Conclusions

Since the 21st century, the hazard intensity of meteorological droughts has been increasing in the South Tibet Valley and the Hengduan Mountains, leading to the expansion of drought hazard zone in the QTP. It is suggested that meteorological monitoring and early warning systems should be further improved in these areas, meanwhile the drought prevention capacity in grain producing areas and pastoral areas should be strengthened.

Rainstorm hazard zone has expanded by 28.4% and the exposed population has increased by 13.1% in the last 20 years. Intensified rainstorm hazard is detectable in the densely populated prefectures of Xining and Lhasa; hence, the technical standard of urban infrastructures should be improved to address climate risks. Rainstorm-induced geological hazards such as landslides, collapses and debris flows should also be taken precautions in mountain areas.

Due to the reduction in snowstorm intensity in the Tanglha Mountains and Qinghai Plateau, snowstorm hazard zone has reduced by 14.5% in the last two decades but the area fluctuates greatly on an inter-annual scale. Therefore, it is still necessary to improve the management of livestock industry in snow seasons of the QTP.

Hailstorm hazard zone distributes in the central part of the QTP, including the prefectures of Nagqu, Haixi and the western part of Yushu. Hailstorms show an overall descending trend since the 21st century but ascend in the western part of Qinghai Plateau. Improvements in weather forecast capacity is the key to mitigate losses of hailstorms.

Based on an integrated assessment model, we quantified the integrated intensity of multiple meteorological hazards, and further developed a multi-hazard regionalization map that could be referenced for territorial planning and disaster risk management in the QTP.

Author Contributions: Conceptualization, S.S. and Q.Z.; methodology, S.S.; validation, Q.Z. and R.Y.; formal analysis, S.S.; investigation, Y.X. and R.Y.; resources, Q.Z.; data curation, S.S. and Y.X.; writing—original draft preparation, S.S. and Y.X.; writing—review and editing, Q.Z. and R.Y.; visualization, S.S.; supervision, Q.Z.; project administration, R.Y.; funding acquisition, Q.Z. All authors have read and agreed to the published version of the manuscript.

Funding: This research has been financially supported by the China National Key R&D Program (Grant No. 2019YFA0606904 and 2018YFC1509003), the National Natural Science Foundation of China (Grant No. 41701103 and 41771536), the Major Research and Development Program of China Railway Group (P2018T006), and the UK-China Cooperation on Climate Change Risk Assessment.

Conflicts of Interest: The authors declare no conflict of interest.

References

1. Xu, W.; Zhuo, L.; Zheng, J.; Ge, Y.; Gu, Z.; Tian, Y. Assessment of the casualty risk of multiple meteorological hazards in China. *Int. J. Environ. Res. Public Health* **2016**, *13*, 222. [CrossRef] [PubMed]
2. Aksha, S.K.; Resler, L.M.; Juran, L.; Carstensen, L. A geospatial analysis of multi-hazard risk in Dharan, Nepal. *Geomat. Nat. Hazards Risk* **2020**, *11*, 88–111. [CrossRef]
3. Duan, A.; Xiao, Z. Does the climate warming hiatus exist over the Tibetan Plateau? *Sci. Rep.* **2015**, *5*, 13711. [CrossRef] [PubMed]
4. Liu, X.; Chen, B. Climatic warming in the Tibetan Plateau during recent decades. *Int. J. Climatol.* **2000**, *20*, 1729–1742. [CrossRef]
5. Song, C.; Pei, T.; Zhou, C. The role of changing multiscale temperature variability in extreme temperature events on the eastern and central Tibetan Plateau during 1960–2008. *Int. J. Climatol.* **2014**, *34*, 3683–3701. [CrossRef]
6. Yang, T.; Hao, X.; Shao, Q.; Xu, C.; Zhao, C.; Chen, X.; Wang, W. Multi-model ensemble projections in temperature and precipitation extremes of the Tibetan Plateau in the 21st century. *Glob. Planet. Chang.* **2012**, *80–81*, 1–13. [CrossRef]
7. Yin, H.; Sun, Y.; Donat, M. Changes in temperature extremes on the Tibetan Plateau and their attribution. *Environ. Res. Lett.* **2019**, *14*, 124015. [CrossRef]
8. You, Q.; Wu, F.; Shen, L.; Pepin, N.; Jiang, Z.; Kang, S. Tibetan Plateau amplification of climate extremes under global warming of 1.5 °C, 2 °C and 3 °C. *Glob. Planet. Chang.* **2020**, *192*, 103261. [CrossRef]
9. Fang, Y.; Zhao, C.; Ding, Y.; Qin, D.; Huang, J. Impacts of snow disaster on meat production and adaptation: An empirical analysis in the yellow river source region. *Sustain. Sci.* **2016**, *11*, 249–260. [CrossRef]
10. Gao, J.; Huang, X.; Ma, X.; Feng, Q.; Liang, T.; Xie, H. Snow disaster early warning in pastoral areas of Qinghai Province, China. *Remote Sens.* **2017**, *9*, 475. [CrossRef]
11. Li, Y.; Ye, T.; Liu, W.; Gao, Y. Linking livestock snow disaster mortality and environmental stressors in the Qinghai-Tibetan Plateau: Quantification based on generalized additive models. *Sci. Total Environ.* **2018**, *625*, 87–95. [CrossRef]
12. Wang, W.; Liang, T.; Huang, X.; Feng, Q.; Xie, H.; Liu, X.; Chen, M.; Wang, X. Early warning of snow-caused disasters in pastoral areas on the Tibetan Plateau. *Nat. Hazards Earth Syst. Sci.* **2013**, *13*, 1411–1425. [CrossRef]
13. IPCC. *Climate Change: Impacts, Adaptation, and Vulnerability*; Cambridge University Press: Cambridge, UK; New York, NY, USA, 2014.
14. Wang, J.; Wang, Y.; Wang, S. Biophysical and socioeconomic drivers of the dynamics in snow hazard impacts across scales and over heterogeneous landscape in Northern Tibet. *Nat. Hazards* **2016**, *81*, 1499–1514. [CrossRef]
15. Anderegg, W.; Kane, J.; Anderegg, L. Consequences of widespread tree mortality triggered by drought and temperature stress. *Nat. Clim. Chang.* **2013**, *3*, 30–36. [CrossRef]
16. Bartletta, M.; Klein, T.; Jansen, S.; Choat, B.; Sack, L. The correlations and sequence of plant stomatal, hydraulic, and wilting responses to drought. *Proc. Natl. Acad. Sci. USA* **2016**, *113*, 13098–13103. [CrossRef]
17. Asner, G.; Brodrick, P.; Anderson, C.; Vaughn, N.; Knapp, D.; Martin, R. Progressive forest canopy water loss during the 2012-2015 California drought. *Proc. Natl. Acad. Sci. USA* **2016**, *113*, 249–255. [CrossRef] [PubMed]
18. Sperry, J.; Love, D. What plant hydraulics can tell us about responses to climate-change droughts. *New Phytol.* **2015**, *207*, 14–27. [CrossRef]
19. Li, S.; Yao, Z.; Liu, Z.; Rui, W.; Adam, J. The spatio-temporal characteristics of drought across Tibet, china: Derived from meteorological and agricultural drought indexes. *Theor. Appl. Climatol.* **2019**, *137*, 1–16. [CrossRef]
20. Yin, J.; Gentine, P.; Zhou, S.; Sullivan, S.; Wang, R.; Zhang, Y.; Guo, S. Large increase in global storm runoff extremes driven by climate and anthropogenic changes. *Nat. Commun.* **2018**, *9*, 4389. [CrossRef]
21. Hohl, R.; Schiesser, H.; Aller, D. Hailfall: The relationship between radar-derived hail kinetic energy and hail damage to buildings. *Atmos. Res.* **2002**, *63*, 177–207. [CrossRef]

22. Li, M.; Zhang, Q.; Zhang, F. Hail day frequency trends and associated atmospheric circulation patterns over China during 1960-2012. *J. Clim.* **2016**, *29*, 7027–7044. [CrossRef]
23. Zou, T.; Zhang, Q.; Li, W.; Li, J. Responses of Hail and Storm Days to Climate Chang. in the Tibetan Plateau. *Geophys. Res. Lett.* **2018**, *45*, 4485–4493. [CrossRef]
24. Changnon, S. Increasing major hail losses in the U.S. *Clim. Chang.* **2009**, *96*, 161–166. [CrossRef]
25. Púcik, T.; Castellano, C.; Groenemeijer, P.; Kühne, T.; Rädler, A.; Antonescu, B.; Faust, E. Large hail incidence and its economic and societal impacts across Europe. *Mon. Weather Rev.* **2019**, *147*, 3901–3916. [CrossRef]
26. Zscheischler, J.; Westra, S.; Van Den Hurk, B.; Seneviratne, S.; Ward, P.; Pitman, A.; AghaKouchak, A.; Bresch, D.; Leonard, M.; Wahl, T.; et al. Future climate risk from compound events. *Nat. Clim. Chang.* **2018**, *8*, 469–477. [CrossRef]
27. Liu, B.; Siu, Y.; Mitchell, G. A quantitative model for estimating risk from multiple interacting natural hazards: An application to northeast Zhejiang, China. *Stoch. Environ. Res. Risk Assess.* **2017**, *31*, 1319–1340. [CrossRef]
28. Liu, B.; Siu, Y.; Mitchell, G.; Xu, W. Exceedance probability of multiple natural hazards: Risk assessment in China's Yangtze River Delta. *Nat. Hazards* **2013**, *69*, 2039–2055. [CrossRef]
29. Kong, J.; Simonovic, S. Probabilistic Multiple Hazard Resilience Model of an Interdependent Infrastructure System. *Risk Anal.* **2019**, *39*, 1843–1863. [CrossRef]
30. Yin, H.; Cao, C.; Xu, M.; Chen, W.; Ni, X.; Chen, X. Long-term snow disasters during 1982–2012 in the Tibetan Plateau using satellite data. *Geomat. Nat. Hazards Risk* **2017**, *8*, 466–477. [CrossRef]
31. Huang, W.; Qiu, T.; Yang, Z.; Lin, D.; Wright, J.; Wang, B.; He, X. On the Formation Mechanism for Wintertime Extreme Precipitation Events Over the Southeastern Tibetan Plateau. *J. Geophys. Res. Atmos.* **2018**, *123*, 12692–12714. [CrossRef]
32. Yao, T.; Xue, Y.; Chen, D.; Chen, F.; Thompson, L.; Cui, P.; Li, Q. Recent third pole's rapid warming accompanies cryospheric melt and water cycle intensification and interactions between monsoon and environment: Multidisciplinary approach with observations, modeling, and analysis. *Bull. Am. Meteorol. Soc.* **2019**, *100*, 423–444. [CrossRef]
33. Wang, S.; Zhou, L.; Wei, Y. Integrated risk assessment of snow disaster over the Qinghai-Tibet Plateau. *Geomat. Nat. Hazards Risk* **2019**, *10*, 740–757.
34. Chen, Q.; Liu, F.; Chen, R.; Zhao, Z.; Zhang, Y.; Cui, P.; Zheng, D. Trends and risk evolution of drought disasters in Tibet Region, China. *J. Geogr. Sci.* **2019**, *29*, 1859–1875. [CrossRef]
35. Zhang, A.; Wang, J.; Jiang, Y.; Chen, Y.; Shi, P. Spatiotemporal changes of hazard intensity-adjusted population exposure to multiple hazards in Tibet during 1982–2015. *Int. J. Dis. Risk Sci.* **2018**, *9*, 541–554. [CrossRef]

sustainability

Article

Locomotion of Slope Geohazards Responding to Climate Change in the Qinghai-Tibetan Plateau and Its Adjacent Regions

Yiru Jia [1,2], Jifu Liu [1,2], Lanlan Guo [2,3,*], Zhifei Deng [2], Jiaoyang Li [2] and Hao Zheng [2]

1 Key Laboratory of Environmental Change and Natural Disaster, Academy of Disaster Reduction and Emergence Management, Faculty of Geographical Science, Beijing Normal University, Beijing 100875, China; yirujia@mail.bnu.edu.cn (Y.J.); liujifu@bnu.edu.cn (J.L.)
2 Faculty of Geographical Science, Beijing Normal University, Beijing 100875, China; 201531480026@mail.bnu.edu.cn (Z.D.); 201821051182@mail.bnu.edu.cn (J.L.); zhenghao@mail.bnu.edu.cn (H.Z.)
3 State Key Laboratory of Earth Surface Processes and Resource Ecology, Beijing Normal University, Beijing 100875, China
* Correspondence: guolanlan@bnu.edu.cn; Tel.: +010-58804674

Abstract: Slope geohazards, which cause significant social, economic and environmental losses, have been increasing worldwide over the last few decades. Climate change-induced higher temperatures and shifted precipitation patterns enhance the slope geohazard risks. This study traced the spatial transference of slope geohazards in the Qinghai-Tibet Plateau (QTP) and investigated the potential climatic factors. The results show that 93% of slope geohazards occurred in seasonally frozen regions, 2.6% of which were located in permafrost regions, with an average altitude of 3818 m. The slope geohazards are mainly concentrated at 1493–1988 m. Over time, the altitude of the slope geohazards was gradually increased, and the mean altitude tended to spread from 1984 m to 2562 m by 2009, while the slope gradient varied only slightly. The number of slope geohazards increased with time and was most obvious in spring, especially in the areas above an altitude of 3000 m. The increase in temperature and precipitation in spring may be an important reason for this phenomenon, because the results suggest that the rate of air warming and precipitation at geohazard sites increased gradually. Based on the observation of the spatial location, altitude and temperature growth rate of slope geohazards, it is noted that new geohazard clusters (NGCs) appear in the study area, and there is still a possibility of migration under the future climate conditions. Based on future climate forecast data, we estimate that the low-, moderate- and high-sensitivity areas of the QTP will be mainly south of 30° N in 2030, will extend to the south of 33° N in 2060 and will continue to expand to the south of 35° N in 2099; we also estimate that the proportion of high-sensitivity areas will increase from 10.93% in 2030 to 14.17% in 2060 and 17.48% in 2099.

Keywords: Qinghai-Tibet Plateau; climate change; slope geohazards; new geohazard clusters

Citation: Jia, Y.; Liu, J.; Guo, L.; Deng, Z.; Li, J.; Zheng, H. Locomotion of Slope Geohazards Responding to Climate Change in the Qinghai-Tibetan Plateau and Its Adjacent Regions. *Sustainability* **2021**, *13*, 10488. https://doi.org/10.3390/su131910488

Academic Editors: Xiaodong Yan, Jia Yang and Shaofei Jin

Received: 20 July 2021
Accepted: 2 September 2021
Published: 22 September 2021

1. Introduction

Slope geohazards are widespread worldwide [1], and cause significant social, economic and environmental losses [2,3]. Slope geohazards in high-altitude areas have become more frequent and are closely related to global climate change [4–6], which has attracted increasing attention from researchers. The global climate has warmed significantly in recent decades, mainly at high altitudes and latitudes [7]. The shrinkage of glaciers and the degradation of permafrost are expected to significantly worsen the geotechnical and mechanical properties of rocks, debris and soils in high mountain areas [8,9], e.g., by changing active bed thickness and fracture conditions by reducing shear strength to affect slope or wall-rock stability [10]. Permafrost and general cryospheric degradation may have played a role in increasing slope failure, thereby resulting in slope geohazards at high elevations since the beginning of the 21st century [11].

The spatial and temporal distribution patterns of most landslides are affected by rainfall events and earthquakes [12]. The NASA research team [13] has suggested that warmer temperatures will cause more rain in the High Mountain Asia region of China, Tibet and Nepal, which could lead to increased landslide activity along the China–Nepal border. Heavy rainfall causes the increase in pore water pressure and the reduction in cohesion and friction coefficients [14,15]. The cumulative effect of the freeze–thaw cycles weaken the rock and propagate fissures. The current changes in permafrost conditions caused by atmospheric warming also affect the stability of steep rock surfaces in high mountain areas [9]. Climate change may increase the frequency of slope geohazards [16,17] and alter the spatial extent of slope geohazards [18].

As the third pole in the world, the Qinghai-Tibet Plateau (QTP) has the largest cryosphere system at low- and mid-latitudes, and is a sensitive area for climate change [19,20]. The QTP is prone to slope geohazards due to strong tectonic activity, complex geomorphology and climate change [21]. Additionally, in the QTP, the frequency and scale of slope geohazards have increased in recent years [22]. With the continuous warming of the climate, the thaw slump activity in the Qilian Mountain area is increasing, and the growth rate of thaw slump activity is accelerating [23]. In El Niño–Niña years, the frequency of debris flows in the mountainous areas of southeastern Tibet and the Hengduan Mountains has an obvious increasing trend due to increasing heavy rain events [24]. The accelerated melting of glaciers on the QTP and loose moraine deposits may form mudslides and dammed lakes [19,25]. Despite a good understanding of the spatial relationships between slope geohazards and their causative factors on the QTP [23,24], research on the distribution and evolution of the overall slope geohazards in the plateau in time series is insufficient. More attention should be given to the future evolution of slope geohazards in the QTP.

In this study, we collected data on slope geohazards from 1905 to 2015 in the study area, coupled with meteorological data and topographic data to assess the variation in spatial and temporal slope geohazards. The primary objectives of this study are to (1) obtain slope geohazard distribution patterns and location shifts, (2) identify factors affecting changes in slope geohazards over the study area and (3) predict the future occurrence of slope geohazards in the study area under global climate change. The contribution of this study is the proposed shift in slope geohazards in the study area and the distribution of slope geohazard risks under future climate conditions.

2. Study Area

The QTP is the youngest, highest and largest plateau in the world with an average height of over 4000 m and an area of approximately 2.57×10^6 km^2. It stretches from the Pamirs to the Hengduan Mountains and from south of the Himalayas to north of the Kunlun-Qilian Mountains [26]. In this study, the study area included not only the main body of the Tibet Plateau, but also the surrounding areas bordering on the Tibet Plateau, namely western Xinjiang, central and southern Gansu, central Sichuan and northern and central Yunnan (Figure 1a). The study area ranges in slope from 0 to 54° (Figure 1a). The land cover of the study area mainly consists of grassland, bare land, forest, cultivated land and permanent snow and ice (Figure 1c). The distribution of permafrost in the QTP is shown in Figure S1, and the areas of permafrost and seasonally frozen regions are 1.06×10^6 km^2 and 1.46×10^6 km^2, respectively [27].

The average annual air temperature of the study area ranges from −21.23 °C to 24.56 °C, and increases from northwest to southeast (Figure 1d). The spatial distribution of soil temperature is similar to that of the air temperature, with average annual temperatures within the range of −11.72 to 23.93 °C (Figure 1e). From 1980 to 2015, the annual average air temperature in the study area presented an increasing trend and the rate of temperature rise was 0.48 °C/decade, which was more than twice the global temperature rise in the same period [28]. The soil temperature increased at a rate of 0.067 °C/decade (Figure S2).

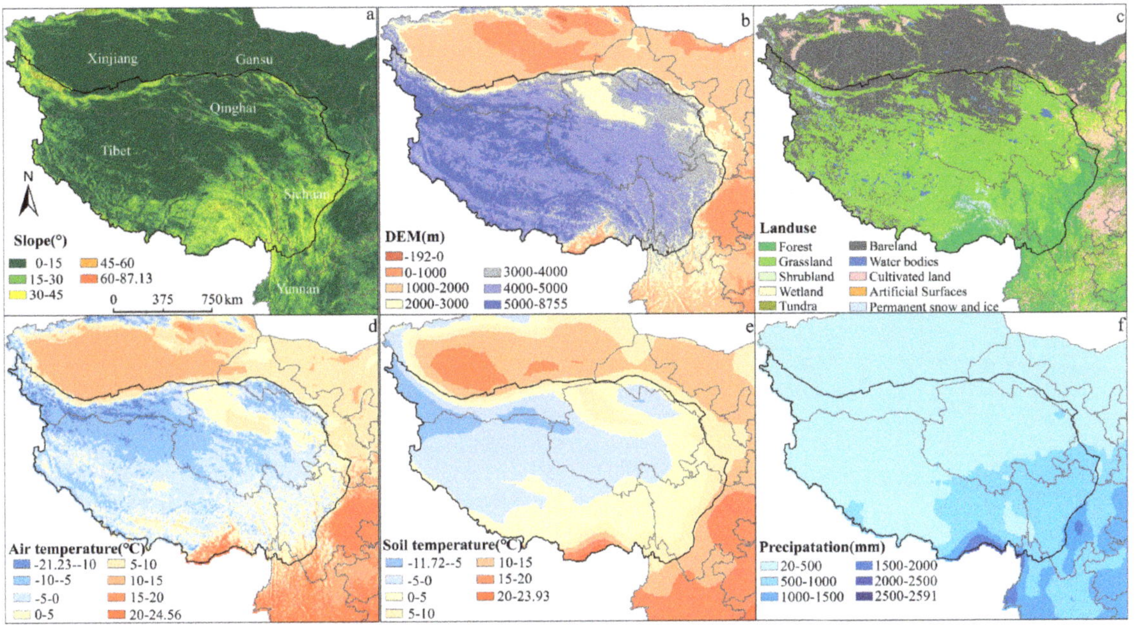

Figure 1. Overview of the study area. (**a**) Slope gradient; (**b**) elevation; (**c**) land use classification; (**d**) mean annual air temperature; (**e**) mean annual soil temperature; (**f**) mean annual precipitation.

The distribution of annual precipitation showed a distinct ladder, ranging from 20–2161 mm and decreasing from the southeast to the northwest (Figure 1f). Precipitation has increased in most regions during past decades in the central, northern and western parts of the study area, while it decreased in the eastern, southern and southeastern parts of the study area. The rate of increase in precipitation for the whole QTP was 32 mm/decade from 1980 to 2015 (Figure S2).

Slope geohazards in the study area are frequent and diverse. In general, slope geohazards occur frequently at the eastern edge, northeastern part and southern mountains, and the prevalence at the eastern edge is higher than that in the western part [29–31]. Since the 1930s, 27 outbursts have occurred in 18 glacial moraine lakes in Tibet due to ice avalanches and landslides, causing heavy casualties and destroying a large number of villages, farmland and infrastructure [32]. The two Baige landslides in October and November 2018 caused direct economic losses of more than CNY 10 billion [33]. Global warming has caused an increase in slope geohazards, and the severity makes the study of geohazards in the QTP and its surrounding areas more important.

3. Data and Methods

3.1. Data

The slope geohazards were compiled from the Centre for Research on the Epidemiology of Disasters (CRED) International Disaster Database (EM-DAT; http://www.em-dat.net, accessed on 13 June 2020)), the Durham Fatal Landslide Database (DFLD; http://www.landslidecentre.org/database.htm, accessed on 13 June 2020) and some previous publications [34–39]. We obtained the detailed administrative zoning locations of the geohazard sites and located them using Google Earth. The dataset includes 897 slope geohazards, including landslides, debris flows and rockfalls and does not include events directly caused by earthquakes.

A 90 × 90 m digital elevation model (DEM) from the Google Earth engine was collected for use as topographic information. The 90 × 90 m slope data were obtained through DEM conversion. A low DEM resolution will reduce the accuracy of the location attributes of slope geohazards and lead to a small distinction between the altitude and slope of geohazards. When studying slope failure over a long period of time, perfect DEM resolution may not exist, because there is no resolution that can display the scale of all different slope failures distributed across different times and locations [40]. Land cover data consist of GlobeLand30's 30 × 30 m resolution dataset obtained from the National Geomatics Center of China (NGCC; http://www.ngcc.cn/ngcc/, accessed on20 June 2020). The GlobeLand dataset 30 includes 10 type groups and truly reflects the actual land cover of the study area.

Monthly precipitation and air temperature data (0.1° × 0.1°) for the period 1980 to 2015 were obtained from the China Meteorological Forcing Dataset (CMFD, https://data.tpdc.ac.cn/zh-hans/, accessed on 20 June 2020) and developed by the Cold and Arid Regions Science Data Center (CARSDC) of the Chinese Academy of Science. Daily soil temperature data (0.75° × 0.75°) were obtained from ERA-Interim data provided by the European Centre for Medium-Range Weather Forecasts (ECMWF, https://www.ecmwf.int/en/forecasts/datasets, accessed on 25 June 2020). The soil data were divided into four layers: 0–7 cm, 7–28 cm, 28–100 cm and 100–255 cm. The soil temperature in the study is the average temperature of the four layers of soil.

All the model outputs used to assess the risk of slope geohazards in this study were sourced from CMIP5 (i.e., the fifth phase of the Coupled Model Mutual Comparison Project; https://esgf-node.llnl.gov/search/cmip5/, accessed on 20 July 2020), which is the most widely used and highly coordinated project in the international climate model project [41,42]. Eight climate models were selected under the highest emission scenarios (representative concentration pathway 8.5, RCP 8.5) from the CMIP5 data archive (Table S1). For each model, monthly precipitation data, surface air temperature and soil temperature of 2030, 2060 and 2099 were used for the study.

3.2. *Methods*

3.2.1. Analysis of Impact Factors

The linear trend [43] over the period 1980 to 2015 for the precipitation and the temperature was calculated for the entire period and each season using the non-parametric Mann-Kendall statistic.

Spatial distribution analysis was performed in a GIS environment through spatial statistical analysis functions [44], as spatial analysis can be used to reveal the patterns of slope geohazard spatial distribution. A grid with a resolution of 0.1° × 0.1° was established in the study area and the area of the study area was calculated accordingly. Slope geohazard coverage areas were defined as the sum of pixel grid areas with slope geohazards identified. We interpolated the spatial meteorological and topographic data with a resolution of 0.1° × 0.1° by using the bilinear interpolation method to generate the spatial distribution of factors such as slope, altitude, rainfall and temperature. The frequency of slope geohazard occurrence in different scopes of each factor was counted, and the correlation between factors and slope geohazards was established.

To explore the distribution variation of slope geohazards with different factors, we divided 36 years into 4 time periods (1980–1989, 1990–1999, 2000–2009, 2010–2015) and compared slope geohazard-covered areas and occurrence seasons in different periods. The correlation statistics of slope geohazards and factors were carried out in different periods.

3.2.2. Subsection

Considering the lag effect of climate on slope geohazards, the meteorological conditions of the season when slope geohazards occur and the season before the slope geohazards occur were analyzed. For example, for summer slope geohazards, the precipitation in spring and summer in the region where the slope geohazards are located should be taken into account; for spring slope geohazards, the precipitation in the preceding winter and spring of the year of slope geohazards should be taken into account. The average precipitation values of June, July and August in the region of a certain summer geohazard were taken as the summer precipitation of the geohazard site. The average precipitation value of March, April and May in the location of the summer geohazards was taken as the spring precipitation of the geohazard site. Then, the mean precipitation of each year's summer and spring geohazards was obtained, and the precipitation series of the time series were established, to obtain the change trend of summer precipitation and spring precipitation corresponding to the summer geohazards.

In the same way, the summer air temperature, soil temperature, spring air temperature and soil temperature of the summer geohazards were obtained, and the spring precipitation, air temperature, soil temperature and winter precipitation, air temperature and soil temperature of the spring geohazards were obtained. Then, we obtained the trend of them over the time series.

3.2.3. Geohazard Susceptibility Analysis

The random forest model is an integrated learning model based on a decision tree, also known as a simple and efficient artificial intelligence algorithm. Beyond the previously reported geohazard clusters (GCs) distributed on the eastern and southeastern margins of the QTP, slope geohazards occurred above 3000 m in the interior of the QTP and were defined as new geohazard clusters (NGCs). A total of 188 geohazards were collected from the NGCs, and 188 non-geohazard sample points were randomly selected from the QTP, which together constituted 376 total samples. Eighty percent of 376 sample points were randomly selected as the training set of the random forest training calculation, and the remaining 20% were selected as the validation dataset to verify the prediction rate of the results. Altitude, slope and historical meteorological data were selected as the influencing factors in the model. The number of decision trees was set as 500, and the number of parallel simulations was set as 200. Altitude, slope and future meteorological data were brought into the constructed model to obtain the distribution of geohazard susceptibility in 2030, 2060 and 2099. The historical and future meteorological data all included soil temperature in spring, precipitation and soil temperature in summer, and precipitation and soil temperature in winter in 2030, 2060 and 2099.

4. Results

4.1. Disaster-Prone Regions of Slope Geohazards

A total of 897 slope geohazard-covered areas within approximately 73.80×10^3 km^2 were recorded from 1905–2015. A total of 775 slope geohazards were recorded from 1980–2015 with an average frequency of approximately 21.5 times/year (Figure 2a). The numbers and covered areas of slope geohazards have grown rapidly over the past few decades. The slope of the cumulative frequency curve and the cumulative covered area curve after 1980 was larger than that before 1980 and was partly derived from the incomplete records in early decades; furthermore, the frequency and covered area increment in geohazards were the highest in 1998 and 1999 (Figure 2a, b). Notice here that the dataset may slightly underestimate the occurrence of the incidents, for which there are two main reasons:, i.e., the dataset inevitably fails to capture some smaller events, especially in remote mountainous areas [45], and previous event records are not systematic, and data are more difficult to obtain.

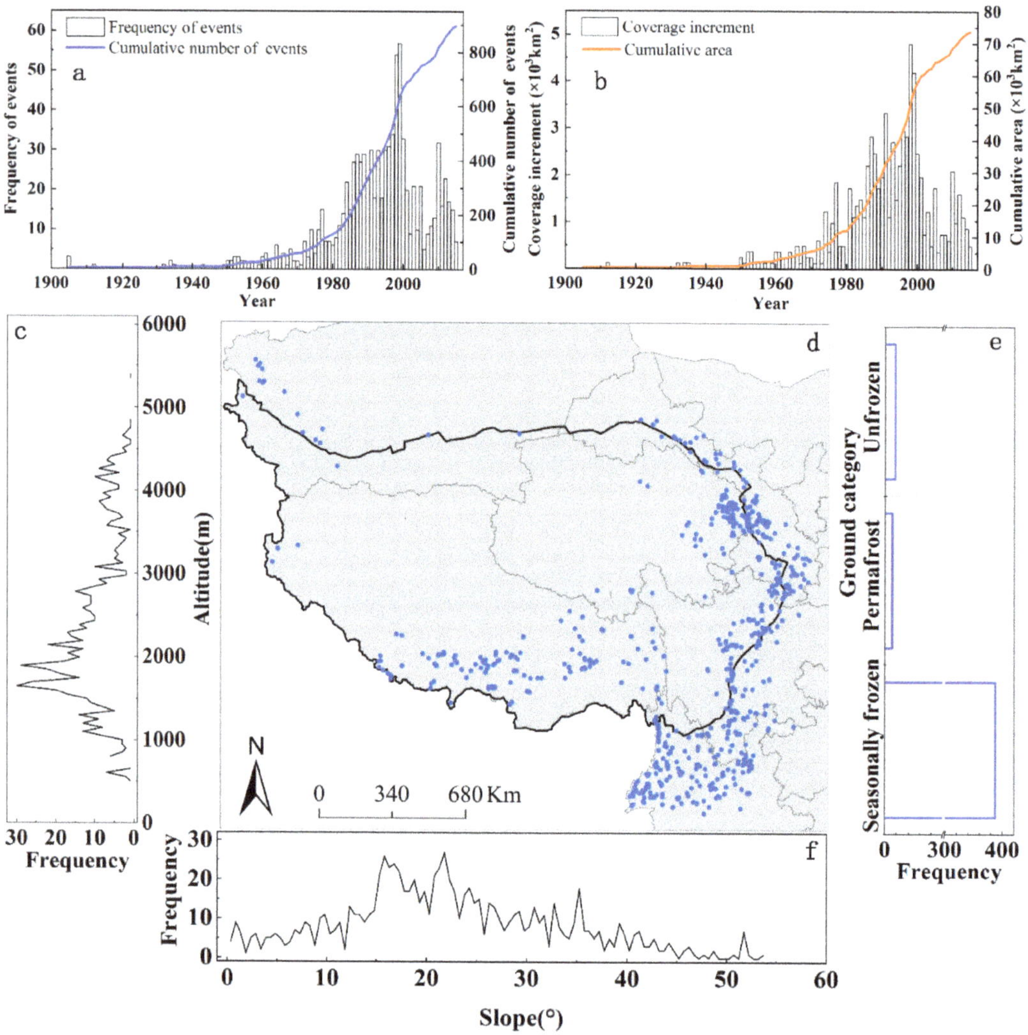

Figure 2. Slope geohazard distributions in the study area, 1905-2015. (**a**) Slope geohazard frequency per year and cumulative frequency (blue curve); (**b**) covered area of the slope geohazards per year and cumulative covered area (orange curve); (**c**) frequency of slope geohazards at different altitudes; (**d**) spatial distribution of slope geohazards in the study area; (**e**) frequency of slope geohazards in permafrost regions; (**f**) frequency of slope geohazards at different slopes.

Slope geohazards in the study area occurred in slope ranges of 0.36°–53.70° (Figure 2f), and their peak frequency was 13.82°–36.25°. The altitudinal gradient for slope geohazards ranged from 381 m to 5673 m, with peak frequency existing at 1493–1988 m (Figure 2c). The slope geohazard concentrated highly corresponded to the southeastern margins of the QTP, which are the slope geohazard-prone areas with high rates of tectonic processes and intense rainfall in mountain regions [46–49]. Meanwhile, the slope geohazard distribution also appeared to favor areas of freezing and thawing activities [23,50]. Ninety-three percent of slope geohazards occurred in seasonally frozen regions (Figure 2e), 2.6% were located in permafrost regions and the average altitude was 3818 m. The observed increased temperature and shifted precipitation distribution on the QTP are expected to impact freezing and thawing activities and related slope geohazards.

The occurrence of recorded slope geohazards throughout the year was unevenly distributed, with more events occurring in summer (June–August), followed by autumn (September–November) (Figure 3a). During the period between winter and spring (December–May), the recorded slope geohazard occurrence was low. This pattern reflects the dominant global trigger of slope geohazards, namely, the occurrence of precipitation associated with the Northern Hemisphere summer monsoon [51]. The annual precipitation of slope geohazards presented a bimodal distribution, with the first peak at 254–353 mm and the other peak at 510–550 mm (Figure 3b). In the region with abundant precipitation, slope geohazards are relatively more distributed in the region with annual precipitation of 790–930 mm (Figure 3b). The areas with precipitation in the months of slope geohazards of 50–150 mm are most prone to slope geohazards (Figure S3a). The areas with mean annual air temperatures and soil temperatures ranging from 3.82–9.74 °C (Figure 3c) and 4.67–9.09 °C (Figure 3d) were found to be the most prone to slope geohazards. Overall, slope geohazards, in accordance with these meteorological conditions, were mainly distributed at the eastern edge and southeast of the study area.

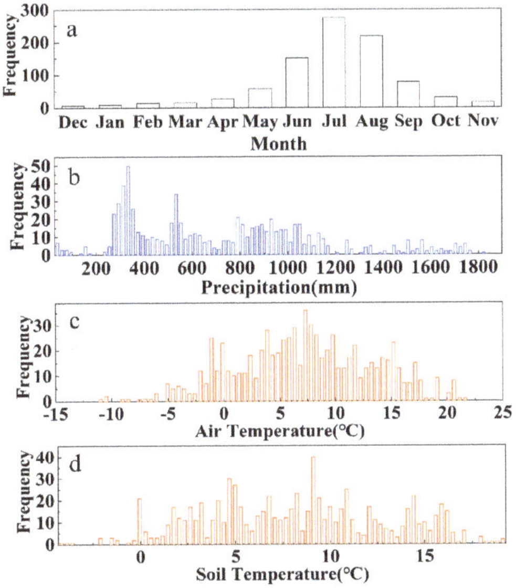

Figure 3. The temporal and meteorological characteristics of slope geohazard distributions in the study area. (**a**) Monthly frequency of slope geohazards; (**b**) annual precipitation at slope geohazard sites; (**c**) average annual air temperature at slope geohazard sites; (**d**) average annual soil temperature at slope geohazard sites.

4.2. *The Temporal Evolution of Slope Geohazards*

We further analyzed the distribution characteristics of slope geohazards in the study area from 1905 to 2015, and explored the changes in slope geohazards with time. For the convenience of comparison, the slope geohazards were divided into five groups according to the time of occurrence (Figure 4a), before 1979 (122 events), 1980–1989 (188 events), 1990–1999 (325 events), 2000–2009 (152 events) and 2010–2015 (110 events). The covered area average annual growth rates of slope geohazards were approximately 1601.7 km^2/year for 1980–1989, 2759.9 km^2/year for 1990–1999, 1108.9 km^2/year for 2000–2009 and 714.6 km^2/year for 2010–2015 (Figure 4b). The areal growth rate of slope geohazards from 1990–1999 was 1.7 times as much as from 1980–1989; after 2000, the areal growth rate of slope geohazards was less than from 1990–1999. Slope geohazards spread more widely over time and are likely to become more widespread.

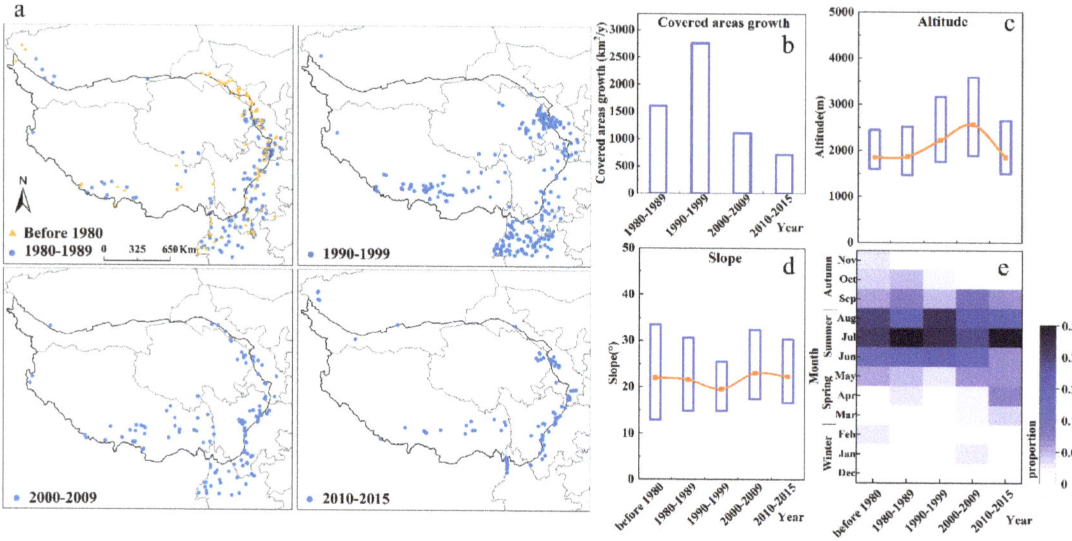

Figure 4. Slope geohazard change in different periods. (**a**) Spatial distribution of slope geohazards in different periods (the orange triangles represent geohazard sites before 1980, and the blue circles represent geohazard sites from 1980–1989, 1990–1999, 2000–2009, and 2010-2015); (**b**) the growth rate of the covered area; (**c**) the altitude of slope geohazards in different periods (the blue box represents 25%–75% of altitude range in geohazard sites, and the orange curve represents the variation in average altitude); (**d**) the slope gradient of slope geohazards in different periods (the blue box represents 25%–75% of slope range in geohazard sites, and the orange curve represents the variation in average slope); (**e**) the monthly proportion of slope geohazards in different periods.

Topographically, slope geohazards vary with time. By 2009, the mean altitude of slope geohazards had increased from 1984 m to 2562 m, with a growth rate of 238 m/10 year (Figure 4c). Then, the mean altitude of the slope geohazards decreased to 1857 m from 2010–2015 because the recorded events in this period were generally located at the lower altitude of the eastern boundary. In contrast, the slope range of slope geohazards remained stable during the study period, with the average slope maintained at 19.53°–22.96° from 1905–2015 (Figure 4b), which indicates that the influence of slope on geohazards in the study area has no obvious change. The altitude of the slope geohazards gradually increases, while the slope gradient varies only slightly. Therefore, the slope is still the most important condition to determine the occurrence of slope geohazards in the study area.

Slope geohazards spread to the spring, although summer and autumn were still the frequent seasons for slope geohazards from 1905–2015 (Figure 4d). Nine percent of slope geohazards occurred in spring before 1980, and 10%, 6%, 14% and 25% of the slope geohazards occurred in spring in the following four periods. The increase in slope geohazards in spring is related to climate change. It was further found that the seasonal variation is more prominent in the area above 2000 m (Figure S4). Before 1980, slope geohazards above 2000 m mostly occurred from May–October and in February, increased to April and December from 1980–1989, expanded to February–November from 1990–1999, expanded to January–November from 2000–2009 and occurred from March–September and in December from 2010–2015. The increase in slope geohazards in high-altitude areas in spring indicates that the driving factors of some slope geohazards in the study area have changed and that the impact of temperature rise is more obvious for slope geohazards.

The increasing frequency of slope geohazards is not only related to precipitation, but also affected by soil temperature. Considering the lagging influence of climate on geohazards, the precipitation and temperature of the previous season should also be taken into account in addition to the precipitation and temperature of the season in which the geohazards occur. The summer monthly precipitation corresponding to the summer geohazards decreased (Figure 5a), while the spring monthly precipitation increased (Figure 5b). The summer and spring soil monthly temperature corresponding to the summer geohazards both decreased (Figure 5c,d). This result suggested that summer geohazards are affected by summer precipitation, spring precipitation, summer soil temperature and spring soil temperature and tend to develop in regions with lower soil temperature in spring, lower soil temperature and less precipitation in summer and that the increase in spring precipitation may lead to geohazards in some areas during summer. The spring and winter soil monthly temperatures corresponding to spring geohazards decreased at a rate of 1.3 °C/10a (Figure 5e,f). This result suggested that spring geohazards are affected by winter and spring soil temperatures and tend to develop in regions with lower soil temperatures in winter and spring. There was no obvious influence on spring geohazards with spring, winter precipitation and air temperature (Figure S5a,b,e,f) and no obvious influence on summer geohazards with spring or summer air temperature (Figure S5c,d).

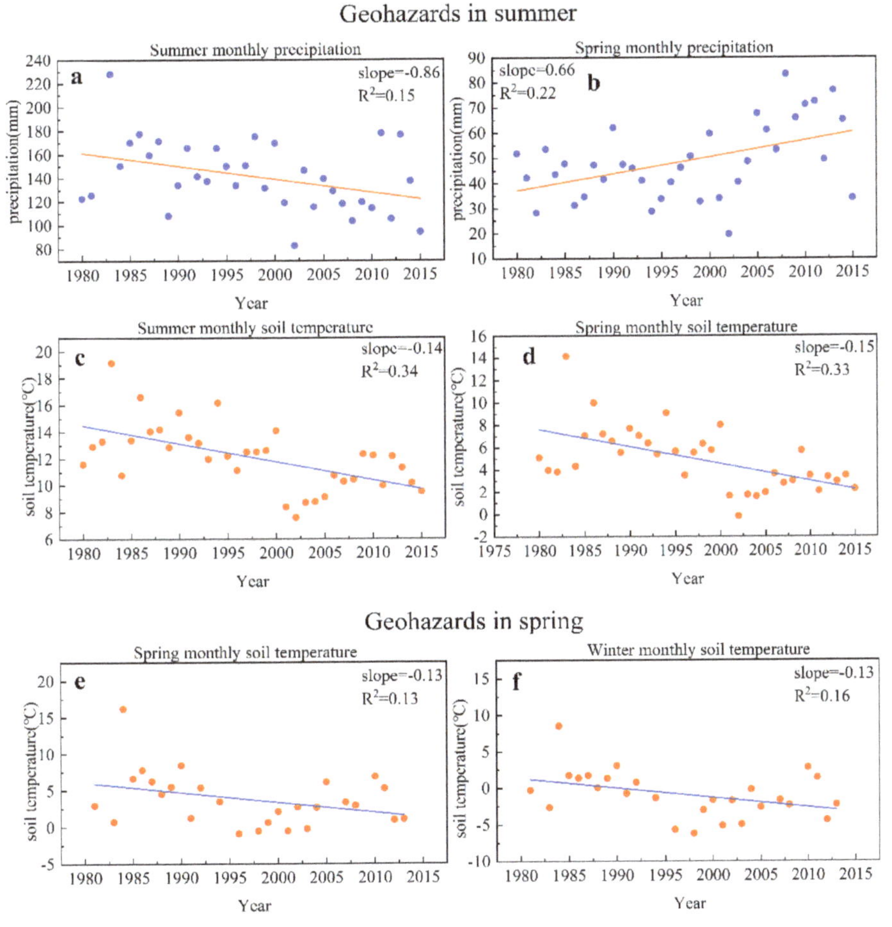

Figure 5. Changes in precipitation and temperature in different seasons of geohazards in different years. (**a**) Summer monthly precipitation in the area of summer geohazards; (**b**) spring monthly precipitation in the area of summer geohazards; (**c**) summer monthly soil temperature in the area of the summer geohazards; (**d**) spring monthly soil temperature in the area of the summer geohazards; (**e**) spring monthly soil temperature in the area of the spring geohazards; (**f**) winter monthly soil temperature in the area of the spring geohazards.

5. Discussion

5.1. Locomotion of Slope Geohazards in the QTP and Its Adjacent Regions

The QTP is one of the regions with the most frequent slope geohazards because of its complex geological environments, strong tectonic activity, poor stability of rock slopes and frequent rainstorms [52]. We observed not only the significant growth of slope geohazards but also an accelerating growth rate of scale in past decades, which is consistent with reports that the scale and frequency of slope geohazards in the QTP corridor are increasing [53]. Although such a trend was associated with improving skill in data collection, other field observations also reported an increase in slope failure on the QTP. In the permafrost region of the Qilian Mountains [23] and Fenghuo Mountain [50], slope failure has accelerated, and in the Beiluhe Region on the QTP, the total number of thaw slumps and the total surface area increased significantly [54]. Considering the increased human and economic vulnerability with plateau development, the slope geohazard risk is likely to be higher and requires more attention.

The spatial distribution displayed within the slope geohazard dataset is instructive, with the occurrence of slope geohazards separately concentrated in two altitude ranges. On the one hand, the well-known slope geohazard-prone area along the Himalayan Arc Regions in southeastern edges of the plateau [1,46] and, on the other hand, regions in the central plateau with higher altitudes (above 3000 m) have been suggested to be ever-increasing slope geohazards. The slope geohazards at higher altitudes showed particularity in the occurrence season. Different from traditional slope geohazards, which mainly occur in summer and autumn, the frequency of slope geohazards in high-altitude areas increases significantly in spring [55]. The spatial and temporal migration of slope geohazards both suggested that thaw slump activity is rapidly accelerating in the alpine permafrost regions on the central QTP [23].

It is interesting to find that the mean slope ranges remained steady at 19°–23° within all slope geohazards of different decades. Slope is still the key factor in predicting the slope geohazard risk, and slope size affects the stress distribution inside the slope body and controls the characteristics of surface runoff and groundwater [56]. Statistical results show that 70% of the area is within slope ranges of 13.82°–36.25° on the QTP, covering an area of 7.1×10^5 km^2. In higher elevation stands, rising spring temperatures and increasing precipitation will magnify slope geohazard risk [5]. This hypothesis is supported by our results and by short- and long-term studies in the Arctic and other alpine regions. In addition, the prolonged season of high-altitude geological disasters has been affected by climate warming [57].

5.2. Comparison of Two Slope Geohazard Clusters

Observed from the slope geohazard spatial distribution location and elevation variation, we could easily find the migration of slope geohazards. Beyond the previously reported geohazard clusters (GCs) distributed on the eastern and southeastern margins of the QTP, we found that slope geohazards occurred above 3000 m in the interior of the QTP and were defined as new geohazard clusters (NGCs) (Figure 6a). The distribution of GCs and NGCs is shown in Figure 6a. We compared the altitude, slope, season, precipitation, air temperature and soil temperature of the geohazards to illustrate the different driving factors of these two clusters.

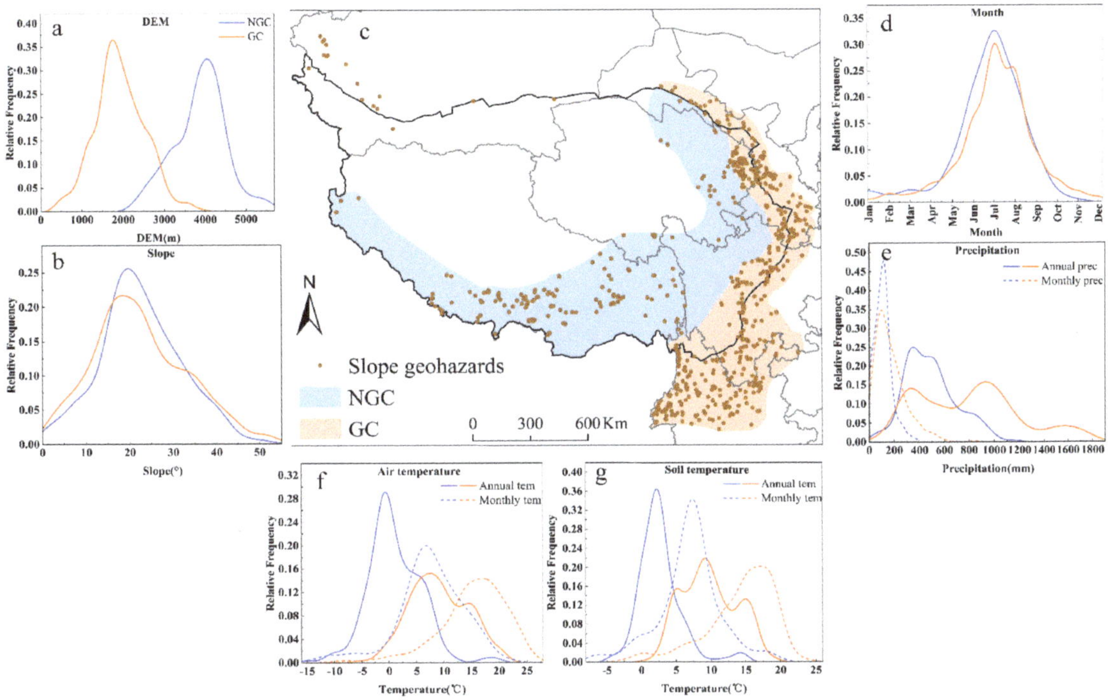

Figure 6. Comparison of the characteristics of new slope geohazard clusters (NGCs) and previously reported geohazard clusters (GCs). (**a**) Elevation-geohazard frequency curve comparison (blue curve refers to NGC, orange curve refers to GC); (**b**) slope-geohazard frequency curve comparison; (**c**) the comparison of the two clusters spatial distributions; (**d**) month-geohazard frequency curve comparison; (**e**) precipitation-geohazard frequency curve comparison (the solid line refers to the annual precipitation of geohazards, while the dotted line refers to the monthly precipitation of geohazards); (**f**) air temperature-geohazard frequency curve comparison (the solid line refers to the annual temperature of geohazards, while the dotted line refers to the monthly temperature of geohazards); (**g**) soil temperature-geohazard frequency curve comparison.

In terms of topography, the peak altitude is 1750 m in the GC while it is 4050 m in the NGC, but the frequent slope of the two disaster clusters is 13.82°–36.25°. The influence of slope on geohazards does not vary greatly with region. Summer and autumn are the rainy seasons on the QTP, so June–August is the peak period for the two disaster clusters. However, the proportion of NGCs occurring from May to July is higher than that of GCs, which may be related to the melting of snow and ice in high-altitude areas. In the seasonally frozen area of China, the occurrence of a landslide in summer and autumn is usually induced by rainfall, while in spring, it is caused by the freeze–thaw process [58]. There are two annual rainfall peaks in the GC, one of which is 300–400 mm in the northeastern part of Qinghai Province, which is in the transition zone from the QTP to Loess Plateau with complex terrain [59,60], and the other is 800–1100 mm in Yunnan and the northern part of Sichuan where both terrain and precipitation are conducive to slope geohazard development [29,61,62]. The annual precipitation of the NGC is mostly 300–600 mm. However, there is little difference in the monthly precipitation of the two disaster clusters, most of which are between 70 mm and 120 mm, which suggests that slope geohazards are more common in areas within this monthly precipitation threshold. The different spatial distributions of the two geohazard clusters also mean that there are significant differences in temperature; for example, slope geohazards occur most frequently in NGCs with an average annual temperature of −0.74 °C, while the peak temperature of the frequent geohazard clusters is 7.68 °C in the GC. The air temperature in the months of the geohazards was 6.66 °C in the NGC and 16.91 °C in the GC. Similarly, this difference is also reflected in soil temperature.

We further adopted the random forest model and future climatic data simulated by CMIP5 (RCP 8.5) to obtain the distribution of geohazard susceptibility on the QTP in 2030, 2060 and 2099. Altitude, slope and simulated future air temperature, soil temperature and precipitation were brought into the constructed model. The receiver operating characteristic (ROC) curve of the random forest model in this study is shown in Figure S6, and the area under the curve (AUC) is 0.9702. As the climate changes gradually, slope geohazards will spread over a wider area of the QTP. The proportion of very low-sensitivity areas decreased from 51.44% in 2030 to 42.95% in 2060 and 30.95% in 2099 (Figure 7). The areas of low-sensitivity areas, moderate-sensitivity areas and high-sensitivity areas are increasing. The proportion of highly sensitive areas increased from 10.93% in 2030 to 14.17% in 2060 and 17.48% in 2099. The moderate- and high-sensitivity areas are mainly distributed in the east and south of the QTP. The low-, moderate- and high-sensitivity areas of the QTP are mainly south of 30° N in 2030, extend to the south of 33° N in 2060 and continue to expand to the south of 35° N in 2099. The low-sensitivity region in the northern plateau gradually expands from 37° N inward to 35° N. Geohazards on the QTP will spread from the east and south to the central region.

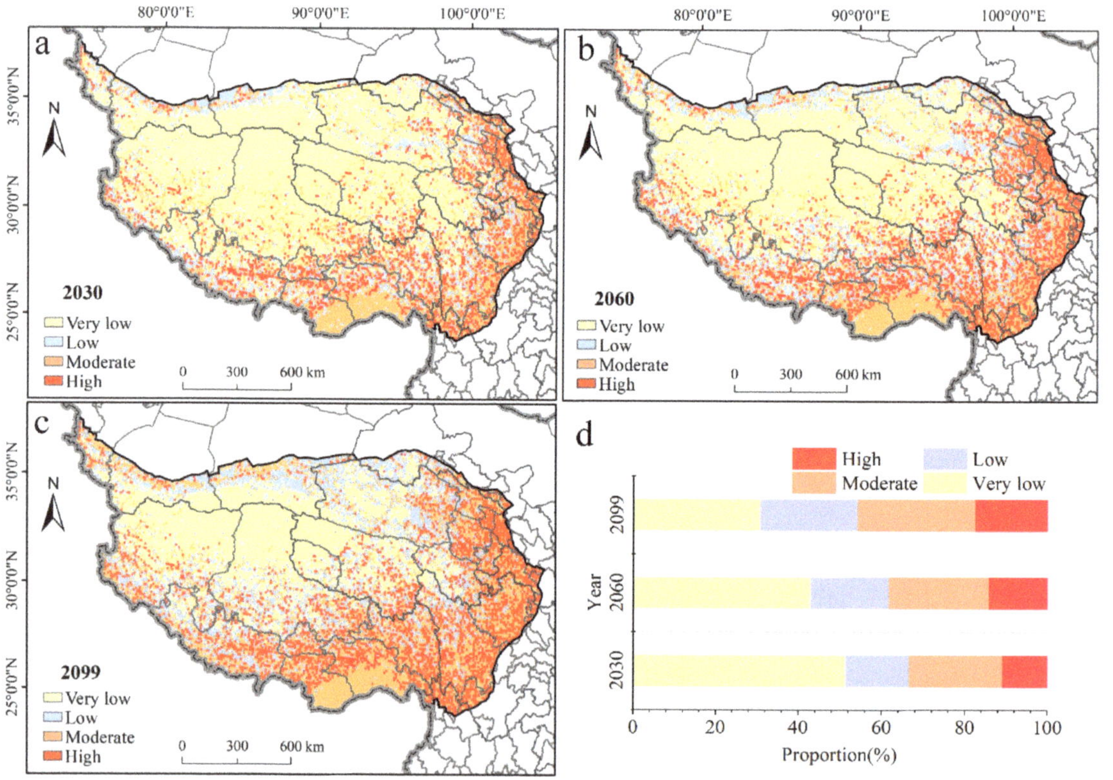

Figure 7. The slope geohazard sensitivity map area proportion of different grades. (**a**) Slope geohazard sensitivity map in 2030; (**b**) slope geohazard sensitivity map in 2060; (**c**) slope geohazard sensitivity map in 2099; (**d**) area variation across the sensitivity grades in different periods.

6. Conclusions

In this study on the QTP, the variation in the spatial and temporal distribution of 897 slope geohazards was assessed. Geographical and meteorological element correlations with the occurrence of slope geohazards were evaluated. By judging the factors that affect changes in slope geohazards, unknown risk areas were forecast. The main findings of our study are as follows.

Ninety-three percent of slope geohazards occurred in seasonally frozen regions, 2.6% were located in permafrost regions and the average altitude was 3818 m. The slope geohazards are mainly concentrated at approximately 2000 m, and the mean altitude spreads from 1984 m to 2562 m. Over time, the altitude of the slope geohazards gradually increases, while the slope gradient varies only slightly. Slope geohazards increased in the spring, especially in areas above 3000 m. The air warming rate and precipitation growth rate at geohazard sites increased gradually, indicating that the increase in spring air temperature and precipitation is an important reason for the increase in spring slope geohazards. Based on the observation of the spatial location, altitude and temperature growth rate of slope geohazards, an NGC appears in the study area, and there is still a possibility of migration under future climate conditions. We estimate that the low-, moderate- and high-sensitivity areas of the QTP are mainly south of 30° N in 2030, extend to the south of 33° N in 2060, continue to expand to the south of 35° N in 2099 and the proportion of high-sensitivity areas increases from 10.93% in 2030 to 14.17% in 2060 and 17.48% in 2099.

However, to explore the specific impact factors of slope geohazards in high-altitude mountain areas, we will further analyze the formation mechanism of slope geohazards with high-resolution datasets of slope geohazards and rigorous long-term records of daily temperature and precipitation in the future [57].

Supplementary Materials: The following are available online at https://www.mdpi.com/article/10.3390/su131910488/s1, Figure S1. Spatial distribution of the field survey regions of the QTP; Figure S2. Changes of temperature and precipitation over the QTP, 1980–2015; Figure S3. (a) precipitation, (b) air temperature and (c) soil temperature in the month when the slope geohazards occurred; Figure S4. The proportion of slope geohazards occurring at different elevations in 4 periods; Figure S5. Changes of precipitation and temperature in different seasons of geohazards in different years. (a) spring monthly precipitation in the area of spring geohazards; (b) winter monthly precipitation in the area of spring geohazards; (c) summer monthly air temperature in the area of the summer geohazards; (d) spring monthly air temperature in the area of the summer geohazards; (e) spring monthly air temperature in the area of the spring geohazards; (f) winter monthly air temperature in the area of the spring geohazards; Figure S6. The ROC curve of the random forest model in this study; Table S1. Eight CMIP5 models were used in this study.

Author Contributions: Conceptualization, J.L. (Jifu Liu), L.G.; methodology, Y.J., Z.D.; software, Y.J., Z.D.; validation, all authors; formal analysis, Y.J., L.G., H.Z.; investigation, Y.J., J.L. (Jiaoyang Li); resources, J.L. (Jifu Liu), L.G.; data curation, Y.J., J.L. (Jiaoyang Li), H.Z.; writing—original draft preparation, Y.J.; writing—review and editing, Y.J., L.G.; visualization, Z.D., H.Z.; supervision, L.G.; project administration, J.L. (Jifu Liu), L.G.; funding acquisition, J.L. (Jifu Liu). All authors have read and agreed to the published version of the manuscript.

Funding: This research was funded by the Second Tibetan Plateau Scientific Expedition and Research Program (STEP), Grant No. 2019QZKK0906, and the National Key R&D Program of China, Grant No. 2018YFC1509003.

Institutional Review Board Statement: Not applicable.

Informed Consent Statement: Not applicable.

Data Availability Statement: The data supporting reported results may be made available upon request.

Acknowledgments: We would like to thank the high-performance computing support from the Center for Geodata and Analysis, Faculty of Geographical Science, Beijing Normal University (https://gda.bnu.edu.cn/, accessed on 20 June 2020).

Conflicts of Interest: The authors declare no conflict of interest.

References

1. Petley, D. Global patterns of loss of life from landslides. *Geology* **2012**, *40*, 927–930. [CrossRef]
2. Andres, N.; Badoux, A. The Swiss flood and landslide damage database: Normalisation and trends. *J. Flood Risk Manag.* **2018**, *12*, e12510. [CrossRef]
3. Haque, U.; Blum, P.; da Silva, P.F.; Andersen, P.; Pilz, J.; Chalov, S.R.; Malet, J.-P.; Auflič, M.J.; Andres, N.; Poyiadji, E.; et al. Fatal landslides in Europe. *Landslides* **2016**, *13*, 1545–1554. [CrossRef]
4. Saez, J.L.; Corona, C.; Stoffel, M.; Berger, F. Climate change increases frequency of shallow spring landslides in the French Alps. *Geology* **2013**, *41*, 619–622. [CrossRef]
5. Stoffel, M.; Tiranti, D.; Huggel, C. Climate change impacts on mass movements—Case studies from the European Alps. *Sci. Total Environ.* **2014**, *493*, 1255–1266. [CrossRef]
6. Vergara Dal Pont, I.; Moreiras, S.M.; Santibañez Ossa, F.; Araneo, D.; Ferrando, F. Debris flows triggered from melt of seasonal snow and ice within the active layer in the semi-arid Andes. *Permafr. Periglac. Process.* **2019**, *31*, 57–68. [CrossRef]
7. Schar, C.; Vidale, P.L.; Luthi, D.; Haberli, C.F.; Liniger, M.A.; Appenzeller, C. The role of increasing temperature variability in European summer heatwaves. *Nature* **2004**, *427*, 332–336. [CrossRef]
8. Kääb, A.; Chiarle, M.; Raup, B.; Schneider, C. Climate change impacts on mountain glaciers and permafrost. *Glob. Planet. Chang.* **2007**, *56*, vii–ix. [CrossRef]
9. Noetzli, J.; Huggel, C.; Hoelzle, M.; Haeberli, W. GIS-based modelling of rock-ice avalanches from Alpine permafrost areas. *Comput. Geosci.* **2006**, *10*, 161–178. [CrossRef]
10. Davies, M.C.R.; Hamza, O.; Harris, C. The effect of rise in mean annual temperature on the stability of rock slopes containing ice-filled discontinuities. *Permafr. Periglac. Process.* **2001**, *12*, 137–144. [CrossRef]
11. Chiarle, M.; Mortara, G. Geomorphological impact of climate change on alpine glacial and periglacialareas. *Conf. Proc.* **2008**, *2*, 111–122.
12. Qiu, H.; Cui, Y.; Hu, S.; Yang, D.; Pei, Y.; Yang, W. Temporal and spatial distributions of landslides in the Qinba Mountains, Shaanxi Province, China. *Geomat. Nat. Hazards Risk* **2019**, *10*, 599–621. [CrossRef]
13. Climate Change Could Trigger More Landslides in High Mountain Asia. Available online: https://climate.nasa.gov/news/2951/climate-change-could-trigger-more-landslides-in-high-mountain-asia/ (accessed on 5 July 2021).
14. Tokashiki, N.; Aydan, O. Kita-Uebaru natural rock slope failure and its back analysis. *Environ. Earth Sci* **2011**, *62*, 25–31. [CrossRef]
15. Tonini, M.; Cama, M. Spatio-temporal pattern distribution of landslides causing damage in Switzerland. *Landslides* **2019**, *16*, 2103–2113. [CrossRef]
16. Khan, M.S.; Hossain, S.; Ahmed, A.; Faysal, M. Investigation of a shallow slope failure on expansive clay in Texas. *Eng. Geol.* **2017**, *219*, 118–129. [CrossRef]
17. Winter, M.G.; Dixon, N.; Wasowski, J.; Dijkstra, T.A. Introduction to land-use and climate change impacts on landslides. *Q. J. Eng. Geol. Hydrogeol.* **2010**, *43*, 367–370. [CrossRef]
18. Papathoma-Köhle, M. Vulnerability curves vs. vulnerability indicators: Application of an indicator-based methodology for debris-flow hazards. *Nat. Hazards Earth Syst. Sci.* **2016**, *16*, 1771–1790. [CrossRef]
19. Jia, H.; Chen, F.; Pan, D. Disaster Chain Analysis of Avalanche and Landslide and the River Blocking Dam of the Yarlung Zangbo River in Milin County of Tibet on 17 and 29 October 2018. *Int. J. Environ. Res. Public Health* **2019**, *16*, 4707. [CrossRef] [PubMed]
20. Li, X.; Cheng, G.; Jin, H.; Kang, E.; Che, T.; Jin, R.; Wu, L.; Nan, Z.; Wang, J.; Shen, Y. Cryospheric change in China. *Glob. Planet. Chang.* **2008**, *62*, 210–218. [CrossRef]
21. Yao, T.; Xue, Y.; Chen, D.; Chen, F.; Thompson, L.; Cui, P.; Koike, T.; Lau, W.K.-M.; Lettenmaier, D.; Mosbrugger, V.; et al. Recent Third Pole's Rapid Warming Accompanies Cryospheric Melt and Water Cycle Intensification and Interactions between Monsoon and Environment: Multidisciplinary Approach with Observations, Modeling, and Analysis. *Bull. Am. Meteorol. Soc.* **2019**, *100*, 423–444. [CrossRef]
22. Hao, J.; Zhao, L.; Li, R.; Xie, C.; Wu, T.; Wu, X.; Hu, G.; Zou, D.; Zhu, X.; Ni, J.; et al. Investigation of a Small Landslide in the Qinghai-Tibet Plateau by InSAR and Absolute Deformation Model. *Remote Sens.* **2019**, *11*, 2126. [CrossRef]
23. Mu, C.; Shang, J.; Zhang, T.; Fan, C.; Wang, S.; Peng, X.; Zhong, W.; Zhang, F.; Mu, M.; Jia, L. Acceleration of thaw slump during 1997–2017 in the Qilian Mountains of the northern Qinghai-Tibetan plateau. *Landslides* **2020**, *17*, 1051–1062. [CrossRef]
24. Gao, Y.J.; Cheng, N.S.; Hu, G.S.; Deng, M.S. Temporal and Spatial Coupling Relationship Between Debris Flow and El Nino-La Nina Event in Southwest China. *J. Yangtze River Sci. Res. Inst.* **2019**, *36*, 43–48. [CrossRef]
25. Wei, R.; Zeng, Q.; Davies, T.; Yuan, G.; Wang, K.; Xue, X.; Yin, Q. Geohazard cascade and mechanism of large debris flows in Tianmo gully, SE Tibetan Plateau and implications to hazard monitoring. *Eng. Geol.* **2018**, *233*, 172–182. [CrossRef]
26. Yang, Y.; Wu, Q.; Jin, H.; Wang, Q.; Huang, Y.; Luo, D.; Gao, S.; Jin, X. Delineating the hydrological processes and hydraulic connectivities under permafrost degradation on Northeastern Qinghai-Tibet Plateau, China. *J. Hydrol.* **2019**, *569*, 359–372. [CrossRef]

27. Zou, D.; Zhao, L.; Sheng, Y.; Chen, J.; Hu, G.; Wu, T.; Wu, J.; Xie, C.; Wu, X.; Pang, Q.; et al. A new map of permafrost distribution on the Tibetan Plateau. *Cryosphere* **2017**, *11*, 2527–2542. [CrossRef]
28. Chen, D.; Xu, B.; Yao, T.; Guo, Z.; Cui, P.; Chen, F.; Zhang, R.; Zhang, X.; Zhang, Y.; Fan, J.; et al. Assessment of past, present and future environmental changes on the Tibetan Plateau. *Chin. Sci Bull.* **2015**, *60*, 3025–3035. [CrossRef]
29. Yang, F.; Wang, W.; Wang, H.; Li, X.; Jiang, W. The Distribution Characters of the Stratospheric Zonal Mean Temperature and Water Vapor during 1979–2011. *Plateau Mt. Meteorol. Res.* **2016**, *36*, 63–67.
30. Zhang, Y.; Guo, C.; Yao, X.; Yang, Z.; Wu, R.; Du, G. Research on the Geohazard Effect of Active Fault on the Eastern Margin of the Tibetan Plateau. *Acta Geosci. Sin.* **2016**, *37*, 277–286. [CrossRef]
31. Zhou, J.; Cui, P.; Hao, M. Comprehensive analyses of the initiation and entrainment processes of the 2000 Yigong catastrophic landslide in Tibet, China. *Landslides* **2016**, *13*, 39–54. [CrossRef]
32. Cui, P.; Chen, R.; Xiang, L.; Su, F. Risk analysis of mountain hazards in Tibetan Plateau under the global warming. *Progress. Inquisitiones De Mutat. Clim.* **2014**, *10*, 103–109. [CrossRef]
33. Deng, J.; Dai, F.; Wen, B. Catastrophic mechanisms and risk control of major landslides in Tibetan Plateau. *Adv. Eng. Sci.* **2019**, *52*, 1–8. [CrossRef]
34. Dong, A.; Wen, K. *Chinese Meteorological Disasters Ceremony (Gansu Volume)*; China Meteorological Press: Beijing, China, 2005. (In Chinese)
35. Liu, G.; Wen, K. *Chinese Meteorological Disasters Ceremony (Tibet Volume)*; China Meteorological Press: Beijing, China, 2008. (In Chinese)
36. Liu, J.; Wen, K. *Chinese Meteorological Disasters Ceremony (Yunnan Volume)*; China Meteorological Press: Beijing, China, 2006. (In Chinese)
37. Shi, Y.; Wen, K. *Chinese Meteorological Disasters Ceremony (Xinjiang Volume)*; China Meteorological Press: Beijing, China, 2006. (In Chinese)
38. Wang, X.; Wen, K. *Chinese Meteorological Disasters Ceremony (Qinghai Volume)*; China Meteorological Press: Beijing, China, 2007. (In Chinese)
39. Zhan, Z.; Wen, K. *Chinese Meteorological Disasters Ceremony (Sichuan Volume) (in Chinese)*; China Meteorological Press: Beijing, China, 2006. (In Chinese)
40. Liu, C.; Liu, X.; Chen, Q.; Gao, W. Impact of DEM spatial resolution on landslide extraction using object-oriented methods. *Remote Sens. Technol. Appl.* **2014**, *29*, 631–638.
41. Taylor, K.E.; Stouffer, R.J.; Meehl, G.A. An Overview of CMIP5 and the Experiment Design. *Bull. Am. Meteorol. Soc.* **2012**, *93*, 485–498. [CrossRef]
42. You, Q.; Min, J.; Kang, S. Rapid warming in the Tibetan Plateau from observations and CMIP5 models in recent decades. *Int. J. Climatol.* **2016**, *36*, 2660–2670. [CrossRef]
43. Sen, P.K. Estimates of the Regression Coefficient Based on Kendall's Tau. *J. Am. Stat. Assoc.* **1968**, *63*, 1379–1389. [CrossRef]
44. Zhao, B.; Wang, Y.; Chen, M.; Luo, Y.; Liang, R.; Li, J. Typical characteristics of large-scale landslides in the transition belt between the Qinghai-Tibet Plateau and the Loess Plateau. *Arab. J. Geosci.* **2019**, *12*, 470. [CrossRef]
45. Sepúlveda, S.A.; Petley, D.N. Regional trends and controlling factors of fatal landslides in Latin America and the Caribbean. *Nat. Hazards Earth Syst. Sci.* **2015**, *15*, 1821–1833. [CrossRef]
46. Bai, S.; Lu, P.; Thiebes, B. Comparing characteristics of rainfall- and earthquake-triggered landslides in the Upper Minjiang catchment, China. *Eng. Geol.* **2020**, *268*, 105518. [CrossRef]
47. Huang, R. Some catastrophic landslides since the twentieth century in the southwest of China. *Landslides* **2009**, *6*, 69–81. [CrossRef]
48. Bai, Y.; Ni, H.; Ge, H. Advances research progress on geohazard effect of active faults on the southeastern margin of the Tibetan Plateau. *J. Geomech.* **2019**, *25*, 1116–2019. [CrossRef]
49. Luo, F.; Ren, G.; Li, H.; Li, Y. Development rules of landslides and collapses along provincial highway S216 in Sichuan. *J. Yangtze River Sci. Res. Inst.* **2019**, *36*, 37–41.
50. Ma, W.; Niu, F.; Satoshi, A.; Jin, D. Slope instability phenomena in permafrost regions of Qinghai-Tibet Plateau, China. *Landslides* **2006**, *3*, 260–264. [CrossRef]
51. Petley, D.N. On the impact of climate change and population growth on the occurrence of fatal landslides in South, East and SE Asia. *Q. J. Eng. Geol. Hydrogeol.* **2010**, *43*, 487–496. [CrossRef]
52. Cao, J.; Zhang, Z.; Du, J.; Zhang, L.; Song, Y.; Sun, G. Multi-geohazards susceptibility mapping based on machine learning—A case study in Jiuzhaigou, China. *Nat. Hazards* **2020**, *102*, 851–871. [CrossRef]
53. Niu, F.; Luo, J.; Lin, Z.; Fang, J.; Liu, M. Thaw-induced slope failures and stability analyses in permafrost regions of the Qinghai-Tibet Plateau, China. *Landslides* **2016**, *13*, 55–65. [CrossRef]
54. Luo, J.; Niu, F.; Lin, Z.; Liu, M.; Yin, G. Recent acceleration of thaw slumping in permafrost terrain of Qinghai-Tibet Plateau: An example from the Beiluhe Region. *Geomorphology* **2019**, *341*, 79–85. [CrossRef]
55. Lin, Q.; Wang, Y.; Glade, T.; Zhang, J.; Zhang, Y. Assessing the spatiotemporal impact of climate change on event rainfall characteristics influencing landslide occurrences based on multiple GCM projections in China. *Clim. Chang.* **2020**, *162*, 761–779. [CrossRef]
56. Kui, Z. The study of geological emergence of susceptible division in Yunan district-based on Arcgis. *J. Geol. Hazards Environ. Preserv.* **2020**, *31*, 38–42.

57. Stoffel, M.; Mendlik, T.; Schneuwly-Bollschweiler, M.; Gobiet, A. Possible impacts of climate change on debris-flow activity in the Swiss Alps. *Clim. Chang.* **2013**, *122*, 141–155. [CrossRef]
58. Li, T.; Li, P.; Wang, H. Forming Mechanism of Landslides in the Seasonal Frozen Loess Region in China. In *Landslides in Cold Regions in the Context of Climate Change*; Springer International Publishing: New York, NY, USA, 2014; pp. 41–51.
59. Zhao, D.; Lancuo, Z.; Hou, G.; Xu, C.; Li, W. Assessment of geological disaster susceptibility in the Hehuang Valley of Qinghai Province. *J. Geomech.* **2021**, *27*, 84–95.
60. Wei, G.; Yin, Z.; Shi, L.; Ma, W.; Cui, X. Zoning of geological disasters of Hualong county in Qinghai province. *Chin. J. Geol. Hazard. Control.* **2013**, *24*, 86–92. [CrossRef]
61. Wang, J.; Tian, T.; Zou, Y.; Wang, J.; Fu, X. Mechanical Mechanism of the Rainfall-inducing Deposit Slope Failure in Yunnan Region. *J. Water Resour. Archit. Eng.* **2020**, *18*, 204–227.
62. He, R.; Lin, Q.; Wang, Y.; Song, C. Factors and high risk area analysis of geological hazards in Yunnan. *J. Catastrophology* **2015**, *30*, 208–213. [CrossRef]

sustainability

MDPI

Article

Changes in the Frequency of Extreme Cooling Events in Winter over China and Their Relationship with Arctic Oscillation

Shuaifeng Song and Xiaodong Yan *

State Key Laboratory of Earth Surface Processes and Resource Ecology, Faculty of Geographical Science, Beijing Normal University, Beijing 100875, China; ssf@mail.bnu.edu.cn
* Correspondence: yxd@bnu.edu.cn

Abstract: Extreme weather and climate events are becoming increasingly frequent and have gained an increasing amount of attention. Extreme cooling (EC) events are a major challenge to socioeconomic sustainability and human health. Based on meteorological stations and NCEP/NCAR reanalysis data, this study analyzed the temporal and spatial distributions of EC events in winter in China by using the relative threshold and the relationship between EC events and the Arctic Oscillation (AO) index during the period of 1961–2017. The results show that the frequency of EC events in China decreased by 0.730 d in these 57 years, with a trend of −0.1 d/10 y. Northeast China had the highest frequency of EC events in winter, with an average of 4 d. In addition, EC events are significantly negatively correlated with the AO index in China, with a correlation coefficient of −0.5, and the AO index accounts for approximately 21% of the EC event variance. The strongest correlations are mainly located in Northwest China. Our research shows that significant changes in the mid–high latitude atmospheric circulation anomalies, which are associated with the AO, are responsible for EC events. These findings provide theoretical guidance for the prediction and simulation of EC events.

Keywords: extreme cooling events; Arctic Oscillation; winter in China; atmospheric circulation

Citation: Song, S.; Yan, X. Changes in the Frequency of Extreme Cooling Events in Winter over China and Their Relationship with Arctic Oscillation. *Sustainability* **2021**, *13*, 11491. https://doi.org/10.3390/su132011491

Academic Editor: Adriana Del Borghi

Received: 22 August 2021
Accepted: 13 October 2021
Published: 18 October 2021

1. Introduction

According to the sixth assessment report of the IPCC, the global surface temperature has shown an upward linear trend, increasing by 0.99 °C since the 21st century compared to the preindustrial period [1]. Changes in extreme weather and climate events, which have caused serious impacts on society, the ecological system, and public health [2–5], are more sensitive to global warming than the mean climate [6,7]. The frequency and intensity of extreme weather and climate events are also increasing [8,9]. However, the impact of these changes is directly felt by people in the form of day-to-day temperature changes. Extreme cooling (EC) events represent a sharp decrease in temperature between contiguous days, and such events may be a major challenge to socioeconomic sustainability and human health. Several studies have found that EC events can easily lead to the onset of disease [10]. EC events are strongly associated with nonaccidental deaths, cardiovascular deaths, and respiratory deaths, especially for elderly individuals [11,12]. For example, Guo et al. [11] showed that for every drop of 3 °C on two consecutive days, there was a 15.7% increase in nonaccidental deaths in the population. Furthermore, studies have shown that EC events are associated with the onset of infectious diseases. EC events can significantly increase the number of infectious diseases, such as hand, foot, and mouth disease [12,13], respiratory tract infections [14], and pneumonia [15]. In addition, EC events affect industrial and agricultural production and transportation conditions. In 2008, an extreme cold surge invaded central and southern China, causing economic losses of more than USD 22 billion, and 129 people lost their lives [16]. China has complex climatic conditions and a large population. Therefore, it is of great practical significance to clarify the spatial and temporal characteristics of EC events in China and their possible mechanisms and thereby improve the prediction of these events and reduce human casualties and property losses.

The spatial and temporal patterns of EC events in winter have become the focus for many researchers. Studies have shown that global cooling events above 10 °C are decreasing [17]. As one of the more influential EC events in winter in China, cold waves have also received widespread attention due to their temporal and spatial variability. In recent decades, the frequency of cold surges in China has shown a decreasing trend [18]. However, most definitions are based on absolute thresholds [19–21]. China is a vast country, and the climate varies greatly from region to region. Considering that people in different regions have different adaptive capacities and emergency measures for EC events, the definition of EC events for different regions should be defined by relative thresholds. Xu et al. [22] used daily minimum temperature data and the rotating empirical orthogonal function (REOF) method to divide China into seven regions, and different thresholds were attached to each region. The results showed that the frequency of EC events is higher in the north and lower in the south, and the overall trend of the change in the frequency of EC events is decreasing. Zhai et al. [23] proposed using a certain percentile value as the threshold for extreme weather events, and exceeding this threshold is considered to be the definition of an extreme weather event. Cai et al. [24] defined EC events in eastern China using the 90% quantile. The results showed that EC events are decreasing in eastern China. Therefore, most scholars have concluded that EC events have decreased in China using different definitions of EC events. However, the intensity and weakening trends of EC event reduction in different regions still need further attention.

A dominant pattern of the Northern Hemisphere in winter is the Arctic Oscillation (AO) [25,26]. The AO has a significant effect on climate variability and air temperatures in the Northern Hemisphere at middle and high latitudes [27–29]. Therefore, quantifying the relationship between EC events and the AO index can effectively improve EC event prediction. Recent studies have shown a positive correlation between the AO and winter temperatures in most parts of China [30]. The AO phase shift usually leads to weather and climate anomalies at middle and high latitudes in the North Atlantic, North Africa, and East Asia [31–33]. In addition, cold events occur more frequently in East Asia during the negative phase of the AO, and the East Asian trough deepens with stronger East Asian winter winds [34]. When El Niño and the positive-phase AO, or La Niña and the negative-phase AO, are combined, the temperature anomaly in northern China accelerates [35].

Therefore, studies on the effects of the AO on winter temperatures in China have been more extensive. However, few studies have been conducted to investigate the relationship between the AO index and EC events in China; in particular, the effects of the AO index on EC events in different regions are unclear. Given this situation, this study defined EC events using observed and reanalyzed data, combined with the relative threshold method, and analyzed the relationships between EC events and the AO index. The spatial and temporal variation characteristics of EC events and the influencing mechanisms of the AO index in China were also examined. The results of this study deepen our understanding of cold-related extreme events and provide theoretical guidance for the prediction of EC events.

2. Materials and Methods

2.1. Data

Daily mean surface air temperature data were obtained from the National Meteorological Information Center of the China Meteorological Administration. We mainly considered the average state of the daily temperature cycle and variations as the condition of EC events, and the daily mean temperature was used instead of the maximum and minimum temperatures. These data were subjected to strict quality control, and a total of 1115 meteorological stations were finally selected. The winter months during the period of 1961–2017 were selected. To highlight the influence of different natural geographic conditions on EC events, China was divided into seven regions: Northeast China, Northwest China, North China, Central China, East China, South China, and Southwest China (Figure 1).

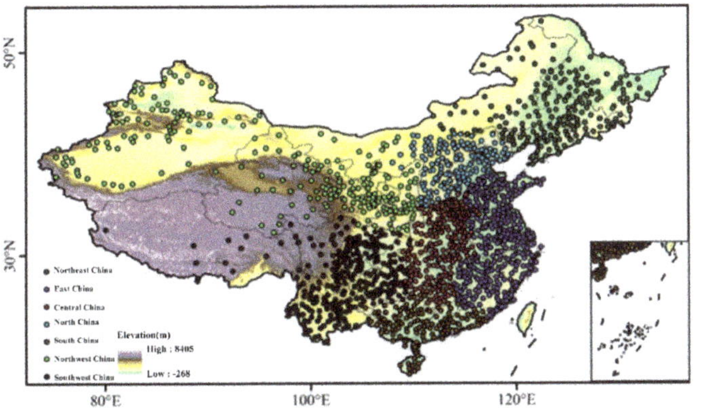

Figure 1. Spatial distribution of the 1115 meteorological stations and elevation of the topography over mainland China.

Monthly mean geopotential height, sea level pressure, and zonal and meridional wind speed data were obtained from the National Centers for Environmental Prediction/National Center for Atmospheric Research (NCEP/NCAR) reanalysis product [36]. The spatial resolution was 2.5° × 2.5°, and a total of 144 × 73 grid points were obtained globally. This study applied AO index data from the Climate Prediction Center (CPC) of the National Oceanic and Atmospheric Administrator (NOAA); the data can be downloaded from https://www.cpc.ncep.noaa.gov/products/precip/CWlink/daily_ao_index/ao.shtml (accessed on 14 November 2020) [37,38]. The AO index was defined according to Thompson and Wallace [25].

2.2. EC Event Definitions and Calculations

A day-to-day temperature change is estimated as the temperature change between neighboring days (TCN) [39]:

$$TCN = T_i - T_{i-1} \quad (i = 1,\ 2,\ 3,\ \ldots,\ n) \tag{1}$$

where TCN denotes a change in the average daily temperature for day i, and T_i (T_{i-1}) denotes the average daily temperature for day i (the previous day is denoted as $i-1$). The term n is the total number of days in winter. $TCN < 0$ indicates a cooling event.

The TCN data of a single station in winter were sorted from highest to lowest (except for the positive TCN values), and the value of the 90th percentile was taken as the threshold for EC events. Only TCN values exceeding this threshold were considered to be indicative of an EC event. According to this method, the cooling thresholds of 1115 stations in China were obtained as the criteria for EC events.

2.3. Correlation Analysis

The relationship between EC events and the AO index is calculated by correlation coefficients and is often expressed as R:

$$R = \frac{\sum_{i=1}^{n}(x_i - x_m)(y_i - y_m)}{\sqrt{\sum_{i=1}^{n}(x_i - x_m)^2} \cdot \sqrt{\sum_{i=1}^{n}(y_i - y_m)^2}} \tag{2}$$

where n denotes the number of years, and x_m and y_m are the average values of x and y, respectively. R assumes values in the range of $[-1, 1]$. Positive R values indicate a positive correlation between x and y; negative R values indicate a negative correlation between x and y. Significance levels of the correlation coefficient were estimated according to the two-tailed Student t-test [40].

2.4. Synthetic Analysis

In this paper, the conventional synthetic analysis method was used to analyze the changes in each element when an AO event occurs. The results were tested for statistical significance, and t-tests were used for two overall means, x, y.

$$t = \frac{x_m - y_m}{\sqrt{\frac{(n_1-1)s_1{}^2+(n_2-1)s_2{}^2}{n_1+n_2-2}}\sqrt{\frac{1}{n_1}+\frac{1}{n_2}}} \quad (3)$$

where n_1 and n_2 denote the sequence lengths of samples x and y, respectively; x_m and y_m are the means; s_1 and s_2 are the variances; and $n_1 + n_2 - 2$ is the overall degree of freedom. The t-distribution table was queried to determine if the results were significant.

2.5. Linear Trend Analysis

The long-term trend of the air temperature time series data was analyzed using the linear tendency estimate method [41]. A simple linear regression was performed between the temperature variable (y) and the corresponding time (x):

$$y = ax + b \quad (4)$$

where a is a linear regression coefficient, which represents the rate of change in EC events. A positive or negative value indicates an increasing or decreasing trend of EC events, respectively. The trend results were tested for significance using the t-test at the 95% confidence level.

3. Results

3.1. Temporal Variation and Spatial Pattern of EC Events

Figure 2h shows the time series of EC events in China from 1961 to 2017. The EC events show an obvious interannual and interdecadal variability, but the trend is not significant, with a decline of 0.730 d in these 57 years, and a trend of −0.128 d/10 y ($p > 0.05$). The largest EC event occurred in 1965 at 7.803 d, and the smallest occurred in 2006 at 2.643 d, with an average of 4.265 d. The interdecadal variability shows that the EC events in China had a rapidly increasing trend from 1961 to 1970 and a decreasing trend from the 1970s to the present, with the fastest decreasing trend in the 1990s at −0.038 d/10 y ($p > 0.05$).

To facilitate the analysis of the regional characteristics of the EC events, we calculated the frequency and trend of winter EC events for each station in seven regions of China. The results show that the frequency of EC events displays a spatial pattern of more in the north and less in the south (Figure 3). Most EC events occurred in Northeast China, with an average of 4.373 d, followed by Southwest China and Northwest China, with average values of 4.335 d and 4.334 d, respectively. The lowest number of EC events occurred in South China, at 4.059 d (Table 1).

In terms of the trend of winter EC events, obvious decreases were observed (Figure 2). The statistical results of all meteorological stations show that 858 of the stations had a decreasing trend of winter EC events. Among them, the trend of 240 stations reached the 90% confidence level (Figure 4). The trends of the EC events in the seven regions show that the rate of decrease in Northeast China was the fastest, with an average of −0.196 d/10 y, and approximately 31.1% of the stations showed a significant change. The regional average trends of the frequency of EC events in North China, East China, and Northwest China had values of −0.194 d/10 y, −0.162 d/10 y, and −0.127 d/10 y, respectively. Central China, South China, and Southwest China had smaller decreasing rates, with averages of −0.115 d/10 y, −0.075 d/10 y, and −0.040 d/10 y, respectively, and these values did not pass the 90% confidence level. This result shows that Northeast China and North China are the regions with the most EC events and the fastest decreasing trends.

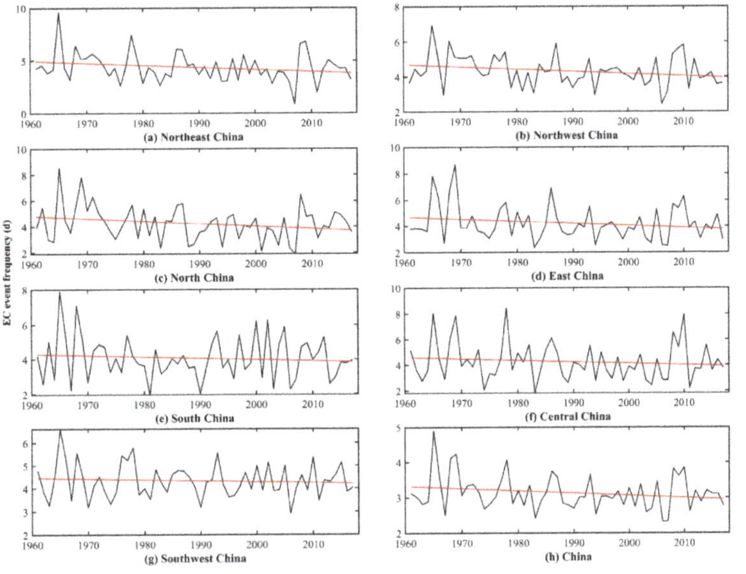

Figure 2. Time series changes of EC events in China and its subregions.

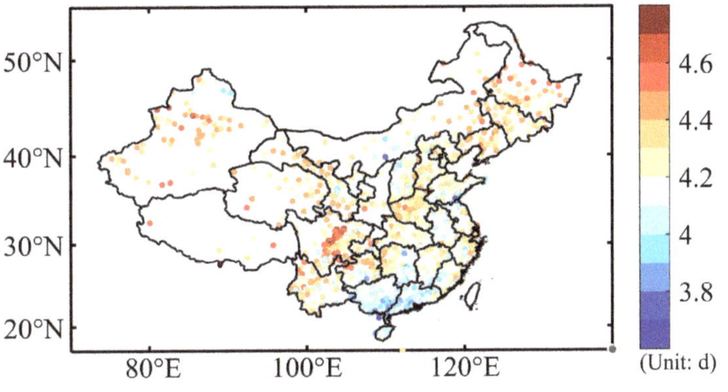

Figure 3. Climatological distribution of winter EC events in China.

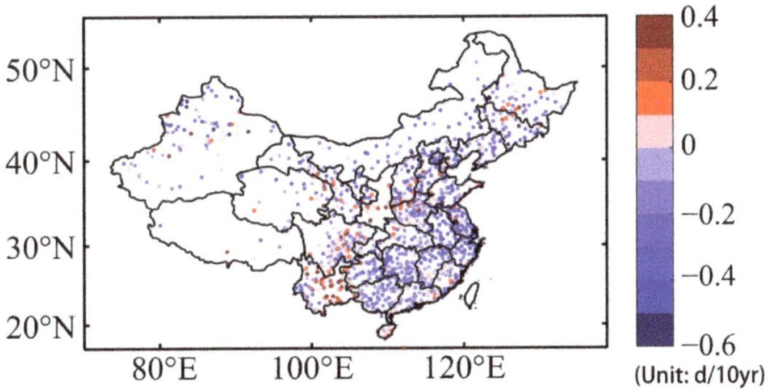

Figure 4. Same as Figure 3 but for the trend.

Table 1. Frequency and trends of EC events in different regions of China (Column 2 and Column 3 represent the range; in Column 4 and Column 5, the number in parentheses indicates the number of meteorological stations that passed the 90% significance test).

	Mean (d)	Trends (d/10 y)	Positive	Negative
Northeast China	4.373 (4.105~4.632)	−0.196 * (−0.720~0.354)	18 (1)	117 (42)
Northwest China	4.334 (3.930~4.667)	−0.127 * (−1.081~0.450)	48 (6)	120 (43)
North China	4.240 (3.719~4.456)	−0.194 * (−0.691~0.310)	19 (2)	94 (37)
East China	4.221 (3.789~4.579)	−0.162 (−0.544~0.231)	32 (0)	207 (64)
South China	4.059 (3.702~4.333)	−0.075 (−0.373~0.338)	30 (1)	81 (7)
Central China	4.250 (3.860~4.509)	−0.115 (−0.384~0.419)	26 (2)	130 (17)
Southwest China	4.335 (3.895~4.719)	−0.040 (−0.785~0.480)	82 (17)	109 (30)
China	4.265 (3.702~4.719)	−0.128 (−1.081~0.480)	255 (29)	858 (240)

* indicates significance at the 0.1 level.

3.2. Correlation between EC Events and the AO Index

As shown in Figure 5, the AO and EC events were significantly negatively correlated, with a correlation coefficient of −0.459 ($p < 0.01$). When the AO was strong, the number of EC events in China was low, and the AO can explain 21.1% of the variation in EC events in China. To detach EC events from the global warming environment, this study detrended EC events and the AO index. The results show that regardless of whether EC events and the AO index were detrended separately or simultaneously, EC events and the AO index in China still had a highly significant negative correlation (Table 2).

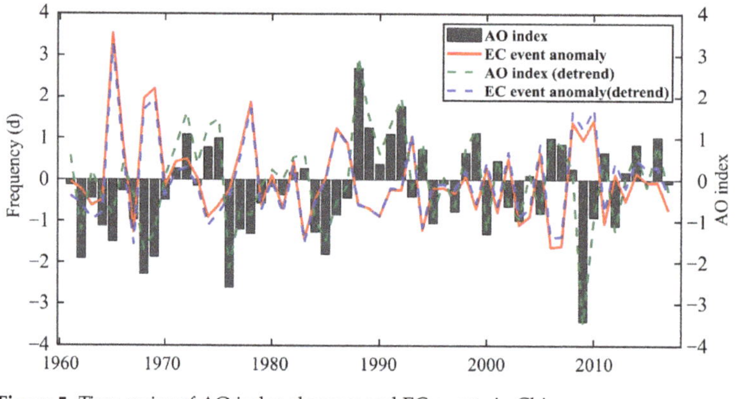

Figure 5. Time series of AO index changes and EC events in China.

Table 2. The correlation between EC events and the AO index in different regions.

	Northeast China	Northwest China	North China	East China	South China	Central China	Southwest China	China
$R(AOI_{detrend}, EC)$	−0.237	−0.471 ***	−0.374 ***	−0.390 ***	−0.285 **	−0.389 ***	−0.302 **	−0.421 ***
$R(AOI, EC_{detrend})$	−0.237	−0.471 ***	−0.374 ***	−0.386 ***	−0.278 **	−0.381 ***	−0.295 **	−0.418 ***
$R(AOI_{detrend}, EC_{detrend})$	−0.244	−0.485 ***	−0.385 ***	−0.398 ***	−0.286 **	−0.392 ***	−0.303 **	−0.431 ***
$R(AOI, EC)$	−0.286 **	−0.514 ***	−0.421 ***	−0.427 ***	−0.301 **	−0.408 ***	−0.315 **	−0.459 ***

** and *** indicate significance at the 0.05 and 0.01 levels, respectively.

Spatially, the correlation coefficient between EC events and the simultaneous AO index was mainly negative, and the correlation was positive only in Southwest China (Figure 6). From the regional perspective, the frequency of EC events in the seven regions was significantly negatively correlated with the AO index. The Northwest China region had the strongest correlation, and the correlation coefficient exceeded −0.5.

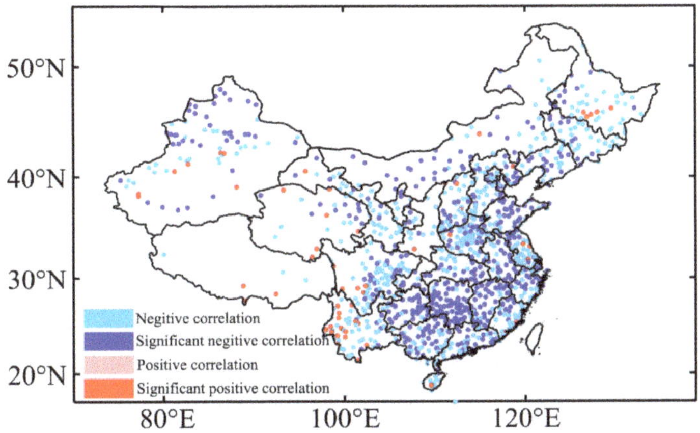

Figure 6. Climatological distribution of the correlation coefficient between EC events and the simultaneous AO index.

In addition, the relationship between the AO and EC events was closely related to the variation in atmospheric circulation anomalies associated with the AO in different climate contexts. Therefore, further analysis of the variability of atmospheric circulation anomalies associated with the AO is needed.

3.3. *Possible Mechanisms of AO Affecting EC Events*

The above results illustrate that EC events in China have a significant negative correlation with the AO. To further study the mechanisms by which the AO influences EC events in China, we selected the first eight years with the highest and lowest anomaly values from the winter AO index change series from 1961 to 2017 to define strong and weak AO index years, respectively. Among them, the strong AO index years were 2006, 2016, 1972, 1991, 1999, 1989, 1992, and 1988, and the weak AO index years were 2009, 1976, 1968, 1962, 1969, 1985, 1965, and 2000. As shown in Figure 7, the frequency of EC events was consistently lower when the AO index was strong, except for the southwestern station and some other stations. A total of 24.9% of these meteorological stations passed the 0.05 significance test, indicating that the typical years selected were representative of the anomalies of EC events in China.

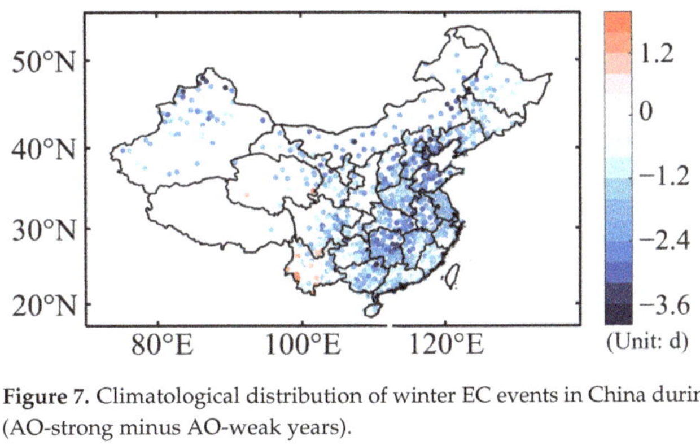

Figure 7. Climatological distribution of winter EC events in China during AO index anomaly years (AO-strong minus AO-weak years).

Figure 8 shows the spatial distribution of the difference between the sea level pressure, 850 hPa and 500 hPa geopotential heights, and wind fields in the strong and weak years. From the spatial distribution of the sea level pressure and wind fields (Figure 8a), the air pressure was significantly lower at high latitudes and higher in China, indicating that the Siberian high and Aleutian low were weak at this time, and that the near-surface northerly wind became weaker, which was not conducive to the occurrence of EC events in China. The 850 hPa geopotential height and wind field in the positive anomaly year (Figure 8b) were further analyzed. When the AO index was strong, the geopotential height in the high-latitude region was higher, and the geopotential height in the middle- and low-latitude regions was lower. The colder airflow activity range was limited to the polar region, and China was affected by warm airflow from the Pacific Ocean in the south. From the 500 hPa geopotential height and wind field (Figure 8c), the geopotential height was significantly higher in China when the AO index was strong. At this time, the East Asian trough was weak, and China was controlled by anomalous anticyclonic circulation. Therefore, there were fewer EC events in China. Significant changes in the mid–high latitude atmospheric circulation anomalies associated with the AO led to the occurrence of the EC events.

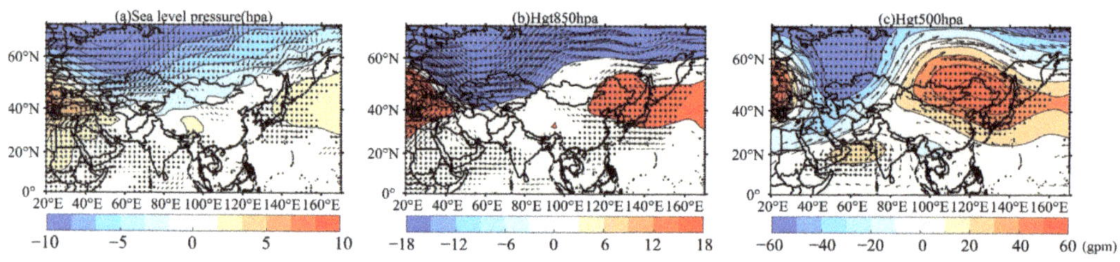

Figure 8. The difference in (**a**) sea level pressure, (**b**) 850 hPa and (**c**) 500 hPa geopotential heights, and wind fields between strong and weak AO index years (dotted areas indicate the areas that passed the 0.05 significance test; the wind field grid passes the significance test when u, v, and w pass 0.05).

4. Discussion

The purpose of this study was to use observational data to calculate the temporal and spatial variation in EC events in China under the 90th percentile threshold and to determine the relationship between EC events and the AO index in different regions. To assess the consistency of different EC event definitions with the conclusions of this study, the following discussion is presented.

EC events are defined using three relative threshold methods (90th, 95th, and 99th percentiles) and an absolute threshold (cold surge) method. A cold surge is defined as a temperature decrease within a 24-h period greater than 8 °C, a 48-h period greater than 10 °C, or a 72-h period greater than 12 °C, where the minimum temperature is less than 4 °C [42]. The results show that the frequency of EC events in China based on all four methods from 1961 to 2017 showed a decreasing trend of −0.128 d/10 y (90th percentile), −0.089 d/10 y (95th percentile), −0.028 d/10 y (99th percentile), and −0.132 d/10 y (cold surge) (Figure 9). Both the relative and absolute threshold methods using different percentile definitions showed a decreasing trend of EC events in China, which is in line with the findings of most studies [21]. In addition, the decreasing trend in the frequency of EC events is defined using the 90th percentile, and the cold surge is the closest.

To test the consistency of the correlation between different EC event definitions and the AO index, the correlation between the AO index and EC events before and after detrending was calculated. The results show that EC events, as defined by all four methods, were highly significantly correlated with the AO index, regardless of the detrending treatment. Among them, the EC events defined according to the 90th percentile were above the highest correlation coefficient of −0.459 to −0.418 (Figure 10).

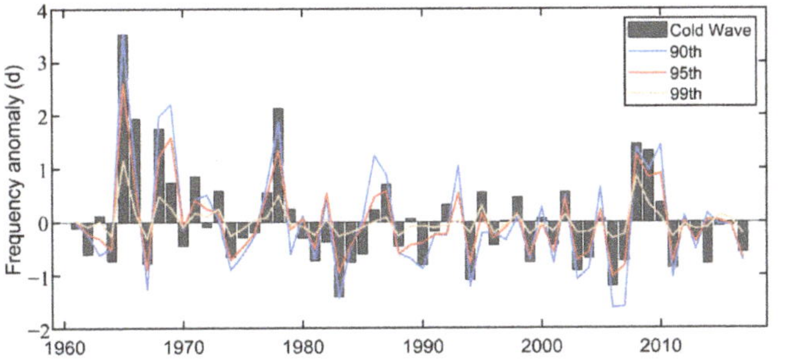

Figure 9. Time series of winter EC events with different definitions in China from 1961 to 2017.

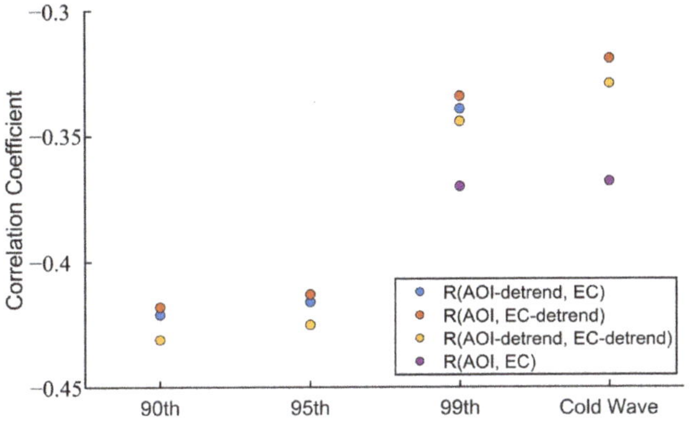

Figure 10. The correlation between winter EC events with different definitions and the AO index (AOI) in China.

It should be noted that the analysis of the AO mechanisms influencing EC events mainly began when the AO in the main circulation systems affected the winter temperature in China. However, the entire weather process by which the AO affects EC events in China is not discussed in detail here. Furthermore, in addition to the AO, the factors affecting the average winter temperature in China include ENSO [43], Arctic sea ice [44], and the East Asian winter monsoon [45]. The relationship between these factors and EC events and their mechanisms will be the focus of our future work.

5. Conclusions

The results of this study show that EC events in China exhibit an obvious interannual and interdecadal variability. EC events decreased by 0.730 d in the 57 years from 1961 to 2017, with a rate of −0.128 d/10 y. Most winter EC events occurred in Northeast China, with an average of 4.373 d.

The AO was significantly negatively correlated with EC events, with a correlation coefficient of –0.459. After detrending EC events and the AO index, EC events still had a highly significant negative correlation with the AO index in China.

When the AO was strong, the Siberian high and Aleutian low were weak, and the near-surface northerly winds became weaker. The East Asian trough in the middle troposphere obviously weakened. Thus, the propagation of cold polar air was limited by the atmospheric circulation situation in the middle and high latitudes. China has had fewer EC events influenced by warmer Pacific Ocean currents. In addition, the occurrence of EC events due to other factors should be considered in future research.

Author Contributions: Conceptualization, resources, methodology, writing—review and editing, project administration, and funding acquisition, X.Y.; software, formal analysis, data curation, writing—original draft preparation, and investigation, S.S. All authors have read and agreed to the published version of the manuscript.

Funding: This research was funded by the National Key R&D Program of China, grant numbers 2018YFC509003 and 2019YFA0606904.

Institutional Review Board Statement: Not applicable.

Informed Consent Statement: Not applicable.

Data Availability Statement: The data are not publicly available due to the confidentiality of the research projects.

Acknowledgments: We would like to thank Daoyi Gong for help in this study. We also acknowledge the China Meteorological Administration for temperature data and information.

Conflicts of Interest: The authors declare no conflict of interest.

References

1. IPCC. Summary for Policymakers. In *Climate Change 2021: The Physical Science Basis. Contribution of Working Group I to the Sixth Assessment Report of the Intergovernmental Panel on Climate Change*; Cambridge University Press: Cambridge, UK, 2021.
2. Easterling, D.R.; Meehl, G.A.; Parmesan, C.; Changnon, S.A.; Karl, T.R.; Mearns, L.O. Climate Extremes: Observations, Modeling, and Impacts. *Science* **2000**, *289*, 2068–2074. [CrossRef]
3. Orlowsky, B.; Seneviratne, S.I. Global changes in extreme events: Regional and seasonal dimension. *Clim. Chang.* **2011**, *110*, 669–696. [CrossRef]
4. Martinich, J.; Crimmins, A. Climate damages and adaptation potential across diverse sectors of the United States. *Nat. Clim. Chang.* **2019**, *9*, 397–404. [CrossRef]
5. Török, I.; Croitoru, A.-E.; Man, T.-C. Assessing the Impact of Extreme Temperature Conditions on Social Vulnerability. *Sustainability* **2021**, *13*, 8510. [CrossRef]
6. Schoetter, R.; Cattiaux, J.; Douville, H. Changes of western European heat wave characteristics projected by the CMIP5 ensemble. *Clim. Dyn.* **2014**, *45*, 1601–1616. [CrossRef]
7. Lewis, S.C.; King, A.D.; Perkins-Kirkpatrick, S.E. Defining a New Normal for Extremes in a Warming World. *Bull. Am. Meteorol. Soc.* **2017**, *98*, 1139–1151. [CrossRef]
8. Sui, Y.; Lang, X.; Jiang, D. Projected signals in climate extremes over China associated with a 2 °C global warming under two RCP scenarios. *Int. J. Clim.* **2018**, *38*, e678–e697. [CrossRef]
9. Zhao, N.; Chen, M. A Comprehensive Study of Spatiotemporal Variations in Temperature Extremes across China during 1960–2018. *Sustainability* **2021**, *13*, 3807. [CrossRef]
10. McMichael, A.J. Globalization, Climate Change, and Human Health. *N. Engl. J. Med.* **2013**, *368*, 1335–1343. [CrossRef]
11. Guo, Y.; Barnett, A.; Yu, W.; Pan, X.; Ye, X.; Huang, C.; Tong, S. A Large Change in Temperature between Neighbouring Days Increases the Risk of Mortality. *PLoS ONE* **2011**, *6*, e16511. [CrossRef]
12. Cheng, J.; Zhu, R.; Xu, Z.; Xu, X.; Wang, X.; Li, K.; Su, H. Temperature variation between neighboring days and mortality: A distributed lag non-linear analysis. *Int. J. Public Health* **2014**, *59*, 923–931. [CrossRef]
13. Xu, J.; Zhao, D.; Su, H.; Xie, M.; Cheng, J.; Wang, X.; Li, K.; Yang, H.; Wen, L.; Wang, B. Impact of temperature variability on childhood hand, foot and mouth disease in Huainan, China. *Public Health* **2016**, *134*, 86–94. [CrossRef]
14. Liu, Y.; Guo, Y.; Wang, C.; Li, W.; Lu, J.; Shen, S.; Xia, H.; He, J.; Qiu, X. Association between Temperature Change and Outpatient Visits for Respiratory Tract Infections among Children in Guangzhou, China. *Int. J. Environ. Res. Public Health* **2015**, *12*, 439–454. [CrossRef]
15. Xu, Z.; Hu, W.; Tong, S. Temperature variability and childhood pneumonia: An ecological study. *Environ. Health* **2014**, *13*, 51. [CrossRef] [PubMed]
16. Zhou, B.; Gu, L.; Ding, Y.; Shao, L.; Wu, Z.; Yang, X.; Li, C.; Li, Z.; Wang, X.; Cao, Y.; et al. The Great 2008 Chinese Ice Storm: Its Socioeconomic–Ecological Impact and Sustainability Lessons Learned. *Bull. Am. Meteorol. Soc.* **2011**, *92*, 47–60. [CrossRef]
17. Zhou, X.; Wang, Q.; Yang, T. Decreases in days with sudden day-to-day temperature change in the warming world. *Glob. Planet. Chang.* **2020**, *192*, 103239. [CrossRef]
18. Gao, W.; Duan, K.; Li, S. Spatial-temporal variations in cold surge events in northern China during the period 1960–2016. *J. Geogr. Sci.* **2019**, *29*, 971–983. [CrossRef]
19. Song, L.; Wu, R. Processes for Occurrence of Strong Cold Events over Eastern China. *J. Clim.* **2017**, *30*, 9247–9266. [CrossRef]
20. Shi, J.; Cui, L.; Ma, Y.; Du, H.; Wen, K. Trends in temperature extremes and their association with circulation patterns in China during 1961–2015. *Atmos. Res.* **2018**, *212*, 259–272. [CrossRef]
21. Ting, D.; Wei-Hong, Q.; Zhong-Wei, Y. Characteristics and Changes of Cold Surge Events over China during 1960–2015. *Atmos. Ocean. Sci. Lett.* **2009**, *2*, 339–344. [CrossRef]

22. Xu, M.; Guan, Z.; Cai, Q. Spatial and temporal evolution features of cooling extremes in China during winter half year from 1960 to 2015. *J. Meteorol. Sci.* **2020**, *40*, 733–743. (In Chinese)
23. Zhai, P.; Pan, X. Change in extreme temperature and precipitation over northern China during the second half of the 20th century. *Acta Geogr. Sin.* **2003**, *58*, 1–10. (In Chinese)
24. Cai, Q.; Guan, Z.; Xu, M. Temporal and spatial characteristics of extreme cooling events in eastern China in winter monsoon period during 1960 and 2012. *Trans. Atmos. Sci.* **2020**, *43*, 458–468. (In Chinese)
25. Thompson, D.W.J.; Wallace, J.M. The Arctic oscillation signature in the wintertime geopotential height and temperature fields. *Geophys. Res. Lett.* **1998**, *25*, 1297–1300. [CrossRef]
26. Ambaum, M.H.P.; Hoskins, B.J.; Stephenson, D.B. Arctic Oscillation or North Atlantic Oscillation? *J. Clim.* **2001**, *14*, 3495–3507. [CrossRef]
27. Wang, L.; Gong, H.; Lan, X. Interdecadal variation of the Arctic Oscillation and its influence on climate. *Trans. Atmos. Sci.* **2021**, *44*, 50–60. (In Chinese)
28. Ghasemi, A.R.; Khalili, D. The influence of the Arctic Oscillation on winter temperatures in Iran. *Theor. Appl. Clim.* **2006**, *85*, 149–164. [CrossRef]
29. He, S.; Wang, H. Impact of the November/December Arctic Oscillation on the following January temperature in East Asia. *J. Geophys. Res. Atmos.* **2013**, *118*, 12981–12998. [CrossRef]
30. Gong, D.-Y.; Wang, S.-W.; Zhu, J.-H. East Asian Winter Monsoon and Arctic Oscillation. *Geophys. Res. Lett.* **2001**, *28*, 2073–2076. [CrossRef]
31. Wettstein, J.J.; Mearns, L.O. The Influence of the North Atlantic–Arctic Oscillation on Mean, Variance, and Extremes of Temperature in the Northeastern United States and Canada. *J. Clim.* **2002**, *15*, 3586–3600. [CrossRef]
32. Li, F.; Wang, H. Autumn Sea Ice Cover, Winter Northern Hemisphere Annular Mode, and Winter Precipitation in Eurasia. *J. Clim.* **2012**, *26*, 3968–3981. [CrossRef]
33. Park, T.-W.; Ho, C.-H.; Yang, S. Relationship between the Arctic Oscillation and Cold Surges over East Asia. *J. Clim.* **2011**, *24*, 68–83. [CrossRef]
34. Thompson, D.W.J. Regional Climate Impacts of the Northern Hemisphere Annular Mode. *Science* **2001**, *293*, 85–89. [CrossRef]
35. Chen, S.; Chen, W.; Yu, B. The influence of boreal spring Arctic Oscillation on the subsequent winter ENSO in CMIP5 models. *Clim. Dyn.* **2017**, *48*, 2949–2965. [CrossRef]
36. Kalnay, E.; Kanamitsu, M.; Kistler, R.; Collins, W.; Deaven, D.; Gandin, L.; Iredell, M.; Saha, S.; White, G.; Woollen, J. The NCEP/NCAR 40-year reanalysis project. *Bull. Am. Meteorol. Soc.* **1996**, *77*, 437–471. [CrossRef]
37. Zhou, S.; Miller, A.J.; Wang, J.; Angell, J.K. Trends of NAO and AO and their associations with stratospheric processes. *Geophys. Res. Lett.* **2001**, *28*, 4107–4110. [CrossRef]
38. Peings, Y.; Brun, E.; Mauvais, V.; Douville, H. How stationary is the relationship between Siberian snow and Arctic Oscillation over the 20th century? *Geophys. Res. Lett.* **2013**, *40*, 183–188. [CrossRef]
39. Lei, X.; Liu, L.; Chen, R.; Liu, C.; Hong, J.; Cao, L.; Lu, Y.; Dong, X.; Chen, X.; Qiu, X.; et al. Temperature changes between neighboring days and childhood asthma: A seasonal analysis in Shanghai, China. *Int. J. Biometeorol.* **2021**, *65*, 827–836. [CrossRef] [PubMed]
40. Chen, S.; Song, L. Recent Strengthened Impact of the Winter Arctic Oscillation on the Southeast Asian Surface Air Temperature Variation. *Atmosphere* **2019**, *10*, 164. [CrossRef]
41. Tokarska, K.B.; Stolpe, M.B.; Sippel, S.; Fischer, E.M.; Smith, C.J.; Lehner, F.; Knutti, R. Past warming trend constrains future warming in CMIP6 models. *Sci. Adv.* **2020**, *6*, eaaz9549. [CrossRef] [PubMed]
42. Gao, W.; Duan, K.; Li, S. A spatial–temporal analysis of cold surge days in northern China during 1960–2016. *Nat. Hazards* **2021**, *108*, 147–162. [CrossRef]
43. Chen, S.; Chen, W.; Wei, K. Recent trends in winter temperature extremes in eastern China and their relationship with the Arctic Oscillation and ENSO. *Adv. Atmos. Sci.* **2013**, *30*, 1712–1724. [CrossRef]
44. Zuo, J.; Ren, H.-L.; Wu, B.; Li, W. Predictability of winter temperature in China from previous autumn Arctic sea ice. *Clim. Dyn.* **2016**, *47*, 2331–2343. [CrossRef]
45. Jia, Z.; Bollasina, M.A.; Li, C.; Doherty, R.; Wild, O. Changes in the relationship between ENSO and the East Asian winter monsoon under global warming. *Environ. Res. Lett.* **2020**, *15*, 124056. [CrossRef]

sustainability

MDPI

Article

Impact of Model Resolution on the Simulation of Precipitation Extremes over China

Neng Luo and Yan Guo *

State Key Laboratory of Earth Surface Processes and Resource Ecology, Faculty of Geographical Science, Beijing Normal University, Beijing 100875, China; luoneng@mail.bnu.edu.cn
* Correspondence: guoyan@bnu.edu.cn

Abstract: Climate models tend to overestimate light precipitation and underestimate heavy precipitation due to low model resolution. This work investigated the impact of model resolution on simulating the precipitation extremes over China during 1995–2014, based on five models from Coupled Model Intercomparison Project 6 (CMIP6), each having low- and high-resolution versions. Six extreme indices were employed: simple daily intensity index (SDII), wet days (WD), total precipitation (PRCPTOT), extreme precipitation amount (R95p), heavy precipitation days (R20mm), and consecutive dry days (CDD). Models with high resolution demonstrated better performance in reproducing the pattern of climatological precipitation extremes over China, especially in the western Sichuan Basin along the eastern side of the Tibetan Plateau (D1), South China (D2), and the Yangtze-Yellow River basins (D3). Decreased biases of precipitation exist in all high-resolution models over D1, with the largest decease in root mean square error (RMSE) being 48.4% in CNRM-CM6. The improvement could be attributed to fewer weak precipitation events (0 mm/day–10 mm/day) in high-resolution models in comparison with their counterparts with low resolutions. In addition, high-resolution models also show smaller biases over D2, which is associated with better capturing of the distribution of daily precipitation frequency and improvement of the simulation of the vertical distribution of moisture content.

Keywords: precipitation; China; CMIP6; model resolutions

Citation: Luo, N.; Guo, Y. Impact of Model Resolution on the Simulation of Precipitation Extremes over China. *Sustainability* **2022**, *14*, 25. https://doi.org/10.3390/su14010025

Academic Editor: Ozgur Kisi

Received: 1 December 2021
Accepted: 19 December 2021
Published: 21 December 2021

1. Introduction

Increased climate extremes in terms of frequency and intensity, such as extreme precipitation events, have been observed in the context of global warming, which have immensely adverse impacts on economic production and human life. The simulation in precipitation extremes over complex terrain is challenging but crucial for investing the performance of models. For example, the Pacific Northwest of the United States is characterized by complex topography, including land-sea contrasts and mountains ranges, which is challenging for global models [1]. China is strongly affected by precipitation extremes due to complex terrain, vast areas, and the influence of the topographically complex Tibetan Plateau and East Asian monsoon system. Many extreme precipitation events have been reported in China. On 6–7 May 2017, the coastal city of Guangzhou experienced severe extreme precipitation with maximum rainfall above 500 mm in one day, eventually causing enormous economic losses [2]. Severe flooding occurred in the Yangtze River in 2020 and caused at least 141 deaths, with direct economic losses of 82.23 billion yuan by 12 July 2020 [3]. On 20 July 2021, Zhengzhou experienced a record-breaking torrential rainfall of above 200 mm within one hour, eventually causing direct economic losses amounting to 53.2 billion yuan [4]. Therefore, the best way to improve the ability of climate models to simulate or project precipitation extremes has received a lot of attention from the climate community.

Previous studies have demonstrated the advantage of high-resolution simulation, including improvements in large-scale atmospheric and oceanic circulation [5,6], tropical cyclones [7,8], El Niño-Southern Oscillation [9], and precipitation [10–12]. For example, the authors of [13] found that high-resolution models obtain better performance in simulating the frequency and percentage of extreme precipitation in eastern China. Using two atmospheric general circulation models from the UK Met Office Hadley Centre, the authors of [14] demonstrated that the model resolution has an effect on the hydrological cycle by increasing (decreasing) precipitation over land (ocean), eventually making high-resolution simulations closer to observations over the ocean but further away over land. The authors of [15] indicated that a model with a horizontal grid spacing less than 4 km (convection-permitting model), better representing orography and variations in surface fields, can improve precipitation simulation in mountainous regions. The authors of [16] indicated that models in higher resolution better simulate the precipitation extremes in the mountainous regions over northern China and Sichuan but that their skill in simulating precipitation over the Maritime Continent is low.

Currently, the World Climate Research Programme (WCRP) launched the latest Coupled Model Intercomparison Project (CMIP6), which aimed to answer new scientific questions in the field of climate change and provided data available to achieve the scientific goal set by the WCRP. Most CMIP6 models have updated parameterization schemes and higher resolutions compared with CMIP5 models [17]. Studies have demonstrated that CMIP6 models generally help simulate the mean and extreme precipitation compared with CMIP5 models [18–22]. For example, the authors of [23] evaluated the performance of six CMIP6 and CMIP5 models in simulating extreme precipitation in China and indicated that heavy precipitation events (>20 mm/day) over southern China are well captured by CMIP6 models, which might be related to model resolutions and precipitation-related parameterization schemes. The authors of [24] indicated that the southeast-northwest gradient of rainfall over China is underestimated in CMIP6 models but less than that in CMIP5 models, which is possibly associated with model resolution. The authors of [25] assessed the performance of 18 CMIP6 models participating in the High-Resolution Model Inter-comparison Project (HighResMIP) in simulating the climatological precipitation over China and indicated that models in higher resolution show better skill in simulating the spatial distribution, and the precipitation bias over the Tibetan Plateau, northern China, and southern China has been obviously reduced.

Several factors, such as physical parameterization, affect the model simulation of precipitation extremes besides model resolution. Thus, the impact of model resolution should be investigated using identical models, which only differ with respect to resolution. Based on five state-of-the-art CMIP6 models, with each having a pair of high-resolution and low-resolution configurations, we aim to study the impact of model resolution on their fidelity in simulating the precipitation extremes over China in this study. We used five CMIP6 models, with each having high-resolution and low-resolution configurations. We found that models with improved resolutions show common improvements in the simulation of precipitation extremes in China compared with their counterparts with low resolutions. In addition, the vertical distribution of moisture content is better represented in high-resolution models. Our results indicate high-resolution models are crucial for improving projection in precipitation extremes events.

2. Materials and Methods

2.1. Observational and Model Data

The gridded observational precipitation data were from CN05.1 on a 0.25 × 0.25 grid reference [26]. The dataset was produced by the National Climate Center of the China Meteorological Administration from over 2400 observation stations over China. The reliability of this dataset has been proven in previous studies [27,28]. Relative humidity data were from ERA5 reanalysis dataset in horizontal resolution of 0.25° × 0.25° [29].

In this study, five CMIP6 models, each having high-resolution and low-resolution configurations (totally ten models), were obtained from data portals of the Earth System Grid Federation (https://esgf-node.llnl.gov/projects/esgf-llnl/, accessed on 31 August

2021). Detailed information about the ten models is provided in Table 1. Only the first realization was used for each model. To achieve a convenient comparison, all model outputs were re-gridded into $1^\circ \times 1^\circ$grid using the bilinear interpolation method. The period of 1995–2014 was chosen as the climatology to perform evaluation.

Table 1. CMIP6 models used in the study.

Institute	High	Resolution	Low	Resolution
National Centre for Meteorological Research, France	CNRM-CM6-1-HR	$0.5^\circ \times 0.5^\circ$	CNRM-CM6-1	$1.4^\circ \times 1.4^\circ$
EC-Earth consortium	EC-Earth3-Veg	$0.7^\circ \times 0.7^\circ$	EC-Earth3-Veg-LR	$1.125^\circ \times 1.25^\circ$
Met Office Hadley Centre, UK	HadGEM3-GC31-MM	$0.556^\circ \times 0.833^\circ$	HadGEM3-GC31-LL	$1.25^\circ \times 1.875^\circ$
Max Planck Institute for Meteorology, Germany	MPI-ESM1-2-HR	$0.94^\circ \times 0.94^\circ$	MPI-ESM1-2-LR	$1.9^\circ \times 1.9^\circ$
NorESM Climate modelling Consortium consisting of CICERO	NorESM2-MM	$0.94^\circ \times 1.25^\circ$	NorESM2-LM	$1.9^\circ \times 2.5^\circ$

2.2. Methods

Six precipitation extreme indices were employed (Table 2): simple daily intensity index (SDII), wet days (WD), total precipitation (PRCPTOT), extreme precipitation amount (R95p), heavy precipitation days (R20mm), and consecutive dry days (CDD). These indices have been commonly used in previous studies [30,31].

Table 2. Definitions of precipitation indices used in this study. RR is daily precipitation.

Label	Index Definition	Units
PRCPTOT	Annual total precipitation on wet days (RR $\geq$ 1 mm)	mm
WD	Annual mean count of wet days (RR $\geq$ 1 mm)	days
SDII	Mean precipitation on wet days (RR $\geq$ 1 mm)	mm/day
CDD	Annual count of maximum number of consecutive dry days (RR < 1 mm)	days
R95p	Accumulated precipitation amounts when RR > 95th percentile	mm
R20mm	Annual count of days when RR $\geq$ 20 mm	days

Taylor diagrams are commonly used to evaluate two spatial patterns and to provide a pathway of how well the simulation matches the observation in terms of the standard deviation, correlation coefficient, and root mean square error (RMSE) [32].

3. Results

3.1. Comparison between Models in High-Resolution and Low-Resolution

The spatial pattern of model biases in climatological PRCPTOT over China from five individual models and their ensemble mean (MME) are shown in Figure 1. Compared with models in low resolution, high-resolution models show obvious improvement over southern China. To facilitate quantitative description of these improvements, three regions were determined as follows: western Sichuan Basin along the east side of the Tibetan Plateau (D1), South China (D2), and the Yangtze-Yellow River basins (D3) (black boxes in Figure 1). The common wet biases over D1 are reduced, with RMSE decreased by 48.4% in CNRM-CM6, 11.5% in EC-Earth3, 34.1% in HadGEM3, 32.1% in MPI-ESM1, 18.0% in NorESM2, and 34.3% in MME, respectively. This improvement has also been shown in previous studies [33,34], and it has been attributed to better representing of complex terrain [35]. Most models present decreased biases over D2, with RMSE reductions reaching 29.9%, 24.0%, 25.4%, 27.2%, and 26.8% in CNRM-CM6, EC-Earth3, HadGEM3, NorESM2, and MME, respectively. However, it should be noted that the high-resolution model is not implicitly perfect, such as MPI-ESM1, with RMSE increasing by 17.7% over D2 compared with the low-resolution versions. In addition, the biases over D3 are also reduced in the high-resolution models compared with the low-resolution models, with the largest reduction in bias existing in HadGEM3 at a magnitude exceeding 52.2%.

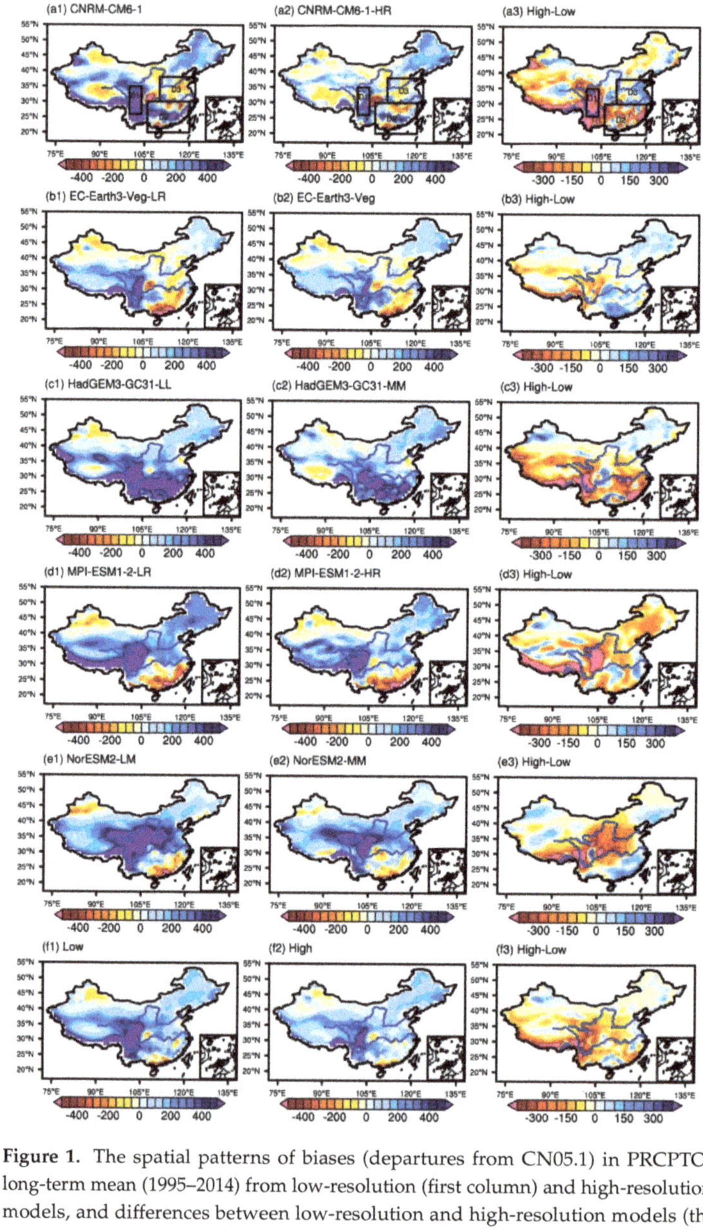

Figure 1. The spatial patterns of biases (departures from CN05.1) in PRCPTOT (unit: mm) for long-term mean (1995–2014) from low-resolution (first column) and high-resolution (second column) models, and differences between low-resolution and high-resolution models (third column). The last row shows multimodels with low-resolution ensemble means (**f1**) and multimodels with high-resolution ensemble means (**f2**), respectively.

The spatial pattern of model bias in WD index over China is shown in Figure 2. The biases over D1 are obviously lower in the high-resolution models, with RMSE reductions reaching 12.5%, 15.1%, 32.8%, 54.3%, 31.4%, and 42.2% in CNRM-CM6, EC-Earth3, HadGEM3, NorESM2, and MME, respectively. In addition, the biases over D2 and D3 in most models, with improved resolutions, are also lower. For example, biases over D2 (D3) in MPI-ESM1 are reduced by 20.4% (50.0%).

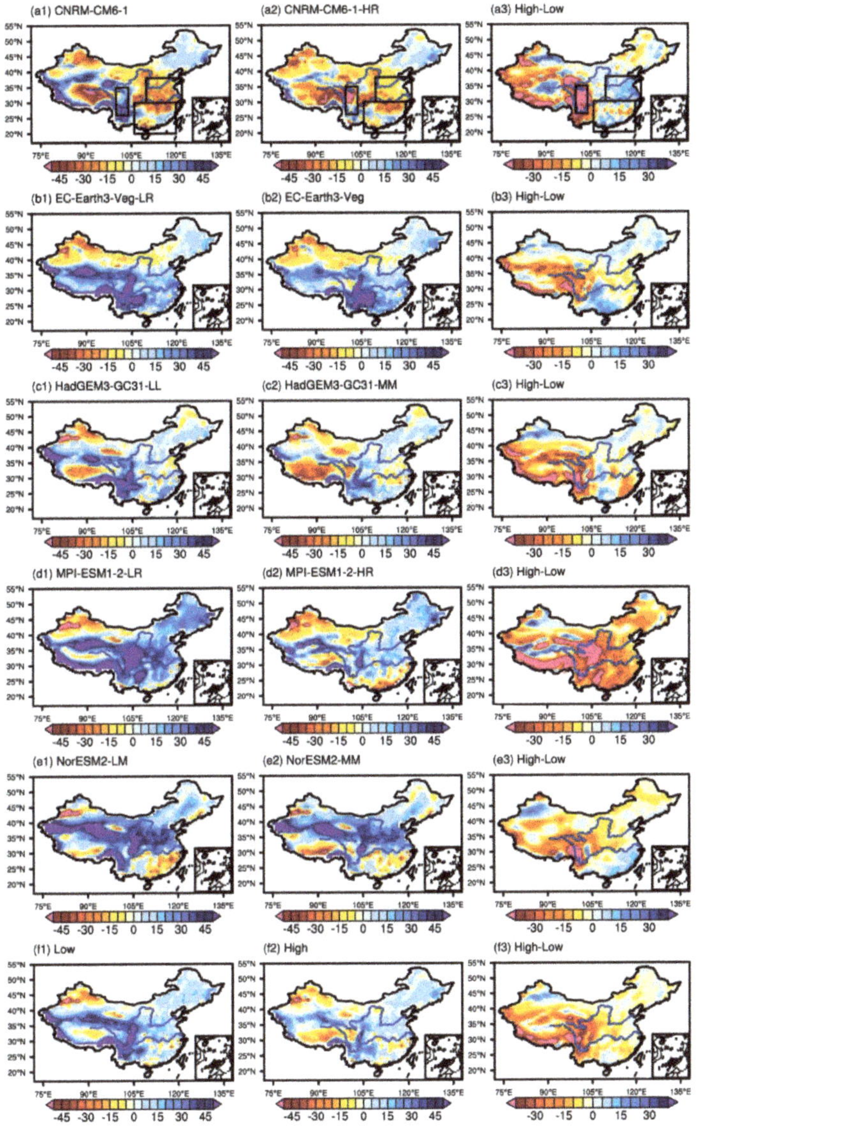

Figure 2. The spatial patterns of biases (departures from CN05.1) in WD (unit: days) for long-term mean (1995–2014) from low-resolution (first column) and high-resolution (second column) models, and differences between low-resolution and high-resolution models (third column). The last row shows multimodels with low-resolution ensemble means (**f1**) and multimodels with high-resolution ensemble means (**f2**), respectively.

The spatial pattern of model biases in representing the SDII over China are shown in Figure 3. The biases over D2 and D3 are remarkably lower in nearly all models with high resolutions, in comparison with low resolution models. For example, the RMSE decreased by 30.7% (22.1%) in CNRM-CM6, 14.6% (21.1%) in EC-Earth3, 19.7% (37.3%) in HadGEM3, 36.6% (20.1%) in MPI-ESM1, and 10.8% (18.9%) in MME. Regional averaged biases in PRCPTOT, WD, and SDII over D1, D2, and D3 from the individual models and MME are shown in Table 3. The reduction in wet biases over D1 in CNRM-CM6, HadGEM3, NorESM2, and MME could be due to the decreased precipitation frequency and the weakened precipitation intensity. However, for EC-Earth3 and MPI-ESM1, the contribution of precipitation frequency reduction is dominant. The reduction in biases over D2 in NorESM2 is associated with precipitation frequency. In EC-Earth3, HadGEM3, and MME, the reduction biases over D2 is related to precipitation frequency and precipitation intensity. In addition, the decreased biases over D3 in CNRM-CM6 and MPI-ESM1 could result from precipitation frequency. In HadGEM3, NorESM2, and MME, the decreased biases could be associated with precipitation frequency and precipitation intensity.

Table 3. The area-weighted average biases (departures from CN05.1) and differences between high-resolution and low-resolution models in PRCPTOT (unit: mm), WD (unit: days), and SDII (unit: mm/day) over D1, D2, and D3 from individual model, multimodels with low-resolution ensemble means, and multimodels with high-resolution ensemble means.

Model	D1			D2			D3		
	PRCPTOT	WD	SDII	PRCPTOT	WD	SDII	PRCPTOT	WD	SDII
CNRM-CM6-1-HR	+292.7	−15.6	+2.3	+53.1	−8.9	+0.8	+17.1	−11.7	+1.4
CNRM-CM6-1	+814.0	+31.6	+3.1	+159.4	−11.0	+1.7	−38.2	−19.5	+1.8
High–Low	−521.3	−47.2	−0.8	−106.3	+2.1	−0.9	+55.3	+7.8	−0.4
EC-Earth3-Veg	+351.7	+31.9	+0.75	−47.1	+25.4	−1.7	−6.0	+9.8	−1.0
EC-Earth3-Veg-LR	+433.7	+46.9	+0.65	−145.7	+20.2	−2.0	−4.2	+14.3	−1.3
High–Low	−82.0	−15	+0.1	+98.6	+5.2	+0.3	−1.8	−4.5	+0.3
HadGEM3-GC31-MM	+406.0	+14.9	+1.8	+383.6	+5.9	+2.0	+77.2	−0.9	+0.9
HadGEM3-GC31-LL	+667.6	+40.6	+2.1	+532.8	+8.8	+2.7	+191.9	+2.0	+1.7
High–Low	−261.6	−25.7	−0.3	−149.2	−2.9	−0.7	−114.7	−2.9	−0.8
MPI-ESM1-2-HR	+520.1	+14.5	+2.5	−220.4	−6.6	−1.0	+113.9	+9.5	+0.2
MPI-ESM1-2-LR	+798.5	+55.3	+2.1	−186.0	+17.8	−2.1	+151.2	+24.4	−0.6
High–Low	−278.4	−40.8	+0.4	−34.4	−24.4	+1.1	−37.3	−14.9	+0.8
NorESM2-MM	+520.7	+16.5	+2.4	−60.6	−4.4	−0.1	+192.1	+24.2	−0.4
NorESM2-LM	+685.4	+34.0	+2.5	−85.7	−8.1	−0.02	+321.8	+27.4	+0.5
High–Low	−164.7	−17.5	−0.1	+25.1	+3.7	−0.08	−129.7	−3.2	−0.9
High	+418.2	+12.4	+1.9	+21.7	+2.2	+0.01	+78.9	+6.2	+0.2
Low	+679.8	+41.7	+2.1	+55.0	+5.5	+0.07	+124.5	+9.7	+0.4
High–Low	−261.6	−29.3	−0.2	−33.3	−3.3	−0.06	−45.6	−3.5	−0.2

Simulation for the climatological extreme indices, R95p, R20mm, and CDD, were evaluated. In observation, the pattern of R95p and R20mm (Figures S1 and S2) shows a transition from small values in northwestern China to large values in southeastern China, consistent with the total precipitation (PRCPTOT) (Figure S3). The pattern of CDD shows an opposite distribution (Figure S4), which has been indicated in previous studies [36,37].

Figure 4 shows the spatial patterns of model biases in R95p over China. The biases in the high-resolution models over D2 (D3) are significantly reduced compared with those in the low-resolution models, with the reduction in RMSE reaching 22.8% (30.7%) in CNRM-CM6, 11.2% (15.6%) in HadGEM3, 31.2% (29.3%) in NorESM2, and 26.9% (6.6%) in MME. For the simulation over D1, obvious decreasing biases are found in CNRM-CM6, MPI-ESM1, and MME, with RMSE reductions of approximately 45.8%, 29.6%, and 26.8%, respectively.

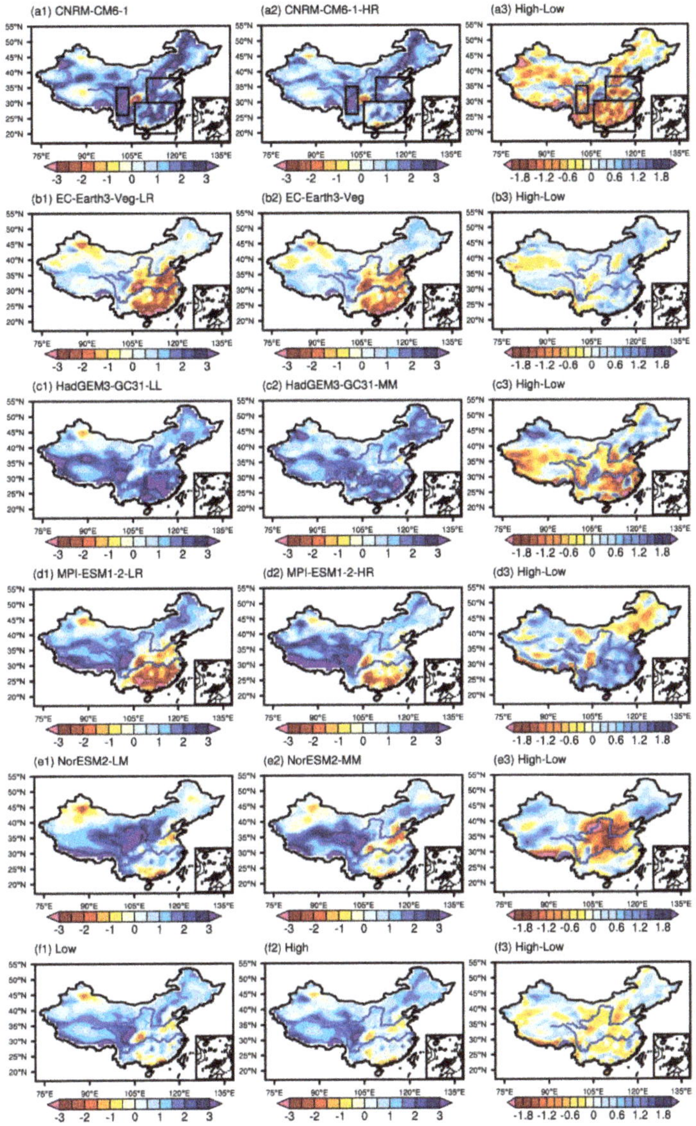

Figure 3. The spatial patterns of biases (departures from CN05.1) in SDII (unit: mm/day) for long-term mean (1995–2014) from low-resolution (first column) and high-resolution (second column) models, and differences between low-resolution and high-resolution models (third column). The last row shows multimodels with low-resolution ensemble means (**f1**) and multimodels with high-resolution ensemble means (**f2**), respectively.

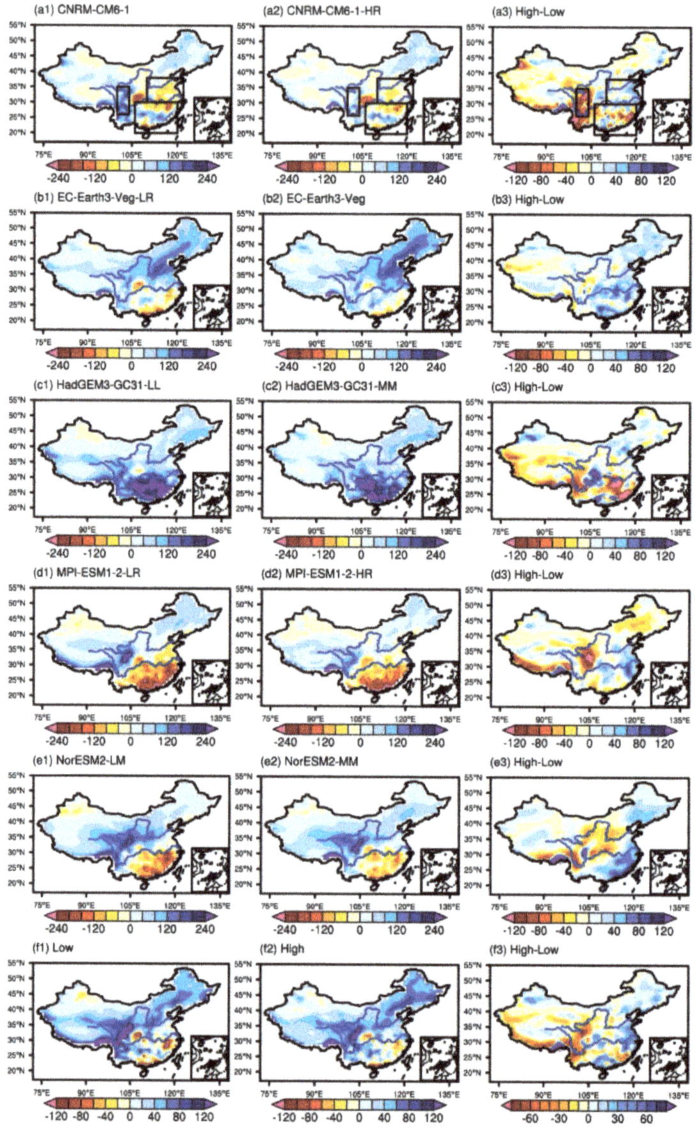

Figure 4. The spatial patterns of biases (departures from CN05.1) in R95p (unit: mm) for long-term mean (1995–2014) from low-resolution (first column) and high-resolution (second column) models, and differences between low-resolution and high-resolution models (third column). The last row shows multimodels with low-resolution ensemble means (**f1**) and multimodels with high-resolution ensemble means (**f2**), respectively.

The spatial distribution of model biases in R20mm over China is shown in Figure 5. We found that all models with high resolution show a uniform reduction in biases over D1, D2, and D3. For example, the obvious reduction in RMSE is 33.3% (2.0%) in CNRM-CM6, 31.6% (56.4%) in HadGEM3, 15.6% (53.1%) in NorESM2, and 18.1% (29.8%) in MME over D2 (D3). Figure 6 shows the spatial pattern of model biases in CDD over China. When compared with low resolution models, the biases with high-resolution models over D1,

D2 and D3 are reduced to some extent. For example, the reduction in RMSE is 29.0% (43.5%) in CNRM-CM6 and 20.0% (21.8%) in NorESM2 over D1 (D2). Furthermore, biases in the western part of Northwest China (36–46° N, 75–100° E) are also lower in models with high resolutions, with the reduction of RMSE achieving 23.9% in CNRM-CM6, 16.5% in HadGEM3, and 11.0% in MME. The better simulation of the CDD index could be associated with the simulation of the WD index (Figure 2). That is, the reduction in wet days may contribute to increased dry days.

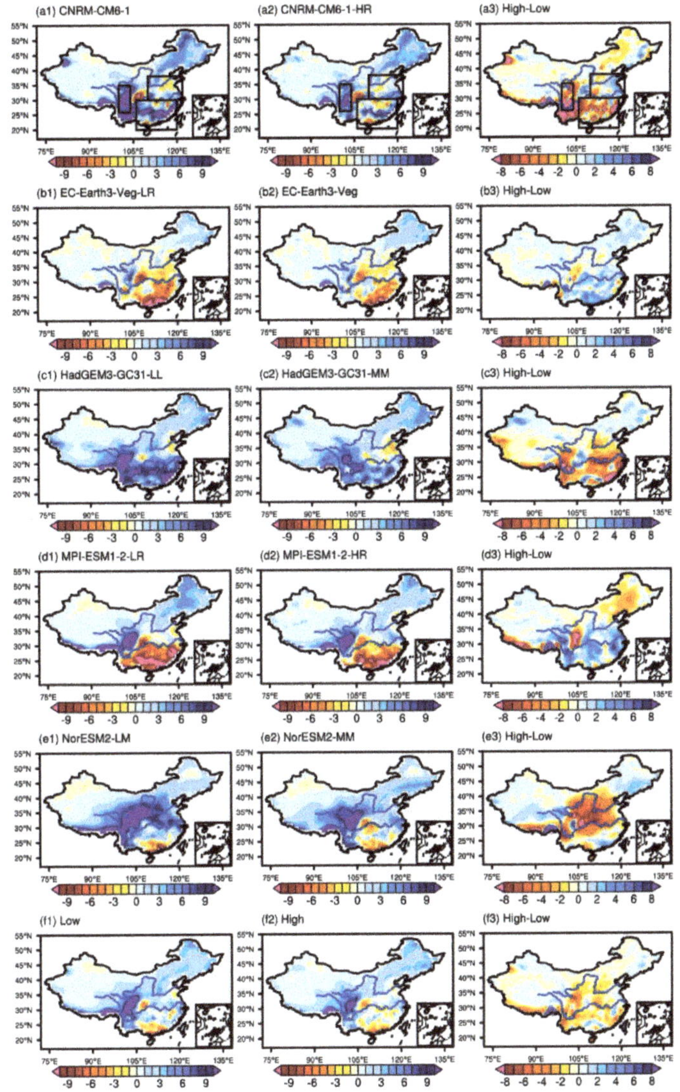

Figure 5. The spatial patterns of biases (departures from CN05.1) in R20mm (unit: days) for long-term mean (1995–2014) from low-resolution (first column) and high-resolution (second column) models, and differences between low-resolutions and high-resolutions models (third column). The last row shows multimodels with low-resolution ensemble means (**f1**) and multimodels with high-resolution ensemble means (**f2**), respectively.

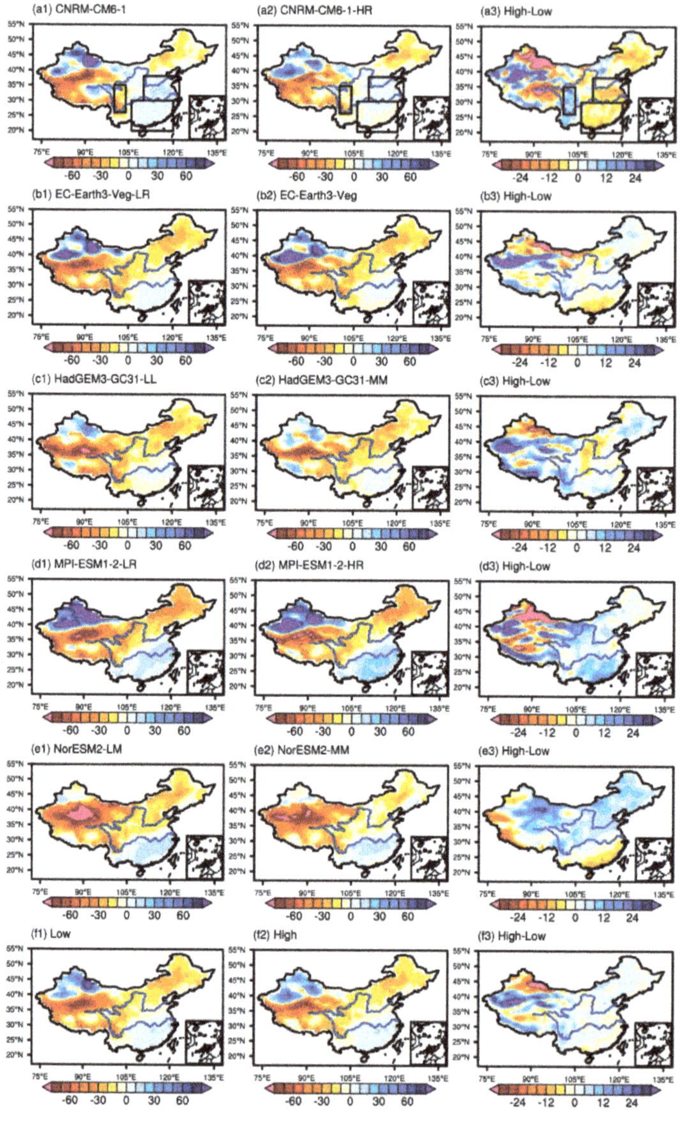

Figure 6. The spatial patterns of biases (departures from CN05.1) in CDD (unit: days) for long-term mean (1995–2014) from low-resolution (first column) and high-resolution (second column) models, and differences between low-resolution and high-resolution models (third column). The last row shows multimodels with low-resolution ensemble means (**f1**) and multimodels with high-resolution ensemble means (**f2**), respectively.

Evaluation of models using Taylor diagrams (Figure 7) for six extreme precipitation indices shows that most models in high resolution show improvements in simulating WD over China, compared with their counterparts in low resolution, such as the CNRM-CM6, EC-Earth3, and MPI-ESM1 models. Most models with improved resolutions show no obvious improvement in representing the features of R95p and in particular for R20mm over China. However, for the CDD index, all models with high resolutions show better performance compared with their counterparts with low resolutions. It is worth noting that models show a wide range of skill in reproducing the spatial distribution of CDD over China. That is, the inter-model distance is obviously large (Figure 7f).

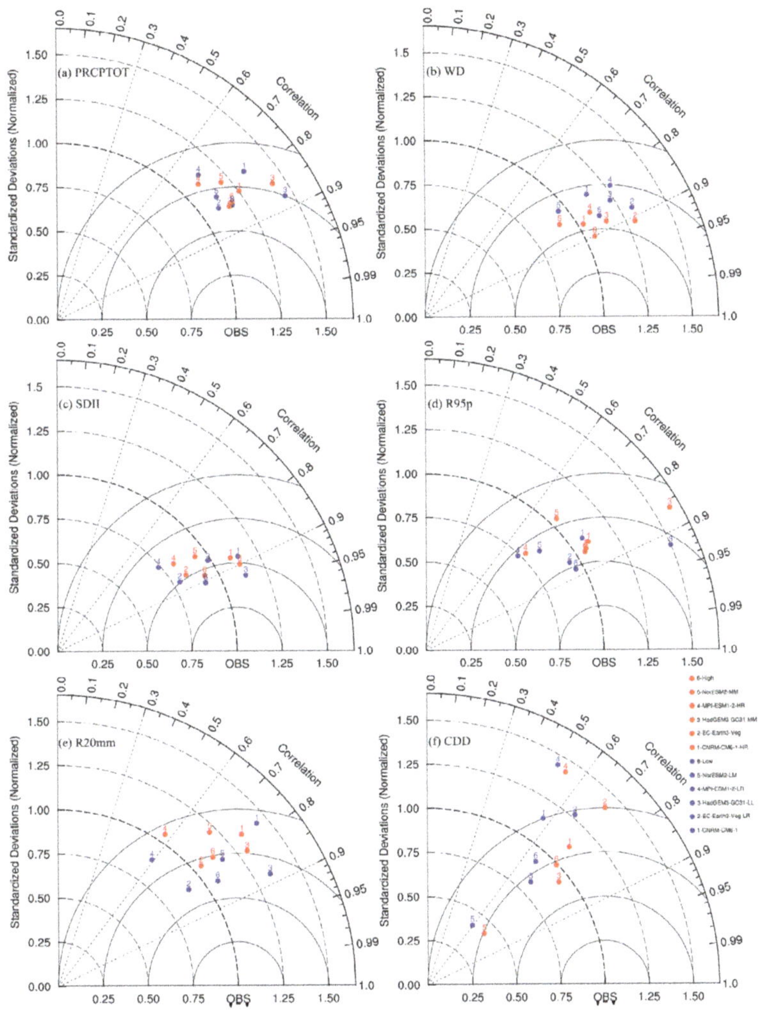

Figure 7. The Taylor diagram for three precipitation indices (PRCPTOT (**a**), WD (**b**), SDII (**c**), R95p (**d**), R20mm (**e**), and CDD (**f**)) over China from individual models with high and low resolutions, multimodels with low-resolution ensemble means, and multimodels with high-resolution ensemble means.

3.2. Possible Reasons for the Improved Precipitation Extremes Simulation in High-Resolution Models

Previous studies have reported that model resolutions might have non-negligible effect in simulations on the distribution of daily precipitation frequency [38]. Figure 8 shows the distributions of daily precipitation frequency as a function of daily precipitation intensity from 0 to 100 mm/day regionally averaged over D1, D2, and D3. Fewer weak precipitation events (0 mm/day–10 mm/day) could be found in nearly all models with updated resolutions over D1 in comparison with their counterparts with low resolutions. Therefore, the decreased biases of precipitation over D1 are associated with fewer weak precipitation events. The so-called "drizzling too much" phenomenon presented by climate models in previous studies [39] is improved with updated model resolutions to some extent. For CNRM-CM6 and HadGEM3 (EC-Earth3 and NorESM2), these models with high resolutions capture decreasing (increasing) 10 mm/day–60 mm/day (>30 mm/day) precipitation compared with their counterparts with low resolutions, which is associated with decreased (increased) precipitation over D2 (Figures 1 and 4). Comparing MME in low-resolution and high resolution versions simulates less 0 mm/day–30 mm/day precipitation, thus causing the decreased precipitation over D2 (Figure 1). In addition, for HadGEM3 and NorESM2, high-resolution versions show decreased precipitation over D3 (Figures 1 and 4), which is also related to the simulated precipitation of less than 10 mm/day–50 mm/day in these models in comparison with low resolutions.

Previous studies have reported that the vertical profile of the moisture content of the atmosphere [40,41] plays an important role in the simulation of convective rainfall. Therefore, the vertical distribution of moisture with simulated extreme precipitation was studied. Figure 9 shows the structure of moisture with precipitation in summer over D2 from models and observations. The combination of ERA5 relative humidity with CN05.1 precipitation (CN05.1/ERA5) was seen as the observation. Extreme precipitation related to strong deep convection is more easily triggered when the lower atmosphere above the boundary layer experiences sufficient moistening and a deep layer of the air column is nearly saturated [40]. We noticed that all models with low resolutions show excessive moisture in the upper atmosphere (500 hPa–250 hPa) in comparison with high-resolution versions. The simulation of the vertical distribution of moisture in high-resolution versions is closer to the observation, in comparison with low-resolution versions, thus possibly contributing to the better performance in the simulation of extreme precipitation with high-resolution versions.

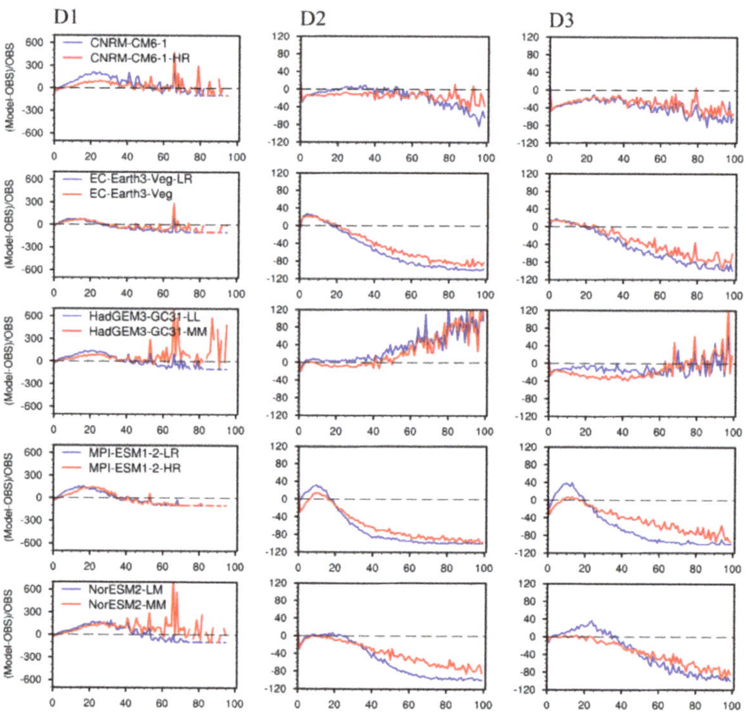

Figure 8. *Cont.*

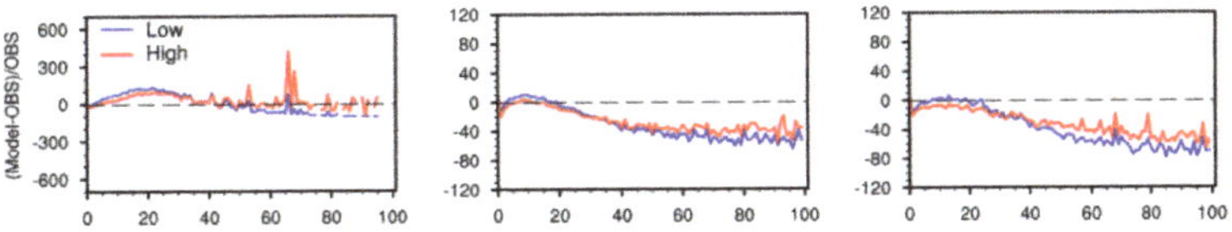

Figure 8. Relative bias curve of precipitation frequency distribution as a function of daily precipitation intensity (bin size 1 mm day^{-1}) over three subregions from individual models with low and high resolutions, multimodels with low-resolution ensemble means, and multimodels with high-resolution ensemble means (left y-axis, units: %). The unit of the x-axis is mm day^{-1}. Biases are shown as percentages relative to the observations. The first to third column corresponds to (**D1**), (**D2**), and (**D3**), respectively.

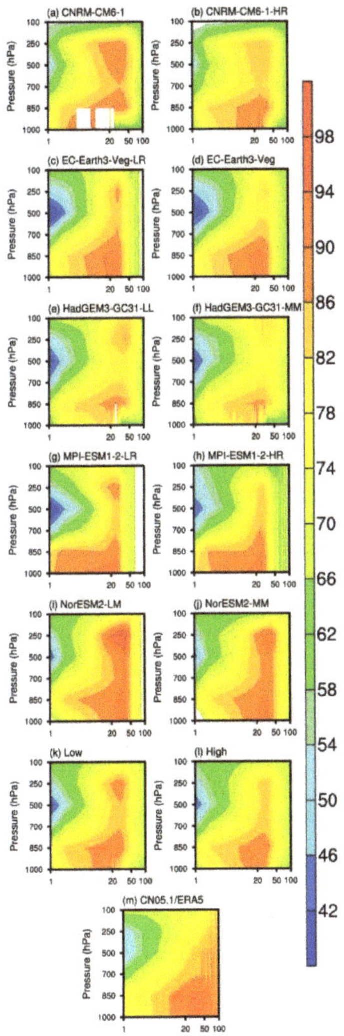

Figure 9. Composite vertical profiles of relative humidity (unit: %) based on precipitation intensity in summer over D2 from individual models (**a**–**j**), multimodels with low-resolution ensemble means (**k**), multimodels with high-resolution ensemble means (**l**), and CN05.1/ERA5 (**m**).

4. Discussion

This work provides a preliminary evaluation of precipitation extremes in terms of the effect of model resolution. Although consistent improvements in light rainfall can be found with updated resolutions, large model differences still exist in the five models. For example, the impact of higher resolutions on the simulation of R95p varies greatly in the five models. The above differences are possibly associated with the different reactions of some physical processes to the resolution. Thus, some key physical processes are also important for better simulation in high-resolution models.

The number of models used in this study is limited and model resolutions are not high enough. To fully investigate the impact of model resolution on the simulation of precipitation extremes, more models with higher resolutions from CMIP6 are required, and further assessment of the impact of model resolution in other ways is also needed.

5. Conclusions

Based on five CMIP6 models, each having a pair of high-resolution and low-resolution configurations, the difference between low-resolution models and high-resolution models in representing the spatial pattern of climatological precipitation extremes over China during 1995–2014 was compared. Generally, models in high resolution show better performance over China, especially over D1, D2, and D3. The decreased biases of total precipitation (PRCPTOT) over D1 were found in all models with updated resolutions, with the largest decease in RMSE being 48.4% in CNRM-CM6, which is associated with fewer weak precipitation events (0 mm/day–10 mm/day). Models with updated resolutions also show smaller biases than their counterparts with low resolutions for the simulation of precipitation over D2. For instance, the RMSE is reduced by 29.9% in CNRM-CM6 for PRCPTOT index and 31.2% in NorESM2 for extreme precipitation amount (R95p), respectively. These improvements could be attributed to better representation of the distribution of daily precipitation frequency. In addition, the vertical distribution of moisture content in models with updated resolutions is closer to the observation in comparison with models with low resolutions, thus showing better performance for extreme precipitation over D2.

Overall, our results indicate that models with improved resolution show an obvious advantage in the simulation of precipitation extremes, in particular for some regions with complex terrain. In addition, models with updated resolution also demonstrate an improvement in the simulation for light precipitation and heavy precipitation. To some extent, high-resolution models increase our confidence in precipitation extremes simulation. However, high-resolution simulation is very expensive, especially for global simulation. Therefore, higher demands are placed on the development of computers. In addition, some key physical processes are also crucial for simulation in extreme precipitation. Better physical parameterizations are also needed.

Supplementary Materials: The following are available online at https://www.mdpi.com/article/10.3390/su14010025/s1, Figure S1: Spatial distribution of R95p (unit: mm) for long-term mean (1995–2014) from individual models (a–j) and observation (k), Figure S2: Spatial distribution of R20mm (unit: days) for long-term mean (1995–2014) from individual models (a–j) and observation (k), Figure S3: Spatial distribution of PRCPTOT (unit: mm) for long-term mean (1995–2014) from individual models (a–j) and observation (k). Figure S4: Spatial distribution of CDD (unit: days) for long-term mean (1995–2014) from individual models (a–j) and observation (k).

Author Contributions: Conceptualization, supervision, and writing—review and editing, Y.G.; writing—original draft preparation and visualization, N.L. All authors have read and agreed to the published version of the manuscript.

Funding: This study is jointly supported by the National Key Research and Development Program of China, grant numbers 2020YFA0608201 and 2018YFC1509002.

Institutional Review Board Statement: Not applicable.

Informed Consent Statement: Not applicable.

Data Availability Statement: Not applicable.

Acknowledgments: We thank the three anonymous reviewers for their helpful comments and suggestions. We also appreciate Rui He for helpful advice for our manuscript.

Conflicts of Interest: The authors declare no conflict of interest.

References

1. Duliè, V.R.; Zhang, Y.; Salathé, E.P. Extreme precipitation and temperature over the U.S. Pacific Northwest: A comparison between observations, reanalysis data, and regional models. *J. Clim.* **2011**, *24*, 1950–1964. [CrossRef]
2. Gao, Z.; Zhu, J.; Guo, Y.; Luo, N.; Fu, Y.; Wang, T. Impact of Land Surface Processes on a Record-Breaking Rainfall Event on May 06–07, 2017, in Guangzhou, China. *J. Geophys. Res. Atmos.* **2021**, *126*, e2020JD032997. [CrossRef]
3. Zhou, Z.Q.; Xie, S.P.; Zhang, R. Historic Yangtze flooding of 2020 tied to extreme Indian Ocean conditions. *Proc. Natl. Acad. Sci. USA* **2021**, *118*, e2022255118. [CrossRef] [PubMed]
4. Su, A.; Lü, X.; Cui, L.; Li, Z.; Xi, L.; Li, H. The Basic Observational Analysis of "7.20" Extreme Rainstorm in Zhengzhou. *Torrential Rain Disasters* **2021**, *40*, 445–454.
5. Duffy, P.B.; Govindasamy, B.; Lorio, J.P.; Milovich, J.; Sperber, K.R.; Taylor, K.E.; Wehner, M.F.; Thompson, S.L. High-resolution simulations of global climate, Part 1: Present climate. *Clim. Dyn.* **2003**, *21*, 371–390. [CrossRef]
6. Greenberg, D.A.; Dupont, F.; Lyard, F.H.; Lynch, D.R.; Werner, F.E. Resolution issues in numerical models of oceanic and coastal circulation. *Cont. Shelf Res.* **2007**, *27*, 1317–1343. [CrossRef]
7. Knutson, T.R.; McBride, J.L.; Chan, J.; Emanuel, K.; Holland, G.; Landsea, C.; Held, I.; Kossin, J.P.; Srivastava, A.K.; Sugi, M. Tropical cyclones and climate change. *Nat. Geosci.* **2010**, *3*, 157–163. [CrossRef]
8. Vidale, P.L.; Hodges, K.; Vannière, B.; Davini, P.; Roberts, M.J.; Strommen, K.; Weisheimer, A.; Plesca, E.; Corti, S. Impact of stochastic physics and model resolution on the simulation of tropical cyclones in climate GCMs. *J. Clim.* **2021**, *34*, 4315–4341. [CrossRef]
9. Guilyardi, E.; Gualdi, S.; Slingo, J.; Navarra, A.; Delecluse, P.; Cole, J.; Madec, G.; Roberts, M.; Latif, M.; Terray, L. Representing El Niño in coupled ocean-atmosphere GCMs: The dominant role of the atmospheric component. *J. Clim.* **2004**, *17*, 4623–4629. [CrossRef]
10. Rauscher, S.A.; Coppola, E.; Piani, C.; Giorgi, F. Resolution effects on regional climate model simulations of seasonal precipitation over Europe. *Clim. Dyn.* **2010**, *35*, 685–711. [CrossRef]
11. Huang, D.Q.; Zhu, J.; Zhang, Y.C.; Huang, A.N. Uncertainties on the simulated summer precipitation over Eastern China from the CMIP5 models. *J. Geophys. Res. Atmos.* **2013**, *118*, 9035–9047. [CrossRef]
12. Zhang, L.; Wu, P.; Zhou, T.; Roberts, M.J.; Schiemann, R. Added value of high resolution models in simulating global precipitation characteristics. *Atmos. Sci. Lett.* **2016**, *17*, 646–657. [CrossRef]
13. He, S.; Yang, J.; Bao, Q.; Wang, L.; Wang, B. Fidelity of the observational/reanalysis datasets and global climate models in representation of extreme precipitation in East China. *J. Clim.* **2019**, *32*, 195–212. [CrossRef]
14. Demory, M.E.; Vidale, P.L.; Roberts, M.J.; Berrisford, P.; Strachan, J.; Schiemann, R.; Mizielinski, M.S. The role of horizontal resolution in simulating drivers of the global hydrological cycle. *Clim. Dyn.* **2014**, *42*, 2201–2225. [CrossRef]
15. Prein, A.F.; Langhans, W.; Fosser, G.; Ferrone, A.; Ban, N.; Goergen, K.; Keller, M.; Tölle, M.; Gutjahr, O.; Feser, F.; et al. A review on regional convection-permitting climate modeling: Demonstrations, prospects, and challenges. *Rev. Geophys.* **2015**, *53*, 323–361. [CrossRef] [PubMed]
16. Lin, L.; Gettelman, A.; Xu, Y.; Wu, C.; Wang, Z.; Rosenbloom, N.; Bates, S.C.; Dong, W. CAM6 simulation of mean and extreme precipitation over Asia: Sensitivity to upgraded physical parameterizations and higher horizontal resolution. *Geosci. Model Dev.* **2019**, *12*, 3773–3793. [CrossRef]
17. Eyring, V.; Bony, S.; Meehl, G.A.; Ronald, S.; Oceanic, N. Overview of the Coupled Model Intercomparison Project Phase 6 (CMIP6) experimental design and organisation. *Geosci. Model Dev. Discuss.* **2015**, *8*, 10539–10583. [CrossRef]
18. Xin, X.; Wu, T.; Zhang, J.; Yao, J.; Fang, Y. Comparison of CMIP6 and CMIP5 simulations of precipitation in China and the East Asian summer monsoon. *Int. J. Climatol.* **2020**, *40*, 6423–6440. [CrossRef]
19. Chen, H.; Sun, J.; Lin, W.; Xu, H. Comparison of CMIP6 and CMIP5 models in simulating climate extremes. *Sci. Bull.* **2020**, *65*, 1415–1418. [CrossRef]
20. Zhu, X.; Ji, Z.; Wen, X.; Lee, S.Y.; Wei, Z.; Zheng, Z.; Dong, W. Historical and projected climate change over three major river basins in China from Fifth and Sixth Coupled Model Intercomparison Project models. *Int. J. Climatol.* **2021**, 1–19. [CrossRef]
21. Zhu, Y.Y.; Yang, S. Evaluation of CMIP6 for historical temperature and precipitation over the Tibetan Plateau and its comparison with CMIP5. *Adv. Clim. Chang. Res.* **2020**, *11*, 239–251. [CrossRef]
22. Dong, T.; Dong, W. Evaluation of extreme precipitation over Asia in CMIP6 models. *Clim. Dyn.* **2021**, *57*, 1751–1769. [CrossRef]
23. Luo, N.; Guo, Y.; Chou, J.; Gao, Z. Added value of CMIP6 models over CMIP5 models in simulating the climatological precipitation extremes in. *Int. J. Climatol.* **2021**, 1–17. [CrossRef]
24. Jiang, D.; Hu, D.; Tian, Z.; Lang, X. Differences between CMIP6 and CMIP5 Models in Simulating Climate over China and the East Asian Monsoon. *Adv. Atmos. Sci.* **2020**, *37*, 1102–1118. [CrossRef]

25. Hu, Y.-Y.; Xu, Y.; Li, J.-J.; Han, Z.-Y. Evaluation on the performance of CMIP6 global climate models with different horizontal resolution in simulating the precipitation over China. *Clim. Chang. Res.* **2021**, *17*, 730–743.
26. Wu, J.; Gao, X.J. A gridded daily observation dataset over China region and comparison with the other datasets. *Acta Geophys. Sin.* **2013**, *56*, 1102–1111. [CrossRef]
27. Yin, H.; Donat, M.G.; Alexander, L.V.; Sun, Y. Multi-dataset comparison of gridded observed temperature and precipitation extremes over China. *Int. J. Climatol.* **2015**, *35*, 2809–2827. [CrossRef]
28. Zhou, B.; Xu, Y.; Wu, J.; Dong, S.; Shi, Y. Changes in temperature and precipitation extreme indices over China: Analysis of a high-resolution grid dataset. *Int. J. Climatol.* **2016**, *36*, 1051–1066. [CrossRef]
29. Hersbach, H.; Bell, B.; Berrisford, P.; Hirahara, S.; Horányi, A.; Muñoz-Sabater, J.; Nicolas, J.; Peubey, C.; Radu, R.; Schepers, D.; et al. The ERA5 Global Reanalysis. *Q. J. R. Meteorol. Soc.* **2020**, *146*, 1999–2049. [CrossRef]
30. Xu, C.; Luo, Y.; Xu, Y. Projected changes of precipitation extremes in river basins over China. *Quat. Int.* **2011**, *244*, 149–158. [CrossRef]
31. Zhou, B.; Wen, Q.H.; Xu, Y.; Song, L.; Zhang, X. Projected changes in temperature and precipitation extremes in China by the CMIP5 multimodel ensembles. *J. Clim.* **2014**, *27*, 6591–6611. [CrossRef]
32. Taylor, K.E. Summarizing multiple aspects of model performance in a single diagram. *J. Geophys. Res. Atmos.* **2001**, *106*, 7183–7192. [CrossRef]
33. Gao, X.; Shi, Y.; Song, R.; Giorgi, F.; Wang, Y.; Zhang, D. Reduction of future monsoon precipitation over China: Comparison between a high resolution RCM simulation and the driving GCM. *Meteorol. Atmos. Phys.* **2008**, *100*, 73–86. [CrossRef]
34. Zou, L.; Zhou, T. Near future (2016-40) summer precipitation changes over China as projected by a regional climate model (RCM) under the RCP8.5 emissions scenario: Comparison between RCM downscaling and the driving GCM. *Adv. Atmos. Sci.* **2013**, *30*, 806–818. [CrossRef]
35. Feng, L.; Zhou, T.; Wu, B.; Li, T.; Luo, J.J. Projection of future precipitation change over China with a high-resolution global atmospheric model. *Adv. Atmos. Sci.* **2011**, *28*, 464–476. [CrossRef]
36. Bao, J.; Feng, J.; Wang, Y. Dynamical downscaling simulation and future projection of precipitation over China. *J. Geophys. Res. Atmos.* **2015**, *120*, 8227–8243. [CrossRef]
37. Wang, H.J.; Sun, J.Q.; Chen, H.P.; Zhu, Y.L.; Zhang, Y.; Jiang, D.B.; Lang, X.M.; Fan, K.; Yu, E.T.; Yang, S. Extreme climate in China: Facts, simulation and projection. *Meteorol. Z.* **2012**, *21*, 279–304. [CrossRef]
38. Ganai, M.; Krishna, R.P.M.; Mukhopadhyay, P.; Mahakur, M. The impact of revised simplified Arakawa-Schubert scheme on the simulation of mean and diurnal variability associated with active and break phases of Indian summer monsoon using CFSv2. *J. Geophys. Res.* **2016**, *121*, 9301–9323. [CrossRef]
39. Stephens, G.L.; L'Ecuyer, T.; Forbes, R.; Gettlemen, A.; Golaz, J.C.; Bodas-Salcedo, A.; Suzuki, K.; Gabriel, P.; Haynes, J. Dreary state of precipitation in global models. *J. Geophys. Res. Atmos.* **2010**, *115*, 1–14. [CrossRef]
40. Kim, D.; Xavier, P.; Maloney, E.; Wheeler, M.; Waliser, D.; Sperber, K.; Hendon, H.; Zhang, C.; Neale, R.; Hwang, Y.T.; et al. Process-oriented MJO simulation diagnostic: Moisture sensitivity of simulated convection. *J. Clim.* **2014**, *27*, 5379–5395. [CrossRef]
41. Holloway, C.E.; Neelin, D.J. Moisture vertical structure, column water vapor, and tropical deep convection. *J. Atmos. Sci.* **2009**, *66*, 1665–1683. [CrossRef]

sustainability

MDPI

Article

Global Forest Types Based on Climatic and Vegetation Data

Chen Xu [1], Xianliang Zhang [2,*], Rocío Hernandez-Clemente [3], Wei Lu [2] and Rubén D. Manzanedo [4]

1 College of Landscape Architecture and Tourism, Hebei Agricultural University, Baoding 071001, China; xuchen2016@163.com
2 College of Forestry, Hebei Agricultural University, Baoding 071001, China; sanpangzi1228@126.com
3 Department of Geography, Swansea University, Singleton Park, Swansea SA2 8PP, UK; r.hernandez-clemente@swansea.ac.uk
4 Department of Environmental Systems Science, Institute of Integrative Biology, D-USYS, ETH-Zürich, 8006 Zurich, Switzerland; ruben.delgadomanzanedo@usys.ethz.ch
* Correspondence: zhxianliang85@gmail.com

Abstract: Forest types are generally identified using vegetation or land-use types. However, vegetation classifications less frequently consider the actual forest attributes within each type. To address this in an objective way across different regions and to link forest attributes with their climate, we aimed to improve the distribution of forest types to be more realistic and useful for biodiversity preservation, forest management, and ecological and forestry research. The forest types were classified using an unsupervised cluster analysis method by combining climate variables with normalized difference vegetation index (NDVI) data. Unforested regions were masked out to constrict our study to forest type distributions, using a 20% tree cover threshold. Descriptive names were given to the defined forest types based on annual temperature, precipitation, and NDVI values. Forest types had distinct climate and vegetation characteristics. Regions with similar NDVI values, but with different climate characteristics, which would be merged in previous classifications, could be clearly distinguished. However, small-range forest types, such as montane forests, were challenging to differentiate. At macroscale, the resulting forest types are largely consistent with land-cover types or vegetation types defined in previous studies. However, considering both potential and current vegetation data allowed us to create a more realistic type distribution that differentiates actual vegetation types and thus can be more informative for forest managers, conservationists, and forest ecologists. The newly generated forest type distribution is freely available to download and use for non-commercial purposes as a GeoTIFF file via doi: 10.13140/RG.2.2.19197.90082).

Keywords: forest types; NDVI; AVHRR GIMMS; temperature range; precipitation range

Citation: Xu, C.; Zhang, X.; Hernandez-Clemente, R.; Lu, W.; Manzanedo, R.D. Global Forest Types Based on Climatic and Vegetation Data. *Sustainability* **2022**, *14*, 634. https://doi.org/10.3390/su14020634

Academic Editor: Ivo Machar

Received: 27 November 2021
Accepted: 4 January 2022
Published: 7 January 2022

1. Introduction

Forests vary in structure and function across the world. The broad-scale vegetation units with common formation characteristics, due to similar climates, are known as vegetation types [1]. Generally, forest types are derived from vegetation or land use types. In fact, the first way to define forest types was based on a vegetation classification. Vegetation types were originally developed based on the idea that similar climates select for similar plant forms [2], and therefore the resulting types were mostly climate-based. The first formal climate classification system was defined by Köppen and was also used to predict the global vegetation distribution [3]. Other systems for delineating types based on climate variables include Holdridge life zones [4,5], Box models [6,7] and Whittaker's biome types, though in the case of Whittaker's biomes, predefined vegetation unites were mapped onto a climate space [8,9]. The predicted vegetation produced by biogeographical models (e.g., BIOME3 [10], dynamic global vegetation models, and bioclimatic maps [11]), is also mainly derived from an assumed relationship between functional types and climate variables. The ecoregions defined by Olson et al. [12] relied on climate data, expert judgment, and species assemblages to differentiate certain forest types. By mainly considering climate data, these methods define forest types that better describe the potential vegetation of an area. They may, however, fail to correspond with the actual vegetation, since this is defined by the interaction between the potential vegetation and multiple factors, such as human influence, species interactions, and biogeographical history.

Forest types were also defined in land-cover classifications. Land-cover classifications delineate vegetation types based on satellite imagery [13–18]. Functional biomes have also been defined using vegetation information [19]. Climatic vegetation types, on the other hand, try to reflect the regional vegetation characteristics in terms of climate by merging climate and NDVI data, and reveal the actual vegetation distribution, taking advantage of the positive aspects of both approaches [20]. Although this method showed great promise in Zhang et al. [20], the coarse resolution of the data, and the limited number of vegetation types used, resulted in a forest type distribution with relatively low accuracy. The forest attributes and their linkage to climate were also not well investigated in that study. A reanalysis and improved definition of the method in Zhang et al. [20] is timely, and it can provide a useful global forest type cartography that can more accurately represent actual vegetation distributions.

Different vegetation classifications are useful for different purposes. For example, climate-based vegetation classifications emphasize the distribution of vegetation types, while land use classifications highlight the role of land cover and human activity. It is important, consequently, to clarify the intent of new classifications. Our classification focuses on forest types, as they have been shown to be reliably characterized using satellite data [14,16]. An accurate definition of forest types is fundamental for preserving biodiversity [11] and forest ecological research (e.g., for studies that compare and explore the drivers of large-scale forest productivity [21]). To this end, forest types should reflect the actual main forest types present in different regions, but this is not guaranteed when using forest type classification based only on climate. For instance, the main forest type in the Northeast China Plain is temperate sub-humid broadleaf forest, and it is generally classified as cropland in land use and vegetation classifications [22], which does not properly capture the actual characteristics of the forested ecosystem. Classifications based only on normalized difference vegetation index (NDVI) values, on the other hand, would not separate different forest types with similar NDVI values but very different functional compositions [20]. We argue that forest types delineated to reflect the actual forest distribution using both vegetation and climate data will be more useful for multiple uses, from management to research [23].

We aim to create regionally and locally coherent maps of the global forest types that emphasize the forest role in every region, that could be useful for research and applied forest studies, and flexible enough to be easily updated as climate change or human actions alter the characteristics or distribution of forest types. To this end, we combined vegetation and climate data to refine the definition of forest types. Then, we compared NDVI values between different forest types to investigate how well these forest types were separated in our method.

2. Materials and Methods

2.1. Vegetation Data

NDVI is the most commonly used vegetation index to represent vegetation greenness. Monthly NDVI data at 8 km resolution were retrieved from the Advanced Very High Resolution Radiometer-based Global Inventory Modelling and Mapping Studies (AVHRR-GIMMS) dataset (https://ecocast.arc.nasa.gov/data/pub/gimms/3g.v1/, accessed on 1 July 2021) for 1982–2013 [24]. The data have been processed to reduce effects of navigation errors, major volcanic eruptions, and orbital drift of older satellites [25]. AVHRR-GIMMS NDVI product was found to be useful for studying linkage between land surface phenology and climate over wide range of vegetation [24]. NDVI values range from -1 to 1, where negative values represent an absence of vegetation and positive values indicate vegetated land. The average monthly NDVI values from 1982 to 2013 were used to reflect the mean state of vegetation dynamics.

Enhanced vegetation index (EVI) enhances the vegetation signal with improved sensitivity in high biomass regions. Monthly EVI data at 8 km resolution were retrieved from the website (https://www.usgs.gov/land-resources/nli/landsat/, accessed on 1 July 2021). EVI data were tested as a replacement for NDVI data.

2.2. Climate Data

We used a global gridded climate dataset with a spatial resolution of 1–18 km^2 [26]. We used the average monthly values of mean temperature and total precipitation for 1970–2000 at a resolution of 8 km^2, to match the resolution of NDVI data [26] (http://worldclim.org/version2, accessed on 1 July 2021).

Although the timespan of the climate data (1970–2000) slightly differs from that of the NDVI data (1983–2013), this climate dataset has been widely used as a reliable high-resolution climate dataset in bioclimatic studies [27,28]. In addition, the main purpose of this climate data is to reflect the long-term mean state of the climate for a comparable period to that of the forest type classification, and thus we do not expect time coverage differences between NDVI and climate data to affect our results.

2.3. Forest Type Classification

K-means method [29] had been proven to be an effective method in defining climate types, climatic vegetation types, and different forest types (e.g., [30–34]). The K-means method is an unsupervised clustering method that separates multivariable data into a given number of clusters according to their distances to the center of the clusters. Four steps are involved in implementing the algorithm. First, k centroids are randomly selected as initial centroids if the data are supposed to be classified into k groups. Second, each point is allocated to its nearest centroid. Third, the k centroids are recalculated and assigned as new centroids. Finally, steps two and three are iterated until the centroids are stable. Supervised classification methods, such as random forest, require training data to classify the forest types. However, the outcome of classification was not prior known before classification. Therefore, K-means method is more suitable to be used in forest type classification than supervised methods because it performs without training data.

NDVI is widely used to define vegetation types and land use types (e.g., [14–17]). Forest types with similar NDVI values were inspected using climate data to check climate differences in vegetation with similar NDVI. Consequently, our final forest type classification is based on monthly mean temperature, total precipitation, and NDVI data. The main advantage of this combinative approach is that types with similar NDVI values under different climate conditions can be separated.

Monthly mean temperatures in all the grid cells were listed in one row and arranged by 12 months to be an $n \times 12$ matrix, as well as monthly total precipitation and NDVI. The precipitation was log-transformed to reduce the influence of unit as precipitation ranges from 0 to over 8000 cm. Hence, monthly mean temperature, total precipitation, and NDVI values were combined in an $n \times 36$ matrix, X:

$$\begin{bmatrix} T_{11} & \cdots & T_{1m} & P_{11} & \cdots & P_{1m} & \text{NDVI}_{11} & \cdots & \text{NDVI}_{1m} \\ \vdots & \vdots & \vdots & \vdots & \vdots & \vdots & \vdots & \vdots & \vdots \\ T_{n1} & \cdots & T_{nm} & P_{n1} & \cdots & P_{nm} & \text{NDVI}_{n1} & \cdots & \text{NDVI}_{nm} \end{bmatrix} \quad (1)$$

where T is monthly mean temperature, P is monthly mean precipitation after being log-transformed, NDVI is monthly NDVI value, m corresponds to the month number, and n is the number of all the grid cells in the global land area, except Antarctica. As the seasonal cycle of climate and vegetation in the Southern Hemisphere is opposite to that in the Northern Hemisphere, T, P, and NDVI in the Southern Hemisphere were adjusted to their corresponding months in the Northern Hemisphere, to make them comparable (see [20]).

Then, the *n* columns were classified into k clusters based on the monthly multivariable attributes. The classification was implemented using stats package in R software [35].

Climate variables were rescaled to eliminate the potential influence of combining multiple units, using the following formula:

$$Z_i = \frac{x_i - min(X)}{\max(X) - min(X)} \tag{2}$$

where Z_i is the standard index, with a scale from 0 to 1, X is a variable, x_i is every value in X, and max(X) and min(X) are the maximum and minimum X values, respectively.

A shortcoming of the K-means method is that the number of clusters should be defined beforehand. In our case, we chose to follow the number of forest types used by the GlobCover land-cover classification, created by the European Space Agency, which identified 12 different types related to forest. The resulting forest types were given descriptive names based on their annual temperature, precipitation, and NDVI values, trying to reflect the forest attribute and corresponding climatic conditions in each of them. The climatic modifiers in the names were designated following the criteria described in Table 1. Whether the tree cover was evergreen or deciduous was determined by the variation in monthly NDVI values and the related references [14,36,37]. The NDVI was low in winter if the type is dominated by deciduous trees.

Table 1. Criteria used to assign forest type names.

Climate Variables	Label	Criteria
Annual Temperature (°C)	Tropical	>20
	Sub-Tropical	(10, 20]
	Temperate	(0, 10]
	Sub-frigid	(−5, 0]
	Frigid	(−15, −5]
Annual Precipitation (mm)	Rain	>2000
	Moist	(1200, 2000]
	Humid	(800, 1200]
	Sub-humid	(400, 800]
	Semiarid	(200, 400]

We masked out 'no forest' regions to constrict our forest type distributions, using forest cover data [38]. Regions with lower than 20% tree cover were considered 'no forest'. This matches the forest definition of the National Forest Inventory for forest inventory [39] and had been used previously in nearby studies [40,41]. The FAO forest definitions entail some flexibility regarding the tree cover threshold between 10% and 30%, creating a non-uniform definition for forest types [42]. While the 10% canopy-cover threshold in FAO's definition has been used in some studies [43,44], it has also been criticized for including wooded grassland ecosystems [45], which have markedly different ecosystem dynamics. We decided to use a 20% minimum tree cover threshold, as it has been used for defining forest regions before [43,44]. It is important to note that the threshold used for defining forests will likely influence the definition and classification of forest types [42]. Thus, the research or management goal should be considered carefully when using a type classification, paying particular attention to the forest definition used.

The sensitivity of the global forest classification to the selected variables and the spatial resolution of selected variables were tested by changing the selected variable and changing the spatial resolution of selected variables.

We compared the resulting type distribution with the GlobCover land use maps (http://due.esrin.esa.int/, accessed on 1 July 2021). We also classified the forest type based on climate and EVI data to test whether EVI was a better vegetation indication than NDVI.

2.4. Climate Conditions in Every Forest Type

After classifying the forest types, we calculated the annual mean temperature, total annual precipitation, and mean NDVI value for every forest type to highlight their differences.

3. Results

3.1. Distribution of Forest Types

Twelve forest types were classified based on 8 km monthly NDVI, temperature, and precipitation data using the K-means method (Figure 1). Both vegetation and climate characteristics differed between different forest types (Table 2). The forest types were named based on their annual temperature, precipitation, and NDVI values (Table 2). We derived eight satellite images from Google Earth to compare our defined forest types in eight sites (Figure 2). The site pictures reflected the forest conditions in their defined forest types well. The other four types were not shown for the lack of photos.

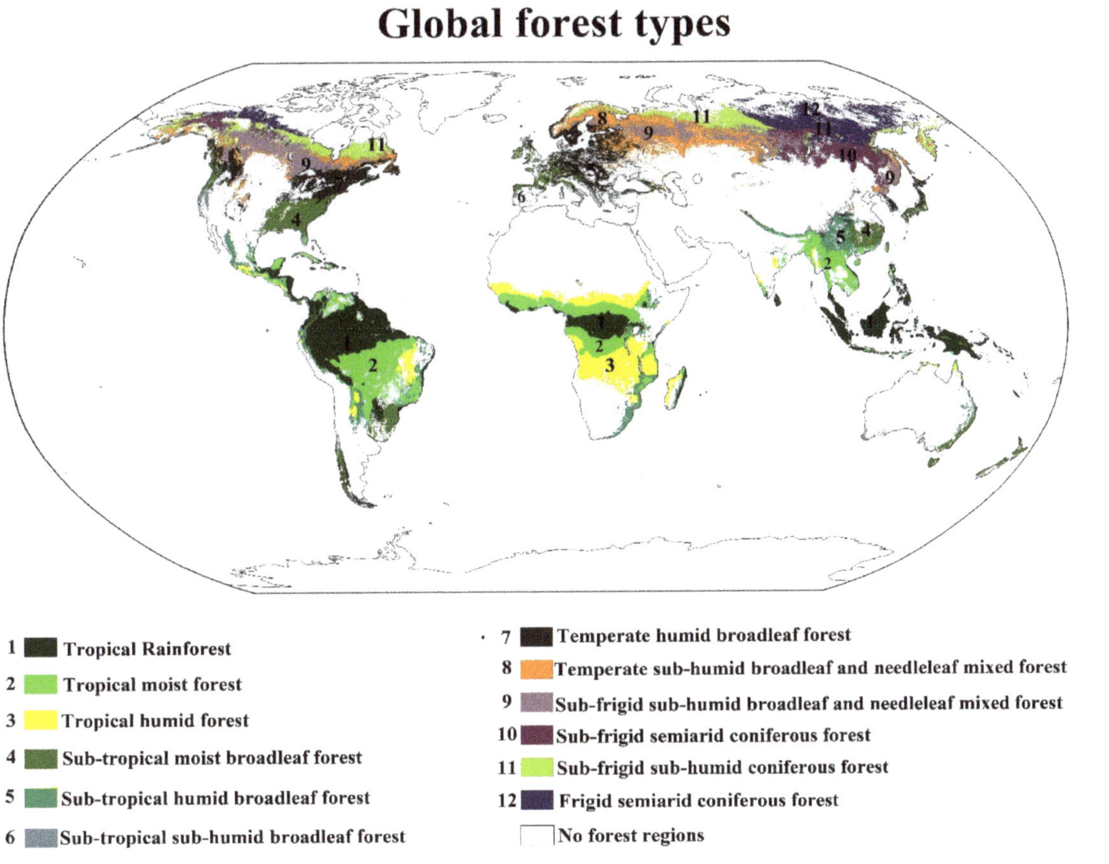

Figure 1. Geographic distribution of forest types classified based on 8 km resolution monthly vegetation and climate data. Forest was defined as tree cover over 20% threshold value.

Table 2. Forest types and their annual NDVI values, temperature (°C), and precipitation (mm).

Code	Type	Mean Annual NDVI	Mean Annual Temperature (°C)	Total Annual Precipitation (mm)
1	Tropical rainforest	0.72 ± 0.09	25.2 ± 1.9	2404.6 ± 677
2	Tropical moist forest	0.69 ± 0.08	24.5 ± 2.0	1644.5 ± 512
3	Tropical humid forest	0.59 ± 0.07	23.5 ± 2.7	1096.9 ± 394
4	Sub-tropical moist broadleaf forest	0.70 ± 0.08	14.0 ± 3.8	1490.9 ± 577
5	Sub-tropical humid broadleaf forest	0.58 ± 0.10	19.2 ± 4.3	1094.9 ± 390
6	Sub-tropical sub-humid broadleaf forest	0.60 ± 0.12	13.5 ± 3.6	732.3 ± 296
7	Temperate humid broadleaf forest	0.58 ± 0.07	6.3 ± 2.7	913.2 ± 402
8	Temperate sub-humid broadleaf and needleleaf mixed forest	0.40 ± 0.05	1.2 ± 2.5	639.9 ± 234
9	Sub-frigid sub-humid broadleaf and needleleaf mixed forest	0.56 ± 0.04	−0.1 ± 2.3	567.9 ± 144
10	Sub-frigid semiarid coniferous forest	0.42 ± 0.05	−3.6 ± 2.7	437.4 ± 200
11	Sub-frigid sub-humid coniferous forest	0.28 ± 0.06	−4.4 ± 2.2	547.3 ± 145
12	Frigid semiarid coniferous forest	0.27 ± 0.05	−10.1 ± 2.4	323.7 ± 91

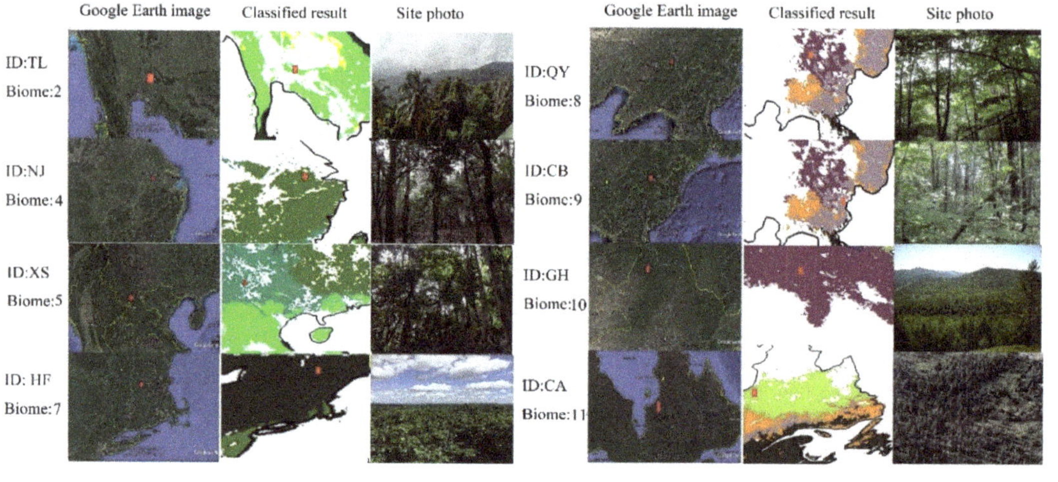

Figure 2. Comparison of our forest types with Google Earth images in eight ground-truth points. ID represents the photo site TL (14.5° N, 101.5° E), NJ (31.1° N, 118.8° E), (XS (21.9° N, 101.2° E), HF (42.5° N, 72.15° S), QY (41.8° N, 124.9° E), CB (41.9° N, 127.8° E), GH (50.8° N, 121.5° E), CA (57.8° N, 76.1° S). Type ID was same as that in Table 2. The red color marked the location of photo site. The other four types were not listed due to our lack of photos for them.

3.1.1. Tropical Forests

Forest types 1 to 3 are tropical forests. Type 1 is along the equator. The local vegetation is rainforest; therefore, it is designated as a tropical rainforest. Type 2 is generally adjacent to type 1; however, the precipitation has stronger seasonal variations (Figure 3). The precipitation in type 2 is lower than that in type 1, so we designated it as a tropical moist forest. The vegetation in type 3 is sparse forest, and it is a tropical humid forest.

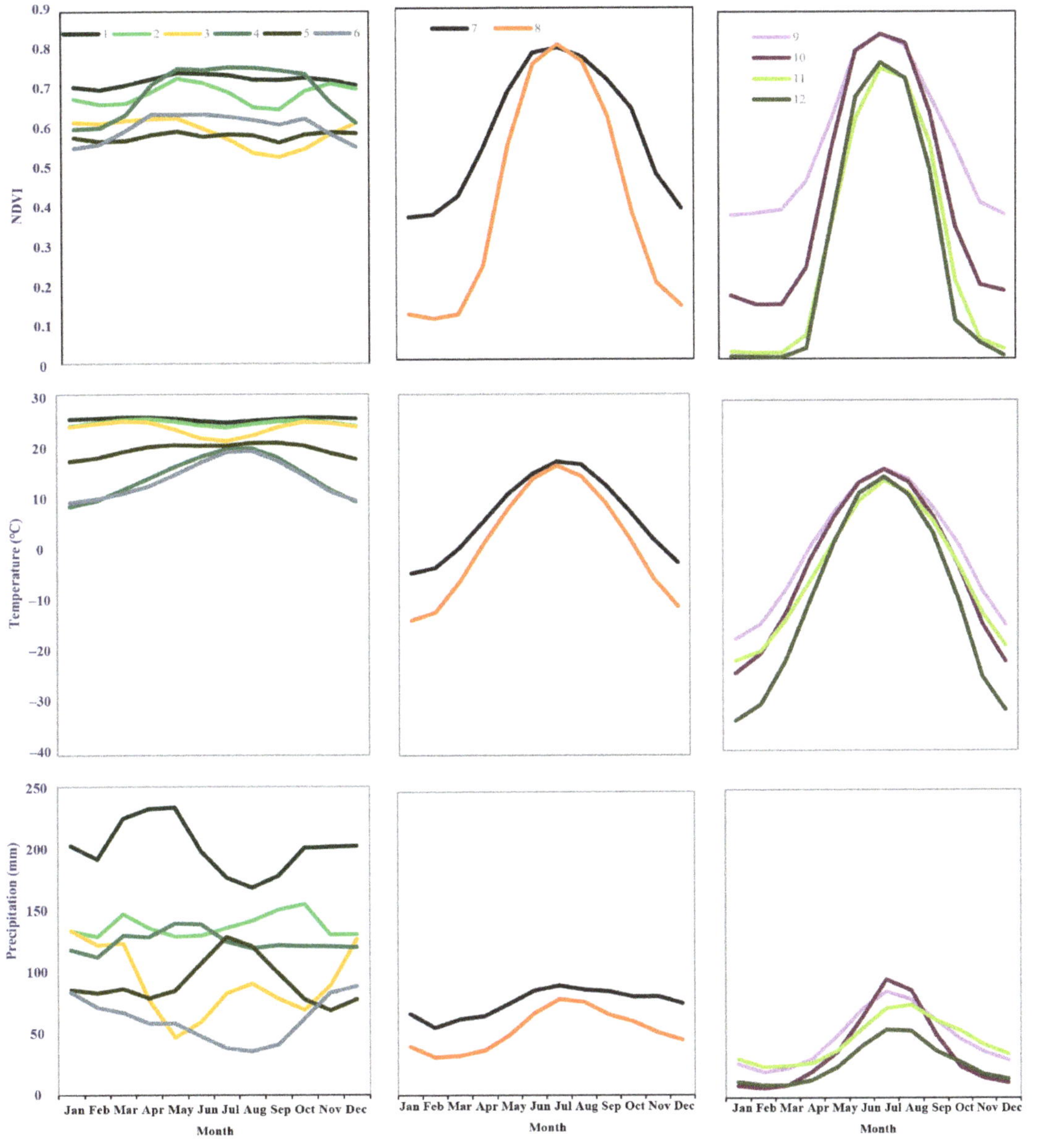

Figure 3. Monthly NDVI, temperature, and precipitation data for each type. See type names in Figure 1. The type colors are same as those in Figure 1.

3.1.2. Subtropical Forests

Forest types 4 to 6 are subtropical forests. Broadleaf forests are the main landscapes. Type 4 includes mainly broadleaf forests in southeastern China and southeastern America and is termed a subtropical moist broadleaf forest. Type 5 includes broadleaf forests in southwestern China and it is a subtropical humid broadleaf forest. Type 6 is referred to as a subtropical sub-humid broadleaf forest and is located in the Mediterranean and western North America.

3.1.3. Temperate Forests

Forest types 7 to 8 are temperate forests. The types include a broadleaf forest and mixed forest. Type 7 includes a temperate forest in northeastern America and western Europe, termed a temperate humid broadleaf forest. Type 8 is a temperate sub-humid broadleaf and needleleaf mixed forest and is located in central Europe and Canada.

3.1.4. Sub-Frigid Forests

Forest types 9 to 11 are sub-frigid forests. Type 9 is mixed with Type 8, but had a lower temperature than type 8, and it is designated as a sub-frigid sub-humid broadleaf and needleleaf mixed forest. Type 10, a sub-frigid semiarid coniferous forest, includes the coniferous forest in southern Russia and northwestern Canada. Type 11, a sub-frigid sub-humid coniferous forest, is located in northern Eurasia and northern Canada.

3.1.5. Frigid Forests

Forest type 12 is a frigid forest with coniferous trees. A deciduous coniferous forest is the typical vegetation in northern Russia, so type 12 is a denominated frigid semiarid coniferous forest.

3.2. Robust of Forest Type Classification

Global forest classifications based on bioclimatic variables, including annual mean temperature, mean diurnal range, isothermality, temperature seasonality, max temperature of the warmest month, min temperature of the coldest month, the temperature's annual range, mean temperature of the wettest quarter, mean temperature of the driest quarter, mean temperature of the warmest quarter, mean temperature of the coldest quarter, annual precipitation, precipitation of the wettest month, precipitation of the driest month, precipitation seasonality, precipitation of the wettest quarter, precipitation of the driest quarter, precipitation of the warmest quarter, and precipitation of the coldest quarter, produced forest types that had distinct climate and NDVI characteristics with above classification (Figure S1). Global forest types, classified based on maximum and minimum temperatures, total precipitation, and NDVI, are similar to those classified based on mean temperature, total precipitation and NDVI. Global forest types based on a $0.17 \times 0.17^{\circ}$ resolution mean temperature, total precipitation, and NDVI had similar results with those based on a $0.08 \times 0.08^{\circ}$ resolution dataset. However, global forest classification could not be well conducted based on a $0.5 \times 0.5^{\circ}$ resolution dataset. Therefore, our classification produced robust results on global forest classification.

3.3. Climatic Control on Global Distribution of Forest Types

Forest types sharply differed in their monthly NDVI signatures (Figure 3). Rainforests and moist forests had high NDVI values throughout the year. Evergreen broadleaf forests had NDVI values greater than 0.4. Broadleaf forests and mixed broadleaf forests had low NDVI values in winter because of fallen leaves from the broadleaf trees. NDVI values in evergreen broadleaf forests ranged from 0.4 to 0.8. The winter NDVI values were lower than 0.05 in the frigid types, even in evergreen needleleaf forest.

In general, forest types showed good differentiation and clustering in regard to temperature, precipitation, and vegetation parameters (Figure 4). Including the NDVI resulted in tighter and more compact forest type definitions than those of classifications based uniquely in precipitation and temperature, which showed blurred limits between forest types and discontinuous climatic envelopes (Figure 4). The temperature range of every forest type was roughly consistent with the criteria which we defined to describe the name of the forest type, and most forest types corresponded with wide precipitation ranges (Figure 4).

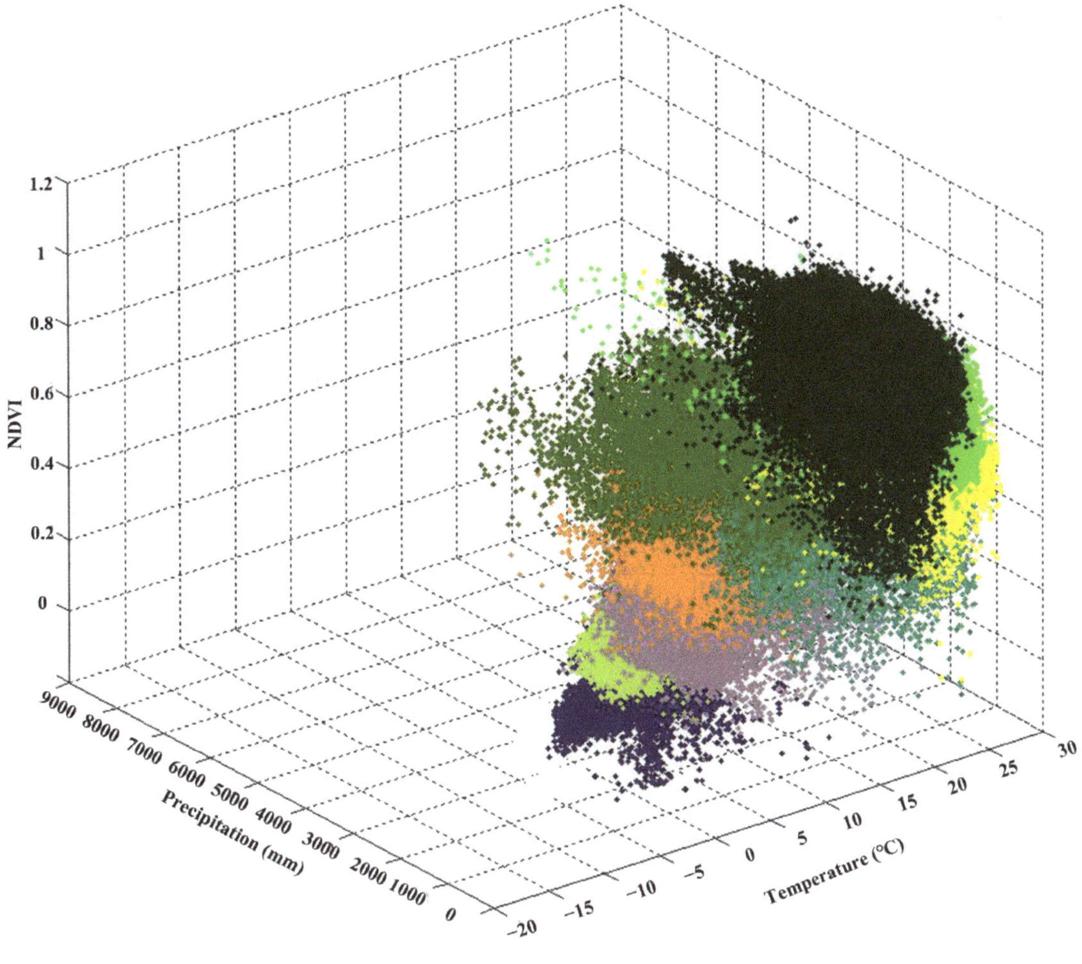

Figure 4. The corresponding annual mean temperature, total precipitation and NDVI for every forest type. The type colors are same as those in Figure 1.

How NDVI values and climate attributes vary between forest types is illustrated using the example of the location of tropical and tropical monsoon rainforests (Figure 5). This example shows that the annual mean NDVI value of the tropical rainforest is much higher than that of the tropical monsoon rainforest (Figure 5B), as well as differing in the distribution patterns of annual mean temperature and total precipitation (Figure 5C,D).

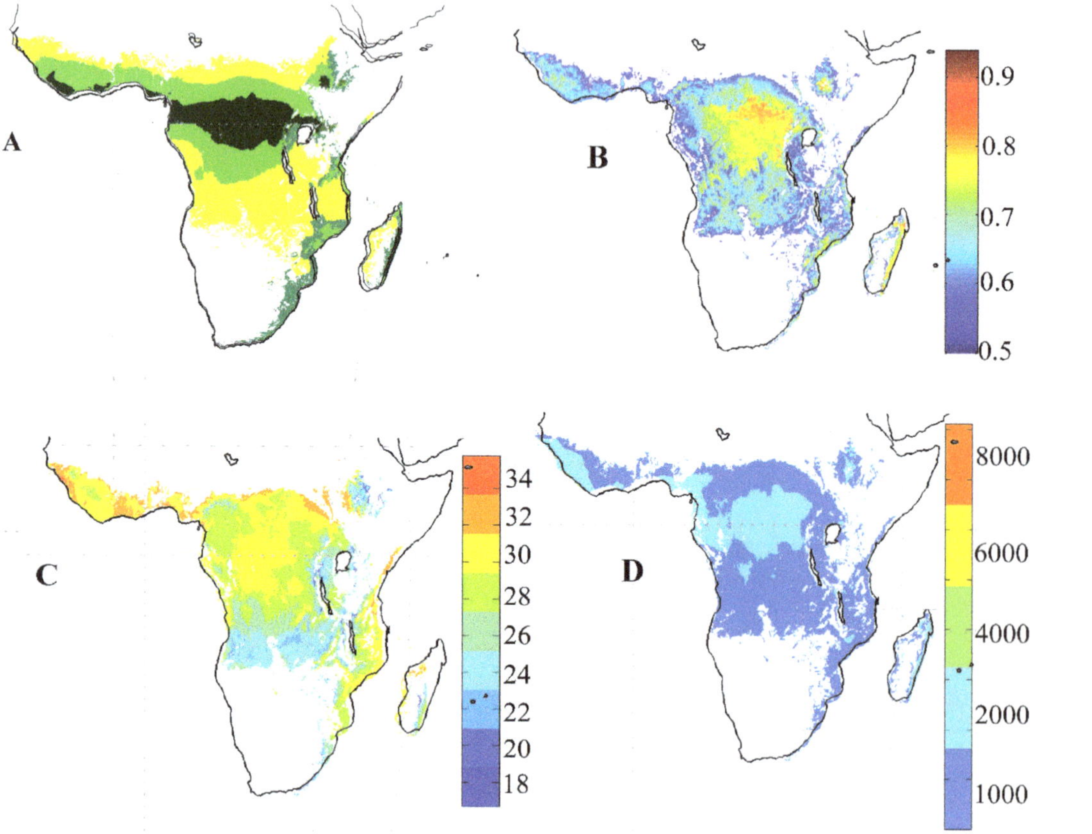

Figure 5. The boundary between tropical rainforest and tropical monsoon rainforest defined in our study (**A**), and the distributions of NDVI values (**B**), annual temperature (°C, **C**), and annual total precipitation (mm, **D**) in these regions. The type colors are same as those in Figure 1.

Our method correctly identified both forest types. The comparison of our defined forest types with the GlobCover land-cover maps showed that the corresponding types were roughly distributed in similar regions (Figure S2). However, there are important differences between our classification and previous classifications at fine scales. The forest types defined based on climate and EVI data were not classified well, compared with those defined based on climate and NDVI data (Figure S3). For instance, the tropical rainforest was divided into two different types when compared with other vegetation classifications. Given the less spatially-consistent grouping when using EVI data, we considered this a less suitable option for the classification of forest types than NDVI data.

4. Discussion

We produced a high-resolution global forest type cartography, delineated using the K-means clustering method based on monthly NDVI, temperature, and precipitation data. In previous studies, forest types have been identified as potential vegetation based mainly on climate data (e.g., [3,6,10]), or as actual vegetation based on satellite imagery (e.g., [14,16]). By contrast, here we used a forest type definition that we believe is closer to the modern concept of type (discussed in [23,46]) by considering not only climate variation, but also the patterns of monthly changes in the actual vegetation. There were clear differences in the monthly variations of NDVI values, temperature, and precipitation between forest types. Compared to a climate-based classification, also considering NDVI values has the advantage of reflecting the realized rather than the potential types. The example of Africa in Figure 5 stresses this, showing how two tropical forest types could be well separated when we also consider NDVI as vegetation data. However, overall, our classification is still highly consistent with climate-based forest types and should be seen as a refinement of them, rather than a challenge to previous work.

There are clear differences in interpretation between our forest type classification and other vegetation classifications. Vegetation or land-cover-based classifications highlight the vegetation or land physical attributes, while our forest types emphasize the forest attributes. In addition, land cover classifications include human-transformed vegetation, such as pasture and urban buildings [47], while we masked out human-created types when identifying the forest regions. There is no detailed information on the forest types defined using land-cover and vegetation classification. Our classification improves the detail in forest types and contains more information on forest types than other vegetation classifications.

Our classification seems to accurately identify known forest ecosystems, whose climatic definitions were well separated. This is particularly important for distinct forest types that share similar NDVI values. For instance, we could clearly identify known sub-divisions of the boreal forest that could not be differentiated in an objective way using only climate, i.e., sub-frigid sub-humid deciduous coniferous forests, frigid semiarid coniferous forests, and sub-frigid sub-humid broadleaf and needleleaf mixed forests, clearly distinguishing sub-frigid from frigid coniferous ecosystems. The main difference between sub-frigid semiarid coniferous forests and frigid semiarid coniferous forests is that the main forest ecosystem in the former type is the larch-dominated bright coniferous forest while the sparse larch trees with shrub dominate in the latter forest type.

An important feature of our model is that it includes a dynamic definition of forest types. Since the NDVI is a characteristic of the vegetation and changes every year, the forest types defined by the NDVI can change with time, as the main vegetation evolves, or as a response to changing climate (e.g., [48]). That way, it is possible to regularly update the forest type classification, to keep the forest types accurate and to study the effect that a changing climate has on forested landscapes. However, in the short- and medium-timescale, we expect these forest types to be quite stable due to the rather stable signal of the long-term climate data (compared with a NDVI-only model).

It should be noted that there are some forest types that are not only determined by climate but also by edaphic and/or hydrological conditions, such as dry forests (vs. wooded grassland), riverine forests and swamp forests, which our approach is not able to differentiate. Another main limitation of our classification is that forest types occupying a small region (i.e., with an extension lower than that provided by the macroclimate data products we used), and those not having a distinct NDVI and climate to nearby forest types would be merged into nearby forest types. Tropical and temperate montane forests were not picked up by our clustering analysis, likely due to their limited spatial extension and lack of climate data resolution. Increasing data resolution across the world will alleviate this problem. Alternatively, including a proxy for 'montane conditions', such as, perhaps, relative elevation or a combination of solar radiation and exposition, would aid in this goal.

It is not easy to make a one-on-one comparison between forest type classifications because they differ in the number of types they consider, they are based on different datasets, and are designed for different purposes [19]. While there are important differences between our classification and previous classifications on fine scales, on a large scale our forest types tended to largely agree with corresponding land-cover types or vegetation types defined in previous studies [14,16]. GlobCover forest types had a high large scale agreement with ours (Figure S2). However, using better datasets, updating our methods, and including a larger number of types improved our ability to correctly identify the boundaries between different forest types at a finer scale. We also included self-descriptive forest labels, rather than types based on climate notation. These labels are easier and more intuitive to interpret for non-scientists. The result is a more accurate distribution of forest types, which will hopefully be more suitable for forestry and biogeographic studies.

5. Conclusions

We developed here a high-resolution global forest type classification cartography by integrating NDVI, temperature, and precipitation data. These forest types are largely consistent with previous type definitions, but also take explicitly into account the actual vegetation and their growth patterns. The distribution of forest types is freely available in (doi: 10.13140/RG.2.2.19197.90082, currently available in Figure S4).

Supplementary Materials: The following are available online at https://www.mdpi.com/article/10.3390/su14020634/s1, Figure S1: Global forest classifications with changing selected variables and the spatial resolution of selected variables. Global forest classification based on NDVI and bioclimatic variables (annual mean temperature, mean diurnal range, isothermality, temperature seasonality, max temperature of the warmest month, min temperature of the coldest month, temperature annual range, mean temperature of the wettest quarter, mean temperature of the driest quarter, mean temperature of the warmest quarter, mean temperature of the coldest quarter, annual precipitation, precipitation of the wettest month, precipitation of the driest month, precipitation seasonality, precipitation of the wettest quarter, precipitation of the driest quarter, precipitation of the warmest quarter and precipitation of the coldest quarter) (**A**), and classification based on maximum and minimum temperatures, precipitation, water vapor pressure, and NDVI data (**B**), and global forest classifications based on mean temperature, total precipitation, and the NDVI from $0.17 \times 0.17°$ (**C**) and $0.5 \times 0.5°$ (**D**) resolution dataset. Figure S2: Comparison between our forest types with the GlobCover land-cover map. Figure S3: Forest types defined based on climate and EVI data. The type colors are the same as those in Figure 1, but some types do not overlap with those in Figure 1, and they are marked with different colors. Figure S4: The GeoTIFF file for our forest types.

Author Contributions: C.X. and X.Z. processed the data, analyzed the results and wrote the majority of the manuscript. X.Z., R.D.M. designed the experiment and methodology. R.D.M., R.H.-C. participated in the manuscript writing, and reviewed and edited the earlier version. C.X., and W.L. participated in data processing and data collection. All authors have read and agreed to the published version of the manuscript.

Funding: This work was funded by the Education Department of Hebei Province (BJ2020025), the National Natural Science Foundation of China (41601045), and talents introduction program in Hebei Agricultural University (YJ201918). R.D.M. is supported by the Swiss National Science Foundation, through the Post.Doc Mobility program.

Institutional Review Board Statement: Not applicable.

Informed Consent Statement: Not applicable.

Data Availability Statement: The datasets used during the current study are available from the websites in the paper.

Conflicts of Interest: The authors declare no conflict of interest.

References

1. Schimper, A.F.W.; Fisher, W.R.; Groom, P.; Balfour, I.B. Plant-Geography upon a Physiological Basis. *Nature* **1960**, *70*, 573–574.
2. Schimper, A.F.W. *Pflanzengeographie auf Physiologischer Grundlage*; Engelmann, H.R., Ed.; Gustav Fischer: Jena, Germany, 1908; pp. 747–749.
3. Köppen, W.P. *Die Klimate der Erde*; De Gruyter: Berlin, Germany, 1923.
4. Holdridge, L.R. *Life Zone Ecology*; Tropical Science Center: San Jose, Costa Rica, 1967.
5. Holdridge, L.R. Determination of World Plant Formations from Simple Climatic Data. *Science* **1947**, *105*, 367–368. [CrossRef] [PubMed]
6. Shimwell, D.W.; Box, E.D.; Lieth, H. Macroclimate and Plant Forms: An Introdution to Predictive Modeling in Phytogeography. *J. Appl. Ecol.* **1982**, *19*, 993. [CrossRef]
7. Box, E.O. Plant functional types and climate at the global scale. *J. Veg. Sci.* **1996**, *7*, 309–320. [CrossRef]
8. Whittaker, R.H. Classification of natural communities. *Bot. Rev.* **1962**, *28*, 1–239. [CrossRef]

9. Whittaker, R.H. *Communities and Ecosystems*; Princeton University Press: Princeton, NJ, USA, 1970; pp. 465–477.
10. Haxeltine, A.; Prentice, I.C. BIOME3: An equilibrium terrestrial biosphere model based on ecophysiological constraints, resource availability, and competition among plant functional types. *Glob. Biogeochem. Cycles* **1996**, *10*, 693–709. [CrossRef]
11. Metzger, M.J.; Bunce, R.G.H.; Jongman, R.H.G.; Sayre, R.; Trabucco, A.; Zomer, R. A high-resolution bioclimate map of the world: A unifying framework for global biodiversity research and monitoring. *Glob. Ecol. Biogeogr.* **2013**, *22*, 630–638. [CrossRef]
12. Olson, D.M.; Dinerstein, E.; Wikramanayake, E.D.; Burgess, N.D.; Powell, G.V.N.; Underwood, E.C.; D'Amico, J.A.; Itoua, I.; Strand, H.E.; Morrison, J.C. Terrestrial Ecoregions of the World: A New Map of Life on Earth: A new global map of terrestrial ecoregions provides an innovative tool for conserving biodiversity. *BioScience* **2001**, *51*, 933–938. [CrossRef]
13. DeFries, R.S.; Townshend, J.R.G. NDVI-derived land cover classifications at a global scale. *Int. J. Remote Sens.* **1994**, *15*, 3567–3586. [CrossRef]
14. DeFries, R.S.; Hansen, M.C.; Townshend, J.R.G.; Janetos, A.C.; Loveland, T.R. A new global 1-km dataset of percentage tree cover derived from remote sensing. *Glob. Change Biol.* **2000**, *6*, 247–254. [CrossRef]
15. De Fries, R.S.; Hansen, M.; Townshend, J.R.G.; Sohlberg, R. Global land cover classifications at 8 km spatial resolution: The use of training data derived from Landsat imagery in decision tree classifiers. *Int. J. Remote Sens.* **1998**, *19*, 3141–3168. [CrossRef]
16. Loveland, T.R.; Reed, B.C.; Brown, J.F.; Ohlen, D.O.; Zhu, Z.; Yang, L.; Merchant, J.W. Development of a global land cover characteristics database and IGBP DISCover from 1 km AVHRR data. *Int. J. Remote Sens.* **2000**, *21*, 1303–1330. [CrossRef]
17. Friedl, M.A.; McIver, D.K.; Baccini, A.; Gao, F.; Schaaf, C.; Hodges, J.C.F.; Zhang, X.Y.; Muchoney, D.; Strahler, A.H.; Woodcock, C.E.; et al. Global land cover mapping from MODIS: Algorithms and early results. *Remote Sens. Environ.* **2002**, *83*, 287–302. [CrossRef]
18. Bartholomé, E.; Belward, A.S. GLC2000: A new approach to global land cover mapping from Earth observation data. *Int. J. Remote Sens.* **2005**, *26*, 1959–1977. [CrossRef]
19. Higgins, S.I.; Buitenwerf, R.; Moncrieff, G. Defining functional biomes and monitoring their change globally. *Glob. Change Biol.* **2016**, *22*, 3583–3593. [CrossRef] [PubMed]
20. Zhang, X.; Wu, S.; Yan, X.; Chen, Z. A global classification of vegetation based on NDVI, rainfall and temperature. *Int. J. Clim.* **2017**, *37*, 2318–2324. [CrossRef]
21. Beck, P.S.A.; Juday, G.P.; Alix, C.; Barber, V.A.; Winslow, S.E.; Sousa, E.E.; Heiser, P.; Herriges, J.D.; Goetz, S. Changes in forest productivity across Alaska consistent with biome shift. *Ecol. Lett.* **2011**, *14*, 373–379. [CrossRef] [PubMed]
22. Liu, J.; Melillo, J.M.; Tian, H.; Zhuang, D.; Zhang, Z. China's changing landscape during the 1990s: Large-scale land transformations estimated with satellite data. *Geophys. Res. Lett.* **2005**, *32*. [CrossRef]
23. Moncrieff, G.R.; Bond, W.J.; Higgins, S.I. Revising the biome concept for understanding and predicting global change impacts. *J. Biogeogr.* **2016**, *43*, 863–873. [CrossRef]
24. Araghi, A.; Martinez, C.J.; Adamowski, J.; Olesen, J.E. Associations between large-scale climate oscillations and land surface phenology in Iran. *Agric. For. Meteorol.* **2019**, *278*, 107682. [CrossRef]
25. Pinzon, J.E.; Tucker, C.J. A non-stationary 1981–2012 AVHRR NDVI3g time series. *Remote Sens.* **2014**, *6*, 6929–6960. [CrossRef]
26. Fick, S.E.; Hijmans, R.J. WorldClim 2: New 1-km spatial resolution climate surfaces for global land areas. *Int. J. Climatol.* **2017**, *37*, 4302–4315. [CrossRef]
27. Mammola, S.; Leroy, B. Applying species distribution models to caves and other subterranean habitats. *Ecography* **2018**, *41*, 1194–1208. [CrossRef]
28. Thorne, J.H.; Choe, H.; Stine, P.A.; Chambers, J.C.; Holguin, A.; Kerr, A.C.; Schwartz, M.W. Climate change vulnerability assessment of forests in the Southwest USA. *Clim. Change* **2018**, *148*, 387–402. [CrossRef]
29. Lloyd, S.P. Least squares quantization in PCM. *IEEE Trans. Inf. Theory* **1982**, *28*, 129–137. [CrossRef]
30. Zhang, X.; Yan, X. Temporal change of climate zones in China in the context of climate warming. *Arch. Meteorol. Geophys. Bioclimatol. Ser. B* **2013**, *115*, 167–175. [CrossRef]
31. Zhang, X.; Yan, X. Spatiotemporal change in geographical distribution of global climate types in the context of climate warming. *Clim. Dyn.* **2014**, *43*, 595–605. [CrossRef]
32. Zhang, X.; Yan, X. Deficiencies in the simulation of the geographic distribution of climate types by global climate models. *Clim. Dyn.* **2015**, *46*, 2749–2757. [CrossRef]
33. Zhang, X.; Yan, X.; Chen, Z. Geographic distribution of global climate zones under future scenarios. *Int. J. Clim.* **2017**, *37*, 4327–4334. [CrossRef]
34. Zhang, X.; Huang, X. Human disturbance caused stronger influences on global vegetation change than climate change. *PeerJ* **2019**, *7*, e7763. [CrossRef]
35. R Core Team. *R: A Language and Environment for Statistical Computing*; R Foundation for Statistical Computing: Vienna, Austria, 2018. Available online: http://www.R-project.org/ (accessed on 1 January 2022).
36. Hansen, M.C.; DeFries, R.S.; Townshend, J.R.G.; Carroll, M.; Dimiceli, C.; Sohlberg, R.A. Global percent tree cover at a spatial resolution of 500 meters: First results of the MODIS vegetation continuous fields algorithm. *Earth Interact.* **2003**, *7*, 1–15. [CrossRef]
37. Chen, J.; Chen, J.; Liao, A.; Cao, X.; Chen, L.; Chen, X.; He, C.; Han, G.; Peng, S.; Lu, M.; et al. Global land cover mapping at 30 m resolution: A POK-based operational approach. *ISPRS J. Photogramm. Remote Sens.* **2015**, *103*, 7–27. [CrossRef]
38. Hansen, M.C.; Potapov, P.V.; Moore, R.; Hancher, M.; Turubanova, S.A.; Tyukavina, A.; Thau, D.; Stehman, S.V.; Goetz, S.J.; Loveland, T.R.; et al. High-resolution global maps of 21st-century forest cover change. *Science* **2013**, *342*, 850–853. [CrossRef]

39. Lawrence, M.; McRoberts, R.E.; Tomppo, E.; Gschwantner, T.; Gabler, K. Comparisons of national forest inventories. In *National Forest Inventories*; Tomppo, E., Gschwantner, T., Lawrence, M., McRoberts, R., Eds.; Springer: Dordrecht, Germany, 2010; pp. 19–32.
40. Bastin, J.-F.; Berrahmouni, N.; Grainger, A.; Maniatis, D.; Mollicone, D.; Moore, R.; Patriarca, C.; Picard, N.; Sparrow, B.; Abraham, E.M.; et al. The extent of forest in dryland biomes. *Science* **2017**, *356*, 635–638. [CrossRef]
41. Schmitt, C.B. Global tropical forest types as support for the consideration of biodiversity under REDD+. *Carbon Manag.* **2013**, *4*, 501–517. [CrossRef]
42. Schmitt, C.B.; Burgess, N.D.; Coad, L.; Belokurov, A.; Besançon, C.; Boisrobert, L.; Campbell, A.; Fish, L.; Gliddon, D.; Humphries, K.; et al. Global analysis of the protection status of the world's forests. *Biol. Conserv.* **2009**, *142*, 2122–2130. [CrossRef]
43. Zhang, D.; Wang, H.; Wang, X.; Lü, Z. Accuracy assessment of the global forest watch tree cover 2000 in China. *Int. J. Appl. Earth Obs. Geoinf.* **2020**, *87*, 102033. [CrossRef]
44. Song, D.-X.; Huang, C.; Noojipady, P.; Channan, S.; Townshend, J. Comparison of remote sensing based forest area and change estimation with national forestry inventory between 2000 and 2005 in China. Proceedings of 2014 IEEE Geoscience and Remote Sensing Symposium, Quebec City, QC, Canada, 13–18 July 2014; pp. 4268–4271.
45. Griffith, D.M.; Lehmann, C.E.R.; Strömberg, C.A.E.; Parr, C.L.; Pennington, R.T.; Sankaran, M.; Ratnam, J.; Still, C.J.; Powell, R.L.; Hanan, N.P.; et al. Comment on "The extent of forest in dryland biomes". *Science* **2017**, *358*, 1309. [CrossRef] [PubMed]
46. Mucina, L. Biome: Evolution of a crucial ecological and biogeographical concept. *New Phytol.* **2019**, *222*, 97–114. [CrossRef] [PubMed]
47. Ellis, E.; Goldewijk, K.K.; Siebert, S.; Lightman, D.; Ramankutty, N. Anthropogenic transformation of the biomes, 1700 to 2000. *Glob. Ecol. Biogeogr.* **2010**, *19*, 589–606. [CrossRef]
48. Scheffer, M.; Hirota, M.; Holmgren, M.; van Nes, E.; Chapin, F.S. Thresholds for boreal biome transitions. *Proc. Natl. Acad. Sci. USA* **2012**, *109*, 21384–21389. [CrossRef] [PubMed]

sustainability

MDPI

Article

Quality Evaluation of the 0.01° Multi-Source Fusion Precipitation Product and Its Application in Extreme Precipitation Event

Zheng Wang [1], Yang Pan [1], Junxia Gu [1,*], Yu Zhang [2] and Jianrong Wang [3]

[1] National Meteorological Information Center, Beijing 100081, China; wangz@cma.gov.cn (Z.W.); pany@cma.gov.cn (Y.P.)
[2] Henan Meteorological Observation Data Center, Zhengzhou 450000, China; zhangyuc477201@cma.cn
[3] Anhui Meteorological Information Center, Hefei 230000, China; faithwakin@sina.com
* Correspondence: gujx@cma.gov.cn

Abstract: High-resolution and high-quality precipitation data play an important role in Numerical Weather Prediction Model testing, mountain flood geological disaster monitoring, hydrological monitoring and prediction and have become an urgent need for the development of modern meteorological business. The 0.01° multi-source fusion precipitation product is the latest precipitation product developed by the National Meteorological Information Center to meet the above needs. Taking the hourly precipitation observation data of 2400 national automatic stations as the evaluation base, independent and non-independent test methods are used to evaluate the 0.01° multi-source fusion precipitation product in 2020. The product quality differences between the 0.01° precipitation product and the 0.05° precipitation product are compared, and their application in extreme precipitation events are analyzed. The results show that, in the independent test, the product quality of the 0.01° precipitation product and the 0.05° precipitation product are basically the same, which is better than that of each single input data source, and the product quality in winter and spring is slightly lower than that in summer, and both products have better quality in the east in China. The evaluation results of the 0.01° precipitation product in the non-independent test are far better than that of the 0.05° product. The root mean square error and the correlation coefficient of the 0.01° multi-source fusion precipitation product are 0.169 mm/h and 0.995, respectively. In the extreme precipitation case analysis, the 0.01° precipitation product, which is more consistent with the station observation values, effectively improves the problem that the extreme value of the 0.05° product is lower than that of station observation values and greatly improves the accuracy of the precipitation extreme value in the product. The 0.01° multi-source fusion precipitation product has better spatial continuity, a more detailed description of precipitation spatial distribution and a more accurate reflection of precipitation extreme values, which will better provide precipitation data support for refined meteorological services, major activity support, disaster prevention and reduction, etc.

Keywords: high-resolution and high-quality precipitation data; independent and non-independent test; the 0.01° multi-source fusion precipitation product; extreme precipitation event

Citation: Wang, Z.; Pan, Y.; Gu, J.; Zhang, Y.; Wang, J. Quality Evaluation of the 0.01° Multi-Source Fusion Precipitation Product and Its Application in Extreme Precipitation Event. *Sustainability* **2022**, *14*, 616. https://doi.org/10.3390/su14020616

Academic Editors: Xiaodong Yan, Jia Yang and Shaofei Jin

Received: 24 November 2021
Accepted: 29 December 2021
Published: 6 January 2022

1. Introduction

Precipitation data are the basis of weather and climate monitoring, climate change research, model prediction tests and meteorological and hydrological prediction. They play an extremely important role in flood season prediction, meteorological prediction, agricultural guidance and disaster prevention and reduction. With the rapid development of meteorological observation systems, more and more observation data and numerical model simulation data such as ground automatic weather station data, radar data and satellite data can be used, and various industries have higher and higher requirements for grid precipitation products. High-resolution and high-quality precipitation data have gradually become an urgent need for the development of modern meteorological business [1]. Multi-source fusion precipitation products can combine the advantages of precipitation data from different sources, and have gradually become the mainstream trend in the development of high-quality precipitation products in the world in recent decades [2–11].

In recent years, many meteorological institutions in China have committed themselves to research the multi-source fusion technology, develop different kinds of 0.05° precipitation products, and significantly improve the quality of precipitation products in China. The National Meteorological Satellite Center of China has developed FY series satellite precipitation products based on the revision of the intelligent objective analysis method considering station distance and angle [12]. The National Meteorological Information Center of China, using the Probability Density Function matching (PDF) + Optimal Interpolation (OI) method of the U.S. Climate Prediction Center (CPC), has developed two-source fusion precipitation products based on ground station data and FY satellites data and two-source fusion precipitation products based on ground station data and CMORPH satellites data [12–19]. On the basis of two-source fusion precipitation products, using Probability Density Function matching (PDF) + Bayesian Multi-model Average (BMA) + Optimal Interpolation (OI) method, the National Meteorological Information Center has developed a series of three-source fusion precipitation products based on ground station data, radar data and satellite data [20,21]. As grid products with high precision, high quality and continuous time and space [1], multi-source fusion precipitation products have been widely used in the fields of model forecast testing [22,23], hydrological forecast [24] and meteorological live monitoring in provinces of China.

In order to further meet the needs of high-resolution and high-quality grid precipitation data in the fields of intelligent grid forecasting business development, refined meteorological services and disaster prevention and reduction, the National Meteorological Information Center has developed a 0.01° multi-source fusion precipitation product based on ground station data, radar data and satellite data by using Probability Density Function matching (PDF) + Bayesian Multi-model Average (BMA) + Spatial Downscaling (DS) + Optimal Interpolation (OI) method [15]. The product has further improved the quality and resolution of the fusion precipitation product, including adding more precipitation data sources, multi-source quality control of station observation data, quality control of fusion precipitation products, optimization of fusion parameters of multi-grid variational analysis, etc. Now the precipitation product has completed the business distribution to all provinces, and the data are connected to the China Integrated Meteorological Information Service System (CIMISS) data environment to provide services through the Meteorological Unified Service Interface Community (MUSIC) interface. In this study, to evaluate the quality and application effect of the product in China, independent and non-independent tests are used to evaluate the 0.01° multi-source fusion precipitation products in 2020, and the effect of the product on the characterization of extreme precipitation is analyzed.

2. Data and Processing

There are seven data sources of multi-source fusion precipitation products: (1) Observation data of ground automatic weather stations with hourly precipitation data of more than 60,000 automatic weather stations in China after quality control. (2) Satellite inversion precipitation products: The FY2 satellite inversion precipitation product developed by the National Satellite Meteorological Center, nominal projection, spatial resolution of about 4 km and time resolution of 1 h; The CMORPH satellite inversion precipitation product with global resolution of about 7 km and 30 min developed by the US Climate Prediction Center; The GsMAP satellite inversion precipitation product with global resolution of 1hour and 10 km developed by Japan Aerospace Exploration Agency (JAXA); The IMERG satellite inversion precipitation product with 30 min and 10 km resolution developed by NASA. (3) Radar precipitation estimation data: The national radar quantitative precipitation estimation product with hourly and 0.01° resolution developed by the Meteorological Observation Center of China Meteorological Administration; The national radar quantitative precipitation estimation product with hourly and 0.01° resolution developed by the National Meteorological Information Center of China Meteorological Administration.

In view of the missing detection of erroneous precipitation data observed by automatic stations, after real-time quality control, the automatic weather stations hourly precipitation data also adopted the collaborative quality control technology of multi-source meteorological data, including consistency check between weather phenomena and precipitation and consistency check with radar estimated precipitation. In addition, the threshold lookup table of precipitation data and blacklist of ground automatic weather stations were established. The optimized quality control algorithm strengthens the screening of micro precipitation, the screening of false precipitation data and extreme outliers of products and improves the product quality in many aspects.

The optimal selection method is used in the fusion application of radar data and satellite data, and the weight coefficient of the linear fusion method that is adjusted by season and region is used in the Bayesian Multi-model Average of satellite data. The main development process of the 0.01° multi-source fusion precipitation product is shown in Figure 1.

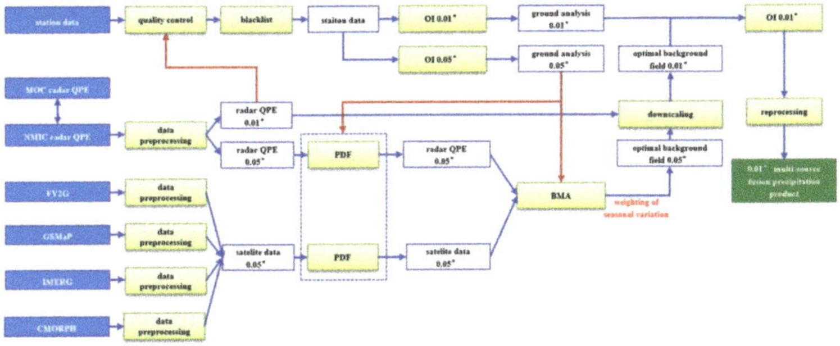

Figure 1. The main development process of the 0.01° multi-source fusion precipitation product.

The 0.01° and 0.05° resolution fusion analysis products can be downloaded through MUSIC, National Meteorological business intranet (http://idata.cma/) (accessed on 3 November 2021) and China Meteorological Data Network (http://data.cma.cn/) (accessed on 3 November 2021).

In addition, in the reprocessing of multi-source fusion precipitation products, some grid values will be replaced by the station precipitation value. In order to ensure the objectivity of the evaluation results, the evaluated product data are not replaced.

3. Product Evaluation Research

In this study, the hourly precipitation observation data of 2400 national automatic stations are used as the evaluation base, and the independent and non-independent test methods are used to evaluate the 0.01° multi-source fusion precipitation product in 2020. In the independent test, the observed precipitation data of 2400 national stations do not participate in the ground grid analysis and fusion and are used as the "truth" data of the test. In the non-independent test, the observed precipitation data of 2400 national stations participate in the ground grid analysis and fusion and are still used as the "truth" data of the test.

The statistical evaluation indexes in this paper are mean error (ME), root mean square error (RMSE) and Pearson correlation coefficient (R):

$$\mathrm{ME} = \frac{1}{\mathrm{n}}\sum_{i=1}^{n}(pi - gi) \tag{1}$$

$$\text{RMSE} = \sqrt{\frac{1}{\text{n}}\sum_{i=1}^{n}(pi - gi)^2} \tag{2}$$

$$\text{R} = \frac{\sum_{i=1}^{n}(pi - \overline{p})(gi - \overline{g})}{\sqrt{\sum_{i=1}^{n}(pi - \overline{p})^2}\sqrt{\sum_{i=1}^{n}(gi - \overline{g})^2}} \tag{3}$$

In the formula, gi is the precipitation observation data of 2400 national stations which is regarded as the "true value" and pi is the precipitation value interpolated from each tested precipitation product to 2400 national stations.

3.1. Evaluation of Data from Different Sources and Time Series of Product Evaluation Results

The overall independent quality evaluation results of different resolution fusion precipitation products and the input data used in the products are shown in Table 1. The independent testing results in 2020 show that the quality of the 0.01° fusion precipitation product is basically the same as that of the 0.05° product, the Correlation Coefficients of both products are higher than 0.85, the Root Mean Square Error is less than 0.6 mm/h, and they are better than the quality of each single input datum. In the comparison of various input data, the quality of ground analysis is the best, followed by radar, and both are better than all kinds of satellite precipitation data. Among all kinds of satellite retrieved precipitation input data, the quality of the IMERG-L is the best, followed by the CMORPH, and both are better than other kinds of satellite precipitation products.

Table 1. The overall independent quality evaluation results of the fusion precipitation products and input data sources (2020).

Number	Data	ME (mm/h)		RMSE (mm/h)		R	
		0.01°	**0.05°**	**0.01°**	**0.05°**	**0.01°**	**0.05°**
1	Ground analysis	−0.013	−0.015	0.576	0.597	0.857	0.845
2	Radar QPE	−0.016	−0.016	0.815	0.813	0.720	0.704
3	FY2G	/	0.017	/	1.478	/	0.245
4	GSMaP-now	/	0.018	/	1.288	/	0.244
5	GSMaP-nrt	/	0.004	/	1.284	/	0.299
6	CMORPH	/	−0.034	/	1.102	/	0.366
7	IMERG-L	/	0.015	/	1.070	/	0.451
8	Precipitation products	−0.010	−0.007	0.524	0.519	0.851	0.854

Figure 2 shows the evaluation result time series of independent tests of fusion precipitation products from January to December 2020. In the whole year, there are little differences in the quality of the 0.01° fusion precipitation product and the 0.05° fusion precipitation product. In winter and spring, due to the suspension of the Tipper rain gauge in northern China (about 60,000 stations across the country are reduced to about 40,000 stations, as shown in Figure 3), and the insufficient inversion and estimation ability of satellite and radar detection means for solid precipitation, the quality of the products is affected, and the correlation coefficient is unstable. The root mean square errors of the 0.01° product and the 0.05° product are relatively consistent, and because the root mean square error is closely related to precipitation, the root mean square error in summer is higher than in winter and spring. Compared with the 0.05° precipitation product, the mean error of the 0.01° precipitation product in summer is slightly lower, which is closely related to the analysis radius of ground analysis of the two resolution products.

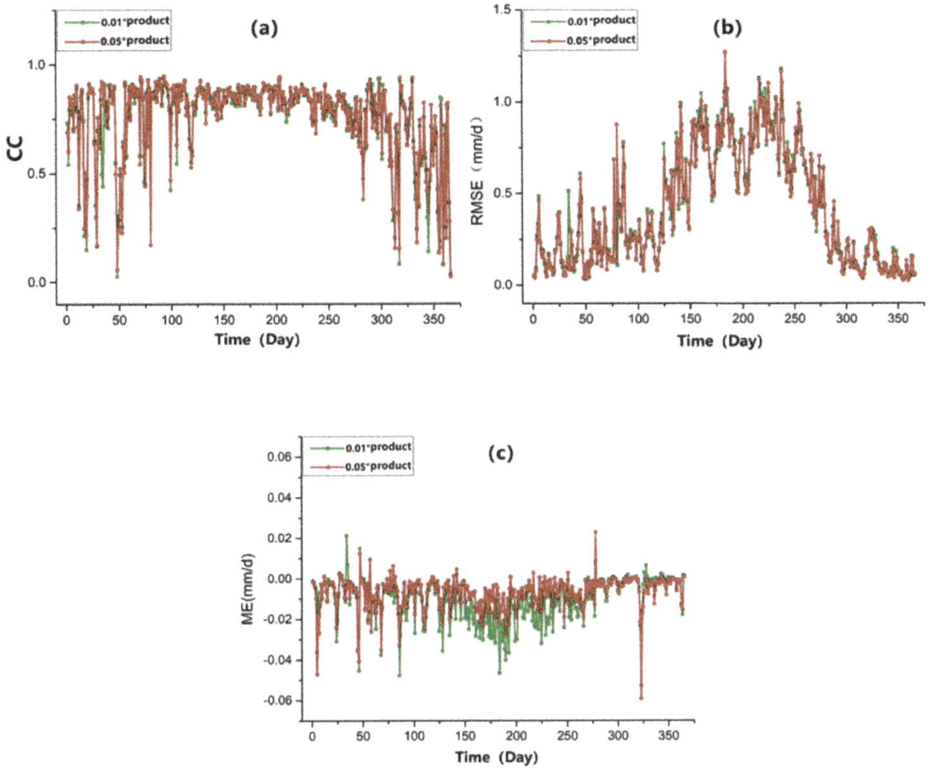

Figure 2. The evaluation results time series of independent test of fusion precipitation products in 2020 ((**a**): correlation coefficient, (**b**): root mean square error, (**c**): mean error, green line represents the 0.01° precipitation product, green line represents the 0.05° precipitation product).

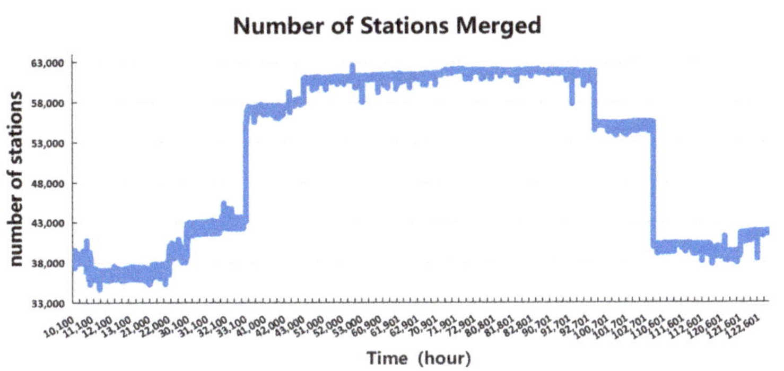

Figure 3. Number of automatic stations merged in the 0.01° precipitation product every hour in 2020. (10,100 represents UTC 00:00 on 1 January, and 122,601 represents UTC 01:00, 26 December).

3.2. Spatial Distribution of Product Evaluation Index

The spatial distribution of independent test evaluation results of fusion precipitation products in 2020 is shown in Figure 4. The spatial distribution of independent test evaluation results of the 0.01° fusion precipitation product and the 0.05° fusion precipitation product is basically the same. In terms of the correlation coefficient (a1 and b1), the correlation coefficient in the dense area of stations in the east can basically exceed 0.8, and the correlation coefficient in the north and west is relatively low, which is related to the sparse distribution of ground stations and relatively few precipitation data in this area. The spatial distribution of the root mean square error is affected by precipitation. The root mean square error in the southeast is greater than that in the west and north with less precipitation (a2 and b2). The proportion of stations with root mean square error of 0~0.8 mm/h for 0.01° and 0.05° products is 90.41% and 89.56%, respectively. In the spatial distribution of mean error (a3 and b3), stations with absolute deviation greater than 0.05 mm/h are mainly distributed in areas with large annual precipitation in 2020. The proportion of stations with mean error of the 0.01° product and the 0.05° product between −0.05~0.05 mm/h are 96.08% and 95.15%, respectively. In addition, the 0.05° product has more positive mean error values. The proportion of stations with mean error of the 0.01° product and the 0.05° product greater than 0.05 mm/h is 0.7% and 2.0%, respectively, which is consistent with the small overall mean error of the 0.01° product in Figure 2 and Table 1. As can be seen from the spatial distribution of independent test results (Figure 4), the quality of the 0.01° product is slightly better than that of the 0.05° product, and both products have better quality in the east.

3.3. Comparison of Independent and Non-Independent Tests

Taking July 2020 as an example, the independent test and non-independent test results of the 0.01° fusion precipitation product and the 0.05° fusion precipitation product are analyzed and compared, as shown in Table 2. Test results in July 2020 show that independent test results of the 0.01° multi-source fusion precipitation product are basically consistent with those of the 0.05° product, and the evaluation results of the non-independent test indicate that the 0.01° multi-source fusion precipitation product is obviously better than the 0.05° product. In the non-independent test, the product quality of the 0.01° product and the 0.05° product was improved, and the product quality of the 0.01° product was improved more significantly.

Table 2. Comparison of independent test and non-independent test results of fusion precipitation products (July 2020).

Number	Data	ME (mm/h)	RMSE (mm/h)	R
1	0.01° (non-independent)	0.0001	0.169	0.995
2	0.05° (non-independent)	−0.0091	0.681	0.907
3	0.01° (independent)	−0.0253	0.825	0.863
4	0.05° (independent)	−0.0134	0.799	0.869

Correlation coefficient, root mean square error and mean error time series of the independent test and the non-independent test of the 0.01° fusion precipitation product and the 0.05° fusion precipitation product in July 2020 are shown in Figure 5. It can be seen from the time series, the independent test results of the 0.01° multi-source fusion precipitation product are basically consistent with those of the 0.05° product, and non-independent test results of the 0.01° product are slightly worse than those of the 0.05° product. In the non-independent test, the 0.01° precipitation product is much better than the 0.05° product, which means the precipitation of the 0.01° product is more consistent with the station observation values. For example, the root mean square error and the correlation coefficient of the non-independent test of the 0.01° multi-source fusion precipitation product are 0.169 mm/h and 0.995, respectively, and the root mean square error and the correlation coefficient of the non-independent test of the 0.05° product are 0.681 mm/h and 0.907, respectively. In addition, compared with other evaluation results, non-independent test results of the 0.01° product are much more stable.

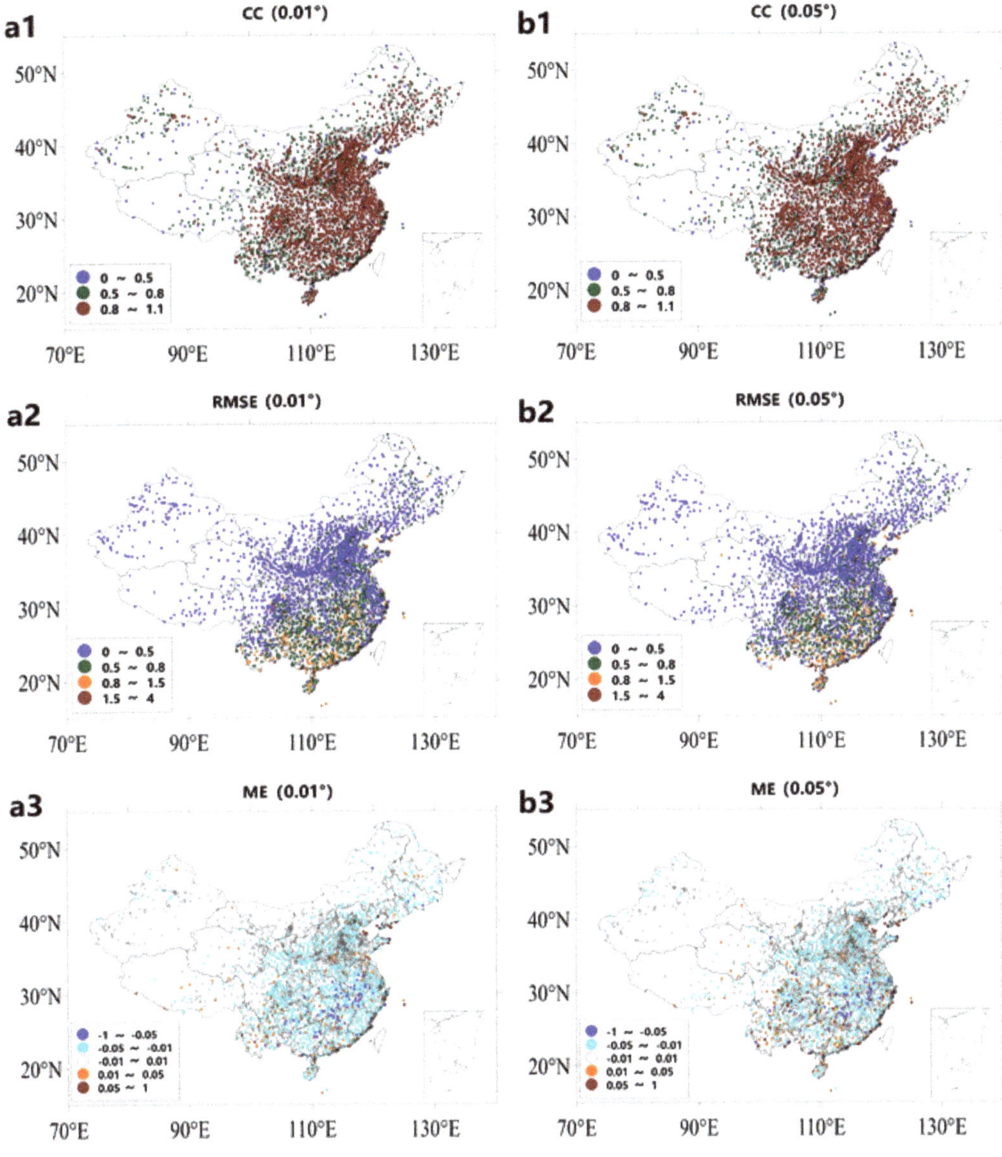

Figure 4. The spatial distribution of independent test evaluation results of fusion precipitation products in 2020. ((**a1**,**b1**): correlation coefficient, (**a2**,**b2**): root mean square error, (**a3**,**b3**): mean error).

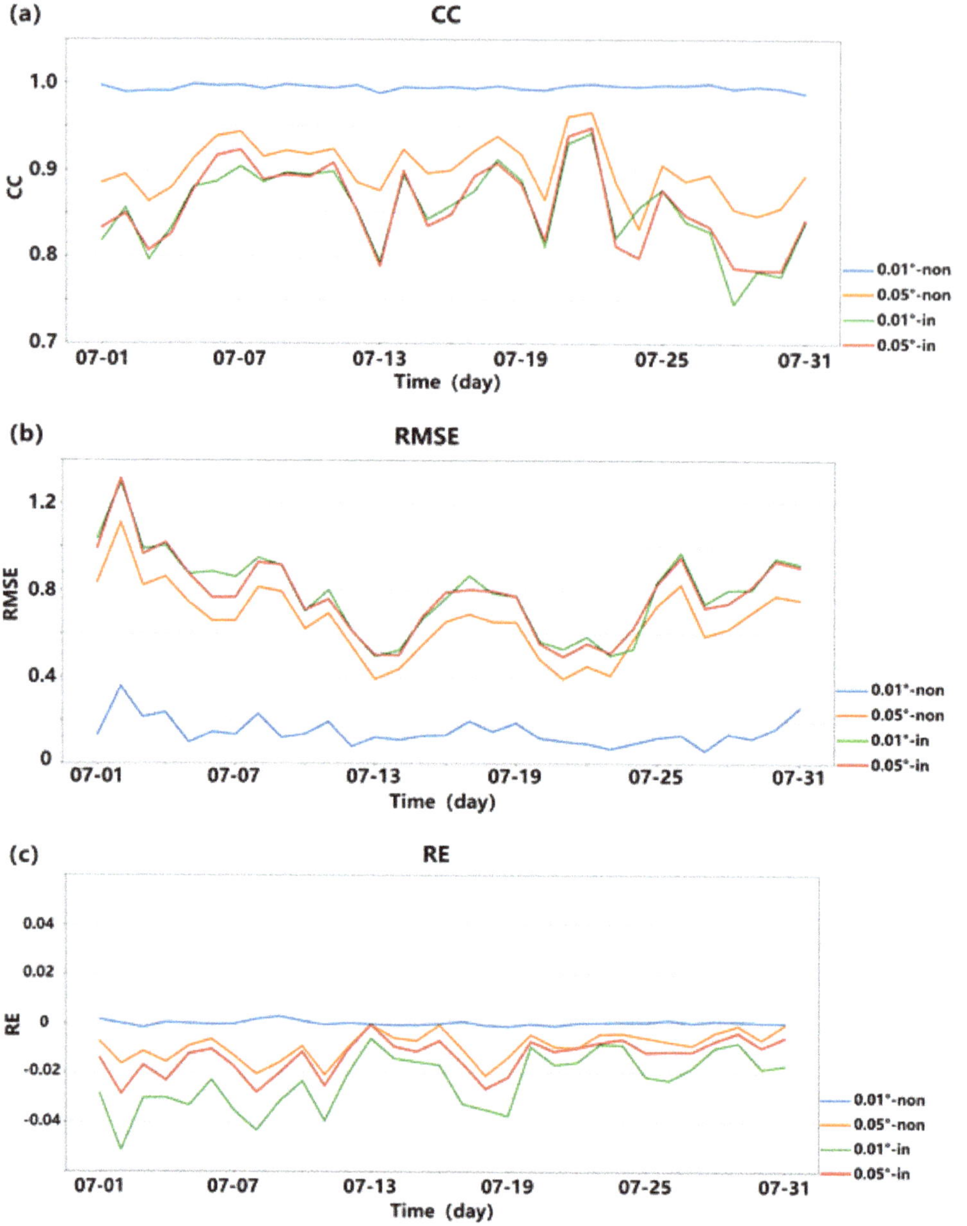

Figure 5. Time series of independent and non-independent test result of fusion precipitation products ((**a**): correlation coefficient, (**b**): root mean square error, (**c**): mean error, blue line: 0.01° non-independent, orange line: 0.05° non-independent, green line: 0.01° independent, red line: 0.05° independent).

4. Application of Fusion Precipitation Products in Extreme Precipitation Event

Taking the 7.20 rainstorm event in Henan in 2021 as an example, the application effect of the 0.01° precipitation product and the 0.05° precipitation product in this event is analyzed. Figure 6 shows the spatial distribution of the 0.01° precipitation product and the 0.05° precipitation product in Henan on 20 July 2021. It can be seen from the figure that the spatial distribution of the 0.01° precipitation product and the 0.05° precipitation product is similar on the whole, and both have better spatial continuity than that of station data. The 0.01° multi-source precipitation product describes the spatial distribution of precipitation more finely and reflects the extreme precipitation more continuously and accurately. In the figure, on 20 July, the 24 h cumulative extreme precipitation of the station and the 0.01° precipitation product was 687.9 mm, which of the 0.05° precipitation product was only 580.2 mm. The spatial location of the 0.01° product extreme precipitation is more matched with that of station extreme precipitation. It can be seen that the 0.01° multi-source fusion precipitation product effectively improves the problem that the extreme precipitation of the 0.05° product is smaller, and greatly improves the accuracy of extreme precipitation of fusion precipitation products.

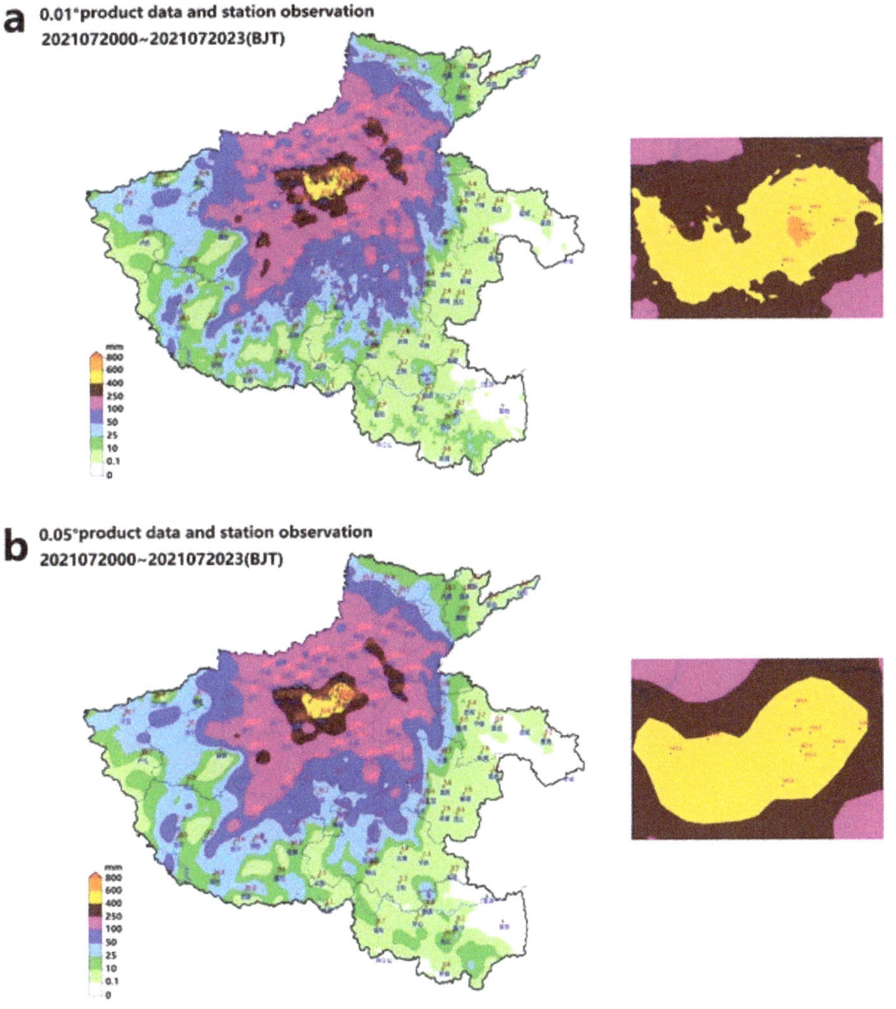

Figure 6. The spatial distribution of 0.01° and 0.05° precipitation products in Henan on 20 July 2021 (the picture on the right is an enlarged picture of the extreme precipitation region).

5. Conclusions

On the basis of the 0.05° three-source fusion precipitation product based on ground station data, radar data and FY2G satellite data, the National Meteorological Information Center has developed the 0.01° multi-source fusion precipitation product by using Probability Density Function matching (PDF) + Bayesian Multi-model Average (BMA) + Spatial Downscaling (DS) + Optimal Interpolation (OI). Taking the hourly precipitation observation data of 2400 national automatic stations as the evaluation base, independent and non-independent test methods are used to evaluate the 0.01° multi-source fusion precipitation product in 2020, the differences between the 0.01° multi-source fusion precipitation product and the 0.05° three-source fusion precipitation product are compared, and the spatial fineness and extreme value accuracy of extreme precipitation portrayed by precipitation products are analyzed. The main conclusions are as follows:

(1) From the overall independent test results in 2020, the quality of the 0.01° fusion precipitation product is basically the same as that of the 0.05° product. Both products are better than that of each single input data source. Among all data sources, the ground analysis quality is the best, followed by radar data, and the IMERG satellite precipitation data is the best among all satellite data sources. Both products have better quality in summer than that in winter and spring, and better quality in the east in China than that in the west.

(2) Independent test results of the 0.01° fusion precipitation product are basically consistent with those of the 0.05° product, which are slightly worse than the non-independent test results of the 0.05° product. The evaluation results of the 0.01° fusion precipitation product in the non-independent test are far better than those of the 0.05° product, which means the precipitation of 0.01° product is more consistent with the station observation values.

(3) The 0.01° multi-source fusion precipitation product has better spatial continuity, more detailed description of precipitation spatial distribution and more accurate embodiment of precipitation extreme value, which effectively improves the problem of the small extreme value of the 0.05° product and greatly improves the accuracy of precipitation extreme value.

High-quality and high-timeliness of the 0.01° multi-source fusion precipitation product in China will provide real-time precipitation data support for the upgrading of China's intelligent grid forecasting business to 0.01° resolution, and will play a significant economic and social benefit in disaster prevention and mitigation such as flood control and drought relief, refined meteorological services and guarantee of major activities. At present, the 0.01° multi-source fusion precipitation real-time product has been commercialized and is available for download. Although the effect of the 0.01° multi-source precipitation product in extreme precipitation events is preliminarily analyzed in this paper, there is still a lack of more detailed research, which can further analyze the applicability of the 0.01°multi-source precipitation product in extreme precipitation in different seasons and different precipitation levels. These contents will be reflected in future research, and this research is already in progress.

Author Contributions: J.G., Y.P. and Z.W. designed the study; Z.W. and J.W. performed the precipitation data preparation and analysis; Y.Z., J.W. and Z.W. draw the figures and pictures; the manuscript was written by Z.W. with significant contributions from all coauthors; the manuscript was revised by Z.W., J.G. and Y.P. All authors have read and agreed to the published version of the manuscript.

Funding: This research was funded by the National Key Research and Development Project, grant number 2018YFC1506604, and the Special Program for Innovation and Development of China Meteorological Administration, grant number CXFZ2021Z007.

Institutional Review Board Statement: Not applicable.

Informed Consent Statement: Not applicable.

Data Availability Statement: Not applicable.

Acknowledgments: This work was supported by the National Key Research and Development Project (2018YFC1506604) and the Special Program for Innovation and Development of China Meteorological Administration (CXFZ2021Z007).

Conflicts of Interest: The authors declare no conflict of interest.

References

1. Shi, C.; Pan, Y.; Gu, J.; Xu, B.; Han, S.; Zhu, Z.; Zhang, L.; Sun, S.; Jiang, Z. A review of multi-source meteorological data fusion product. *Acta Meteorol. Sin.* **2019**, *77*, 774–783.
2. Hong, Y.; Hsu, K.L.; Sorooshian, S.; Gao, X. Precipitation estimation from remotely sensed imagery using an artificial neural network cloud classification system. *J. Appl. Meteorol.* **2004**, *43*, 1834–1853. [CrossRef]
3. Huffman, G.J.; Adler, R.F.; Arkin, P.; Chang, A.; Ferraro, R.; Gruber, A.; Janowiak, J.; McNab, A.; Rudolf, B.; Schneider, U. The global precipitation climatology project (GPCP)combined precipitation dataset. *Bull. Am. Meteorol. Soc.* **1997**, *78*, 5–20. [CrossRef]
4. Huffman, G.J.; Adler, R.F.; Morrissey, M.M.; Bolvin, D.T.; Curtis, S.; Joyce, R.; McGavock, B.; Susskind, J. Global precipitation at one-degree daily resolution from multisatellite observations. *J. Hydrometeorol.* **2001**, *2*, 36–50. [CrossRef]
5. Huffman, G.J.; Bolvin, D.T.; Nelkin, E.J.; Wolff, D.B.; Adler, R.F.; Gu, G.; Hong, Y.; Bowman, K.P.; Stocker, E.F. The TRMM multisatellite precipitation analysis (TMPA): Quasi-global, multiyear, combined-sensor precipitation estimates at fine scales. *J. Hydrometeorol.* **2007**, *8*, 38–55. [CrossRef]
6. Seo, D.J. Real-time estimation of rainfall fields using radar rainfall and rain guage data. *J. Hydrol.* **1998**, *208*, 37–52. [CrossRef]
7. Chen, T.; Ren, L.; Yuan, F.; Tang, T. Merging ground and satellite-based precipitation data sets for improved hydrological simulations in the Xijiang River basin of China. *Stoch. Environ. Res. Risk Assess.* **2019**, *33*, 1893–1905. [CrossRef]
8. Zhang, L.; Li, X.; Zheng, D. Merging multiple satellite-based precipitation products and gauge observations using a novel double machine learning approach. *J. Hydrol.* **2021**, *594*, 125969. [CrossRef]
9. Lyu, F.; Tang, G.; Behrangi, A.; Wang, T.; Tan, X.; Ma, Z.; Xiong, W. Precipitation Merging Based on the Triple Collocation Method Across Mainland China. *IEEE Trans. Geosci. Remote Sens.* **2020**, *59*, 3161–3176. [CrossRef]
10. Hazra, A.; Maggioni, V.; Houser, P. A Monte Carlo-Based Multi-Objective Optimization Approach to Merge Different Precipitation Estimates for Land Surface Modeling. *J. Hydrol.* **2019**, *570*, 454–462. [CrossRef]
11. Cassalho, F.; Rennó, C.D.; dos Reis, J.B.C. Hydrologic Validation of MERGE Precipitation Products over Anthropogenic Watersheds. *Water* **2020**, *12*, 1268. [CrossRef]
12. Lu, N.M.; You, R.; Zhang, W.J. A fusing technique with satellite precipitation estimate and raingauge data. *Acta Meteorol. Sin.* **2004**, *18*, 141–146.
13. Shen, Y.; Zhao, P.; Pan, Y.; Yu, J. A high spatiotemporal gauge satellite merged precipitation analysis over China. *J. Geophys. Res.* **2014**, *119*, 3063–3075. [CrossRef]
14. Xie, P.P.; Xiong, A.Y. A conceptual model for constructing high-resolution gauge-satellite merged precipitation analyses. *J. Geophys. Res.* **2011**, *116*, D21106. [CrossRef]
15. Pan, Y.; Gu, J.; Yu, J.; Shen, Y.; Shi, C.; Zhou, Z. Test of merging methods for multi-source observed precipitation products at high resolution over China. *Acta Meteorol. Sin.* **2018**, *76*, 755–766.
16. Pan, Y.; Shen, Y.; Yu, J.; Zhao, P. Analysis of the combined gauge-satellite hourly precipitation over China based on the OI technique. *Acta Meteorol. Sin.* **2012**, *70*, 1381–1389.
17. Shen, Y.; Pan, Y.; Yu, J.J.; Zhao, P.; Zhou, Z.J. Quality assessment of hourly merged precipitation product over China. *Trans. Atmos. Sci.* **2013**, *36*, 37–46.
18. Yu, J.J.; Shen, Y.; Pan, Y.; Zhao, P.; Zhou, Z. Improvement of satellite-based precipi- tation estimates over China based on probability density function matching method. *J. Appl. Meteorol. Sci.* **2013**, *24*, 544–553.
19. Yu, J.; Shen, Y.; Pan, Y.; Xiong, A. Comparative assessment between the daily merged precipitation dataset over China and the world's popular counterparts. *Acta Meteorol. Sin.* **2015**, *73*, 394–410.
20. Pan, Y.; Shen, Y.; Yu, J.; Xiong, A. An experiment of high- resolution gauge-radar-satellite combined precipitation retrieval based on the Bayesian merging method. *Acta Meteorol. Sin.* **2015**, *73*, 177–186.
21. Sun, S.; Shi, C.; Pan, Y.; Gu, J.; Bai, L.; Su, C.; Han, S.; Sun, J. The Improved Effects Evaluation of Three-Source Merged of Precipitation Products in China. *J. China Hydrol.* **2020**, *40*, 10–23.
22. Gu, C.; Cheng, W.; Xu, Y.; Deng, Z.; Cheng, R. Analysis on impacts of radar data integration and cold or hot start on heavy rainfall prediction of AREM model. *Torrential Rain Disasters* **2018**, *37*, 574–581.
23. Li, H.L.; Cui, C.G.; Wang, Z.B. Scientific Designs, Functions and Applications of LAPS. *Torrential Rain Disasters* **2009**, *28*, 64–70.
24. Li, X.; Zhou, Z.; Li, Z.; Pan, Y.; Pan, Y.; Shen, Y.; Xu, B.; Gu, J. Quality Assessment of China Merged Precipitation Product Using Hydrological Data in Jiangxi Province. *Meteorol. Mon.* **2017**, *43*, 1534–1546.

MDPI
St. Alban-Anlage 66
4052 Basel
Switzerland
Tel. +41 61 683 77 34
Fax +41 61 302 89 18
www.mdpi.com

Sustainability Editorial Office
E-mail: sustainability@mdpi.com
www.mdpi.com/journal/sustainability

www.ingramcontent.com/pod-product-compliance
Lightning Source LLC
Chambersburg PA
CBHW061453220926
51850CB00004B/85
9783036558493